Modern Arylation Methods

Edited by
Lutz Ackermann

Further Reading

de Vries, J. G., Elsevier, C. J. (Eds.)

The Handbook of Homogeneous Hydrogenation

3 Volumes
2007
ISBN: 978-3-527-31161-3

Yamamoto, H., Ishihara, K. (Eds.)

Acid Catalysis in Modern Organic Synthesis

2 Volumes
2008
ISBN: 978-3-527-31724-0

Oro, L. A., Claver, C. (Eds.)

Iridium Complexes in Organic Synthesis

2009
ISBN: 978-3-527-31996-1

Hiersemann, M., Nubbemeyer, U. (Eds.)

The Claisen Rearrangement
Methods and Applications

2007
ISBN: 978-3-527-30825-5

Bolm, C., Hahn, F. E. (Eds.)

Activating Unreactive Substrates
The Role of Secondary Interactions

2009
ISBN: 978-3-527-31823-0

Sheldon, R. A., Arends, I., Hanefeld, U.

Green Chemistry and Catalysis

2007
ISBN: 978-3-527-30715-9

Modern Arylation Methods

Edited by
Lutz Ackermann

WILEY-VCH Verlag GmbH & Co. KGaA

The Editors

Prof. Dr. Lutz Ackermann
Georg-August-Universität Göttingen
Institut für Organische Chemie und
Biomolekulare Chemie
Tammannstr. 2
37077 Göttingen
Germany

Library of Congress Card No.: applied for

British Library Cataloguing-in-Publication Data
A catalogue record for this book is available from the British Library.

Bibliographic information published by the Deutsche Nationalbibliothek
The Deutsche Nationalbibliothek lists this publication in the Deutsche Nationalbibliografie; detailed bibliographic data are available on the Internet at <http://dnb.d-nb.de>.

Composition SNP Best-set Typesetter Ltd., Hong Kong
Printing betz-druck GmbH, Darmstadt
Bookbinding Litges & Dopf GmbH, Heppenheim

Printed in the Federal Republic of Germany
Printed on acid-free paper

ISBN: 978-3-527-31937-4

Contents

Modern Arylation Methods. Edited by Lutz Ackermann

ISBN: 978-3-527-31937-4

Preface

Arenes and heteroarenes are essential substructures of numerous compounds with activities that are relevant to a variety of important areas of research, ranging *inter alia* from medicinal chemistry and biology to materials sciences. As a result, the selective preparation of these omnipresent moieties is of the utmost relevance to synthetic chemists, both in industry and academia. The introduction of already existing aryl- or heteroaryl-groups – which we recognize today as 'arylation chemistry' – arguably constitutes the most generally applicable approach to accomplish this task. Thus, the recent growing impact of – and also interest in – arylation chemistry is reflected by the increasing numbers of references that contain the term 'arylation' {SciFinder Scholar (October 2008): 41 (1968), 124 (1978), 178 (1988), 188 (1998), 755 (2007)}. Hence, *Modern Arylation Methods* summarizes the diverse aspects of arylation reactions, with a particular focus on recent developments in this area. Within the book, following a brief introduction, industrial practitioners review important transition metal-catalyzed cross-coupling reactions, as well as carbon–heteroatom bond-forming processes. The influence of catalytic strategies – and particularly of those that employ transition metal complexes – on arylation reactions with haloalkanes, alkenes, alkynes and carbonyl compounds as substrates is subsequently described. The next three chapters detail not only the experimental observations but also the mechanistic considerations of ecologically benign C–H bond functionalization reactions. Finally, the book concludes with two chapters on arylations, which involve arynes or radicals as key intermediates, and a summary of photochemically initiated transformations. I hope that this topical selection will be useful to the reader, and that it will serve as a stimulus for further exciting developments in this rapidly evolving research area.

The chapters of this book were written by internationally renowned authorities, to whom I am very thankful for such outstanding contributions. I also wish to express my gratitude to Elke Maase, Rainer Münz, Hans-Jochen Schmitt and the staff of the editorial team of Wiley-VCH for their continuous help during this project. Further, I gratefully acknowledge the assistance of my coworkers during

Modern Arylation Methods. Edited by Lutz Ackermann

ISBN: 978-3-527-31937-4

the editorial process, particularly of Sergei I. Kozhushkov, Stefan Beußhausen and Gabriele Keil-Knepel. Most importantly, I am deeply thankful to Daniela Rais for her invaluable advice, encouragement and support.

Göttingen, December 2008 *Lutz Ackermann*

List of Contributors

Lutz Ackermann
Institut für Organische und Biomolekulare Chemie
Georg-August-Universität Göttingen
Tammannstrasse 2
37077 Göttingen
Germany

Angelo Albini
University of Pavia
Department of Organic Chemistry
Viale Taramelli 10
27100 Pavia
Italy

Erick M. Carreira
ETH Zürich
Laboratorium für Organische Chemie
HCI H 335
Wolfgang-Pauli-Strasse 10
8093 Zürich
Switzerland

Yu Chen
Department of Chemistry
Iowa State University
Ames, Iowa 50011
USA

Christian Defieber
ETH Zürich
Laboratorium für Organische Chemie
HCI H 335
Wolfgang-Pauli-Strasse 10
8093 Zürich
Switzerland

Paula de Mendoza
Institute of Chemical Research of Catalonia (ICIQ)
Av. Països Catalans 16
43007 Tarragona
Spain

Valentina Dichiarante
University of Pavia
Department of Organic Chemistry
Viale Taramelli 10
27100 Pavia
Italy

Antonio M. Echavarren
Institute of Chemical Research of Catalonia (ICIQ)
Av. Països Catalans 16
43007 Tarragona
Spain

Modern Arylation Methods. Edited by Lutz Ackermann

ISBN: 978-3-527-31937-4

Maurizio Fagnoni
University of Pavia
Department of Organic Chemistry
Viale Taramelli 10
27100 Pavia
Italy

Shingo Ito
International Research Center for Elements Science
Institute for Chemical Research
Kyoto University, Uji
Kyoto 611-0011
Japan

Richard C. Larock
Department of Chemistry
Iowa State University
Ames, Iowa 50011
USA

Steven V. Ley
Department of Chemistry
University of Cambridge
Cambridge CB2 1EW
United Kingdom

Adam Littke
Biogen Idec
Department of Process Chemistry
14 Cambridge Center
Cambridge, MA 02142
USA

Masahiro Miura
Osaka University
Department of Applied Chemistry
Faculty of Engineering
Suita
Osaka 565-0871
Japan

Masaharu Nakamura
International Research Center for Elements Science
Institute for Chemical Research
Kyoto University, Uji
Kyoto 611-0011
Japan

Steven P. Nolan
Institute of Chemical Research of Catalonia (ICIQ)
Av. Països Catalans 16
43007 Tarragona
Spain

Martin Oestreich
Organisch-Chemisches Institut
Westfälische Wilhelms-Universität Münster
Corrensstrasse 40
48149 Münster
Germany

Tetsuya Satoh
Osaka University
Department of Applied Chemistry
Faculty of Engineering
Suita
Osaka 565-0871
Japan

Björn Schlummer
Saltigo GmbH
51369 Leverkusen
Germany

Ulrich Scholz
Boehringer Ingelheim Pharma
GmbH & Co. KG
55216 Ingelheim am Rhein
Germany

Birte Schulte
Organisch-Chemisches Institut
Westfälische Wilhelms-Universität
Münster
Corrensstrasse 40
48149 Münster
Germany

Armido Studer
Organisch-Chemisches Institut
Westfälische Wilhelms-Universität
Münster
Corrensstrasse 40
48149 Münster
Germany

Andrew W. Thomas
Discovery Chemistry
F. Hoffmann-La Roche, Ltd
4070 Basel
Switzerland

Verena T. Trepohl
Organisch-Chemisches Institut
Westfälische Wilhelms-Universität
Münster
Corrensstrasse 40
48149 Münster
Germany

Santiago E. Vaillard
Organisch-Chemisches Institut
Westfälische Wilhelms-Universität
Münster
Corrensstrasse 40
48149 Münster
Germany

Rubén Vicente
Institut für Organische und
Biomolekulare Chemie
Georg-August-Universität
Göttingen
Tammannstrasse 2
37077 Göttingen
Germany

Mihai S. Viciu
ETH Zürich
Department of Chemistry and
Applied Biosciences
Swiss Federal Institute of Technology
HCI H 232
8093 Zürich
Switzerland

1
Arylation Reactions: A Historical Perspective

Lutz Ackermann

1.1
Structure and Bonding of Benzene

Substituted arenes and heteroarenes are ubiquitous substructures of both naturally occurring and synthetic organic compounds which have activities that are relevant to a variety of research areas, ranging from biology to the materials sciences. Representative examples of historically important synthetic arenes include the analgesic aspirin, acetylsalicylic acid (**1**), or the first natural pigment to be duplicated synthetically in an industrial environment, alizarin (**2**) (Figure 1.1).

The specific activities displayed by arene-containing compounds are due largely to the remarkable relative stability and, thus, the unique chemical reactivity of arenes. As a result, their structure and bonding were one of the most fascinating problems facing chemists during the mid-nineteenth century. Interestingly, a cyclic planar structure was first described for benzene and its derivatives by Johann Josef Loschmidt in 1861 (Figure 1.2), a pioneer of nineteenth-century physics and chemistry [1].

Loschmidt proposed circular planar structures for 121 arenes, including those of benzene (**3**) and aniline (**4**) (Figure 1.3) [1, 2]. Whilst the chemical formulae for further important structural motifs, such as cyclopropanes, as well as multiple carbon–carbon bonds, were also proposed in this prescient report, it was largely ignored by the chemical community. Indeed, it was not until 1913 that Loschmidts' work was recognized by Richard Anschütz who, interestingly, was one of August Friedrich Kekulé von Stradonitz's students [3].

In 1865, August Friedrich Kekulé von Stradonitz (Figure 1.4) suggested a planar structure for benzene, in which carbon–carbon single and carbon–carbon double bonds alternated within a six-membered ring [4]. However, this proposal did not account for the observation of only one isomer of *ortho*-disubstituted benzenes. Therefore, Kekulé revised his model in 1872, postulating a rapid equilibrium between two structures (Scheme 1.1). This beautiful D_{6h}-symmetrical structure of benzene was later confirmed by its X-ray crystal structure and its electron-diffraction data, which highlighted the equivalence of all six carbon–carbon bonds

Modern Arylation Methods. Edited by Lutz Ackermann

ISBN: 978-3-527-31937-4

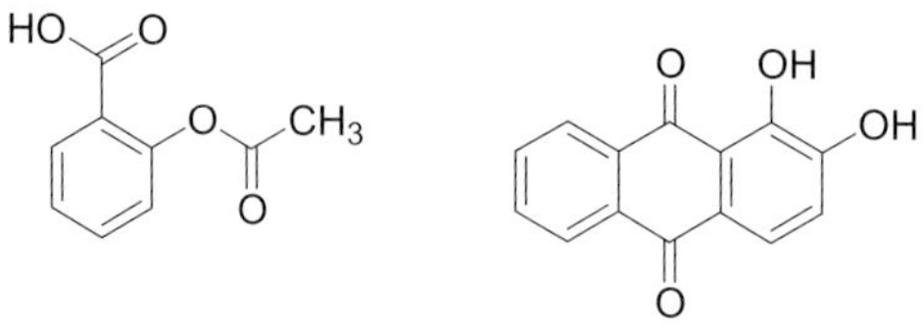

HO O O CH_3 O

O OH OH O

1 2

Figure 1.1 Selected historically relevant substituted arenes.

Figure 1.2 Johann Josef Loschmidt (15 March 1821–8 July 1895).

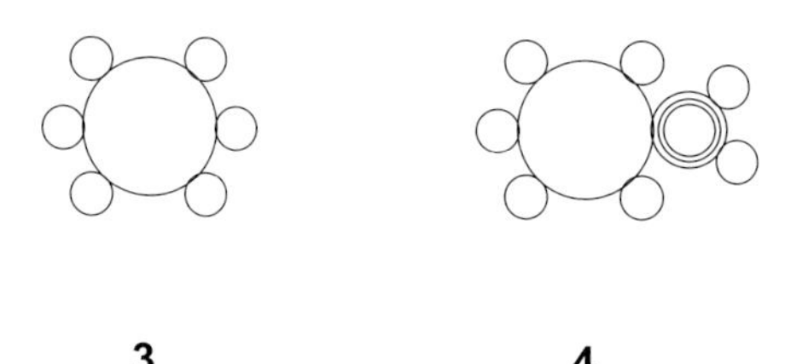

Figure 1.3 Loschmidt's molecular structural formulae for benzene (**3**) and aniline (**4**).

[5]. According to R. Robinson (1925), these experimentally obtained data are represented by the commonly used symbol **5**, in which delocalized π-electrons are represented by a circle within a hexagon (Scheme 1.1).

The term 'aromatic' was coined in 1855 by August Wilhelm von Hofmann (1818–1892), before the physical mechanism determining so-called *aromaticity* was unraveled. Originally, this definition was derived from a characteristic property, namely a sweet scent that was associated with some 'aromatic' compounds. Subsequently, it was shown that arenes are unsaturated compounds, but showed a chemical reactivity which differed from that of both alkenes or alkynes. This unique reactivity pattern of arenes was used at the end of the nineteenth century as the only criterion of aromaticity. At the start of the twentieth century, additional

Figure 1.4 August Friedrich Kekulé von Stradonitz (7 September 1829–13 July 1896).

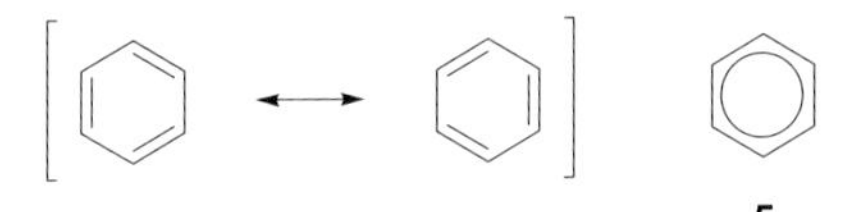

Scheme 1.1 Structural representations of D_{6h}-symmetrical benzene.

criteria of aromaticity were elaborated, such as the relative stability of arenes with respect to the corresponding hypothetical cyclooligoenes [6]. Later, the development of quantum mechanics set the stage for the molecular orbital theory, the application of which enabled Hückel to devise his well-known ($4n + 2$) rule as a criterion for aromaticity in 1931 [7, 8]. A definition of aromaticity based on carbon–carbon bond lengths in arenes was also delineated during the twentieth century. While this model proved useful where accurate bond lengths were available, it was, unfortunately, found not to be generally applicable [9]. Hence, today one of the most reliable criteria of aromaticity is based on the magnetic properties of arenes. The anomalous chemical shifts of resonances in ^{1}H NMR spectra led to less satisfactory analyses for some compounds. In contrast, the diamagnetic susceptibility exaltation [6] or the nucleus-independent chemical shift (NICS) [10] were found often to be more reliable criteria of aromaticity [5, 11].

1.2 Syntheses of Substituted (Hetero)Arenes, and the Contents of this Book

Historically, the name 'benzene' derives from benzoin, which was known to European pharmacists and perfumers since the fifteenth century. Thus, pure benzene

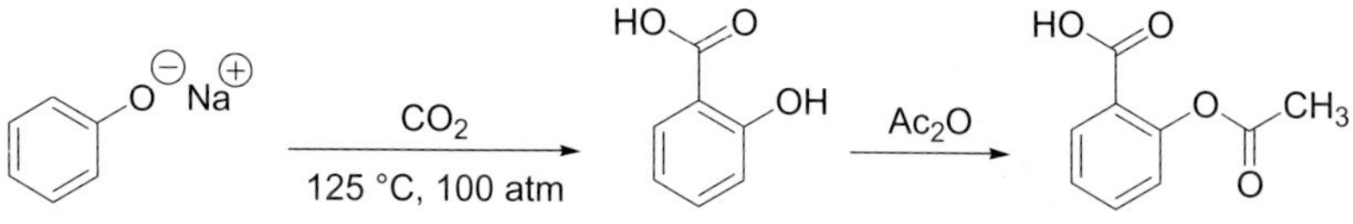

Scheme 1.2 Industrial synthesis of acetylsalicylic acid (**1**) (ASA, aspirin, 1897).

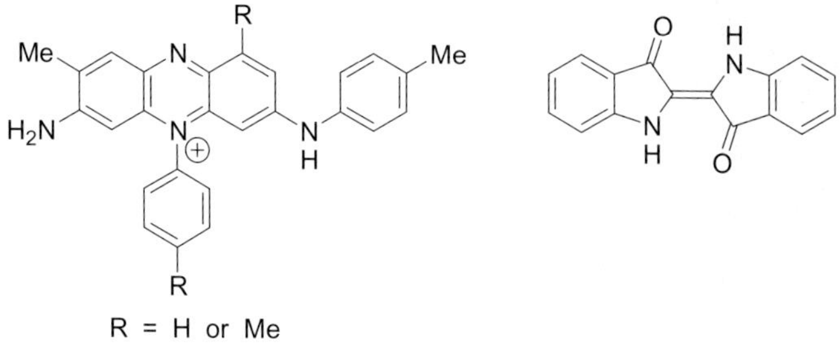

Figure 1.5 Selected historically relevant substituted arenes.

could be prepared via the decarboxylation of benzoic acid, *flowers of benzoin* (E. Mitscherlich, 1833), which in turn was obtained from gum benzoin [5]. Benzene was initially isolated and identified as a chemical compound from the condensed phase of pyrolyzed whale oil in 1825 by Michael Faraday [12]. However, it was not until the 1840s that benzene would be produced in an industrial environment based on its isolation from coal tar by Charles Mansfield in 1836.

Since their discovery during the 1860s, electrophilic aromatic substitution reactions played the dominant role for functionalizations of arenes, and were often the method of choice for the synthesis of substituted arenes. For example, Hermann Kolbe, a student of Friedrich Wöhler, devised a synthesis for salicylic acid (**7**) [13, 14], which set the stage for the industrial preparation of acetylsalicylic acid (ASA, aspirin) (**1**) by Arthur Eichengrün and Felix Hoffmann at Bayer in 1897 (Scheme 1.2).

This representative historically relevant example, as with the industrial preparation of important dyes such as mauveine (Tyrian purple, **8**) (1856, Perkin), alizarin (**2**) (1869, Perkin, Graebe and Liebermann) or indigo (**9**) (1890, Heumann) (Figure 1.5), illustrates the significant economical impact of aromatic substitution chemistry [15].

Modern aspects of electrophilic aromatic substitution chemistry address the development of enantioselective variants of these direct (hetero)arene functionalization reactions. For example, enantiomerically enriched metal catalysts, as well as organocatalysts, allowed for the asymmetric addition reactions of (hetero)arenes onto (α,β-unsaturated) carbonyl compounds. Additionally, highly enantioselective arylations of carbonyl compounds were accomplished with organometallic reagents

Figure 1.6 Fritz Ullmann (2 July 1875–17 March 1939).

as nucleophilic aryl donors. These exciting new methodologies of modern arylation chemistry are comprehensively reviewed by Christian Defieber and Erick M. Carreira in Chapter 8.

In addition to electrophilic aromatic substitution reactions, a variety of valuable approaches to the synthesis of substituted (hetero)arenes was developed [16]. Consequently, a fully comprehensive review on the entire history of arylation methodologies would call for an encyclopedia, and is, thus, beyond the scope of this introductory chapter. However, both metal-mediated and metal-catalyzed reactions [17, 18] had, arguably, the strongest impact on the development of modern arene syntheses [19, 20]. While olefin metatheses reactions catalyzed by homogeneous catalysts proved to be particularly valuable for ring-closing reactions [21], late transition metal-catalyzed (cross)coupling reactions were among the most important chemical innovations of the twentieth century. Thus, the palladium-catalyzed Tsuji–Trost reaction allowed for novel strategies in C—C bond formations, which had a major impact on asymmetric syntheses [22]. The relevance of catalytic (cross)coupling chemistry is arguably most evident in (hetero)aryl–(hetero)aryl and (hetero)aryl–heteroatom bond-forming reactions [17, 18, 23], and therefore the majority of the chapters in this book discuss recent innovations in this rapidly developing area of research. As a result, the following personal historical perspective focuses rather on the evolution of metal-catalyzed coupling chemistry.

While electrophilic substitution reactions enabled the direct functionalizations of (hetero)arenes, Fittig observed, during the nineteenth century, that halogenated arenes could be employed for the preparation of alkylated arenes in the presence of sodium under Wurtz's reaction conditions [24–26].

In 1901, Fritz Ullmann (Figure 1.6) showed that stoichiometric amounts of copper enabled the reductive coupling of haloarenes for the synthesis of symmetrically substituted biaryls (Scheme 1.3) [27].

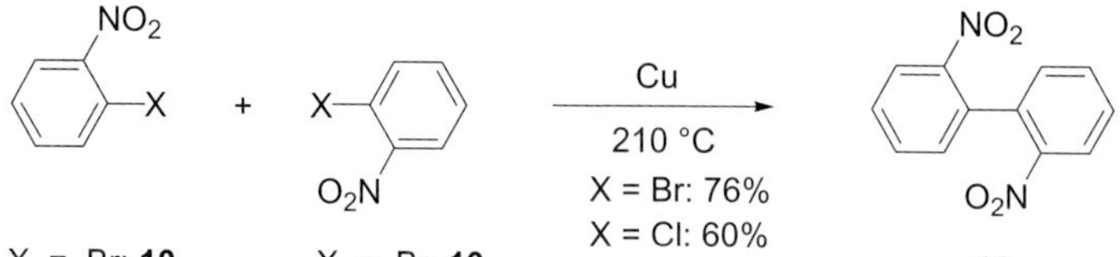

X = Br: **10**
X = Cl: **11**

Cu, 210 °C, X = Br: 76%, X = Cl: 60%

12

Scheme 1.3 Copper-mediated reductive coupling of haloarenes (Ullmann, 1901).

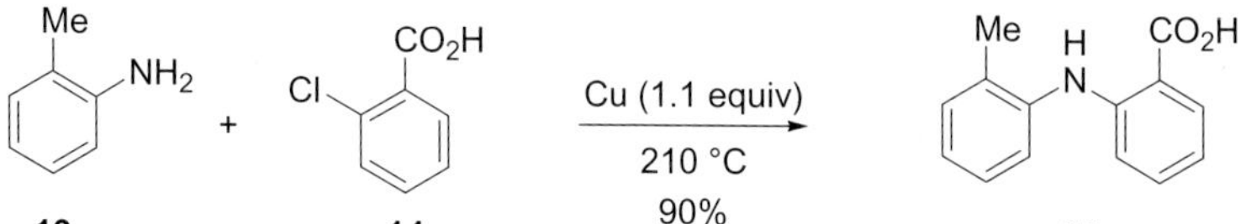

13 + **14** → Cu (1.1 equiv), 210 °C, 90% → **15**

Scheme 1.4 Copper-mediated arylation of amine **13** (Ullmann, 1903).

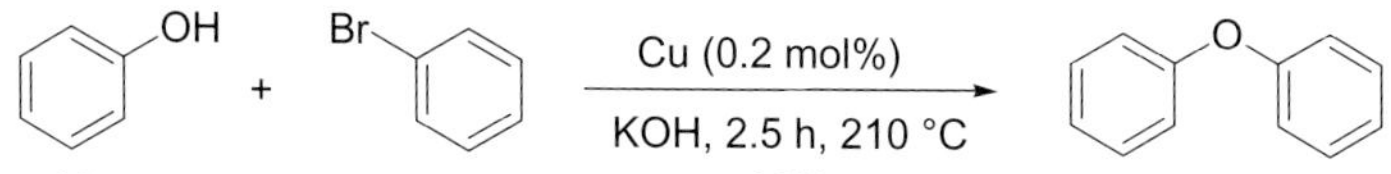

16 + **17** → Cu (0.2 mol%), KOH, 2.5 h, 210 °C, 90% → **18**

Scheme 1.5 Copper-catalyzed arylation of phenol (**16**) (Ullmann, 1905).

Based on these regioselective copper-mediated C—C bond-forming reactions, Ullmann found in 1903 that arylations of amines were also viable when using haloarenes as coupling partners. Notably, this report illustrated at an early stage that chlorides could be employed as leaving groups on an arene, even for the conversion of more sterically hindered *ortho*-substituted starting materials (Scheme 1.4) [28].

These reports on copper-mediated coupling reactions set the stage for pioneering reports by Ullmann and Irma Goldberg (his later wife) on the use of catalytic amounts of copper for coupling reactions of haloarenes. Thus, the arylation of phenol (**16**) and its derivatives with low catalyst loadings of copper enabled the selective syntheses of diarylethers (Scheme 1.5) [29].

Importantly, Irma Goldberg showed in 1906 that copper-catalyzed arylation reactions of, potentially chelating, aniline derivatives could be achieved with bromoarenes as coupling partners (Scheme 1.6) [30].

It is noteworthy that this arylation protocol proved also applicable to catalytic transformations of amides as nucleophilic substrates (Scheme 1.7) [30].

Although these reports on copper-mediated and copper-catalyzed coupling reactions highlighted the potential of transition metal compounds for regioselective C—C, C—O and C—N bond formations, rather few further developments of these outstanding methodologies were disclosed during the following decades [31]. However, an improved understanding of organometallic reagents with *umgepolte*

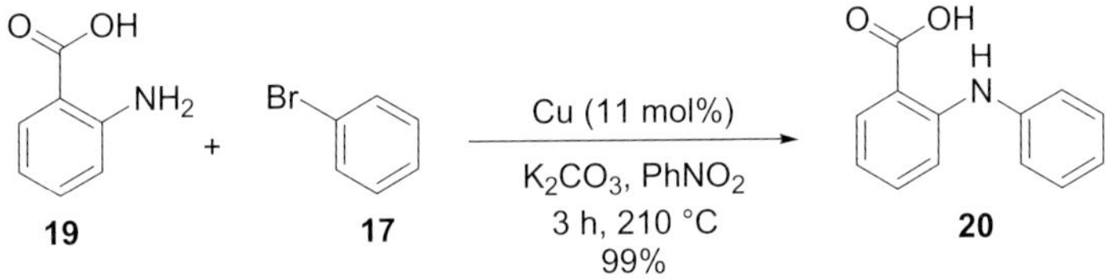

O OH NH$_2$ + Br

Cu (11 mol%)

K_2CO_3, $PhNO_2$
3 h, 210 °C
99%

O OH H N

19 **17** **20**

Scheme 1.6 Catalytic arylation of an aniline derivative (Goldberg, 1906).

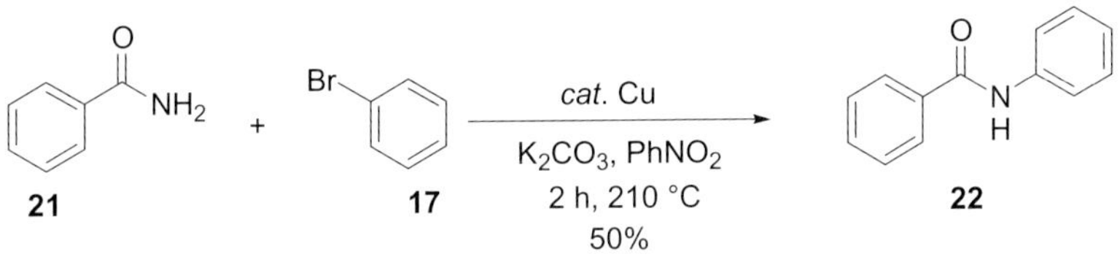

O NH$_2$ + Br

cat. Cu

K_2CO_3, $PhNO_2$
2 h, 210 °C
50%

O N H

21 **17** **22**

Scheme 1.7 Copper-catalyzed arylation of amide **21** (Goldberg, 1906).

reactivity enabled André Job to disclose a catalytic effect of [$NiCl_2$] on the conversion of PhMgBr with ethylene and CO [32, 33].

During the first half of the twentieth century, alternative methodologies for the preparation of arenes were discovered by Walter Reppe at the Badische Anilin-&-Soda-Fabrik (BASF), employing homogeneous metal catalysts. Thus, a scalable nickel-catalyzed synthesis of cyclooctatetraene from acetylene was devised [34]. While remarkable progress has been made in transition metal-catalyzed cycloaddition chemistry since these early reports [35], the achievement of satisfactory regioselectivities in the syntheses of unsymmetrically substituted (hetero)arenes still represents a major challenge when applying these methodologies [36].

During the 1940s, Kharasch and coworkers observed a dramatic influence of transition metal salts on the formation and reactivity of Grignard reagents [37]. It was found that catalytic amounts of [$CoCl_2$], [$MnCl_2$], [$FeCl_3$] or [$NiCl_2$] allowed for efficient homocouplings of organomagnesium reagents in the presence of organic halides, such as bromo- or chlorobenzene, which were thought to act as terminal oxidants (Scheme 1.8) [38].

Also during the 1940s, the formation of biphenyl (**24**) from fluorobenzene (**26**) and phenyl lithium (**25**) in an uncatalyzed reaction (Scheme 1.9) led Wittig to propose a dehydrobenzene, specifically an *ortho*-benzyne, as intermediate [39, 40], the formation of which was confirmed by Roberts in 1953 [41]. Modern arylation methodologies based on arynes as key intermediates are reviewed by Yu Chen and Richard Larock in Chapter 12.

Later, during the 1960s, selective cross-coupling reactions between alkyl, alkenyl or aryl halides and stoichiometric amounts of allyl nickel or alkyl copper reagents were developed (Scheme 1.10). These methodologies displayed a remarkable functional group tolerance and proved to be among the most effective protocols for $C(sp^3)$–$C(sp^3)$ cross-coupling reactions [42, 43]. A comprehensive review on metal-catalyzed cross-coupling reactions for challenging arylations of nonactivated

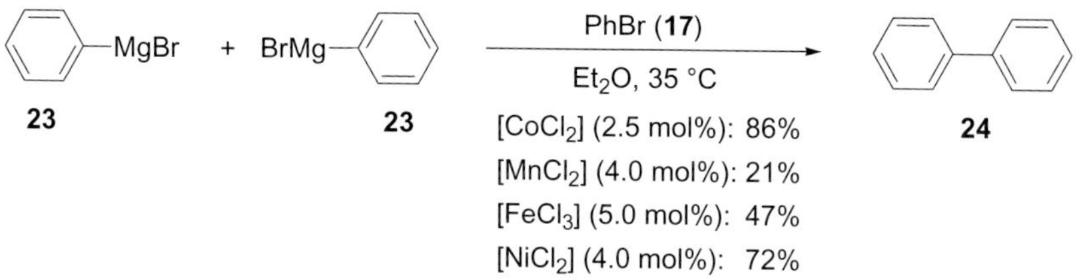

PhBr (**17**)
Et_2O, 35 °C
23 + **23** → **24**
$[CoCl_2]$ (2.5 mol%): 86%
$[MnCl_2]$ (4.0 mol%): 21%
$[FeCl_3]$ (5.0 mol%): 47%
$[NiCl_2]$ (4.0 mol%): 72%

Scheme 1.8 Metal-catalyzed homocouplings of Grignard reagent **23** (Kharasch, 1941).

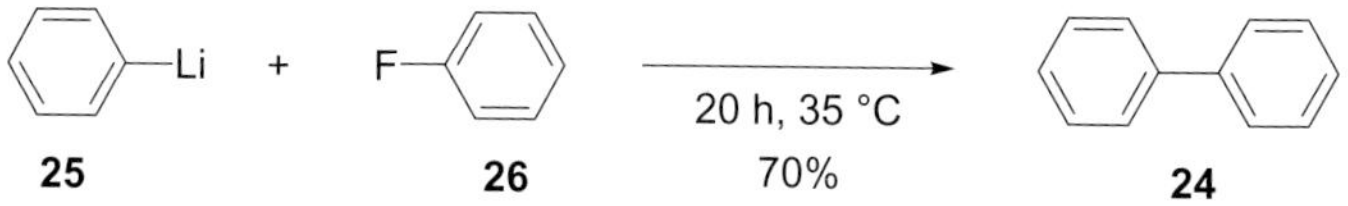

25 + **26** → **24**
20 h, 35 °C
70%

Scheme 1.9 Biphenyl synthesis through an aryne intermediate (Wittig, 1942).

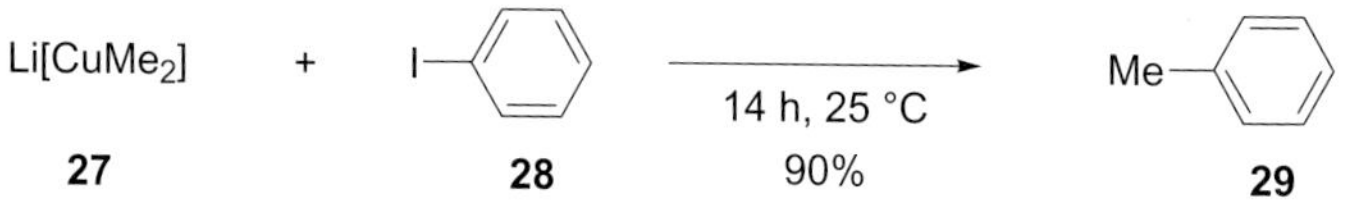

$Li[CuMe_2]$ (**27**) + **28** → **29**
14 h, 25 °C
90%

Scheme 1.10 Methylation of iodobenzene (**28**) with stoichiometric amounts of copper reagent **27** (Corey, 1967).

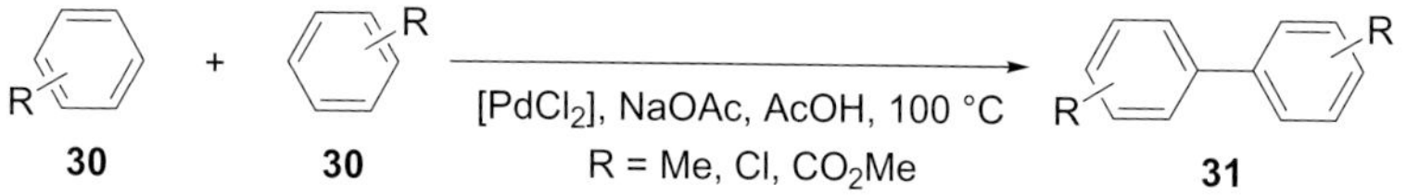

30 + **30** → **31**
$[PdCl_2]$, NaOAc, AcOH, 100 °C
R = Me, Cl, CO_2Me

Scheme 1.11 Intermolecular oxidative homo-coupling (van Helden, Verberg, 1965).

alkyl-substituted electrophiles is provided by Masaharu Nakamura and Shingo Ito in Chapter 5.

In 1965, a palladium-mediated oxidative homo-coupling of arenes was reported. Through the use of $[PdCl_2]$, along with sodium acetate as additive in acetic acid as solvent, biaryls could be obtained, with a regioselectivity being indicative of an electrophilic substitution-type mechanism (Scheme 1.11) [44]. Subsequently, a palladium-catalyzed dehydrogenative coupling of arenes was accomplished with oxygen as terminal oxidant [45, 46]. Despite remarkable progress in oxidative biaryl syntheses, the achievement of chemoselectivity and regioselectivity in intermolecular cross-coupling reactions continues to offer a major challenge [47, 48].

In 1968, palladium-catalyzed oxidative arylations of alkenes using stoichiometric amounts of organometallic arylating agents, predominantly mercury-derived compounds, were disclosed by Heck [49]. These studies indicated that economical terminal oxidants, such as molecular oxygen, could be employed (Scheme 1.12) [49, 50]. Unfortunately, however, a significant limitation of this arylation protocol was represented by the use of highly toxic organomercury compounds [51]. In contrast, Fujiwara, Moritani and coworkers probed the use of stoichiometric amounts of

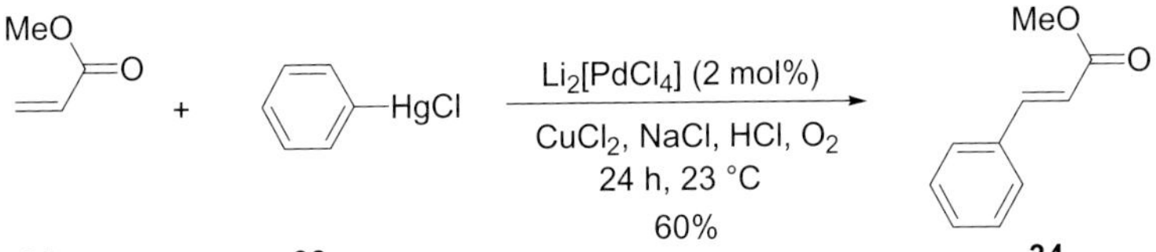

MeO, O, + HgCl, $Li_2[PdCl_4]$ (2 mol%), $CuCl_2$, NaCl, HCl, O_2, 24 h, 23 °C, 60%, MeO, O

32 **33** **34**

Scheme 1.12 Palladium-catalyzed oxidative arylation of alkenes (Heck, 1968).

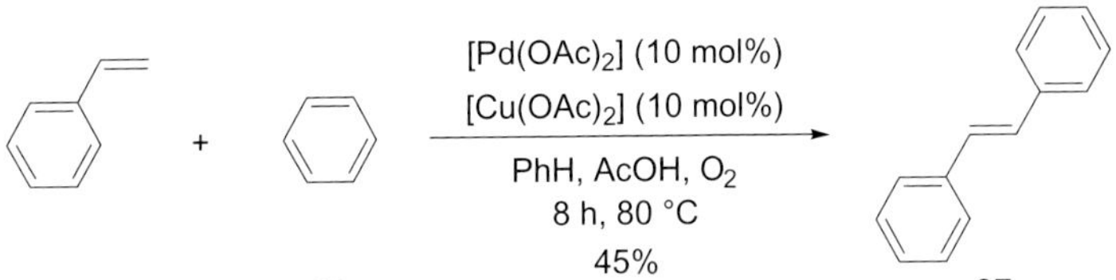

+, $[Pd(OAc)_2]$ (10 mol%), $[Cu(OAc)_2]$ (10 mol%), PhH, AcOH, O_2, 8 h, 80 °C, 45%

35 **36** **37**

Scheme 1.13 Palladium-catalyzed oxidative arylation of alkenes with arenes (Fujiwara, Moritani, 1969).

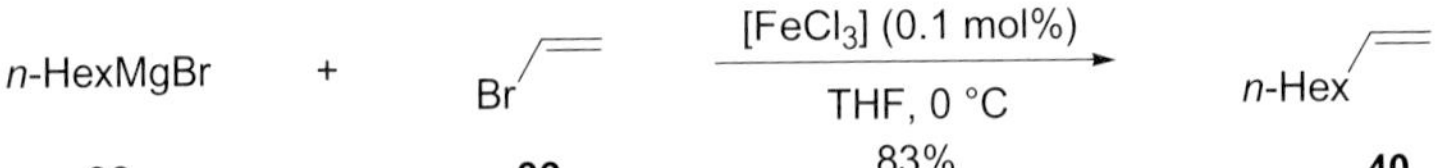

n-HexMgBr + Br, $[FeCl_3]$ (0.1 mol%), THF, 0 °C, 83%, *n*-Hex

38 **39** **40**

Scheme 1.14 Iron-catalyzed cross-coupling with alkenyl halide **39** (Kochi, 1971).

palladium compounds for a direct coupling between arenes and alkenes in 1967 [52, 53]. Based on these stoichiometric transformations, a catalytic oxidative arylation of alkenes using simple arenes as arylating reagents was devised. Again, this intermolecular [54] catalytic C–H bond functionalization could be achieved with *inter alias* molecular oxygen as terminal oxidant, and likely proceeded via an electrophilic aromatic substitution-type mechanism (Scheme 1.13) [55–57].

Another breakthrough in modern cross-coupling chemistry was accomplished during the early 1970s, when Kochi and coworkers disclosed a pioneering report in which electrophiles displayed $C(sp^2)$–Br bonds as functional groups. Interestingly, this iron-catalyzed cross-coupling was used for the functionalization of alkenyl bromides (Scheme 1.14) [58, 59], but was not applied to haloarenes as electrophiles until recently [60–63]. Nonetheless, Kochi's report highlighted the remarkable prospective applications of late transition metal complexes to $C(sp^2)$–$C(sp^2)$ bond formations through catalytic cross-coupling chemistry.

A remarkable catalytic conversion of aryl halides was disclosed in 1971 – namely the arylation of alkenes [64] – which can be considered as being related to Heck's arylations of alkenes with arylmercury, -tin or -lead compounds [49, 50]. Here, Mizoroki and coworkers found that catalytic amounts of $[PdCl_2]$ or heterogeneous palladium black (**41**) enabled transformations of ethylene (**42**) or monosubstituted alkenes when using iodobenzene (**28**) as electrophile and KOAc as inorganic base (Scheme 1.15) [64, 65].

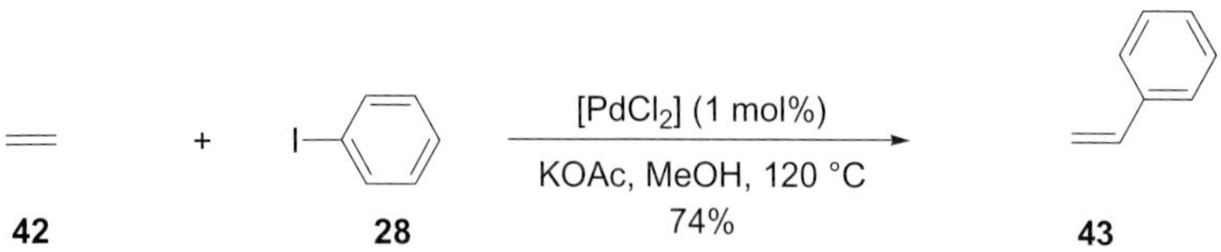

[$PdCl_2$] (1 mol%)
KOAc, MeOH, 120 °C
74%
42 **28** **43**

Scheme 1.15 Palladium-catalyzed arylation of alkene **42** (Mizoroki, 1971).

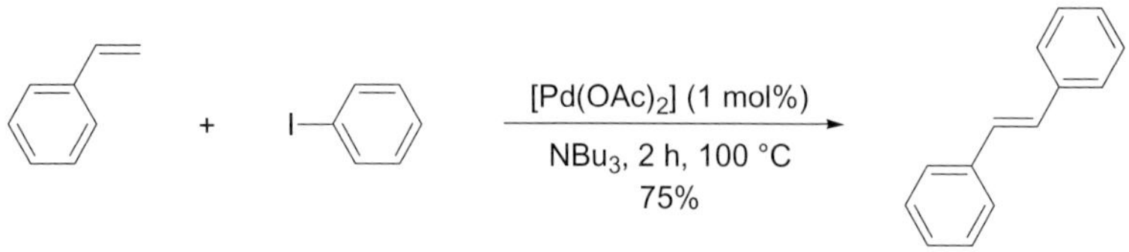

[$Pd(OAc)_2$] (1 mol%)
NBu_3, 2 h, 100 °C
75%
35 **28** **37**

Scheme 1.16 Palladium-catalyzed arylation of styrene (**35**) (Heck, 1972).

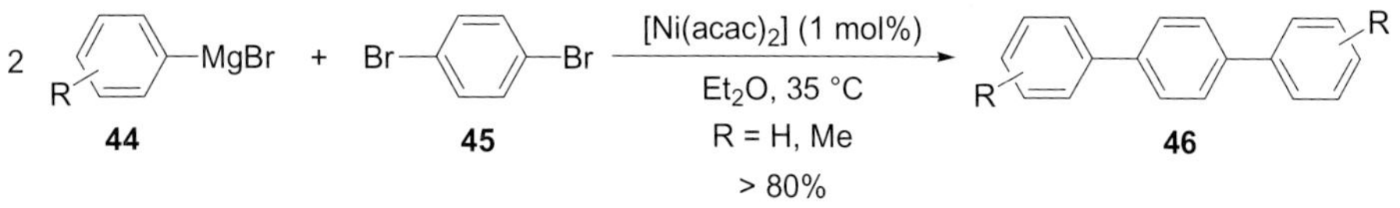

2 R–C6H4–MgBr + Br–C6H4–Br
[$Ni(acac)_2$] (1 mol%)
Et_2O, 35 °C
R = H, Me
> 80%
44 **45** **46**

Scheme 1.17 Nickel-catalyzed C_{aryl}–C_{aryl} cross-coupling (Corriu, 1972).

In 1972, Heck showed, independently, that aryl, benzyl and styryl halides would react with alkenes in the presence of a hindered amine as organic base at a lower reaction temperature of 100 °C (Scheme 1.16) [51, 66]. A detailed overview on the modern aspects of Mizoroki–Heck reactions, as well as their oxidative variants (Fujiwara reactions), is provided by Martin Oestreich and Verena Trepohl in Chapter 7.

A catalytic C_{aryl}–C_{aryl} cross-coupling was reported by Corriu and Masse in early 1972, when different nickel and cobalt compounds were probed in the cross-coupling between alkenyl or aryl bromides and aryl Grignard reagents, with [$Ni(acac)_2$], giving rise to superior results. For example, the catalytic cross-coupling of dibromide **45** provided high yields of the corresponding *para*-terphenyls **46** (Scheme 1.17) [33, 67].

Based on studies by Yamamoto on stoichiometric oxidative addition reactions of chlorobenzene to nickel complexes [68, 69], Kumada, Tamao and coworkers reported independently on highly efficient nickel-catalyzed cross-coupling reactions between Grignard reagents and aryl or alkenyl chlorides. Importantly, the beneficial effect of phosphine ligands, allowing for the effective cross-coupling reactions of chloroarenes, was demonstrated (Scheme 1.18) [70].

Furthermore, a catalytic cycle for this cross-coupling reaction was proposed that served as a blueprint for many subsequently developed transformations. The working mode of the catalytically competent species [TM] was based on: (a) an

2 *n*-BuMgBr + 48 → 49; $[NiCl_2(dppe)_2]$ (0.7 mol%), Et_2O, 35 °C, 94%

47 **48** **49**

Scheme 1.18 Nickel-catalyzed cross-coupling of dichloride **48** (Kumada, Tamao, 1972).

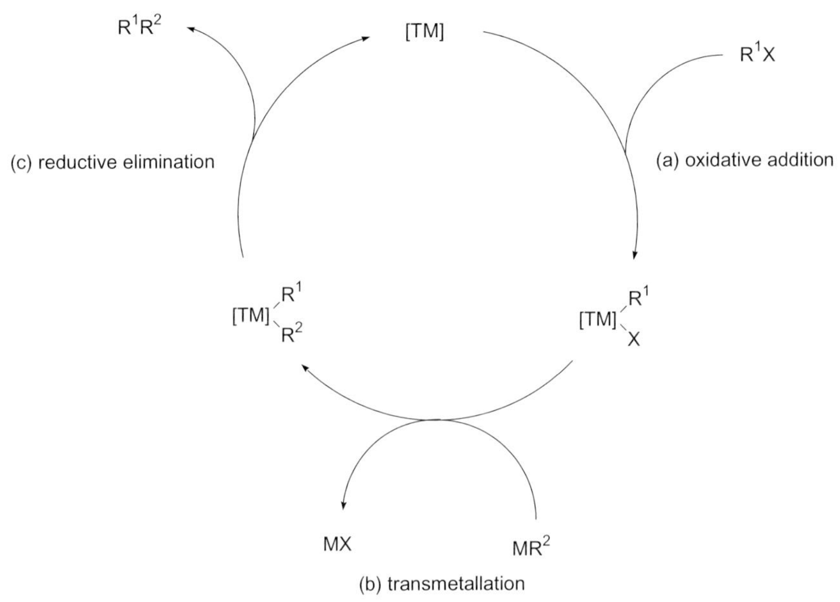

Scheme 1.19 Generalized catalytic cycle for metal-catalyzed cross-coupling reactions.

initial oxidative addition; (b) a subsequent transmetallation; and (c) a final reductive elimination, as illustrated in Scheme 1.19 [70, 71].

In 1975, both Cassar [72] and Murahashi and coworkers [73] reported on the applications of palladium complexes to catalytic cross-coupling chemistry [74]. Initially, Cassar compared nickel and palladium complexes in the catalytic cross-coupling between aryl or alkenyl halides and terminal alkynes, and found that nickel-based compounds exhibited no catalytic activity. In contrast, $[Pd(PPh_3)_4]$ enabled catalytic coupling reactions of various organic halides bearing $C(sp^2)$–X bonds as reactive functional groups, notably in the absence of any copper [75] additives (Scheme 1.20) [72].

At about the same time, Heck disclosed an independent report on a comparable palladium-catalyzed arylation of terminal alkynes. Here, the use of amines as bases allowed for the catalytic transformations of alkenyl and aryl halides, again in the absence of additional copper compounds (Scheme 1.21) [76].

Subsequently, Sonogashira and coworkers observed that the addition of catalytic amounts of [CuI] enabled arylations of terminal alkynes to proceed at significantly

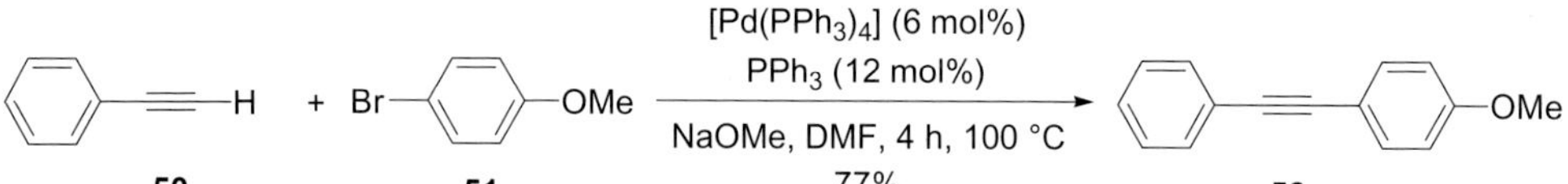

$[Pd(PPh_3)_4]$ (6 mol%)
PPh_3 (12 mol%)
NaOMe, DMF, 4 h, 100 °C
77%

50 51 52

Scheme 1.20 Palladium-catalyzed synthesis of internal alkyne **52** (Cassar, 1975).

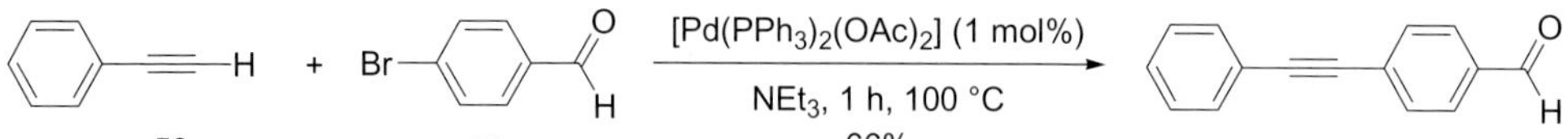

$[Pd(PPh_3)_2(OAc)_2]$ (1 mol%)
NEt_3, 1 h, 100 °C
66%

50 53 54

Scheme 1.21 Palladium-catalyzed arylation of terminal alkyne **50** (Heck, 1975).

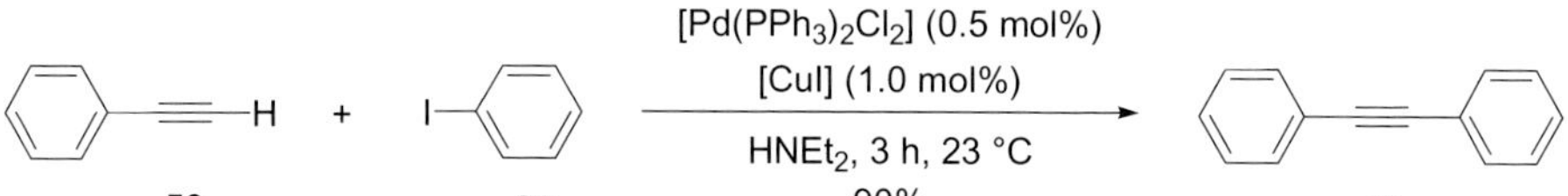

$[Pd(PPh_3)_2Cl_2]$ (0.5 mol%)
[CuI] (1.0 mol%)
$HNEt_2$, 3 h, 23 °C
90%

50 28 55

Scheme 1.22 Palladium-catalyzed arylation of terminal alkyne **50** in the presence of [CuI] (Sonogashira, 1975).

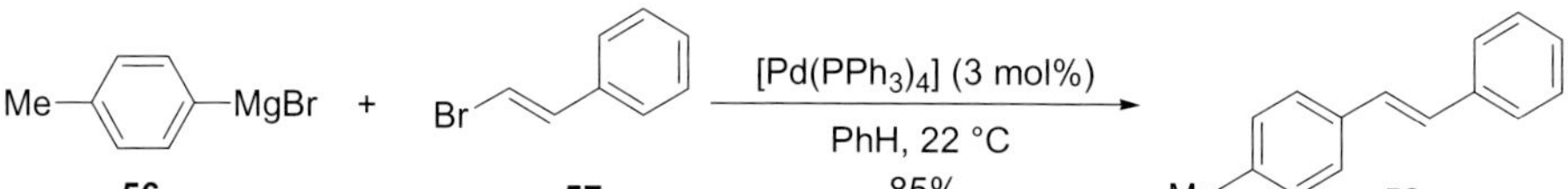

$[Pd(PPh_3)_4]$ (3 mol%)
PhH, 22 °C
85%

56 57 58

Scheme 1.23 Palladium-catalyzed cross-coupling of Grignard reagent **56** (Murahashi, 1975).

milder reaction conditions (Scheme 1.22) [77, 78]. Modern aspects of arylation reactions of alkynes, along with a detailed discussion on their mechanisms, are presented by Steven P. Nolan and Mihai S. Viciu in Chapter 6.

Also in 1975, Murahashi probed the use of palladium catalysts in cross-coupling reactions of alkenyl halides with lithium- or magnesium-based nucleophiles. While stoichiometric amounts of $[Pd(PPh_3)_4]$ were required for reactions with lithium derivatives as nucleophiles [79], the corresponding Grignard [80] reagents could be coupled with catalytic amounts of palladium complexes (Scheme 1.23) [73, 74].

Thus far, catalytic coupling reactions had focused on the use of magnesium-based organometallics as nucleophiles. However, in 1976, Negishi revealed the use of alkenylalanes as valuable coupling partners for nickel- [81] (Scheme 1.24) or palladium-catalyzed [82] diastereoselective syntheses of substituted alkenes [83, 84]. These results showcased that organometallics other than Grignard reagents could be employed for efficient cross-coupling chemistry. Furthermore, a comparison between nickel- and palladium-based catalysts indicated the latter to be superior with respect to diastereoselective syntheses [83]. These observations set the stage for evaluating the scope of further organometallic reagents in palladium-

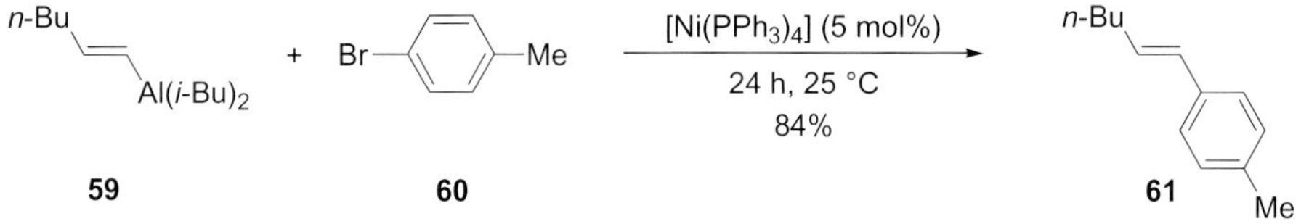

Scheme 1.24 Nickel-catalyzed cross-coupling of alane **59** (Negishi, 1976).

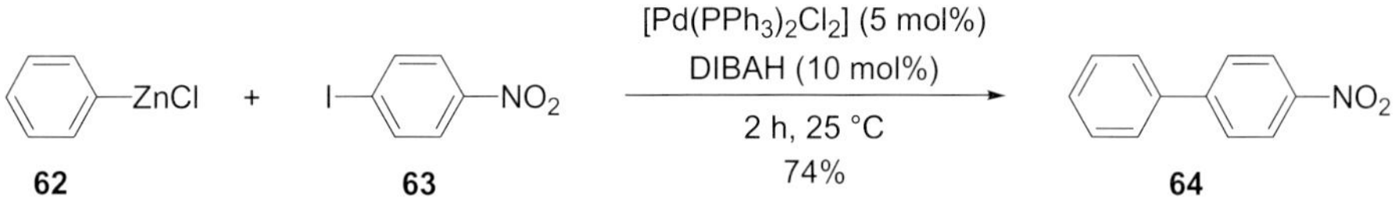

Scheme 1.25 Palladium-catalyzed cross-coupling with arylzinc **62** (Negishi, 1977).

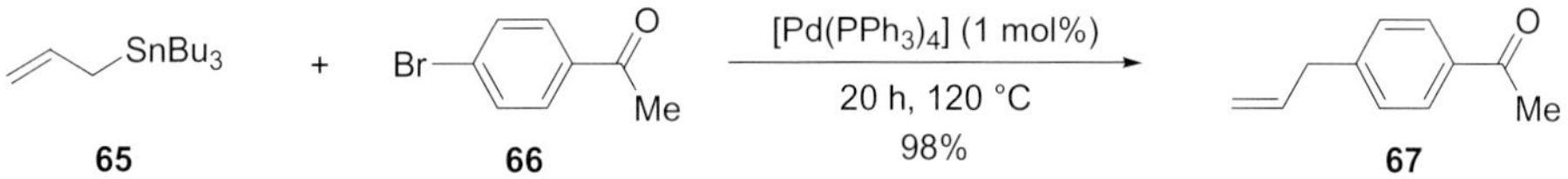

Scheme 1.26 Palladium-catalyzed cross-coupling with allyltin **65** (Kosugi, Migita, 1977).

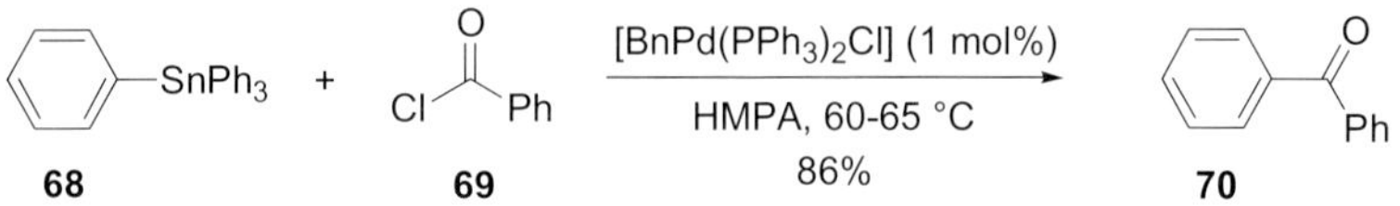

Scheme 1.27 Palladium-catalyzed arylation of acyl chloride **69** (Stille, 1978).

catalyzed cross-coupling chemistry. As a consequence, Negishi and coworkers disclosed catalytic arylations and alkenylations [85] of aryl halides using arylzinc reagents. The decreased nucleophilicity of organozinc compounds resulted in an excellent functional group tolerance (Scheme 1.25), particularly when compared to couplings with Grignard reagents [86–88].

Also in 1977, Kosugi, Migita and coworkers reported on the use of organotin compounds for the palladium-catalyzed cross-couplings of aryl [89] and acyl halides [90]. The allylation of haloarenes and the arylation of acyl halides displayed again a remarkable wide substrate scope (Scheme 1.26) [91].

Stille and Milstein showed, in 1978, that palladium-catalyzed arylations of acyl halides with aryltins occurred under exceedingly mild reactions conditions when using $[BnPd(PPh_3)_2Cl]$ as catalyst (Scheme 1.27) [92]. Furthermore, Negishi developed *inter alia* palladium-catalyzed cross-coupling reactions between lithium 1-alkynyltributylborates and haloarenes [93], while Murahashi employed a transmetallation strategy for the use of trialkylboranes [94]. Suzuki, Miyaura and coworkers, in contrast, unraveled the beneficial effect of added bases in palladium-catalyzed cross-coupling reactions between alkenylboranes and organic halides [95]. Importantly, mild bases (such as Na_2CO_3) enabled catalytic C_{aryl}–C_{aryl}

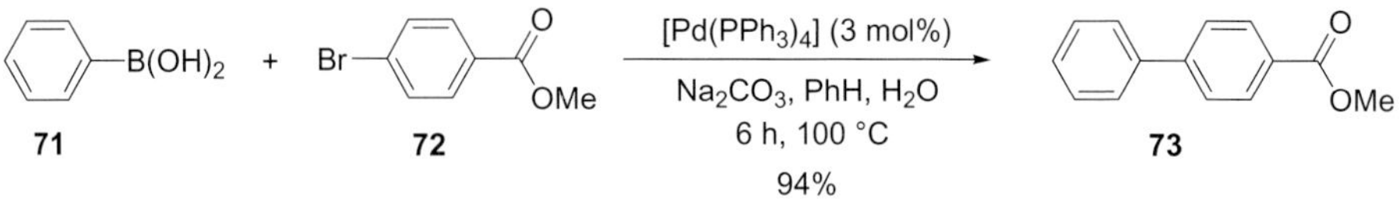

Scheme 1.28 Palladium-catalyzed arylation with phenyl boronic acid (71) (Suzuki, 1981).

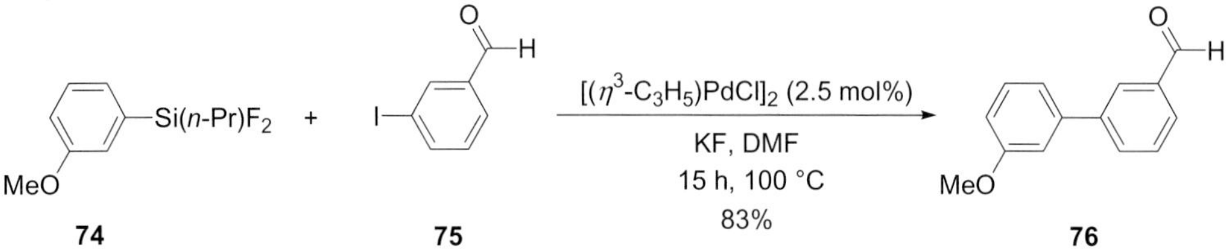

Scheme 1.29 Palladium-catalyzed arylation with silicon-based nucleophile 74 (Hiyama, 1989).

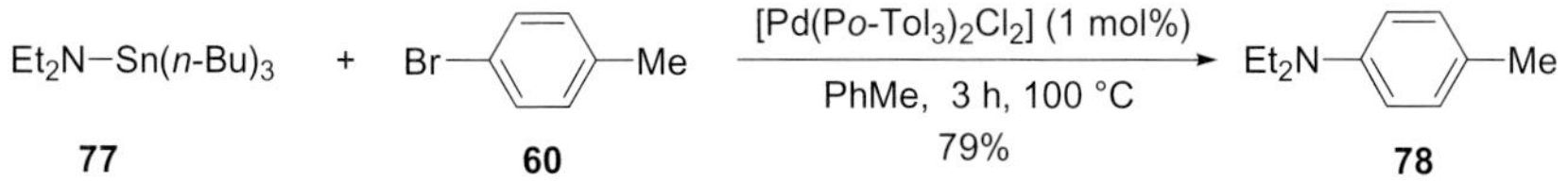

Scheme 1.30 Palladium-catalyzed arylation of tin amide 77 (Kosugi, Migita, 1983).

bond formations to proceed efficiently in aqueous reaction media (Scheme 1.28) [96, 97].

Cross-coupling reactions of alkenyl-substituted silicon-based nucleophiles were reported by Kumada [98], and by Hallberg in 1982 [99], as well as by Kikukawa subsequently [100]. However, efficient cross-couplings of aryl-substituted silicon-based nucleophiles were accomplished through the use of fluoride-containing additives, as reported by Hiyama in 1989. This *in situ* generation of pentacoordinated silicates gave rise to cross-couplings with a useful functional group tolerance (Scheme 1.29) [101, 102].

Further valuable catalytic cross-coupling reactions between organometallic reagents and aryl halides were reported, which matured to being indispensable tools in modern organic synthesis. In Chapter 2, Adam F. Littke summarizes the modern aspects of catalytic cross-coupling reactions, with particular focus on the use of challenging organic electrophiles, such as aryl chlorides, fluorides and tosylates, for $C(sp^2)$–$C(sp^2)$ bond formations.

In 1983, Kosugi, Migita and coworkers found that palladium-catalyzed aminations of bromoarenes could be achieved when tin amides were used as nucleophiles (Scheme 1.30) [103, 104].

In the 1990s, a breakthrough was independently accomplished by Buchwald [105] (Scheme 1.31) and Hartwig [106] (Scheme 1.32), establishing broadly applicable palladium-catalyzed Ullmann-type aminations of haloarenes, employing amines directly as nucleophiles. Thereby, the preparation and use of tin amides could be circumvented, and intramolecular [105] aminations of haloarenes were shown also to occur efficiently.

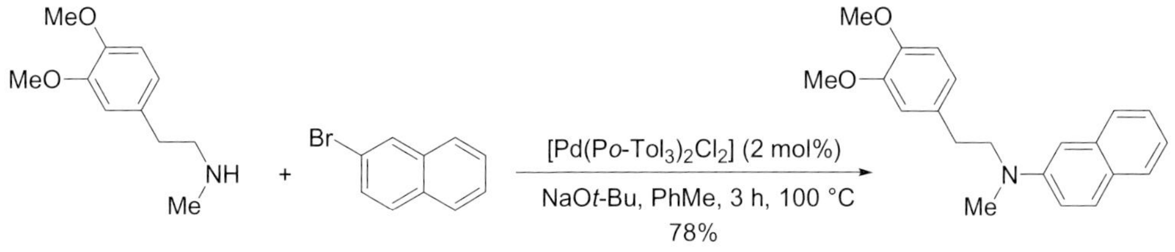

[Pd(Po-Tol$_3$)$_2$Cl$_2$] (2 mol%)
NaO*t*-Bu, PhMe, 3 h, 100 °C
78%

79 **80** **81**

Scheme 1.31 Palladium-catalyzed arylation of amine **79** (Buchwald, 1995).

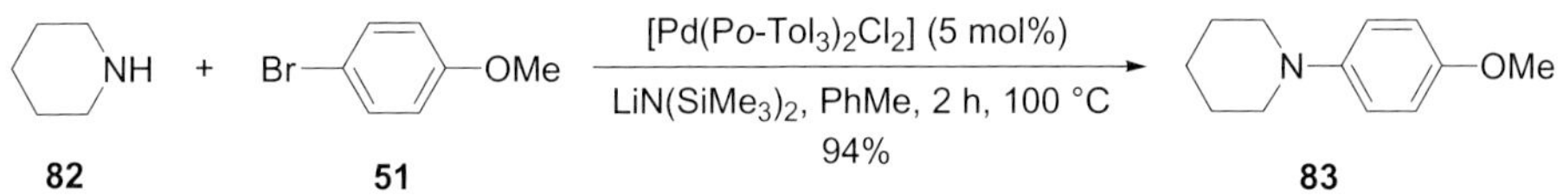

[Pd(Po-Tol$_3$)$_2$Cl$_2$] (5 mol%)
LiN(SiMe$_3$)$_2$, PhMe, 2 h, 100 °C
94%

82 **51** **83**

Scheme 1.32 Palladium-catalyzed arylation of amine **82** (Hartwig, 1995).

These palladium-catalyzed amination reactions of haloarenes are among the most popular modern methodologies for C(sp^2)–N bond formations. A detailed summary on the state of the art of these (also industrially relevant) reactions, as well as of related arylation reactions of α-C–H acidic compounds, is provided in Chapter 3 by Björn Schlummer and Ulrich Scholz.

In addition to Ullmann's copper-catalyzed [107] C(sp^2)–N bond formations using haloarenes, a remarkable complementary approach was developed by Chan and Lam during the 1990s using boron-based arylating reagents [108, 109]. A detailed review by Andrew W. Thomas and Steven V. Ley on these oxidative copper-mediated and copper-catalyzed arylations, as well as their extension to alcohols [108, 110], is to be found in Chapter 4.

The presented methodologies for selective intermolecular C_{aryl}–C_{aryl} bond formations largely relied on the use of prefunctionalized coupling partners as substrates. Predominantly, these transformations made use of organometallic reagents as nucleophilic coupling partners. An ecologically benign and economically attractive alternative is represented by the direct functionalization of C–H bonds in simple arenes [111]. An early example of a metal-catalyzed direct arylation was disclosed in 1982 by Ames. During attempts directed towards palladium-catalyzed intermolecular arylation of alkene **85** following Heck's protocol, cyclization product **86** was obtained through an intramolecular direct arylation reaction (Scheme 1.33) [112].

An extension of this protocol showed that the alkene was not essential for achieving catalytic turnover. This allowed for syntheses of various important heterocycles, as illustrated for the preparation of dibenzofuran **88** in Scheme 1.34 [113–115].

In 1985, Ohta and coworkers reported on intermolecular direct arylations, and probed the palladium-catalyzed direct arylations of indole derivatives with electron-

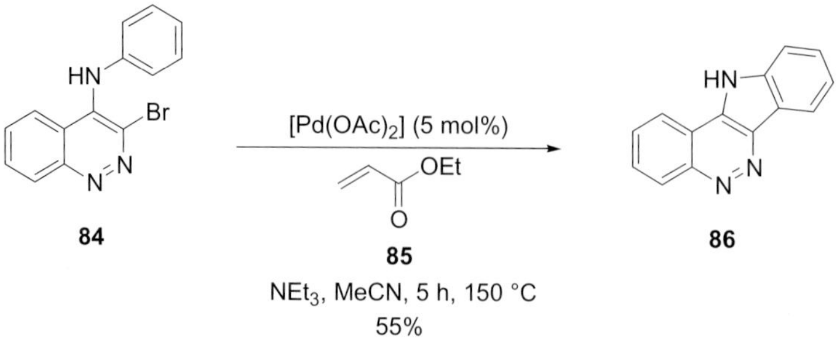

HN Br 84 [Pd(OAc)$_2$] (5 mol%) OEt O 85 NEt$_3$, MeCN, 5 h, 150 °C 55% HN N=N 86

Scheme 1.33 Palladium-catalyzed intramolecular direct arylation (Ames, 1982).

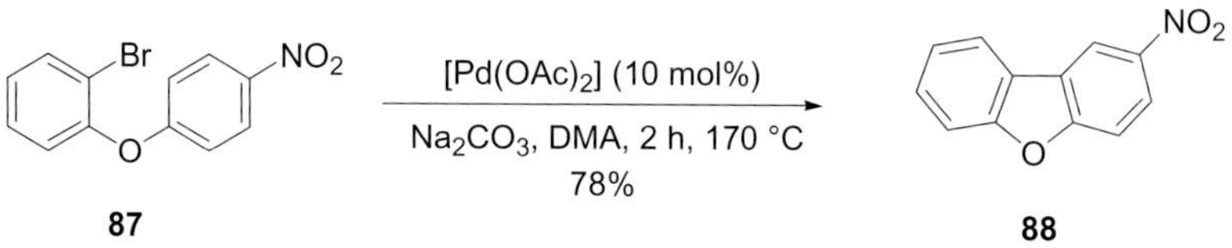

Br NO$_2$ O 87 [Pd(OAc)$_2$] (10 mol%) Na$_2$CO$_3$, DMA, 2 h, 170 °C 78% NO$_2$ O 88

Scheme 1.34 Palladium-catalyzed intramolecular direct arylation for a dibenzofuran synthesis (Ames, 1983).

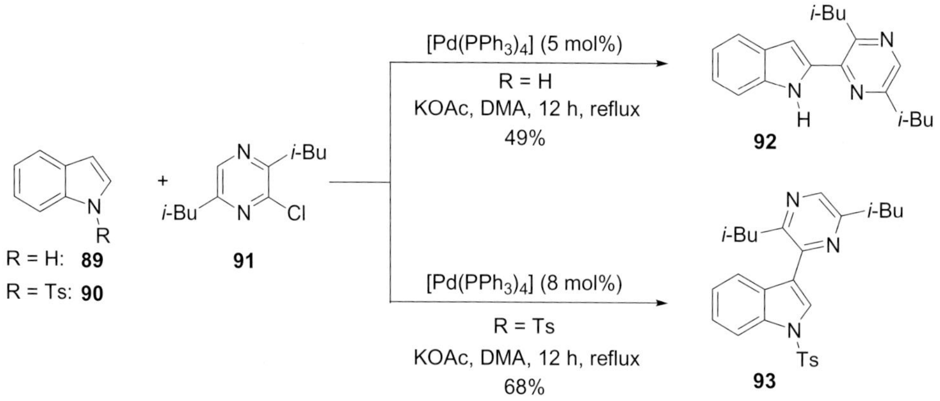

N R R = H: 89 R = Ts: 90 + *i*-Bu N N Cl *i*-Bu 91

[Pd(PPh$_3$)$_4$] (5 mol%) R = H KOAc, DMA, 12 h, reflux 49% → 92

[Pd(PPh$_3$)$_4$] (8 mol%) R = Ts KOAc, DMA, 12 h, reflux 68% → 93

Scheme 1.35 Palladium-catalyzed intermolecular direct arylation (Ohta, 1985 and 1989).

deficient heteroaryl chloride **91**. Interestingly, the regioselectivity of this intermolecular reaction was found to be strongly influenced by the substitution pattern of the indole on position N-1. Hence, indole (**89**) [116], as well as its N-alkylated derivatives [117], yielded the C-2 arylated products. In contrast, *N*-tosyl indole (**90**) gave rise to functionalization at position C-3, with good to excellent regioselectivities (Scheme 1.35) [117].

During the following decades, palladium-catalyzed direct arylations proved to be valuable tools for organic synthesis, and a comprehensive review on these transformations is provided by Masahiro Miura and Tetsuya Satoh in Chapter 10.

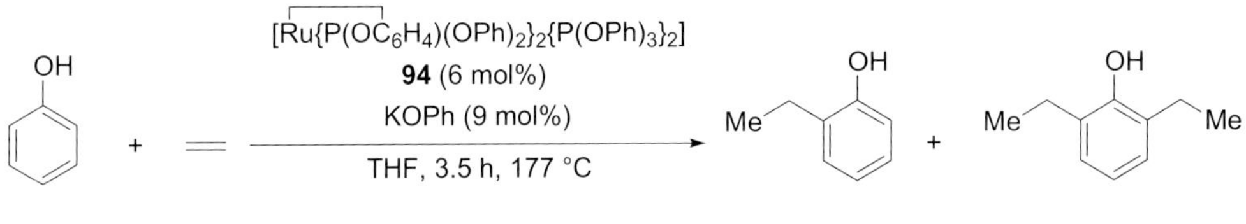

$[Ru\{P(OC_6H_4)(OPh)_2\}_2\{P(OPh)_3\}_2]$
94 (6 mol%)
KOPh (9 mol%)
THF, 3.5 h, 177 °C

16 + **42** → **95** (13%) + **96** (75%)

Scheme 1.36 Ruthenium-catalyzed alkylation of phenol (16) (Lewis, 1986).

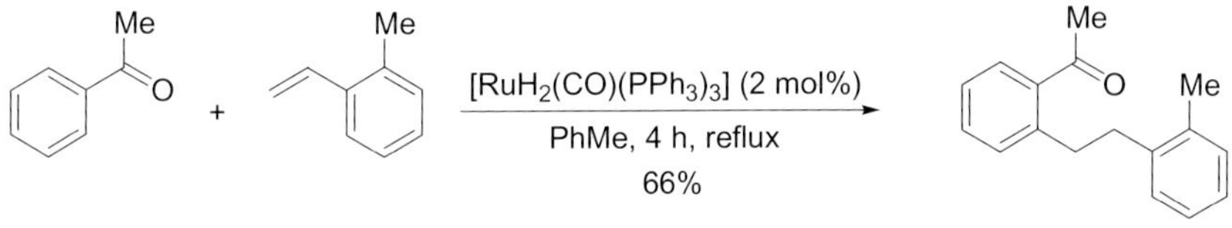

97 + **98** → **99**

$[RuH_2(CO)(PPh_3)_3]$ (2 mol%)
PhMe, 4 h, reflux
66%

Scheme 1.37 Ruthenium-catalyzed intermolecular C—H bond functionalization (Murai, 1993).

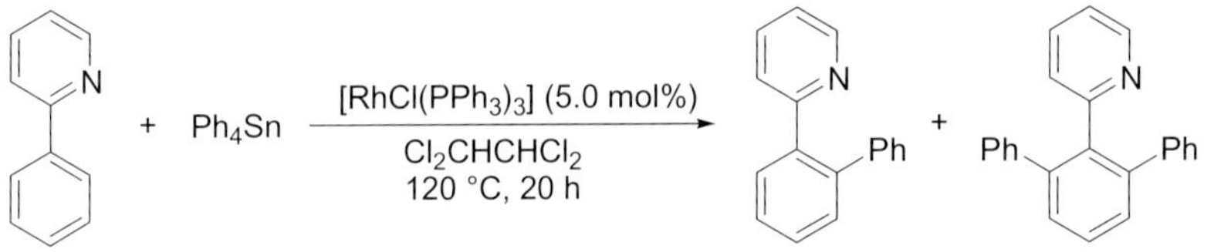

100 + Ph_4Sn **101** → **102** (56%) + **103** (20%)

$[RhCl(PPh_3)_3]$ (5.0 mol%)
$Cl_2CHCHCl_2$
120 °C, 20 h

Scheme 1.38 Rhodium-catalyzed direct arylation of arene **100** (Oi, Inoue, 1998).

In 1986, a highly regioselective catalytic hydroarylation [118] of alkenes was disclosed by Lewis, whereby a cyclometallated ruthenium catalyst **94** allowed for a directed (hence *ortho*-selective) alkylation of phenols through the *in situ* formation of the corresponding phosphites (Scheme 1.36) [119].

In 1993, Murai reported on highly efficient alkylations of aromatic ketones with olefins, where $[RuH_2(CO)PPh_3)_3]$ served as the catalyst for regioselective C—H bond functionalizations (Scheme 1.37) [120–123].

These examples illustrated the potential of ruthenium complexes for selective intermolecular C—H bond functionalization reactions. Furthermore, they sowed the seeds of ruthenium-catalyzed direct arylation reactions, which proved to be rather broadly applicable [124–127].

A comparable concept was applied in 1998 to rhodium-catalyzed oxidative direct arylations, which employed organometallic compounds for regioselective direct arylations (Scheme 1.38) [128].

A comprehensive review on direct arylations with ruthenium and rhodium complexes is provided in Chapter 9 by Lutz Ackermann and Rubén Vicente, where recently reported iridium-, copper- and iron-catalyzed processes are also discussed. The current mechanistic understanding of transition metal-catalyzed direct arylation reactions is summarized in Chapter 11 by Paula de Mendoza and Antonio M. Echavarren.

A notable alternative to metal-catalyzed direct arylation reactions is represented by methodologies that involve the formation of radicals as intermediates. The modern aspects of these transformations, along with further synthetically useful radical-based arylation methods, such as the Meerwein arylation of alkenes [129], are reviewed in Chapter 13 by Santiago E. Vaillard, Birte Schulte and Armido Studer. In some cases, radical intermediates can be generated through photochemically initiated electron-transfer reactions and, given their ecologically and economically attractive features, these photochemical arylations are highly promising for future developments [130]. Significant progress was accomplished in this important research area [131], which is reviewed by Valentina Dichiarante, Maurizio Fagnoni and Angelo Albini in Chapter 14.

Abbreviations

acac	acetylacetonate
Ac	acetyl
atm	atmosphere
Bn	benzyl
Bu	butyl
cat	catalytic
DIBAH	diisobutylaluminum hydride
DMA	*N,N*-dimethylacetamide
DMF	*N,N*-dimethylformamide
dppe	1,2-bis(diphenylphosphino)ethane
Hex	hexyl
HMPA	hexamethylphosphoramide
NICS	nucleus-independent chemical shift
NMR	nuclear magnetic resonance
Ph	phenyl
Pr	propyl
THF	tetrahydrofuran
Tol	tolyl

References

1 Loschmidt, J. (1861) *Chemische Studien, A. Constitutions-Formeln der organischen Chemie in geographischer Darstellung, B. Das Mariotte'sche Gesetz*, Wien.

2 Wiswesser, W.J. (1989) *Aldrichim. Acta*, 22, 17–19.

3 Anschütz, R. (1913) *Konstitutions-Formeln der organischen Chemie in graphischer*

Darstellung, J. Loschmidt, republished in Ostwald's *Klassiker der exakten Wissenschaft*, Verlag von Wilhelm Engelmann, Leipzig.

4 (a) Kekulé, A. (1865) *Bull. Soc. Chim. Fr.*, **3**, 98–110;
(b) Kekulé, A. (1866) *Liebigs Ann. Chem.*, **137**, 129–96.

5 For alternative benzene structures, such as bicyclo[2.2.0]hexa-2,5-diene, originally reported by Sir James Dewar in 1867, as well as for a detailed historical review on benzene structures, see: (a) Astruc, D. (2002) Arene chemistry: from historical notes to the state of the art, in *Modern Arene Chemistry* (ed. D. Astruc), Wiley-VCH Verlag GmbH, Weinheim, pp. 1–19;
(b) Rocke, A.J. (1985) *Ann. Sci.*, **42**, 355–81.

6 Selected reviews on aromaticity and antiaromaticity: (a) Badger, G.M. (1969) *Aromatic Character and Aromaticity*, University Press, Cambridge;
(b) Breslow, R. (1973) *Acc. Chem. Res.*, **6**, 393–8;
(c) Agranat, I. (1973) Theoretical aromatic chemistry, in *MTP International Review of Science, Organic Chemistry, Series One, Aromatic Compounds*, Vol. 3 (ed. H. Zollinger), Butterworths, London, pp. 139–77;
(d) Agranat, I. and Barak, A. (1976) The controversial notion of aromaticity, in *MTP International Review of Science, Organic Chemistry, Series One, Aromatic Compounds*, Vol. 3 (ed. H. Zollinger), Butterworths, London, pp. 1–177;
(e) Garratt, P.J. (1986) *Aromaticity*, John Wiley & Sons, Inc., New York;
(f) Minkin, V.I., Glukhovtsev, M.N. and Simkin, B.Y. (1994) *Aromaticity and Antiaromaticity: Electronic and Structural Aspects*, John Wiley & Sons, Inc., New York;
(g) Schleyer, P.v.R. and Jiao, H.J. (1996) *Pure Appl. Chem.*, **68**, 209–18;
(h) special issue on aromaticity: Schleyer, P.v.R. (ed.) (2001) *Chem. Rev.*, **101**, 1115 ff;
(i) Balaban, A.T., Oniciu, D.C. and Katritzky, A.R. (2004) *Chem. Rev.*, **104**, 2777–812;
(j) Special issue on delocalization – pi and sigma: Schleyer, P.v.R. (ed.) (2005) *Chem. Rev.*, **105**, 3433 ff.

7 Hückel, E. (1931) *Z. Phys.*, **70**, 204–86.

8 For a recent report on the remarkable synthesis of a Möbius [16] annulene, see: Ajami, D., Oeckler, O., Simon, A., Herges, R. (2003) *Nature*, **426**, 819–21.

9 Krygowski, T.M. and Cyranski, M.K. (2001) *Chem. Rev.*, **101**, 1385–419.

10 For Nucleus-Independent Chemical Shifts (NICS) as a simple and efficient aromaticity probe, see: (a) Schleyer, P.v.R., Maerker, C., Dransfeld, A., Jiao, H. and Hommes, N.J.v.E. (1996) *J. Am. Chem. Soc.*, **118**, 6317–18;
(b) Chen, Z., Wannere, C.S., Corminboeuf, C., Puchta, R. and Schleyer, P.v.R. (2005) *Chem. Rev.*, **105**, 3842–88.

11 Note that the concept of aromaticity continues to be a valuable stimulus for the development of chemical bond theory, and that some aspects of aromaticity are still controversially discussed: (a) Pogodin, S. and Agranat, I. (2007) *J. Org. Chem.*, **72**, 10096–107;
(b) Mills, N.S. and Llagostera, K.B. (2007) *J. Org. Chem.*, **72**, 9163–9;
(c) Wannere, C.S., Sattelmeyer, K.W., Schaefer, H.F., III and Schleyer, P.v.R. (2004) *Angew. Chem. Int. Ed.*, **43**, 4200–6;
(d) Seal, P. and Chakrabarti, S. (2007) *J. Phys. Chem. A*, **111**, 9988–94.

12 Faraday, M. (1825) *Philos. Trans. R. Soc. Lond.*, **115**, 440–66.

13 Kolbe, H. (1860) *Justus Liebigs Ann. Chem.*, **113**, 125–7.

14 Charles Frédéric Gerhardt prepared acetylsalicylic acid in 1853, though not in pure form: (a) Gerhardt, C.F. (1853) *Ann. Chem. Pharm.*, **87**, 149–79;
(b) Gerhardt, C.F. (1853) *Justus Liebigs Ann. Chem.*, **87**, 57–84.

15 For further illustrative examples, see for example: Nicolaou, K.C. and Mantagnon, T. (2008) *Molecules that Changed the World*, Wiley-VCH Verlag GmbH, Weinheim.

16 Astruc, D. (ed.) (2002) *Modern Arene Chemistry*, Wiley-VCH Verlag GmbH, Weinheim.

17 A valuable access to substituted arenes was also provided with the development of the chromium-templated Dötz-reaction: Dötz, K.H. and Stendal, J., Jr (2002) The chromium-templated carbene benzannulation approach to densely functionalized arenes (Dötz reaction), in, *Modern Arene Chemistry* (ed. D. Astruc), Wiley-VCH Verlag GmbH, pp. 250–96.

18 For regioselective functionalizations of substituted arenes through directed *ortho*-metallation reactions with stoichiometric amounts of organometallic bases, see: (a) Snieckus, V. and Macklin, T. (2005) Metallation of arenes. Directed ortho and remote metallation (DoM and DreM), in *Handbook of C–H Transformations* (ed. G. Dyker), Wiley-VCH Verlag GmbH, Weinheim, pp. 106–18;
(b) Hartung, C.G. and Snieckus, V. (2002) The directed ortho metallation reaction – a point of departure of new synthetic aromatic chemistry, in *Modern Arene Chemistry* (ed. D. Astruc), Wiley-VCH Verlag GmbH, pp. 330–67.

19 For reviews, see: (a) de Meijere, A. and Diederich, F. (eds) (2004) *Metal-Catalyzed Cross-Coupling Reactions*, 2nd edn, Wiley-VCH Verlag GmbH, Weinheim;
(b) Beller, M. and Bolm, C. (eds) (2004) *Transition Metals for Organic Synthesis*, 2nd edn, Wiley-VCH Verlag GmbH, Weinheim.

20 The potential of cross-coupling chemistry was among others illustrated by their applications to various syntheses of complex natural products: Nicolaou, K.C., Bulger, P.G. and Sarlah, D. (2005) *Angew. Chem. Int. Ed.*, **44**, 4442–89.

21 A review on syntheses of aromatic compounds based on ring-closing metathesis: Donohoe, T.J., Orr, A.J. and Bingham, M. (2006) *Angew. Chem. Int. Ed.*, **45**, 2664–70.

22 Tsuji, J. (2004) *Palladium Reagents and Catalysts*, 2nd edn, John Wiley & Sons, Ltd, Chichester.

23 Hassan, J., Sevignon, M., Gozzi, C., Schulz, E. and Lemaire, M. (2002) *Chem. Rev.*, **102**, 1359–469.

24 Wurtz, A. (1855) *Justus Liebigs Ann. Chem.*, **96**, 364–75.

25 Tollens, B. and Fittig, R. (1864) *Justus Liebigs Ann. Chem.*, **131**, 303–23.

26 Fittig, R. (1862) *Justus Liebigs Ann. Chem.*, **121**, 361–5.

27 Ullmann, F. and Bielecki, J. (1901) *Chem. Ber.*, **34**, 2174–85.

28 Ullmann, F. (1903) *Chem. Ber.*, **36**, 2382–4.

29 Ullmann, F. and Sponagel, P. (1905) *Chem. Ber.*, **38**, 2211–12.

30 Goldberg, I. (1906) *Chem. Ber.*, **39**, 1691–2.

31 For an early example of palladium-catalyzed reductive homocouplings of haloarenes in the presence of hydrazine, see: Busch, M. and Schmidt, W. (1929) *Chem. Ber.*, **62**, 2612–20.

32 Job, A. and Reich, R. (1924) *C. R. Hebd. Seances Acad. Sci.*, **179**, 330–2.

33 Corriu, R.J.P. (2002) *J. Organomet. Chem.*, **653**, 20–2.

34 (a) Reppe, W. (1949) *Neue Entwicklungen auf dem Gebiet der Chemie des Acetylen und Kohlenoxyds*, Springer, Berlin-Göttingen-Heidelberg;
(b) For a recent computational study, see Straub, B.F. and Gollub, C. (2004) *Chem. Eur. J.*, **10**, 3081–90.

35 (a) For representative recent examples, see: Nishida, G., Noguchi, K., Hirano, M. and Tanaka, K. (2008) *Angew. Chem. Int. Ed.*, **47**, 3410–13;
(b) Hilt, G. and Janikowski, J. (2008) *Angew. Chem. Int. Ed.*, **47**, 5243–5;
(c) Barluenga, J., Fernández-Rodríguez, M.A., García-García, P. and Aguilar, E. (2008), *J. Am. Chem. Soc.*, **130**, 2764–5;
(d) Wender, P.A. and Christy, J.P. (2007) *J. Am. Chem. Soc.*, **129**, 13402–3; and references cited therein.

36 For reviews, see: (a) Grotjahn, D.B. (1995) *Comprehensive Organometallic Chemistry II*, Vol. 12 (eds L.S. Hegedus, E.W. Abel, F.G.A. Stone and G. Wilkinson), Pergamon, Oxford, pp. 741–70;
(b) Bönnemann, H. and Brijoux, W. (2004) *Transition Metals for Organic Synthesis*, 2nd edn (eds M. Beller and C. Bolm), Wiley-VCH Verlag GmbH, Weinheim, pp. 171–97;
(c) Nakamura, I. and Yamamoto, Y. (2004) *Chem. Rev.*, **104**, 2127–98.

37 Kharasch, M.S. and Reinmuth, O. (1954) *Grignard Reagents of Nonmetallic Substances*, Prentice-Hall, Inc., New York.

38 Kharasch, M.S. and Fields, E.K. (1941) *J. Am. Chem. Soc.*, **63**, 2316–20.

39 Wittig, G. (1942) *Naturwissenschaften*, **30**, 696–703.

40 For the identification of 1,4-diradicals in the Bergman cyclization, see: Jones, R.R. and Bergman, R.G. (1972) *J. Am. Chem. Soc.*, **94**, 660–1.

41 Roberts, J.D., Simmons, J., Carlsmith, L.A. and Vaughan, C.W. (1953) *J. Am. Chem. Soc.*, **75**, 3290–1.

42 For selective cross-coupling reactions between haloarenes and stoichiometric amounts of allyl nickel reagents, see: (a) Corey, E.J. and Semmelhack, M.F. (1967) *J. Am. Chem. Soc.*, **89**, 2755–7;
(b) For cross-couplings with alkyl copper compounds, see: Corey, E.J. and Posner, G.H. (1967) *J. Am. Chem. Soc.*, **89**, 3911–12;
(c) Corey, E.J. and Posner, G.H. (1968) *J. Am. Chem. Soc.*, **90**, 5615.

43 Posner, G.H. (1975) *Org. React.*, **22**, 253–400.

44 van Helden, R. and Verberg, G. (1965) *Recl. Trav. Chim. Pays-Bas*, **84**, 1263–73.

45 Iataaki, H. and Yoshimoto, H. (1973) *J. Org. Chem.*, **38**, 76–9.

46 See also: (a) Yoshimoto, H. and Itatani, H. (1973) *J. Catal.*, **31**, 8–12;
(b) Clark, F.R.S., Norman, R.O.C., Thomas, C.B. and Willson, J.S. (1974) *J. Chem. Soc., Perkin Trans.*, **11**, 1289–94.

47 Beccalli, E.M., Broggini, G., Martinelli, M. and Sottocornola, S. (2007) *Chem. Rev.*, **107**, 5318–65.

48 For a recent example of selective intermolecular oxidative palladium-catalyzed cross-coupling between indoles and arenes, see: Stuart, D.R. and Fagnou, K. (2007) *Science*, **316**, 1172–5.

49 Heck, R.F. (1968) *J. Am. Chem. Soc.*, **90**, 5518–26.

50 Heck, R.F. (1969) *J. Am. Chem. Soc.*, **91**, 6707–14.

51 Heck, R.F. (2006) *Synlett*, 2855–60.

52 Moritani, I. and Fujiwara, Y. (1967) *Tetrahedron Lett.*, **8**, 1119–22.

53 Fujiwara, Y., Moritani, I. and Matsuda, M. (1968) *Tetrahedron*, **24**, 4819–24.

54 For selected examples of intramolecular variants, see: (a) Iida, H., Yuasa, Y. and Kibayashi, C. (1980) *J. Org. Chem.*, **45**, 2938–42;
(b) Knölker, H.-J. and O'Sullivan, N. (1994) *Tetrahedron*, **50**, 10893–908.;
(c) Hagelin, H., Oslob, J.D. and Akermark, B. (1999) *Chem. Eur. J.*, **5**, 2413–16; and cited references.

55 Fujiwara, Y., Moritani, I., Danno, S., Asano, R. and Teranishi, S. (1969) *J. Am. Chem. Soc.*, **91**, 7166–9.

56 Jia, C., Kitamura, T. and Fujiwara, Y. (2001) *Acc. Chem. Res.*, **34**, 633–9.

57 Fujiwara, Y. and Kitamura, T. (2005) Fujiwara reaction: palladium-catalyzed hydroarylations of alkynes and alkenes, in *Handbook of C-H Transformations* (ed. G. Dyker), Wiley-VCH Verlag GmbH, pp. 194–202.

58 Tamura, M. and Kochi, J. (1971) *J. Am. Chem. Soc.*, **93**, 1487–9.

59 Tamura, M. and Kochi, J. (1971) *Synthesis*, 303–5.

60 Fürstner, A. and Leitner, A. (2002) *Angew. Chem. Int. Ed.*, **41**, 609–12.

61 Sherry, B.D. and Fürstner, A. (2008) *Acc. Chem. Res.*, **41**, DOI: 10.1021/ar 800039X

62 (a) Correa, A., García Mancheño, O. and Bolm, C. (2008) *Chem. Soc. Rev.*, **37**, 1108–17;
(b) Bolm, C., Legros, J. and Zani, L. (2004) *Chem. Rev.*, **104**, 6217–54.

63 Note that iron-catalyzed C_{aryl}–C_{aryl} cross-coupling reactions are still challenging. For recent reports on such coupling reactions, see: (a) Sapountzis, I., Lin, W., Kofink, C.C., Despotopoulou, C. and Knochel, P. (2005) *Angew. Chem. Int. Ed.*, **44**, 1654–8;
(b) Hatakeyama, T. and Nakamura, M. (2007) *J. Am. Chem. Soc.*, **129**, 9844–5.

64 Mizoroki, T., Mori, K. and Ozaki, A. (1971) *Bull. Chem. Soc. Jpn*, **44**, 581.

65 Mori, K., Mizoroki, T. and Ozaki, A. (1973) *Bull. Chem. Soc. Jpn*, **46**, 1505–8.

66 Heck, R.F. and Nolley, J.P., Jr (1972) *J. Org. Chem.*, **37**, 2320–2.

67 Corriu, R.J.P. and Masse, J.P. (1972) *J. Chem. Soc., Chem. Commun.*, 144.

68 Uchino, M., Yamamoto, A. and Ikeda, S. (1970) *J. Organomet. Chem.*, **24**, C63–4.

69 Yamamoto, A. (2002) *J. Organomet. Chem.*, **653**, 5–10.

70 Tamao, K., Sumitani, K. and Kumada, M. (1972) *J. Am. Chem. Soc.*, **94**, 4374–6.

71 Tamao, K. (2002) *J. Organomet. Chem.*, **653**, 23–6.

72 Cassar, L. (1975) *J. Organomet. Chem.*, **93**, 253–7.

73 Yamamura, M., Moritani, I. and Murahashi, S.-I. (1975) *J. Organomet. Chem.*, **91**, C39–42.

74 Murahashi, S.-I. (2002) *J. Organomet. Chem.*, **653**, 27–33.

75 For cross-coupling reactions between iodoarenes and stoichiometric amounts of copper acetylides in the absence of palladium compounds, see: (a) Castro, C.E. and Stephens, R.D. (1963) *J. Org. Chem.*, **28**, 2163;
(b) Castro, C.E., Gaughan, E.J. and Owsley, D.C. (1966) *J. Org. Chem.*, **31**, 4071–8.

76 Dieck, H.A. and Heck, R.F. (1975) *J. Organomet. Chem.*, **93**, 259–63.

77 Sonogashira, K., Tohda, Y. and Hagihara, N. (1975) *Tetrahedron Lett.*, **50**, 4467–70.

78 Sonogashira, K. (2002) *J. Organomet. Chem.*, **653**, 46–9.

79 In 1979, palladium-catalyzed cross-couplings with organolithium compounds were reported: Murahashi, S.-I., Yamamura, M., Yanagisawa, K.-I., Mita, N. and Kondo, K. (1979) *J. Org. Chem.*, **44**, 2408–17.

80 For further early examples of palladium-catalyzed cross-coupling reactions of Grignard reagents, see: (a) Fauvarque, J.F. and Jutand, A. (1976) *Bull. Soc. Chem. Fr.*, 765–70;
(b) Sekiya, A. and Ishikawa, N. (1976) *J. Organomet. Chem.*, **118**, 349–54.

81 Negishi, E.-i. and Baba, S. (1976) *J. Chem. Soc. Chem. Commun.*, 596–7.

82 Baba, S. and Negishi, E.-i. (1976) *J. Am. Chem. Soc.*, **98**, 6729–31.

83 Negishi, E.-i. (2002) *J. Organomet. Chem.*, **653**, 34–40.

84 Negishi, E.-i. (2007) *Bull. Chem. Soc. Jpn*, **80**, 233–57.

85 King, A.O., Okukado, N., Negishi, E.-i. (1977) *J. Chem. Soc. Chem. Commun.*, **19**, 683–4.

86 Negishi, E.-i. and King, A.O. (1977) *J. Org. Chem.*, **42**, 1821–3.

87 See also: Fauvarque, J.F. and Jutand, A. (1977) *J. Organomet. Chem.*, **132**, C17–19.

88 For cross-coupling reactions with zirconium-based nucleophiles, see: Negishi, E.-i. and Van Horn, D.E. (1977) *J. Am. Chem. Soc.*, **99**, 3168–70.

89 Kosugi, M., Sasazawa, K., Shimizu, Y. and Migita, T. (1977) *Chem. Lett.*, 301–2.

90 Kosugi, M., Shimizu, Y. and Migita, T. (1977) *Chem. Lett.*, 1423–4.

91 Kosugi, M. and Fugami, K. (2002) *J. Organomet. Chem.*, **653**, 50–3.

92 Milstein, D. and Stille, J.K. (1978) *J. Am. Chem. Soc.*, **100**, 3636–8.

93 Negishi, E.-i. (1982) *Acc. Chem. Res.*, **15**, 340–8.

94 Kondo, K. and Murahashi, S.-I. (1979) *Tetrahedron Lett.*, **20**, 1237–40.

95 Miyaura, N., Yamada, K. and Suzuki, A. (1979) *Tetrahedron Lett.*, **20**, 3437–40.

96 Miyaura, N., Yanagi, T. and Suzuki, A. (1981) *Synth. Commun.*, **11**, 513–19.

97 Miyaura, N. (2002) *J. Organomet. Chem.*, **653**, 54–7.

98 Yoshida, J.-i., Tamao, K., Yamamoto, H., Kakui, T., Uchida, T. and Kumada, M. (1982) *Organometallics*, **1**, 542–9.

99 Hallberg, A. and Westerlund, C. (1982) *Chem. Lett.*, 1993–4.

100 Kikukawa, K., Ikenaga, K., Wada, F. and Matsuda, T. (1983) *Chem. Lett.*, 1337–40.

101 Hatanaka, Y., Fukushima, S. and Hiyama, T. (1989) *Chem. Lett.*, 1711–14.

102 Hiyama, T. (2002) *J. Organomet. Chem.*, **653**, 58–61.

103 Kosugi, M., Kameyama, M. and Migita, T. (1983) *Chem. Lett.*, 927–8.

104 For palladium-catalyzed amidations of haloarenes using CO and anilines, see: Schoenberg, A. and Heck, R.F. (1974) *J. Org. Chem.*, **39**, 3327–31.

105 Guram, A.S., Rennels, R.A. and Buchwald, S.L. (1995) *Angew. Chem. Int. Ed. Engl.*, **34**, 1348–50.

106 Louie, J. and Hartwig, J.F. (1995) *Tetrahedron Lett.*, **36**, 3609–12.

107 For selected recent examples of nickel-catalyzed aminations of haloarenes, see:

(a) Gao, C.-Y. and Yang, L.-M. (2008) *J. Org. Chem.*, **73**, 1624–7;
(b) Matsubara, K., Ueno, K., Koga, Y. and Hara, K. (2007) *J. Org. Chem.*, **72**, 5069–76;
(c) Wolfe, J.P. and Buchwald, S.L. (1997) *J. Am. Chem. Soc.*, **119**, 6054–8;
(d) Lipshutz, B.H. and Ueada, H. (2000) *Angew. Chem. Int. Ed.*, **39**, 4492–4;
(e) Brenner, E., Schneider, R. and Fort, Y. (1999) *Tetrahedron*, **55**, 12829–42;
(f) Desmarets, C., Schneider, R. and Fort, Y. (2002) *J. Org. Chem.*, **67**, 3029–36; and references cited therein.

108 Chan, D.M.T., Monaco, K.L., Wang, R.-P. and Winters, M.P. (1998) *Tetrahedron Lett.*, **39**, 2933–6.

109 Lam, P.Y.S., Clark, C.G., Saubern, S., Adams, J., Winters, M.P., Chan, D.M.T. and Combs, A. (1998) *Tetrahedron Lett.*, **39**, 2941–4.

110 Evans, D.A., Katz, J.L. and West, T.R. (1998) *Tetrahedron Lett.*, **39**, 2937–40.

111 For recent reviews, see: (a) Li, B.-J., Yang, S.-D. and Shi, Z.-J. (2008) *Synlett*, 949–57;
(b) Ackermann, L. (2007) *Top. Organomet. Chem.*, **24**, 35–60;
(c) Satoh, T. and Miura, M. (2007) *Top. Organomet. Chem.*, **24**, 61–84;
(d) Alberico, D., Scott, M.E. and Lautens, M. (2007) *Chem. Rev.*, **107**, 174–238;
(e) Bergman, R.G. (2007) *Nature*, **446**, 391–3;
(f) Campeau, L.-C., Stuart, D.R. and Fagnou, K. (2007) *Aldrichim. Acta*, **40**, 35–41;
(g) Seregin, I.V. and Gevorgyan, V. (2007) *Chem. Soc. Rev.*, **36**, 1173–93;
(h) Daugulis, O., Zaitsev, V.G., Shabashov, D., Pham, Q.N. and Lazareva, A. (2006) *Synlett*, 3382–8;
(i) Yu, J.-Q., Giri, R. and Chen, X. (2006) *Org. Biomol. Chem.*, **4**, 4041–7.

112 Ames, D.E. and Bull, D. (1982) *Tetrahedron*, **38**, 383–7.

113 Ames, D.E. and Opalko, A. (1983) *Synthesis*, 234–5.

114 Ames, D.E. and Opalko, A. (1984) *Tetrahedron*, **40**, 1919–25.

115 Interestingly, an intramolecular palladium-catalyzed *oxidative* arylation for a dibenzofuran synthesis was reported earlier: Shiotani, A. and Itatani, H. (1974), *Angew. Chem. Int. Ed.*, **13**, 471–2.

116 Akita, Y., Inoue, A., Yamamoto, K., Ohta, A., Kurihara, T. and Shimizu, M. (1985) *Heterocycles*, **23**, 2327–33.

117 Akita, Y., Itagaki, Y., Takizawa, S. and Ohta, A. (1989) *Chem. Pharm. Bull.*, **37**, 1477–80.

118 For recent reviews on catalytic hydroarylation reactions, see: (a) Bandini, M., Emer, E., Tommasi, S. and Umani-Ronchi, A. (2006) *Eur. J. Org. Chem.*, 3527–44;
(b) Nevado, C. and Echavarren, A.M. (2005) *Synthesis*, 167–82;
(c) Liu, C., Bender, C.F., Han, X. and Widenhoefer, R.A. (2007) *Chem. Commun.*, 3607–18;
(d) Kakiuchi, F. and Chatani, N. (2003) *Adv. Synth. Catal.*, **345**, 1077–101, and references cited therein.

119 Lewis, L.N. and Smith, J.F. (1986) *J. Am. Chem. Soc.*, **108**, 2728–35.

120 Murai, S., Kakiuchi, F., Sekine, S., Tanaka, Y., Kamatani, A., Sonoda, M. and Chatani, N. (1993) *Nature*, **366**, 529–31.

121 Kakiuchi, F. and Murai, S. (2002) *Acc. Chem. Res.*, **35**, 826–34.

122 Kakiuchi, F. and Chatani, N. (2004) *Ruthenium Catalysts and Fine Chemistry* (ed. C. Bruneau and P.H. Dixneuf), Springer, Berlin, Heidelberg, pp. 45–79.

123 Kakiuchi, F. and Chatani, N. (2004) *Ruthenium in Organic Synthesis* (ed. S.-I. Murahashi), Wiley-VCH Verlag GmbH, Weinheim, pp. 219–55.

124 Oi, S., Fukita, S., Hirata, N., Watanuki, N., Miyano, S. and Inoue, Y. (2001) *Org. Lett.*, **3**, 2579–81.

125 Ackermann, L. (2005) *Org. Lett.*, **7**, 3123–5.

126 Oi, S., Sakai, K. and Inoue, Y. (2005) *Org. Lett.*, **7**, 4009–11.

127 Ackermann, L., Althammer, A. and Born, R. (2006) *Angew. Chem. Int. Ed.*, **45**, 2619–22.

128 Oi, S., Fukita, S. and Inoue, Y. (1998) *Chem. Commun.*, 2439–40.

129 Meerwein, H., Büchner, E. and van Emster, K. (1939) *J. Prakt. Chem.*, **152**, 237–66.

130 Dichiarante, V., Fagnoni, M. and Albini, A. (2007) *Angew. Chem. Int. Ed.*, **46**, 6495–8.

131 Dichiarante, V. and Fagnoni, M. (2008) *Synlett*, 787–800.

2
Metal-Catalyzed Coupling Reactions with Aryl Chlorides, Tosylates and Fluorides

Adam Littke

2.1
Introduction

Palladium- and nickel-catalyzed coupling reactions are recognized to be among the most powerful carbon–carbon bond-forming reactions available to the synthetic organic chemist (Equation 2.1) [1]:

$$RX + MR^1 \xrightarrow[\text{(additive)}]{\text{Pd(0) or Ni(0) catalyst}} R\text{-}R^1 \quad (2.1)$$

R, R^1 = aryl, alkenyl; X = Br, I, OTf
M = B (Suzuki), Sn (Stille), Si (Hiyama), Zn (Negishi), Mg (Kumada), etc.

where R, R^1 = aryl, alkenyl, X = Br, I, OTf, and M = B (Suzuki), Sn (Stille), Si (Hiyama), Zn (Negishi) and Mg (Kumada).

Depending on the nature of the coupling partners, the conditions can be relatively mild, with particularly reactive substrates (e.g. alkenyl iodides) often undergoing reactions at ambient temperature. With the notable exception of the Stille reaction, which utilizes organotin compounds, the byproducts of many palladium- and nickel-catalyzed coupling reactions are relatively benign. In addition, many of these reactions exhibit high functional group tolerance. All of these factors have led to coupling reactions becoming powerful tools in the areas of total synthesis, materials and polymer chemistry, combinatorial and solid-phase chemistry, medicinal chemistry, as well as more recently, process chemistry.

Until the late 1990s, one major limitation of all palladium-catalyzed coupling reactions was represented by the poor reactivity of aryl chlorides and aryl tosylates

Modern Arylation Methods. Edited by Lutz Ackermann

ISBN: 978-3-527-31937-4

compared to the more traditionally employed and more reactive aryl bromides, iodides and triflates. At the time, it was well understood that if the reactivity issues could be overcome, then aryl chlorides would become the substrates of choice for large-scale pharmaceutical and industrial coupling processes, due to their lower cost and ready availability as compared to their more reactive counterparts [2]. Although further success has been observed with nickel-based catalysts, there are certain drawbacks associated with the use of nickel versus palladium complexes, such as lower functional group tolerance [3], lower selectivity [4] and greater toxicity [5].

Aryl tosylates, although less readily available than aryl chlorides, would be a welcome substrate class addition due to their more easy handling (particularly compared with aryl triflates) and access from phenols using relatively inexpensive reagents. Finally, from a historic standpoint, aryl fluorides have been seen as a challenging and underutilized class of compounds for palladium- and nickel-catalyzed coupling reactions. Spurned by the success of activating less-reactive aryl chlorides and tosylates, many new methods for activating aryl fluorides have also been developed.

2.2 Coupling Reactions of Aryl Chlorides

The low reactivity of aryl chlorides is usually attributed to the strength of the C—Cl bond (bond dissociation energies for Ph–X: Cl, 96 kcal mol^{-1}; Br, 81 kcal mol^{-1}; I, 65 kcal mol^{-1}), which leads to a reluctance by aryl chlorides to oxidatively add to either Pd(0) or Ni(0), which is a critical initial step in palladium- and nickel-catalyzed coupling reactions (Scheme 2.1) [6, 7].

In 2002, Littke and Fu produced a fairly comprehensive review of palladium-catalyzed coupling reactions of aryl chlorides [8]. Although in this chapter we will attempt to encompass many of the key and seminal developments that occurred

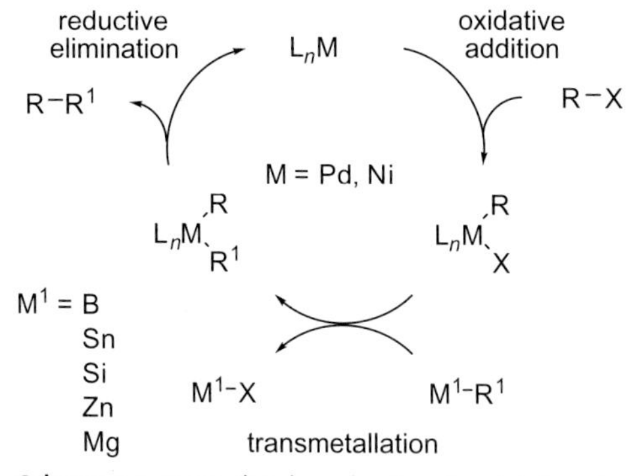

Scheme 2.1 Generalized mechanism for nickel- and palladium-catalyzed cross-coupling reactions.

prior to 2002, the focus will be on developments since 2002. Due to the amount of material available, however, the chapter will be divided into sections relating to nickel- and palladium-based reactions of aryl chlorides, with the palladium section being further subdivided by the nucleophilic substrate.

2.2.1 Nickel-Catalyzed Cross-Couplings of Aryl Chlorides

As early as 1972, Kumada and coworkers reported that aryl chlorides could undergo cross-couplings with Grignard reagents catalyzed by a nickel-1,2-bis(diphenylphosphino)ethane (dppe) complex (Equation 2.2) [9, 10].

Cl, Cl + BrMg–n-Bu → [$NiCl_2$(dppe)] (0.7 mol%), Et_2O, reflux, 94% → n-Bu, n-Bu (2.2)

More than 20 years later, Miyaura reported the first nickel-catalyzed Suzuki coupling of aryl chlorides. Initially, *n*-BuLi was used to reduce nickel(II) to nickel(0) (Equation 2.3) [11].

OMe, MeO, Cl, H_2N + $(HO)_2B$, Me → [$NiCl_2$(dppf)] (3 mol%), dppf (3 mol%), n-BuLi, K_3PO_4, dioxane, 80 °C, 92% → OMe, MeO, H_2N, Me (2.3)

Subsequent studies by Miyaura [12] and others [13] demonstrated that the presence of a strong reducing agent was not required (Equation 2.4).

N, Cl + $(HO)_2B$, Me → [$NiCl_2(PPh_3)_2$] (3 mol%), PPh_3 (6 mol%), K_3PO_4, toluene, 100 °C, 88% → N, Me (2.4)

Even water can be used as a cosolvent, as was shown by Genet using the water-soluble phosphine ligand sodium triphenylphosphinotrimetasulfonate (TPPTS) (Equation 2.5) [14].

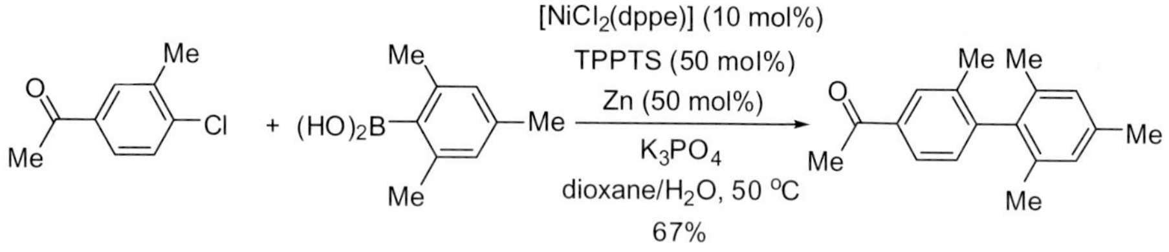
Me
O
Me
Cl
+ (HO)2B
Me
Me
Me
[NiCl2(dppe)] (10 mol%)
TPPTS (50 mol%)
Zn (50 mol%)
K3PO4
dioxane/H2O, 50 °C
67%
O
Me
Me Me
Me
Me

(2.5)

Heterogeneous-based nickel on carbon may also be used for Suzuki couplings of aryl chlorides (Equation 2.6) [15].

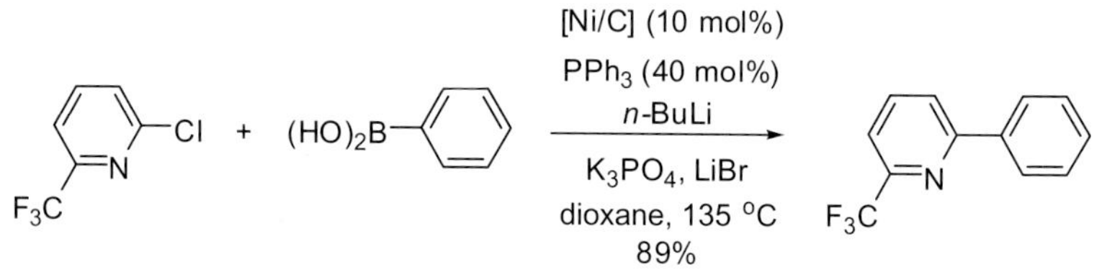
Cl
N
F3C
+
(HO)2B
[Ni/C] (10 mol%)
PPh3 (40 mol%)
n-BuLi
K3PO4, LiBr
dioxane, 135 °C
89%
N
F3C

(2.6)

Hu has shown that Suzuki couplings can take place at ambient temperature. Here, air-sensitive $Ni(COD)_2$ can be used with either PPh_3 or PCy_3 (Equation 2.7). Alternatively, air-stable $[Ni(PPh_3)_2Cl_2]$ may be employed as catalyst, although the use of *n*-BuLi as reducing agent is required [16].

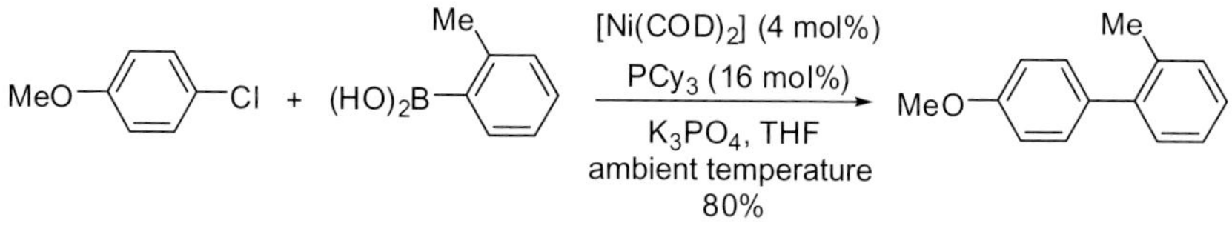
MeO
Cl
+
(HO)2B
Me
[Ni(COD)2] (4 mol%)
PCy3 (16 mol%)
K3PO4, THF
ambient temperature
80%
MeO
Me

(2.7)

A PPh_3-based nickel catalyst can be used for Stille couplings of aryl chlorides with alkenyl-, alkynyl- and allyl-based tin reagents. The nickel(0) catalyst is generated from $Ni(acac)_2$ and diisobutylaluminum hydride (DIBAL-H) (Equation 2.8) [17].

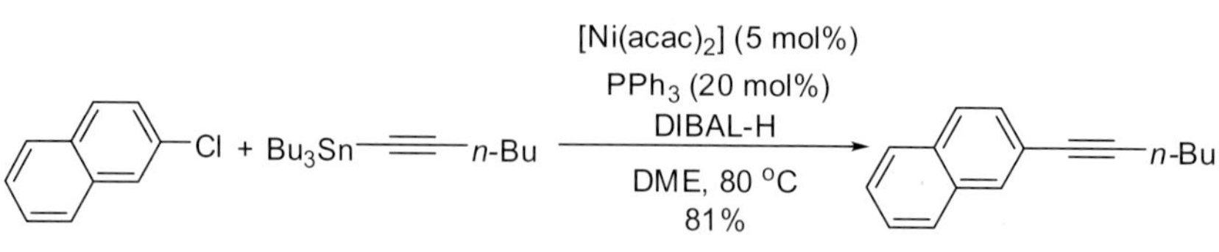
Cl + Bu3Sn
n-Bu
[Ni(acac)2] (5 mol%)
PPh3 (20 mol%)
DIBAL-H
DME, 80 °C
81%
n-Bu

(2.8)

Lipshutz has demonstrated heterogeneous nickel on charcoal as catalyst for Negishi couplings of aryl chlorides, utilizing highly functionalized zinc reagents (Equation 2.9) [18, 19].

$$\text{MeO–}C_6H_3(CO_2Et)\text{–Cl} + \text{IZn–}(CH_2)_4CN \xrightarrow[\text{THF, 67 °C; 89\%}]{\text{[Ni/C] (5 mol\%), } PPh_3 \text{ (20 mol\%), } n\text{-BuLi}} \text{MeO–}C_6H_3(CO_2Et)\text{–}(CH_2)_4CN \quad (2.9)$$

More recently, Schneider and coworkers have shown that a nickel/*N*-heterocylic carbene-based catalyst can be used for the coupling of aryl chlorides with organomanganese reagents (Equation 2.10) although, unfortunately, electron-rich aryl chlorides give low yields under these reaction conditions [20].

$$\text{1-chloronaphthalene} + \text{ClMn–}C_6H_4\text{–OMe} \xrightarrow[\text{THF, ambient temperature; 90\%}]{[Ni(acac)_2] \text{ (5 mol\%), IPrHCl (10 mol\%)}} \text{1-(4-MeOC}_6H_4\text{)naphthalene} \quad (2.10)$$

IPrHCl: Ar–N⁺(imidazolium)N–Ar Cl⁻, Ar = 2,6-(i-Pr)$_2C_6H_3$

2.2.2 Palladium-Catalyzed Cross-Coupling Reactions

2.2.2.1 Suzuki Reaction

One of the earliest examples of a Suzuki reaction of a heteroaryl chloride was provided by Gronowitz, who examined the coupling of 2,4-dichloropyrimidine with 2-thienylboronic acid and established that the 4-chloro group is more reactive than the 2-chloro group (Equation 2.11) [21].

$$\text{2,4-dichloropyrimidine} + (HO)_2B\text{–(2-thienyl)} \xrightarrow[\text{DME, reflux; 59\%}]{[Pd(PPh_3)_4] \text{ (3 mol\%), aq. } Na_2CO_3} \text{2-chloro-4-(2-thienyl)pyrimidine} \quad (2.11)$$

Caron has recently reported the use of a Suzuki reaction of a heteroaryl chloride to produce 2-phenyl-3-aminopyridine, a key intermediate in the synthesis of 2-phenyl-3-aminopiperidine, an important pharmacophore present in potent nonpeptidic NK1 receptor antagonists. Although the direct coupling of 2-chloro-3-aminopyridine with phenylboronic acid was unsuccessful, a one-pot protection–Suzuki–deprotection sequence furnished the target compound in excellent yield on a >100 g scale (Equation 2.12) [22].

PhCHO, $PhB(OH)_2$
$[Pd(PPh_3)_2Cl_2]$ (0.4 mol%)
aq. Na_2CO_3
toluene, 85 °C
99% overall
$H_3O^{\oplus}$

(2.12)

Uemura has shown that aryl chlorides which are η^6-bound to $Cr(CO)_3$ are remarkably reactive coupling partners in Suzuki reactions [23, 24]. Even in the presence of the electron-donating, deactivating *ortho*-methoxy substituent, the aryl chloride couples with an arylboronic acid (Equation 2.13). Furthermore, no homocoupled 4′-bromobiphenylboronic acid is observed, establishing that highly selective activation of a C—Cl bond is occurring in the presence of a typically more reactive C—Br bond.

$[Pd(PPh_3)_4]$ (10 mol%)
Na_2CO_3
$MeOH/H_2O$
75 °C
60%

no $(HO)_2B$–C_6H_4–C_6H_4–Br

homocoupled 4'-bromobiphenylboronic acid

(2.13)

An important observation regarding Suzuki couplings of aryl chlorides was reported in 1997 by Shen, who established that palladium complexes that include a bulky, electron-rich trialkylphosphine, such as tricyclohexylphosphine (PCy_3), catalyze Suzuki couplings of activated aryl chlorides at 100 °C (Equation 2.14) [25].

$$3\text{-}O_2N\text{-}C_6H_4\text{-}Cl + (HO)_2B\text{-}C_6H_5 \xrightarrow[\text{NMP, 100 °C, 97\%}]{[Pd(PCy_3)_2Cl_2]\ (5\ mol\%),\ CsF} 3\text{-}O_2N\text{-}C_6H_4\text{-}C_6H_5 \qquad (2.14)$$

Bidentate 1,3-bis(diphenylphosphino)propane (dppp) is also useful in certain cases, whereas 1,4-bis(diphenylphosphino)butane (dppb), PPh_3, P(2-furyl)$_3$ and $AsPh_3$ are generally ineffective. If, however, the aryl chloride is highly activated – that is, it bears two electron-withdrawing groups – $[Pd(PPh_3)_2Cl_2]$ can be employed [26]. Shen speculated that the electron-rich nature of PCy_3 relative to PPh_3 might facilitate oxidative addition of the C(sp^2)—Cl bond to palladium(0), and that the steric demand of PCy_3 might favor ligand dissociation to afford an active monophosphine-palladium catalyst.

Subsequently, Littke and Fu demonstrated a versatile method for palladium-catalyzed Suzuki cross-couplings of aryl chlorides, using the sterically demanding and electron-rich trialkylphosphine, tri-*tert*-butylphosphine, P(*t*-Bu)$_3$ (Equation 2.15) [27], with a P(*t*-Bu)$_3$ to palladium ratio of between 1.0 and 1.5 being most effective. Electronically deactivated and hindered aryl chlorides proved to be suitable substrates for this catalyst system. In this initial study, $[Pd_2(dba)_3]$ (dba = dibenzylideneacetone), Cs_2CO_3, and 1,4-dioxane were employed as the palladium source, activator and solvent, respectively. Although a variety of other commercially available phosphines were probed, including PPh_3 and BINAP, the only phosphine to show any useful reactivity was the similarly sterically demanding and electron-rich PCy_3.

$$X\text{-}C_6H_4\text{-}Cl + (HO)_2B\text{-}C_6H_4\text{-}Y \xrightarrow[\text{Cs}_2\text{CO}_3,\ \text{dioxane},\ 80\text{-}90\ °C,\ 82\text{-}92\%\ \text{yield}]{[Pd_2(dba)_3]\ (1.5\ mol\%),\ P(t\text{-}Bu)_3\ (3.6\ mol\%)} X\text{-}C_6H_4\text{-}C_6H_4\text{-}Y \qquad (2.15)$$

X = 4-C(O)Me, Me, OMe, NH_2, 2-Me

Y = 4-CF_3, H, OMe, 2-Me

Fu subsequently determined that KF is a more effective additive than Cs_2CO_3, allowing Suzuki cross-couplings of activated aryl chlorides, including heteroaryl chlorides, to proceed at ambient temperature (P(*t*-Bu)$_3$: Pd = 1 : 1) [28]. As this P(*t*-Bu)$_3$-based palladium catalyst system exhibited a highly unusual reactivity profile, an unprecedented selectivity for the coupling of an aryl chloride in preference to that of an aryl triflate was observed (Equation 2.16).

TfO–C$_6$H$_4$–Cl + (HO)$_2$B–(2-MeC$_6$H$_4$) → [Pd$_2$(dba)$_3$] (1.5 mol%), P(*t*-Bu)$_3$ (3 mol%), KF, THF, ambient temperature, 95% → TfO–C$_6$H$_4$–(2-MeC$_6$H$_4$)

(2.16)

Alkylboronic acids, which are often less reactive in Suzuki reactions than arylboronic acids, are suitable substrates for this catalyst, with turnover numbers (TONs) as high as 9700 being achieved with this system. In addition, Fu was able to demonstrate that the air-stable and crystalline tetrafluoroborate salt, P(*t*-Bu)$_3$HBF$_4$ was equally efficient as the low-melting and air-sensitive free phosphine P(*t*-Bu)$_3$ [29].

The observation that using a 1:1 ratio of P(*t*-Bu)$_3$ to palladium is important to obtain high reactivity in Suzuki reactions of aryl chlorides at ambient temperature, in addition to mechanistic studies involving ^{31}P NMR experiments, strongly suggests that a monophosphine-palladium adduct may play a key role in Pd/P(*t*-Bu)$_3$-catalyzed couplings, as Shen had speculated for the similarly bulky and electron-rich PCy$_3$ [30]. Thus, P(*t*-Bu)$_3$ may be a particularly effective ligand for the couplings of aryl chlorides, because the steric demand of P(*t*-Bu)$_3$ facilitates dissociation to a monophosphine adduct, to which the aryl chloride rapidly oxidatively adds, due to the electron-rich nature of P(*t*-Bu)$_3$. Recent density functional theory (DFT) studies conducted by Marder and coworkers support a monophosphine pathway for oxidative addition of aryl chlorides to palladium(0) versus a bisphosphine pathway, which is usually invoked for PPh$_3$ [31].

Ishiyama and Miyaura have disclosed that [Pd(dba)$_2$] and PCy$_3$ yields an effective catalyst for the borylation of aryl chlorides using bis(pinacolato)diboron (Equation 2.17) [32].

X–C$_6$H$_4$–Cl + bis(pinacolato)diboron → [Pd(dba)$_2$] (3-6 mol%), PCy$_3$ (7-14 mol%), KOAc, dioxane, 80 °C, 70-94% → X–C$_6$H$_4$–B(pin)

X = 4-CHO, CO$_2$Me, NMe$_2$
2-NO$_2$, CN, Me, OMe

(2.17)

In addition, a broad spectrum of functional groups are tolerated. Highly electron-rich chlorides, such as 4-chloro-*N*,*N*-dimethylaniline, are suitable substrates, as are heteroaryl chlorides, such as 3-chloropyridine and 2-chlorobenzo[*b*]furan. PCy$_3$ is more efficient than dialkylarylphosphines, triarylphosphines and, interestingly, P(*t*-Bu)$_3$.

PCy_3 also proved to be the ligand of choice for a wide-ranging study of Suzuki couplings of heteroaryl chlorides and heteroarylboronic acids (Equation 2.18) where aryl boronate esters and aryl trifluoroborates could also be employed [33].

[$Pd_2(dba)_3$] (1 mol%), PCy_3 (2.4 mol%), K_3PO_4, dioxane/H_2O, 100 °C, 97% (2.18)

Beller has demonstrated that a new bulky, electron-rich trialkylphosphine, di(1-adamantyl)-*n*-butylphosphine, *n*-$BuPAd_2$, can afford excellent TONs in palladium-catalyzed Suzuki reactions of aryl chlorides [34]. For the coupling of 4-chlorotoluene, this phosphine achieves a TON of 17400 (0.005 mol% Pd, 87% yield) (Equation 2.19), compared with a TON of 9200 with commercially available P(*t*-Bu)$_3$ (0.01 mol% Pd, 92% yield). High TONs (>10000) can also be obtained with challenging aryl chlorides, such as 2-chloro-*m*-xylene and 4-chloroanisole.

Me–C$_6$H$_4$–Cl + $(HO)_2B$–Ph → [Pd(OAc)$_2$] (0.005 mol%), *n*-$BuPAd_2$ (0.01 mol%), K_3PO_4, toluene, 100 °C, 87% → Me–C$_6$H$_4$–Ph (2.19)

9-Fluorenyldialkylphosphines represent another interesting class of trialkylphosphines that can be used for Suzuki couplings of aryl chlorides. Sulfonylation of the aromatic moiety of the fluorenyl group results in a water-soluble ligand that can be used for couplings in aqueous solvent (Equation 2.20) [35–37].

H_2N–SO_2–C$_6$H$_4$–Cl + $(HO)_2B$–(1-naphthyl) → [Na_2PdCl_4] (0.5 mol%), **1** (1 mol%), K_2CO_3, H_2O, 90 °C, 99% → H_2N–SO_2–C$_6$H$_4$–(1-naphthyl)

1: HO_3S-fluorenyl with Et and $P(Cy)_2H^{\oplus}$ $HSO_4^{\ominus}$ (2.20)

Concurrent with Fu's studies using P(*t*-Bu)$_3$, Buchwald reported that aminophosphine **2** is a very effective ligand for palladium-catalyzed Suzuki reactions of aryl chlorides [38]. Remarkable at the time, this catalyst system couples a broad spectrum of aryl chlorides, including electron-neutral and electron-rich substrates, at ambient temperature. CsF was the base of choice, although at a temperature of 100 °C the less-expensive K_3PO_4 could be employed. Buchwald subsequently determined that biphenyl ligands **3** and **4** could be even more effective than **2** in palladium-catalyzed Suzuki reactions of aryl chlorides, thereby establishing that the amino group of **2** is not essential for high activity [39]. With ligand **4**, Suzuki couplings of a wide array of partners can be achieved with 0.5–1.5% Pd and KF as the activator at ambient temperature (Equation 2.21).

Cl + $(HO)_2B$ — [Pd(OAc)$_2$] (0.5-1.5 mol%), **4** (1-3 mol%), KF, THF, ambient temperature, 88-98%

X = 4-NO_2, CN, CO_2Me, Me, OMe, 2-C(O)Me, CH_2CN, OMe, 3,5-$(OMe)_2$

Y = 3-C(O)Me, H, 2-OMe

R^1, PR$_2$

R = Cy, R^1 = NMe$_2$ **2**
R = Cy, R^1 = H **3**
R = *t*-Bu, R^1 = H **4**

(2.21)

A further improvement was the development of ligands **5** and **6**, the use of which allowed for the synthesis of tetra-*ortho*-substituted biaryls, as well as the use of heterocyclic aryl chlorides and heteroarylboronic acids and esters (Equation 2.22) [40]

O, Me, N, Cl + $(HO)_2B$, NH — [Pd$_2$(dba)$_3$] (1 mol%), **6** (1 mol%), K_3PO_4, *n*-BuOH, 120 °C, 91%

MeO, PCy$_2$, OMe — **5**

i-Pr, PCy$_2$, *i*-Pr, *i*-Pr — **6**

(2.22)

Ligands **5** and **6** are particularly efficacious for the borylation of aryl chlorides to pinacol boronate esters, thus allowing the direct 'one-pot' synthesis of unsymmetrical biaryls from two aryl chlorides (Equation 2.23) [41]. Due to the wide breadth and scope that this family of biaryldialkylphosphine ligands offers for palladium-catalyzed couplings of aryl chlorides, many are commercially available and all are highly crystalline, air-stable solids, and hence easy to use.

1. $[Pd_2(dba)_3]$ (1 mol%), **5** (8 mol%), KOAc, dioxane, 110 °C; 2. 2-chloropyridine, K_3PO_4, dioxane/H_2O, 110 °C; 92%

(2.23)

Subsequent to Buchwald's initial communication in 1998, a number of other groups have developed related variants of bulky, electron-rich aryldialkylphosphine ligands for Suzuki (and related) cross-couplings of aryl chlorides [42]. Beller and coworkers have developed readily synthesized and tunable *N*-aryl-2-(dialkylphosp hino)imidazole and -benzimidazole as ligands for palladium-catalyzed Suzuki couplings of aryl chlorides (Equation 2.24), and a number of these are now available commercially [43].

$[Pd(OAc)_2]$ (0.05 mol%), **7** (0.5 mol%), K_3PO_4, toluene, 100 °C; 94%

$P(t\text{-}Bu)_2$

7

(2.24)

Triazole-based monophosphine ligands, ClickPhos, such as **8**, are particularly efficient for Suzuki couplings of highly hindered substrates (Equation 2.25) [44].

$$\text{2-Cl-1,3-}Me_2C_6H_3 + (HO)_2B\text{-}C_6H_4\text{-2-OMe} \xrightarrow[\substack{K_3PO_4 \\ \text{toluene, } 120\,^{\circ}C \\ 72\%}]{\substack{[Pd(dba)_2]\ (1\ mol\%) \\ \mathbf{8}\ (2\ mol\%)}} \text{2,6-}Me_2C_6H_3\text{-}C_6H_4\text{-2-OMe} \tag{2.25}$$

8

Guram and coworkers at Amgen have developed air-stable palladium–phosphine complexes for the Suzuki coupling of heteroaryl chlorides with a diverse array of aryl and heteroaryl boronic acids (Equation 2.26), with TONs of up to 10 000 possible in certain cases [45].

$$\text{MeO-pyridazinyl-Cl} + (HO)_2B\text{-(2-fluoropyridin-3-yl)} \xrightarrow[\substack{KOAc \\ CH_3CN/H_2O \\ 99\%}]{\mathbf{9}\ (1\ mol\%)} \text{MeO-pyridazinyl-(2-fluoropyridin-3-yl)} \tag{2.26}$$

$$\mathbf{9}: [PdCl_2\{P(t\text{-}Bu)_2(C_6H_4\text{-}4\text{-}NMe_2)\}_2]$$

Although traditional triarylphosphine-based palladium catalysts, such as $[Pd(PPh_3)_4]$, are typically not sufficiently reactive for aryl chlorides, several groups have identified novel triarylphosphine ligands that do allow couplings of aryl chlorides. In 2001, both Richards [46] and Fu [47] demonstrated, independently, that ferrocenyl-based phosphines could be used for Suzuki cross-couplings of aryl chlorides, including unactivated and deactivated aryl chlorides (Equation 2.27).

$$\text{2-MeC}_6\text{H}_4\text{Cl} + (HO)_2B\text{-C}_6\text{H}_4\text{-4-OMe} \xrightarrow[\substack{K_3PO_4.H_2O \\ \text{toluene, 70 °C} \\ 82\%}]{\substack{[Pd_2(dba)_3)]\ (1.5\ mol\%) \\ \mathbf{10}\ (6\ mol\%)}} \text{2-Me-4'-OMe-biphenyl} \quad (2.27)$$

10: Cp^*Fe (η-C₅H₃(SiMe₃)(PPh₂))

As PPh_3 is ineffective under these conditions, the unusual reactivity of the diphenylferrocenylphosphine can be attributed to the greater electron-donating ability [48] and the increased bulk of the ferrocenyl group, relative to a phenyl substituent.

More recently, Kwong and Chan disclosed a very similar diphenylferrocene ligand for Suzuki couplings of aryl chlorides. Activated aryl chlorides can undergo reactions at ambient temperature, while alkylboronic acids can be used in addition to arylboronic acids (Equation 2.28) [49, 50].

$$\text{2-MeOC}_6\text{H}_4\text{Cl} + (HO)_2B{-}n\text{-Bu} \xrightarrow[\substack{K_3PO_4 \\ \text{toluene, 110 °C} \\ 88\%}]{\substack{[Pd_2(dba)_3)]\ (0.5\ mol\%) \\ \mathbf{11}\ (1.2\ mol\%)}} \text{2-MeOC}_6\text{H}_4{-}n\text{-Bu} \quad (2.28)$$

11: CpFe (η-C₅H₄PPh₂)

Phosphinous acids represent a promising new class of phosphorus ligands for Suzuki cross-couplings of unactivated aryl chlorides. The hydrolysis of diorganophosphorus halides generates secondary phosphine oxides in an equilibrium with their less-stable phosphinous acid tautomers (Equation 2.29).

R
P−X
R
H_2O
X = Cl, Br
R
R−P=O
H
R
P−OH
R
a phosphinous acid
$[Pd(COD)Cl_2]$
Cl
Cl
Pd
R
P
R
OH
2

(2.29)

The binding of a phosphinous acid to palladium can provide a phosphorus-bound adduct, which may be deprotonated to yield an electron-rich, anionic palladium–phosphine complex suitable as a catalyst for coupling processes. Li has recently demonstrated the viability of this strategy, establishing that catalytic systems consisting of $[Pd_2(dba)_3]$ and $(t\text{-}Bu)_2P(O)H$, as well as $[Pd_2(dba)_3]$, $(t\text{-}Bu)_2PCl$ and H_2O, effect Suzuki reactions of hindered and electron-rich aryl chlorides (Equation 2.30) [51, 52].

OMe
Cl
+
$(HO)_2B$
$[Pd_2(dba)_3)]$ (2.5 mol%)
$(t\text{-}Bu)_2PCl/H_2O$ (5 mol%)
CsF
dioxane, 100 °C
91%
OMe

(2.30)

Heteroatom-substituted secondary phosphine oxide (HASPO) preligands, such as *H*-phosphonates and their derivatives, display significantly different steric and electronic properties. These preligands (e.g. compound **12**), as well as the corresponding phosphine chlorides, were found to be highly effective for Suzuki reactions of aryl chlorides (Equation 2.31) [53].

Me
Cl
+
$(HO)_2B$
Me
$[Pd(dba)_2]$ (2 mol%)
12 (4 mol%)
KO*t*-Bu
THF, 60 °C
95%
Me
Me

(2.31)

O
H
P
t-Bu
N
N
t-Bu
12

Bulky, electron-rich imino-proazaphosphatranes are another relatively new and novel ligand class for coupling reactions of aryl chlorides (Equation 2.32) [54, 55].

$[Pd(OAc)_2]$ (2 mol%)
13 (4 mol%)
Cs_2CO_3
toluene, 80 °C
93%

$(i\text{-Bu})_2P$
t-Bu
t-Bu
t-Bu
13

(2.32)

Carbon-based ligands, specifically, nucleophilic *N*-heterocyclic carbenes (NHCs), are very promising alternatives to bulky, electron-rich phosphines for coupling reactions of aryl chlorides, and other electrophiles [56]. Herrmann was the first to apply this family of ligands to palladium-catalyzed coupling reactions, showing that a first-generation catalyst can achieve the cross-coupling of the activated 4′-chloroacetophenone with phenylboronic acid in 60% yield at 120 °C [57]. He subsequently determined that mixed complexes, bearing PCy_3 and NHCs, can couple unactivated aryl chlorides at elevated temperature [58].

Trudell and Nolan have reported that palladium adducts of Arduengo's carbene ligand, 1,3-bis(2,4,6-trimethylphenyl)imidazol-2-ylidene (IMes) [59], are active catalysts for Suzuki cross-couplings of a wide variety of aryl chlorides with arylboronic acids (Equation 2.33) [60].

$[Pd_2(dba)_3]$ (1.5 mol%)
IMesHCl (3 mol%)
Cs_2CO_3
dioxane, 80 °C
93%

Ar = 2,4,6-$(Me)_3C_6H_2$
IMesHCl

(2.33)

Handling of the air- and moisture-sensitive carbene can be avoided through use of the easier-to-handle, commercially available chloride salt (IMesHCl), which can be deprotonated *in situ* to generate the free carbene ligand. Independent studies conducted by Herrmann confirmed the observed high activity of IMes-based palladium complexes for the coupling of aryl chlorides [61].

Subsequently, Nolan has shown that NHC-bearing palladacycles are even more active catalysts, allowing for the coupling of deactivated aryl chlorides at ambient temperature, the synthesis of tri-substituted biaryls [62], and couplings of heteroaromatic substrates at catalyst loadings as low as 50 ppm palladium (Equation 2.34) [63].

Ar−N N−Ar
PdCl(allyl) (50 ppm)
KO*t*-Bu
i-PrOH, 80 °C
91%
Ar = 2,6-(*i*-Pr)$_2$C$_6$H$_3$

(2.34)

By employing 9-alkyl-9-BBN derivatives as coupling partners, Fürstner has expanded the scope of NHC-palladium catalysts in Suzuki reactions of aryl chlorides (Equation 2.35) [64].

[Pd(OAc)$_2$] (2 mol%)
IPrHCl (4 mol%)
KOMe
THF, 67 °C
83-84%
R = allyl, *n*-C$_{14}$H$_{30}$

(2.35)

The most effective ligand for these processes is not IMes, but instead the more bulky 2,6-diisopropylphenyl-substituted carbene, IPr, which was generated *in situ* from commercially available IPrHCl. Use of the preformed carbene, rather than *in situ* generation from the chloride salt, leads to significantly worse results. Both, alkenyl- and alkynyl-9-BBN derivatives are also suitable partners in these cross-couplings.

Other groups have followed the investigations of Nolan and Herrmann with novel NHC-based palladium catalysts. Carbenes derived from bisoxazolines are suitable for the synthesis of highly hindered tetra-*ortho*-substituted biaryls (Equation 2.36) [65].

Me, Cl, Me + $(HO)_2B$, Et, Et

$[Pd(OAc)_2]$ (3 mol%)
14 (3.6 mol%)
KO*t*-Bu, KH
K_3PO_4
toluene/THF
100 °C
75%

Me Et, Me Et

O, O, N, +, N, OTf$^-$, 8, 8
14

(2.36)

Bis-phenanthryl *N*-heterocyclic carbenes can also be used for the synthesis of tetra-*ortho*-substituted biaryls. In addition, other coupling partners, such as pinacolboranates and alkenylboronic acids, may be used (Equation 2.37) [66, 67].

H_2N, Cl + $(HO)_2B$, Ph

$[Pd(OAc)_2]$ (2 mol%)
15 (4 mol%)
KF/18-crown 6
toluene, 50 °C
87%

H_2N, Ph

Cy, Cy, Cl^-, N, +, N, Cy, Cy
15

(2.37)

Isolated examples of Suzuki couplings of aryl chlorides performed under phosphine- and NHC-ligand-free or heterogeneous conditions have been reported, although most of these are limited to activated, electron-deficient aryl chlorides [68]. For example, Sun and Sowa have recently described the application of heterogeneous palladium on charcoal to Suzuki reactions of activated aryl chlorides with phenylboronic (Equation 2.38) [69].

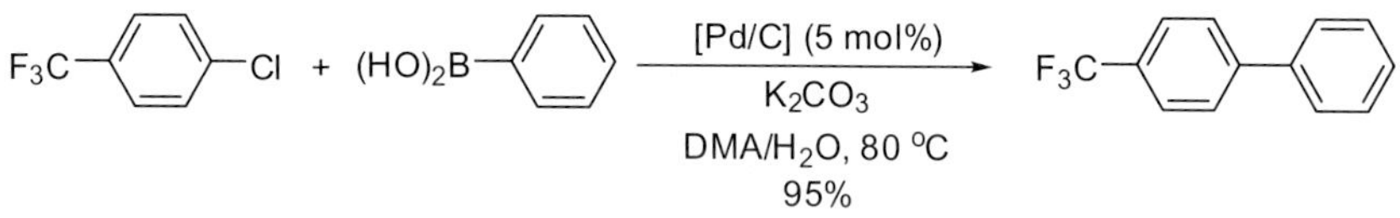

(2.38)

Couplings of electron-neutral and electron-rich aryl chlorides proceed in only modest yields (32–54%). Here, the choice of solvent (DMA : H_2O, 20 : 1) is important, as additional water leads to a significant amount of homocoupling, while a lower proportion of water results in severely retarded reaction rates. Separation of the product from the catalyst can be achieved through simple filtration, which leaves less than 1.0 ppm palladium in the product (<0.10% loss of Pd, based on the initial quantity). It should be noted that coupling does not occur in the presence of an equivalent of PPh_3, which is an interesting contrast to the analogous nickel on charcoal-based catalyst for Suzuki reactions of aryl chlorides, for which PPh_3 appears to be required for activity [70].

2.2.2.2 **Stille Reaction**

Stille cross-couplings of heteroaryl chlorides, mainly nitrogen-containing heterocycles, have often employed traditional triarylphosphine-based palladium catalysts. In one of the earliest examples, Yamanaka showed that 2- and 4-chloro-3-nitropyridines react with (*Z*)-1-ethoxy-2-tributylstannylethylene in good yield, with the products possibly being converted to 1*H*-pyrrolopyridines (Equation 2.39) [71].

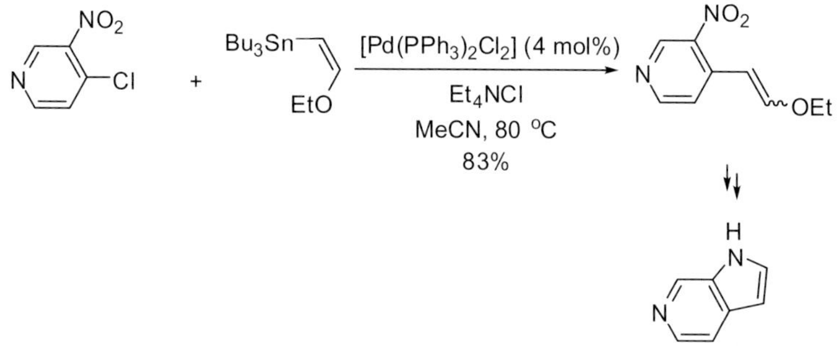

(2.39)

For halopyrimidines, the 4-position is the most activated, followed by the 2-position, and then the 5-position. One can therefore achieve selective couplings of the 4-chloro group of 2,4- [72] and 4,5-dichloropyrimidines [73], as well as the 2-chloro substituent of 2,5-dichloropyrimidines [74]. Interestingly, a 4-chloro group can react in preference to a 5-bromo group, allowing the stepwise functionalization of 2,4-dichloro-5-bromopyrimidine (Scheme 2.2) [70].

The first example of a palladium-catalyzed Stille couplings of a nonheteroaryl chloride was provided by Migita, who established that $[Pd(PPh_3)_4]$ catalyzes the

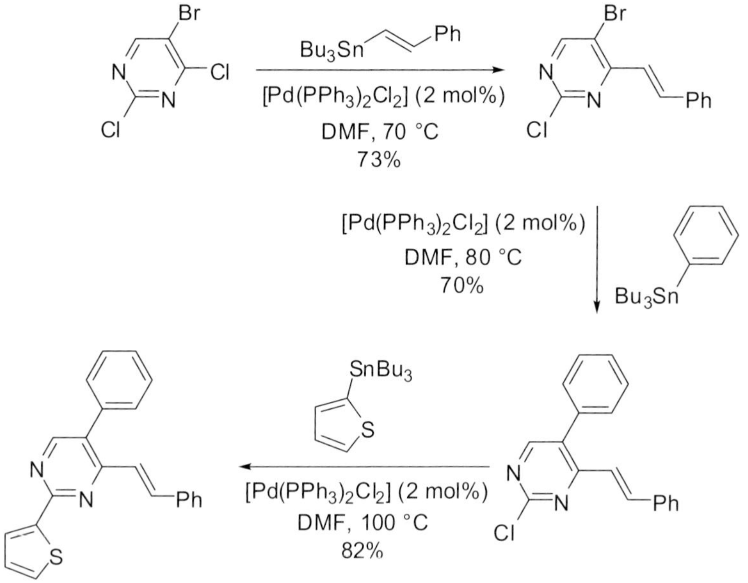

Scheme 2.2 Stepwise coupling of 2,4-dichloro-5-bromopyrimidine.

cross-coupling of allyltributyltin with a highly activated aryl chloride, 1-chloro-4-nitrobenzene, in moderate yield (Equation 2.40) [75, 76].

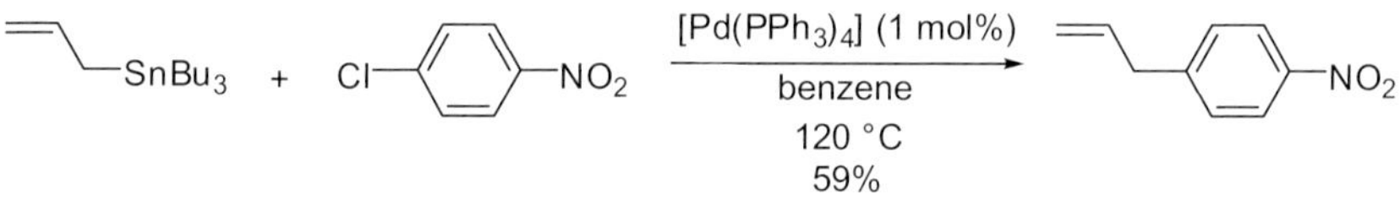

(2.40)

In 1999, Fu reported the first general method for achieving Stille cross-couplings of unactivated aryl chlorides using a palladium precursor and $P(t\text{-Bu})_3$, along with CsF as base [77]. This catalyst works well for electron-deficient, electron-rich and hindered aryl chlorides (Equation 2.41).

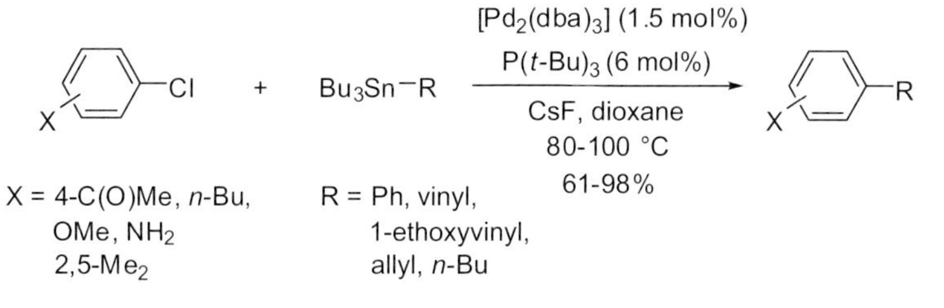

(2.41)

A variety of substituents R can be transferred from tin, including typically unreactive alkyl substituents. The coupling proceeds much more slowly in the absence of fluoride, which appears to play at least two useful roles in this chemistry: (i) it activates the organotin reagent for transmetallation; and (ii) it produces insoluble Bu_3SnF, which can be easily separated from the product. Difficulties in eliminating tin impurities are commonly encountered during purifications of Stille reactions [78].

Fu subsequently demonstrated that commercially available, crystalline [Pd(P(*t*-Bu)$_3$)$_2$], can be employed as a preformed catalyst for the Stille couplings of aryl chlorides [79]. This method is effective even for very hindered substrates, as evidenced by the synthesis of a tetra-*ortho*-substituted biaryl (Equation 2.42).

[Pd(P(*t*-Bu)$_3$)$_2$] (3 mol%)
CsF, dioxane
100 °C
89%

(2.42)

In addition, [Pd(P(*t*-Bu)$_3$)$_2$] accomplishes the selective coupling of a chloride in the presence of a triflate and, for certain activated aryl chlorides, the reactions can be carried out at ambient temperature. A TON as high as 920 was achieved for the Stille cross-coupling of an unactivated aryl chloride. The air-stable tetrafluoroborate salt P(*t*-Bu)$_3$HBF$_4$ may also be used for many of these couplings (Equation 2.43) [29].

[Pd$_2$(dba)$_3$] (1.5 mol%)
P(*t*-Bu)$_3$HBF$_4$ (3 mol%)
CsF, dioxane
ambient temperature
89%

(2.43)

In 2001, Nolan reported that NHC–palladium complexes, in combination with a fluoride additive, catalyze the Stille cross-coupling of electron-deficient aryl chlorides (Equation 2.44) [80].

[Pd(OAc)$_2$] (3 mol%)
IPrHCl (3 mol%)
n-Bu$_4$NF
dioxane/THF
100 °C
91%

(2.44)

2,6-Diisopropylphenyl-substituted carbene IPr and *N,N′*-di(1-adamantyl)-substituted carbene IAd were found to be superior to other carbenes, including mesityl-substituted IMes. Poor to moderate yields (15–54%) were obtained for the reactions of unactivated aryl chlorides.

Palladium–phosphinous acid complexes can be used for Stille couplings of aryl chlorides in water. In the case of 4,7-dichloroquinoline, a highly selective coupling at the 4-position is observed, with no products from coupling at the 7-position being detected (Equation 2.45) [81, 82].

$[(t\text{-}Bu_2P(OH))_2PdCl_2]$ (6 mol%), Cy_2NMe, H_2O, 135-140 °C, 76%

(2.45)

Verkade has performed a very thorough study on the use of bulky proazaphosphatrane ligands for Stille couplings of aryl chlorides. The scope and mildness of the reaction conditions are quite impressive, with a wide range of aryl and heteroaryl chlorides participating with an assortment of tin reagents (Equation 2.46) [83].

$[Pd_2(dba)_3]$ (1.5 mol%), **16** (6 mol%), CsF, dioxane, 110°C, 69%

16

(2.46)

2.2.2.3 **Hiyama Coupling**

The first examples of aryl chlorides participating in this type of process were provided by Matsumoto, who observed that $[Pd(PPh_3)_4]$ catalyzed the cross-coupling of nitro-substituted aryl chlorides with hexamethyldisilane to furnish arylsilanes [84]. In the case of 2,5-dichloronitrobenzene, a high level of selectivity for substitution at the *ortho*-position to the nitro functionality can be obtained (Equation 2.47) [85, 86].

$$\text{Cl-C}_6\text{H}_3(\text{NO}_2)\text{-Cl} + \text{Me}_3\text{Si-SiMe}_3 \xrightarrow[\text{HMPA, reflux; 62\%}]{[\text{Pd(PPh}_3)_4] \text{ (1 mol\%)}} \text{Cl-C}_6\text{H}_3(\text{NO}_2)\text{-SiMe}_3 \quad (2.47)$$

In 1996, Hatanaka considerably expanded the scope of this process with $[Pd(Pi\text{-}Pr_3)_2Cl_2]$ as the catalyst in the presence of KF, establishing that a wide variety of electron-deficient aryl chlorides cross-couple with arylchlorosilanes (Equation 2.48) [87].

$$3\text{-MeC(O)-C}_6\text{H}_4\text{-Cl} + \text{Cl}_2\text{EtSi-C}_6\text{H}_4\text{-OMe} \xrightarrow[\text{KF, DMF, 120 °C; 62\%}]{[\text{Pd(P}i\text{-Pr}_3)_2\text{Cl}_2] \text{ (0.5 mol\%)}} 3\text{-MeC(O)-C}_6\text{H}_4\text{-C}_6\text{H}_4\text{-OMe} \quad (2.48)$$

The bulky, electron-rich bidentate ligand 1,2-bis(dicyclohexylphosphino)ethane is also effective in certain instances, whereas PPh_3-based catalysts are ineffective. Unfortunately, unactivated aryl chlorides do not couple in satisfactory yields under these reaction conditions. However, with $[Pd(PEt_3)_2Cl_2]$ as the catalyst and n-Bu_4NF as the fluoride source, alkenylchlorosilanes also participate in Hiyama cross-couplings with activated aryl chlorides. Subsequently it was determined that inexpensive NaOH could also be used as a promoter [88].

In 2007, Buchwald reported that biaryl ligand **3** was effective for the silylation of both electron-neutral and electron-rich aryl chlorides with hexamethyldisilane (Equation 2.49). Interestingly, the use of electron-deficient aryl chlorides requires slightly modified conditions and a different biaryl ligand (Equation 2.50) [89].

$$\text{5-chloroindole} + \text{Me}_3\text{Si-SiMe}_3 \xrightarrow[\text{KF, dioxane/H}_2\text{O, 100 °C; 86\%}]{[\text{Pd}_2(\text{dba}_3)] \text{ (1 mol\%)}, \mathbf{3} \text{ (6 mol)\%}} \text{5-(trimethylsilyl)indole} \quad (2.49)$$

N Cl + $Me_3Si-SiMe_3$ → N $SiMe_3$

$[Pd_2(dba_3)]$ (1.5 mol%)
17 (9 mol%)
LiOAc
DMF/H_2O
100 °C
82%

Me_2N
Pt-Bu_2
17

(2.50)

Siloxanes also serve as useful partners in Hiyama cross-couplings of aryl chlorides. DeShong has shown that palladium-catalyzed reactions of phenyltrimethylsiloxane with a range of substrates, including challenging electron-neutral and electron-rich aryl chlorides, can be achieved in the presence of Buchwald's 2-(dicyclohexylphosphino)biphenyl ligand (**3**), albeit using 20 mol% palladium (Equation 2.51) [90].

MeO Cl + $(MeO)_3Si$ → MeO

$[Pd_2(dba)_3]$ (10 mol%)
3 (15 mol%)
n-Bu4NF
DMF, 85 °C
71%

(2.51)

In addition, Nolan has determined that use of the NHC IPrHCl allows for couplings of phenyl- and vinyltrimethylsiloxane with activated aryl chlorides (Equation 2.52) [91].

NC Cl + $(MeO)_3Si$ → NC

$[Pd(OAc)_2]$ (3 mol%)
IPrHCl (3 mol%)
n-Bu_4NF
dioxane/THF
80 °C
100% conversion

(2.52)

Ackermann has shown that palladium complexes derived from heteroatom-substituted secondary phosphine oxides are efficient for couplings of siloxanes and activated aryl chlorides (Equation 2.53) [53].

2-chloropyridine + $(MeO)_3Si$–Ph → 2-phenylpyridine

$[Pd(dba)_2]$ (5 mol%)
18 (10 mol%)
n-Bu_4NF
dioxane, 80 °C
63%

18

(2.53)

Murata has shown that a bulky dialkylaryl*bis*phosphine, *i*-Pr-DPEphos, allows for Hiyama couplings of a wide variety of aryl chlorides, including unactivated and deactivated aryl chlorides (Equation 2.54) [92]. Interestingly, replacing the *iso*-propyl group with a *tert*-butyl or phenyl group results in lower yields and significant amounts of the hydrodehalogenated byproducts.

Me–C_6H_4–Cl + $(MeO)_3Si$–Ph → Me–C_6H_4–Ph

$[Pd(dba)_2]$ (3 mol%)
i-Pr-DPEphos (4 mol%)
n-Bu_4NF
toluene, 110 °C
89%

$P(i\text{-}Pr)_2$ $P(i\text{-}Pr)_2$

i-Pr-DPEphos

(2.54)

Diastereospecific couplings of both (*E*)- and (*Z*)-alkenylsilanolates with a wide range of aryl chlorides can be accomplished using Buchwald dialkylbiarylphosphine ligand **5**. Hence, even tri- and tetra-substituted alkenylsilanols can be used allowing access to sterodefined tri- and tetra-substituted olefins (Equation 2.55) [93, 94].

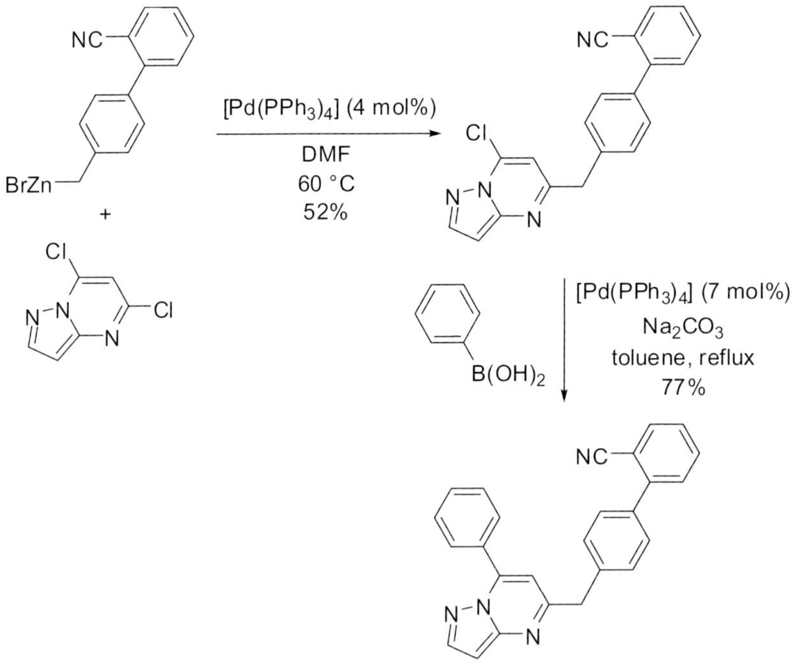

Scheme 2.3 Selective Negishi cross-coupling of a heteroaryl dichloride.

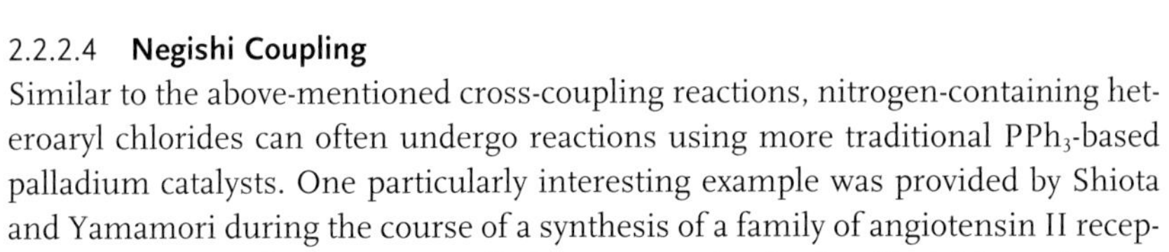

(2.55)

2.2.2.4 **Negishi Coupling**

Similar to the above-mentioned cross-coupling reactions, nitrogen-containing heteroaryl chlorides can often undergo reactions using more traditional PPh_3-based palladium catalysts. One particularly interesting example was provided by Shiota and Yamamori during the course of a synthesis of a family of angiotensin II receptor antagonists. Their strategy relied upon a regioselective Negishi coupling of 5,7-dichloropyrazolo[1,5-*a*]pyrimidine, which proceeds as desired when the reaction is performed in dimethylformamide (DMF), in which only 7% of the unwanted isomer is produced (Scheme 2.3) [95]. In contrast, no selectivity is observed when tetrahydrofuran (THF) is used as the solvent. A second palladium-catalyzed cross-coupling, namely a Suzuki reaction, then furnishes the target compound in good yield. Interestingly, a direct nucleophilic substitution of the starting dichloride by the benzylzinc reagent in the presence of LiCl, but in the absence of a palladium catalyst, leads to a preferential substitution of the chloride at 7-position.

The first example of a palladium-catalyzed Negishi coupling of an unactivated aryl chloride, chlorobenzene, was reported by Herrmann, with a palladacycle as the catalyst (Equation 2.56), although the scope of this method with respect to aryl chlorides was not discussed [96, 97].

Cl + BrZn— ; palladacycle [Pd(OAc)(P(o-Tol)$_2$)]$_2$ (2 mol%), THF, 90 °C, 88%

(2.56)

The first general protocol for accomplishing Negishi couplings of unactivated, including electron-rich, aryl chlorides, was provided by Fu in 2001, using commercially available [Pd{P(t-Bu)$_3$}$_2$] as the catalyst (Equation 2.57) [98].

[Pd(Pt-Bu$_3$)$_2$] (2 mol%), THF/NMP, 100 °C, 87-97%

X = 4-CO$_2$Me, NO$_2$, B(OR)$_2$, nBu, OMe, 2-Me, 2,6-Me$_2$

Y = 4-OMe, 2-Me, 2,6-Me$_2$

(2.57)

Functionalities, such as nitro and ester groups, are tolerated, and heteroaryl chlorides, such as chlorothiophenes and chloropyridines, are suitable substrates. Hindered compounds can be cross-coupled effectively, allowing the synthesis of a tetra-*ortho*-substituted biaryl, and TONs as high as 3000 also being achieved. Finally, primary and secondary *alkyl*zinc reagents can be employed; in the case of *s*-BuZnCl, coupling with 2-chlorotoluene generates predominantly the desired compound, 2-*s*-butyltoluene, along with 8% of the isomerized product, 2-*n*-butyltoluene (Equation 2.58).

ClZn—R, R = n-Bu, s-Bu; [Pd(Pt-Bu$_3$)$_2$] (2 mol%), THF/NMP, 100 °C, 70-83% (2.58)

In 2004, Buchwald reported that biaryldialkyl phosphine ligand **19** is an extremely effective supporting ligand for palladium-catalyzed Negishi couplings of aryl chlorides. Hindered di-, tri- and tetra-*ortho*-substituted biaryls can be generated using only 0.1–1.0 mol% palladium, while a variety of heteroaryl chlorides may also be efficiently coupled with arylzinc reagents (Equation 2.59) [99].

$[Pd_2(dba)_3]$ (0.5 mol%), **19** (2 mol%), THF, 70 °C, 92%

(2.59)

19

Preformed air-stable palladium(II) complexes featuring phosphinous acid ligands are also reasonably effective catalysts for Negishi couplings of aryl chlorides (Equation 2.60) [100].

$[(t\text{-}Bu_2P(OH))_2PdCl_2]$ (5 mol%), THF/NMP, reflux, 57%

(2.60)

2.2.2.5 Kumada Coupling

A few palladium-catalyzed Kumada cross-couplings of heteroaryl chlorides have been reported [101]. Knochel has recently determined that certain 2-chloropyridine derivatives react with functionalized aryl Grignard reagents under mild conditions (Equation 2.61) [102].

$[Pd(dba)_2]$ (5 mol%), dppf (5 mol%), THF, −40 °C, 92%

(2.61)

Because of the low temperature employed, the presence of normally incompatible esters is tolerated.

Katayama described the first examples of palladium-catalyzed Kumada cross-couplings of (activated) nonheteroaryl chlorides. Thus, the selective mono-coupling of dichlorobenzenes with aryl and alkyl Grignard reagents can be achieved with [Pd(dppf)Cl_2] as the catalyst (Equation 2.62) with, in all cases, less than 5% of the dialkylated product being observed [103].

$$\text{X-C}_6\text{H}_4\text{-Cl} + \text{ClMg}{-}n\text{-Pr} \xrightarrow[\text{THF, 85 °C; 68-84\%}]{[\text{Pd(dppf)Cl}_2]\ (0.1\ \text{mol\%}),\ \text{dppf}\ (0.1\text{mol\%})} \text{X-C}_6\text{H}_4{-}n\text{-Pr} \tag{2.62}$$

X = 2-Cl, 3-Cl, 4-Cl

The first report of a palladium-catalyzed Kumada cross-coupling of an unactivated aryl chloride was provided by Herrmann, who showed that palladacycles accomplish the coupling of chlorobenzene with MeMgBr and PhMgCl [104]. Electron-rich aryl chlorides react with arylmagnesium bromides in the presence of [$Pd_2(dba)_3$] and imidazolium salt IPrHCl. Here, a slight excess of Grignard reagent is used to deprotonate the salt and generate the free carbene ligand *in situ* (Equation 2.63) [105].

$$\text{X-C}_6\text{H}_4\text{-Cl} + \text{BrMg-C}_6\text{H}_4\text{-Y} \xrightarrow[\text{dioxane/THF, 80 °C; 83-99\%}]{[\text{Pd}_2(\text{dba})_3]\ (1\ \text{mol\%}),\ \text{IPrHCl}\ (4\ \text{mol\%})} \text{X-C}_6\text{H}_4\text{-C}_6\text{H}_4\text{-Y} \tag{2.63}$$

X = 4-Me, OMe, OH, 2,5-Me_2, 2,6-Me_2

Y = H, 4-Me, 3-Me, 2-F, 2,4,6-Me_3

Hindered substrates, such as 2-chloro-*m*-xylene, can also be coupled efficiently, although the utility of this Kumada cross-coupling process is limited by its relatively low functional-group tolerance. Thus, for example the reaction of methyl 4-chlorobenzoate results in the formation of significant amounts of undesired side products.

The PEPPSI (pyridine-enhanced, precatalyst, preparation, stabilization and initiation) precatalyst **20** is a very versatile catalyst for Kumada couplings of highly hindered substrates and heterocycles at ambient temperature (Equation 2.64) [106].

20 (2 mol%), THF, ambient temperature, 85%

Ar = 2,6-(i-Pr)$_2$C$_6$H$_3$

20

(2.64)

Trialkylphosphines have also been used for Kumada couplings of aryl chlorides. Notably, the air-stable phosphine salt P(t-Bu)$_3$HBF$_4$ has been used for microwave-assisted couplings of aryl chlorides with Grignard reagents (Equation 2.65) [107].

[Pd$_2$(dba)$_3$] (1 mol%), P(t-Bu)$_3$HBF$_4$ (4 mol%), THF, microwaves, 175 °C, 94%

(2.65)

2.3 Coupling Reactions of Aryl Fluorides

Aryl fluorides are more expensive than aryl chlorides, and thus less attractive substrates for many industrial applications. However, activation of C—F bonds contributes to the understanding of the reactivity of very stable bonds and is, therefore, of key importance in organometallic chemistry. In 1999, Widdowson demonstrated the first example of an aryl fluoride participating in a palladium-catalyzed coupling reaction, albeit as an electronically activated tricarbonyl–chromium complex (Equation 2.66) [108].

MeO–C$_6$H$_4$(Cr(CO)$_3$)–F + (HO)$_2$B–C$_6$H$_4$–Me → [Pd$_2$(dba)$_3$] (5 mol%), PMe$_3$ (20 mol%), Cs$_2$CO$_3$, DME, 85 °C, 77% → MeO–C$_6$H$_4$(Cr(CO)$_3$)–C$_6$H$_4$–Me

(2.66)

Sterically small (yet still basic) trimethylphosphine proved to be a superior ligand compared to the bulky tricyclohexylphosphine, presumably due to steric hindrance from the chromiumtricarbonyl moiety. Subsequently, it was shown that uncomplexed aryl fluorides could participate in Suzuki couplings, although the presence of a strongly electron-withdrawing nitro group in the *ortho*-position, which can coordinate an incoming palladium species, along with a second electron-withdrawing group in the *para*-position, was required [109]. Concurrently, a research group at Pfizer was also able to demonstrate both Suzuki and Stille couplings of aryl fluorides that contained an *ortho*-nitro group as well as a second electron-withdrawing group using [Pd(PPh$_3$)$_4$] as the catalyst (Equation 2.67) [110].

OHC–C$_6$H$_3$(NO$_2$)–F + Bu$_3$Sn–C$_6$H$_5$ → [Pd(PPh$_3$)$_4$] (10 mol%), DMF, 65 °C, 65% → OHC–C$_6$H$_3$(NO$_2$)–C$_6$H$_5$

(2.67)

Substrate scope was further enhanced in 2005 to aryl fluorides bearing an *ortho*-carboxylate group and a *para*-electron-withdrawing group. Computational studies revealed that coordination from the oxygen of the *ortho*-group to palladium plays an important role through transition-state stabilization, thereby lowering the activation energy barrier (Equation 2.68) [111, 112].

R–C$_6$H$_3$(CO$_2$H)–F → [PdL$_x$] → [R–C$_6$H$_3$(C(=O)–O···Pd(L$_x$)–F)]‡

(2.68)

Recently, nickel-based catalysts have featured more prominently in coupling reactions of aryl fluorides. In 2001, Herrmann was able to demonstrate Kumada couplings of aryl fluorides, including electron-rich aryl fluorides, using the tetrafluoroborate salt of the *N*-heterocyclic carbene IPr as the supporting ligand for nickel (Equation 2.69) [113].

Me–C$_6$H$_4$–F + BrMg–(2,4,6-Me$_3$C$_6$H$_2$) → [Ni(acac)$_2$] (5 mol%), IPrBF$_4$ (5 mol%), THF, ambient temperature, 65%

(2.69)

A ligand: nickel ratio of 1:1 performed better than a 2:1 mixture, suggesting that the catalytically active species is a zero-valent nickel species coordinated by a single carbene ligand. Hammett studies and the low amounts of terphenyl byproducts observed (which could arise through radical intermediates) suggested a polar mechanism, most likely through oxidative addition, similar to palladium-catalyzed couplings.

The use of phosphines, rather than carbene ligands, for the nickel-catalyzed coupling of aryl fluorides was first demonstrated using heteroaryl fluorides as substrates and organomagnesium reagents as the coupling partners [114]. The bidentate ligands dppe, dppf and dppp were all found to be effective to varying degrees; however, unactivated 4-fluorotoluene could undergo couplings in only moderate yields (59–61%) under similar reaction conditions (Equation 2.70).

Ph–(pyridazinyl)–F + BrMg–C$_6$H$_4$–OMe → [NiCl$_2$(dppe)] (5 mol%), THF, ambient temperature, 59%

(2.70)

The use of air-stable heteroatom-substituted secondary phosphine oxides as preligands for nickel results in a highly active system for Kumada couplings of aryl fluorides at ambient temperature (Equation 2.71) [115]. Most likely, the active nickel catalyst is generated through coordination of a neutral HASPO preligand through phosphorus, and its subsequent deprotonation, yielding a highly reactive electron-rich nickel species for oxidative addition.

1-fluoronaphthalene + ClMg–C$_6$H$_5$ → [Ni(acac)$_2$] (3 mol%), **21** (3 mol%), THF, 20 °C, 84%

i-Pr *i*-Pr O H P N N *i*-Pr *i*-Pr **21**

(2.71)

A study of both nickel- and palladium-catalyzed coupling reactions of polyfluorinated arenes and Grignard reagents revealed that palladium-based catalysts were more effective for the synthesis of monosubstituted products, while nickel complexes could be used for the generation of di- and trisubstituted products (Equation 2.72) [116, 117].

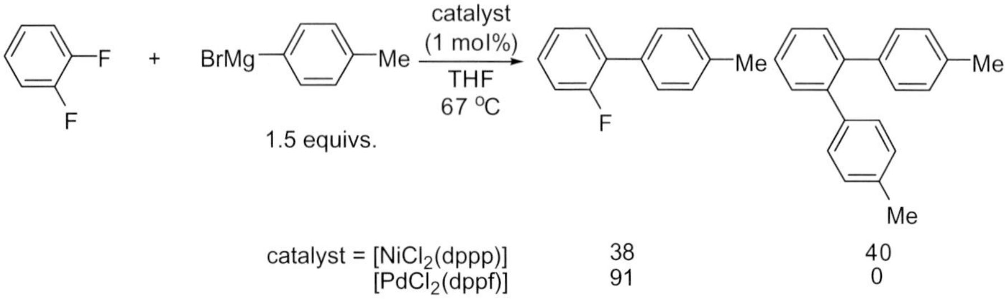

(2.72)

Almost all of the examples of nickel-catalyzed coupling reactions of aryl fluorides have focused on the use of organomagnesium reagents. In 2006, Radius and coworkers demonstrated highly selective Suzuki couplings of perfluorinated arenes using a nickel-carbene based catalyst system (Equation 2.73). However, no products arising from coupling at the *ortho*- or *meta*-positions were observed in the crude reaction mixtures [118, 119].

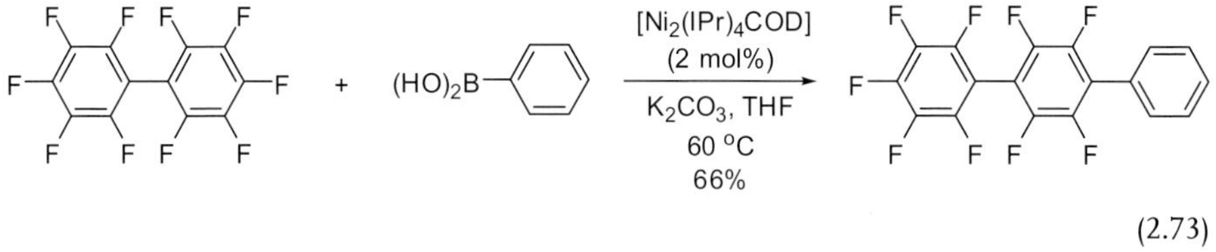

(2.73)

2.4 Coupling Reactions of Aryl Tosylates

Although coupling reactions of aryl triflates are well established, the high cost of triflating agents and the higher hydrolytic stability and crystallinity of aryl tosylates compared to those of the corresponding triflates, makes aryl tosylates a very attractive class of substrates for both nickel- and palladium-catalyzed coupling reactions. Although the lower reactivity of aryl tosylates versus aryl triflates has precluded their wide application, significant progress has recently been made in this area.

As early as 1995, Sasaki and coworkers noted that $[Pd(PPh_3)_4]$ could be used as the catalyst for Stille couplings of guanosine-based tosylates and vinyltributyltin. Interestingly, the analogue's triflate afforded slightly lower yields (Equation 2.74) [120, 121].

TBDMSO, TBDMSO, N, N, N, N, H_2N, OTs + Bu_3Sn — $[Pd(PPh_3)_4]$ (*cat.*), LiCl, dioxane, 100 °C, 94% → TBDMSO, TBDMSO, N, N, N, N, H_2N

(2.74)

In 2003, Buchwald reported the first general method for the palladium-catalyzed Suzuki coupling of aryl tosylates using biaryldialkylphosphine ligand **6** as the ligand. A variety of other tested ligands, including NHCs and P(*t*-Bu)$_3$, were mostly ineffective [122]. Electronically diverse aryl tosylates and aryl boronic acids can be used, as can highly hindered boronic acids (Equation 2.75).

Me, N, S, OTs + $(HO)_2B$, *i*-Pr, *i*-Pr, *i*-Pr — $[Pd(OAc)_2]$ (3 mol%), **6** (7 mol%), $K_3PO_4 \cdot H_2O$, *t*-BuOH, 80 °C, 89% → Me, N, S, *i*-Pr, *i*-Pr, *i*-Pr

(2.75)

In contrast, significant steric bulk on the aryl tosylate is not well tolerated, which suggests that the transmetallation step is not rate-determining.

Concurrently, Roy and Hartwig were able to demonstrate Kumada couplings of aryl tosylates at ambient temperature using P(*o*-Tol)$_3$ and a sterically hindered version of the commercially available Josiphos ligand [123, 124]. Interestingly, for 4-chlorophenyltosylate, reaction occurs at the tosylate (Equation 2.76), in contrast to the palladium-catalyzed Suzuki couplings of 4-chlorophenyltosylates using ligand **6**, where the reaction occurs with the chloride as leaving group [122]. Both, alkyl and alkenylmagnesium halides can be used, which further expands the scope of this chemistry.

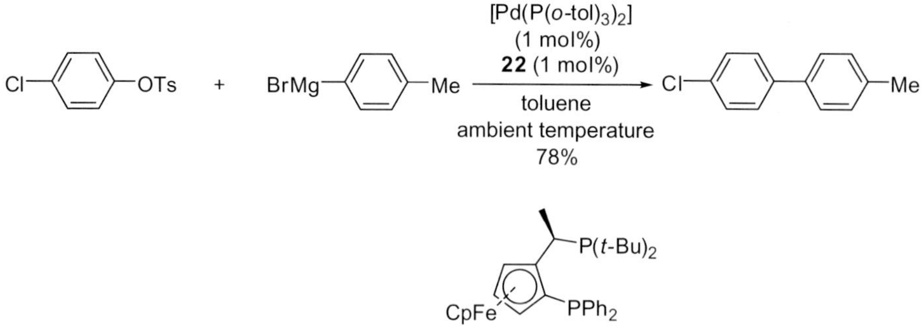

Cl–C6H4–OTs + BrMg–C6H4–Me → Cl–C6H4–C6H4–Me

[Pd(P(o-tol)3)2] (1 mol%), **22** (1 mol%), toluene, ambient temperature, 78%

P(t-Bu)2 PPh2 CpFe

22

(2.76)

More recently, Ackermann has expanded on the use of air-stable *H*-phosphonates as preligands for palladium-catalyzed processes by demonstrating their effectiveness for Kumada couplings of aryl tosylates. Both, chlorides and fluorides are tolerated, and heteroaryl tosylates proved also to be excellent substrates when using commercially available preligand **23** (Equation 2.77) [125].

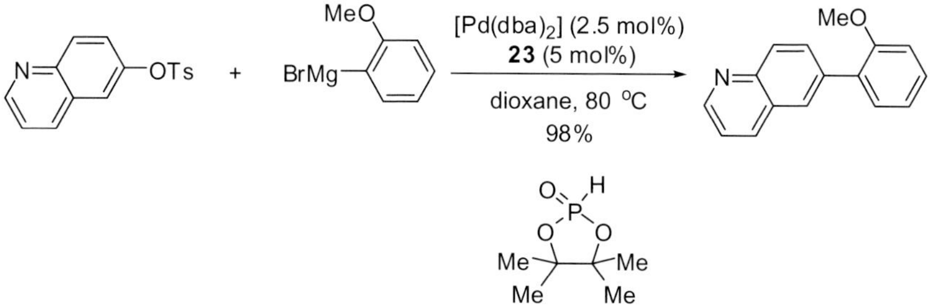

[Pd(dba)2] (2.5 mol%), **23** (5 mol%), dioxane, 80 °C, 98%

MeO OTs BrMg N

23

(2.77)

Nickel-based catalysts have also been used to activate aryl tosylates. In 1996, Kobayshi reported that the activated *para*-methoxycarbonylphenyl tosylate underwent Suzuki coupling with a lithium organoborate using [NiCl$_2$(PPh$_3$)$_2$] as the catalyst (Equation 2.78) [126].

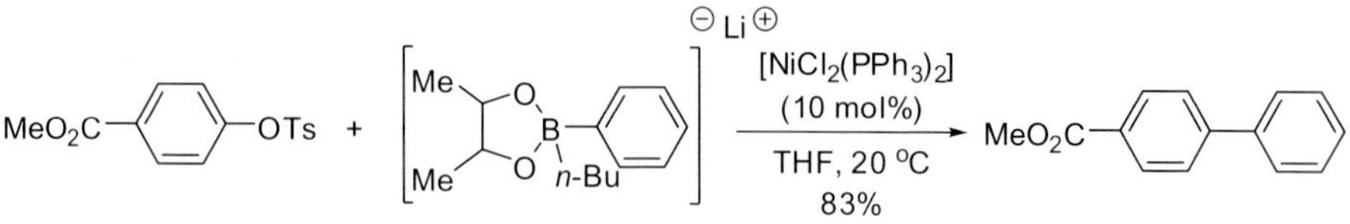

MeO2C–C6H4–OTs + Li⁺ [borate, n-Bu] → MeO2C–C6H4–C6H5

[NiCl2(PPh3)2] (10 mol%), THF, 20 °C, 83%

(2.78)

Monteiro reported a more general method in 2001 for Suzuki couplings of aryl tosylates using a nickel catalyst derived from PCy_3 (Equation 2.79). Mechanistic studies suggest that transmetallation, and not reductive elimination or oxidative addition, is rate-determining [127].

MeO–C6H4–OTs + (HO)2B–C6H5 → MeO–C6H4–C6H5

$[NiCl_2(PCy_3)_2]$ (3 mol%)
PCy_3 (12 mol%)
K_3PO_4
dioxane, 130 °C
89%

(2.79)

The first examples of Suzuki couplings of aryl tosylates at ambient temperature were reported in 2004 by Hu, using a similar PCy_3-based nickel catalyst (Equation 2.80). A variety of ligands were probed, including other bulky and electron-rich trialkylphosphines, such as $P(t\text{-}Bu)_3$ and $P(i\text{-}Pr)_3$, triarylphosphines and dialkylbiarylphosphine ligands, although PCy_3 proved to be the superior ligand [128, 129].

CH_3–C6H4–OTs + (HO)2B–C6H5 → CH_3–C6H4–C6H5

$[Ni(COD)_2]$ (3 mol%)
PCy_3 (12 mol%)
K_3PO_4, THF
ambient temperature
93%

(2.80)

Lipshutz recently reported a Negishi coupling of a highly substituted aryl tosylate utilizing heterogeneous nickel on charcoal as catalyst with PPh_3 as ligand under microwave irradiation (Equation 2.81) [130].

MeO, OMe, Me, CHO substituted aryl–OTs + ClZn–C6H4–Me → MeO, OMe, Me, CHO substituted aryl–C6H4–Me

Ni/C (8 mol%)
PPh_3 (30 mol%)
dioxane
microwaves
150 °C
96%

(2.81)

2.5 Conclusions

A remarkable amount of progress has been reported during the past few years towards solving the longstanding challenge of bringing aryl chlorides, fluorides and tosylates into the family of generally useful substrates for palladium-catalyzed coupling reactions. Through the appropriate choice of ligand, a number of surprisingly mild and general catalyst systems have been developed, with bulky, electron-rich phosphines or phosphine oxides, as well as carbenes, having proved thus far to be the most versatile. The traditional view has been that oxidative addition is the problematic step in the catalytic cycle for these less-reactive substrates. Hence, the facilitation of this step is certainly a necessary – but not necessarily a sufficient – condition for achieving effective catalysis. Consequently, the developed ligands will also have steric and electronic properties appropriate to the subsequent steps in the catalytic cycle.

With regards to aryl chlorides, their lower cost and greater availability, combined with the commercial availability of many of the more versatile ligands discussed herein, has already spurned extensive interest and increased industrial applications of palladium-catalyzed coupling reactions, notably in medicinal and process chemistry. In the case of aryl fluorides and tosylates, considerably less development has been undertaken than for aryl chlorides. However, given the multitude of reported catalytic systems used to activate aryl chlorides with both palladium or nickel complexes, it is simply a matter of time before similar promising and exciting developments concerning fluorides and tosylates are disclosed.

Abbreviations

acac	acetylacetonate
Ac	acetyl
Ad	adamantyl
aq	aqueous
Bn	benzyl
Bu	butyl
BBN	borabicyclo[3.3.1]nonane
COD	1,5-cyclo-octadiene
Cp	cyclopentadienyl
Cp*	1,2,3,4,5-pentamethylcyclopentadienyl
Cy	cyclohexyl
dba	dibenzylideneacetone
DIBAL-H	diisobutylaluminum hydride
DMA	*N,N*-dimethylacetamide
DME	1,2-dimethoxyethane
DMF	*N,N*-dimethylformamide
DPEphos	1,1′-bis(diphenylphosphino)ether
dppb	1,4-bis(diphenylphosphino)butane
dppe	1,2-bis(dipenylphosphino)ethane
dppf	1,1′-bis(dipenylphosphino)ferrocene
dppp	1,3-bis(dipenylphosphino)propane
Et	ethyl

Et_2O	diethyl ether
HMPA	hexamethylphosphoramide
IAd	1,3-bis(adamantyl)imidazol-2-ylidene
IMes	1,3-bis(2,4,6-trimethylphenyl)imidazol-2-ylidene
IPr	1,3-bis(2,6-diisopropylphenyl)imidazol-2-ylidene
Me	methyl
NHC	*N*-heterocyclic carbene
NMP	1-methyl-2-pyrrolidinone
Ph	phenyl
Pr	propyl
TBDMS	*tert*-butyldimethylsilyl
Tf	triflate (trifluoromethanesulfonyl)
THF	tetrahydrofuran
tol	tolyl
TPPTS	sodium triphenylphosphinotrimetasulfonate
Ts	tosyl (*para*-toluenesulfonyl)

References

1 (a) de Meijere, A. and Diedrich, F. (2004) *Metal-Catalyzed Cross-Coupling Reactions*, Vol. 1 and 2, 2nd edn, Wiley-VCH Verlag GmbH, Weinheim;
(b) Tsuji, J. (2004) *Palladium Reagents and Catalysts*, 2nd edn, John Wiley & Sons, Ltd, Chichester.

2 For a discussion, see: (a) Grushin, V.V. and Alper, H. (1999) *Activation of Unreactive Bonds and Organic Synthesis* (ed. S. Murai), Springer-Verlag, Berlin, pp. 193–226;
(b) Grushin, V.V. and Alper, H. (1994) *Chem. Rev.*, **94**, 1047–62.

3 Brandsma, L., Vasilevsky, S.F. and Verkruijsse, H.D. (1998) *Application of Transition Metal Catalysts in Organic Synthesis*, Vol. 150, Springer-Verlag, New York, pp. 228–9.

4 Geissler, H. (1998) *Transition Metals for Organic Synthesis*, Vol. 1 (eds M. Beller and C. Bolm), Wiley-VCH Verlag GmbH, New York, p. 177.

5 (a) Sunderman, F.W., Jr (1988) *Handbook on Toxicity of Inorganic Compounds* (eds H.G. Seiler and H. Sigel), Marcel Dekker, New York, Chap. 37;
(b) Bradford, C.W. and Chase, B.J. (1988) *Handbook on Toxicity of Inorganic Compounds* (eds H.G. Seiler and H. Sigel), Marcel Dekker, New York, Chap. 43;
(c) Carson, B.L. (1986) *Toxicology and Biological Monitoring of Metals in Humans*, Lewis Publishers, Chelsea, MI.

6 Fitton, P. and Rick, E.A. (1971) *J. Organomet. Chem.*, **28**, 287–91.

7 (a) In reality, this is an oversimplification of the problem; see Grushin, V.V. and Alper, H. (1999) in *Activation of Unreactive Bonds and Organic Synthesis* (ed. S. Murai), Springer-Verlag, Berlin, p. 203;
(b) Herrmann, W.A. (1996) *Applied Homogeneous Catalysis with Organometallic Compounds. A Comprehensive Handbook*, Vol. 1 (eds B. Cornils and W.A. Herrmann), Wiley-VCH Verlag GmbH, New York, p. 722.

8 Littke, A.F. and Fu, G.C. (2002) *Angew. Chem. Int. Ed.*, **41**, 4176–211.

9 (a) Tamao, K., Sumitani, K. and Kumada, M. (1972) *J. Am. Chem. Soc.*, **94**, 4374–6;
(b) Kumada, M. (1980) *Pure Appl. Chem.*, **52**, 669–79.

10 For more recent examples of nickel-catalyzed Kumada couplings of aryl chlorides, see: (a) Li, G.Y. (2001) *Angew. Chem. Int. Ed.*, **40**, 1513–16;

(b) Li, G.Y. and Marshall, W.J. (2002) *Organometallics*, **21**, 590–1;
(c) Yoshikai, N., Mashima, H. and Nakamura, E. (2005) *J. Am. Chem. Soc.*, **127**, 17978–9;
(d) Ackermann, L., Born, R., Spatz, J.H. and Meyer, D. (2005) *Angew. Chem. Int. Ed.*, **44**, 7216–19;
(e) Schneider, S.K., Rentzsch, C.F., Kruger, A., Raubenheimer, H.G. and Herrmann, W.A. (2007) *J. Mol. Cat. A*, **265**, 50–8;
(f) Lau, S.Y.W., Hughes, G., O'Shea, P.D. and Davies, I.W. (2007) *Org. Lett.*, **9**, 2239–42.

11 (a) Saito, S., Sakai, M. and Miyaura, N. (1996) *Tetrahedron Lett.*, **37**, 2993–6;
(b) Saito, S., Oh-tani, S. and Miyaura, N. (1997) *J. Org. Chem.*, **62**, 8024–30.

12 Inada, K. and Miyaura, N. (2000) *Tetrahedron*, **56**, 8657–60.

13 (a) Indolese, A.F. (1997) *Tetrahedron Lett.*, **38**, 3513–16;
(b) Percec, V., Golding, G.M., Smidrkal, J. and Weichold, O. (2004) *J. Org. Chem.*, **69**, 3447–52.

14 Galland, J.C., Savignac, M. and Genet, J.P. (1999) *Tetrahedron Lett.*, **40**, 2323–6.

15 (a) Lipshutz, B.H., Sclafini, J.A. and Blomgren, P.A. (2000) *Tetrahedron*, **56**, 2139–44;
(b) Frieman, B.A., Taft, B.R., Lee, C.T., Butler, T. and Lipshutz, B.H. (2005) *Synthesis*, 2989–93;
(c) Lipshutz, B.H., Frieman, B.A., Lee, C., Lower, A., Nihan, D.M. and Taft, B.R. (2006) *Chem. Asian J.*, **1**, 417–29.

16 For other recent examples of nickel-catalyzed Suzuki couplings of aryl chlorides, see: (a) Lu, Y., Plocher, E. and Hu, Q.S. (2006) *Adv. Synth. Catal.*, **348**, 841–5;
(b) Chen, C. and Yang, L.M. (2007) *Tetrahedron Lett.*, **48**, 2427–30;
(c) Lee, C.C., Ke, W.C., Chan, K.T., Lai, C.L., Hu, C.H. and Lee, H.M. (2007) *Chem. Eur. J.*, **13**, 582–91.

17 (a) Shirakawa, E., Yamasaki, K. and Hiyama, T. (1997) *J. Chem. Soc., Perkin Trans.*, **1**, 2449–50;
(b) Shirakawa, E., Yamasaki, K. and Hiyama, T. (1998) *Synthesis*, 1544–9.

18 (a) Lipshutz, B.H. and Blomgren, P.A. (1999) *J. Am. Chem. Soc.*, **121**, 5819–20;
(b) Lipshutz, B.H. and Frieman, B. (2004) *Tetrahedron*, **60**, 1309–16;
(c) Lipshutz, B.H., Frieman, B.A., Lee, C., Lower, A., Nihan, D.M. and Taft, B.R. (2006) *Chem. Asian J.*, **1**, 417–29.

19 For another example of a nickel-catalyzed Negishi coupling of an aryl chloride, see: Walla, P. and Kappe, C.O. (2004) *Chem. Commun.*, 564–5.

20 Leleu, A., Fort, Y. and Schneider, R. (2006) *Adv. Synth. Catal*, **348**, 1086–92.

21 (a) Gronowitz, S., Hornfeldt, A.-B., Kristjansson, V. and Musil, T. (1986) *Chem. Scr.*, **26**, 305–9;
For more recent examples of this selectivity, see:
(b) Cocuzza, A.J., Hobbs, F.W., Arnold, C.R., Chidester, D.R., Yarem, J.A., Culp, S., Fitzgerald, L. and Gilligan, P.J. (1999) *Bioorg. Med. Chem. Lett.*, **9**, 1057–62;
(c) Jiang, B. and Yang, C.-G. (2000) *Heterocycles*, **53**, 1489–98;
(d) Schomaker, J.M. and Delia, T.J. (2001) *J. Org. Chem.*, **66**, 7125–8;
(e) Gong, Y. and Pauls, H.W. (2000) *Synlett*, 829–31.

22 Caron, S., Massett, S.S., Bogle, D.E., Castaldi, M.J. and Braish, T.F. (2001) *Org. Proc. Res. Dev.*, **5**, 254–6.

23 (a) Uemura, M., Nishimura, H., Kamikawa, K., Nakayama, K. and Hayashi, Y. (1994) *Tetrahedron Lett.*, **35**, 1909–12;
(b) Uemura, M., Nishimura, H. and Hayashi, T. (1994) *J. Organomet. Chem.*, **473**, 129–37;
(c) Prim, D., Tranchier, J.-P., Rose-Munch, F., Rose, E. and Vaissermann, J. (2000) *Eur. J. Inorg. Chem.*, 901–5.

24 For a review of palladium-catalyzed carbon–carbon bond formation using tricarbonyl(η^6-chloroarene)chromium complexes, see: Carpentier, J.F., Petit, F., Mortreux, A., Dufand, V., Basset, J.-M. and Thivolle-Cazat, J. (1993) *J. Mol. Cat.*, **81**, 1–15.

25 Shen, W. (1997) *Tetrahedron Lett.*, **38**, 5575–8.

26 Shen, W., Fakhoury, S., Donner, G., Henry, K., Lee, J., Zhang, H., Cohen, J., Warner, R., Saeed, B., Cherian, S., Tahir, S., Kovar, P., Bauch, S.-C., Ng, J., Marsh,

K., Sham, H. and Rosenberg, S. (1999) *Bioorg. Med. Chem. Lett.*, **9**, 703–8.

27 Littke, A.F. and Fu, G.C. (1998) *Angew. Chem. Int. Ed.*, **37**, 3387–8.

28 Littke, A.F., Dai, C. and Fu, G.C. (2000) *J. Am. Chem. Soc.*, **122**, 4020–8.

29 Netherton, M.R. and Fu, G.C. (2001) *Org. Lett.*, **3**, 4295–8.

30 For early studies of P(o-Tol)3-based monophosphine-palladium complexes, see: (a) Hartwig, J.F. and Paul, F. (1995) *J. Am. Chem. Soc.*, **117**, 5373–4;
(b) Hartwig, J.F. (1997) *Synlett*, 329–40.

31 Lam, K.C., Marder, T.B. and Lin, Z. (2007) *Organometallics*, **26**, 758–60.

32 Ishiyama, T., Ishida, K. and Miyaura, N. (2001) *Tetrahedron*, **57**, 9813–16.

33 Kudo, N., Perseghini, M. and Fu, G.C. (2006) *Angew. Chem. Int. Ed.*, **45**, 1282–4.

34 (a) Zapf, A., Ehrentraut, A. and Beller, M. (2000) *Angew. Chem. Int. Ed.*, **39**, 4153–5;
(b) Tewari, A., Hein, M., Zapf, A. and Beller, M. (2004) *Synthesis*, 935–41.

35 Fleckenstein, C.A. and Plenio, H. (2007) *Chem. Eur. J.*, 2701–16.

36 For related work utilizing 1-indenyldialkylphosphines and cyclopentadienyldialkylphopshines for palladium-catalyzed Suzuki coupling of aryl chlorides, see: Fleckenstein, C.A. and Plenio, H. (2007) *Organometallics*, **26**, 2758–67.

37 (a) For other recent examples of Suzuki couplings of aryl chlorides utilizing trialkyl phosphines, see: Bedford, R.B., Hazelwood, S.L., Limmert, M.E., Albisson, D.A., Draper, S.M., Scully, P.N., Coles, S.J. and Hursthouse, M.B. (2003) *Chem. Eur. J.*, 3216–27;
(b) Bedford, R.B., Butts, C.P., Hurst, T.E. and Lidstrom, P. (2004) *Adv. Synth. Catal.*, **346**, 1627–30;
(c) Moreno-Manas, M., Pleixats, R. and Serra-Muns, A. (2006) *Synlett*, 3001–4;
(d) Li, J., Deng, C. and Xie, Y. (2007) *Synth. Commun.*, 2433–48.

38 Old, D.W., Wolfe, J.P. and Buchwald, S.L. (1998) *J. Am. Chem. Soc.*, **120**, 9722–3.

39 (a) Wolfe, J.P. and Buchwald, S.L. (1999) *Angew. Chem. Int. Ed.*, **38**, 2413–16;
(b) Wolfe, J.P., Singer, R.A., Yang, B.H. and Buchwald, S.L. (1999) *J. Am. Chem. Soc.*, **121**, 9550–61.

40 (a) Yin, J., Rainka, M.P., Zhang, X.X. and Buchwald, S.L. (2002) *J. Am. Chem. Soc.*, **124**, 1162–3;
(b) Walker, S.D., Barder, T.E., Martinelli, J.R. and Buchwald, S.L. (2004) *Angew. Chem. Int. Ed.*, **43**, 1871–6;
(c) Barder, T.E., Walker, S.D., Martinelli, J.R. and Buchwald, S.L. (2005) *J. Am. Chem. Soc.*, **127**, 4685–96;
(d) Billingsly, K.L., Anderson, K.W. and Buchwald, S.L. (2006) *Angew. Chem. Int. Ed.*, **45**, 3484–8;
(e) Billingsly, K. and Buchwald, S.L. (2007) *J. Am. Chem. Soc.*, **129**, 3358–66;
(f) Barder, T.E. and Buchwald, S.L. (2007) *Org. Lett.*, **9**, 137–9.

41 Billingsly, K.L., Barder, T.E. and Buchwald, S.L. (2007) *Angew. Chem. Int. Ed.*, **46**, 5539–363.

42 (a) Bei, X., Crevier, T., Guram, A.S., Jandeleit, B., Powers, T.S., Turner, H.W., Uno, T. and Weinberg, W.H. (1999) *Tetrahedron Lett.*, **40**, 3855–8;
(b) Bei, X., Turner, H.W., Weinberg, W.H., Guram, A.S. and Petersen, J.L. (1999) *J. Org. Chem.*, **64**, 6797–803;
(c) Roca, F.X. and Richards, C.J. (2003) *Chem. Commun.*, 3002–3;
(d) Adjabeng, G., Brenstrum, T., Wilson, J., Frampton, C., Robertson, A., Hillhouse, J., McNulty, J. and Capretta, A. (2003) *Org. Lett.*, **5**, 953–5;
(e) Jensen, J.F. and Johannsen, M. (2003) *Org. Lett.*, **5**, 3025–8;
(f) Colacot, T.J. and Shea, H.A. (2004) *Org. Lett.*, **6**, 3731–4;
(g) Baillie, C., Zhang, L. and Xiao, J. (2004) *J. Org. Chem.*, **69**, 7779–82;
(h) Smith, R.C., Woloszynck, R.A., Chen, W., Ren, T. and Protasiewicz, J.D. (2004) *Tetrahedron Lett.*, **45**, 8327–30;
(i) Weng, Z., Teo, S., Koh, L.L. and Hor, T.S.A. (2004) *Organometallics*, **23**, 4342–5;
(j) Dai, W. and Zhang, Y. (2005) *Tetrahedron Lett.*, **46**, 1377–81;
(k) Konovets, A., Penciu, A., Framery, E., Percina, N., Goux-Henry, C. and Sinou, D. (2005) *Tetrahedron Lett.*, **46**, 3205–8;
(l) Teo, S., Weng, Z. and Hor, T.S.A. (2006) *Organometallics*, **25**, 1199–205;

(m) Demchuk, O.M., Yoruk, B., Blackburn, T. and Snieckus, V. (2006) *Synlett*, 2908–13;
(n) So, C.M., Lau, C.P. and Kwong, F.Y. (2007) *Org. Lett.*, **9**, 2795–8;
(o) Thimmaiah, M. and Fang, S. (2007) *Tetrahedron*, **63**, 6879–86.

43 Harkal, S., Rataboul, F., Zapf, A., Fuhrmann, C., Riermeier, T., Monsees, A. and Beller, M. (2004) *Adv. Synth. Catal.*, **346**, 1742–8.

44 Dai, Q., Gao, W., Liu, D., Kapes, L.M. and Zhang, X. (2006) *J. Org. Chem.*, **71**, 3928–34.

45 (a) Guram, A.S., King, A.O., Allen, J.G., Wang, X., Schenkel, L.B., Chan, J., Bunel, E.E., Faul, M.M., Larsen, R.D., Martinelli, M.J. and Reider, P.J. (2006) *Org. Lett.*, **8**, 1787–9;
(b) Guram, A.S., Wang, X., Bunel, E.E., Faul, M.M., Larsen, R.D. and Martinelli, M.J. (2007) *J. Org. Chem.*, **72**, 5104–12.

46 Pickett, T.E. and Richards, C.J. (2001) *Tetrahedron Lett.*, **42**, 3767–9.

47 Liu, S.-Y., Choi, M.J. and Fu, G.C. (2001) *Chem. Commun.*, 2408–9.

48 For example, see: Allenmark, S. (1974) *Tetrahedron Lett.*, **15**, 371–4.

49 Kwong, F.Y., Chan, K.S., Yeung, C.H. and Chan, A.S.C. (2004) *Chem. Commun.*, 2336–7.

50 For other examples of triarylphosphine-based palladium catalysts for Suzuki couplings of aryl chlorides, see: (a) Kocovsky, P., Vyskocil, S., Cisarova, I., Sejbal, J., Tislerova, I., Smrcina, M., Lloyd-Jones, G.C., Stephen, S.C., Butts, C.P., Murray, M. and Langer, V. (1999) *J. Am. Chem. Soc.*, **121**, 7714–15;
(b) Ohta, H., Tokunaga, M., Obora, Y., Iwai, T., Iwasawa, T., Fujihara, T. and Tsuji, Y. (2007) *Org. Lett.*, **9**, 89–92.

51 (a) Li, G.Y. (2001) *Angew. Chem. Int. Ed.*, **40**, 1513–16;
(b) Li, G.Y., Zheng, G. and Noonan, A.F. (2001) *J. Org. Chem.*, **66**, 8677–81;
(c) Li, G.Y. (2002) *J. Org. Chem.*, **67**, 3643–50.

52 See also: Miao, G., Ye, P., Yu, L. and Baldino, C.M. (2005) *J. Org. Chem.*, **70**, 2332–4.

53 (a) Ackermann, L. and Born, R. (2005) *Angew. Chem. Int. Ed.*, **44**, 2444–7;
(b) Ackermann, L., Gschrei, C.J., Althammer, A. and Riederer, M. (2006) *Chem. Commun.*, 1419–21;
(c) Ackermann, L., Born, R., Spatz, J.H., Althammer, A. and Gschrei, C.J. (2006) *Pure Appl. Chem.*, **78**, 209–14.

54 Kingston, J.V. and Verkade, J.G. (2007) *J. Org. Chem.*, **72**, 2816–22.

55 For some related Suzuki couplings of aryl chlorides utilizing *P,N*-type ligands, see: (a) Clarke, M.L., Cole-Hamilton, D.J. and Woollins, J.D. (2001) *J. Chem. Soc. Dalton Trans.*, 2721–3;
(b) Cho, S., Kim, H., Yim, H., Kim, M., Lee, J., Kim, J. and Yoon, Y. (2007) *Tetrahedron Lett.*, **63**, 1345–52.

56 For reviews of carbene ligands and their applications in homogenous catalysis, see: (a) Herrmann, W.A. and Kocher, C. (1997) *Angew. Chem. Int. Ed. Engl.*, **36**, 2162–87;
(b) Bourissou, D., Guerret, O., Gabbai, F. P. and Bertrand, G. (2000) *Chem. Rev.*, **100**, 39–91.
(c) Jafarpour, L. and Nolan, S.P. (2001) *Adv. Organomet. Chem.*, **46**, 181–222;
(d) Kantchev, E.A.B., O'Brien, C.J. and Organ, M.G. (2007) *Angew. Chem. Int. Ed.*, **46**, 2768–813;
(e) Diez-Gonzalez, S. and Nolan, S.P. (2007) *Top. Organomet. Chem.*, **21**, 47–82.

57 Herrmann, W.A., Reisinger, C.-P. and Spiegler, M. (1998) *J. Organomet. Chem.*, **557**, 93–6.

58 (a) Weskamp, T., Bohm, V.P.W. and Herrmann, W.A. (1999) *J. Organomet. Chem.*, **585**, 348–52;
(b) Herrmann, W.A., Bohm, V.P.W., Gstottmayr, C.W.K., Grosche, M., Reisinger, C.-P. and Weskamp, T. (2001) *J. Organomet. Chem.*, **617–618**, 616–28.

59 (a) Arduengo, A.J.III, , Dias, H.V.R., Harlow, R.L. and Kline, M. (1992) *J. Am. Chem. Soc.*, **114**, 5530–4;
(b) Arduengo, A.J.III, , Krafczyk, R., Schmutzler, R., Craig, H.A., Goerlich, J.R., Marshall, W.J. and Unverzagt, M. (1999) *Tetrahedron*, **55**, 14523–34;
(c) Arduengo, A.J., III (1999) *Acc. Chem. Res.*, **32**, 913–21.

60 Zhang, C., Huang, J., Trudell, M.L. and Nolan, S.P. (1999) *J. Org. Chem.*, **64**, 3804–5.

61 Bohm, V.P.W., Gstottmayr, C.W.K., Weskamp, T. and Herrmann, W.A. (2000) *J. Organomet. Chem.*, **595**, 186–90.

62 (a) Navarro, O., Kelly, R.A., III and Nolan, S.P. (2003) *J. Am. Chem. Soc.*, **125**, 16194–5;
(b) Navarro, O., Oonishi, Y., Kelly, R.A., Stevens, E.D., Briel, O. and Nolan, S.P. (2004) *J. Organomet. Chem.*, **689**, 3722–7;
(c) Marion, N., Navarro, O., Mei, J., Stevens, E.D., Scott, N.M. and Nolan, S.P. (2006) *J. Am. Chem. Soc.*, **128**, 4101–11.

63 Navarro, O., Marion, N., Mei, J. and Nolan, S.P. (2006) *Chem. Eur. J.*, **13**, 5142–8.

64 Furstner, A. and Leitner, A. (2001) *Synlett*, 290–2.

65 Altenhoff, G., Goddard, R., Lehmann, C.W. and Glorius, F. (2004) *J. Am. Chem. Soc.*, **126**, 15195–201.

66 Song, C., Ma, Y., Chai, Q., Ma, C., Jiang, W. and Andrus, M.B. (2005) *Tetrahedron*, **61**, 7438–46.

67 For other recent examples of Pd/N-heterocyclic carbene-catalyzed Suzuki couplings of aryl chlorides, see:
(a) Zhang, C. and Trudell, M.L. (2000) *Tetrahedron Lett.*, **41**, 595–8;
(b) Ding, S., Gray, N.S., Ding, Q. and Schultz, P.G. (2001) *Tetrahedron Lett.*, **42**, 8751–5;
(c) Arentsen, K., Caddick, S. and Cloke, F.G.N. (2005) *Tetrahedron*, **61**, 9710–15;
(d) Huang, W., Guo, J., Xiao, Y., Zhu, M., Zou, G. and Tang, J. (2005) *Tetrahedron*, **61**, 9783–90;
(e) Ozdemir, I., Demir, S. and Cetinkaya, B. (2005) *Tetrahedron*, **61**, 9791–8;
(f) Brendgen, T., Frank, M. and Schatz, J. (2006) *Eur. J. Org. Chem.*, 2378–83;
(g) Fleckenstein, C., Roy, S., Leuthauser, S. and Plenio, H. (2007) *Chem. Commun.*, 2870–2.

68 For some representative examples, see:
(a) Reetz, M.T., Breinbauer, R. and Wanninger, K. (1996) *Tetrahedron Lett.*, **37**, 4499–502;
(b) Bumagin, N.A. and Bykov, V.V. (1997) *Tetrahedron*, **53**, 14437–50;
(c) Zim, D., Monteiro, A.L. and Dupont, J. (2000) *Tetrahedron Lett.*, **41**, 8199–202;
(d) Villemin, D., Gomez-Escalonilla, M.J. and Saint-Clair, J.-F. (2001) *Tetrahedron Lett.*, **42**, 635–7;
(e) Grasa, G.A., Hillier, A.C. and Nolan, S.P. (2001) *Org. Lett.*, **3**, 1077–80.

69 LeBlond, C.R., Andrews, A.T., Sun, Y. and Sowa, J.R., Jr (2001) *Org. Lett.*, **3**, 1555–7.

70 Lipshutz, B.H., Sclafani, J.A. and Blomgren, P.A. (2000) *Tetrahedron*, **56**, 2139–44.

71 Sakamoto, T., Satoh, C., Kondo, Y. and Yamanaka, H. (1992) *Heterocycles*, **34**, 2379–84.

72 (a) Solberg, J. and Undheim, K. (1989) *Acta Chem. Scand.*, **43**, 62–8;
(b) Benneche, T. (1990) *Acta Chem. Scand.*, **44**, 927–31.

73 (a) Kondo, Y., Watanabe, R., Sakamoto, T. and Yamanaka, H. (1989) *Chem. Pharm. Bull.*, **37**, 2814–16;
(b) Majeed, A.J., Antonsen, O., Benneche, T. and Undheim, K. (1989) *Tetrahedron*, **45**, 993–1006.

74 Solberg, J. and Undheim, K. (1989) *Acta Chem. Scand.*, **43**, 62–8.

75 (a) Kosugi, M., Sasazawa, K., Shimizu, Y. and Migita, T. (1977) *Chem. Lett.*, 301–2;
See also: (b) Kashin, A.N., Bumagina, I.G., Bumagin, N.A. and Beletskaya, I.P. (1981) *Zh. Org. Khim.*, **17**, 21–8.

76 For examples of (η^6-chloroarene)chromium tricarbonyl complexes participating in palladium-catalyzed Stille couplings, see: (a) Scott, W.J. (1987) *J. Chem. Soc. Chem. Commun.*, 1755–6;
(b) Clough, J.M., Mann, I.S. and Widdowson, D.A. (1987) *Tetrahedron Lett.*, **28**, 2645–8;
(c) Mitchell, T.N., Kwekat, K., Rutschow, D. and Schneider, U. (1989) *Tetrahedron*, **45**, 969–78;
(d) Wright, M.E. (1989) *Organometallics*, **8**, 407–11;
(e) Wright, M.E. (1989) *Macromolecules*, **22**, 3256–9;
(f) Prim, D., Tranchier, J.-P., Rose-Munch, F., Rose, E. and Vaissermann, J. (2000) *Eur. J. Inorg. Chem.*, 901–5.

77 Littke, A.F. and Fu, G.C. (1999) *Angew. Chem. Int. Ed.*, **38**, 2411–13.

78 (a) Farina, V. and Rothin, G.P. (1996) *Advances in Metal-Organic Chemistry*, Vol. 5 (ed. L.S. Liebeskind), JAI, London, pp. 45–6;
(b) Brandsma, L., Vasilevsky, S.F. and Verkruijsse, H.D. (1998) *Application of Transition Metal Catalysts in Organic Synthesis*, Springer-Verlag, New York, p. 246;
(c) Farina, V., Krishnamurthy, V. and Scott, W.J. (1997) *Org. React.*, **50**, 54–5.

79 Littke, A.F., Schwarz, L. and Fu, G.C. (2002) *J. Am. Chem. Soc.*, **124**, 6343–8.

80 Grasa, G.A. and Nolan, S.P. (2001) *Org. Lett.*, **3**, 119–22.

81 (a) Wolf, C. and Lerebours, R. (2003) *J. Org. Chem.*, **68**, 7077–84;
(b) Wolf, C. and Lerebours, R. (2003) *J. Org. Chem.*, **68**, 7551–4.

82 For a Stille coupling of 4-chloronitrobenzene using a heteroatom substituted secondary phosphine oxide, see: Ackermann, L., Gschrei, C.J., Althammer, A. and Riederer, M. (2006) *Chem. Commun.*, 1419–21.

83 (a) Su, W., Urgaonkar, S. and Verkade, J.G. (2004) *Org. Lett.*, **6**, 1421–4;
(b) Su, W., Urgaonkar, S., McLaughlin, P.A. and Verkade, J.G. (2004) *J. Am. Chem. Soc.*, **126**, 16433–9.

84 (a) Matsumoto, H., Nagashima, S., Yoshihiro, K. and Nagai, Y. (1975) *J. Organomet. Chem.*, **85**, C1–3;
(b) Matsumoto, H., Yoshihiro, K., Nagashima, S., Watanabe, H. and Nagai, Y. (1977) *J. Organomet. Chem.*, **128**, 409–13.

85 Matsumoto, H., Shono, K. and Nagai, Y. (1981) *J. Organomet. Chem.*, **208**, 145–52.

86 For other examples of couplings of aryl chlorides with organodisilanes, see: (a) Azarian, D., Dua, S.S., Eaborn, C. and Walton, D.R.M. (1976) *J. Organomet. Chem.*, **117**, C55–7;
(b) Babin, P., Bennetau, B., Theurig, M. and Dunogues, J. (1993) *J. Organomet. Chem.*, **446**, 135–8;
(c) Lachance, N. and Gallant, M. (1998) *Tetrahedron Lett.*, **39**, 171–4.

87 Hagiwara, K.-i., Gouda, E., Hatanaka, Y. and Hiyama, T. (1996) *J. Org. Chem.*, **61**, 7232–3.

88 Hagiwara, E., Hatanaka, K.-i. Gouda, Y. and Hiyama, T. (1997) *Tetrahedron Lett.*, **38**, 439–42.

89 McNeill, E., Barder, T.E. and Buchwald, S.L. (2007) *Org. Lett.*, **9**, 3785–8.

90 Mowery, M.E. and DeShong, P. (1999) *Org. Lett.*, **1**, 2137–40.

91 Lee, H.M. and Nolan, S.P. (2000) *Org. Lett.*, **2**, 2053–5.

92 Murata, M., Yoshida, S., Nirei, S., Watanabe, S. and Masuda, Y. (2006) *Synlett*, 118–20.

93 Denmark, S.E. and Kallemeyn, J.M. (2006) *J. Am. Chem. Soc.*, **128**, 15958–9.

94 For other recent examples of Hiyama couplings of aryl/alkenyl siloxanes with aryl chlorides, see: (a) Wolf, C., Lerebours, R. and Tanzani, E.H. (2003) *Synthesis*, 2069–73;
(b) Ju, J., Nam, H., Jung, H.M. and Lee, S. (2006) *Tetrahedron Lett.*, **47**, 8673–8;
(c) Alacid, E. and Najera, C. (2006) *Adv. Synth. Catal.*, **348**, 2085–91.

95 Shiota, T. and Yamamori, T. (1999) *J. Org. Chem.*, **64**, 453–7.

96 Herrmann, W.A., Bohm, V.P.W. and Reisinger, C.-P. (1999) *J. Organomet. Chem.*, **576**, 23–41.

97 For some representative examples of Negishi couplings of activated aryl chlorides, see: (a) Uemura, M., Nishimura, H., Kamikawa, K., Nakayama, K. and Hayashi, Y. (1994) *Tetrahedron Lett.*, **35**, 1909–12;
(b) Uemura, M., Nishimura, H. and Hayashi, T. (1994) *J. Organomet. Chem.*, **473**, 129–37;
(c) Miller, J.A. and Farrell, R.P. (1998) *Tetrahedron Lett.*, **39**, 6441–4;
(d) Miller, J.A. and Farrell, R.P. (1998) *Tetrahedron Lett.*, **39**, 7275–8;
(e) Sonoda, M., Inaba, A., Itahashi, K. and Tobe, Y. (2001) *Org. Lett.*, **3**, 2419–21.

98 Dai, C. and Fu, G.C. (2001) *J. Am. Chem. Soc.*, **123**, 2719–24.

99 Milne, J.E. and Buchwald, S.L. (2004) *J. Am. Chem. Soc.*, **126**, 13028–32.

100 (a) Li, G.Y. (2002) *J. Org. Chem.*, **67**, 3643–50;
(b) Li, G.Y. (2002) *J. Organomet. Chem.*, **653**, 63–8.

101 For example, see: (a) Isobe, K. and Kawaguchi, S. (1981) *Heterocycles*, **16**, 1603–12;

(b) Minato, A., Suzuki, K., Tamao, K. and Kumada, M. (1984) *J. Chem. Soc. Chem. Commun.*, 511–13.

102 Bonnet, V., Mongin, F., Trecourt, F., Queguiner, G. and Knochel, P. (2001) *Tetrahedron Lett.*, **42**, 5717–19.

103 Katayama, T. and Umeno, M. (1991) *Chem. Lett.*, 2073–6.

104 Herrmann, W.A., Bohm, V.P.W. and Reisinger, C.-P. (1999) *J. Organomet. Chem.*, **576**, 23–41.

105 Huang, J. and Nolan, S.P. (1999) *J. Am. Chem. Soc.*, **121**, 9889–90.

106 Organ, M.G., Abdel-Hadi, M., Avola, S., Hadei, N., Nasielski, J., O'Brien, C.J. and Valente, C. (2007) *Chem. Eur. J.*, 150–7.

107 Walla, P. and Kappe, C.O. (2004) *Chem. Commun.*, 564–5.

108 Widdowson, D.A. and Wilhelm, R. (1999) *Chem. Commun.*, 2211–12.

109 Widdowson, D.A. and Wilhelm, R. (2003) *Chem. Commun.*, 578–9.

110 Kim, Y.M. and Yu, S. (2003) *J. Am. Chem. Soc.*, **125**, 1696–7.

111 Bahmanyar, S., Borer, B.C., Kim, Y.M., Kurtz, D.M. and Yu, S. (2005) *Org. Lett.*, **7**, 1011–14.

112 For other examples of palladium-catalyzed coupling reactions of aryl fluorides, see: (a) Mikami, K., Miyamoto, T. and Hatano, M. (2004) *Chem. Commun.*, 2082–3;
(b) Dankwardt, J.W. (2005) *J. Organomet. Chem.*, **690**, 932–8.

113 Bohm, V.P.W., Gstottmayr, C.W.K., Weskamp, T. and Herrmann, W.A. (2001) *Angew. Chem. Int. Ed.*, **40**, 3387–9.

114 Mongin, F., Mojovic, L., Guillamet, B., Trecourt, F. and Queguiner, G. (2002) *J. Org. Chem.*, **67**, 8991–4.

115 Ackermann, L., Born, R., Spatz, J.H. and Meyer, D. (2005) *Angew. Chem. Int. Ed.*, **44**, 7216–19.

116 Saeki, T., Takashima, Y. and Tamao, K. (2005) *Synlett*, 1771–4.

117 For other examples of nickel-catalyzed Kumada couplings of aryl fluorides, see: (a) Dankwardt, J.W. (2005) *J. Organomet. Chem.*, **690**, 932–8;
(b) Steffen, A., Sladek, M.I., Braun, T., Neumann, B. and Stammler, H.G. (2005) *Organometallics*, **24**, 4057–64;
(c) Yoshikai, N., Mashima, H. and Nakamura, E. (2005) *J. Am. Chem. Soc.*, **127**, 17978–9;
(d) Lu, Y., Plocher, E. and Hu, Q.-S. (2006) *Adv. Synth. Catal.*, **348**, 841–5;
(e) Inamoto, K., Sakamoto, J.-i., Kuroda, T. and Hiroya, K. (2007) *Synthesis*, 2853–61.

118 Schaub, T., Backes, M. and Radius, U. (2006) *J. Am. Chem. Soc.*, **128**, 15964–5.

119 For an example of a nickel-catalyzed Suzuki coupling of a heteroaryl fluoride, see: Liu, J. and Robins, M.J. (2005) *Org. Lett.*, **7**, 1149–51.

120 Nagatsugi, F., Uemura, K., Nakashima, S., Maeda, M. and Sasaki, S. (1995) *Tetrahedron Lett.*, **36**, 421–4.

121 For a related example, see: Lakshman, M.K., Thomson, P.F., Nuqui, M.A., Hilmer, J.H., Sevova, N. and Boggess, B. (2002) *Org. Lett.*, **4**, 1479–82.

122 Nguyen, H.N., Huang, X. and Buchwald, S.L. (2003) *J. Am. Chem. Soc.*, **125**, 11818–19.

123 (a) Roy, A.H. and Hartwig, J.F. (2003) *J. Am. Chem. Soc.*, **125**, 8704–5;
(b) Limmert, M.E., Roy, A.H. and Hartwig, J.F. (2005) *J. Org. Chem.*, **70**, 9364–70.

124 For a related mechanistic study of the oxidative addition of aryl sulfonates to palladium(0), see: Roy, A.H. and Hartwig, J.F. (2004) *Organometallics*, **23**, 194–202.

125 Ackermann, L. and Althammer, A. (2006) *Org. Lett.*, **8**, 3457–60.

126 Kobayashi, Y. and Mizojiri, R. (1996) *Tetrahedron Lett.*, **37**, 8531–4.

127 Zim, D., Lando, V.R., Dupont, J. and Monteiro, A.L. (2001) *Org. Lett.*, **3**, 3049–51.

128 Tang, Z.Y. and Hu, Q.S. (2004) *J. Am. Chem. Soc.*, **126**, 3059.

129 For other examples of nickel-catalyzed Suzuki couplings of aryl tosylates, see: (a) Percec, V., Golding, G.M., Smidrkal, J. and Weichold, O. (2004) *J. Org. Chem.*, **69**, 3447–52;
(b) Tang, Z.Y., Spinella, S. and Hu, Q.S. (2006) *Tetrahedron Lett.*, **47**, 2427–30.

130 Lipshutz, B.H., Frieman, B.A., Lee, C., Lower, A., Nihan, D.M. and Taft, B.R. (2006) *Chem. Asian J.*, **1**, 417–29.

3
Palladium-Catalyzed Arylations of Amines and α-C—H Acidic Compounds

Björn Schlummer and Ulrich Scholz

3.1
Introduction

The development of palladium catalysis as a standard synthetic tool in all areas of organic chemistry over the past decades has been impressive [1]. Indeed, the palladium-catalyzed arylation of amines represents one of the prime examples of an extremely fast transition, from an early research phase into an industrially viable and broadly applied technical process, in only 10 years. This reaction – which was dubbed the Buchwald–Hartwig amination after two key players in the field – can today be considered as a standard established method in the synthesis of arylamines, and has been widely reviewed [2]. In addition to allowing access to important target structures that are of biological or technical relevance in synthetic laboratories in industry or academia, the procedure has also overcome the problems of implementation in large-scale processes in chemical industry, where the demands for a new technology are often even higher and usually governed by substantially different parameters such as costs, space–time efficiency and technical feasibility [3].

During the period that the arylation of amines was being developed, another powerful methodology was discovered via a byproduct identified when carbonyl-containing substrates were employed. In the case of ketones bearing an acidic proton in the α-position, an arylation reaction at that position was observed, and this in turn gave rise to the development of arylations of α-C—H acidic compounds. Subsequently, this methodology matured significantly such that today a wide variety of substrates – including nitriles, esters and amides – can be arylated in the α-position using state-of-the-art catalytic systems [4].

Modern Arylation Methods. Edited by Lutz Ackermann

ISBN: 978-3-527-31937-4

3.2
Palladium-Catalyzed Arylations of Amines

3.2.1
Historical Development

The general concept of forming arylamines by the reaction of aryl halides with amines or their derivatives, can be traced back to the early days of organic chemistry, in the late nineteenth century. The easiest approach to these targets consists of a nucleophilic aromatic substitution reaction [5] using aryl halides and, although this dates back to 1891, it remains an effective and industrially relevant synthetic methodology to this day. Unfortunately, however, the method is limited to a small number of electrophile–nucleophile combinations, usually comprising electron-rich amines and heteroaryl or electron-deficient aryl halides, bearing one or more electron-withdrawing substituents, such as the nitro group [6]. Since, due to the harsh reaction conditions required, isomers can be formed through aryne intermediates [7], several other approaches have reported, including the substitution of nitro groups [8], direct aminations (such as the NASH reaction [9]) or radical processes, as in the Minisci reaction [10].

In 1901, when Ullmann introduced copper as an aid to C—N bond formations, he observed that copper in fact facilitated the substitution of halides at arenes [11]. The same principle was then applied to improve the synthesis of aryl ethers, and in 1903 Ullmann reported the synthesis of diphenyl amine derivatives with the aid of copper [12]. The important milestone publication from Ullmann and Irma Goldberg, who at the time were collaborating closely, was a report on the synthesis of arylanthranilic acid in 1905, whereby anthranilic acid was reacted with bromobenzene [13]. Surprisingly, the reaction conditions (catalyst [Cu], K_2CO_3, 210 °C, Ph-NO_2) still define the general principle that is applied today in modern, copper-catalyzed couplings. The tremendous development which followed, especially during the past decade, has led to significant improvements [14], with copper catalysis now having matured to become a standard technology. There remain, however, several drawbacks associated with its use, including limitations in substrate scope and the need for relatively high reaction temperatures. Perhaps most importantly, there are also environmental aspects such as the toxicity of copper in wastewater streams in industrial processes.

It is one of the many curiosities in the history of organic chemistry that the general approach to producing arylamines by catalytic C—N bond formation entered the arena of palladium-catalyzed transformations only during the late twentieth century. Previously, the major breakthroughs in palladium-catalyzed C—C cross-coupling chemistry had been achieved by key players such as Heck [15], Suzuki [16], Stille [17], Negishi [18] and others. This was accompanied by the mechanistic understanding that palladium is a key transition metal in the oxidative addition of C(sp^2)-halogen bonds, allowing coupling of the resultant intermediates with nucleophiles. Despite this, it was not until 1983 that palladium catalysis was

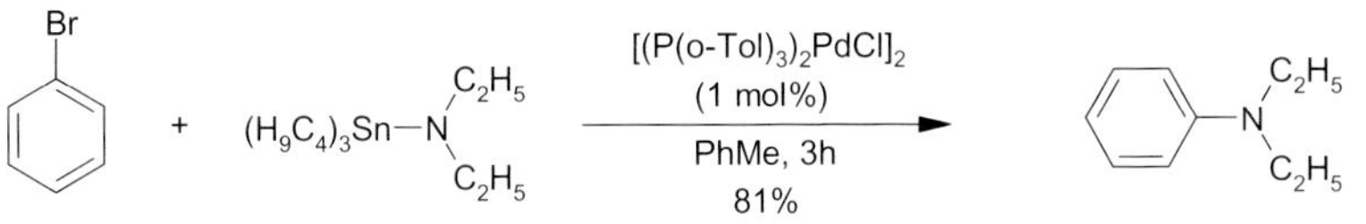

Scheme 3.1 Catalytic synthesis of *N,N*-diethylaniline by Migita (1983).

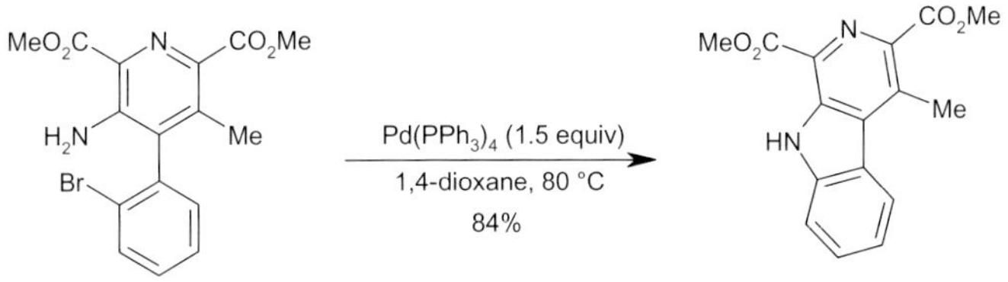

Scheme 3.2 Boger's intramolecular palladium-mediated amination from 1984.

first applied to the synthesis of arylamines, and when Kosugi and Migita first reported the synthesis of *N,N*-diethylaniline (Scheme 3.1) [19].

Of special note was the use of the electronically nonactivated bromobenzene as electrophile, combined with [{P(*o*-Tol)$_3$}$_2$PdCl$_2$] as catalyst for the transformation. A clear disadvantage here was the fact that the nucleophile, a tin amide, was not only toxic but also notoriously difficult to prepare [20].

During the following ten years a scattering of examples was reported which applied this concept. In 1984, Boger described an intramolecular palladium-mediated aryl amination using stoichiometric amounts of [Pd(PPh$_3$)$_4$] [21]. Although the reaction was accomplished by employing a free amine as nucleophile, this transformation was not developed further into an intermolecular variant (Scheme 3.2). Another example was described by Dhzemilev in 1987, in which the tin amide was substituted with a magnesium amide using Pd(OAc)$_2$ with PPh$_3$ in tetrahydrofuran (THF) as the reaction system [22].

Subsequently, this area of research lay almost dormant until Buchwald reported the *in situ* preparation of tin amides in 1994. Here, Bu$_3$Sn-NEt$_2$ was used as a relay for the formation of tin amides derived from the amine starting materials, through the removal of HNEt$_2$ [23].

The *annus mirabilis* of the palladium-catalyzed C—N bond formation was 1995, when Buchwald and Hartwig reported, independently, two milestone events – Buchwald the amination of aryl bromides with different amines using [{P(*o*-Tol)$_3$}$_2$PdCl$_2$] and NaO*t*-Bu as base [24], and Hartwig a similar procedure using LHMDS as base (Scheme 3.3) [25].

Based on these contributions and subsequent reports, the reaction is today often referred to as the Buchwald–Hartwig amination, and is considered as a standard fundamental organic reaction. Although a review of the myriad of exciting developments occurring since 1995 is beyond the scope of this chapter, we will nonetheless attempt to provide details of the current state of the art.

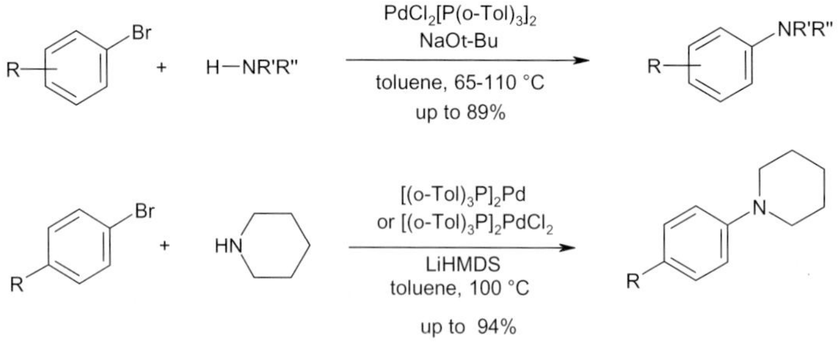

R–C$_6$H$_4$–Br + H–NR'R'' → R–C$_6$H$_4$–NR'R''
$PdCl_2[P(o\text{-}Tol)_3]_2$, NaOt-Bu, toluene, 65-110 °C, up to 89%

R–C$_6$H$_4$–Br + HN(piperidine) → R–C$_6$H$_4$–N(piperidine)
$[(o\text{-}Tol)_3P]_2Pd$ or $[(o\text{-}Tol)_3P]_2PdCl_2$, LiHMDS, toluene, 100 °C, up to 94%

Scheme 3.3 Buchwald's and Hartwig's aminations.

3.2.2
Catalytic Systems

In principle, the main components of the reaction system have not changed significantly since 1995, with palladium(0) or palladium(II) compounds continuing to constitute the main component of the catalytic systems. Although heterogeneous catalytic approaches have been reported [26], the vast majority of aminations are carried out using a suitable palladium precursor that dissolves during the course of the reaction.

The ligand, arguably, has the biggest influence on the reactivity of the catalyst, and the ease of modification of either the ligand backbone or the nature of the coordinating atom leads to the parameter matrix for the ligand being particularly important. Thus, the emphasis of most modern research activities is placed on the development of new or optimized ligand sets.

The base is a necessary component in the current reaction systems since, during palladium-catalyzed aminations an acid is formed that must be trapped by a suitable base. As most bases used in these reactions are commercially available substances (e.g. metal salts of basic anions), the latest developments have focused on tuning the catalytic system in order to allow the use of milder, but readily available, bases.

A solvent, which is usually selected from the standard organic solvent repertoire, is not a necessary prerequisite for the reaction itself, as shown in examples of solventless couplings. Apart from interactions with the palladium complexes, the solvent fulfils the usual role of dissolving all of the reaction components, dissipating the heat of reaction, and facilitating the work-up. A careful approach is suggested when encountering commonly dubbed 'green chemistry approaches' [27] which claim to avoid the use of solvents, or to substitute these with water. Rather, such methods should be judged on the basis of the overall flow of materials over the entire process, including waste streams and solvent usage in work-up. In fact, in most industrial processes the largest solvent and waste streams are normally generated during the work-up, and not during the reaction itself.

3.2.2.1 Palladium Sources

The palladium precursors for the catalytically active complexes can be classified as palladium(0) and palladium(II) compounds.

The most often used palladium(0) precursors are $Pd_2(dba)_3$ or $Pd(dba)_2$ (dba = dibenzylidenacetone; this is released during the formation of the active complex). In a study conducted by Fairlamb [28], it was shown that substituents on the arene of dba have a significant effect on the overall catalytic activity, destabilizing the η^2-interaction between palladium(0) and the dba derivative. Nevertheless, this effect is reported to be of minor importance in palladium-catalyzed aminations using $P(t\text{-}Bu)_3$ as ligand, possibly owing to the fact that dba in general is prone to react with amines, thus reducing the interference of dba itself with the catalyst [29]. The otherwise very popular $Pd(PPh_3)_4$ is less attractive in amination technology, as its activity is usually insufficient for efficient couplings of demanding substrates. Therefore, the use of an additional ligand is necessary, although this potentially creates difficulties with regards to ligand exchange rates and slowing down the catalytic reaction.

Palladium(II) precursors are more abundant as catalysts due to the larger variety of available metal salts. The most prominent palladium(II) precursor is $Pd(OAc)_2$, but in some cases the use of $[(\eta^2\text{-}C_3H_5)PdCl]_2$ or $Pd(acac)_2$ is beneficial. Nevertheless, it must be evaluated – based on the nature of the published application – if the use of a specific palladium precursor resulted from a catalyst screening with a reasonable parameter base, or if more pragmatic or subjective reasons were the basis for its choice. All palladium(II) precursors must be reduced to palladium(0) to enable oxidative addition of the carbon–halogen bond. As the ease of reduction of the palladium(II) starting material can depend on many reaction components – such as the amine, the base, the ligand, the solvent or the functionalized aryl halide – a variety of reaction pathways exists for the formation of the required palladium(0) species. In very challenging cases the use of reducing agents, such as phenylboronic acid [30], water [31] or sodium formate [32], is beneficial.

The application of *immobilized palladium* sources has also been reported recently. For example, Kobayashi described the use of polymer-incarcerated palladium catalysts for the synthesis of acridones [33]. Here, the catalyst was prepared from $Pd(PPh_3)_4$ and a polystyrene-based copolymer. Although the amination yields and scope using additional stabilizing ligands are not comparable to those obtained in homogeneous reaction, the use of this palladium precursor may have its advantages in the area of combinatorial catalysis [34]. The applications of preformed palladium complexes will be discussed in Section 3.2.2.2.6. The recovery or removal of palladium after the reaction is driven by two factors: (i) the minimization of costs; and (ii) the reduction of heavy-metal contamination of the products. A variety of strategies used for this purpose, most of which are based on scavenging reagents, have been described in a review [35].

3.2.2.2 Ligands

A plethora of ligands is available for palladium-catalyzed C–N bond formations, using a seemingly unlimited variety of combinations of amines and organic electrophiles. In general, the vast majority of problems posed by very challenging combinations of coupling partners has been solved. After an initial phase of ligand development, which has resulted in the synthesis and application of today's privileged highly active catalytic systems, ligand design is currently governed by ease of preparation (which mostly determines the costs of ligand production), stability and minimal catalyst loading, as well as nonchemical aspects such as patent protection.

As in many areas of transition-metal catalysis, it is difficult to classify the activity and versatility of ligands objectively, as most of the available data are derived from research groups that are also originators of the respective ligand class. Nevertheless, in the present authors' view, the field has matured to a state in which some privileged ligands have emerged over the past years in a number of independent applications reported by a large variety of different academic, as well as industrial, groups [36, 37].

3.2.2.2.1 **Bisphosphine Ligands** This ligand class was among the first to be used in palladium-catalyzed aminations, with BINAP being the most prominent example (Figure 3.1). Even today, a large number of reactions are accomplished using this privileged ligand, possibly owing to its good commercial availability as well as only slightly limited patent protection. Most reported examples refer to the racemic ligand, which shows a slightly better solubility in organic solvents than its enantiomerically enriched versions. Apart from being the center of various mechanistic studies [38], including a recent re-evaluation of the mechanism [39], this ligand has proven to be broadly applicable. Thus, it was found useful in combination with NaOMe as base [40], transformations of primary [41, 38] and acylic secondary amines [42], of aryl iodides at ambient temperature [43], of chiral amines [44], of aryl triflates [45], selective arylations of primary amines versus secondary amines [46] and also polymerization reactions [47]. The combination of BINAP with inexpensive $PdCl_2$ [48] was reported, as well as special applications to the synthesis of benzothiophenes [49].

Both, XantPhos and DPEPhos are remarkable ligands which were developed by van Leeuwen and coworkers [50]. These ligands are especially useful in the couplings of various amine derivatives, such as amides [51, 52], hydrazines [53], oxazolidinones [54] and ureas [55]. Furthermore, applications of XantPhos to the diphenylamine synthesis [48], and couplings of alkylarylamines [56], as well as

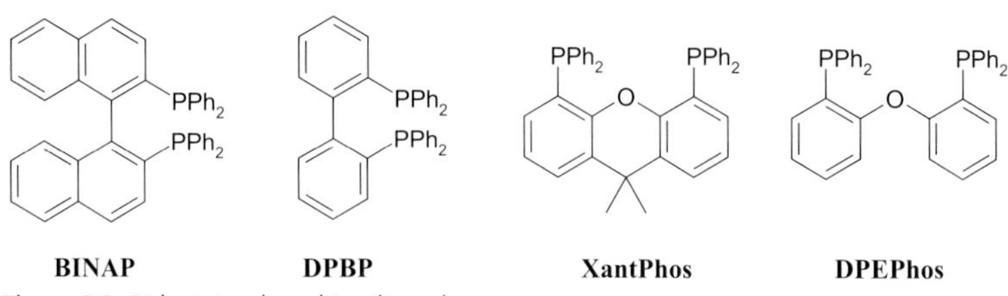

Figure 3.1 Bidentate phosphine ligands.

N-heterocycles [57]. Buchwald used XantPhos as a model ligand for mechanistic studies [51], thereby elucidating its ability to act in a *trans*-chelating mode [58].

DPEPhos was employed relatively early in the development of palladium-catalyzed aminations [59] and constitutes a viable ligand for the use of $PdCl_2$ [48].

DPBP was applied by Hayashi for the amination of aryl bromides [60], but no further applications of this ligand have been reported.

3.2.2.2.2 **Biphenyl-Based Monodentate Phosphine Ligands** This ligand class was used for the palladium-catalyzed amination of organic electrophiles by Buchwald in 1998 (Figure 3.2) [61]. Generally speaking, this privileged ligand set constitutes one of the most active ligand classes in the majority of aryl aminations.

DavePhos showed very high reactivity, in combination with mild bases, in the coupling of secondary amines [61, 62], aryl chlorides [63], even at ambient temperature, as well as in the preparation of aza crown ethers [64], and can be prepared in a surprisingly straightforward manner (Figure 3.2) [65]. Further, it was used for example in the arylation of 3-aminopentyl nitrile with aryl chlorides [66].

JohnPhos was employed for reactions of cyclic amines at ambient temperature [67], couplings of acyclic secondary amines [61] and for triarylamine synthesis [61, 67].

Cy-JohnPhos displayed wide applicability for aminations of aryl chlorides [61, 68].

X-Phos is probably the most active and versatile ligand of the biphenyl monophosphines for C—N bond formations [30]. It is one of the very few ligands that was found applicable to the conversion of nonfluorinated sulfonic esters [30]. These substrates are very attractive from an economical point of view [30].

Recently, the susceptibility of this ligand class to oxidation was the topic of a comprehensive study [69]. Despite their fairly electron-rich nature, these phosphines were found to be only moderately prone to oxidation, a property explained mainly by the unfavorable conformations in a transition-state model due to the interactions between an arene and the lone pair on the phosphorus. The synthesis of this ligand class was successfully scaled up to multi-hundred kilogram level, illustrating the importance of these ligands in chemical industry [70].

3.2.2.2.3 **Non-Biphenyl-Based Monophosphine Ligands** Among monophosphine ligands, P(*t*-Bu)$_3$ is one of the privileged (Figure 3.3). This compound was patented

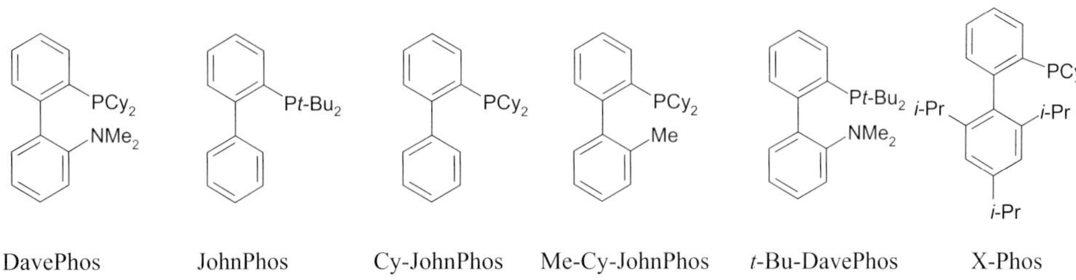

DavePhos JohnPhos Cy-JohnPhos Me-Cy-JohnPhos *t*-Bu-DavePhos X-Phos

Figure 3.2 Buchwald's biphenyl-based monodentate phosphines.

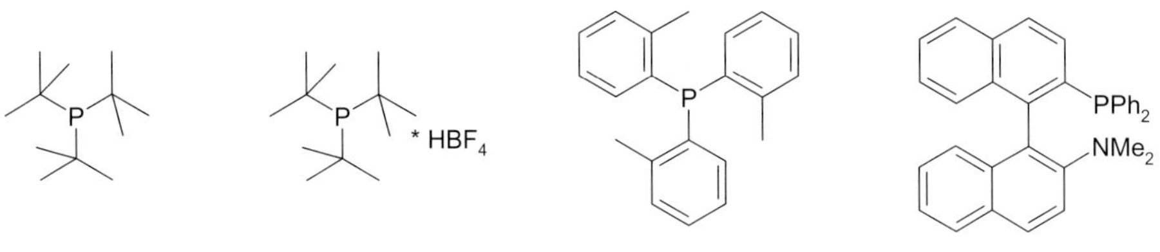

$P(t\text{-}Bu)_3$ $P(t\text{-}Bu)_3\text{-}HBF_4$ $P(o\text{-}Tol)_3$ MAP

Figure 3.3 Non-biphenyl-based monophosphine ligands.

by the Japanese company Tosoh at a very early stage, in 1997 [71]. Due to its very electron-rich nature and its major steric hindrance, $P(t\text{-}Bu)_3$ is among the most active ligands today for arylations of amines, and has been used extensively by the Hartwig research group [2]. Unfortunately, its high activity is accompanied by drawbacks in its handling, such as a high susceptibility to oxidation contributing to its pyrophoric nature, as well as its low melting point (30–36 °C). Nevertheless, $P(t\text{-}Bu)_3$ is commercially available even on a technical scale, and can be handled safely using a glovebox or Schlenk techniques, especially as a solution, for example in toluene [72].

Particularly for small-scale applications, Fu's report on the use of the tetrafluoroborate salt of $P(t\text{-}Bu)_3$ constitutes a viable alternative [73]. $P(t\text{-}Bu)_3$ has been used in many areas of palladium-catalyzed amination, such as the arylations of piperazines [74], couplings of secondary amines at ambient temperature [75] and triarylamine syntheses [75, 76]. PCy_3 was used for the palladium-catalyzed amination of aryl chlorides [77].

$P(o\text{-}Tol)_3$ was among the first ligands to be employed in palladium-catalyzed aminations, and has found applications in the field of triarylamine syntheses [76, 78] and couplings of aryl iodides [79]. However, it is not used extensively today.

MAP, a ligand designed by Kocosvký in 1997 [80], was among the first ligands for C—N bond formations, but is no longer used to any large extent.

3.2.2.2.4 **Ferrocene-Based Phosphine Ligands** Ferrocene-based ligands were among the first to be applied to palladium-catalyzed aminations, with dppf being the most prominent example (Figure 3.4). This is partly due to the early mechanistic assumptions that chelating ligands accelerate the product-forming reductive elimination [2]. Thus, DPPF was used as a viable alternative to BINAP, with comparable results [81]. Several applications in triarylamine syntheses were reported [76, 78], as well as couplings of aryl triflate [82], of nonaflates [76], and of primary amines [78]. Furthermore, selective monoarylations of primary diamines were also disclosed [83].

JosiPhos-type ligands, such as PPF-OMe (R = OMe, Figure 3.4), were particularly effective in promoting the conversion of acyclic secondary amines [84] and allowing the use of mild base Cs_2CO_3 [85]. With PPF-A (R = NMe_2, Figure 3.4) certain couplings of dibutylamine are accompanied by very little dehalogenation

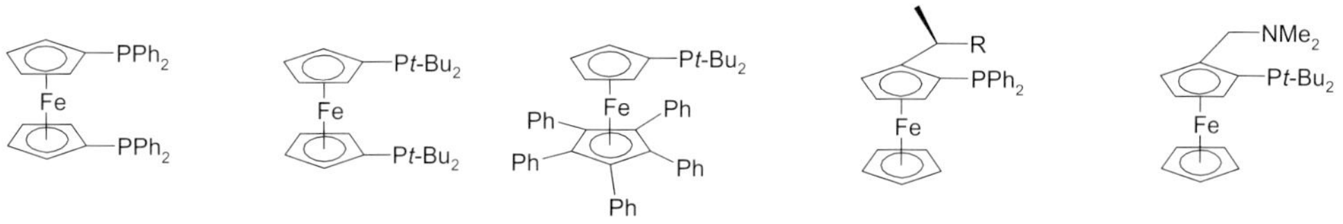

DPPF **D*t*BPF** **Q-Phos** **Josiphos-Type** **(CH_2NMe_2)D*t*BPF**

R = OMe: PPF-OMe

R_2 = NMe: PPF-A

Figure 3.4 Ferrocene-based ligands.

of the aryl halide and enable conversion of electron-rich aryl halides at ambient temperature [84]. Recently, Josiphos was used extensively by the Hartwig research group to couple heteroaryl and functionalized aryl chlorides with primary amines [86].

1-(*N*,*N*-Dimethylaminomethyl)-2-(di-*tert*-butylphosphino)ferrocene ((CH_2NMe_2)D*t*BPF, Figure 3.4) was used extensively for the coupling of halogenated indoles with cyclic amines [87].

Hartwig's Q-Phos is the ligand of choice for C—O bond formations, however performing also well in C—N bond formations [88]. The D*t*BPF ligand (Figure 3.4) is also effective in the coupling of cyclic amines with aryl chlorides [89].

3.2.2.2.5 ***N*-Heterocyclic Carbenes** The rapid development of *N*-heterocyclic carbenes (NHCs) as highly active ligands for transition metal-catalyzed processes during the past 10 years has been impressive [90]. Arguably triggered by their application in the rapid evolution of olefin metathesis [91], research on carbene chemistry led to the early recognition that these nonphosphine ligands combine a high electron density with considerable steric bulk – two factors that significantly influence the performance of transition-metal catalysts. Furthermore, the synthesis of NHCs is rather straightforward, starting from inexpensive raw materials [92, 93]. In most cases, the imidazolium salt is used as precursor for the carbene ligand [94].

Nolan reported several examples of palladium-catalyzed aminations using imidazolium salts as precursors to an unsaturated NHC ligand (Figure 3.5). Applications include arylations of piperidine [63] and couplings of aryl chlorides with additional secondary amines.

The corresponding saturated NHC ligands (Figure 3.5) were used by Hartwig in the coupling of aryl chlorides at ambient temperature [95]. In a recent study, NHC ligands were used with LHMDS as base to achieve couplings of aryl bromides and iodides under mild reaction conditions [96] although, unfortunately, the yields for challenging substrate combinations were found unsatisfactory.

Fort used carbene ligands for intramolecular aminations, yielding five-, six- and seven-membered nitrogen heterocycles [97].

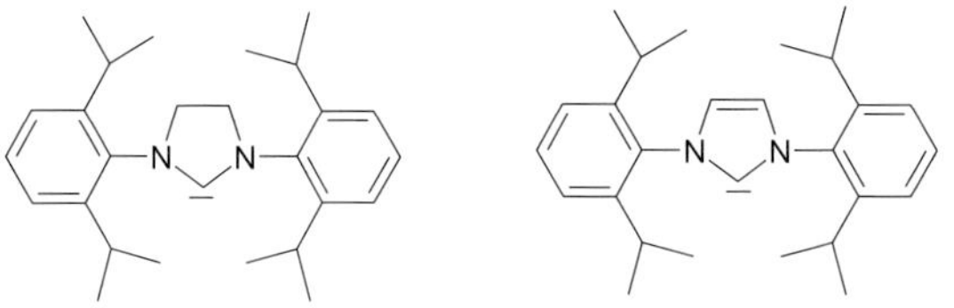

imidazol-2-ylidene imidazolin-2-ylidene

Figure 3.5 Carbene ligands.

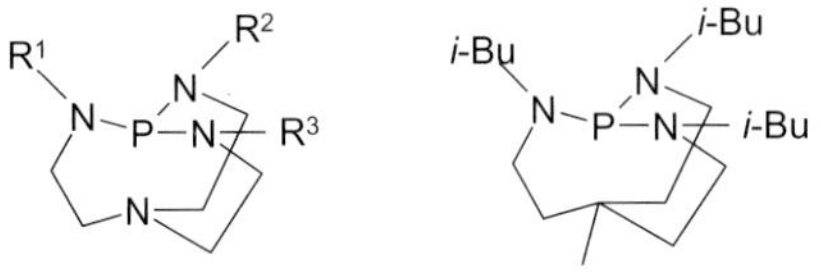

$R^1 = R^2 = R^3$ = Me, *i*-Pr, *i*-Bu, Bn
$R^1 = R^2$ = Bn, R^3 = *i*-Bu
$R^1 = R^2$ = *i*-Bu, R^3 = Bn

Figure 3.6 Verkade's-TAP ligands.

Due to the air-sensitivity of the free carbenes, catalysts derived thereof are often employed as preformed palladium–carbene complexes (these are discussed in Section 3.2.2.2.6).

3.2.2.2.6 **Recent Ligand Developments** Owing to the rapidly growing importance of palladium-catalyzed C—N bond formations in all areas of synthetic chemistry, a variety of different research groups have initiated programs in the quest for new ligands. The key drivers in such programs include the development of simpler and more easily accessible ligands, the identification of more active ligands for challenging substrate combinations, as well as possible nonproprietary ligand systems (this being one of the main motivations for industrial groups), and the discovery of completely new ligand classes.

Verkade: TAP Ligands Verkade introduced a new class of ligands for C—N bond formations, namely bicyclic tri-amino-phosphine (TAP) ligands (Figure 3.6) [98, 99]. The obvious advantage of the TAP ligands is their simple preparation. These ligands possess wide scope in the general coupling of halogenated arenes with various amines, with their catalytic activity in combination with the inexpensive base NaOH in couplings of aryl chlorides [99] being of particular note. Recently, these phosphines were used in a one-pot synthesis of unsymmetrically substituted *trans*-4-*N,N*-diarylaminostilbenes in a sequential coupling procedure [100]. The functional group tolerance could be extended to include alcohols, phenols, acetanilides, amides and enolizable ketones when employing LHMDS as base [101].

Singer: Arylpyrazole- and Arylpyrrole-Based Phosphine Ligands Singer introduced heterocyclic analogues of biphenyl monophosphine ligands (Figure 3.7). Singer's

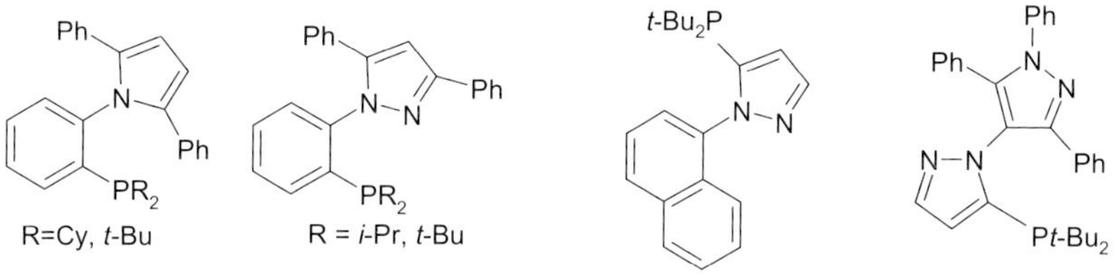

R=Cy, *t*-Bu　　R = *i*-Pr, *t*-Bu

Figure 3.7 Singer's Ligands.

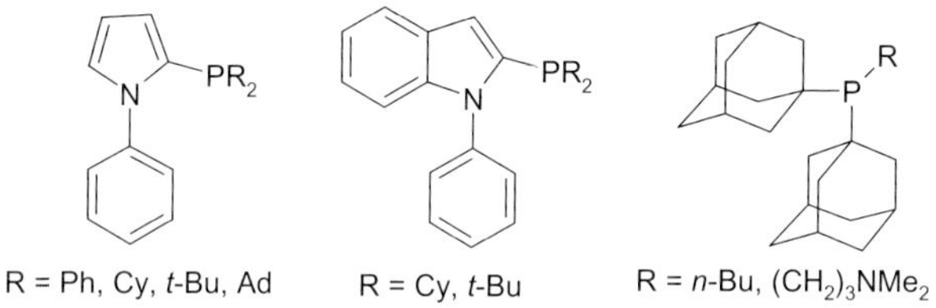

R = Ph, Cy, *t*-Bu, Ad　　R = Cy, *t*-Bu　　R = *n*-Bu, $(CH_2)_3NMe_2$

Figure 3.8 Beller's ligands.

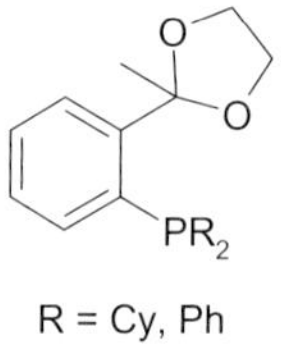

R = Cy, Ph

Figure 3.9 Guram's ligand.

studies work is a prime example of an industrial contribution in the quest for nonproprietary phosphines, with a clear focus on practicality. This ligand system is simple to prepare, and complexes derived thereof show interesting catalytic activity in a variety of applications [102]. Recently, a pyrazolylphosphine and a bipyrazole-based ligand were synthesized and applied to palladium-catalyzed aminations [103] although, interestingly, the latter ligand showed a much higher activity than the former. Thus, the bipyrazole-based ligand was prepared on a 2 kg scale as a crystalline solid [103]. A research group at Pfizer also reported the synthesis of other ligand families, including sulfone-containing phosphines and imidazole-substituted phosphines [104].

Beller: Adamantyl-, Arylpyrrole- and Arylindol-Based Phosphine Ligands The adamantly-, arylpyrrole-/imidazole- and indole-derived ligands [105] (Figure 3.8) presented by Beller are related to P(*t*-Bu)$_3$ and Singer's heterocyclic ligands, respectively. They were applied to aminations of aryl chlorides under mild reaction conditions [105].

Guram: P–O Ligands Guram's P–O ligands (Figure 3.9) were used for example for the coupling of bromoacetophenones with piperidine [106].

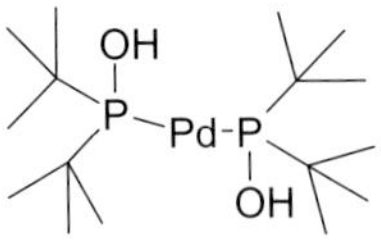

Figure 3.10 Li's dialkylphosphine oxide complex of palladium.

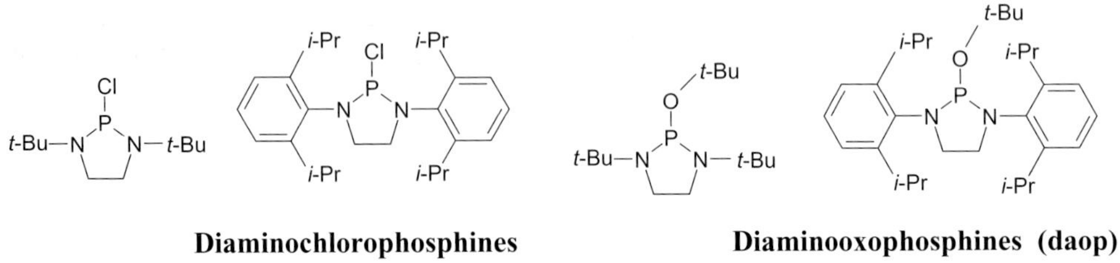

Diaminochlorophosphines

Diaminooxophosphines (daop)

Figure 3.11 Ackermann's ligands.

Ackermann: HASPO Ligands Ackermann presented two new ligand families, named heteroatom-substituted secondary phosphine oxides (HASPO) and the corresponding phosphine chlorides. The development of the former ligand class is related to studies of Li (DuPont), who described the use of *dialkyl*phosphine oxides with electronically distinct properties as active and easily accessible ligands for palladium-catalyzed aminations (Figure 3.10) [107].

Ackermann's diaminochlorophosphines and diamino-oxophosphines (daop) (Figure 3.11) show broad catalytic activity in the functionalization of electronically deactivated aryl chlorides, bromides and iodides [108], especially when using sterically hindered anilines bearing *ortho*-substituents [109]. The attractiveness of this ligand class is found in the modular character of the diaminophosphine derivatives, which are easily accessible and stable compounds [110].

Miscellaneous Several other research groups have presented the development of new ligands (Figure 3.12), the scope and limitations of which are still largely under investigation.

X. Zhang presented a triazole-based monophosphine dubbed Click-Phos [111]. This ligand can be obtained in two steps using a [3+2] cycloaddition as the key step, and used to functionalize aryl chlorides. Nevertheless, NaO*t*-Bu was employed as strong base and only a limited number of examples are available thus far.

Z. Zhang reported the synthesis and application of electron-rich MOP-type ligands in a comprehensive study [112]. The reported scope of these binaphthyl-based ligands is quite impressive, and matches those observed for of the biphenyl analogues. Nevertheless, their synthesis starting from BINOL includes several demanding synthetic transformations.

Shaughnessy presented bulky alkylphosphines (DTBNpP) bearing neopentyl substituents [113]. While being similar to $P(t\text{-}Bu)_3$, their larger cone angle renders

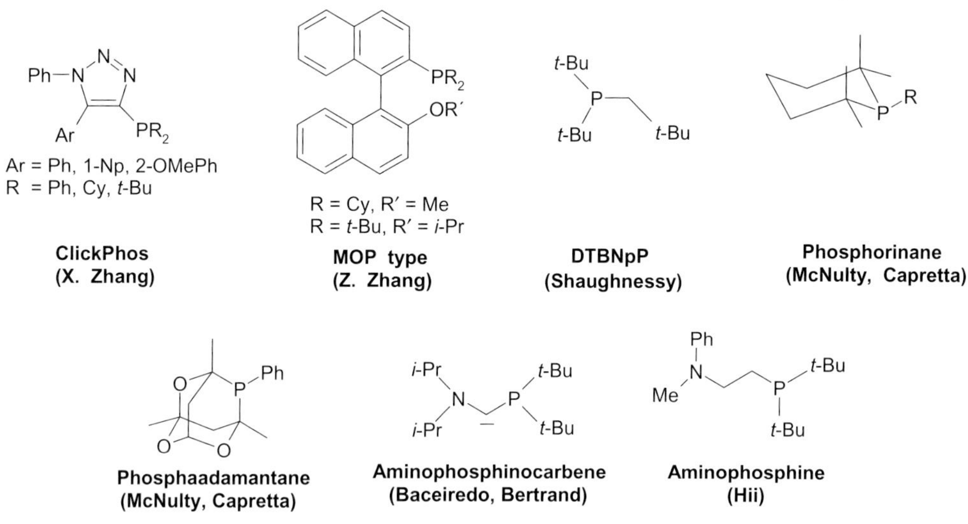

Figure 3.12 Recently developed ligands.

these ligands even more active. They can be synthesized on a bulk scale and are claimed not to be encumbered by licensing issues, although their air-sensitivity seems to be comparable to that of $P(t\text{-Bu})_3$. The scope of this ligand family is rather broad, and includes the coupling of deactivated aryl chlorides at elevated temperatures [113].

Hong reported the use of a cobalt-containing phosphine ligand in palladium-catalyzed aminations [114]. Unfortunately, this system does not offer any improved reactivity, and in addition its complex structure–containing two cobalt atoms–renders the ligand less attractive. Nonetheless, a very interesting palladium complex featuring this coordinated ligand could be isolated and characterized.

McNulty and Capretta disclosed the synthesis and use of phosphorinane ligands [115], and demonstrated catalytic activity in the amination of deactivated aryl chlorides. However, the use of NaO*t*-Bu as a strong base at elevated temperatures limits the scope of these compounds. The ligand scaffold is readily accessed by double conjugate additions of monosubstituted phosphines to phorone, followed by Wolf–Kishner reduction of the remaining ketone.

The same authors reported the use of a phospha-adamantane as a suitable ligand for catalytic aminations [116]. This ligand was found useful for the aminations of aryl iodides, bromides and chlorides in high yields, using NaO*t*-Bu as base. The ligand is crystalline, air-stable and can be recovered using chromatographic procedures.

Baceiredo and Bertrand combined the two privileged coordination moieties to an (amino)(phosphino)carbene ligand [117], after which the palladium complex formed by reaction of this ligand and $PdCl_2$(cod) was isolated and structurally characterized. Whilst the results of the amination experiments were not

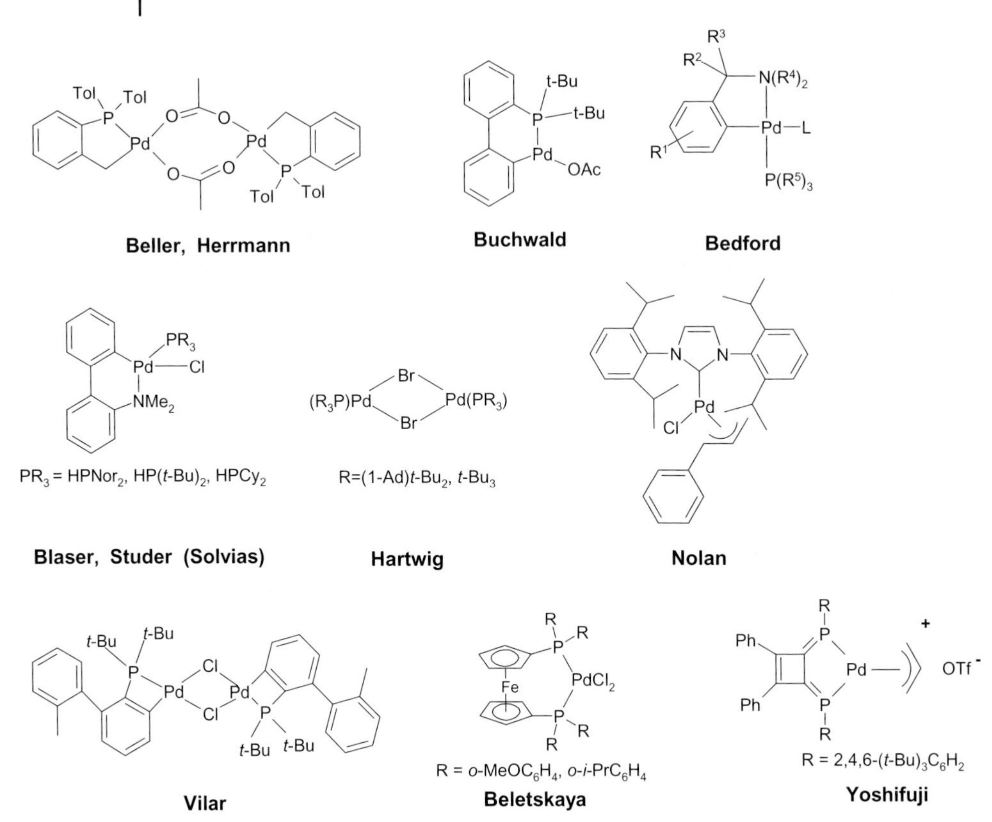

Figure 3.13 Preformed palladium complexes.

particularly impressive, they may lead to the development of cyclic versions of aminophosphinocarbenes that should show improved catalytic performance.

Hii reported an aminophosphine ligand similar to Amphos [118]. This ligand is prepared in four steps, starting from *N*-methylaniline and bromoacetic acid ethyl ester. An elevated temperature of 110 °C was required to achieve aminations of aryl bromides with secondary amines. Surprisingly, the order of reactivity is reported to be reversed to that usually encountered for palladium-catalyzed couplings, and consequently electron-deficient aryl bromides were aminated in lower yields than the corresponding electron-rich electrophiles. The ligand : metal ratio was identified as one of the key parameters for the reaction.

Preformed Palladium Complexes Preformed palladium complexes have been utilized by several research groups and are also produced on a commercial basis by a variety of companies (Figure 3.13). Occasionally, these complexes are more stable towards air or moisture than the isolated ligands, and often no premixing of precursors or initiation periods to form the active complex are necessary. However, as they lack the flexibility for tuning the ligand : metal ratios, they make variation of the catalyst more difficult.

One of the earliest preformed complexes was the *palladacycle*, as reported by Herrmann and Beller. This complex was shown to be stable at high reaction temperatures, and used in the amination of aryl chlorides with secondary amines [119].

Buchwald proposed a preformed palladium complex of his JohnPhos ligand [120]. Solvias even went a step further and used a palladacycle with a similar structure as precatalyst [121], newer variants of which were also reported [34, 122]. Hartwig showed the superior properties of a dimeric palladium complex derived from $P(t\text{-}Bu)_3$ in C–N bond formation at low temperature [123]; this complex is now marketed by Johnson Matthey.

Nolan presented a preformed version of a NHC-based catalyst which showed a high activity for aminations [124], as well as the synthesis and use of (NHC)Pd(allyl)Cl complexes and a structural study thereof elucidating a structure–activity relationship [125]. Nolan extended this concept more recently within a comprehensive study of (NHC)Pd(allyl)Cl complexes [126], a catalyst which allowed for the amination of aryl chlorides at ambient temperature and with (reported) comparably short reaction times. At higher reaction temperatures, a catalyst loading as low as 10 ppm palladium was claimed to be sufficient. A variant of this preformed NHC complex with coordinated acetylacetonate was also reported by Nolan [127], employing $Pd(acac)_2$ and the imidazolium salt precursor of the NHC as starting materials. Consequently, both aryl- and heteroaryl chlorides were coupled efficiently with secondary amines. Gooßen described a group of preformed naphthoquinone-bridged palladium–NHC complexes which are now marketed by Umicore [128], while Özdemir described the synthesis of a palladium–NHC complex and it application to aminations of aryl chlorides with primary aryl and alkyl amines in ionic liquids [129].

Bedford used a palladacycle complex [130], while Vilar investigated a variety of homobimetallic palladium complexes [131]. These species are based on Buchwald's biphenyl-type ligands and were found to be excellent precatalysts for aminations of aryl chlorides.

Beletskaya disclosed the details of metallocene-bridged bidentate phosphine ligands [132], which can be considered as variations of the well-known dppf-derived palladium complexes, but with additional alkoxy substituents. The same group prepared a series of palladium complexes based on the monoxides of dppf, but their performance in catalytic aminations was lower than that of the corresponding diphosphine complexes [133].

Yoshifuji developed a diphosphinidenecyclobutene ligand [134] and used its preformed palladium complex in solventless aminations in the presence of KO*t*-Bu as base.

3.2.2.3 **Bases**

The addition of at least stoichiometric amounts of a suitable base remains a necessary prerequisite for palladium-catalyzed aminations. Initially, palladium-catalyzed aminations required the use of strong bases, and even today many examples are reported which use strong alkoxide bases, such as NaO*t*-Bu, which limits the functional group tolerance. Thus, the main motivations for the optimization of bases were to improve functional group compatibility through the use of milder bases, and to identify further examples of soluble bases, as these are not accompanied by physical parameters, such as the mixing of heterogeneous systems. The use of inexpensive bases, such as alkali metal hydroxides, was another objective.

Bases commonly used in palladium-catalyzed aminations can be categorized as: (i) *soluble alkoxide bases*, such as NaO*t*-Bu, KO*t*-Bu, NaO*t*-Am or NaOMe; (ii) *sparingly soluble metal carbonates and phosphates*, such as Cs_2CO_3 and K_3PO_4 [52, 135], or in some cases K_2CO_3, silyl amides, such as LHMDS or NaHMDS; and (iii) *inexpensive hydroxides*, such as NaOH or KOH [30, 136]. Interestingly, the generation of phenols via C—O bond formations did not constitute a significant problem. It should be noted, however, that a viable method for a halide-OH exchange was reported recently by Buchwald [137].

In detailed recent studies, Maes investigated the influence of the amount and physical form of the weak base Cs_2CO_3 on the aminations of aryl iodides [118]. In particularly, Maes probed the different particle sizes that always come into play when using sparingly soluble bases. This holds especially true for Cs_2CO_3 [138], for which various commercially available specifications were previously employed.

Although the molar price of the base is more important for larger-scale applications [2], the absolute amounts of the required bases can also be of significance, especially for the high-molecular-weight base Cs_2CO_3. This can have a distinct effect on the practicality of the reaction, as the masses of the bases or alkali halide byproducts may exceed those of the desired coupling products, thus rendering the overall transformation rather inefficient in both economical and environmental terms.

For the coupling of ureas, McLaughlin reported on the use of the weak bases K_2CO_3 and $NaHCO_3$. It should be noted that although the latter was rarely used in palladium-catalyzed amination, it definitely represents an attractive alternative [139]

KF on alumina was reported by Basu as base in solventless mono- or polyaminations [140].

Apart from bisarylations and dehalogenations, most side reactions of palladium-catalyzed aminations did not result from the catalyst itself, but were caused by interactions of the base with functional groups of the substrates.

3.2.2.4 **Solvents**

Although solvents are not a necessary prerequisite for palladium-catalyzed aminations, the vast majority of these couplings was performed in organic solvent systems [2] or even in aqueous systems [136]. During recent years, mixed solvent systems have also been investigated in more detail [141–143].

The solvents most often used in catalytic aminations include toluene, THF, 1,4-dioxane, DME, DMF, NMP and DMSO, as well as their aqueous solutions [2, 30, 144].

Recently, the use of supercritical CO_2 as a reaction medium was reported by Holmes and Smith [145]. Here, the formation of carbamic acid from the corresponding amine with CO_2 could be avoided by the use of silyl amines. With X-Phos as ligand and Cs_2CO_3 as base, a range of couplings was accomplished, including the conversion of deactivated aryl chlorides, pyrroles, indoles and sulfonamides.

Solvent-free catalytic reactions were investigated by Yoshifuji [134] and Basu [140]. Yoshifuji used his preformed palladium complex (see Section 3.2.2.2.6) with KO*t*-Bu as base, while Basu reported solvent-free aminations using KF on alumina as base and BINAP as ligand. The advantages of these procedures are not always obvious, as mixing of the viscous or heterogeneous reaction mixture is more difficult in the absence of any solvent. Further, organic solvents are still required during the work-up of the reactions, either for extraction or for chromatography to isolate the desired products [146, 147].

Stauffer investigated the effect of polar solvents on palladium-catalyzed aminations using preformed [$\{P(t\text{-Bu})_3\}_2Pd$] as catalyst, and found a significant increase in reaction rates when using *N,N*-dimethylacetamide (DMA) or DMF as solvents. [148].

The sensitivity of palladium-catalyzed amination reactions towards air is heavily dependent on the properties of the phosphine ligand. The degassing of solvents is of utmost importance when employing electron-rich phosphines, such as P*t*-Bu_3, although for some other ligands the reactions can even be conducted under air, without any significant loss of efficacy. The presence of water is detrimental for only a few selected applications, as evidenced by several reports on C—N bond formation in aqueous media.

3.2.3 Aryl Halides

The aminations of aryl halides were the first to be studied in palladium-catalyzed C—N bond formations, and are thus well developed nowadays. As the oxidative addition of the C—X bond of the aryl halide often constitutes the rate-limiting step in catalytic aminations, a relative reactivity is usually observed that is comparable to that obtained in most transition metal-catalyzed cross-couplings. Thus, due to the decreasing bond dissociation energies, the reactivity order for halides is as follows: Ph-Cl > Ph-Br > Ph-I [149].

Arguably, protocols for aminations of aryl bromides are the best developed thus far, and the coupling of these electrophiles with the majority of amine nucleophiles has been reported [2]. Here, further developments mostly addressed the functional group compatibility, particularly when using milder bases [101, 150].

As in other transition metal-catalyzed cross-coupling reactions (see Chapter 2), the quest for efficient catalytic systems for the coupling of aryl chlorides has been one of the key motivations for further ligand development. Nowadays, different protocols exist for the aminations of aryl chlorides, even for reactions at ambient temperature. In general, highly active electron-rich phosphines, such as $P(t\text{-Bu})_3$

or the biphenyl monophosphine ligand family, as well as NHC-ligands, are privileged ligands for these transformations. However, these challenging reactions were also achieved recently with novel ligand systems [62, 63, 67, 95, 123]. For example, Maes reported the amination of aryl chlorides with Buchwald's ligands under microwave irradiation [151].

Aryl iodides have not been used extensively in palladium-catalyzed aminations, mainly because they are unattractive from an economical point of view, and do not perform exceptionally well in amination reactions, with dehalogenation often being a significant competing side reaction [98, 101, 118, 152, 153]. Rather, a recent report indicated that the greater proportion of aminations performed thus far has used aryl bromides and aryl chlorides as electrophiles [3].

3.2.4 Arylsulfonic Acid Esters

In recent years, significant efforts have been made to extend the scope of palladium-catalyzed aminations with respect to the leaving group on the electrophile, with particularly attention being focused on the esters of various acids [2, 3]. The main motivations for these developments were the good availability of the phenols as precursors, along with their usually low cost when compared to aryl halides. Further, the substitution pattern of the phenols is often complementary to that of available aryl halides.

Here, aryl triflates were most often employed as electrophiles. By using BINAP or dppf as ligands and strong bases (e.g. NaO*t*-Bu), both Buchwald and Hartwig were able to couple aryl triflates in early examples [45, 82]. Although a variety of ligands was found to be capable of coupling aryl triflates efficiently, the number of successful examples remained significantly lower than when using aryl halides. Subsequently, Skjaerbaek and coworkers studied several ligands in the aminations of aryl triflates, and found X-Phos to be the most efficient [143].

Recently, Buchwald reported the coupling of aryl nonaflates [154] with soluble bases under microwave irradiation. With a catalytic system comprising $Pd_2(dba)_3$, and either XantPhos, X-Phos or an X-Phos derivative as ligand, Buchwald efficiently coupled electron-rich and electron-neutral aryl nonaflates in the presence of DBU or MTBD as base, even when bearing functional groups, such as carbonyl-containing groups [155].

W. Zhang published details of the amination of aryl perfluorooctyl sulfonates [156], the system consisting of $Pd(OAc)_2$, BINAP and Cs_2CO_3, using either microwave irradiation or conventional heating. However, the reaction times were relatively long and the catalyst loading was rather high (10 mol%). Although it may be argued that the use of fluorinated leaving groups is less attractive because of their high costs and potential corrosivity, they continue to attract significant synthetic interest.

Much more appealing is the use of inexpensive aryl tosylates, and their use as electrophiles has also been reported. In 1998, Hartwig disclosed details of the amination of aryl tosylates employing D*t*BPF or JosiPhos as ligands [89]. The scope of JosiPhos-type ligands [157] was extended by Hartwig more recently with aminations of aryl tosylates at ambient temperature [158].

Buchwald investigated aminations of aryl tosylates in a detailed study [30]. Among a variety of ligands, X-Phos proved to be superior for a variety of coupling partners, while furthermore a catalytic system generated from $Pd(OAc)_2$, X-Phos and Cs_2CO_3 in a toluene/*t*-BuOH solvent mixture allowed for the efficient coupling of benzenesulfonates. In addition to commonly employed amine nucleophiles, amides could be arylated efficiently.

However, in their study on the amination of aryl tosylates with X-Phos as ligand and Cs_2CO_3 or NaO*t*-Bu as base, Skjaerbaek and colleagues obtained low yields.

3.2.5 Heteroaromatic Electrophiles

Arylations of heteroaryl electrophiles are often complicated by the fact that the heterocycle (e.g. a pyridine) coordinates to palladium and is, therefore, capable of replacing weakly coordinating ligands [159, 160]. However, the substitution patterns of electrophiles play a significant role with regards to the pathway of the reaction. For example, the 2-position in a pyridine nucleus is usually strongly activated for a standard nucleophilic aromatic substitution; in fact, this reaction can become the dominant transformation, especially in polar solvents using strong bases.

Consequently, a significant research effort has focused on the aminations of heterocyclic halides. Most prominent here are the pyridyl halides, for which several catalytically active systems based on ligands, such as biphenyl monophosphines, carbenes, TAPs or Josiphos-type phosphines, have been reported [52, 98, 62, 67, 99, 161].

Recently, Buchwald investigated the aminations of a variety of heteroaryl halides using his biphenyl-based ligand system [162]. Interestingly, for couplings with halothiophenes, S-Phos and *i*-Pr-S-Phos were the most effective ligands (Figure 3.14). Within this study, X-Phos proved to be the ligand of choice for the coupling of 5-bromopyrimidines [162], whereas benzoxazoles and benzothiazoles could be aminated using X-Phos, S-Phos or *i*-Pr-S-Phos. The haloindoles functioned best when using X-Phos or Dave-Phos [162].

XantPhos was also used in the coupling of various heteroaromatic halides [57, 98, 163]. Hartwig investigated not only the amination of halothiophenes and a

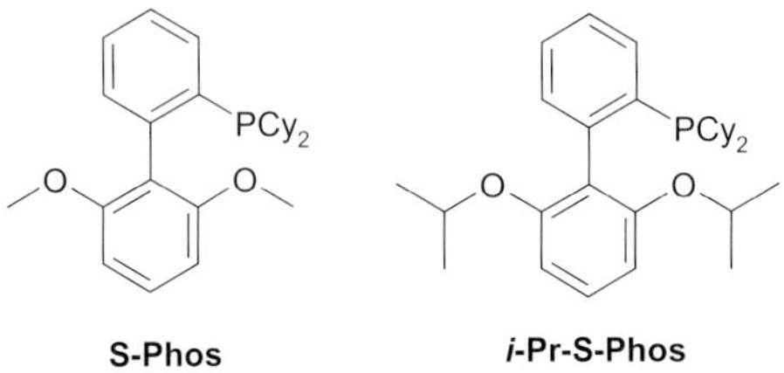

PCy_2 PCy_2

S-Phos ***i*-Pr-S-Phos**

Figure 3.14 S-Phos and *i*-Pr-S-Phos.

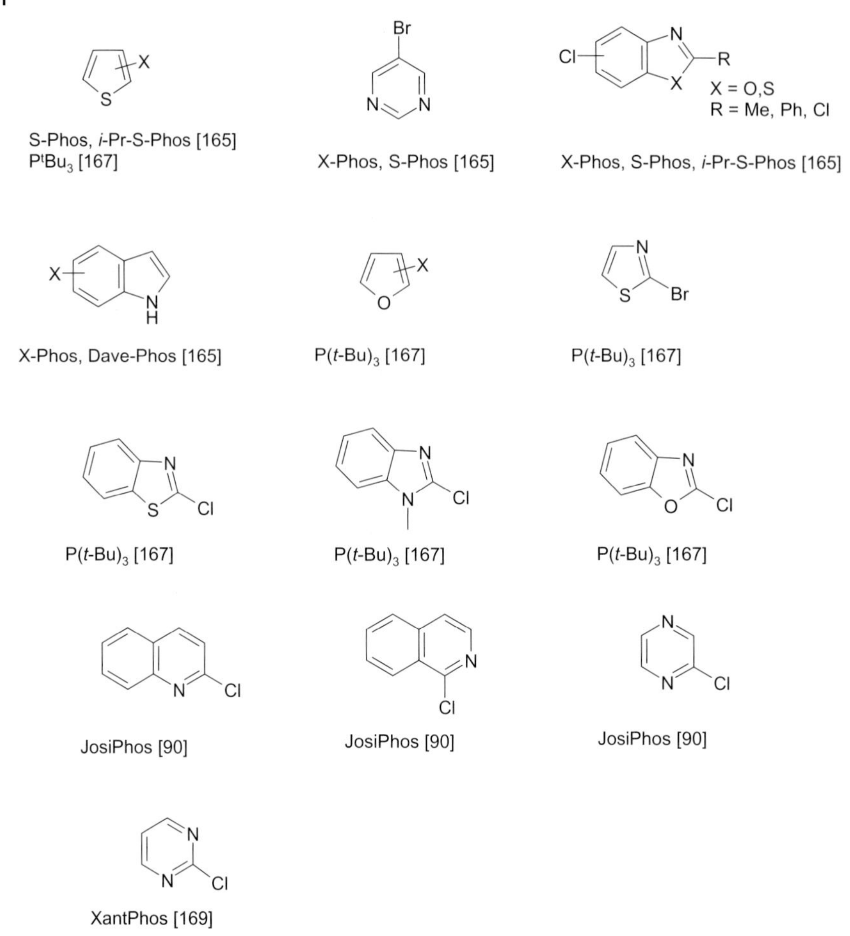

Figure 3.15 Selected recent examples of couplings of heteroaryl halides with electron-rich phosphines.

variety of other heteroaryl halides using P(*t*-Bu)$_3$ as ligand [164] (Figure 3.15), but also the use of JosiPhos for various couplings of heteroaryl halides [86].

Alami reported the use of XantPhos, along with Cs_2CO_3, for the coupling of 3-bromoquinolin-2-(1*H*)-ones with amines and their derivatives [165].

H. Q. Zhang disclosed aminations of chloropyrimidines using XantPhos as ligand and Cs_2CO_3 as base under microwave irradiation [166].

The functionalizations of nucleosides are also of continuous interest when using amination technology, and Lakshman reported several interesting applications in this research area [167].

Beller disclosed details of the coupling of piperazines with various heteroaryl chlorides such as pyridine, pyridazine, quinoline or isoquinoline derivatives, where both adamantyl-based ligands and biphenyl-derived monophosphines proved to be successful [168].

3.2.6 Amines as Nucleophiles

Today, the arylation of aliphatic and aromatic amines is a very mature area of organic chemistry, and both cyclic and acyclic nucleophiles can be coupled efficiently with a variety of electrophiles [2, 3]. As a general rule, the coupling of electron-rich aniline derivatives and of cyclic aliphatic amines are usually the most efficient substrates, and the reactions can be performed under a wide variety of conditions. The transformation of highly electron-deficient aromatic or heteroaromatic amines is more demanding, although more active, electron-rich ligands also offer a wide range of options to achieve these couplings. The coupling of branched aliphatic primary or secondary amines is sometimes hampered by steric effects as well as by competing β-hydride eliminations (the conversion of *N*-heterocycles as nucleophiles will be discussed separately; *vide infra*.) With these substrates, as well as with heteroaryl amines, coordination to palladium may be detrimental for the efficiency of the catalytic system, and consequently only recently significantly improved applications will be discussed at this point.

Recently, the arylation of ammonia as the simplest amine was accomplished by Hartwig [169] (Scheme 3.4). In the same publication, the use of lithium amide for the synthesis of anilines under similar reaction conditions was described; this constitutes a user-friendly protocol for small-scale settings, whereas the direct use of ammonia is attractive within an industrial environment. Subsequently, Buchwald also reported the arylation of ammonia using biphenyl-based ligands [170].

A variety of protocols exists for the coupling of ammonia equivalents that can be converted to the corresponding primary anilines. Hartwig described the use of LHMDS as a potent ammonia surrogate with $P(t\text{-Bu})_3$ as ligand [171, 172], where the silyl groups could be cleaved off easily during the work-up. Buchwald also used this reagent, but in combination with aminotriphenylsilane and biphenyl-based ligands [150].

More recently, Hartwig reported on the use of a zinc variant of LHMDS, namely $Zn(HMDS)_2$, as a mild ammonia equivalent in combination with LiCl or R_4NX as additives and $P(t\text{-Bu})_3$ as ligand [173]. This protocol has the advantage of extended functional group tolerance, allowing the use of substrates with, for example, enolizable groups. Hence, when (*S*)-naproxen methyl ester was deliberately added to the amination reactions no racemization occurred, yet addition of the stronger base LHMDS led to almost complete racemization.

Benzophenone imine was used extensively as an ammonia equivalent by various research groups [49, 174], mostly with BINAP or dppf as ligands [2]. Diallylamine

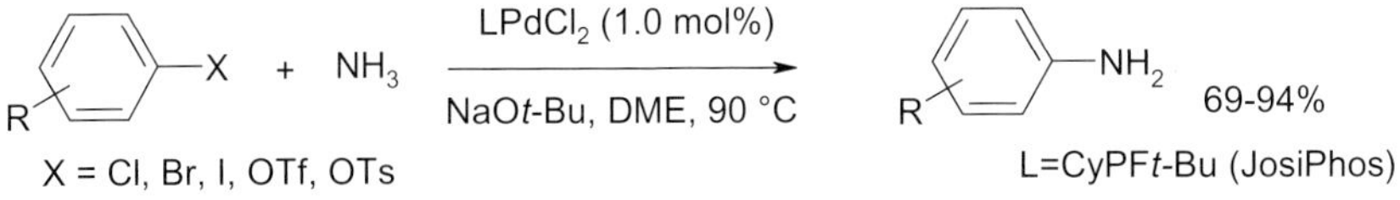

R–C₆H₄–X + NH_3 → (LPdCl₂ (1.0 mol%); NaO*t*-Bu, DME, 90 °C) R–C₆H₄–NH_2 69-94%

X = Cl, Br, I, OTf, OTs   L=CyPF*t*-Bu (JosiPhos)

Scheme 3.4 Palladium-catalyzed arylation of ammonia.

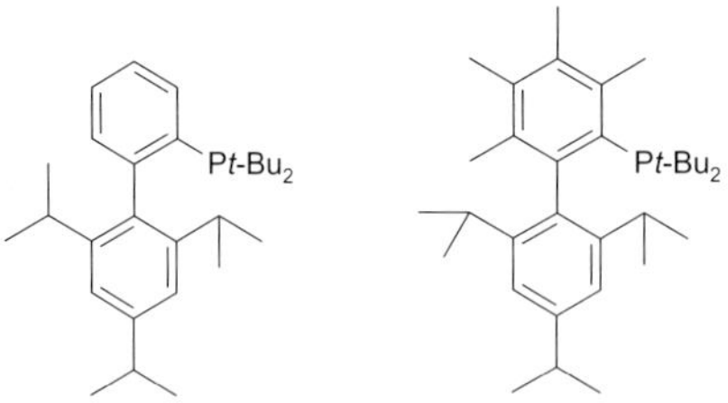

Pt-Bu2 Pt-Bu2

Figure 3.16 *t*-Butyl-X-Phos and permethylated *t*-Butyl-X-Phos.

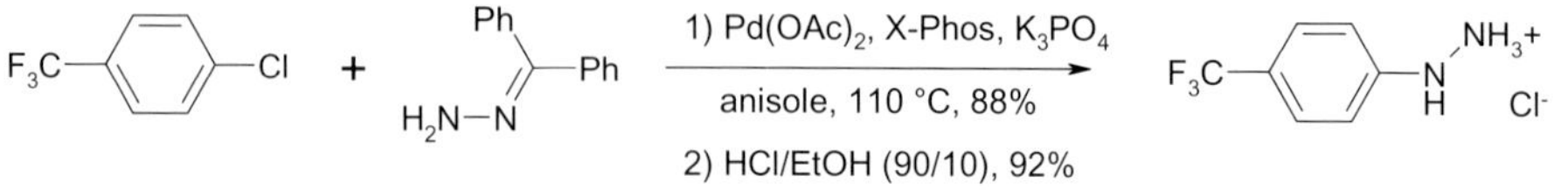

F_3C–C_6H_4–Cl + H_2N–N=C(Ph)Ph → 1) $Pd(OAc)_2$, X-Phos, K_3PO_4, anisole, 110 °C, 88%; 2) HCl/EtOH (90/10), 92% → F_3C–C_6H_4–NH–NH_3^+ Cl^-

Scheme 3.5 Coupling of benzophenone hydrazone on an 18 liter scale.

[175], benzyl- and diphenylmethylamine [176] were each also used to achieve such transformations. A fluorous analogue of benzophenone imine was presented by Herr [177] which facilitated purification by the use of fluorous chromatographic techniques.

The coupling of heteroarylamines and the *N*-arylation of *N*-heterocycles remains a field of intensive research. For example, Buchwald reported the arylation of indazoles and pyrazoles, as well as imidazoles and benzimidazoles [178]. Interestingly, the *t*-Butyl-X-Phos and a permethylated derivative (Figure 3.16) proved to be highly active ligands in these transformations, while NaO*t*-Bu and K_3PO_4 were used as bases in these couplings, in toluene as solvent and at elevated reaction temperatures. However, for many couplings of *N*-heterocycles, the use of copper still represents a viable, often favorable, alternative to palladium-catalyzed processes.

3.2.7
Amine Derivatives as Nucleophiles

The coupling of less-electron-rich N-nucleophiles, such as amides, ureas, carbamates, sulfonamides, guanidines and sulfoximines, can often be accomplished today [2, 3]. The coupling of electron-rich amine derivatives, such as hydrazine, has also been reported, while arylhydrazones were investigated by two industrial groups [65] (Scheme 3.5). However, most research effort was directed to the use of amides that were arylated in intramolecular reactions using ligands, such as MOP, DPEPhos or XantPhos, along with weak bases, such as Cs_2CO_3 or K_2CO_3 [52]. Buchwald reported the coupling of amides using XantPhos as ligand with aryl triflates as electrophiles [52]. In contrast, X-Phos proved to be a good ligand for the amidation of arenesulfonates [30, 144].

Legraverend reported the coupling of a variety of amides with 2,6-dihalopurines [179]; for this, he employed XantPhos as ligand and Cs_2CO_3 as base for arylations,

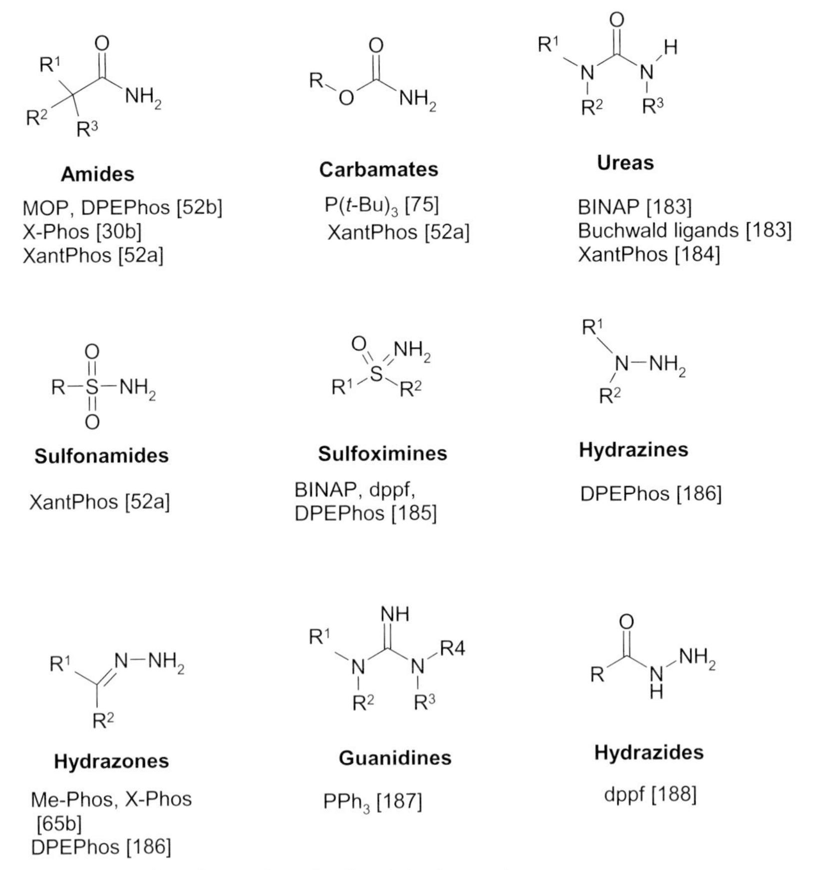

Figure 3.17 Selected examples of amine derivatives and ligands used for their palladium-catalyzed arylations.

and compared its regioselectivity to that of the nucleophilic substitutions using NaH in DMF.

Furuta and Kan reported the cyclization of bromosubstituted amides, yielding phenanthridinone derivatives, using their (phenyl)(naphthyl)phosphine ligands together with Cs_2CO_3 as base [180].

Klapars reported the synthesis of enamides by the coupling of enol tosylates with amides [181]. Here, an *i*-propyl-derivative of dppf was used as ligand together with the weak bases K_3PO_4 or K_2CO_3 in *t*-AmOH as solvent. Previously, XantPhos and Cs_2CO_3 were used for the same purpose in 1,4-dioxane as solvent by Wallace [182].

Carbamates were converted by Hartwig, using P(*t*-Bu)$_3$ as ligand [75]. XantPhos is also a viable ligand for couplings of these substrates [52] (Figure 3.17).

Details of the coupling of closely related ureas and cyclic carbamates were reported using BINAP or biphenyl-based phosphine ligands with weak bases, such as Cs_2CO_3 or K_3PO_4 [183]. McLaughlin reported a palladium-catalyzed urea cyclization with XantPhos as ligand and weak bases, such as Cs_2CO_3, K_3PO_4, K_2CO_3 or $NaHCO_3$ [184].

Sulfoximines were used as coupling partners by Bolm [185], who employed a variety of ligands for this transformation, such as BINAP, dppf or DPEPhos.

The coupling of hydrazones was the focus of a comprehensive investigation by Scholz, Mignani and Buchwald [65] in which both Me-Phos and X-Phos turned out to be the ligands of choice for these amine derivatives (Figure 3.17). With a catalytic system comprising $Pd(OAc)_2$, X-Phos and the inexpensive base K_3PO_4, the coupling of benzophenone hydrazone with 4-chlorobenzotrifluoride could be scaled-up to an 18 liter level, providing the arylhydrazone with 88% yield. Moreover, after treatment with HCl the free hydrazine was obtained in 92% yield (Scheme 3.5).

3.2.8
Applications

Due to an abundance of the arylamine substructure in almost all areas of synthetic organic chemistry–including pharmacological ingredients, crop protection products, natural products or functional materials, such as polyamines–it is not too surprising that the amination technology has been used extensively in a host of research areas.

Interestingly, apart from the plethora of contributions from academic groups, a large number of reports on the application of palladium-catalyzed C—N bond formations have originated from industrial groups, and this is reflected by the rather impressive number of patent applied for. A selection of the most recent examples is provided in Figure 3.18 [2, 3].

The synthesis of Clausine P and its analogues was described by Bedford using $P(t\text{-}Bu)_3$ as ligand and NaO*t*-Bu as base in toluene, with either conventional heating or microwave irradiation [189]. Pujol reported the synthesis of aryl piperazines as scaffolds of biologically active compounds [190], having used $[\{P(o\text{-}Tol)_3\}_2PdCl_2]$ as catalyst, along with either no additional ligand, PPh_3, or BINAP, and Cs_2CO_3 as base to couple aryl bromides with piperazines.

Chattopadhyay reported intramolecular aminations, yielding *cis*-fused furobenzoxazocines as sugar derivatives [191], using a catalytic system composed of Pd_2dba_3, BINAP and a mixture of KO*t*-Bu and K_2CO_3 as base in toluene as solvent. Likewise, Ila described the preparation of benzimidazo[1,2-*a*]quinolines via an intramolecular arylation of an amine, employing $Pd(PPh_3)_4$ as catalyst with K_2CO_3 in DMF [192].

Azacalix[n]arenes were prepared by Yamamoto [193], using Pd_2dba_3 and XantPhos with NaO*t*-Bu as base. Thereby, new azacalix[n]arenes were formed from *meta*-phenylendiamines and aryl dibromides, albeit in low yields.

Trans-4-*N,N*-diaryl aminostilbenes were synthesized by Verkade in a one-pot sequential procedure, consisting of sequential C—N and C—N bond formations [194]. Today, these compounds are widely used as electrophotographic photoconductors and receptors. By using 4-aminostyrene and bromobenzene derivatives as

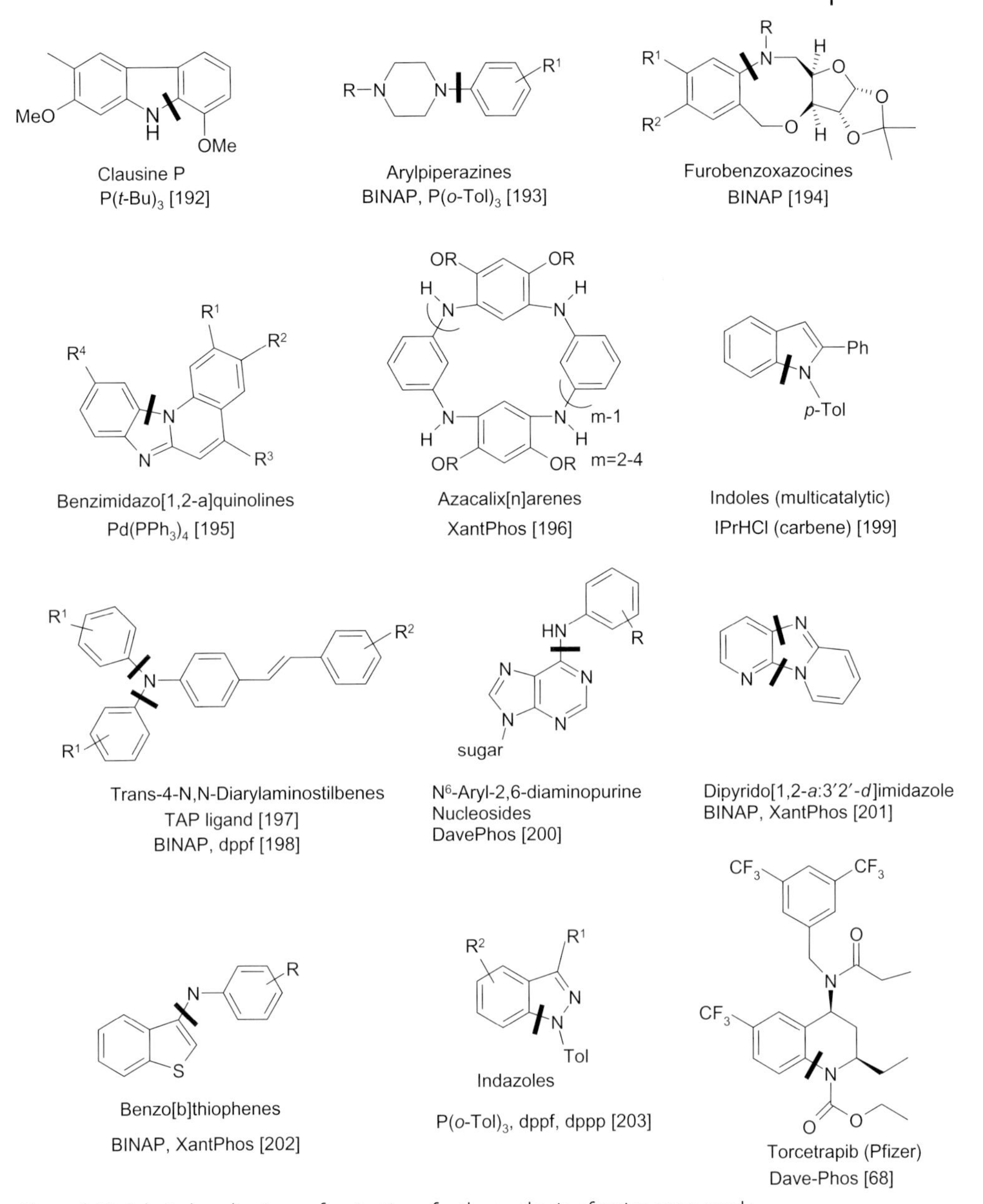

Figure 3.18 Selected applications of aminations for the synthesis of active compounds.

coupling partners, the desired triarylaminostilbenes were formed using a TAP ligand and NaO*t*-Bu as base. Aminostilbenes were also prepared by Yang, using BINAP or DPPF as ligand and NaO*t*-Bu as base [195].

Ackermann reported a multicatalytic one-pot indole synthesis starting from *ortho*-chloro-iodobenzenes and terminal acetylenes. By employing a mixed catalyst system composed of CuI, Cs_2CO_3, $Pd(OAc)_2$ and HIPrCl as carbene ligand, a Sonogashira coupling, an intermolecular amination and an intramolecular hydroamination sequence was accomplished [196].

Sulfonates derived from nucleosides were coupled with aryl amines by Lakshman [197], who used $Pd(OAc)_2$ and Dave-Phos as the catalytic system, along with K_3PO_4 as base, giving rise to a wide range of aryl amines in 1,4-dioxane/*t*-BuOH as solvent mixture.

Dipyrido[1,2-*a*:3′,2′-*d*]imidazole was prepared by Maes using a tandem double palladium-catalyzed amination. Here, $Pd(OAc)_2$ was used as the palladium source, BINAP or XantPhos as ligands, and Cs_2CO_3 as base [198].

Queiroz employed BINAP or XantPhos as ligands in the synthesis of benzo[b]thiophenes [199]. One of the most important drugs based on this structural motif is raloxifene, which is used in the treatment of osteoporosis, and potentially also for Alzheimer's disease.

Sakamoto prepared 3-substituted indazoles with $P(o\text{-Tol})_3$, dppf or dppp as ligand [200]. Here, NaO*t*-Bu, Cs_2CO_3 as well as LHMDS were suitable bases for the intramolecular coupling of arylhydrazone precursors.

A research group from Pfizer reported an extensive process development in the synthesis of their former development candidate torcetrapib [66, 201–203]. An interesting coupling of a primary alkyl amine, which displayed a sterogenic position in the α-position, and with *p*-benzotrifluoride as the electrophile, was described using biphenyl-based phosphine ligands.

3.2.9
Mechanistic Aspects

Since the early days of palladium-catalyzed aminations, the mechanism of the reaction has been the focus of many investigations. In particular, the research group of Hartwig has made seminal contributions during the past decade [2]. Buchwald has also investigated the mechanism of this transformation, focusing on the biphenyl-based ligand system as well as chelating ligands, such as XantPhos.

In a generalized and simplified mechanism, the reaction usually follows the standard catalytic cycle of metal-catalyzed cross-coupling reactions: oxidative addition of the $C(sp^2)$ –X bond to palladium(0), followed by coordination of the amine to the resulting palladium complex, occurring with extrusion of HX that is captured by the base. Finally, reductive elimination yields the coupling product, regenerating the catalytically active palladium(0) species.

In order to elucidate the mechanism involved, a variety of spectroscopic techniques as well as reaction calorimetry have been used for kinetic investigations, along with crystallographic and computational methods [30, 50, 62, 67, 95, 204, 205].

In a recent study, when the mechanism of reactions catalyzed by BINAP-ligated palladium complexes was revisited [39], it was shown that the bromoarene undergoes oxidative addition prior to a reaction of the palladium complex with the

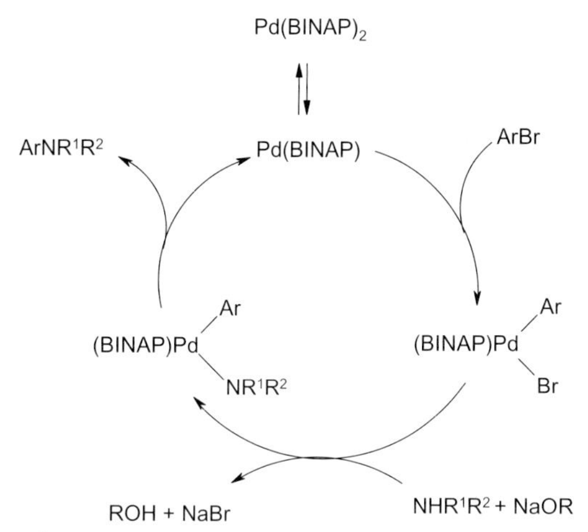

Scheme 3.6 Reaction mechanism with BINAP as ligand.

amine, and not through an initial coordination by the amine coordination (Scheme 3.6).

Subsequently, Green investigated the oxidative addition of aryl chlorides to palladium complexes derived from NHCs and their role in catalytic aminations [206], by using density functional theory (DFT) methods. Here, the most likely transition-state model for oxidative addition is the monocarbene–palladium complex (NHC)Pd(ArCl). For the amination of chlorobenzene, coordination of the T-shaped oxidative addition product by the amine occurs with a rearrangement, placing the amine in a *cis*-position to the aryl group, a prerequisite for reductive elimination [2].

3.2.10
Chirality

The arylations of amines usually do not create a new sterogenic center, except for few cases of hindered rotation in trisubstituted amines. In the past, therefore, chirality has played a minor role in transition metal-catalyzed aminations, and the vast majority of developed ligands were either achiral or were used in racemic form – with the exception of BINAP, for which both enantiomers are easily available.

Nevertheless, in a few selected examples stereochemical considerations do come into play. One very interesting case is that of the kinetic resolution of α-chiral alkyl amines. Ohta reported the use of (*R*)-Tol-BINAP for the reaction of 2-bromotoluene with 1-(2-naphthyl)-ethyl amine with NaOMe as base in the presence of 18-crown-6 [207]. The coupling product could be obtained in 85% yield with an enantiomeric excess (ee) of 74% (Scheme 3.7).

A set of chiral ligands was tested for the conversion of 4-bromobiphenyl with racemic 1-phenylethylamine, including (*R*,*R*)-ChiraPHOS (no product), (*S*,*S*)-

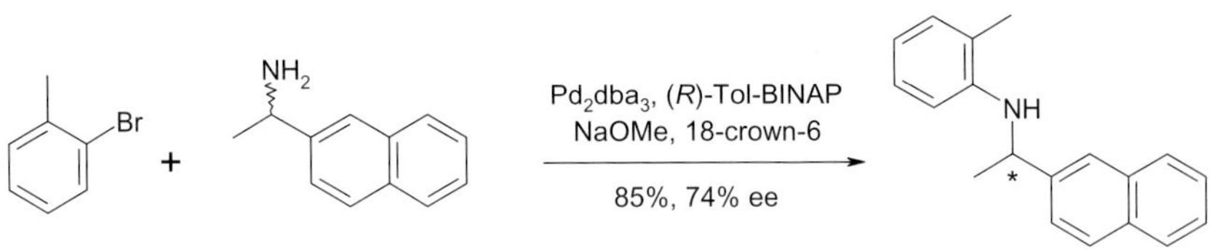

Scheme 3.7 Kinetic resolution of an α-chiral amine using (*R*)-Tol-BINAP [207].

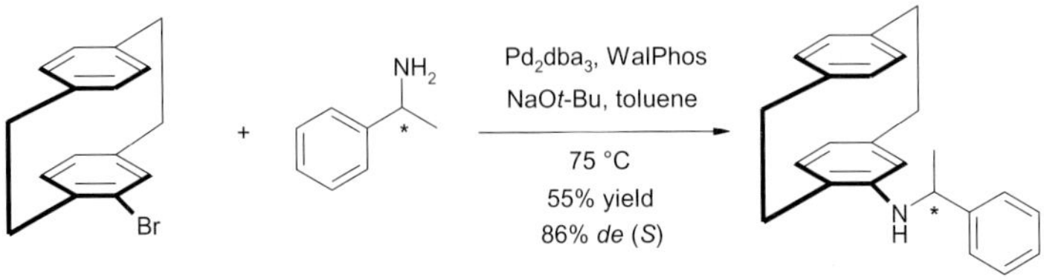

Scheme 3.8 Kinetic resolution of (*rac*)-paracyclophane with chiral ligands [208].

BPPM (no product), (*S,R*)-BPPFA (racemic product), (*S*)-*i*-Pr-MOP (racemic product), (*R,R*)-MOD-DIOP, (*S,S*)-BDPP, (*R*)-MOP and (*S*)-BINAP, the last four of which gave an optically active product.

Bräse reported the kinetic resolution of (*rac*)-4-bromo-[2,2]paracyclophane with (*S*)-phenylethylamine using various chiral ligands, thereby evaluating the effect of matched and mismatched combinations in diastereoselective aminations [208]. A variety of ligands was tested, including enantiomerically-pure or -enriched BINAP, Josiphos, Phanephos, WalPhos-, Mandyphos- and Taniaphos-type ligands. Here, the best results were obtained using the WalPhos ligand class (Scheme 3.8).

In general, the coupling of α-chiral primary alkyl amines is a demanding task due to the facile β-hydride elimination of the corresponding palladium–amide complex, which potentially leads to a loss of stereochemical information. The research group at Pfizer reported the racemization-free coupling of (*R*)-3-amino pentane nitrile with 4-chlorobenzotrifluoride and, after extensive optimization of the reaction conditions, Dave-Phos as ligand and Cs_2CO_3 as base were identified as being the most effective (Scheme 3.9) [66, 201–203].

3.3 Palladium-Catalyzed Arylations of α-C—H Acidic Compounds

3.3.1 Historical Development

The alkylation of α-C—H acidic compounds is, historically, one of the most investigated reactions in organic chemistry. Thus, the results from a plethora of studies on this transformation are available, focusing on the enolate nucleophile or the alkyl electrophile, with asymmetric variants being described in detail. In contrast,

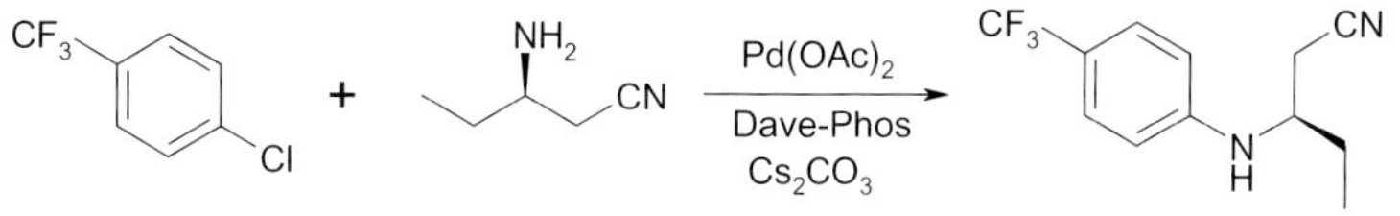

Scheme 3.9 Racemization-free coupling of (*R*)-3-amino pentane nitrile [66, 201–203].

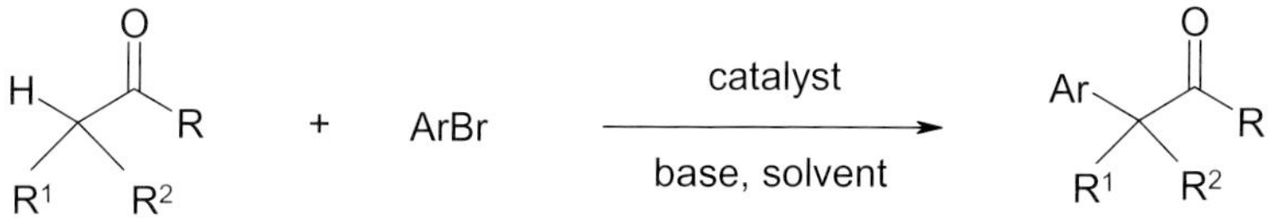

Hartwig: $Pd(dba)_2$, dppf or dtpf, KHMDS or NaO*t*-Bu, THF, reflux
Buchwald: $Pd_2(dba)_3$, BINAP or Tol-BINAP, NaO*t*-Bu, THF, 70 °C
Miura: $PdCl_2$-4LiCl, Cs_2CO_3 (ArI instead of ArBr)

Scheme 3.10 Palladium-catalyzed ketone arylations.

it is somewhat surprising that investigations of the complementary arylation of α-C—H acidic compounds were started rather late [4, 209].

The first attempts to use transition metal-catalyzed cross-coupling methodology to couple enolates with aryl halides were far from being broadly applicable, and involved less-attractive preformed zinc- or tin-enolates [210]. In addition, only ketones or acetates could be employed as enolate precursors. At that time, nickel was the predominant transition metal to mediate these couplings [211]. In an early contribution, Ciufolini arylated soft enolates in an intramolecular fashion with aryl halides using $Pd(PPh_3)_4$, along with NaH, in DMF at elevated temperatures [212]. This represented one of the first palladium-catalyzed arylations of enolates.

The challenges to be met in developing palladium-catalyzed arylations of α-C—H acidic compounds are significant. First, the pK_a-values of the pronucleophile vary over a large range, with the corresponding alkali metal enolates usually being reacted at low temperatures while the cross-couplings are generally performed at elevated reaction temperatures. Condensations of the enolates are to be expected as background reactions. Finally, the nature of the generated arylpalladium-enolate intermediates depend heavily on the substrate used.

During the early stages of development of this new method in 1997, the situation was strikingly similar to that for the palladium-catalyzed arylations of amines. Thus, it is not too surprising that major contributions to this research area originated again not only from the research groups of Buchwald and Hartwig, but also from the group of Miura. These three protagonists and their colleagues reported independently on the palladium-catalyzed couplings of aryl bromides with ketones (Scheme 3.10) [213].

For example, Hartwig used chelating ferrocene-based ligands, such as dppf or dtpf (1,1′-bis(di-*o*-tolylphosphino)ferrocene), while Buchwald employed chelating binaphthyl ligands, such as BINAP or Tol-BINAP. Relatively strong bases were used by both groups, including KHMDS or NaO*t*-Bu.

Miura, in contrast, presented an example of the coupling of iodobenzene with 1,3-diphenyl-2-propanone with $PdCl_2$-4LiCl as catalyst. Initially, it was speculated that chelating ligands were necessary to avoid β-hydride elimination by forming a four-coordinate palladium complex without open coordination sites, which are needed for such an elimination to occur. Later, however, it was found that the coordination of the potentially chelating ligand D*t*-BPF occurred in a η^1-fashion [214].

Consequently, sterically hindered monophosphines, such as $P(t\text{-}Bu)_3$, were tested and provided excellent yields in the α-arylation of C—H-acidic compounds [214]. Today, monodentate phosphine ligands are broadly employed for this palladium-catalyzed arylation technology.

3.3.2
Catalytic Systems

The components of the catalytic systems for these C—C-bond formations are very similar to those used for the corresponding aminations. A palladium precursor in conjunction with a stabilizing and activating ligand define the two key components of the active catalyst. Whilst the focus of palladium-catalyzed amination lies in ligand development, in α-arylation the choice of base plays an equally important role, due mainly to the fact that the base has a decisive influence on the type and reactivity of the enolate formed from the carbonyl compound.

Intensive investigations were performed by Hartwig with regards to the choice of base, which is strongly influenced by the pK_a-values of the employed carbonyl compounds. Among the plethora of bases probed, both strong and sterically hindered bases – such as NaO*t*-Bu, LHMDS, NaHMDS or $LiNCy_2$ – produced very good results. However, the very attractive bases K_3PO_4 and Na_3PO_4 could also be used successfully in the arylation of more acidic substrates, such as malonates, α-cyano esters or protected amino acids [4].

As the catalytic system is heavily dependent on the type of C—H-acidic compound, a general discussion on the catalytic system is not relevant at this point. Rather, in the following sections the reaction systems will be described based on the type of substrate employed.

3.3.3
α-Arylations of Esters

Due to their relatively high acidity, the challenges in the arylation of esters are represented by the disposition of the ester enolates to uncatalyzed condensation reactions, or thermal decompositions. In order to suppress any undesired self-condensation reactions, *t*-butyl-substituted esters are often used as substrates.

For example, Hartwig used $Pd(dba)_2$ in combination with $P(t\text{-}Bu)_3$ or carbene-type ligands as catalysts, whilst LHMDS, NaHMDS or $LiNCy_2$ proved to be effec-

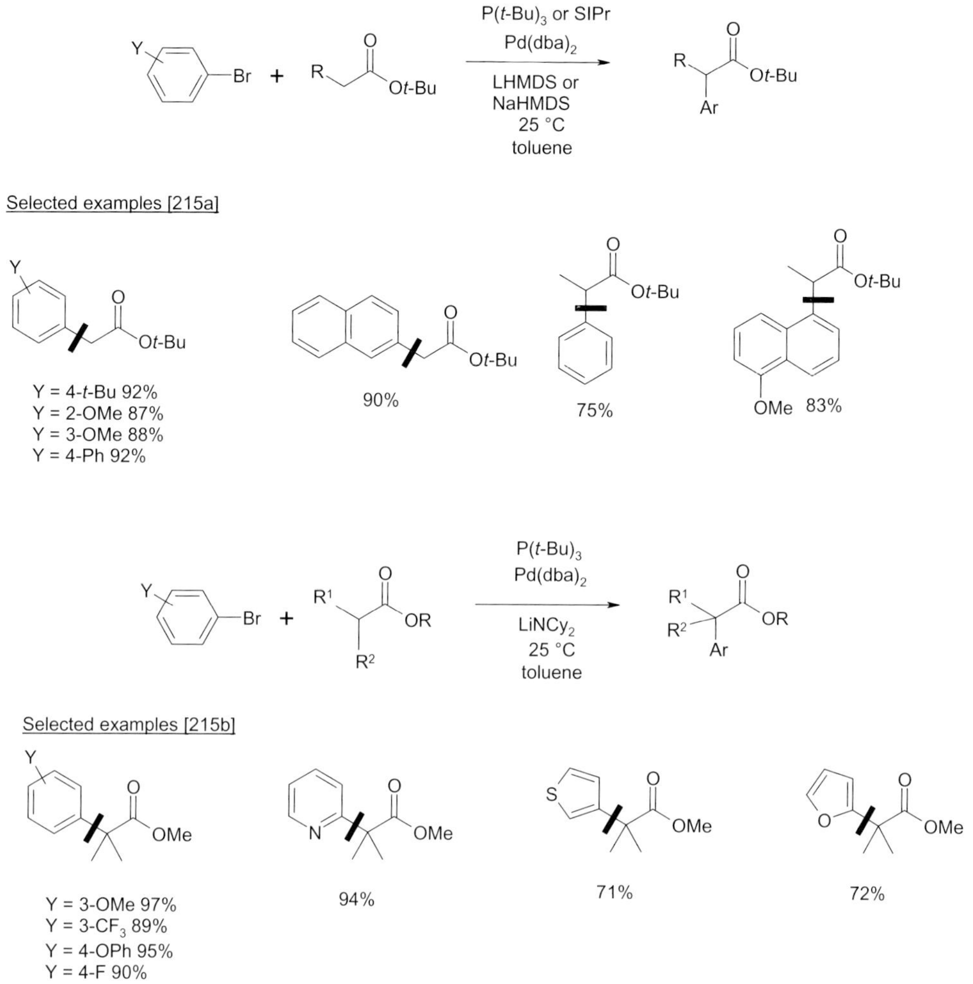

Selected examples [215a]

Y = 4-t-Bu 92%
Y = 2-OMe 87%
Y = 3-OMe 88%
Y = 4-Ph 92%

90%

75%

83%

Selected examples [215b]

Y = 3-OMe 97%
Y = 3-CF_3 89%
Y = 4-OPh 95%
Y = 4-F 90%

94%

71%

72%

Scheme 3.11 α-Arylation of esters.

tive as bases. Moreover, the reactions could be conducted at ambient temperature and delivered the coupling products in excellent yields (Scheme 3.11) [215].

In 2003, Hartwig developed the α-arylation of esters under more neutral reaction conditions [216], and with Q-Phos or its palladium(I) dimer was able to successfully arylate zinc enolates. In contrast, $P(t\text{-}Bu)_3$ enabled arylations with silicon-based enolates, bearing potentially reactive functional groups (Scheme 3.12).

In 2001, Buchwald reported the α-arylation of esters using not only biphenyl-type ligands but also t-Bu-MAP [217]. The t-butyl acetates could be successfully arylated using DavePhos as ligand and LHMDS as base, and the same system was employed for the formation of α-aryl propionic esters and for the α-arylation of ethyl phenyl acetates. Aryl chlorides could also be used as coupling partners for

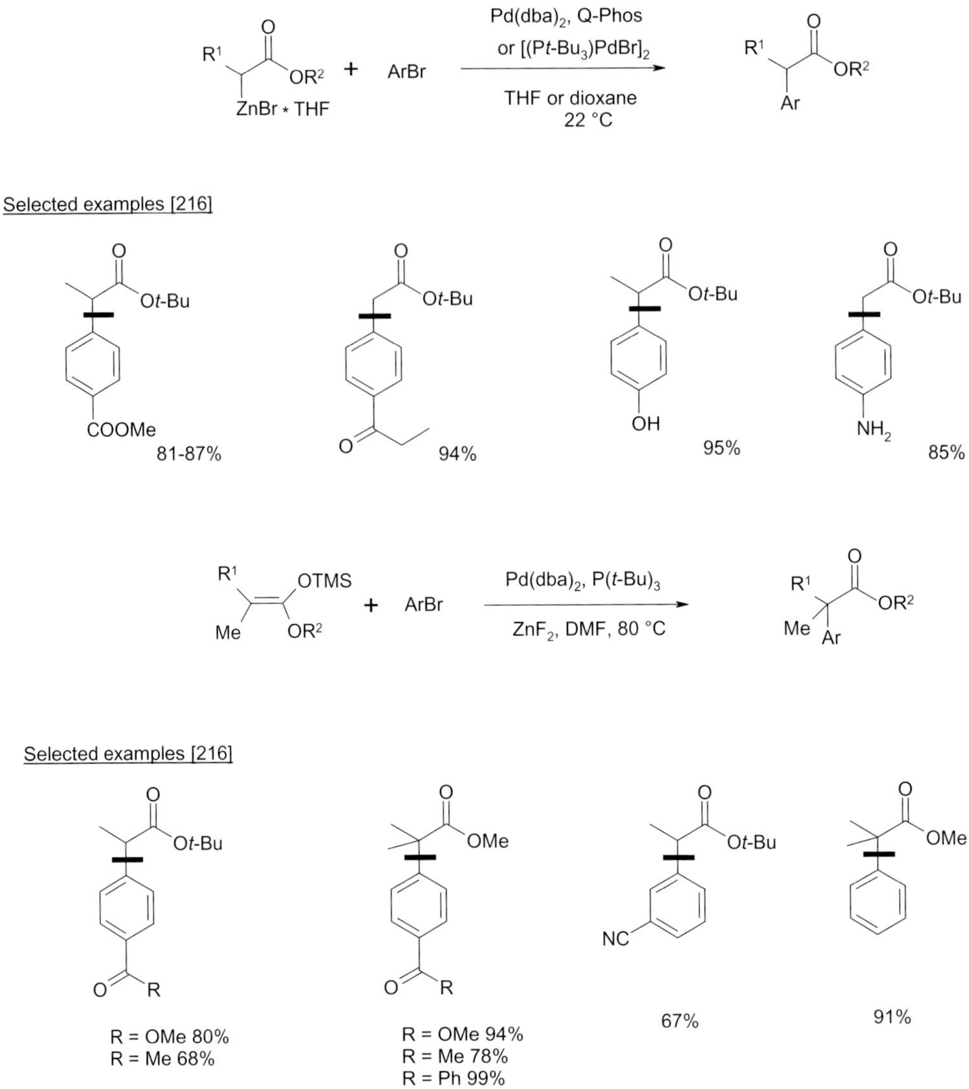
Pd(dba)2, Q-Phos
or [(Pt-Bu3)PdBr]2
THF or dioxane
22 °C
ZnBr * THF
ArBr
Selected examples [216]
81-87%
94%
95%
85%
Pd(dba)2, P(t-Bu)3
ZnF2, DMF, 80 °C
OTMS
Selected examples [216]
R = OMe 80%
R = Me 68%
R = OMe 94%
R = Me 78%
R = Ph 99%
67%
91%

Scheme 3.12 α-Arylation of zinc- or silicon-based enolates.

the first time in the arylation of ethyl phenyl acetates, using a *t*-butyl variant of Dave-Phos. In this methodology, *t*-butyl propionates gave only moderate yields, presumably due to a Claisen condensation occurring as a side reaction (Scheme 3.13).

A synthetically attractive approach is found in the arylation of amino acid esters, because this leads to α-aryl amino acids in a simple, yet modular fashion. This was realized using protected amino acid esters, most often the corresponding imines. Hartwig reported this reaction using P(*t*-Bu)$_3$ as active ligand with K_3PO_4

Ot-Bu + ArBr → ($Pd(OAc)_2$, Dave-Phos; LHMDS, toluene, 22 °C) Ot-Bu, Ar

Selected examples [217]

84% 81% 78% 71%

Ot-Bu + ArBr → ($Pd(OAc)_2$, Dave-Phos; LHMDS, toluene, 80 °C) Ot-Bu, Ar

Selected examples [217]

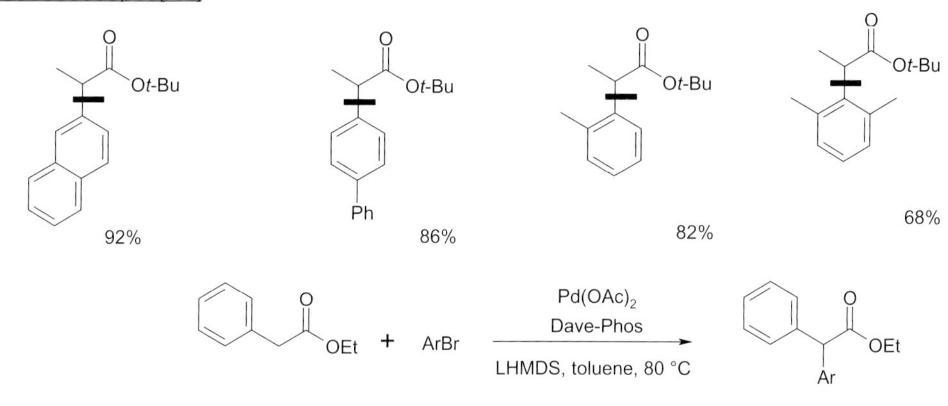

Selected examples [217]

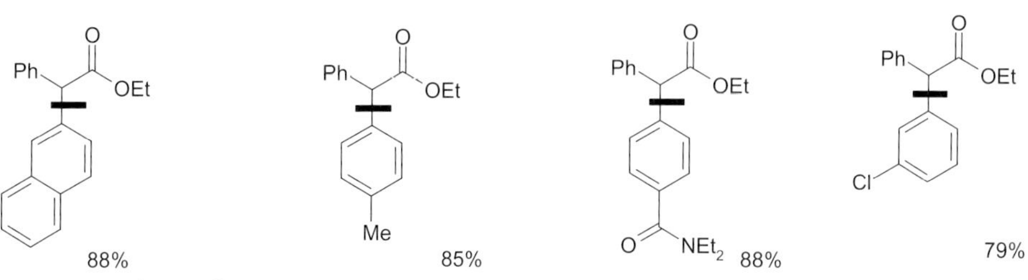

Scheme 3.13 α-Arylation of esters.

as a mild base [215], while Buchwald employed Dave-Phos or its biphenyl variant to arylate amino acid esters in an intramolecular fashion, thereby generating dihydroisoindoles or tetrahydroisoquinolines [218].

3.3.4
α-Arylations of Malonates and α-Cyano Esters

In comparison to the other carbonyl compounds discussed herein, malonates usually have a relatively lower pK_a, thus allowing the use of milder bases for their arylation. Nevertheless, with these nucleophiles, strong binding to palladium (e.g. in a η^2-*O,O*-coordination mode) might be detrimental for reductive elimination.

Hartwig was able successfully to couple di-*t*-butyl and diethyl malonates with a broad variety of aryl bromides and chlorides [4, 209], the catalytic system being composed of $P(t\text{-}Bu)_3$, Q-Phos or $(1\text{-}Ad)P(t\text{-}Bu)_2$ as ligands, NaH, Na_3PO_4 or K_3PO_4 as bases, and $Pd(dba)_2$ as palladium precursor [219] in THF or toluene as solvent. Interestingly, the arylation of alkyl-substituted malonates, potentially to yield arylalkylmalonates, was not possible with any of these systems, although synthetically this problem could be overcome quite simply by reversing the order of reactions – that is, palladium-catalyzed arylation prior to alkylation of the resulting malonates (Scheme 3.14).

Buchwald also reported details of the arylation of malonates using monodentate biaryl phosphine ligands [220]. Here, with $Pd(OAc)_2$, *t*-Bu-DavePhos and the inexpensive base K_3PO_4 diethyl malonate could be arylated successfully. Later, Buchwald also reported on the arylation of malonates employing the attractive aryl benzene sulfonates [221] when, by using $Pd(OAc)_2$ and X-Phos as the catalytic system, together with Cs_2CO_3 as base in toluene, diethyl malonate could be successfully arylated (Scheme 3.15).

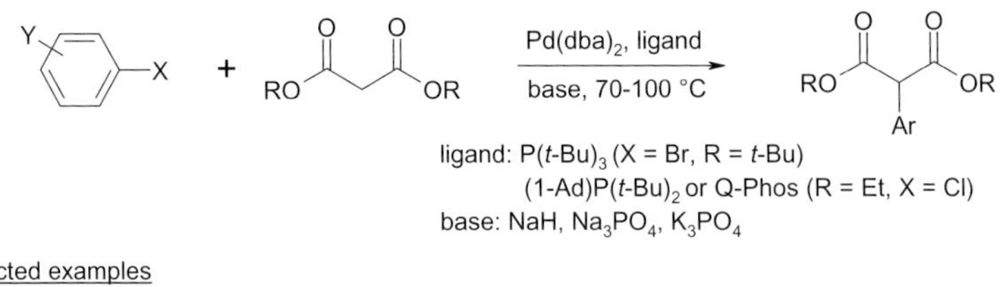

Selected examples

t-BuO Ot-Bu — X = Br 89%, X = Cl 88%

EtO OEt — X = Br 89%, X = Cl 81%

EtO OEt — X = Cl 86%

EtO OEt, F — X = Br 79%

Scheme 3.14 α-Arylations of malonates [219].

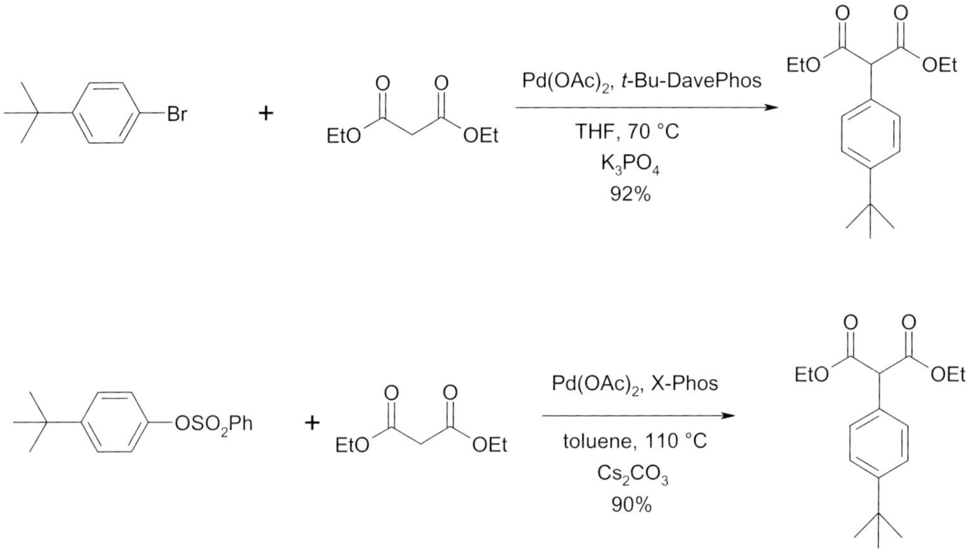

Pd(OAc)$_2$, *t*-Bu-DavePhos
THF, 70 °C
K_3PO_4
92%

Pd(OAc)$_2$, X-Phos
toluene, 110 °C
Cs_2CO_3
90%

Scheme 3.15 α-Arylations of diethyl malonates [221].

Djakovitch reported the arylation of malonates using Pd-exchanged NaY zeolites with *para*-substituted aryl bromides, and compared the heterogeneous with the homogeneous reaction. Obviously, the major advantage of the heterogeneous variant is the easily removable precious metal [222].

A closely related reactivity has been observed for the arylations of cyano esters. For example, with P(*t*-Bu)$_3$, Q-Phos or (1-Ad)P(*t*-Bu)$_2$ as ligand, Hartwig was able successfully to arylate ethyl cyanoacetates using Na_3PO_4 or K_3PO_4 as mild bases [219, 223], while with Q-Phos the selectivity for monoarylation was observed to be higher. Nevertheless, the diarylation of cyano esters could also be developed, using two equivalents of aryl halide with P(*t*-Bu)$_3$ as ligand.

3.3.5 α-Arylations of Ketones

Since the development of palladium-catalyzed α-arylations began using ketones as substrates, this area of research can be considered as having the highest degree of maturity. Whereas, during the early days of ketone arylation, it was speculated that chelating bidentate phosphine ligands were a prerequisite for reductive elimination, it was found later that only one phosphine would bind to the palladium [214]. Consequently, Hartwig developed the palladium-catalyzed arylation of ketones with P(*t*-Bu)$_3$ as ligand, along with NaO*t*-Bu as base (Scheme 3.16) [214].

Buchwald disclosed arylations of a variety of ketones using biphenyl-based ligand systems [220], and also examined phosphine ligand-free reactions, although these proved applicable to only a limited number of substrate combinations.

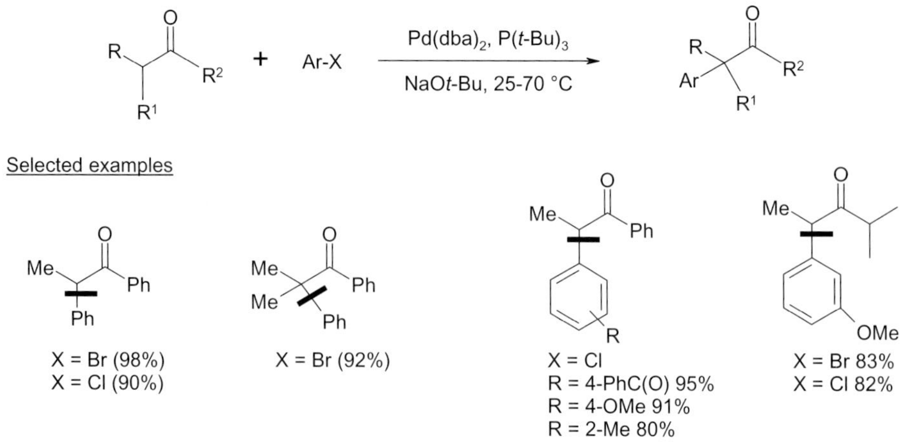

Scheme 3.16 α-Arylations of ketones using P(t-Bu)$_3$ [214].

By using NaOt-Bu as base in combination with Pd(OAc)$_2$ and a biphenylphosphine ligand in toluene, both aliphatic and aromatic ketones could be arylated. Furthermore, K_3PO_4 could be used as base in the arylation of cyclic ketones (Schemes 3.17 and 3.18).

Beller reported details of the arylation of acetophenones with Pd(OAc)$_2$ and n-BuPAd$_2$ as catalytic system using NaOt-Bu as base (Scheme 3.19) [224].

Later, Miura discovered the arylation of α,β-unsaturated carbonyl compounds, and in 1998 reported the arylation of α,β-unsaturated aldehydes and ketones using Pd(OAc)$_2$ together with the inexpensive ligand PPh$_3$ and the weak base Cs_2CO_3 in DMF as solvent [225]. In 2001, Miura also investigated the multiple arylation of alkyl aryl ketones and α,β-unsaturated carbonyl compounds in the presence of Pd(OAc)$_2$, Cs_2CO_3 and monodentate phosphine ligands, such as PPh$_3$, P(o-Tol)$_3$ and P(t-Bu)$_3$ (Scheme 3.20) [226].

Muratake and Natsume reported intramolecular α-arylations of aliphatic ketones using PdCl$_2$(PPh$_3$)$_2$ and Cs_2CO_3, forming a variety of carbocyclic compounds (Scheme 3.21) [227].

Ackermann reported on the arylation of ketones using his diaminochlorophosphine ligands [228]. Here, with Pd(dba)$_2$ as palladium precursor and NaOt-Bu as base, a broad variety of aryl chlorides, bromides and iodides could be employed in the arylation of propiophenones (Scheme 3.22).

3.3.6
α-Arylations of Amides

As the intermolecular arylation of amides is hampered by their relatively high pK_a-values, the double arylation becomes more relevant, since the pK_a of the monoarylated amide is significantly lower than that of the starting material.

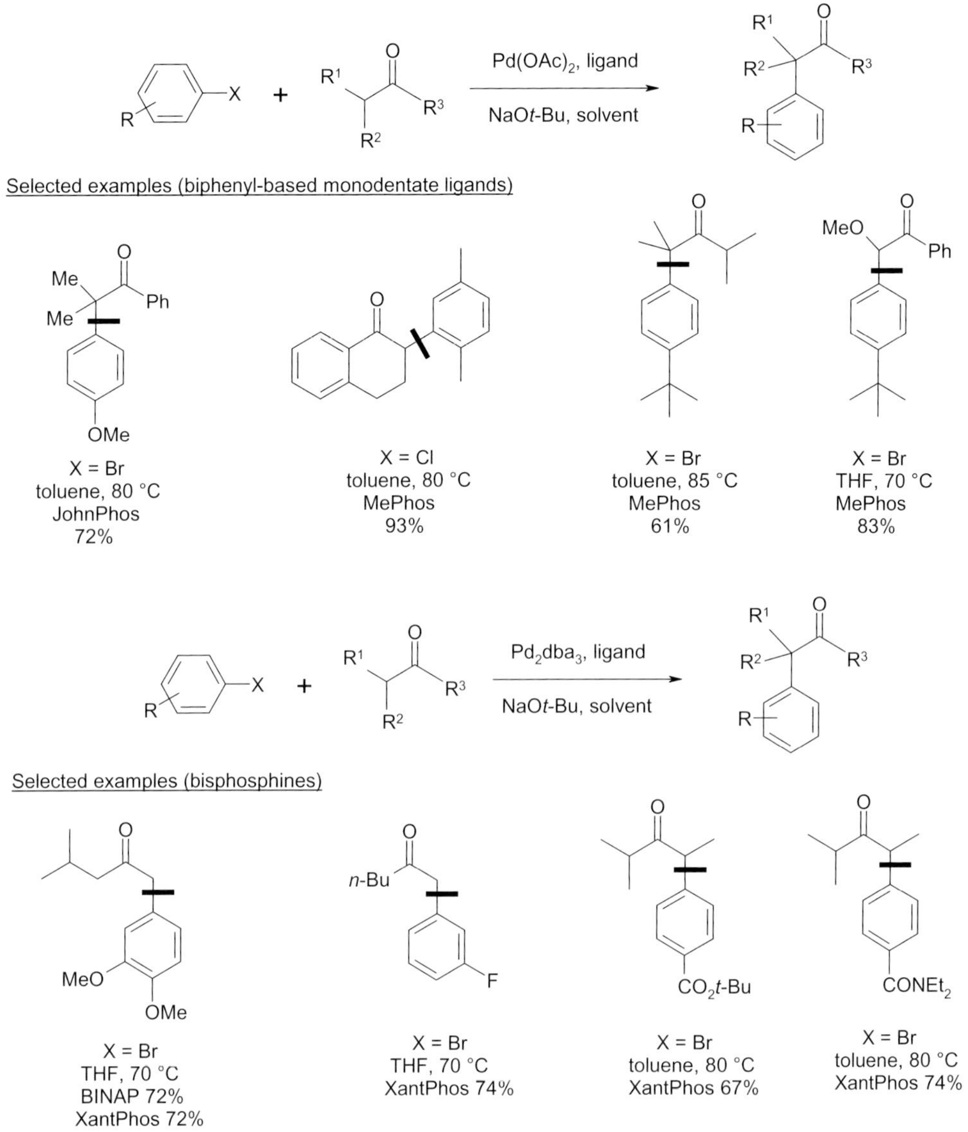

Pd(OAc)$_2$, ligand
NaO*t*-Bu, solvent

Selected examples (biphenyl-based monodentate ligands)

X = Br
toluene, 80 °C
JohnPhos
72%

X = Cl
toluene, 80 °C
MePhos
93%

X = Br
toluene, 85 °C
MePhos
61%

X = Br
THF, 70 °C
MePhos
83%

Pd$_2$dba$_3$, ligand
NaO*t*-Bu, solvent

Selected examples (bisphosphines)

X = Br
THF, 70 °C
BINAP 72%
XantPhos 72%

X = Br
THF, 70 °C
XantPhos 74%

X = Br
toluene, 80 °C
XantPhos 67%

X = Br
toluene, 80 °C
XantPhos 74%

Scheme 3.17 α-Arylations of ketones [220].

Initially, in 1998, Hartwig reported an intermolecular version using BINAP as ligand and KHMDS as base [229], but later reported milder reaction conditions, using zinc enolates as substrates in conjunction with Q-Phos as ligand in 1,4-dioxane as solvent (Scheme 3.23) [216].

Hartwig investigated the intramolecular variant intensively for the synthesis of oxindoles [230], and even succeeded in developing an asymmetric variant using

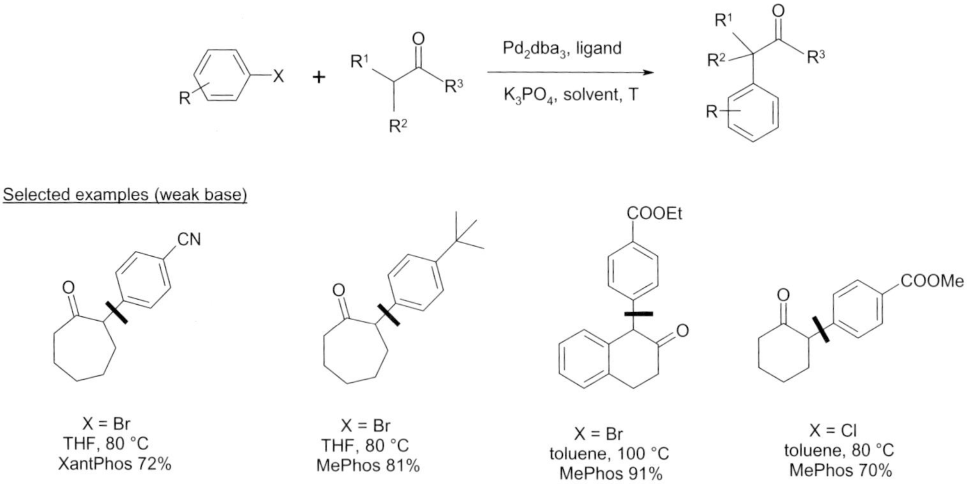
Pd2dba3, ligand
K3PO4, solvent, T
Selected examples (weak base)
X = Br
THF, 80 °C
XantPhos 72%
X = Br
THF, 80 °C
MePhos 81%
X = Br
toluene, 100 °C
MePhos 91%
X = Cl
toluene, 80 °C
MePhos 70%

Scheme 3.18 α-Arylation of cyclic ketones with K_3PO_4 as base [220].

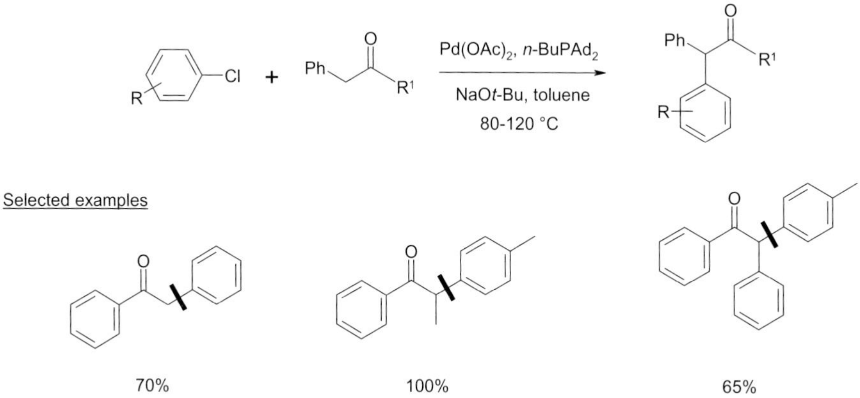
Pd(OAc)2, n-BuPAd2
NaOt-Bu, toluene
80-120 °C
Selected examples
70%
100%
65%

Scheme 3.19 α-Arylation of acetophenones by Beller [224].

chiral NHC-ligands [231]. For the cyclization reaction, both PCy_3 and the carbene ligand SIPr proved to be effective. For the asymmetric transformation, a wide range of enantiomerically pure phosphines was tested, but with unsatisfactory results. Finally, carbene ligands derived from (−)-isopinocampheylamine or (+)-bornylamine constituted the best systems (Scheme 3.24).

3.3.7
α-Arylations of Nitriles

Alkyl nitriles have a relatively high pK_a value compared with that of ketones, for example, and therefore stronger bases are required for their deprotonation. As a

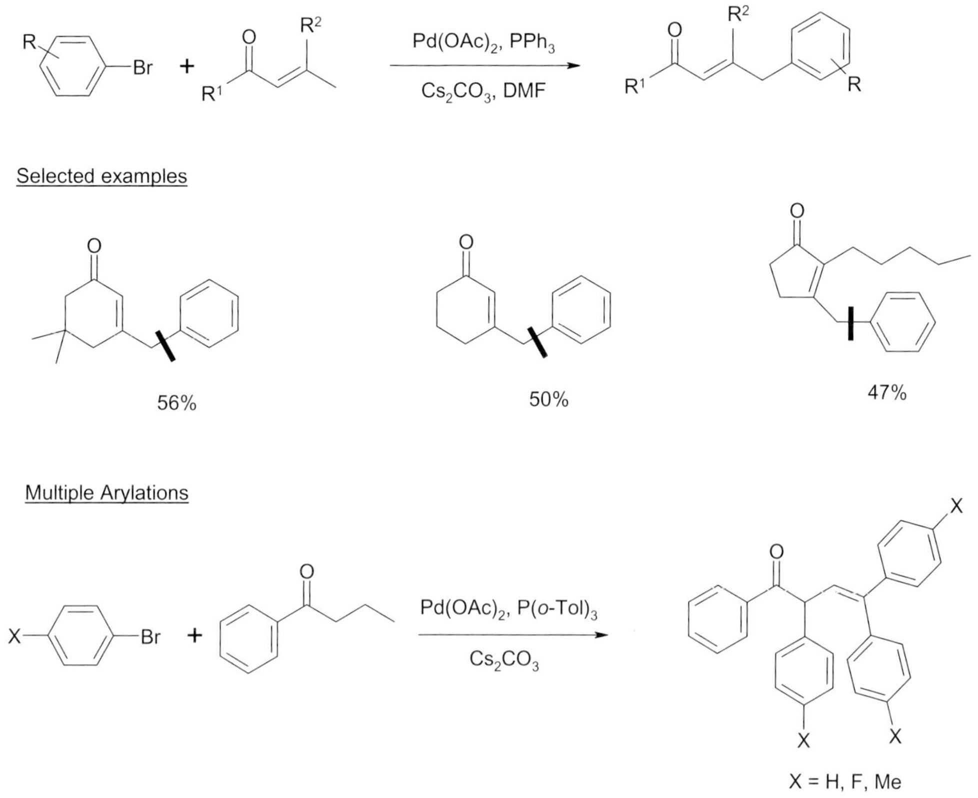

$Pd(OAc)_2$, PPh_3

Cs_2CO_3, DMF

Selected examples

56%

50%

47%

Multiple Arylations

$Pd(OAc)_2$, $P(o\text{-}Tol)_3$

Cs_2CO_3

X = H, F, Me

Scheme 3.20 (Multiple) Arylations of ketones [226].

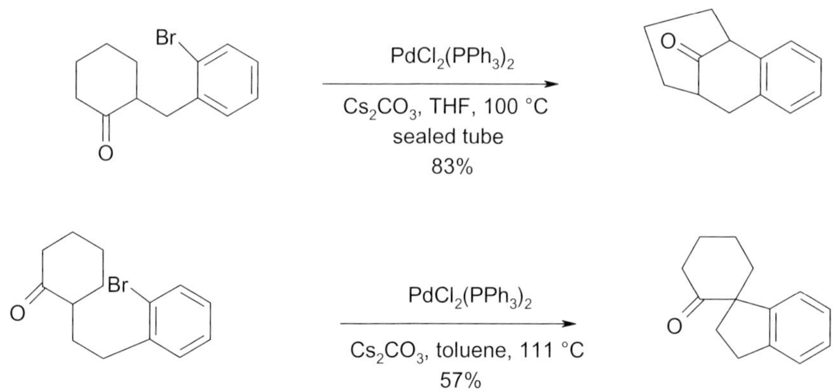

$PdCl_2(PPh_3)_2$

Cs_2CO_3, THF, 100 °C
sealed tube
83%

$PdCl_2(PPh_3)_2$

Cs_2CO_3, toluene, 111 °C
57%

Scheme 3.21 Intramolecular α-arylation of aliphatic ketones by Muratake and Natsume [227].

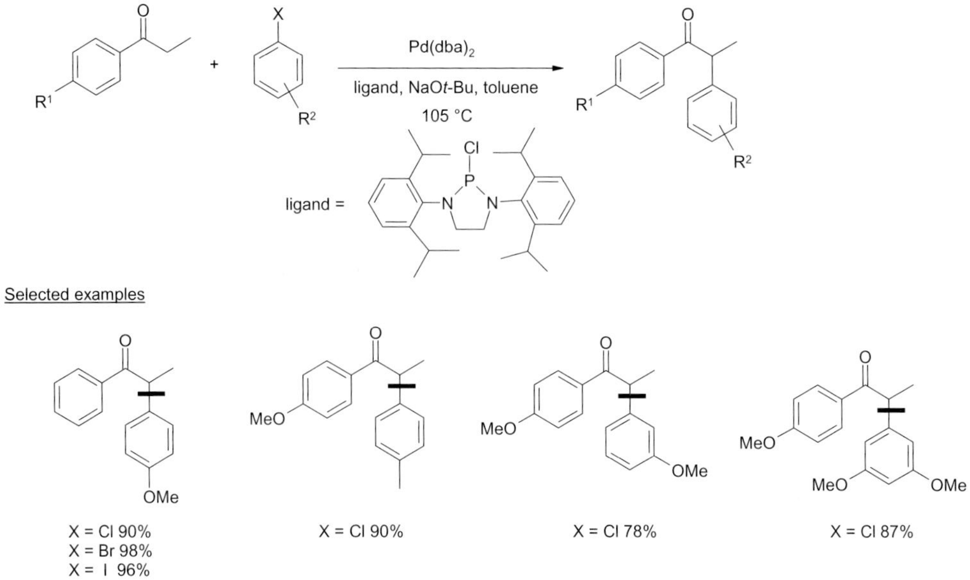

$Pd(dba)_2$
ligand, NaO*t*-Bu, toluene
105 °C

Selected examples

X = Cl 90%
X = Br 98%
X = I 96%

X = Cl 90%

X = Cl 78%

X = Cl 87%

Scheme 3.22 α-Arylations of propiophenones [228].

result, examples of arylation of nitriles were first reported using more acidic substrates such as phenylacetonitrile. In 1998, Miura reported arylations of this substrate with $PdCl_2$, PPh_3 and Cs_2CO_3, although the yields provided were rather low [232].

In 2002, Hartwig reported the arylations of 2-phenylbutyronitrile and butyronitrile using $P(t\text{-}Bu)_3$, but unfortunately this system could not be extended to arylations of other nitriles. BINAP as a potential chelating ligand was used for the arylations of secondary and benzyl-substituted nitriles, whereas when primary nitriles were used diarylation was observed [233]. In 2005, Hartwig reported a mild and selective monoarylation of nitriles [234] when, by using α-silyl nitriles, he was able to perform arylations using XantPhos or $P(t\text{-}Bu)_3$ as ligands together with a half-equivalent of ZnF_2 in DMF at 90 °C. This methodology later proved to be especially useful in the synthesis of verapamil (not shown) (Scheme 3.25).

In 2003, Verkade disclosed details of the arylation of nitriles using his TAP ligands and NaHMDS as base in toluene, and also observed diarylations when using acetonitrile [235].

Klapars and Waldman, of the Merck research group, reported a transition metal-free α-arylation of aliphatic nitriles with heteroaryl halides using KHMDS or NaHMDS as bases in a toluene/THF mixture at 0–25 °C. However, this protocol was strictly limited to the conversion of electronically activated heteroaryl halides [236].

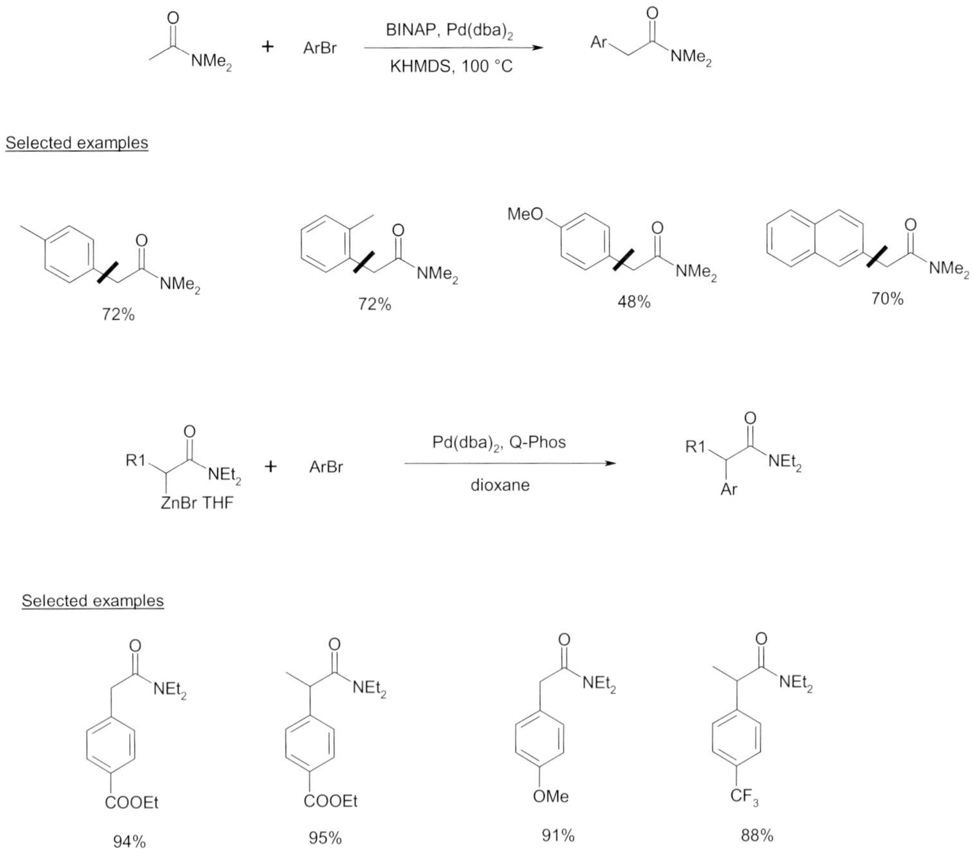

Scheme 3.23 Intermolecular α-arylation of amides [216].

3.4 Summary and Conclusions

Today, the palladium-catalyzed arylation of amines, which often is referred to as the Buchwald–Hartwig reaction, has reached a mature state such that a plethora of protocols is available for the arylation of almost any given amine. Yet, perhaps the most impressive point has been the rapid development of this technology, from a niche application using unattractive tin reagents to becoming one of the basic organic reactions in amine synthesis that is found in every modern organic chemistry textbook. Moreover, this rapid development has been accompanied in parallel by an early implementation of the process into the chemical industry, leading to the first commercially available pharmaceuticals to be synthesized using this reaction as a key step.

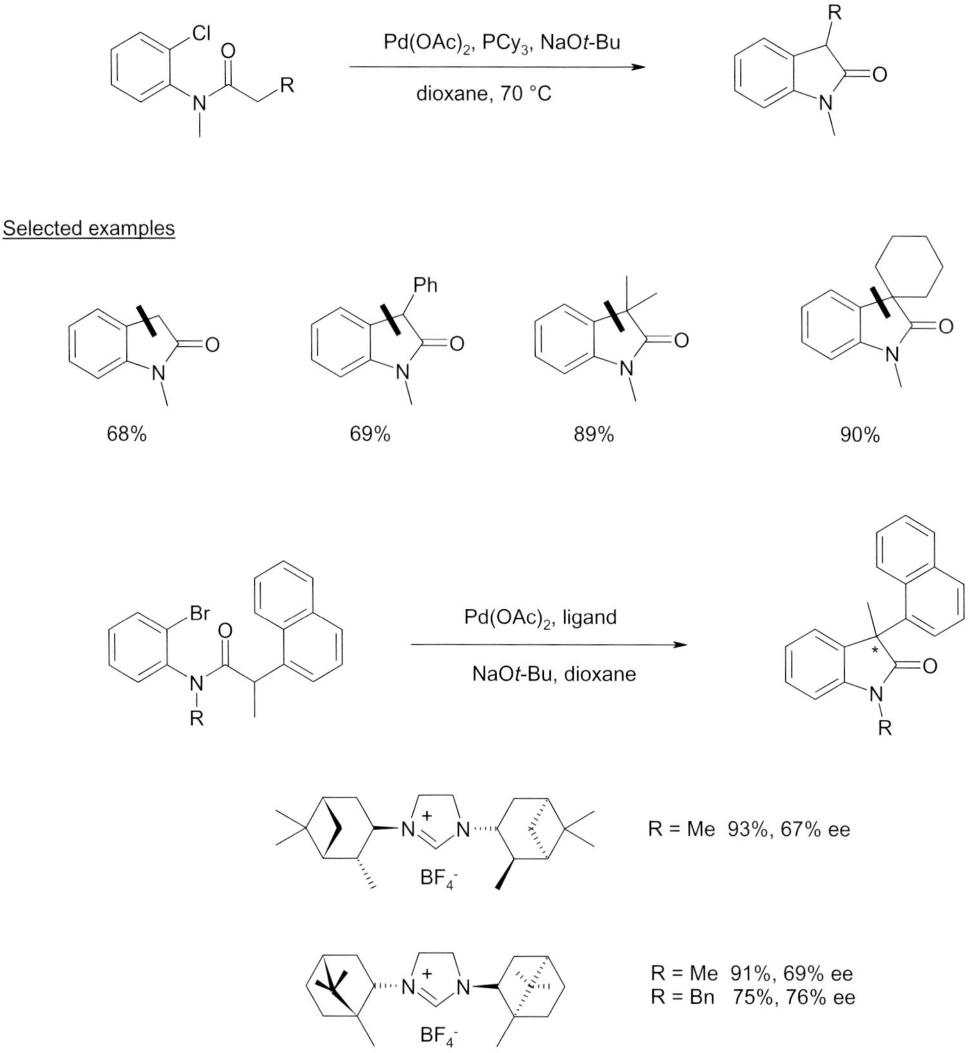

Pd(OAc)$_2$, PCy$_3$, NaO*t*-Bu

dioxane, 70 °C

Selected examples

68% 69% 89% 90%

Pd(OAc)$_2$, ligand

NaO*t*-Bu, dioxane

R = Me 93%, 67% ee

R = Me 91%, 69% ee
R = Bn 75%, 76% ee

Scheme 3.24 Intramolecular α-arylation yielding oxindoles [230, 231].

Based on the observation that α-arylated ketones were obtained as byproducts in the arylation of amines containing carbonyl functionalities, the main potential of this methodology was soon acknowledged and, consequently, it was developed into a very powerful synthetic tool.

Perhaps the most encouraging factor regarding the processes described in this chapter is that, even after more than 100 years of synthetic organic chemistry, fundamental organic methodologies can still be discovered and developed into synthetic tools with far-reaching impact!

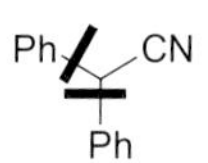

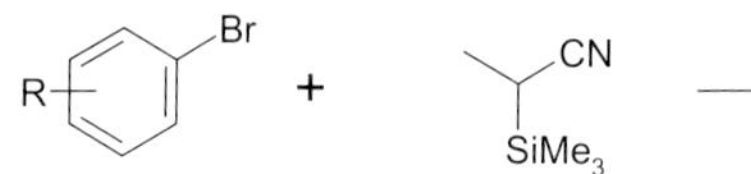

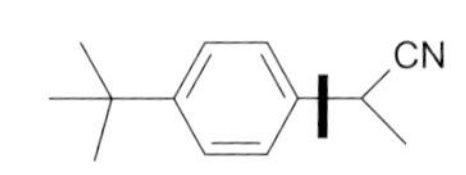

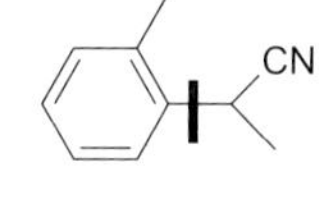

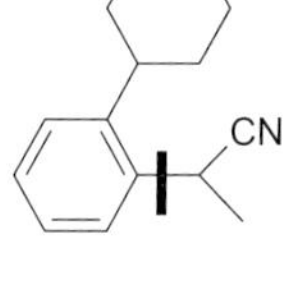

Scheme 3.25 α-Arylation of nitriles by Hartwig [233, 234].

Abbreviations

acac	acetylacetonate
(*S*,*S*)-BDPP	(2*S*,3*S*)-(−)-2,4-bis(diphenylphosphino)pentane
BINAP	[1,1′-binaphthalene]-2,2′-diylbis[diphenylphosphine]
(*S*,*R*)-BPPFA	(*S*)-*N*,*N*-dimethyl-1-[(*R*)-1,2-bis(diphenylphosphino) ferrocenyl]ethylamine
(*S*,*S*)-BPPM	(2*S*,4*S*)-1-(*tert*-butoxycarbonyl)-4-(diphenylphosphino)-2-(diphenylphosphinomethyl)pyrrolidine
daop	diaminooxophosphines
dba	dibenzylideneacetone
DPBP	2,2′-bis(diphenylphosphino)-1,1′-biphenyl
DBU	1,8-diazabicyclo[5.4.0]undec-7-ene
DFT	density functional theory
DME	1,2-dimethoxy ethane

DMF	*N,N*-dimethylformamide
DMSO	dimethyl sulfoxide
DPEPhos	bis(2-diphenylphosphinophenyl)ether
dppf	1,1′-bis(diphenylphosphino)ferrocene
dppp	1,3-bis(diphenylphosphino)propane
D^{t}BPF	1,1′-bis(di-*tert*-butylphosphino)ferrocene
HASPO	Heteroatom-substituted secondary phosphine oxide preligands
HMDS	hexamethyldisilazane
LHMDS	lithium hexamethyldisilazide
MAP	2-amino-2′-diphenylphosphino-1,1′-binaphthyl
(*R,R*)-MOD-DIOP	(4*R*,5*R*)-(1)-4,5-bis[bis(4′-methoxy-3′,5′-dimethylphenyl)-phosphinomethyl]-2,2-dimethyl-1,3-dioxolane
MTBD	7-methyl-1,5,7-triazabicyclo[4.4.0]dec-5-ene
MOP	2-(diphenylphosphino)-2′-methoxy-1,1′-binaphthyl
NaHMDS	sodium hexamethyldisilazide
NHC	*N*-heterocyclic carbene
NMP	*N*-methyl pyrrolidone
SIPr	*N,N*′-bis(2,6-diisopropylphenyl)4,5-dihydroimidazol-2-ylidene
TAP	bicyclic tri-amino-phosphine
THF	tetrahydrofuran
Tol	tolyl
XantPhos	9,9-dimethyl-4,5-bis(diphenylphosphino)xanthene

References

1 (a) de Meijere, A. and Diederich, F. (eds) (2004) *Metal-Catalyzed Cross-Coupling Reactions*, Wiley-VCH Verlag GmbH, Weinheim;
(b) Nicolaou, K.C., Bulger, P.G. and Sarlah, D. (2005) *Angew. Chem. Int. Ed.*, **44**, 4442–89;
(c) Negishi, E.i. (ed.) (2002) *Handbook of Organopalladium Chemistry for Organic Synthesis*, John Wiley & Sons, Inc., Hoboken, N.J.
(d) King, A.O. and Yasuda, N. (2004) *Top. Organomet. Chem.*, **6**, 205–45.

2 (a) Guram, S., Rennels, R.A., Buchwald, S.L., Barta, N.S. and Pearson, W.H. (1996) *Chemtracts Inorg. Chem.*, **8**, 1–5;
(b) Hartwig, J.F. (1997) *Synlett*, 329–40;
(c) Wolfe, J.P., Wagaw, S., Marcoux, J.F. and Buchwald, S.L. (1998) *Acc. Chem. Res.*, **31**, 805–18;
(d) Hartwig, J.F. (1998) *Acc. Chem. Res.*, **31**, 852–60;
(e) Hartwig, J.F. (1998) *Angew. Chem. Int. Ed.*, **37**, 2046–2067;
(f) Yang, B.H. and Buchwald, S.L. (1999) *J. Organomet. Chem.*, **576**, 125–46;
(g) Muci, A.R. and Buchwald, S.L. (2002) *Top. Curr. Chem.*, **219**, 131–209;
(h) Hartwigin, J.F. (2002) *Modern Arene Chemistry* (ed. D. Astruc), Wiley-VCH Verlag GmbH, Weinheim, pp. 107–68;
(i) Prim, D., Campagne, J.M., Joseph, D. and Andrioletti, B. (2002) *Tetrahedron*, **58**, 2041–75;
(j) Littke, A.F. and Fu, G.C. (2002) *Angew. Chem. Int. Ed.*, **41**, 4176–211;
(k) Hartwig, J.F. (2004) *Comprehensive Coordination Chemistry II*, **9**, 369–98;
(l) Hartwig, J.F. (2006) *Synlett*, **9**, 1283–94.

3 Schlummer, B. and Scholz, U. (2004) *Adv. Synth. Catal.*, **346**, 1599–626.

4 (a) Culkin, D.A. and Hartwig, J.F. (2003) *Acc. Chem. Res.*, **36**, 234–45; (b) Lloyd-Jones, G.C. (2002) *Angew. Chem. Int. Ed.*, **41**, 953–6.

5 (a) Crampton, M.R. (1977) Nucleophilic aromatic substitution, in *Organic Reaction Mechanisms*, Vol. **12**, John Wiley & Sons, Ltd, pp. 279–97; (b) Vlasov, V.M. (2003) *Russ. Chem. Rev.*, **72**, 681–703.

6 (a) Turpin, G.S. (1891) *J. Chem. Soc.*, 714; (b) Abegg, R. (1891) *Ber. d. Chem. Ges.*, **24**, 949–58; (c) Pisani, F. (1854) *Compt. Rend.*, **39**, 852.

7 Kym, O. (1895) *J. Prakt. Chem.*, 325–36.

8 (a) Laubenheimer, A. (1876) *Chem. Ber.*, **9**, 1826–9; (b) Laubenheimer, A. (1878) *Chem. Ber.*, **11**, 1155–61; (c) Rudy, H. and Cramer, K.E. (1939) *Chem. Ber.*, **72B**, 227–48.

9 (a) Bergstrom, F.W., Granara, I.M. and Erickson, V. (1942) *J. Org. Chem.*, 98–102; (b) Stern, M.K. and Cheng, B.K. (1993) *J. Org. Chem.*, **58**, 6883–8; (c) Stern, M.K. (1992) US A 5117063, 19920526; (d) De Vera, A.L. (2001) WO A2 2001098252, 20011227.

10 (a) Stella, L., Raynier, B. and Surzur, J.M. (1981) *Tetrahedron*, **37**, 2843–54; (b) Minisci, F. (1976) *Top. Curr. Chem.*, **62**, 1–48; (c) Minisci, F., Galli, R., Cecere, M. and Quilico, A. (1969) FR 93635, 19690425; (d) Minisci, F., Galli, R. and Cecere, M. (1966) *Chem. Ind.*, **48**, 1324–6.

11 Ullmann, F. and Bielecki, J. (1901) *Chem. Ber.*, **34**, 2174–85.

12 Ullmann, F. (1903) *Chem. Ber.*, **36**, 2382–4.

13 Goldberg, I. and Ullmann, D.E. (1905) 173523.

14 (a) Kunz, K., Scholz, U. and Ganzer, D. (2003) *Synlett*, 2428–39; (b) Beletskaya, I.P. and Cheprakov, A.V. (2004) *Coord. Chem. Rev.*, **248**, 2337–64; (c) Nelson, T.D. and Crouch, D.R. (2004) *Org. React.*, **63**, 265–555.

15 Heck, R.F. (2006) *Synlett*, 2855–60.

16 Miyaura, N., Yanagi, T. and Suzuki, A. (1981) *Synth. Commun.*, **11**, 513–19.

17 Milstein, D. and Stille, J.K. (1979) *J. Am. Chem. Soc.*, **101**, 4992–8.

18 Negishi, E.-i., Zeng, X., Tan, Z., Qian, M., Hu, Q. and Huangin, Z. (2004) *Metal-Catalyzed Cross-Coupling Reactions*, 2nd edn (eds A. De Meijere and F. Diederich), Wiley-VCH Verlag GmbH, Weinheim, pp. 815–89.

19 Kosugi, M., Kameyama, M.M. and Migita, T. (1983) *Chem. Lett.*, **12**, 927–8.

20 Singh, A.K. and Bhandari, S. (2003) *Main Group Metal. Chem.*, **26**, 155–211.

21 Boger, D.L., Duff, S.R., Panek, J.S. and Yasuda, M. (1985) *J. Org. Chem.*, **50**, 5782–9.

22 Dzhemilev, U.M., Ibragimov, A.G., Minsker, D.L. and Muslukhov, R.R. (1987) *Izvestiya Akad. Nauk SSSR, Seriya Khim.*, **2**, 406–9.

23 Guram, A.S. and Buchwald, S.L. (1994) *J. Am. Chem. Soc.*, **116**, 7901–2.

24 (a) Buchwald, S.L. (1996), US A 5576460, 19961119; (b) Guram, A.S., Rennels, R.A. and Buchwald, S.L. (1995) *Angew. Chem. Int. Ed.*, **34**, 1348–50.

25 Louie, J. and Hartwig, J.F. (1995) *Tetrahedron Lett.*, **36**, 3609–12.

26 Monguchi, Y., Kitamoto, K., Mori, S., Yanase, T., Mizoguchi, Y., Ikawa, T., Maegawa, T. and Sajiki, H. (2007) Abstracts of Papers. *233rd ACS National Meeting, March 25–29, Chicago, IL, United States.*

27 Matthews, M.A. (2005) Green chemistry, in *Kirk-Othmer Encyclopedia of Chemical Technology*, Vol. **12**, 5th edn (ed. A. Seidel), John Wiley & Sons, Inc., Hoboken, NJ, pp. 799–818.

28 Fairlamb, I.J.S., Kapdi, A.R., Lee, A.F., McGlacken, G.P., Weissburger, F., de Vries, A.H.M. and Schmieder-van de Vondervoort, L. (2006) *Chem. Eur. J.*, **12**, 8750–61.

29 Kozlov, N.S., Pinegina, L. and Tselishchev, B.E. (1968) *Zh. Khim.*, Abstr. No. 9, h230.

30 (a) Huang, X., Anderson, K.W., Zim, D., Jiang, L., Klapars, A. and Buchwald, S.L. (2003) *J. Am. Chem. Soc.*, **125**, 6653–5;

(b) Strieter, E.R., Blackmond, D.G. and Buchwald, S.L. (2003) *J. Am. Chem. Soc.*, **125**, 13978–80.

31 Buchwald, S.L. and Anderson, K.W. (2006) WO 2006074315.

32 Herrmann, W.A., Xfele, K., Preysing, D.v. and Schneider, S.K. (2003) *J. Organomet. Chem.*, **687**, 229–48.

33 Nishio, R., Wessely, S., Sugiura, M. and Kobayashi, S. (2006) *J. Comb. Chem.*, **8**, 459–61.

34 Nettekoven, U., Naud, F., Schnyder, A. and Blaser, H.-U. (2004) *Synlett*, 2549–52.

35 (a) Guino, M. and (Mimi) Hii, K.K. (2005) *Tetrahedron Lett.*, **46**, 6911–13; (b) Rosso, V.W., Lust, D.A., Bernot, P.J., Grosso, J.A., Modi, S.P., Rusowicz, A., Sedergran, T.C., Simpson, J.H., Srivastava, S.K., Humora, M.J. and Anderson, N.G. (1997) *Org. Proc. Res. Dev.*, **1**, 311–14; (c) Garrett, C.E. and Prasad, K. (2004) *Adv. Synth. Catal.*, **346**, 889–900.

36 For an interesting overview of monoligated palladium species as catalysts in cross-coupling, see: Christmann, U. and Vilar, R. (2005) *Angew. Chem. Int. Ed.*, **44**, 366–74.

37 For a comparison of phosphine- and carbene-based ligands, see: Frisch, C., Beller, A., Zapf, A., Briel, O., Kayser, B., Shaikh, N. and Beller, M. (2004) *J. Mol. Catal. A*, **214**, 231–9.

38 (a) Wolfe, J.P., Wagaw, S. and Buchwald, S.L. (1996) *J. Am. Chem. Soc.*, **118**, 7215–16; (b) Alcazar-Roman, L.M., Hartwig, J.F., Rheingold, A.L., Liable-Sands, L.M. and Guzei, I.A. (2000) *J. Am. Chem. Soc.*, **122**, 4618–30; (c) Singh, U.K., Strieter, E.R., Blackmond, D.G. and Buchwald, S.L. (2002) *J. Am. Chem. Soc.*, **124**, 14104–14.

39 Shekhar, S., Ryberg, P., Hartwig, J.F., Mathew, J.S., Blackmond, D.G., Strieter, E.R. and Buchwald, S.L. (2006) *J. Am. Chem. Soc.*, **128**, 3584–91.

40 Prashad, M., Hu, B., Lu, Y.M., Draper, R., Har, D., Repic, O. and Blacklock, T.J. (2000) *J. Org. Chem.*, **68**, 2612–14.

41 Laufer, R.S. and Dmitrienko, G.I. (2002) *J. Am. Chem. Soc.*, **124**, 1854–5.

42 Wolfe, J.P. and Buchwald, S.L. (2000) *J. Org. Chem.*, **65**, 1144–57.

43 Wolfe, J.P. and Buchwald, S.L. (1997) *J. Org. Chem.*, **62**, 6066–8.

44 (a) Marinetti, A., Hubert, P. and Genêt, J.P. (2000) *Eur. J. Org. Chem.*, 1815–20; (b) Wagaw, S., Rennels, R.A. and Buchwald, S.L. (1997) *J. Am. Chem. Soc.*, **119**, 8451–8.

45 (a) Wolfe, J.P. and Buchwald, S.L. (1997) *J. Org. Chem.*, **62**, 1264–7; (b) Ahman, J. and Buchwald, S.L. (1997) *Tetrahedron Lett.*, **38**, 6363–6; (c) Demadrille, R., Moustrou, C., Samat, A. and Guglielmetti, R. (1999) *Heterocyclic Commun.*, **5**, 123; (d) Wentland, M.P., Xu, G., Cioffi, C.L., Ye, Y., Duan, W., Cohen, D.J., Colasurdo, A.M. and Bidlack, J.M. (2000) *Bioorg. Med. Chem. Lett.*, **10**, 183–7.

46 Hong, Y., Senanayake, C.H., Xiang, T., Vandenbossche, C.P., Tanoury, G.J., Bakale, R.P. and Wald, S.A. (1998) *Tetrahedron Lett.*, **39**, 3121–4.

47 Kanbara, T., Izumi, K., Nakadani, Y., Narise, T. and Hasegawa, K. (1997) *Chem. Lett.*, **26**, 1185–6.

48 Zhang, X.X., Harris, M.C., Sadighi, J.P. and Buchwald, S.L. (2001) *Can. J. Chem.*, **79**, 1799–805.

49 Ferreira, I.C.F.R., Queiroz, M.J.R.P. and Kirsch, G. (2003) *Tetrahedron*, **59**, 975–81.

50 Guari, Y., van Strijdonck, G.P.F., Boele, M.D.K., Reek, J.N.H., Kramer, P.C.J. and van Leeuwen, P.W.N.M. (2001) *Chem. Eur. J.*, **7**, 475–82.

51 Yin, J. and Buchwald, S.L. (2002) *J. Am. Chem. Soc.*, **124**, 6043–8.

52 (a) Yin, J. and Buchwald, S.L. (2000) *Org. Lett.*, **2**, 1101–4; (b) Yang, B.H. and Buchwald, S.L. (1999) *Org. Lett.*, **1**, 35–7.

53 Wagaw, S., Yang, B.H. and Buchwald, S.L. (1999) *J. Am. Chem. Soc.*, **121**, 10251–63.

54 Cacchi, S., Fabrizi, G., Goggiamani, A. and Zappia, G. (2001) *Org. Lett.*, **3**, 2539–41.

55 Artamkina, G.A., Sergeev, A.G. and Beletskaya, I.P. (2001) *Tetrahedron Lett.*, **42**, 4381–4.

56 Harris, M.C., Geis, O. and Buchwald, S.L. (1999) *J. Org. Chem.*, **64**, 6019–22.

57 Yin, J., Zhao, M.M., Huffman, M.A. and McNamara, J.M. (2002) *Org. Lett.*, **4**, 3481–4.

58 For recent applications, see: (a) Wang, W., Ding, Q., Fan, R. and Wu, J. (2007) *Tetrahedron Lett.*, **48**, 3647–9;
(b) Begouin, A., Hesse, S., Queiroz, M.-J.R.P. and Kirsch, G. (2007) *Eur. J. Org. Chem.*, **10**, 1678–82;
(c) Hostyn, S., Maes, B.U.W., Van Baelen, G., Gulevskaya, A., Meyers, C. and Smits, K. (2006) *Tetrahedron*, **62**, 4676–84.

59 Sadighi, J.P., Harris, M.C. and Buchwald, S.L. (1998) *Tetrahedron Lett.*, **39**, 5327–30.

60 Ogasawara, M., Yoshehida, K. and Hayashi, T. (2000) *Organometallics*, **19**, 1567–71.

61 Old, D.W., Wolfe, J.P. and Buchwald, S.L. (1998) *J. Am. Chem. Soc.*, **120**, 9722–3.

62 Wolfe, J.P., Tomori, H., Sadighi, J.P., Yin, J. and Buchwald, S.L. (2000) *J. Org. Chem.*, **65**, 1158–74.

63 Old, D.W., Wolfe, J.P. and Buchwald, S.L. (1998) *J. Am. Chem. Soc.*, **120**, 9722–3.

64 Zhang, X.X. and Buchwald, S.L. (2000) *J. Org. Chem.*, **65**, 8027–31.

65 (a) Tomori, H., Fox, J.M. and Buchwald, S.L. (2000) *J. Org. Chem.*, **65**, 5334;
(b) Buchwald, S.L., Mauger, C., Mignani, G. and Scholz, U. (2006) *Adv. Synth. Catal.*, **348**, 23–39.

66 Damon, D.B., Dugger, R.W., Magnus-Aryitey, G., Ruggeri, R.B., Wester, R.T., Tu, M. and Abramov, Y. (2006) *Org. Proc. Res. Dev.*, **10**, 464–71.

67 Wolfe, J.P. and Buchwald, S.L. (1999) *Angew. Chem. Int. Ed.*, **38**, 2413–16.

68 Xu, C., Gong, J.-F. and Wu, Y.-J. (2007) *Tetrahedron Lett.*, **48**, 1619–23.

69 Barder, T.E. and Buchwald, S.L. (2007) *J. Am. Chem. Soc.*, **129**, 5096–101.

70 (a) Schlummer, B. and Scholz, U. (2005) *Spec. Chem. Mag.*, **25**, 22–4;
(b) Schlummer, B. and Scholz, U. (2005) *Chim. Oggi*, **23**, 18–20.

71 (a) Nishiyama, M. and Koie, Y. (1997), EP A1 802173;
(b) Yamamoto, T., Nishiyama, S. and Koie, Y. (1998) JP A2 10310561, 19981124;
(c) Yamamoto, T., Nishiyama, S. and Koie, Y. (1999) JP A2 11080346, 19990326.

72 (a) FMC Lithium (2008) http://www.fmclithium.com/index.html (accessed 12 September 2008);
(b) STREM (2008) http://www.strem.com/code/index.ghc (accessed 12 September 2008).

73 Netherton, M.R. and Fu, G.C. (2001) *Org. Lett.*, **3**, 4295–8.

74 Nishiyama, M., Yamamoto, T. and Koie, Y. (1998) *Tetrahedron Lett.*, **39**, 617–20.

75 Hartwig, J.F., Kawatsura, M., Hauck, S.I., Shaughnessy, K.H. and Alcazar-Roman, L.M. (1999) *J. Org. Chem.*, **64**, 5575–80.

76 (a) Louie, J., Hartwig, J.F. and Fry, A.J. (1997) *J. Am. Chem. Soc.*, **119**, 11695–6;
(b) Louie, J. and Hartwig, J.F. (1998) *Macromolecules*, **31**, 6737–93;
(c) Goodson, F.E. and Hartwig, J.F. (1998) *Macromolecules*, **31**, 1700–3;
(d) Yamamoto, T., Nishiyama, M. and Koie, Y. (1998) *Tetrahedron Lett.*, **39**, 2367–70;
(e) Goodson, F.E., Hauck, S.I. and Hartwig, J.F. (1999) *J. Am. Chem. Soc.*, **121**, 7527–39.

77 Reddy, N.P. and Tanaka, M. (1997) *Tetrahedron Lett.*, **38**, 4807–10.

78 (a) Hauck, S.I., Lakshmi, K.V. and Hartwig, J.F. (1999) *Org. Lett.*, **1**, 2057–60;
(b) Tew, G.N., Pralle, M.U. and Stupp, S.I. (2000) *Angew. Chem. Int. Ed.*, **39**, 517–21;
(c) Braig, T., Müller, D.C., Groß, M., Meerholz, K. and Nuyken, O. (2000) *Macromol. Rapid Commun.* **21**, 583–9;
(d) Thelakkhat, M., Hagen, J., Haarer, D. and Schmidt, H.W. (1999) *Synthetic Methods*, **102**, 1125–8.

79 (a) Wolfe, J.P. and Buchwald, S.L. (1996) *J. Org. Chem.*, **61**, 1133–5;
(b) Zhao, S.H., Miller, A.K., Berger, J. and Flippin, L.A. (1996) *Tetrahedron Lett.*, **37**, 4463–6.

80 (a) Kocovsky, P. (2003) *J. Organomet. Chem.*, **687**, 256–68;
(b) Vysko il, S., Smrčina, M. and Kocovsky, P. (1998) *Tetrahedron Lett.*, **39**, 9289–92.

81 Driver, M.S. and Hartwig, J.F. (1996) *J. Am. Chem. Soc.*, **118**, 7217–18.

82 Louie, J., Driver, M.S., Hamann, B.C. and Hartwig, J.F. (1997) *J. Org. Chem.*, **62**, 1268–73.

83 Beletskaya, I.P., Bessmertnykh, A.G. and Guilard, R. (1999) *Synlett*, 1459–61.

84 Marcoux, J.-F., Wagaw S. and Buchwald, S.L. (1997) *J. Org. Chem.*, **62**, 1568–9.

85 Wolfe, J.P. and Buchwald, S.L. (1997) *Tetrahedron Lett.*, **38**, 6359–62.

86 Shen, Q., Shekhar, S., Stambuli, J.P. and Hartwig, J.F. (2005) *Angew. Chem. Int. Ed.*, **44**, 1371–5.

87 (a) Watanabe, M., Yamamoto, T. and Nishiyama, M. (2000), EP A2 1035114, 20000913;
(b) Watanabe, M., Yamamoto, T. and Nishiyama, M. (2000) *Angew. Chem. Int. Ed.*, **39**(14), 2501–4.

88 Kataoka, N., Shelby, Q., Stambuli, J.P. and Hartwig, J.F. (2002) *J. Org. Chem.*, **67**, 5553–66.

89 Hamann, B.C. and Hartwig, J.F. (1998) *J. Am. Chem. Soc.*, **120**, 7369–70.

90 (a) Kantchev, E.A.B., O'Brien, C.J. and Organ, M.G. (2007) *Angew. Chem. Int. Ed.*, **46**, 2768–813;
(b) Tekavec, T.N. and Louie, J. (2007) *Top. Organomet. Chem.*, **21**, 159–92;
(c) Diez-Gonzalez, S. and Nolan, S.P. (2007) *Top. Organomet. Chem.*, **21**, 47–82;
(d) Glorius, F. (2007) *Top. Organomet. Chem.*, **21**, 1–20;
(e) Herrmann, W.A. (2002) *Angew. Chem. Int. Ed.*, **41**(**8**), 1290–309.

91 Harries-Rees, K., Chauvin, Y., Grubbs, R. and Schrock, R. (2005) *Chem. World*, **2**, 42–4.

92 Krafczyk, A.J., III, Arduengo, R. and Schmuter, R. (1999) *Tetrahedron*, **55**, 14523–34.

93 Fuerstner, A., Alcarazo, M., Cesar, V. and Lehmann, C.W. (2006) *Chem. Commun.*, 2176–8.

94 Caddick, S., Cloke, F.G.N., Clentsmith, G.K.B., Hitchcock, P.B., McKerrecher, D., Titcomb, L.R. and Williams, M.R.V.J. (2001) *J. Organomet. Chem.*, **617**, 635–9.

95 Stauffer, S.R., Lee, S., Stambuli, J.P., Hauck, S.I. and Hartwig, J.F. (2000) *Org. Lett.*, **2**, 1423–6.

96 Lerma, I.C., Cawley, M.J., Cloke, F.G.N., Arentsen, K., Scott, J.S., Pearson, S.E., Hayler, J. and Caddick, S. (2005) *J. Organomet. Chem.*, **690**, 5841–8.

97 Omar-Amrani, R., Schneider, R. and Fort, Y. (2004) *Synthesis*, 2527–34.

98 Urgaonkar, S., Xu, J.H. and Verkade, J.G. (2003) *J. Org. Chem.*, **68**, 8416–23.

99 (a) Urgaonkar, S., Nagarajan, M. and Verkade, J.G. (2003) *Org. Lett.*, **5**, 815–18;
(b) Urgaonkar, S. and Verkade, J.G. (2004) *J. Org. Chem.*, **69**, 9135–42;
(c) Urgaonkar, S., Xu, J.-H. and Verkade, J.G. (2004) *J. Org. Chem.*, **69**(**26**), 9323.

100 Nandakumar, M.V. and Verkade, J.G. (2005) *Tetrahedron*, **61**, 9775–82.

101 Urgaonkar, S. and Verkade, J.G. (2004) *Adv. Synth. Catal.*, **346**, 611–16.

102 Singer, R.A., Caron, S., McDermott, R.E., Arpin, P. and Do, N.M. (2003) *Synthesis*, 1727–31.

103 Singer, R.A., Dore, M., Sieser, J.E. and Berliner, M.A. (2006) *Tetrahedron Lett.*, **47**, 3727–31.

104 Singer, R.A., Tom, N.J., Frost, H.N. and Simon, W.M. (2004) *Tetrahedron Lett.*, **45**, 4715–18.

105 (a) Beller, M., Ehrentraut, A., Fuhrmann, C. and Zapf, A. (2002) WO A1 2002010178, 20020207;
(b) Ehrentraut, A., Zapf, A. and Beller, M. (2002) *J. Mol. Catal. A*, **182-3**, 515–523;
(c) Harkal, S., Rataboul, F., Zapf, A., Fuhrmann, C., Riermeier, T., Monsees, A. and Beller, M. (2004) *Adv. Synth. Catal.*, **346**, 1742–8;
(d) Rataboul, F., Zapf, A., Jackstell, R., Harkal, S., Riermeier, T., Monsees, A., Dingerdissen, U. and Beller, M. (2004) *Chem. Eur. J.*, **10**, 2983–90.

106 Bei, X.H., Uno, T., Norris, J., Turner, H.W., Weinberg, W.H., Guram, A.S. and Peterson, J.L. (1999) *Organometallics*, **18**, 1840–53.

107 Li, G.Y., Zheng, G. and Noonan, A.F. (2001) *J. Org. Chem.*, **66**, 8677–81.

108 Ackermann, L. and Born, R. (2005) *Angew. Chem. Int. Ed.*, **44**, 2444–7.

109 Ackermann, L. (2007) *Synlett*, **4**, 507–26.

110 Ackermann, L., Born, R., Spatz, J.H., Althammer, A. and Gschrei, C.J. (2006) *Pure Appl. Chem.*, **78**, 209–14.

111 Liu, D., Gao, W., Dai, Q. and Zhang, X. (2005) *Org. Lett.*, **7**, 4907–10.

112 Xie, X., Zhang, T.Y. and Zhang, Z. (2006) *J. Org. Chem.*, **71**, 6522–9.

113 Hill, L.L., Moore, L.R., Huang, R., Craciun, R., Vincent, A.J., Dixon, D.A., Chou, J., Woltermann, C.J. and Shaughnessy, K.H. (2006) *J. Org. Chem.*, **71**, 5117–25.

114 Lee, J.-C., Wang, M.-G. and Hong, F.-E. (2005) *Eur. J. Inorg. Chem.*, **24**, 5011–17.

115 Brenstrum, T., Clattenburg, J., Britten, J., Zavorine, S., Dyck, J., Robertson, A.J., McNulty, J. and Capretta, A. (2006) *Org. Lett.*, **8**, 103–5.

116 Gerristma, D., Brenstrum, T., McNulty, J. and Capretta, A. (2004) *Tetrahedron Lett.*, **45**, 8319–21.

117 Teuma, E., Lyon-Saunier, C., Gornitzka, H., Mignani, G., Baceiredo, A. and Bertrand, G. (2005) *J. Organomet. Chem.*, **690**, 5541–5.

118 (a) Meyers, C., Maes, B.U.W., Loones, K.T.J., Bal, G., Lemiere, G.L.F. and Dommisse, R.A. (2004) *J. Org. Chem.*, **69**, 6010–17;
(b) Shaughnessy, K.H. and Booth, R.S. (2001) *Org. Lett.*, **3**, 2757;
(c) DeVasher, R.B., Moore, L.S. and Shaughnessy, K.H. (2004) *J. Org. Chem.*, **69**, 7919–27.

119 Beller, M., Riermeier, T.H., Reisinger, C.P. and Herrmann, W.A. (1997) *Tetrahedron Lett.*, **38**, 2073–4.

120 Zim, D. and Buchwald, S.L. (2003) *Org. Lett.*, **5**, 2413–15.

121 Schnyder, A., Indolese, A.F., Studer, M. and Blaser, H.U. (2002) *Angew. Chem. Int. Ed.*, **41**, 3668–71.

122 Blaser, H.-U., Indolese, A., Naud, F., Nettekoven, U. and Schnyder, A. (2004) *Adv. Synth. Catal.*, **346**, 1583–98.

123 Stambuli, J.P., Kuwano, R. and Hartwig, J.F. (2002) *Angew. Chem. Int. Ed.*, **41**, 4746–8.

124 Viciu, M.S., Kissling, R.M., Stevens, E.D. and Nolan, S.P. (2002) *Org. Lett.*, **4**, 2229–31.

125 Viciu, M.S., Navarro, O., Germaneau, R.F., Kelly, R.A., III, Sommer, W., Marion, N., Stevens, E.D., Cavallo, L. and Nolan, S.P. (2004) *Organometallics*, **23**, 1629–35.

126 Marion, N., Navarro, O., Mei, J., Stevens, E.D., Scott, N.M. and Nolan, S.P. (2006) *J. Am. Chem. Soc.*, **12**, 4101–11.

127 Marion, N., Ecarnot, E.C., Navarro, O., Amoroso, D., Bell, A. and Nolan, S.P. (2006) *J. Org. Chem.*, **71**, 3816–21.

128 Goossen, L.J., Paetzold, J., Briel, O., Rivas-Nass, A., Karch, R. and Kayser, B. (2005) *Synlett*, 275–8.

129 Oezdemir, I., Demir, S., Goek, Y., Cetinkaya, E. and Cetinkaya, B. (2004) *J. Mol. Catal. A*, **222**, 97–102.

130 Bedford, R.B. and Cazin, C.S.J. (2002) GB A 2376946, 20021231.

131 (a) Christmann, U., Pantazis, D.A., Benet-Buchholz, J., McGrady, J.E., Maseras, F. and Vilar, R. (2006) *J. Am. Chem. Soc.*, **128**, 6376–90;
(b) Christmann, U., Vilar, R. and White, A.J.P. (2004) *Chem. Commun.*, **11**, 1294–5.

132 Gusev, O.V., Peganova, T.A., Kalsin, A.M., Vologdin, N.V., Petrovskii, P.V., Lyssenko, K.A., Tsvetkov, A.V. and Beletskaya, I.P. (2006) *Organometallics*, **25**, 2750–60.

133 Gusev, O.V., Peganova, T.A., Kalsin, A.M., Vologdin, N.V., Petrovskii, P.V., Lyssenko, K.A., Tsvetkov, A.V. and Beletskaya, I.P. (2005) *J. Organomet. Chem.*, **690**, 1710–17.

134 Gajare, A.S., Toyota, K., Yoshifuji, M. and Ozawa, F. (2004) *J. Org. Chem.*, **69**, 6504–6.

135 (a) Jiang, L. and Buchwald, S.L. (2004) *Palladium-Catalyzed Aromatic Carbon-Nitrogen Bond Formation in Metal-Catalyzed Cross-Coupling Reactions*, Wiley-VCH Verlag GmbH, pp. 699–760;
(b) Stauffer, S.R. and Hartwig, J.F. (2003) *J. Am. Chem. Soc.*, **125**, 6977–85.

136 Kuwano, R., Utsunomiya, M. and Hartwig, J.F. (2002) *J. Org. Chem.*, **6**, 6479–86.

137 Anderson, K.W., Ikawa, T., Tundel, R.E. and Buchwald, S.L. (2006) *J. Am. Chem. Soc.*, **128**, 10694–5.

138 Flessner, T. and Doye, S. (1999) *J. Prakt. Chem.*, **341**, 186–90.

139 Li, L. and Navasero, N. (2006) *Org. Lett.*, **8**, 3733–6.

140 Basu, B., Das, P., Nanda, A.K., Das, S. and Sarkar, S. (2005) *Synlett*, 1275–8.

141 Schlummer, B. and Scholz, U. (2006) DE 102004056820 A1 20060601.

142 Schlummer, B. and Scholz, U. (2006) I. WO 2006056412 A1 20060601.

143 Jensen, T.A., Liang, X., Tanner, D. and Skjaerbaek, N. (2004) *J. Org. Chem.*, **69**, 4936–47.

144 Buchwald, S.L. and Anderson, K.W. (2006) WO 2006074315 A2 20060713.

145 Smith, C.J., Tsang, M.W.S., Holmes, A.B., Danheiser, R.L. and Tester, J.W. (2005) *Org. Biomol. Chem.*, 3767–81.

146 Varma, R.S. and Ju, Y. (2005) *Green Separation Processes*, 53–87.

147 Cave, G.W.V., Raston, C.L. and Scott, J.L. (2001) *Chem. Commun.*, 2159–69.

148 Stauffer, S.R. and Steinbeiser, M.A. (2005) *Tetrahedron Lett.*, **46**, 2571–5.

149 Grushin, V.V. and Alper, H. (1994) *Chem. Rev.*, **94**, 1047–62.

150 (a) Harris, M.C., Huang, X. and Buchwald, S.L. (2002) *Org. Lett.*, **4**, 2885–8;
(b) Buchwald, S.L. and Huang, X. (2001) *Org. Lett.*, **3**, 3417–19.

151 Maes, B.U.W., Loones, K.T.J., Hostyn, S., Diels, G. and Rombouts, G. (2004) *Tetrahedron*, **60**, 11559–64.

152 (a) Urgaonkar, S., Nagarajan, M. and Verkade, J.G. (2003) *J. Org. Chem.*, **68**, 452–9;
(b) Parrish, C.A. and Buchwald, S.L. (2001) *J. Org. Chem.*, **66**, 3820–7;
(c) Ali, M.H. and Buchwald, S.L. (2001) *J. Org. Chem.*, **66**, 2560–5.

153 Iwaki, T., Yasuhara, A. and Sakamoto, T. (1999) *J. Chem. Soc. Perkin Trans.*, **1**, 1505–10.

154 Anderson, K.W., Mendez-Perez, M., Priego, J. and Buchwald, S.L. (2003) *J. Org. Chem.*, **68**, 9563–73.

155 Tundel, R.E., Anderson, K.W. and Buchwald, S.L. (2006) *J. Org. Chem.*, **71**, 430–3.

156 Zhang, W. and Nagashima, T. (2006) *J. Fluorine Chem.*, **127**, 588–91.

157 Blaser, H.U., Brieden, W., Pugin, B., Spindler, F., Studer, M. and Togni, A. (2002) *Top. Catal.*, **19**, 3–16.

158 Roy, A.H. and Hartwig, J.F. (2003) *J. Am. Chem. Soc.*, **125**, 8704–5.

159 Paul, F., Patt, J. and Hartwig, J.F. (1995) *Organometallics*, **14**, 3030–9.

160 Wagaw, S. and Buchwald, S.L. (1996) *J. Org. Chem.*, **61**, 7240–1.

161 Grasa, G.A., Viciu, M.S., Huang, J. and Nolan, S.P. (2001) *J. Org. Chem.*, **66**, 7729–37.

162 Charles, M.D., Schultz, P. and Buchwald, S.L. (2005) *Org. Lett.*, **7**, 3965–8.

163 Audisio, D., Messaoudi, S., Peyrat, J.-F., Brion, J.-D. and Alami, M. (2007) *Tetrahedron Lett.*, **48**, 6928–32.

164 Hooper, M.W., Utsunomiya, M. and Hartwig, J.F. (2003) *J. Org. Chem.*, **68**, 2861–73.

165 Messaoudi, S., Audisio, D., Brion, J.-D. and Alami, M. (2007) *Tetrahedron*, **63**, 10202–10.

166 Zhang, H.Q., Xia, Z., Vasudevan, A. and Djuric, S.W. (2006) *Tetrahedron Lett.*, **47**, 4881–4.

167 (a) Lakshman, M.K., Keeler, J.C., Hilmer, J.H. and Martin, J.Q. (1999) *J. Am. Chem. Soc.*, **121**, 6090–1;
(b) Lakshman, M.K., Hilmer, J.H., Martin, J.Q., Keeler, J.C., Dinh, Y.Q.V., Ngassa, F.N. and Russon, L.M. (2001) *J. Am. Chem. Soc.*, **123**, 7779–87;
(c) Lakshman, M.K. (2005) *Curr. Org. Synth.*, **2**, 83–112.

168 Michalik, D., Kumar, K., Zapf, A., Tillack, A., Arlt, M., Heinrich, T. and Beller, M. (2004) *Tetrahedron Lett.*, **45**, 2057–61.

169 Shen, Q. and Hartwig, J.F. (2006) *J. Am. Chem. Soc.*, **128**, 10028–9.

170 Surry, D.S. and Buchwald, S.L. (2007) *J. Am. Chem. Soc.*, **129**, 10354–5.

171 Lee, S., Jorgensen, M. and Hartwig, J.F. (2001) *Org. Lett.*, **3**, 2729–32.

172 Hartwig, J.F., Jorgensen, M. and Lee, S. (2001) WO 03/006420 A1, 12.7.2001.

173 Lee, D.-Y. and Hartwig, J.F. (2005) *Org. Lett.*, **7**, 1169–72.

174 (a) Deng, B.-L., Lepoivre, J.A. and Lemiere, G. (1999) *Eur. J. Org. Chem.*, 2683–8;
(b) Mann, G., Hartwig, J.F., Driver, M.S. and Fernandez-Rivas, C. (1998) *J. Am. Chem. Soc.*, **120**, 827–8;
(c) Wolfe, J.P., Ahman, J., Sadighi, J.P., Singer, R.A. and Buchwald, S.L. (1997) *Tetrahedron Lett.*, **38**, 6367–70.

175 Jaime-Figueroa, S., Liu, Y., Muchowski, J.M. and Putman, D.G. (1998) *Tetrahedron Lett.*, **39**, 1313–16.

176 Lim, C.W. and Lee, S.-g. (2000) *Tetrahedron*, **56**, 5131–6.
177 Cioffi, C.L., Berlin, M.L. and Herr, R.J. (2004) *Synlett*, 841–5.
178 Anderson, K.W., Tundel, R.E., Ikawa, T., Altman, R.A. and Buchwald, S.L. (2006) *Angew. Chem. Int. Ed.*, **45**, 6523–7.
179 Piguel, S. and Legraverend, M. (2007) *J. Org. Chem.*, **72**, 7026–9.
180 Furuta, T., Kitamura, Y., Hashimoto, A., Fujii, S., Tanaka, K. and Kan, T. (2007) *Org. Lett.*, **9**, 183–6.
181 Klapars, A., Campos, K.R., Chen, C.-y., and Volante, R.P. (2005) *Org. Lett.*, **7**, 1185–8.
182 Wallace, D.J., Klauber, D.J., Chen, C.-y. and Volante, R.P. (2003) *Org. Lett.*, **5**, 4749–52.
183 (a) Madar, D.J., Kopecka, H., Pireh, D., Pease, J., Pliushchev, M., Sciotti, R.J., Wiedeman, P.E. and Djuric, S.W. (2001) *Tetrahedron Lett.*, **42**, 3681–4;
(b) Ghosh, A., Sieser, J.E., Riou, M., Cai, W. and Rivera-Ruiz, L. (2003) *Org. Lett.*, **5**, 2207–10.
184 McLaughlin, M., Palucki, M. and Davies, I.W. (2006) *Org. Lett.*, **8**, 3311–14.
185 Bolm, C. and Hildebrand, J.P. (2000) *J. Org. Chem.*, **65**, 169–75.
186 (a) Zhu, Y.-M., Kiryu, Y. and Katayama, H. (2002) *Tetrahedron Lett.*, **43**, 3577–80;
(b) Lebedev, Y., Khartulyari, A.S. and Voskoboynikov, A.Z. (2005) *J. Org. Chem.*, **70**, 596–602.
187 Evindar, G. and Batey, R.A. (2003) *Org. Lett.*, **5**, 133–6.
188 Wang, Z., Skerlj, R.T. and Bridger, G.J. (1999) *Tetrahedron Lett.*, **40**, 3543–6.
189 Bedford, R.B. and Betham, M. (2006) *J. Org. Chem.*, **71**, 9403–10.
190 Romero, M., Harrak, Y., Basset, J., Ginet, L., Constans, P. and Pujol, M.D. (2006) *Tetrahedron*, **62**, 9010–16.
191 Neogi, A., Majhi, T.P., Mukhopadhyay, R. and Chattopadhyay, P. (2006) *J. Org. Chem.*, **71**, 3291–4.
192 Venkatesh, C., Sundaram, G.S.M., Ila, H. and Junjappa, H. (2006) *J. Org. Chem.*, **71**, 1280–3.
193 Fukushima, W., Kanbara, T. and Yamamoto, T. (2005) *Synlett*, **19**, 2931–4.
194 Nandakumar, M.V. and Verkade, J.G. (2005) *Angew. Chem. Int. Ed.*, **44**, 3115–18.
195 Yang, J.-S., Liau, K.-L., Wang, C.-M. and Hwang, C.-Y. (2004) *J. Am. Chem. Soc.*, **126**(39), 12325–35.
196 Ackermann, L. (2005) *Org. Lett.*, **7**, 439–42.
197 Gunda, P., Russon, L.M. and Lakshman, M.K. (2004) *Angew. Chem. Int. Ed.*, **43**, 6372–7.
198 Loones, K.T.J., Maes, B.U.W., Dommisse, R.A. and Lemiere, G.L. (2004) *Chem. Commun.*, 2466–7.
199 Queiroz, M.-J.R.P., Begouin, A., Ferreira, I.C.F.R., Kirsch, G., Calhelha, R.C., Barbosa, S. and Estevinho, L.M. (2004) *Eur. J. Org. Chem.*, **17**, 3679–85.
200 Inamoto, K., Katsuno, M., Yoshino, T., Suzuki, I., Hiroya, K. and Sakamoto, T. (2004) *Chem. Lett.*, **33**, 1026–7.
201 Damon, D.B., Dugger, R.W., Hubbs, S.E., Scott, J.M. and Scott, R.W. (2006) *Org. Proc. Res. Dev.*, **10**, 472–80.
202 Damon, D.B., Dugger, R.W. and Scott, R.W. (2002) WO 2002088085 A2, 30.4.2001.
203 Damon, D.B., Dugger, R.W. and Scott, R.W. (2002) WO 2002088069 A2, 30.4.2001.
204 (a) Alcazar-Roman, L.M. and Hartwig, J.F. (2001) *J. Am. Chem. Soc.*, **123**, 12905–6;
(b) Stambuli, J.P., Bühl, M. and Hartwig, J.F. (2002) *J. Am. Chem. Soc.*, **124**, 9346–7.
205 Yin, J., Rainka, M.P., Zhang, X.X. and Buchwald, S.L. (2002) *J. Am. Chem. Soc.*, **124**, 1162–3.
206 Green, J.C., Herbert, B.J. and Lonsdale, R. (2005) *J. Organomet. Chem.*, **690**, 6054–67.
207 Tagashira, J., Imao, D., Yamamoto, T., Ohta, T., Furukawa, I. and Ito, Y. (2005) *Tetrahedron: Asymmetry*, **16**, 2307–14.
208 Kreis, M., Friedmann, C.J. and Bräse, S. (2005) *Chem. Eur. J.*, **11**, 7387–94.
209 Satoh, T., Miura, M. and Nomura, M. (2002) *J. Organomet. Chem.*, **653**, 161–6.
210 (a) Fauvarque, J.F. and Jutand, A. (1979) *J. Organomet. Chem.*, **177**, 273–81;

(b) Galarini, R., Musco, A., Pontellini, R. and Santi, R. (1992) *J. Mol. Catal.*, **72**, L11–13;
(c) Kuwajima, I. and Nakamura, E. (1985) *Acc. Chem. Res.*, **18**, 181–7;
(d) Kosugi, M., Hagiwara, I., Sumiya, T. and Migita, T. (1984) *Bull. Chem. Soc. Jpn*, **57**, 242–6;
(e) Kosugi, M., Negishi, Y., Kameyama, M. and Migita, T. (1985) *Bull. Chem. Soc. Jpn*, **58**, 3383–4.

211 (a) Semmelhack, M.F., Stauffer, R.D. and Rogerson, T.D. (1973) *Tetrahedron Lett.*, 4519–22;
(b) Millard, A.A. and Rathke, M.W. (1977) *J. Am. Chem. Soc.*, **99**, 4833–5.

212 Ciufolini, M.A., Qi, H.B. and Browne, M.E. (1988) *J. Org. Chem.*, **53**(17), 4149–51.

213 (a) Hamann, B.C. and Hartwig, J.F. (1997) *J. Am. Chem. Soc.*, **119**, 12382–3;
(b) Palucki, M. and Buchwald, S.L. (1997) *J. Am. Chem. Soc.*, **119**, 11108–9;
(c) Satoh, T., Kawamura, Y., Miura, M. and Nomura, M. (1997) *Angew. Chem. Int. Ed.*, **36**, 1740–2.

214 Kawatsura, M. and Hartwig, J.F. (1999) *J. Am. Chem. Soc.*, **121**, 1473–8.

215 (a) Lee, S., Beare, N.A. and Hartwig, J.F. (2001) *J. Am. Chem. Soc.*, **123**, 8410–11;
(b) Jørgensen, M., Lee, S., Liu, X., Wolkowski, J.P. and Hartwig, J.F. (2002) *J. Am. Chem. Soc.*, **124**, 12557–65.

216 Hama, T., Liu, X. and Culkin, D.A. (2003) *J. Am. Chem. Soc.*, **125**, 11176–7.

217 Moradi, W.A. and Buchwald, S.L. (2001) *J. Am. Chem. Soc.*, **123**, 7996–8002.

218 Gaertzen, O. and Buchwald, S.L. (2002) *J. Org. Chem.*, **67**, 465–75.

219 (a) Beare, N.A. and Hartwig, J.F. (2002) *J. Org. Chem.*, **67**, 541–55;
(b) Shelby, Q., Kataoka, N., Mann, G. and Hartwig, J.F. (2000) *J. Am. Chem. Soc.*, **122**, 10718–19;
(c) Stambuli, J.P., Stauffer, S.R., Shaughnessy, K.H. and Hartwig, J.F. (2001) *J. Am. Chem. Soc.*, **123**, 2677–8.

220 Fox, J.M., Huang, X., Chieffi, A. and Buchwald, S.L. (2000) *J. Am. Chem. Soc.*, **122**, 1360–70.

221 Nguyen, H.N., Huang, X. and Buchwald, S.L. (2003) *J. Am. Chem. Soc.*, **125**, 11818–19.

222 Djakovitch, L. and Kohler, K. (2000) *J. Organomet. Chem.*, **606**, 101–7.

223 Stauffer, S.R., Beare, N.A., Stambuli, J.P. and Hartwig, J.F. (2001) *J. Am. Chem. Soc.*, **123**, 4641–2.

224 Ehrentraut, A., Zapf, A. and Beller, M. (2002) *Adv. Synth. Catal.*, **344**, 209–17.

225 Terao, Y., Satoh, T., Miura, M. and Nomura, M. (1998) *Tetrahedron Lett.*, **39**, 6203–6.

226 Terao, Y., Kametani, Y., Wakui, H., Satoh, T., Miura, M. and Nomura, M. (2001) *Tetrahedron*, **57**, 5967–74.

227 Muratake, H., Natsume, M. and Nakai, H. (2004) *Tetrahedron*, **60**, 11783–803.

228 Ackermann, L., Spatz, J.H., Gschrei, C.J., Born, R. and Althammer, A. (2006) *Angew. Chem. Int. Ed.*, **45**, 7627–30.

229 Shaughnessy, K.H., Hamann, B.C. and Hartwig, J.F. (1998) *J. Org. Chem.*, **63**, 6546–53.

230 Porcs-Makkay, M., Volk, B., Kapiller-Dezsofi, R., Mezei, T. and Simig, G. (2004) *Monatsh. Chem.*, **135**, 697–711.

231 Lee, S. and Hartwig, J.F. (2001) *J. Org. Chem.*, **66**, 3402–15.

232 Satoh, T., Inoh, J.-i., Kawamura, Y., Kawamura, Y., Miura, M. and Nomura, M. (1998) *Bull. Chem. Soc. Jpn*, **71**, 2239–46.

233 Culkin, D.A. and Hartwig, J.F. (2002) *J. Am. Chem. Soc.*, **124**, 9330–1.

234 Wu, L. and Hartwig, J.F. (2005) *J. Am. Chem. Soc.*, **127**, 15824–32.

235 You, J. and Verkade, J.G. (2003) *J. Org. Chem.*, **68**, 8003–7.

236 Klapars, A., Waldman, J.H., Campos, K.R., Jensen, M.S., McLaughlin, M., Chung, J.Y.L., Cvetovich, R.J. and Chen, C.-yi (2005) *J. Org. Chem.*, **70**, 10186–9.

4
Copper-Catalyzed Arylations of Amines and Alcohols with Boron-Based Arylating Reagents

Andrew W. Thomas and Steven V. Ley

4.1
Introduction

Recent years have witnessed a revolution in the synthetic methods available to transfer an aryl moiety to both oxygen and nitrogen, as well as to sulfur, selenium and phosphorus heteroatoms, mediated by transition metals. Indeed, a number of detailed reviews summarizing key developments in contemporary organic synthesis and many elegant solutions for both simple and complex target molecules of interest have appeared [1], and will continue to do so, as this field of research is maturing considerably each year. A common feature of most of these reviews is the general agreement that the pioneering studies of Ullmann [2] and Goldberg [3] still have importance today, albeit significantly superseded by the modern methods that clearly were not available to these early pioneers of arylation methodologies. The principal advances involve the use of lesser amounts of copper or copper salts, and lower reaction temperatures, to effect transformations that now have evolved to encompass a very broad range of substrates. Hence, the complexity of possible target molecules has now reached new heights. These advances were made possible by specific seminal modern contributions, principally by Chan and Lam through the use of copper(II) acetate and boron-based arylating reagents [4], and Buchwald, Ma and others via the use of copper(I) salts, along with cesium carbonate as base and chelating ligands [5]. Many of the hundreds of subsequent reports in the field of copper-mediated arylations of O—H and N—H bonds may be generally categorized as being rather incremental innovation based on these modern pioneering discoveries. The overall copper-mediated cross-coupling strategies that have developed into mature processes are shown in Figure 4.1 for O—H bonds and in Figure 4.2 for N—H bonds. Although this chapter deals only with boron-based aryl donors, it is clearly important to put into perspective the alternative aryl donors available for copper-mediated arylations of O—H and N—H bonds.

Modern Arylation Methods. Edited by Lutz Ackermann

ISBN: 978-3-527-31937-4

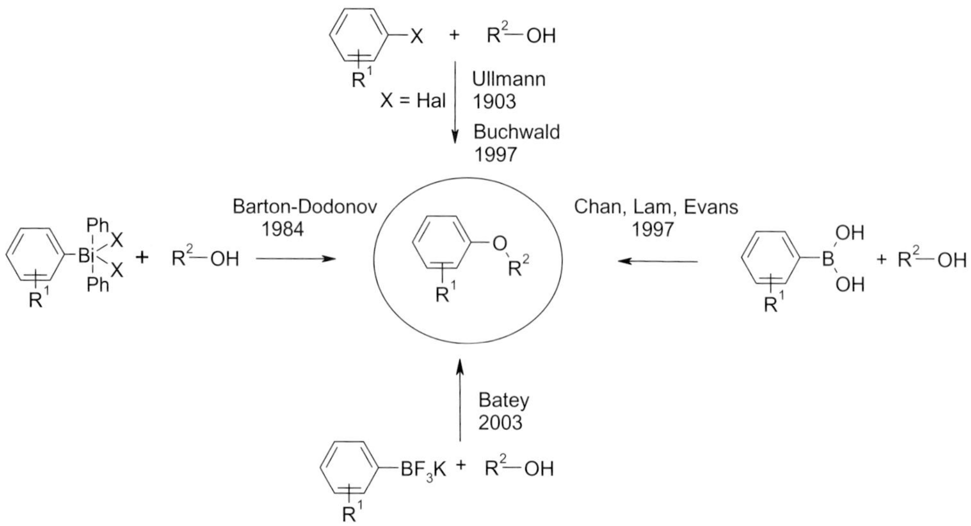

Figure 4.1 Copper-mediated cross-coupling concepts for O–H bond arylation.

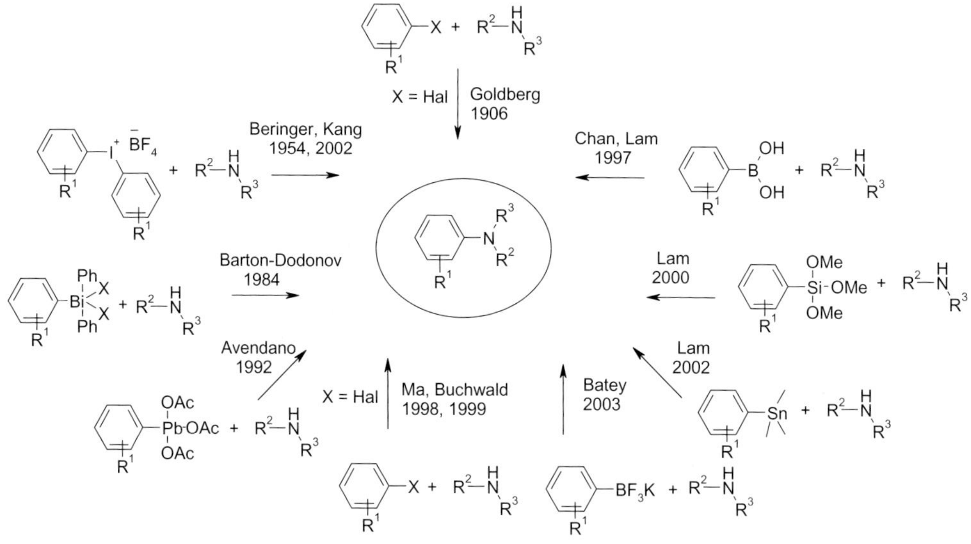

Figure 4.2 Copper-mediated cross-coupling concepts for N–H bond arylation.

This chapter will attempt to summarize selected key learnings and trends, and will also highlight recent developments of these methods solely devoted to the copper-catalyzed oxidative arylations of O–H and N–H bonds with boron-based arylating reagents. Further, it will detail some specifics that have brought the method to its current high levels of popularity within the synthetic organic chemistry community.

4.2 Discovery and Development of a New O—H Bond Arylation Reaction: From Stoichiometric to Catalytic in Copper

The first report by Chan [6], in June 1997, of copper-mediated arylations utilizing boronic acids detailed a new and efficient method that had an impressive scope of both N—H and O—H-containing substrates. This discovery can be traced back [1c] as an evolution of previous powerful methods involving triarylbismuth reagents [7]. This report was quickly followed by two short, but important, communications [4, 8] which can be considered as creating a new mind-set on investigating further this much improved method for carbon–heteroatom bond formations. These initial protocols required stoichiometric amounts of copper(II) acetate for best efficiency and, as such, left much room for improvement. However, it is worth detailing some of the key features of these copper-mediated methodologies by Chan and Evans (Scheme 4.1, Table 4.1). Although the substrate scope was minimal in these first communications, the amount of data published in the scientific literature using these exact methods subsequently has been staggering. Interestingly, the scientific community seems to have no clear preference in the choice of base for this transformation. Triethylamine or pyridine and, in some cases, even mixed-base systems (as shown originally by Evans) have been–and continue to be–used routinely. Evans, importantly, was the first to demonstrate that the use of substoichiometric (10 mol%) amounts of copper salts was viable. Indeed, this was to become the case as exemplified, and in practice these conditions can be used for a wide range of substrates. Here, vigorous stirring of the reaction mixture is necessary to provide the best outcomes. Heating the reaction mixture at 50 °C in dimethylformamide (DMF), instead of the normally employed CH_2Cl_2, allowed the catalytic in copper salt version, this time with 20 mol% $Cu(OAc)_2$, to proceed efficiently with pyridine as base and molecular oxygen as the oxidant. Alternative oxidants, such as TEMPO or pyridine *N*-oxide, were less efficacious [9]. It was realized at an early stage that $Cu(OAc)_2$ seemed to be an optimal copper(II) source, as the use of other copper salts, such as $Cu(OPiv)_2$, $Cu(NO_3)_2$, $Cu(acac)_2$ and $Cu(OCOCF_3)_2$, $CuSO_4$, $CuCl_2$, or $Cu(ClO_4)_2$, gave inferior results and often did not lead to the desired products of an arylation reaction. The use of $Cu(OTf)_2$ resulted in significant C—C bond formation.

The use of molecular sieves resulted in a significant enhancement in the product yield, and excess amounts of the aryl donor were always necessary for higher-yielding reactions. Presumably, the minimum quantity of molecular sieves to be

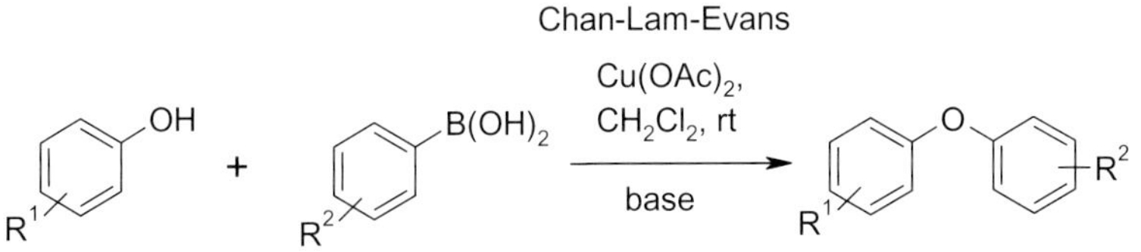

Chan-Lam-Evans

$Cu(OAc)_2$, CH_2Cl_2, rt

base

OH + B(OH)₂ → O

Scheme 4.1 Copper-mediated cross-couplings of phenols with boronic acids (see Table 4.1).

Table 4.1 Stoichiometric and the first catalytic O—H bond arylation.

Copper source	Remarks	Example of conditions	Reference
$Cu(OAc)_2$	Good to high yields, wide substrate scope, *ortho*-halophenols react well, Et_3N preferred base	$Cu(OAc)_2$ (1 equiv), $ArB(OH)_2$ (2 equiv), Et_3N (2 equiv), 24 h	[4]
$Cu(OAc)_2$	Good to high yields, very wide substrate scope, *ortho*-substituted phenols and *ortho*-alkyl-substituted boronic acids tolerated	$Cu(OAc)_2$ (1 equiv), $ArB(OH)_2$ (1–3 equiv), Et_3N (5 equiv), 4 Å MS, 18 h, ambient air	[8]
$Cu(OAc)_2$	Catalytic in copper, 30% yield	$Cu(OAc)_2$ **(0.1 equiv)**, $ArB(OH)_2$ (1–3 equiv), Et_3N (5 equiv), 4 Å MS, 18 h, $\mathbf{O_2}$	[8]
$Cu(OAc)_2$	79%	$Cu(OAc)_2$ **(0.2 equiv)**, pyridine (2 equiv), DMF, 4 Å MS, $\mathbf{O_2}$, 50 °C	[9]
$Cu(OAc)_2$	16%	$Cu(OAc)_2$ **(0.2 equiv)**, pyridine (2 equiv), DMF, 4 Å MS, **TEMPO (1.1 equiv), air**, 50 °C	[9]
$Cu(OAc)_2$	9%	$Cu(OAc)_2$ **(0.2 equiv)**, pyridine (2 equiv), DMF, 4 Å MS, **pyridine *N*-oxide, air**, 50 °C	[9]
$[Cu(OH) \cdot TMEDA]_2Cl_2$	3%	$Cu(OAc)_2$ **(0.2 equiv)**, pyridine (2 equiv), DMF, 4 Å MS, $\mathbf{O_2}$, 50 °C	[9]

MS = molecular sieve.

added should correspond to at least the amount of water that can be produced by the hydrolysis of the boronic acid during the reaction. To allow for the synthesis of products with higher molecular complexity, milder reaction conditions were sought, and it was subsequently shown that, during the reaction, significant amounts of phenol and diphenyl ether were produced which may be due to the formation of water during the reaction. This water could be formed from the boronic acid, which forms the trimeric triaryl boroxine that may then participate in the reaction. Indeed, this was found to be the case in independent experiments when using 0.33 equiv. of the boroxines as aryl donors. Investigations into the need for an inert gas atmosphere during the reaction also led to the observation that ambient atmospheric air, or even molecular oxygen, offered a better reaction outcome, leading to more than double the yields observed under an argon atmosphere. Moreover, the addition of excess base to the reaction medium did not impede the reaction, and may infer that this component plays a dual role as a base and as a participating ligand with an organocopper species.

4.3 Mechanistic Considerations

It has been speculated that the mechanism of copper-mediated (stoichiometric in copper) arylations of phenols involves the following elemental steps: (i) transmetallation of Cu(II) with the arylboronic acid; (ii) coordination of the phenol nucleophile to copper(II); and (iii) reductive elimination, slowly via the Cu(II) species or via air oxidation to the Cu(III) species which can be expected to undergo reductive elimination more rapidly, thereby regenerating a potentially catalytically active copper(I) species (Scheme 4.2). A plausible catalytic mechanism (not illustrating the potential role of substrates as copper ligands) is also shown.

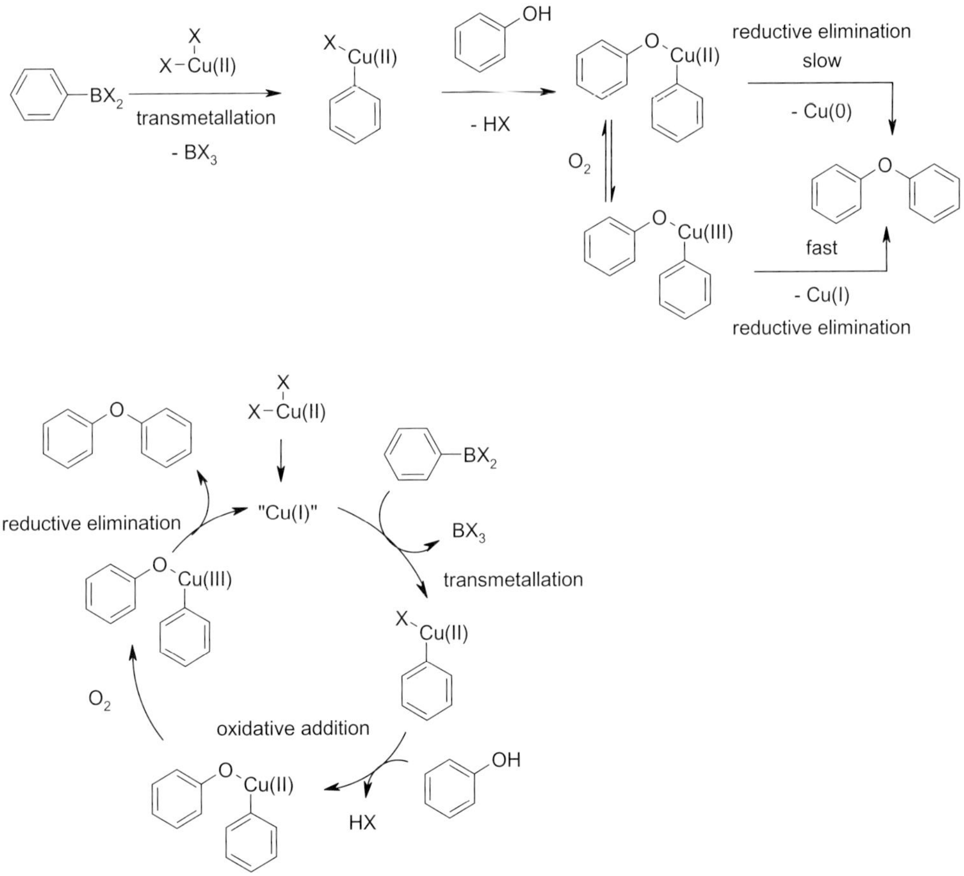

Scheme 4.2 Plausible mechanism for copper-catalyzed cross-couplings of aryl boronic acids with phenols.

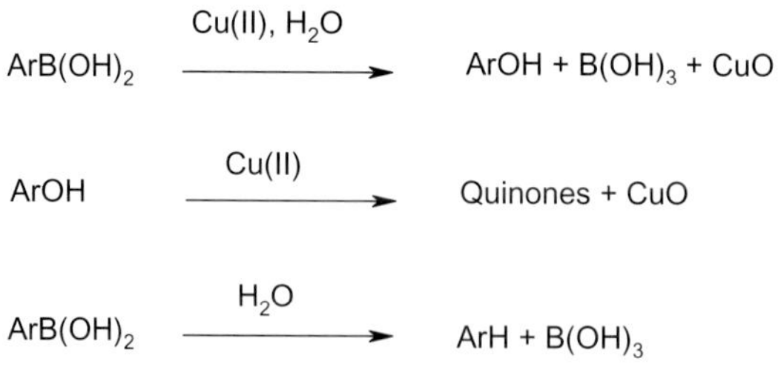

$$ArB(OH)_2 \xrightarrow{Cu(II),\ H_2O} ArOH + B(OH)_3 + CuO$$

$$ArOH \xrightarrow{Cu(II)} Quinones + CuO$$

$$ArB(OH)_2 \xrightarrow{H_2O} ArH + B(OH)_3$$

Scheme 4.3 Possible side reactions of aryl boronic acids.

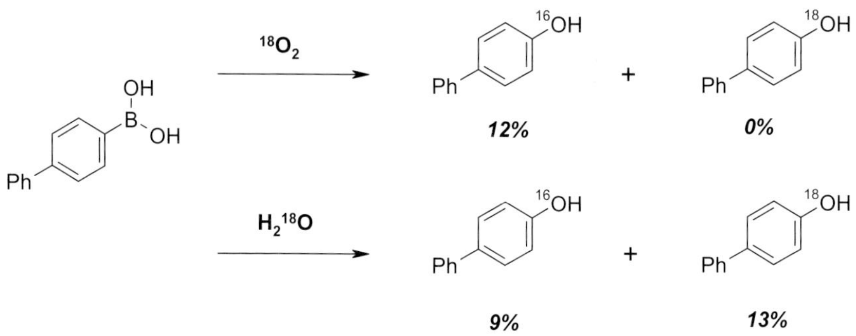

Scheme 4.4 Oxygen isotope-incorporation studies.

During these catalytic or stoichiometric processes, possible side reactions do occur which are shown in Scheme 4.3. These explain why the use of more than 1 equiv. of boronic acid is necessary. Furthermore, during the oxidation of copper(II) to copper(III), hydrogen peroxide is produced, which can decrease the yields of the reaction as a result of its reaction with the aryl boronic acid. This may also explain why more than 1 equiv. of aryl boronic acid provides enhanced yields [10]. Moreover, the aryl boronic acid can form triaryl boroxines [11], and in doing so forms water, which can be removed from the reaction by molecular sieves. Evans and coworkers postulated that phenolic products would be formed as a result of the competitive arylation of water formed during the reaction process.

Lam and coworkers conducted oxygen isotope incorporation studies to validate Evans' hypotheses [12]. By using $^{18}O_2$, it could be shown that the phenols were not formed from atmospheric oxygen, which in turn suggests that the phenol is formed from water produced during the reaction. Indeed, when $H_2{}^{18}O$ was used as additive, the isotope label was incorporated into the biphenylphenol product, as shown in Scheme 4.4.

4.4 Miscellaneous Applications

4.4.1 Additional Applications with $ArB(OH)_2$

Generally, a common feature of the arylation methods is that long reaction times are required, although this may in some cases result in a poor reaction outcomes, especially when using open vessels that allow the solvent to evaporate. Whilst the use of a balloon offers a good compromise, the better approach is to bubble oxygen directly into the reaction mixture. The use of a solid supported reagent, which facilitates reaction product purification, has also been suggested as a good approach, employing a supported form of immobilized copper(II) that is easily prepared in three steps from a Wang resin (Table 4.2) [13]. A key intramolecular phenolic arylation reaction with a boronic acid was used in an efficient synthesis of the cycloisodityrosine moiety of RP 66453 [1b, 14]. This method relied on DMAP as an additive to minimize the formation of an otherwise-formed protodeborylated product and, notably, it was preferable to perform the reaction in the absence of another base under dilute reaction conditions. Symmetrical diarylethers can be obtained by two methods:

Table 4.2 Additional applications.

Copper source	Remarks	Example of conditions	Reference
$Cu(OAc)_2$	Solid-supported catalyst, good yields, good substrate scope, including *ortho*-substitution on phenol	Supported Cu (1.5 equiv), $ArB(OH)_2$ (3 equiv), Et_3N (2.6 equiv), 4 Å MS, CH_2Cl_2, rt, 24 h	[13]
$Cu(OAc)_2$	Use of **DMAP** as an additive, **intramolecular arylation,** absence of pyridine or Et_3N	$Cu(OAc)_2$ (1 equiv), CH_2Cl_2, 4 Å MS, rt, 48 h	[14]
$Cu(OAc)_2$	Symmetrical diarylether formation, 3 equiv $ArB(OH)_2$	H_2O_2 (0.25 equiv), $Cu(OAc)_2$ (0.5 equiv), Et_3N (3 equiv), CH_2Cl_2, 4 Å MS, rt, 48 h	[15]
$Cu(OAc)_2$	Symmetrical diarylether formation, 1 equiv $ArB(OH)_2$	H_2O (10 equiv), $Cu(OAc)_2$ (1 equiv), Et_3N (5 equiv), CH_2Cl_2 : MeCN, rt, 6 h	[16]

MS = molecular sieve; rt = room temperature.

- By using a sequential one-pot method where 0.25 equiv. of H_2O_2 and 0.5 equiv. $Cu(OAc)_2$ is optimal, and yields of up to 90% can be expected for electron-rich substrates [15].
- By using 1 equiv. of $Cu(OAc)_2$ and 10 equiv. of H_2O as an additive, such that the good yields obtained over a wide substrate range are based on 1 equiv. of aryl boronic acid [16].

4.4.2 Alternatives to $ArB(OH)_2$

Boronic acids are generally the most common – and, in most cases studied to date – the superior aryl donors in the Chan–Lam–Evans modified Ullmann condensation reaction. However, other boron-containing sources have been examined and provided significantly improved results [17]. This study detailed a comparison with 1 equiv. of $PhB(OH)_2$ where yields of 17% or 35% were obtained using triethylamine or pyridine as the base, respectively. It is interesting to note that in all cases, except for the boron ester derived from pinacol, improved yields were obtained. It has been postulated correctly by Evans (*vide supra*) that the role of molecular sieves can facilitate the formation of the triphenylboroxine, the cyclic anhydride of phenylboronic acid. The formation of this arylating agent has also been observed spontaneously in the absence of molecular sieves by simply stirring the boronic acid in CH_2Cl_2 at ambient temperature. A suggested protocol is to stir the boronic acids in CH_2Cl_2 at ambient temperature overnight before its use in the Chan–Evans–Lam modified Ullmann condensation reaction. Thereby, only 0.3 equiv. rather than 2.0 or 3.0 equiv. of boronic acid are needed for the best yields. It should be noted that these studies all utilize stoichiometric amounts of copper acetate (Table 4.3) [17]. The synthesis of oxybispyridines using copper-mediated cross-coupling is challenging, and the scope of the methods has been established [18]. The cross-coupling of pyridine-containing boronic acids with bromopyridinols is a poor reaction, and typically produces no product. However, when using a pinacol boronate ester the cross-coupling product could be obtained in 11–16% yield under standard Chan–Lam conditions. The failure is postulated to be due to the lowering of the pK_a (7.7–9.2) of the N-containing 'phenolic moiety', creating a better efficiency in the coordination step of the mechanism due to a greater stability of the anion. This hypothesis is clearly substantiated by the efficiency of the phenolic substrates (pK_a=9.3–9.6) in 63% and 79%, respectively. An interesting development of this method shows that the use of cesium carbonate, instead of an organic base, is also valid, giving a 60% yield of the desired product [19]. Further, the triphenylboroxine–pyridine complex is easily prepared and provides high yields of the arylation product in both the absence or presence of cesium carbonate. Although the exact role of the enhancing effects of the pyridine complex is still not known, it can be speculated that this system might stabilize a heterobimetallic tetracoordinated copper species, and thus facilitate transmetallation

Table 4.3 Additional boron-containing aryl donors.

$ArBX_2$ (+phenol)	Yield (%)	Example of conditions	Reference
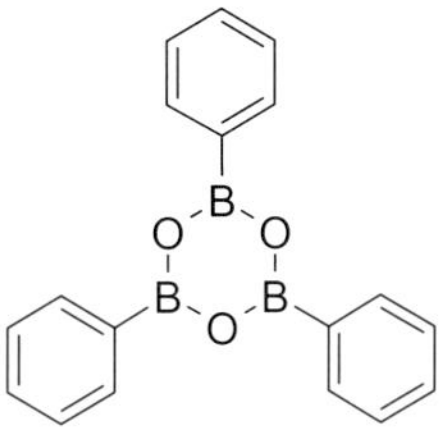	43 (Et_3N) 46 (Pyridine)	$Cu(OAc)_2$ (1 equiv), Et_3N or pyridine (2 equiv), CH_2Cl_2, ambient air	[17]
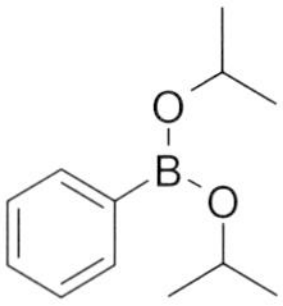	39	$Cu(OAc)_2$ (1 equiv), Et_3N (2 equiv), CH_2Cl_2, ambient air	[17]
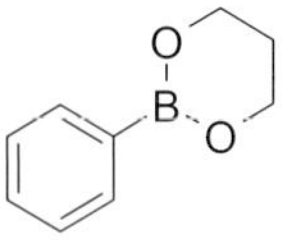	29	$Cu(OAc)_2$ (1 equiv), Et_3N (2 equiv), CH_2Cl_2, ambient air	[17]
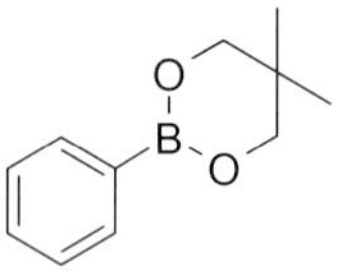	21	$Cu(OAc)_2$ (1 equiv), Et_3N (2 equiv), CH_2Cl_2, ambient air	[17]
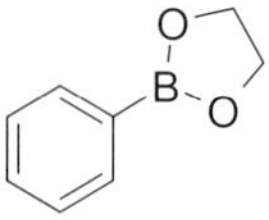	32	$Cu(OAc)_2$ (1 equiv), Et_3N (2 equiv), CH_2Cl_2, ambient air	[17]
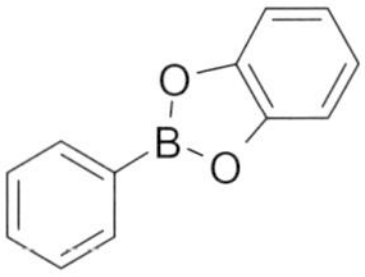	18	$Cu(OAc)_2$ (1 equiv), Et_3N (2 equiv), CH_2Cl_2, ambient air	[17]
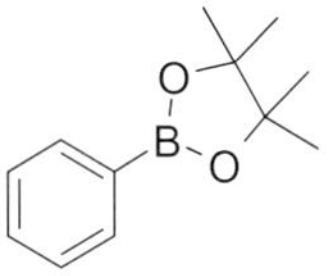	4	Cu( )$_2$ (1 equiv), Et_3N (2 equiv), CH_2Cl_2, ambient air	[17]
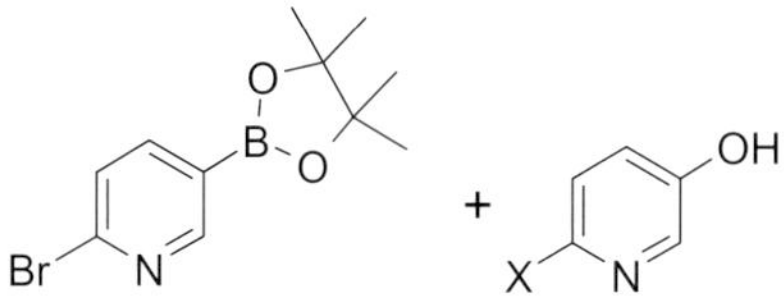	0–11% (X = Br) 16% (X = OBn)	$Cu(OAc)_2$ (1 equiv), Et_3N (2 equiv), CH_2Cl_2, ambient air, rt, 4 days	[18]

Table 4.3 *Continued*

$ArBX_2$ (+phenol)	Yield (%)	Example of conditions	Reference
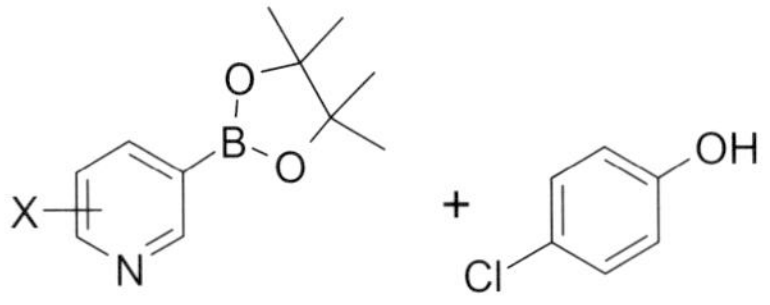	63% (X = 2-Br) 67% (X = 3-Br) 57% (X = 6-Cl)	$Cu(OAc)_2$ (1 equiv), Et_3N (2 equiv), CH_2Cl_2, ambient air, rt, 4 days	[18]
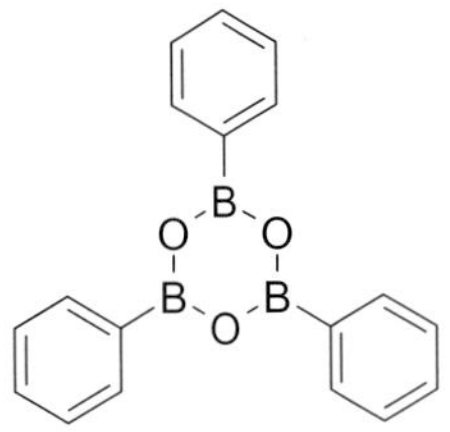 0.66 equiv	5 (no base) 60 (Cs_2CO_3)	$Cu(OAc)_2$ (1 equiv), Cs_2CO_3 (1 equiv), CH_2Cl_2, rt, 24 h	[19]
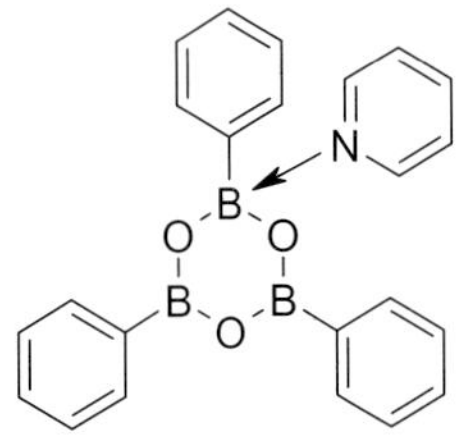 0.66 equiv	93 (no base) 99 (Cs_2CO_3)	$Cu(OAc)_2$ (1 equiv), Cs_2CO_3 (1 equiv), CH_2Cl_2, rt, 24 h	[19]
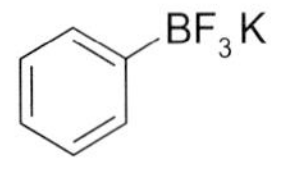 2 equiv	Generally high yields, wide scope, electron-poor arenes not tolerated	$Cu(OAc)_2 \cdot H_2O$ (**0.1 equiv**), DMAP (0.2 equiv), CH_2Cl_2, 4 Å MS, rt, 5 min, then R-OH, $\mathbf{O_2}$, rt, 24 h	[20]

MS = molecular sieve; rt = room temperature.

(Scheme 4.5). Potassium organotrifluoroborate salts are easily accessible, have attracted increased commercial availability, are air- and moisture-stable alternatives to boronic acids as well as their esters, and can be stored for long periods of time without decomposition under ambient reaction conditions. These aryl donors function brilliantly in arylations of phenols, albeit with the requirement for 2 equiv. of the potassium organotrifluoroborate salt being used to gain the best yields. However, the reactions function even at ambient temperature with molecular oxygen as a co-oxidant [20].

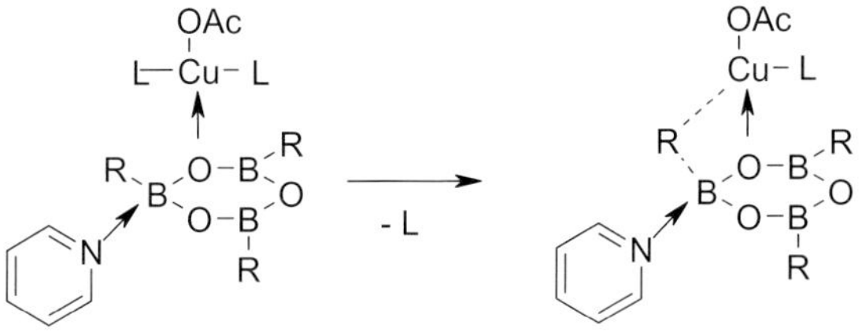

Scheme 4.5 Transmetallation facilitated by pyridine boron complex.

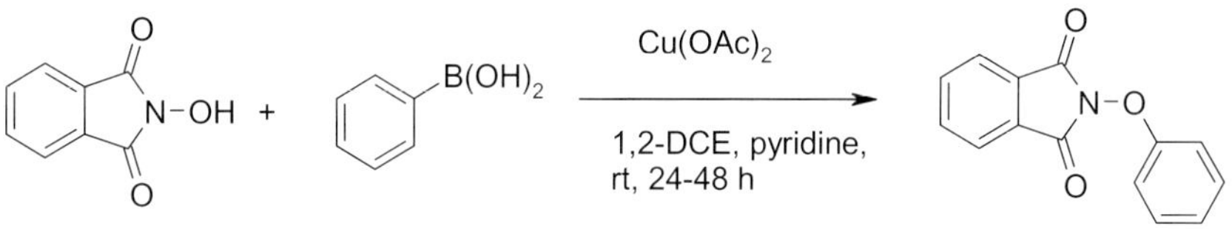

Scheme 4.6 Cross-coupling of *N*-hydroxyphthalimides with arylboronic acids.

4.4.3
Alternatives to Phenols

Most studies conducted on the transfer of aryl groups from boron-based aryl donors to an O—H-containing substrate have dealt specifically with diarylether synthesis, as all successful cross-coupling conditions have been limited to phenols. The protocol of Batey is perhaps special in that it also functions equally well with a wide range of aliphatic alcohols via the use of potassium organotrifluoroborate salts, and can perform efficiently with only catalytic amounts of $Cu(OAc)_2$ in the presence of molecular oxygen and DMAP as a promoting ligand [20]. An alternative protocol exists for the arylation of *N*-hydroxyphthalimides under mild reaction conditions, giving rise to the formation of aryloxyamines in the presence of 4 Å molecular sieves and ambient air in 1,2-dichloroethane (DCE) as solvent (Scheme 4.6) [21]. Again, the addition of a base is beneficial and 1 equiv. of pyridine is preferable. Interestingly, in these transformations – in contrast to the use of phenols as substrates – various examined copper(I) and copper(II) precursors, such as $Cu(OAc)_2$, $Cu(OTf)_2$, CuCl, and $CuBr(SMe_2)$, have resulted in the formation of the desired coupling products in up to 98% yield. Moreover, there appears to be no direct correlation between the effectiveness of a specific copper salt and its oxidation state, and notably Cu, Cu(OAc), $CuCl_2$, and $Cu(CF_3COCHCOCH_3)_2$ were not suitable in effecting the transformation. The use of Et_3N as the base resulted in lower yields (13–37%), and neither DMAP and DABCO nor Cs_2CO_3 were effective bases in the reactions, but rather inhibited the reaction to a significant degree. In line with earlier reports on the cross-coupling of phenols with arylboronic acids, electronically and structurally diverse substituents on the boronic acid were tolerated, and the only recorded exceptions were *ortho*-fluoro substituents and *para*-boronic acid precursors, both of which can undergo oligomerization. The main limitations of the method are, thus, the requirement for stoichiometric amounts of the copper salt, the use of 2 equiv. of the boronic acids, and the incompatibility with an *ortho*-heteroatom substituent or a second boronic acid functionality in the substrate.

4.5 Development of a New N—H Bond Arylation Reaction

4.5.1 Stoichiometric in Copper

This section will begin to deal with examples of the copper-mediated arylation of N—H bonds where stoichiometric amounts of copper salts are used, followed by the development of the truly catalytic in copper versions. This development parallels that of the copper-catalyzed arylation of O—H bonds (*vide supra*). The first reports detailed arylations of a very wide range of N—H bonds, demonstrating the broad substrate compatibility, which includes primary and secondary aliphatic amines, anilines, amides, imides, ureas, sulfonamides, carbamates, diazoles, triazoles and tetrazoles (Scheme 4.7) in both solution and solid phase [4, 22, 23]. There is no base-derived substrate trend for efficiency of these processes, with either triethylamine or pyridine being superior [1c]. As yet, a study of the acidity or basicity of the substrates has not been systematically performed, and in general the practitioners of these processes will attempt to use both bases and, in some cases, mixed triethylamine/pyridine base systems, for example, in the selective N-arylation of 2-pyridones and 3-pyridazinones, seem optimal [24]. In recent years, the range of N—H-containing substrates has been extended considerably, as shown in Table 4.4. Although bifunctional O—H- and N—H-containing substrates have not been examined in detail, it has been shown in competition experiments that anilines react approximately tenfold faster than phenols [1c].

2(1*H*)-Pyrazinones can be efficiently N-arylated with a range of arylboronic acids by using a special protocol which involves the use of a mixed-base system, where a triethylamine : pyridine ratio of 1 : 2 offers the best yields (Table 4.5) [29]. Interestingly, when comparing these processes performed at ambient temperature with a reaction under microwave irradiation (300 W for 1 h) with simultaneous cooling (0 °C), the latter method provides significantly superior results. Copper-mediated, microwave-assisted N-arylation using a KF-Al_2O_3 heterogeneous base in the absence of solvent is an efficient process, and displays a wide substrate scope in relation to both amine and boronic acid components [30]. Typically, the presence of the correct base in N-arylations gives enhanced yields, although a heterogeneous method has also been developed using a basic copper-exchanged fluorapatite (Cu–FAP) at ambient temperature in methanol as solvent within a few hours to effect N—H bond arylations [31]. This process tolerates a very broad range of boronic acid substrates, giving good to excellent yields with both aromatic, heteroaromatic and aliphatic N—H-containing substrates. In addition, the heterogeneous Cu–FAP can be recycled and reused efficiently to give

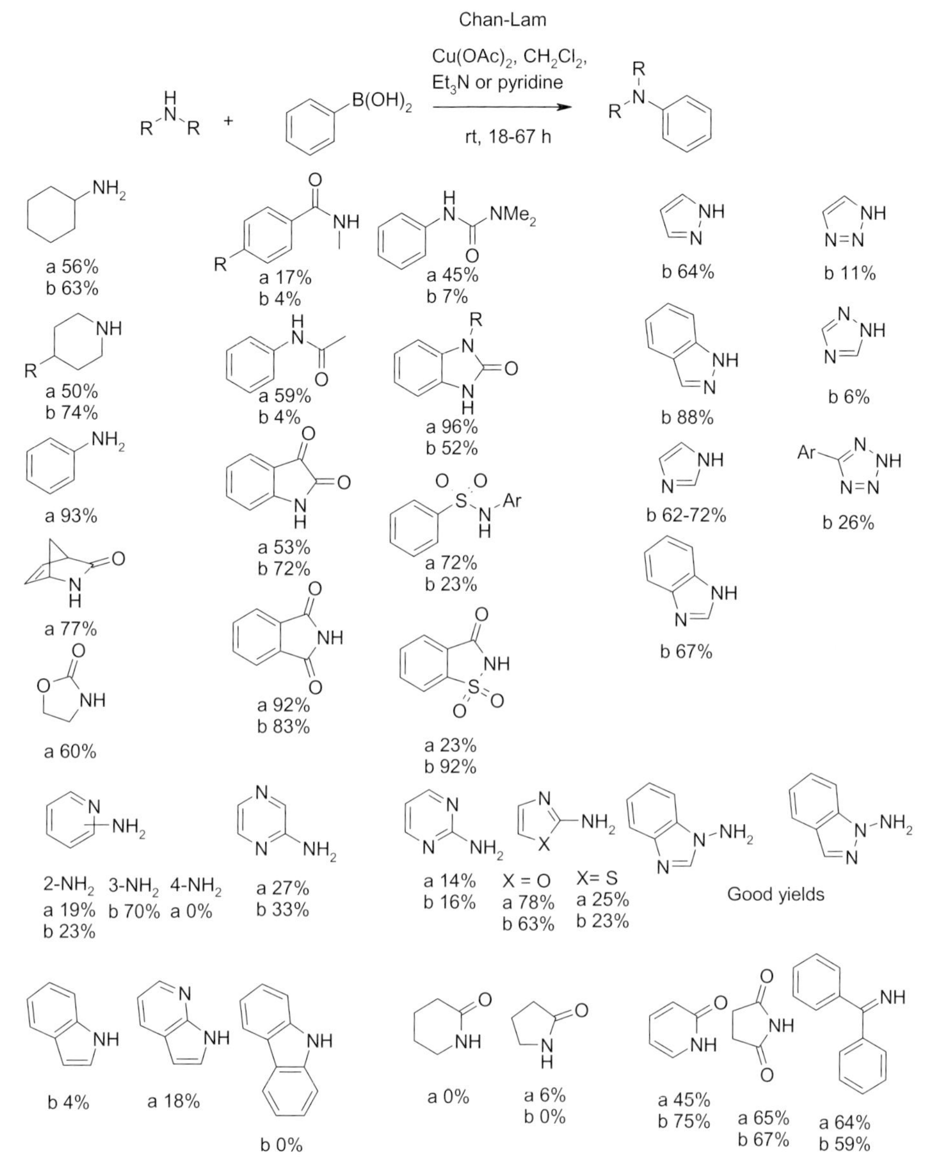

Scheme 4.7 Chan–Lam arylation of N–H bonds with arylboronic acids (Yields: a = Et_3N; b = pyridine).

comparable yields after four use-cycles. Microwave-induced arylations also proceed well in the absence of a solid support and, in this case, DBU was shown to be preferred over triethylamine, pyridine or KO*t*-Bu [32]. In a study exploring structure–activity relationships within an azetidinone structural class, a key N–H arylation reaction using Chan–Lam conditions was used to good effect, allowing

Table 4.4 Copper-mediated N—H bond arylations.

Substrate	Remarks	Example of conditions	Reference
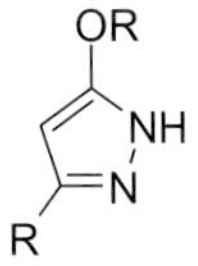 OR NH N R	Mixture of regioisomers, 0–47% yield	$Cu(OAc)_2$ (1.5 equiv), $ArB(OH)_2$ (1.1 equiv), pyridine (2 equiv), CH_2Cl_2, 4 Å MS, ambient air, 72 h	[25]
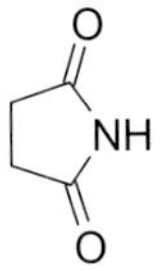 O NH O	Electron-rich aryl boronic acids fail, 36–72% yield	$Cu(OAc)_2$ (1.5 equiv), $ArB(OH)_2$ (1.1 equiv), Et_3N (2 equiv), CH_2Cl_2, 48–72 h	[26]
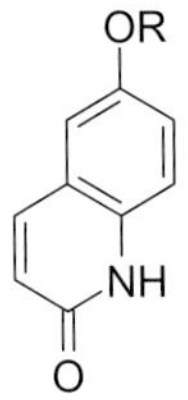 OR NH O	Good yields (76%), mixed-base system	$Cu(OAc)_2$ (2 equiv), $ArB(OH)_2$ (2 equiv), Et_3N (2 equiv) and pyridine (2 equiv), CH_2Cl_2, 48 h	[24]
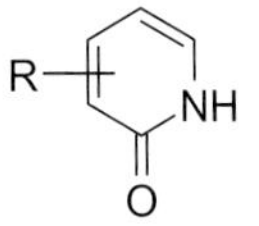 R NH O *multi-subst.* N NH O 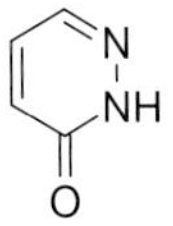	Moderate yields (10–54%)	$Cu(OAc)_2$ (2 equiv), $ArB(OH)_2$ (2 equiv), Et_3N (2 equiv) and pyridine (2 equiv), CH_2Cl_2, 48 h	[24]
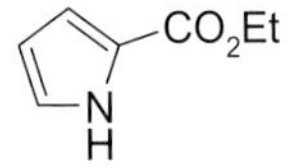 CO_2Et N H	50% yield	$Cu(OAc)_2$ (2 equiv), $ArB(OH)_2$ (3 equiv), pyridine (4 equiv), CH_2Cl_2, 18 h	[27]
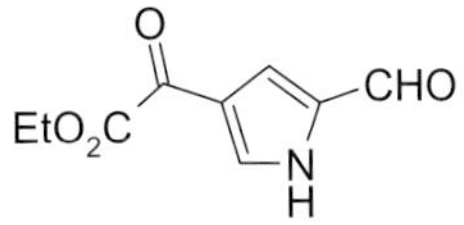 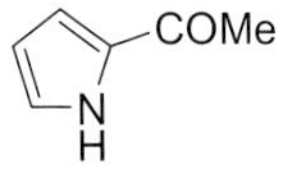 O EtO_2C CHO N H COMe N H	Moderate to good yields, good substrate scope	$Cu(OAc)_2$ (1.5 equiv), $ArB(OH)_2$ (2 equiv), pyridine (2 equiv), CH_2Cl_2, rt, air, **2–13 days**	[27]
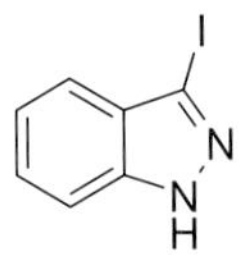 I N N H	Regioselective N-arylation	$Cu(OAc)_2$ (1.5 equiv), $ArB(OH)_2$ (2 equiv), Et_3N (2 equiv), CH_2Cl_2, rt, 1–2 h	[28]
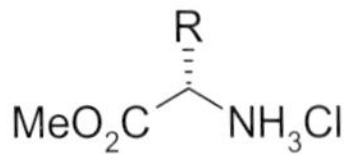 R MeO_2C NH_3Cl	No loss of enantiomeric purity	$Cu(OAc)_2$ (1.1 equiv), *p*-Tol-$B(OH)_2$ (2.0 equiv), Et_3N (2.0 equiv), CH_2Cl_2, 4 Å MS, 1–2 days	[12]

MS = molecular sieve; rt = room temperature.

Table 4.5 Additional special applications.

N-H substrate	Remarks	Example of conditions	Reference
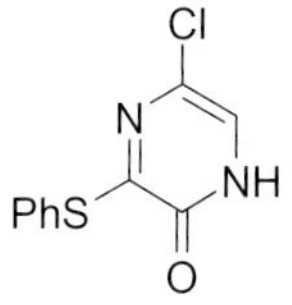 Cl, N, NH, PhS, O	Ambient temperature (64%); **Microwave** 300 W, 0 °C (87%)	$Cu(OAc)_2$ (2 equiv), $PhB(OH)_2$ (2 equiv), Et_3N (2 equiv), CH_2Cl_2, 1 h	[29]
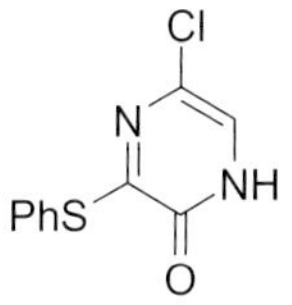 Cl, N, NH, PhS, O	**Microwave** 300 W, 0 °C (94%)	$Cu(OAc)_2$ (2 equiv), 3-CF_3-$ArB(OH)_2$ (2 equiv), Et_3N (1 equiv), pyridine (2 equiv), CH_2Cl_2, 1 h	[29]
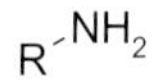 $R-NH_2$ R = (het)aryl, alkyl	**Microwave** (80–240 W), KF-Al_2O_3, good yields	$Cu(OAc)_2$ (1 equiv), $ArB(OH)_2$ (2 equiv), **KF-Al_2O_3** (1 equiv), 10 min	[30]
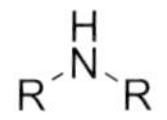 R-NH-R R = (het)aryl, alkyl	**Base-free** method, good yields (78–93%), good substrate scope	**Cu-FAP** (100 mg), $ArB(OH)_2$ (1 mmol), substrate (1.2 mmol), MeOH, rt, 2–12 h	[31]
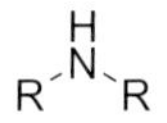 R-NH-R R = (het)aryl, alkyl	**Microwave** (100 °C), yields good to excellent, broad substrate scope	$Cu(OAc)_2$ (2 equiv), amine (1 equiv), $ArB(OH)_2$ (1.5 equiv), **DBU** (2 equiv), 30 min	[32]
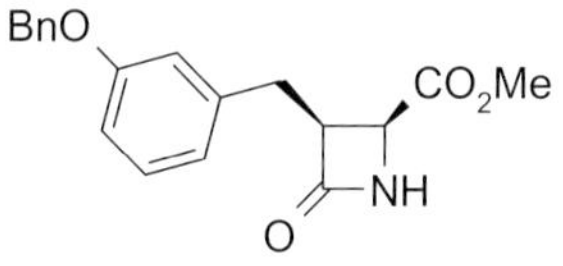 BnO, CO_2Me, NH, O	Yields 93–99%, broad range of $ArB(OH)_2$ (3 equiv)	$Cu(OAc)_2$ (1.2 equiv), Et_3N (5 equiv), pyridine (5 equiv), 1,2-DCE, 4 Å MS, rt, ambient air, 5 h	[33]
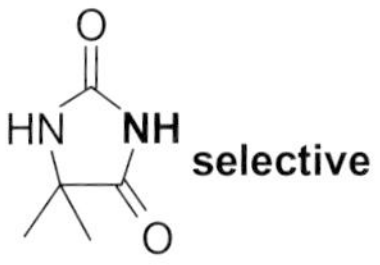 O, HN, **NH** **selective**, O	Yields 38–79%, comparison study between arylbismuthates and aryl boronic acids (1.25 equiv)	$Cu(OAc)_2$ (1 equiv), pyridine (5 equiv), 1,2-DCM, 4 Å MS, rt, O_2, 7 days	[34]
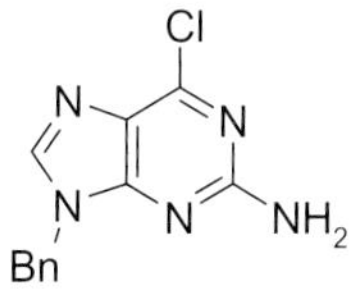 Cl, N, N, N, N, NH_2, Bn	Yields 49–67%, electron-poor, electron-neutral substrates work equally well	$Cu(OAc)_2$ (2 equiv), $PhB(OH)_2$ (3 equiv), Et_3N (2 equiv), **DMAP (1.6 equiv)**, $CHCl_3$, 24 h	[35]

Table 4.5 *Continued*

N-H substrate	Remarks	Example of conditions	Reference
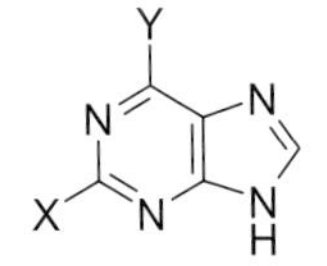 Y, N, N, X, N, N, H X, Y = H, Cl, SMe, SH, SPh, 2-thienyl	Yields 42–81%, good substrate scope of purine and boronic acid	$Cu(OAc)_2$ (1 equiv), $PhB(OH)_2$ (3 equiv), Et_3N (2 equiv), **1,10-phenanthroline (2 equiv)**, 4 Å MS, CH_2Cl_2, rt, 4 days	[36]
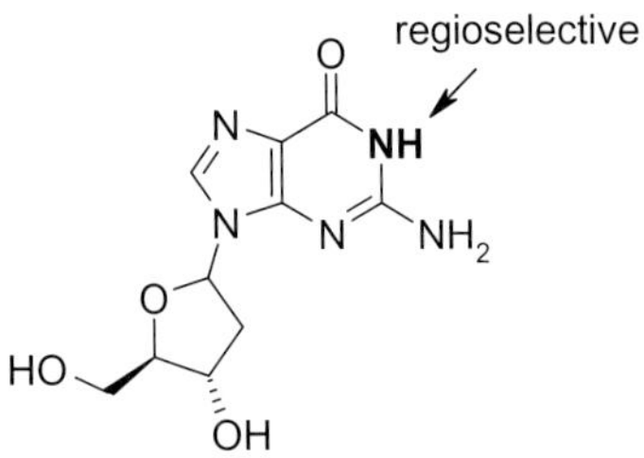 regioselective, O, N, **NH**, N, N, NH_2, O, HO, OH	Generally good yields in absence of chelating ligands, generally good substrate scope of boronic acid (3-pyridinylboronic acid fails)	$Cu(OAc)_2$ (1 equiv), $PhB(OH)_2$ (2 equiv), Et_3N (2 equiv), DMSO, 60 °C, 16 h	[37]
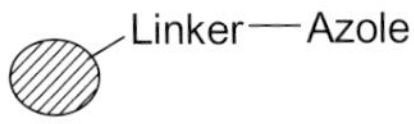 Linker—Azole PEG PS resin support	**Solid-supported chemistry**, yields 56–64%, imidazole, benzimidazole, pyrazole, benzotriazole	$Cu(OAc)_2$ (5 equiv), $ArB(OH)_2$ (3 equiv), pyridine: NMP (1 : 1), 4 Å MS, 3 × 10 s (5 cycles)	[23]
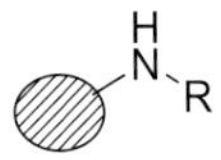 H, N, R Bal-PEG-PS(HL) resin support	**Solid-supported chemistry**, primary or secondary amine immobilized, yields 21–75%	$Cu(OAc)_2$ (2 equiv), $ArB(OH)_2$ (4 equiv), Et_3N (4 equiv), **THF**, 4 Å MS, 3 h (3 cycles)	[38]
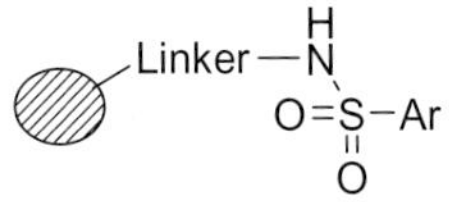 Linker—N(H), O=S(=O)−Ar Fmoc PAL resin support	Solid-supported chemistry, yields 48–81%	$Cu(OAc)_2$ (2 equiv), $ArB(OH)_2$ (4 equiv), Et_3N (4 equiv), **THF**, 4 Å MS, 3 h (2 cycles)	[39]
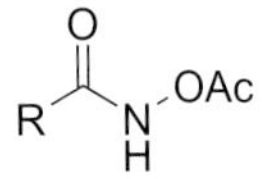 O, R, N, H, OAc R = alykl, (het)aryl	Good yields, wide substrate scope, Ac is the best hydroxamic acid substituent, although others also function well	CuTC (1 equiv), *p*-Tol-$B(OH)_2$ (1.1 equiv), THF or THF : hexanes (1 : 1). 60 °C, 10–18 h	[40]

MS = molecular sieve; rt = room temperature.

a near-quantitative yield of the target products [33]. A study comparing the efficiency of cross-coupling between 5,5-dimethylhydantoin and either arylboronic acids or triarylbismuthates clearly demonstrated the superiority of the boron-based aryl donors [34]. Both, aminopurines and aminopyridines were conveniently N-arylated using DMAP as an additive, and the process was tolerant of a thioether moiety [35]. 1,10-Phenanthroline was a key additive used in the regioselective synthesis of 9-arylpurines, and proved to be superior to triethylamine, pyridine, 2,2′-bipyridine, TMEDA and *N,N′*-diarylethanediimines [36]. A closely related study reached similar conclusions on the prioritization of 1,10-phenanthroline as a ligand in the selective *N*-arylation of 2′-deoxyguanosine, which is best conducted at a slightly elevated temperature (60 °C). Interestingly, it is also possible to perform the reaction in the absence of a chelating ligand with 0.5 equiv. of $Cu(OAc)_2$ to give good yields of the desired product, using triethylamine as base [37]. Today, the generation of libraries of products continues to attract interest, and this new arylation methodology has been demonstrated to be applicable in this respect. In order to speed up the reactions, the use of microwave-induced heating (3×10 s at 1000 W) was successfully employed using open reaction vessels on solid-supported substrates [23]. The overall mild reaction conditions required were notable, and in general chemoselective processes dominated for primary and secondary amines [38], as well as for immobilized sulfonamides [39] and azoles [22]. Although not being a direct N—H bond arylation, a related new non-oxidative protocol for the arylation of *N*-amides with arylboronic acids has also been developed. This methodology shows a broad substrate scope, and utilizes copper(I) rather than the copper(II) sources which have dominated arylation methodologies (Scheme 4.8) [40]. This new protocol utilizes copper(I) thiophene-2-carboxylate (CuTC), although CuOAc offers similar results. Oxygen-free copper(I) sources, such as CuCl, CuI and CuCN, were shown to be ineffective. In terms of solvent choice, DMF, DMA and toluene proved to be inferior to tetrahydrofuran (THF) or THF/hexane solvent mixtures. It is speculated that the mechanism of this copper-mediated reaction proceeds via the oxidative addition of copper(I) to the hydroxamic acid, thereby generating a copper(III) amido species which slowly undergoes transmetallation.

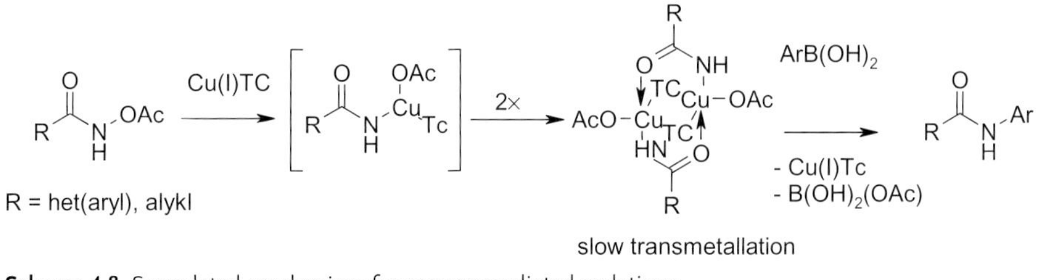

Scheme 4.8 Speculated mechanism for copper-mediated arylations.

4.5.2 Alternatives to Boronic Acids

A systematic study for boronic acid replacements for N—H bond arylations has been investigated, and the outcome compares favorably to the O—H bond arylation reaction (Table 4.6) [17, 41]. However, to date no catalytic in copper variations have been reported with these alternative aryl donors. The best donors are triarylboroxines and boronic acid esters derived from 2,2′-dimethyl-1,3-propanediol and 1,3-propanediol and, as in the case of the O—H bond arylation, the pinacolate esters are poor aryl donors. Sodium tetraphenylborate is also an alternative phenyl donor in the copper(II)-mediated, microwave-assisted N-arylation using KF-Al_2O_3 as heterogeneous base in the absence of solvent [30].

4.6 Development of a New N—H Bond Arylation Reaction: Catalytic in Copper

The first efficient catalytic variation of the N—H bond arylation reaction was reported by Collmann, and was dedicated solely to the synthesis of *N*-aryl imidazoles and *N*-arylbenzimidazoles (Scheme 4.9) [42]. The protocol involves the use of a commercially available dimer $[\{Cu(OH) \cdot TMEDA\}_2]Cl_2$, which was found to be a potent catalyst. Several reaction parameters were optimized, such as the amount of catalyst (aryl boronic acid and imidazole), as well as solvent, reaction time, atmosphere and the addition of molecular sieves, which resulted in an optimized general reaction procedure. Thus, a mixture of 2 equiv. of aryl boronic acid, 1 equiv. of imidazole and 10 mol% $[\{Cu(OH) \cdot TMEDA\}_2]Cl_2$ are stirred in dichloromethane (DCM) at ambient temperature overnight under molecular oxygen. A large array of electronic and structural diversity is tolerated in the boronic acids selected for this study. However, in almost all cases reaction conditions were found that led to good to excellent yields (52–98%).

Environmentally friendly reaction procedures are important and especially attractive when performed in water [43]. Hence, Collman studied exactly the above reaction (Scheme 4.9) with the same substrates and catalyst species, and reported the first examples of N-arylations of imidazoles in water [44]. This reaction system differs from the previously developed reaction procedure only in the replacement of DCM with water. Notably, the addition of phase-transfer catalysts, such as Bu_4NBr, Bu_4NCl, $Hept_4NBr$, $BzEt_3NCl$, $BzBu_3NCl$ and $BzMe_2$tetradecylNCl, offered no substantial benefit in terms of yield or reaction time. It was also shown that a

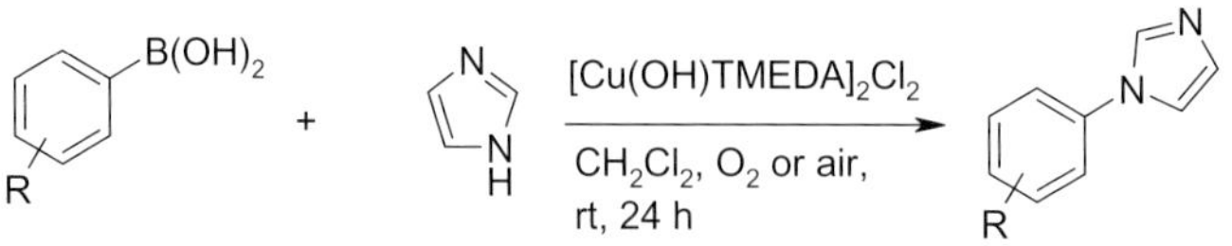

$B(OH)_2$ R + N N H $\xrightarrow[\text{CH}_2\text{Cl}_2,\ \text{O}_2 \text{ or air, rt, 24 h}]{[\text{Cu(OH)TMEDA}]_2\text{Cl}_2}$ N N R

Scheme 4.9 First copper-catalyzed N—H bond arylation with boronic acids.

Table 4.6 Additional boron-based aryl donors.

$ArBX_2$	Yield	Example of conditions	Reference
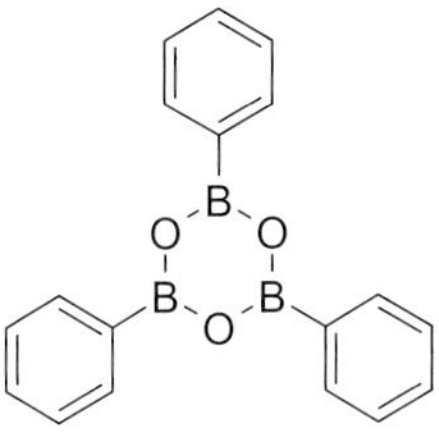 (triphenylboroxine structure)	40–62% (Et_3N or pyridine), 4-ethylpiperidine and 2-benzimidazolinone	$Cu(OAc)_2$ (1 equiv), Et_3N or pyridine (2 equiv), CH_2Cl_2, ambient air	[17]
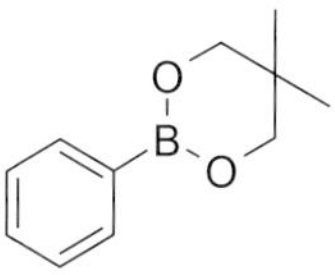 (5,5-dimethyl-2-phenyl-1,3,2-dioxaborinane structure) 2 equiv	46–76% (Et_3N), 4-ethylpiperidine and 2-benzimidazolinone	$Cu(OAc)_2$ (1 equiv), Et_3N (2 equiv), CH_2Cl_2, ambient air	[17]
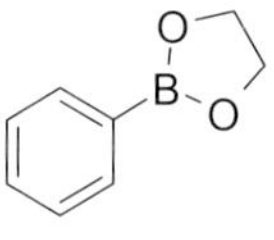 (2-phenyl-1,3,2-dioxaborolane structure) 3 equiv	Cyclic imides, 71%	$Cu(OAc)_2$ (3 equiv), Et_3N (3 equiv), CH_2Cl_2, 4 Å MS, 55 °C, O_2, 20 h	[41]
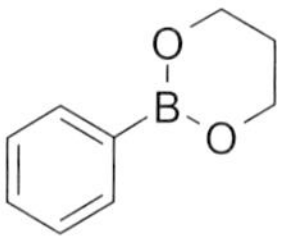 (2-phenyl-1,3,2-dioxaborinane structure) 3 equiv	Cyclic, acyclic imides, **43–97%**	$Cu(OAc)_2$ **(2 equiv)**, Et_3N (3 equiv), CH_2Cl_2, 4 Å MS, 55 °C, O_2, 20 h	[41]
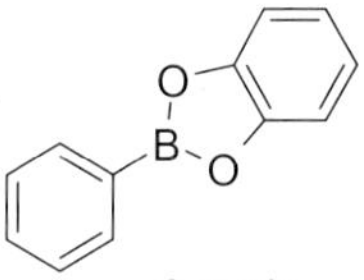 (2-phenyl-1,3,2-benzodioxaborole structure) 2 equiv	Cyclic imides, 32%	$Cu(OAc)_2$ **(2 equiv)**, Et_3N (3 equiv), CH_2Cl_2, 4 Å MS, 47 °C, O_2, 20 h	[41]
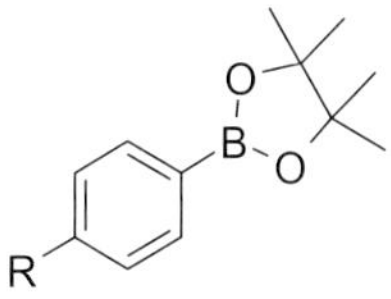 (4-R-phenyl pinacol boronate structure)	R = H: 0% R = NO_2: 43%	$Cu(OAc)_2$ **(2 equiv)**, Et_3N (3 equiv), CH_2Cl_2, 4 Å MS, rt, O_2, 20 h	[41]
Ph_4BNa	**Microwave** (80–240 W), KF-Al_2O_3, good yields	$Cu(OAc)_2$ (1 equiv), RRNH (1 equiv), **KF-Al_2O_3** (1 equiv), 10 min	[30]

MS = molecular sieve; rt = room temperature.

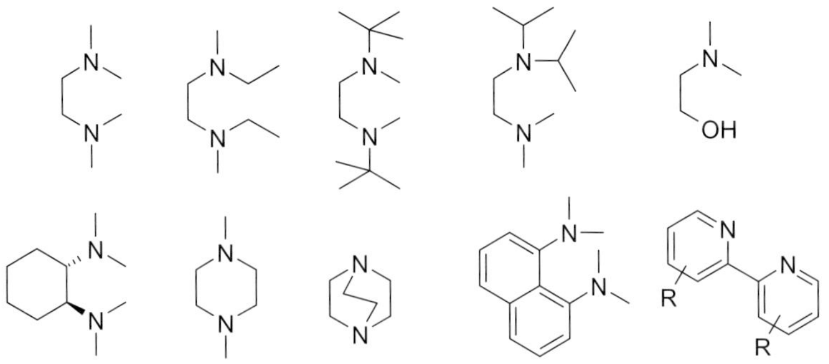

Scheme 4.10 Bidentate ligands examined in the N—H bond arylation of imidazole.

neutral pH value was optimal for the reaction, as in both slightly acidic or basic media the reaction yields were lower. The increased potential for water to intercept the putative reaction intermediates led to slightly lower yields for these transformations. Nevertheless, the products could be prepared in overall acceptable yields of 26–63%. Attempts were made, with limited success, to improve this method by closely investigating other copper catalysts and ligands [45]. The use of a commercially available dimeric μ-hydroxo copper(II) complex with the bidentate TMEDA ligand was investigated first. This was followed by studies with other bidentate nitrogen ligands and different counterions. In this report, additional potential bidentate ligands (Scheme 4.10) and copper(I) salts, such as CuCl, CuBr, CuI or CuOTf, in the presence of molecular oxygen and water, were used to generate a copper(II) species that efficiently catalyzed the cross-coupling of imidazoles, and benzimidazoles with aryl boronic acids. Although all ligands led to acceptable yields, the most efficient catalysis was accomplished with the initially studied TMEDA. Interestingly, it is possible to perform these reactions in the absence of a base under anaerobic reaction conditions, although the use of water in the reaction remains an essential feature [46]. Under these reaction conditions, a β-amino alcohol as a bidentate ligand gave a poor yield of the desired N-arylated imidazole. In contrast, a range of 2,2′-bipyridine ligands in fact proved to be useful alternative bidentate ligands.

4.6.1 Proposed Mechanism

Collman proposed a catalytic cycle for the coupling of imidazole with phenyl boronic acids, as depicted in Scheme 4.11. Although, as yet, no experimental evidence exists for any of the putative elementary reaction steps, the hypothesis can be supported [47]. The individual steps may include: (i) transmetallation of the aryl boronic acid with the catalyst **I**, generating the Cu(II) species **II**; (ii) coordination of the imidazole nucleophile to give adduct **III**; (iii) oxidation of complex **III** to species **IV** in the presence of molecular oxygen or air; (iv) reductive elimination

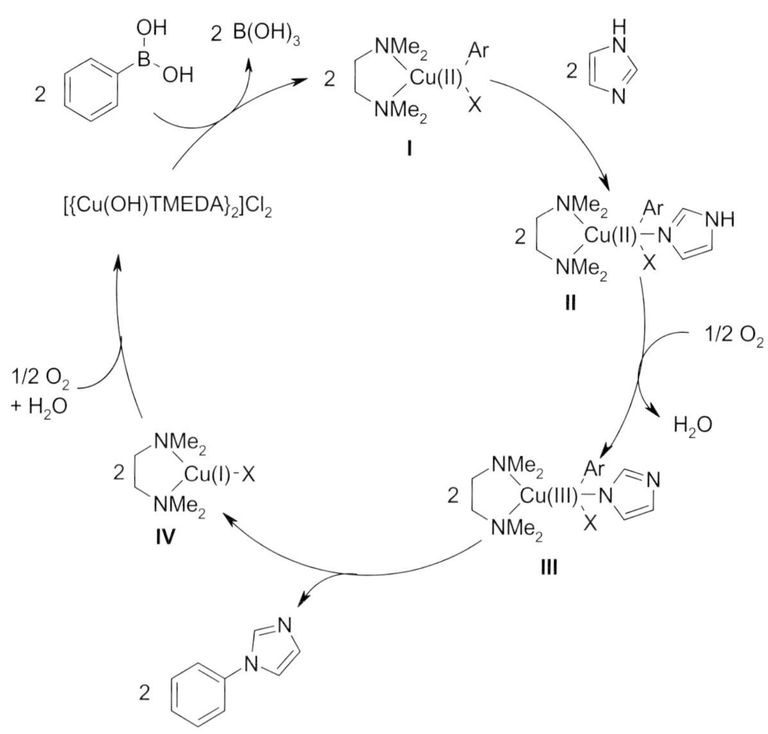

Scheme 4.11 Proposed catalytic cycle for arylations of imidazole with phenylboronic acid and $[\{Cu(OH) \cdot TMEDA\}_2]Cl_2$ as precatalyst.

of the desired product; and, finally (v) oxidation to regenerate the catalytically active homobimetallic copper(II) species **I** [45].

Additional improvements to the novel catalytic variation of the boronic acid cross-coupling have been reported, this time going back to the use of $Cu(OAc)_2$ as the copper(II) source. Buchwald developed a general reaction, involving the use of catalytic amounts (5–20 mol%) of $Cu(OAc)_2$ and myristic acid (n-$C_{13}H_{27}COOH$, 10–40 mol%) as an additive, along with stoichiometric amounts of 2,6-lutidine as base, at ambient temperature [48]. The reason for adding a small amount of myristic acid was to enhance the speed of the reactions, perhaps by rendering the coordinated copper species more soluble in the reaction medium. Substituted anilines as well as primary and secondary aliphatic amines reacted with a series of aryl boronic acids (Scheme 4.12). *p*-Tolylboronic acid was used in the optimization of the protocol, and in early observations it was shown by screening different copper salts that $Cu(OAc)_2$, CuOAc and copper(II) isobutyrate all efficiently catalyzed the reaction with aniline, allowing conversions up to 55%. Vigorous stirring of the reaction mixture allowing good oxygen uptake into the reaction mixture enabled quantitative cross-coupling of aniline derivatives. The reactions were best

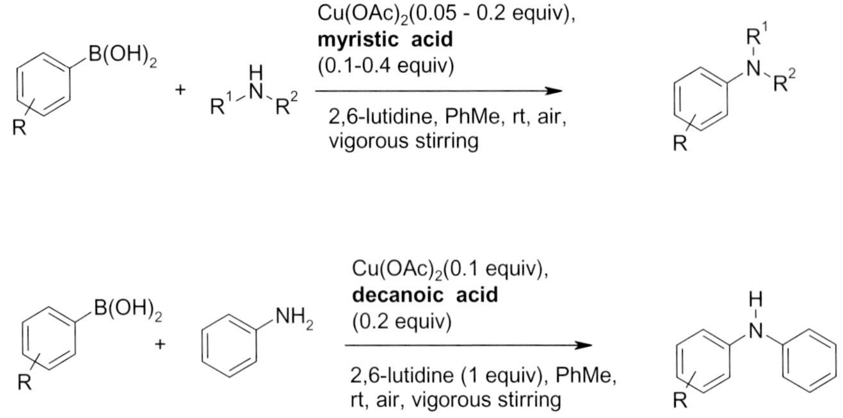

$B(OH)_2$ + R^1–NH–R^2 → Cu(OAc)$_2$(0.05 - 0.2 equiv), **myristic acid** (0.1-0.4 equiv); 2,6-lutidine, PhMe, rt, air, vigorous stirring

$B(OH)_2$ + NH_2 → Cu(OAc)$_2$(0.1 equiv), **decanoic acid** (0.2 equiv); 2,6-lutidine (1 equiv), PhMe, rt, air, vigorous stirring

Scheme 4.12 Copper-catalyzed arylations of aryl boronic acids using carboxylic acid additives.

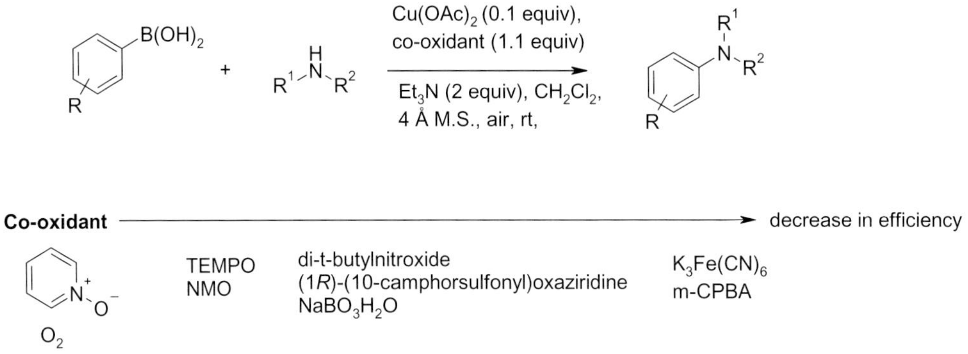

$B(OH)_2$ + R^1–NH–R^2 → Cu(OAc)$_2$ (0.1 equiv), co-oxidant (1.1 equiv); Et_3N (2 equiv), CH_2Cl_2, 4 Å M.S., air, rt,

Co-oxidant → decrease in efficiency

O_2; TEMPO, NMO; di-t-butylnitroxide, (1*R*)-(10-camphorsulfonyl)oxaziridine, $NaBO_3H_2O$; $K_3Fe(CN)_6$, m-CPBA

Scheme 4.13 N—H bond arylation using catalytic amounts of Cu(OAc)$_2$ and a co-oxidant.

performed in flasks with large volumes (100 mL) relative to the amount of used solvent (2 mL). The reaction was seen to tolerate a wide range of electronically and structurally diverse substrates with various functionalities already incorporated into the products, including alkene, ester, alcohol and acetamides, none of which interfered with the desired reaction pathway. Of note here in the reaction with primary amines, was that no diarylated products were observed. Decanoic acid has also been successfully used in the high-yielding cross-coupling of anilines [49].

A general catalytic version of the original Chan–Lam arylation based on using catalytic amounts of Cu(OAc)$_2$ without an additive, but instead employing a co-oxidant has successfully evolved (Scheme 4.13) [9]. Thus, 10 mol% Cu(OAc)$_2$ was found optimal, giving equal results to 0.2 equiv. of Cu(OAc)$_2$, and significantly superior results to 0.01 or 0.05 equiv. when using *N*-ethylbenzimidazolone as substrate and *p*-Tolyl-boronic acid as the aryl donor. A range of co-oxidants was examined with molecular oxygen, pyridine *N*-oxide and TEMPO being the best for

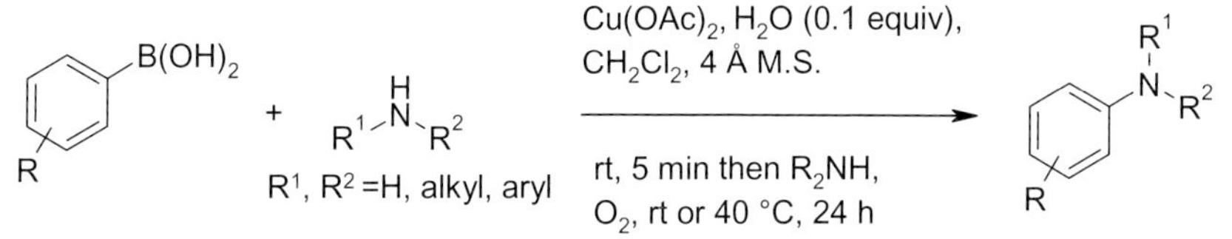

Scheme 4.14 Ligand- and base-free N—H bond arylation using catalytic amounts of $Cu(OAc)_2$ and O_2.

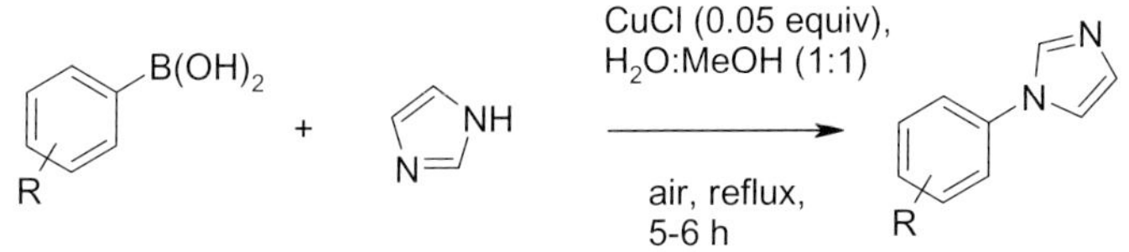

Scheme 4.15 CuCl-catalyzed arylation of imidazole in a mixed protic solvent system.

a wide range of N—H-containing substrates at ambient temperature, or for benzimidazole or sultames as substrates at elevated temperature of 50–65 °C.

These catalytic methods all rely on the use of ligands or a base. A ligand-less, base-free method was subsequently developed to effect high-yielding cross-couplings of anilines and a wide range of aliphatic amines with various aryl boronic acids (Scheme 4.14) [50]. Relative to the myristic acid-based method, this protocol was shown to be superior for aliphatic amines, but inferior in the transformation of aniline derivatives.

Additional studies using simple copper sources, such as CuCl, CuBr, CuI, $CuClO_4$, $CuCl_2 \cdot 2H_2O$, $Cu(OAc)_2 \cdot H_2O$ and $Cu(NO_3)_2 \cdot 3H_2O$, revealed that the arylation of imidazole with phenyl boronic acid can be efficiently conducted when using protic solvents, such as methanol or water [51]. Copper loadings as low as 1 mol%, but typically 5 mol%, were found to be suitable, and optimally 1.2 equiv. of imidazole were used (Scheme 4.15). A mixed water/methanol solvent system gave the best outcome, although methanol could be replaced by either ethanol, acetone or THF to provide similarly high to excellent yields.

An interesting study investigating further the potential for larger-scale processes revealed the important role that the addition of an optimal amount of water has on the reaction of benzimidazole derivatives with arylboronic acids, arylboronates and triarylboroxines (Scheme 4.16) [52]. Whilst the accelerating effect of water is still not fully understood, the reaction times were generally twice as fast when compared to anhydrous methods using molecular sieves. Interestingly, the formation of phenols as side products appears not to depend on the additional amounts of water present in the reaction medium, and therefore did not diminish the yield of the target N-arylated benzimidazole products.

Typically, boronic acids are used in an excess, and a more efficient protocol has been developed which not only proceeds in the absence of a base, but also uses equimolar amounts of imidazole and boronic acid. These processes are best

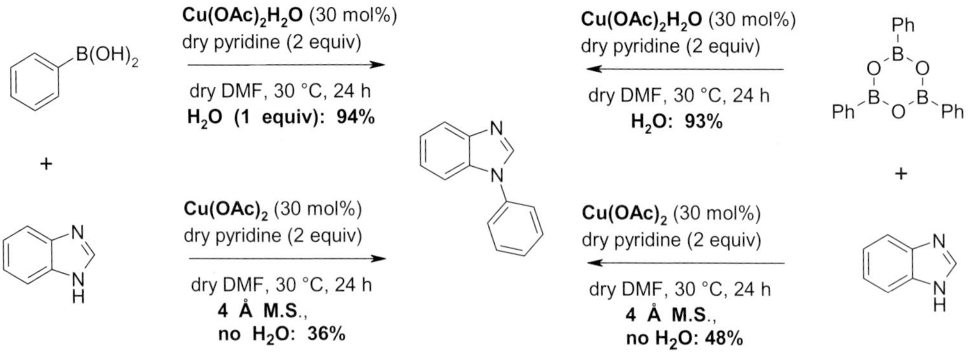

$Cu(OAc)_2H_2O$ (30 mol%), dry pyridine (2 equiv), dry DMF, 30 °C, 24 h, **H_2O (1 equiv): 94%**

$Cu(OAc)_2H_2O$ (30 mol%), dry pyridine (2 equiv), dry DMF, 30 °C, 24 h, **H_2O: 93%**

$Cu(OAc)_2$ (30 mol%), dry pyridine (2 equiv), dry DMF, 30 °C, 24 h, **4 Å M.S., no H_2O: 36%**

$Cu(OAc)_2$ (30 mol%), dry pyridine (2 equiv), dry DMF, 30 °C, 24 h, **4 Å M.S., no H_2O: 48%**

Scheme 4.16 Effect of water on the coupling reaction of benzimidazole with aryl boron donors.

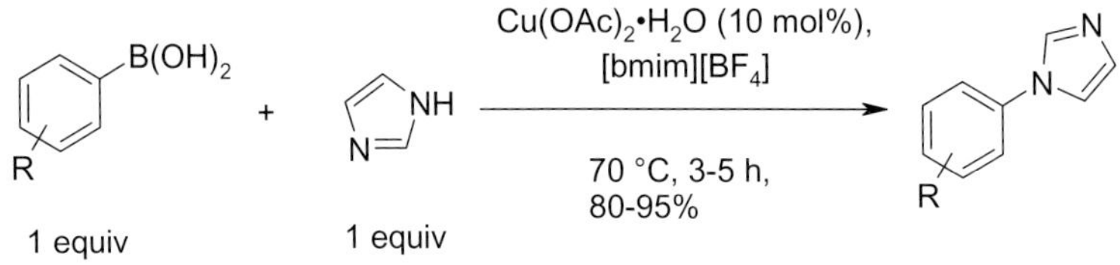

$Cu(OAc)_2{\cdot}H_2O$ (10 mol%), [bmim][BF_4], 70 °C, 3-5 h, 80-95%

1 equiv 1 equiv

Scheme 4.17 Arylation of N—H bonds using a recyclable $Cu(OAc)_2 \cdot H_2O$/[bmim][BF_4] system.

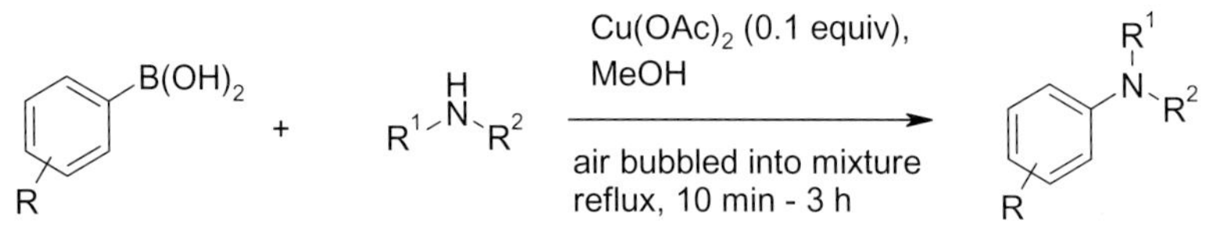

$Cu(OAc)_2$ (0.1 equiv), MeOH, air bubbled into mixture, reflux, 10 min - 3 h

Scheme 4.18 $Cu(OAc)_2 \cdot H_2O$-catalyzed arylations of amines with boronic acids as limiting reagents.

conducted using an ionic liquid [bmim][BF_4] as an ecobenevolent alternative to volatile organic solvents, and can be recycled after use (Scheme 4.17). A range of additional N—H-containing nucleophiles, such as anilines, aliphatic primary and secondary amines, amides, sulfonamides and cyclic imides, all perform well to give the target N-aryl products in good to excellent yield with broad functional group compatibility [53].

When boronic acids are used as the limiting reagent with 1.1 equiv. of the N—H-containing nucleophiles, such as imides, anilines and aliphatic amines, then 10 mol% of $Cu(OAc)_2 \cdot H_2O$ is the preferred catalyst of choice to give good yields of the target N-arylated products [54]. For benzamides, acetamides and sulfonamides, higher catalysts loadings of 40–100 mol% $Cu(OAc)_2 \cdot H_2O$ are required (Scheme 4.18).

The use of potassium aryltrifluoroborate salts as useful bench-stable alternatives to arylboronic acids has been demonstrated in O—H bond arylations (*vide supra*),

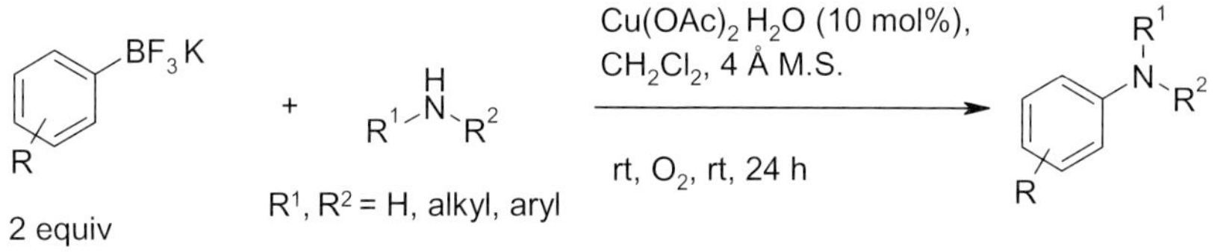

$$\text{R-C}_6\text{H}_4\text{-BF}_3\text{K} \;(2\ \text{equiv}) + \text{R}^1\text{-NH-R}^2 \xrightarrow[\text{rt, O}_2\text{, rt, 24 h}]{\text{Cu(OAc)}_2\ \text{H}_2\text{O (10 mol\%), CH}_2\text{Cl}_2\text{, 4 Å M.S.}} \text{R-C}_6\text{H}_4\text{-NR}^1\text{R}^2$$

R^1, R^2 = H, alkyl, aryl

Scheme 4.19 Additional boron-containing aryl donors.

and required stoichiometric amounts of copper acetate, as well as DMAP as an additive (Scheme 4.19). Interestingly, the N–H bond arylation reactions proceed with much greater levels of ease, and can be conducted without DMAP and with only 10 mol% $Cu(OAc)_2 \cdot H_2O$ [51]. Importantly, here a major side reaction – namely N-dealkylation – is minimized when using aliphatic amines.

Additional interesting N–H-containing substrates have been utilized in copper-catalyzed arylations (Table 4.7). The use of Buchwalds reaction conditions (Scheme 4.13) represents the best procedure for the arylation of substituted aziridines [55]. Sulfoximes are arylated regioselectively on nitrogen at ambient temperature under base-free reaction conditions to give a wide range of *N*-arylsulfoximes [56]. Among a range of copper(I) and copper(II) sources examined, CuI, CuCl, $CuSO_4$, $Cu(OAc)_2 \cdot H_2O$ and $Cu(OAc)_2$ all gave high yields (55–93%) of the desired product, although $Cu(OAc)_2$ was selected as the preferred catalyst for further optimization of the process. With the addition of 2.3 equiv. of the boronic acid aryl donor, good to excellent yields were obtained at ambient temperature in methanol as solvent. As this protocol is so mild, substrates containing potentially reactive groups (e.g. bromides) were arylated chemoselectively. Moreover, no trace of reaction on the methyl group of the sulfoxime was observed, and further substitution on this group was also tolerated.

An oxidative homocoupling of *tert*-butylcarbazate is speculated to form di-*tert*-butyl hydrazodiformate, which in turn smoothly undergoes selective N-arylation at ambient temperature to give the bis-Boc protected-monosubstituted hydrazine as sole product in high yields (Scheme 4.20) [57]. The process is best catalyzed by CuCl, as other copper(I) and copper(II) sources, such as $CuBr \cdot SMe_2$ (51%), CuI (42%), $CuCl_2$ (45%) and $Cu(OAc)_2$ (22%), give only modest yields.

The reaction of arylboronic acids with copper(II) acetate is speculated to form an arylcopper species which can add in a 1,2-fashion to an azodicarboxylate. Subsequent transmetallation with the arylboronic acid and hydrolysis would give the desired product selectively (Scheme 4.21). Indeed, it was found that several copper(I) and copper(II) salts would facilitate this reaction. While CuI (55%), CuCl (73%) and $Cu(OTf)_2$ (86%) could be used – in contrast to the reaction depicted in Scheme 4.19 – $Cu(OAc)_2$ (88%) proved to be superior [58]. Additional azodicarboxylate derivatives function equally well under the optimized reaction conditions. Non-symmetrical azodicarboxylates are also useful substrates and can be arylated in a reproducibly regioselective manner in excellent to quantitative yields at ambient temperature. In this case, methanol was the preferred reaction solvent [59].

Table 4.7 Additional N—H bond-containing substrates.

Substrate	Remarks	Example of conditions	Reference
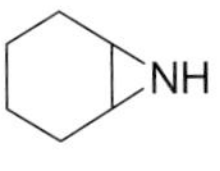 NH	Generally high yields (40–87%)	$Cu(OAc)_2$ **(0.1 equiv)**, $ArB(OH)_2$ (1.5 equiv), myristic acid (0.2 equiv), 2,6-lutidine (1 equiv), toluene, rt	[55]
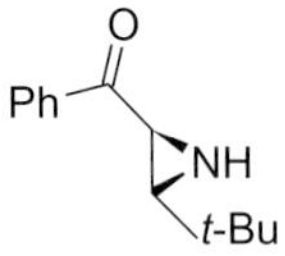 O Ph NH *t*-Bu	Yield 40%	$Cu(OAc)_2$ **(0.1 equiv)**, *p*-F-$ArB(OH)_2$ (1.5 equiv), myristic acid (0.2 equiv), 2,6-lutidine (1 equiv), toluene, 50 °C	[55]
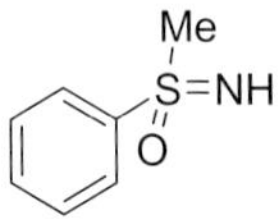 Me S=NH O	Base-free, generally high yields (62–93%), wide substrate scope in aryl boronic acid	$Cu(OAc)_2$ **(0.1 equiv)**, $ArB(OH)_2$ (2.3 equiv), MeOH (0.3 *M*), rt, 12 h	[56]
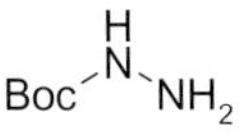 H Boc N NH_2 2 equiv	Generally high yields, selective, O_2 atmosphere decreases the yield, wide substrate scope in aryl boronic acid, hindered substrates react slowly	CuCl **(0.1 equiv)**, $ArB(OH)_2$ (1.1 equiv), pyridine, 1,2-DCE, 3 Å MS, dry air, rt, 6–96 h	[57]
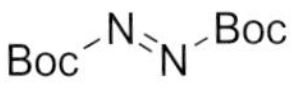 Boc N=N Boc 1 equiv	Range of copper catalysts function well, generally excellent yields, electron-rich or *ortho*-substituted boronic acids work worse	$Cu(OAc)_2$ **(0.1 equiv)**, $ArB(OH)_2$ (2 equiv), THF, N_2, rt, 20 h	[58]
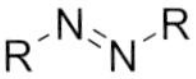 R N=N R 1 equiv	R = CO_2Et: 99% R = CO_2i-Pr: 93% R = $CONR_2$: 99%	$Cu(OAc)_2$ **(0.1 equiv)**, $PhB(OH)_2$ (2 equiv), THF, N_2, rt, 20 h	[58]
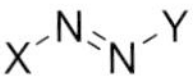 X N=N Y	X = Ph, Y = Boc: 100% X = Y = Boc: 91% X = 4-$NO_2C_6H_4$, Y = Boc: 100%	$Cu(OAc)_2$ **(0.1 equiv)**, $PhB(OH)_2$ (2 equiv), MeOH, rt or reflux, 0.5–1.5 h	[59]

MS = molecular sieve; rt = room temperature.

In the past, copper acetate has been the most popular choice of catalyst for many O—H and N—H bond arylation reactions, although it is not exclusive. The use of catalytic amounts of copper(I) oxide offers a valuable alternative and a range of azoles, anilines and aliphatic amines, all used in slight excess, can each be efficiently functionalized using the optimized protocol (Scheme 4.22), where air as a co-oxidant is essential [60].

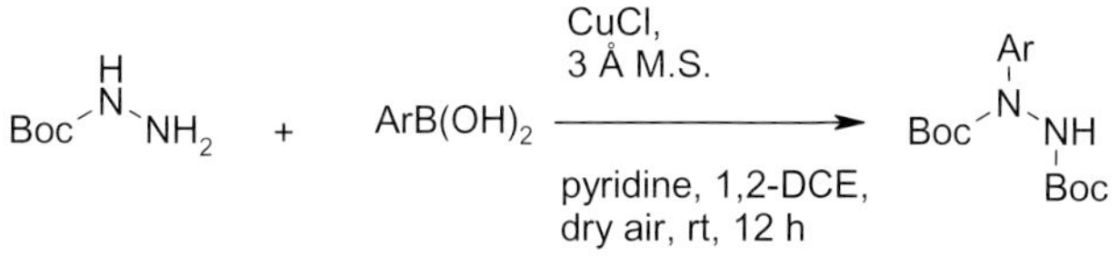

Scheme 4.20 CuCl-catalyzed synthesis of bis-Boc protected-monoarylated hydrazine.

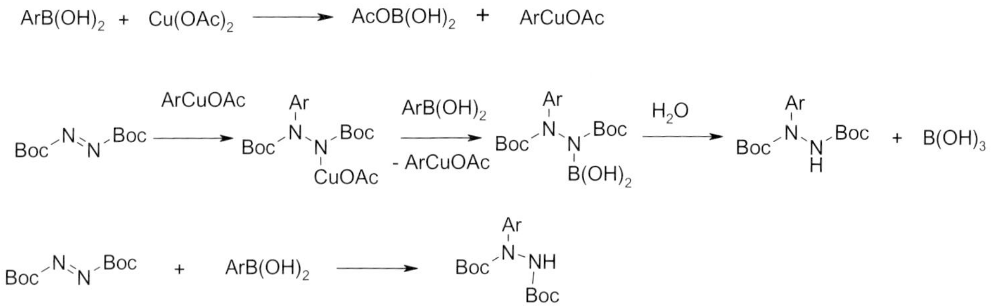

Conditions: $Cu(OAc)_2$, THF, N_2, rt, 20 h or $Cu(OAc)_2$, MeOH, rt or reflux, 0-5-1.5 h

Scheme 4.21 Proposed mechanism for CuCl-catalyzed synthesis of bis-Boc protected-monoarylated hydrazine from azodicarboxylates.

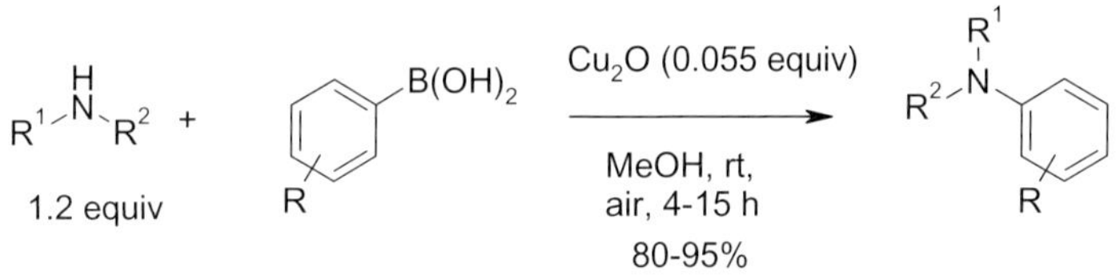

Scheme 4.22 Cu_2O-catalyzed arylations of boronic acids.

Interestingly, in contrast to the numerous intramolecular couplings between alcohols and aryl boronic acids [1b], there are no reports of intramolecular copper-catalyzed N—H bond arylations with boronic acids.

4.6.2
Additional Important Non-N—H Arylation Examples

Although not a direct N—H bond arylation, a related new reductive protocol for the amination of arylboronic acids has been developed which utilizes nitroso arenes. This reaction is certainly worthy of inclusion here, as it broadens our understanding of the cross-coupling processes of heteroatom–carbon bonds via N—O bonded species [61]. These new protocols rely on either stoichiometric amounts of CuCl (Scheme 4.23) or on catalytic amounts of copper(II) methylsalicylate (CuMeSal) (Scheme 4.24). The need for stoichiometric amounts of a copper(I)

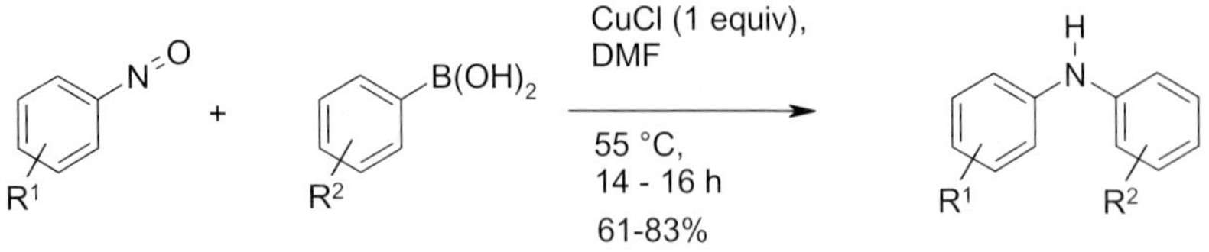

CuCl (1 equiv), DMF; 55 °C, 14 - 16 h; 61-83%

Scheme 4.23 CuCl-mediated reductive aminations of nitroso arenes.

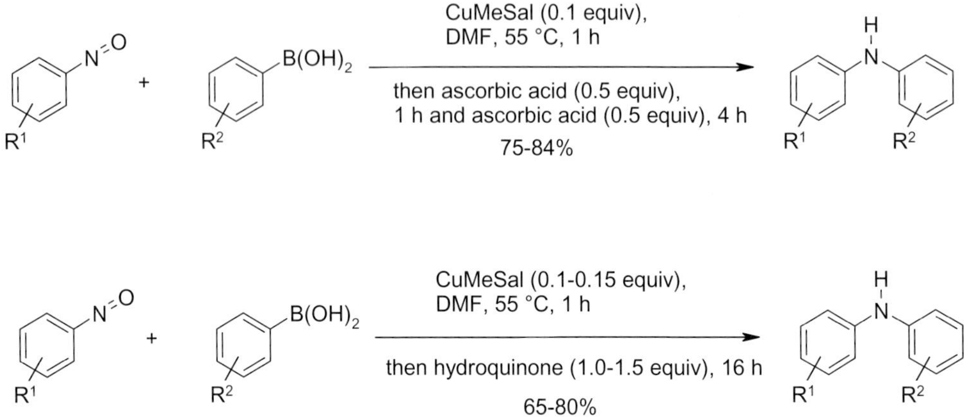

CuMeSal (0.1 equiv), DMF, 55 °C, 1 h; then ascorbic acid (0.5 equiv), 1 h and ascorbic acid (0.5 equiv), 4 h; 75-84%

CuMeSal (0.1-0.15 equiv), DMF, 55 °C, 1 h; then hydroquinone (1.0-1.5 equiv), 16 h; 65-80%

Scheme 4.24 CuMeSal-catalyzed reductive aminations of nitroso aromatics.

source, such as CuCl, in the transformation is necessary due to the required reduction of the N—O bond, as well as its function in mediating the aryl transfer from the boronic acid component. A wide range of copper sources was examined, with CuCl and $[Cu(OTf)_2] \cdot PhH$ and DMF being the best choice of catalyst and solvent, respectively. Under the optimized conditions a wide range of substituents and functional groups are tolerated in the formation of non-symmetrical diarylamines.

A method which would be catalytic in copper and would not rely on excess boronic acid would clearly offer distinct advantages over current protocols. A careful study of potential terminal reducing agents, in addition to the phenylboronic acid, which leads to the formation of phenols, showed that the portion-wise addition of ascorbic acid resulted in an efficient transformation to the desired product, when using even catalytic quantities of CuMeSal (Scheme 4.24). An alternative mild reductant was hydroquinone, which could be added in one batch to the reaction mixture. Although this method requires longer reaction times and affords slightly lower yields than the ascorbic acid-based methods, more importantly the reactions were devoid of aniline or azoxy side products. A mechanism to support these findings has been proposed (Scheme 4.25) which involves the formation of a copper(III) species via complexation of the nitroso arene to the copper(I) salt. Transmetallation with the arylboronic acid is then expected to be facile, followed by reductive elimination of the *N,N*-diarylhydroxylamine as a

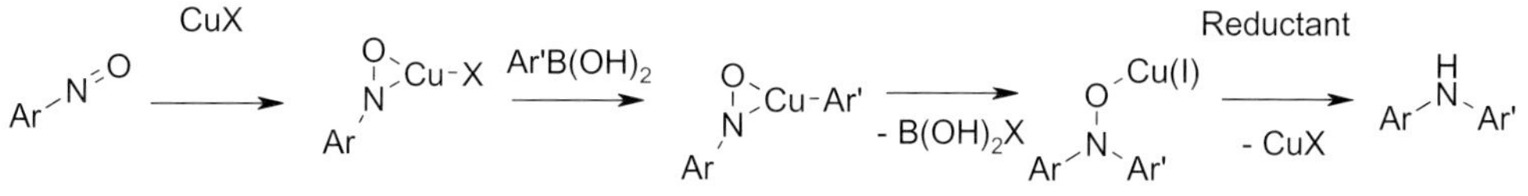
CuX
Ar'B(OH)2
- B(OH)2X
Reductant
- CuX
Cu-X
Cu-Ar'
Cu(I)
Ar
Ar'

X = Cl, MeSal

Reductant = ascorbic acid or hydroquinone

Scheme 4.25 Suggested mechanism of copper-mediated reductive aminations of arylboronic acids with nitroso arenes.

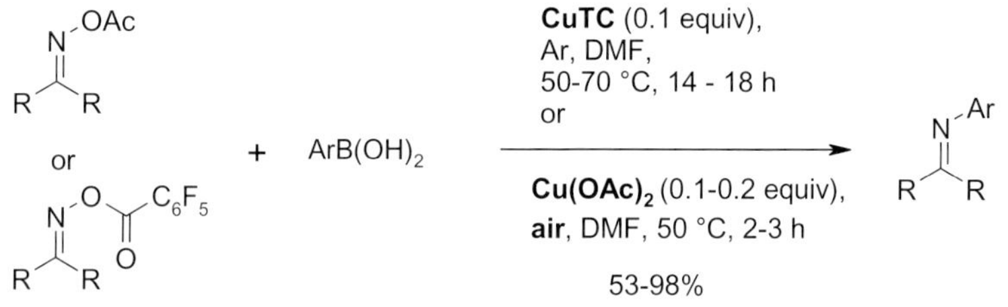
OAc
or
+
ArB(OH)2
CuTC (0.1 equiv),
Ar, DMF,
50-70 °C, 14 - 18 h
or
Cu(OAc)2 (0.1-0.2 equiv),
air, DMF, 50 °C, 2-3 h
53-98%
C6F5
Ar
R

R = (het)aryl, alkyl

Scheme 4.26 Copper-catalyzed N-imination of boronic acids with *O*-acylketoximes.

copper(I) alkoxide. This would be reduced by the reductant to regenerate the copper(I) catalyst and would give the desired diarylamine. In the stoichiometric version of the reaction, this penultimate intermediate would be expected, via a redox process, to generate the diarylamine final product and a copper(II) species.

Scheme 4.26 shows the development of two general reactions based on copper(I) thiophene-2-carboxylate (CuTC) or $Cu(OAc)_2$ for the formal arylation of imines with arylboronic acids [62]. The *O*-acetyl or *O*-pentafluorophenyl oximes react with a wide range of electron-rich, electron-poor and electron-neutral boronic acids, and even *ortho*-substituted substrates react well. The mechanism is proposed to proceed, as shown in Scheme 4.27, via an initial oxidative addition of the copper(I) species, either CuTC or through reduction of $Cu(OAc)_2$ by the boronic acid, to the ketoxime *O*-carboxylate. This is followed by transmetallation and reductive elimination to generate the final product and regenerate the catalytically active copper(I) species.

4.7 Summary and Conclusions

Today, there is a truly wide range of choices available for synthetic chemists requiring to transfer an aryl moiety to an O—H or N—H bond by utilizing copper salts.

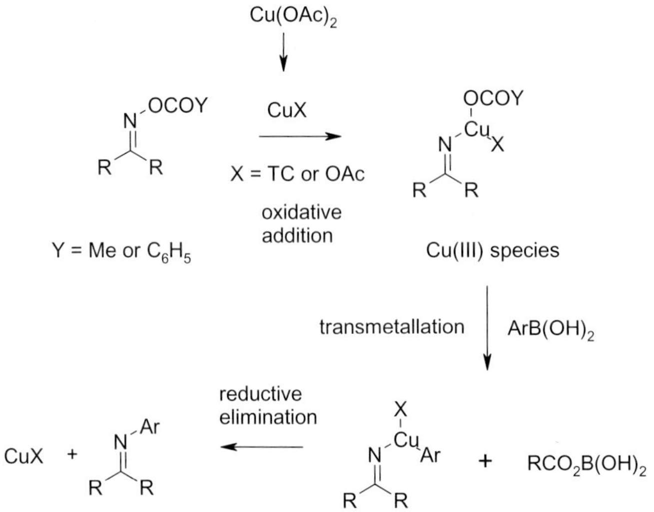

Scheme 4.27 Mechanism of the copper-catalyzed N-imination of boronic acids with *O*-acylketoximes.

To date, however, no one method or protocol is considered to be effective for every possible combination of the two reaction partners, the aryl donor and the O—H- or N—H-containing nucleophile. Clearly, the original seminal studies of Chan and Lam, initially further developed by Evans, have stood the test of time as both a general method of choice that is tolerant of a range of common functional groups. However, severe limitations of such a method persist, including: (i) the need for an excess of copper(II) acetate; (ii) the use of an excess amount of an organic base, usually triethylamine or pyridine; (iii) normally long reaction times; (iv) a requirement for an excess of the arylboronic acid; and (v) in general only modest to good yields of the targeted O- or N-arylated products. Nevertheless, the Chan–Lam–Evans modified Ullmann condensation reaction remains very popular today, especially in medicinal chemistry programs in the pharmaceutical industry, due to its broad efficiency and simplicity of operation for parallel synthesis applications. For O—H bond arylations, the copper salts $Cu(OAc)_2$ and $Cu(OAc)_2 \cdot H_2O$ have dominated to become the most commonly used in arylation transformations as either the catalyst or precatalyst of choice. However, in almost all cases, in order to achieve the best reaction outcome then stoichiometric or even excess amounts of the copper(II) salts are required. The use of molecular sieves has been central to improving reaction efficacy, and also has the effect of minimizing the required amounts of the aryl boronic acid donor. A further crucial advance has been the option to preform the cyclic triarylboroxines (0.33 equiv.), or preferably to use pyridine complexes of cyclic triarylboroxines (0.66 equiv.) with Cs_2CO_3 as base, which function typically in a superior fashion to aryl boronic acids. A more detailed and systematic study of the tolerated aryl groups is merited, and would be useful to confirm this claim, although stoichiometric amounts of $Cu(OAc)_2$ are still required. The above-described developments are all specific to the intermolecular arylation of O—H bonds, while the addition of DMAP as an additive has positive effects on reaction efficiencies and proceeds in the absence of a additional base. Interestingly, when combining the use of DMAP and molecular sieves, a whole new process utilizing 2.0 equiv. of potassium aryltrifluoroborates has evolved into perhaps the method of choice for the copper-catalyzed arylations of the broadest set of O—H bonds, including phenols as well as primary and secondary alcohols. The same process, utilizing 2.0 equiv. of potassium aryltrifluoroborates, also works wonderfully well for arylation of the broadest set of N—H bonds, including anilines as well as primary and secondary amines. In this case, the transformations proceed best in the absence of DMAP as a ligand. Additional key and significant advances in N—H bond arylation have involved the change of the copper(II) source from $Cu(OAc)_2$ to $[Cu(OH) \cdot TMEDA]_2Cl_2$, which allowed a truly catalytic version to be established that is particularly useful for imidazole substrates. It has long been believed – and now has also been demonstrated by the use of myristic and decanoic acid – that copper-solubilizing ligands can improve the efficiency of arylation reactions. In these processes with aryl boronic acids, a wide range of amines can be arylated with catalytic amounts of $Cu(OAc)_2$. Recent developments using Chan–Lam-inspired methods include the N-cyclopropanation of a range of N—H bonds through the use of cyclopropylboronic acids, using both catalytic and stoichiometric amount of copper acetate and a sodium base in the presence of a chelating ligand [63, 64]. Significant important advances have also been made in the area of copper-mediated O—H and N—H bond formation to alkenyl-substituted boron-based reagents [9, 65–68]. Overall, the field of arylation of O—H and N—H bonds utilizing boronic acids and their derivatives has matured significantly since the pioneering efforts of Lam and Chan. The aim of this chapter has not been to provide a critical comparison of other transition metal-mediated processes, especially palladium-catalyzed reactions, and in fact many of the methods are complementary to one another. The current low cost of copper (US$ 4 per lb) compared to palladium (US$ 470 per lb) (based on June 2008 prices) clearly offers a financial advantage to the industrial application of these methods (http://www.metalprices.com/FreeSite/metals/cu/cu.asp).

Abbreviations

Ac	acetyl
acac	acetylacetone
Bn	benzyl
Boc	*tert*-butoxycarbonyl
cat.	catalytic
DABCO	1,4-diazabicyclo[2.2.2]octane
DBU	1,8-diazabicyclo[5.4.0]undec-7-ene
DCE	1,2-dichloroethane
DMAP	4-dimethylaminopyridine

DMF	*N,N*-dimethylformamide
DMSO	dimethyl sulfoxide
mCPBA	*m*-chloroperoxybenzoic acid
NMO	4-methylmorpholine-*N*-oxide
NMP	*N*-methylpyrolidinone
PEG	poly(ethylene glycol)
phen	phenanthroline
Piv	pivaloyl
py	pyridine
Sal	salicylate
TEMPO	tetramethylpiperidinyloxy
tmeda	tetramethylethylenediamine

References

1 (a) Ley, S.V. and Thomas, A.W. (2003) *Angew. Chem. Int. Ed. Engl.*, **42**, 5400–9;
(b) Thomas, A.W. (2007) Product subclass 1: diaryl ethers. *Sci. Synth.*, **31**a, 469–543;
(c) Chan, D.M.T. and Lam, P.Y.S. (2005) *Boronic Acids: Preparation and Applications in Organic Synthesis and Medicine* (ed. D.G. Hall), Wiley-VCH Verlag GmbH, Weinheim, Germany, pp. 205–40;
(d) Theil, F. (1999) *Angew. Chem. Int. Ed. Engl.*, **38**, 2345–7;
(e) Muci, A.R. and Buchwald, S.L. (2002) *Top. Curr. Chem.*, **219**, 131–209;
(f) Wolfe, J.P., Wagaw, S., Marcoux, J.F. and Buchwald, S.L. (1998) *Acc. Chem. Res.*, **31**, 805–18;
(g) Yang, B.Y. and Buchwald, S.L. (1999) *J. Organomet. Chem.*, **576**, 125–46;
(h) Hartwig, J.F. (1998) *Angew. Chem. Int. Ed. Engl.*, **37**, 2046–67;
(i) Hartwig, J.F. (2002) *Handbook of Organopalladium Chemistry for Organic Synthesis*, Vol. 1 (eds E.-i. Negishi and A. de Meijere), John Wiley & Sons, Inc., New York, pp. 1051–96;
(j) Hartwig, J.F. (1999) *Pure Appl. Chem.*, **71**, 1417–23;
(k) Hartwig, J.F. (1998) *Acc. Chem. Res.*, **31**, 852–60;
(l) Hartwig, J.F. (1997) *Synlett*, 329–40;
(m) Littke, A.F. and Fu, G.C. (2002) *Angew. Chem. Int. Ed. Engl.*, **41**, 4177–211;
(n) Lindley, J. (1984) *Tetrahedron*, **40**, 1433–56;
(o) Finet, J.-P., Federov, A.Y., Combes, S. and Boyer, G. (2002) *Curr. Org. Chem.*, **6**, 597–626;
(p) Theil, F. (2003) *Organic Synthesis Highlights*, Vol. **5** (eds H.-G. Schmalz and T. Wirth), Wiley-VCH Verlag GmbH, Weinheim, Germany, pp. 15–21;
(q) Sawyer, J.S. (2000) *Tetrahedron*, **56**, 5045–65;
(r) Kunz, K., Scholz, U. and Ganzer, D. (2003) *Synlett*, 2428–39;
(s) Frlan, R. and Kikelj, D. (2006) *Synthesis*, 2271–85;
(t) Beletskaya, I.P. and Cheprakov, A.V. (2004) *Coord. Chem. Rev.*, **248**, 2337–64;
(u) Finet, J.-P. (1989) *Chem. Rev.*, **89**, 1487–501;
(v) Cai, Q., Zhu, W., Zhang, H., Zhang, Y. and Ma, D. (2005) *Synthesis*, 496–9.

2 (a) Ullmann, F. (1903) *Ber. Deutsch. Chem. Ges.*, **36**, 2382–91;
(b) Ullmann, F. (1904) *Ber. Deutsch. Chem. Ges.*, **37**, 853–7.

3 Goldberg, I. (1906) *Ber. Deutsch. Chem. Ges.*, **39**, 1691–6.

4 Chan, D.M.T., Monaco, K.L., Wang, R.-P. and Winters, M.P. (1998) *Tetrahedron Lett.*, **39**, 2933–6.

5 (a) Marcoux, J.F., Doye, S. and Buchwald, S.L. (1997) *J. Am. Chem. Soc.*, **119**, 10539–40;

(b) Ma, D., Zhang, Y., Yao, J., Wu, S. and Tao, F. (1998) *J. Am. Chem. Soc.*, **120**, 12459–67;
(c) Cristau, H.-J., Cellier, P.P., Hamada, S., Spindler, J.-F. and Taillefer, M. (2004) *Org. Lett.*, **6**, 913–16.

6 Chan, D.M.T. (1997) Abstract M92, 35th National Organic Symposium, June 22–26, 1997, TX, USA.

7 Chan, D.M.T. (1996) *Tetrahedron Lett.*, **37**, 9013–16.

8 Evans, D.A., Katz, J.L. and West, T.R. (1998) *Tetrahedron Lett.*, **39**, 2937–40.

9 Lam, P.Y.S., Vincent, G., Clark, C.G., Deudon, S. and Jadhav, P.K. (2001) *Tetrahedron Lett.*, **42**, 3415–18.

10 Lappert, M.F. (1956) *Chem. Rev.*, **56**, 959–1064.

11 Santucci, L. and Triboulet, C. (1969) *J. Chem. Soc. Chem. Commun.*, 392–6.

12 Lam, P.Y.S., Bonne, D., Vincent, G., Clark, C.G. and Combs, A.P. (2003) *Tetrahedron Lett.*, **44**, 1691–4.

13 Chiang, G.C.H. and Olsson, T. (2004) *Org. Lett.*, **6**, 3079–82.

14 Hitotsuyanagi, Y., Ishikawa, H., Naito, S. and Takeya, K. (2003) *Tetrahedron Lett.*, **44**, 5901–3.

15 Simon, J., Salzbrunn, S., Prakash, G.K.S., Petasis, N.A. and Olah, G.A. (2001) *J. Org. Chem.*, **66**, 633–4.

16 Sagar, A.D., Tale, R.H. and Adude, R.N. (2003) *Tetrahedron Lett.*, **44**, 7061–3.

17 Chan, D.M.T., Monaco, K.L., Li, R. Bonne, D., Clark, C.G. and Lam, P.Y.S. (2003) *Tetrahedron Lett.*, **44**, 3863–5.

18 Voison, A.S., Bouillon, A., Lancelot, J.-C., Lesnard, A. and Rault, S. (2006) *Tetrahedron*, **62**, 6000–5.

19 McKinley, N.F. and O'Shea, D.F. (2004) *J. Org. Chem.*, **69**, 5087–92.

20 Quach, T.D. and Batey, R.A. (2003) *Org. Lett.*, **5**, 1381–4.

21 Petrassi, H.M., Sharpless, K.B. and Kelly, J.W. (2001) *Org. Lett.*, **3**, 139–42.

22 Lam, P.Y.S., Clark, C.G., Saubern, S., Adams, J., Winters, M.P., Chan, D.M.T. and Combs, A. (1998) *Tetrahedron Lett.*, **39**, 2941–4.

23 Combs, A.P., Saubern, S., Rafalski, M. and Lam, P.Y.S. (1999) *Tetrahedron Lett.*, **40**, 1623–6.

24 Mederski, W.W.K.R., Lefort, M., Germann, M. and Kux, D. (1999) *Tetrahedron*, **55**, 12757–70.

25 Rossiter, S., Woo, C.K., Hartzoulakis, B., Wishart, G., Stanyer, L., Labadie, J.W. and Selwood, D.L. (2004) *J. Comb. Chem.*, **6**, 385–90.

26 Cundy, D.J. and Forsyth, S.A. (1998) *Tetrahedron Lett.*, **39**, 7979–82.

27 Yu, S., Saenz, J. and Srirangam, J.K. (2002) *J. Org. Chem.*, **67**, 1699–702.

28 Collot, V., Bovy, P.R. and Rault, S. (2000) *Tetrahedron Lett.*, **41**, 9053–7.

29 Singh, B.K., Appukkuttan, P., Claerhout, S., Parmar, V.S. and Van der Eycken, E. (2006) *Org. Lett.*, **8**, 1863–6.

30 Das, P. and Baus, B. (2004) *Synth. Commun*, **34**, 2177–84.

31 Kantam, M.L., Venkanna, G.T., Sridhar, C., Sreedhar, B. and Choudray, B.M. (2006) *J. Org. Chem.*, **71**, 9522–4.

32 Chen, S., Huang, H., Liu, X., Shen, J., Jiang, H. and Liu, H. (2008) *J. Comb. Chem.*, **10**, 358–60.

33 Wang, W., Devasthale, P., Farrelly, D., Gu, L., Harrity, T., Cap, M., Chu, C., Kunselman, L., Morgan, N., Ponticiello, R., Zebo, R., Zhang, L., Locke, K., Lippy, J., O'Malley, K., Hosagrahara, V., Zhang, L., Kadiyala, P., Chang, C., Muckelbauer, J., Doweyko, A.M., Zahler, R., Ryono, D., Hariharan, N. and Cheng, P.T.W. (2008) *Bioorg. Med. Chem. Lett.*, **18**, 1939–44.

34 Hügel, H.H., Rix, C.J. and Fleck, K. (2006) *Synlett*, 2290–2.

35 Joshi, R.A., Patil, P.S., Mthukrishnan, M., Ramana, C.V. and Gurjar, M.K. (2004) *Tetrahedron Lett.*, **45**, 195–7.

36 Bakkestuen, A.K. and Gundersen, L.-L. (2003) *Tetrahedron Lett.*, **44**, 3359–62.

37 Dai, Q., Ran, C. and Harvey, R.G. (2006) *Tetrahedron*, **62**, 1764–71.

38 Combs, A.P., Tadesse, S., Rafalski, M., Haque, T.S. and Lam, P.Y.S. (2002) *J. Comb. Chem.*, **4**, 179–82.

39 Combs, A.P. and Rafalski, M. (2000) *J. Comb. Chem.*, **2**, 29–32.

40 Zhang, Z., Yu, Y. and Liebeskind, L. (2008) *Org. Lett.*, **10**, 3005–8.

41 Chernick, E.T., Aherns, M.J., Scheidt, K.A. and Waseilewski, M.R. (2005) *J. Org. Chem.*, **70**, 1486–9.

42 Collman, J.P. and Zhong, M. (2000) *Org. Lett.*, **2**, 1233–6.

43 Grieco, P.A. (ed.) (1998) *Organic Synthesis in Water*, Blackie Academic & Professional, London.

44 Collman, J.P., Zhong, M., Zeng, L. and Costanzo, S. (2001) *J. Org. Chem.*, **66**, 1528–31.

45 Collman, J.P., Zhong, M., Zhang, C. and Costanzo, S. (2001) *J. Org. Chem.*, **66**, 7892–7.

46 van Berkel, S.S., Van den Hoogenband, A., Terpstra, J.W., Tromp, M., Van Leeuwen, P.W.N.M. and Van Strijdonck, G.P.F. (2004) *Tetrahedron Lett.*, **45**, 7659–62.

47 Hay, A.S. (1962) *J. Org. Chem.*, **27**, 3320–1.

48 Antilla, J.C. and Buchwald, S.L. (2001) *Org. Lett.*, **3**, 2077–9.

49 Tzschucke, C.C., Murphy, J.M. and Hartwig, J.F. (2007) *Org. Lett.*, **9**, 761–4.

50 Quach, T.D. and Batey, R.A. (2003) *Org. Lett.*, **5**, 4397–400.

51 Lan, J.-B., Chen, L., Yu, X.-Q., You, J.-S. and Xie, R.-G. (2004) *Chem. Commun.*, 188–9.

52 Nishiura, K., Urawa, Y. and Soda, S. (2004) *Adv. Synth. Catal.*, **346**, 1679–84.

53 Kantam, M.L., Neelima, B., Reddy, Ch.V. and Neeraja, V. (2006) *J. Mol. Catal. A*, **249**, 201–6.

54 Lam, J.-B., Zhang, G.-L., Yu, X.-Q., You, J.-S., Chen, L., Yan, M. and Xie, R.-G. (2004) *Synlett*, 1095–7.

55 Sasaki, M., Dalili, S. and Yudin, A.K. (2003) *J. Org. Chem.*, **68**, 2045–7.

56 Moessner, C. and Bolm, C. (2005) *Org. Lett.*, **7**, 2667–9.

57 Kabalka, G.W. and Guchhait, S.K. (2003) *Org. Lett.*, **5**, 4129–31.

58 Uemura, T. and Chatani, N. (2005) *J. Org. Chem.*, **70**, 8631–4.

59 Kisseljova, K., Tsubrik, O., Sillard, R., Maeorg, S. and Maeorg, U. (2006) *Org. Lett.*, **8**, 43–5.

60 Sreedhar, B., Venkanna, G.T., Kumar, K.B.S. and Balasubrahmanyam, V. (2008) *Synthesis*, 795–9.

61 Yu, Y., Srogl, J. and Liebeskind, L.S. (2004) *Org. Lett.*, **6**, 2631–4.

62 Liu, S., Yu, Y. and Liebeskind, L.S. (2007) *Org. Lett.*, **9**, 1947–50.

63 Tsuritani, T., Strotman, N.A., Yamamoto, Y., Kawasaki, M., Yasuda, N. and Mase, T. (2008) *Org. Lett.*, **10**, 1653–5.

64 Bernard, S., Neuville, L. and Zhu, Z. (2008) *J. Org. Chem.*, **73**, 6441–4.

65 Lam, P.Y.S., Vincent, G., Bonne, D. and Clark, C.G. (2003) *Tetrahedron Lett.*, **44**, 4927–31.

66 Dalili, S. and Yudin, A.K. (2005) *Org. Lett.*, **7**, 1161–4.

67 Deagostino, A., Prandi, C., Zavattaro, C. and Venturello, P. (2008) *Eur. J. Org. Chem.*, 1313–23.

68 Bolshan, Y. and Batey, R.A. (2008) *Angew. Chem. Int. Ed. Engl.*, **47**, 2109–12.

5
Metal-Catalyzed Arylations of Nonactivated Alkyl (Pseudo)Halides via Cross-Coupling Reactions

Masaharu Nakamura and Shingo Ito

5.1
Introduction

The introduction of alkyl chains to aromatic rings is a fundamental synthetic transformation for the production of a vast range of chemical materials. The Friedel–Crafts alkylation or related electrophilic substitution of electron-rich aromatics has long served as an indispensable synthetic method for the preparation of alkylated arenes in industrial settings, as well as in academic laboratories. Likewise, the metal-catalyzed cross-coupling reaction has attracted much attention as a versatile synthetic tool for connecting a variety of alkyl groups and aromatic moieties since its discovery during the early 1970s. The discovery and subsequent development of catalytic coupling methods using aryl halides has revolutionized the approach to the synthesis of aromatic compounds. Thus, it is possible to introduce structurally diverse alkyl chains via the alkyl metal species, thereby significantly expanding the structural variation of aromatic compounds [1]. The obvious advantages of these catalytic arylation methods are: (i) the control of regioselectivities; (ii) the use of mild reaction conditions; and (iii) broad functional group compatibility. It is quite natural to expect that the coupling reactions of the inversed coupling partners – namely, reactions between alkyl (pseudo)halides as electrophiles and arylmetals as nucleophiles – constitute another powerful method for the selective syntheses of aromatic compounds. Besides the desired cross-coupling reaction (pathway *a* in Scheme 5.1), however, the reactions of alkyl halides or pseudohalides with arylmetals suffer from several side reactions, including halogen–metal exchange reactions (pathway *b*), β-elimination reactions (pathway *c*), and the generation of radical species (pathway *d*). Even when transition-metal catalysis is employed, efficient cross-coupling using *alkyl* electrophiles is still difficult due to the competing β-hydride elimination via the unstable σ-alkylmetal intermediates involved in the catalytic cycle. However, recent developments in transition metal-catalyzed cross-coupling reactions have provided certain solutions for these problems.

Modern Arylation Methods. Edited by Lutz Ackermann

ISBN: 978-3-527-31937-4

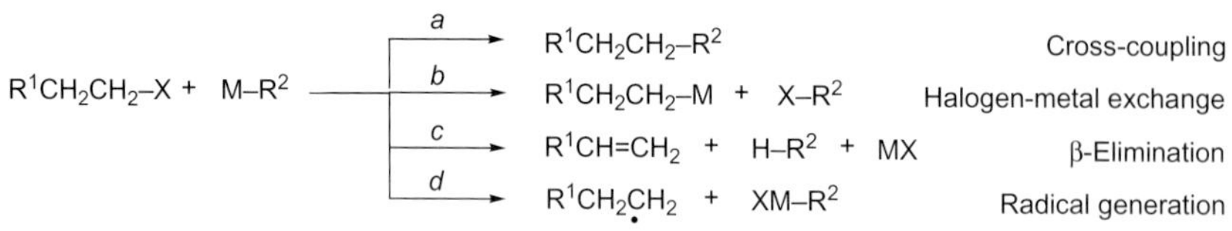

$$R^1CH_2CH_2\text{–}X + M\text{–}R^2 \longrightarrow \begin{cases} \xrightarrow{a} R^1CH_2CH_2\text{–}R^2 & \text{Cross-coupling} \\ \xrightarrow{b} R^1CH_2CH_2\text{–}M + X\text{–}R^2 & \text{Halogen-metal exchange} \\ \xrightarrow{c} R^1CH{=}CH_2 + H\text{–}R^2 + MX & \beta\text{-Elimination} \\ \xrightarrow{d} R^1CH_2\dot{C}H_2 + XM\text{–}R^2 & \text{Radical generation} \end{cases}$$

Scheme 5.1 Possible reaction courses between an alkyl electrophile and an organometallic nucleophile.

In this chapter we describe some modern synthetic methods for connecting various alkyl electrophiles and aromatic nucleophiles, all of which share transition metal catalysis as a common feature. The classical stoichiometric coupling reactions using aryl copper or cuprate reagents [2] are hence omitted. Since the publication of a comprehensive review by Beller and coworkers [3], significant progress has been made in the field of catalytic cross-coupling of alkyl electrophiles. Herein, arylation methods are categorized by the transition metal catalysts employed. Thus, palladium-catalyzed arylations of unactivated alkyl halides are described in Section 5.2, showing that a variety of aryl groups with diverse substitution patterns can be prepared. In Section 5.3, new nickel-catalyzed arylation methods are summarized, which have been developed during the past five years. Special attention will be paid to systems that could serve well for bulky secondary alkyl halides; hence, iron catalysts, which represent a promising substitute for the rare-metal catalysts, are reviewed in Section 5.4. Finally, in Section 5.5 we describe some recent examples of copper- and cobalt-catalyzed arylations of unactivated alkyl halides and sulfonates.

5.2
Palladium-Catalyzed Arylations of Alkyl (Pseudo)Halides

A few notable examples were reported during the early 1990s, where palladium catalysts were employed in cross-coupling reactions of alkyl halides with arylmetal. Widdowson reported that alkyl iodides possessing β-hydrogens could undergo successful cross-couplings with alkyl or aryl Grignard reagents in the presence of a preformed [Pd(0)dppf] catalyst (Equation 5.1) [4]. However, a re-examination by Scott showed that the reported coupling reaction was not reproducible, and that the major product obtained was, in fact, the corresponding reduced product R–H. For example, according to Scott's report, the arylation of decyl iodide with phenylmagnesium bromide proceeded nonselectively to give a 57 : 19 : 24 mixture of decane, decene and decylbenzene under the Widdowson's reaction conditions (Equation 5.2) [5].

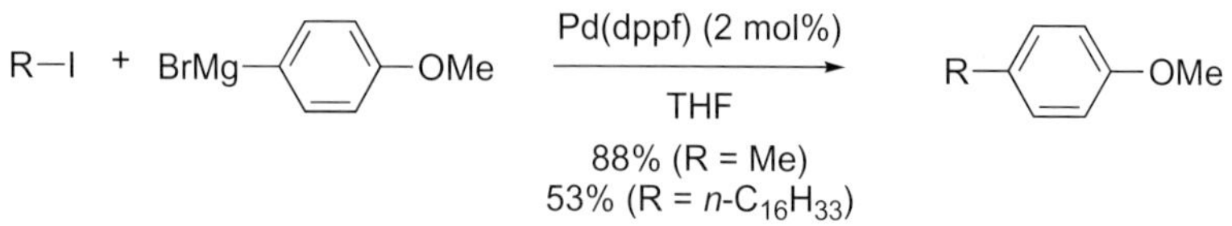

$$R\text{–}I + BrMg\text{–}C_6H_4\text{–}OMe \xrightarrow[\text{THF}]{\text{Pd(dppf) (2 mol\%)}} R\text{–}C_6H_4\text{–}OMe \tag{5.1}$$

88% (R = Me)
53% (R = n-$C_{16}H_{33}$)

$$\text{Dec–I} + \text{BrMg–Ph} \xrightarrow[\text{THF}]{\text{Pd(dppf) (5 mol\%)}} \text{Dec–Ph} \quad \textit{ca.}\ 24\% \tag{5.2}$$

In 1992, Suzuki reported that the cross-coupling between decyl iodide and 9-phenyl-9-borabicyclo[3.3.1]nonane (9-Phenyl-9-BBN) was catalyzed by [$Pd(PPh_3)_4$] in the presence of an excess amount of K_3PO_4 as a base to give the corresponding alkylbenzene in moderate yield (Equation 5.3) [6].

$$\underset{(1.5\ \text{equiv})}{\text{Dec–I}} + \underset{(1.0\ \text{equiv})}{\text{9-BBN–Ph}} \xrightarrow[\text{dioxane, 60 °C}]{\text{Pd(PPh}_3)_4\ \text{(3 mol\%)},\ K_3PO_4\ \text{(3.0 equiv)}} \text{Dec–Ph} \quad 55\% \tag{5.3}$$

B = 9-borabicyclo[3.3.1]nonyl

Charette reported high-yielding arylations of cyclopropyl iodide with arylboronic acids in 1996 [7]. In this case, formation of the elimination byproducts was largely suppressed by virtue of the ring-strain of the alkyl halide. The C–C bond formation took place with retention of configuration at the stereogenic center. In this way, a variety of aromatic groups, including 2- and 3-thienyl groups, can be installed on the cyclopropane ring in a diastereoselective manner (Equation 5.4).

$$\text{BnO–cyclopropyl–I} + \underset{(1.5\ \text{equiv})}{(HO)_2B\text{–Ar}} \xrightarrow[\text{DMF/H}_2\text{O, DME/H}_2\text{O or DMF, 90 °C}]{\text{Pd(OAc)}_2\ \text{(10 mol\%)},\ \text{PPh}_3\ \text{(50 mol\%)},\ \text{base (excess)}} \text{BnO–cyclopropyl–Ar} \tag{5.4}$$

Ar = Ph (80%), p-MeOC_6H_4 (85%), p-ClC_6H_4 (83%), o-Tol (80%), 2-thienyl (70%), 3-thienyl (78%)

Several breakthroughs for the arylation of alkyl halides were reported in 2002. For example, Beller reported that cross-couplings between primary alkyl chlorides possessing β-hydrogens and aryl Grignard reagents could be accomplished using a catalytic system consisting of [$Pd(OAc)_2$] and PCy_3 in a mixed solvent system of NMP and THF (5.5) [8]. Although the studied alkyl halides were limited to primary electrophiles, this arylation reaction can be regarded as one of the first examples demonstrating the synthetic potential of alkyl (pseudo)halides as electrophiles in catalytic arylations.

$$\text{Hex–Cl} + \underset{(1.5\ \text{equiv})}{\text{BrMg–Ar}} \xrightarrow[\text{NMP/THF, rt}]{\text{Pd(OAc)}_2\ (4\ \text{mol\%}),\ \text{PCy}_3\ (4\ \text{mol\%})} \text{Hex–Ar} \quad (5.5)$$

Ar = Ph (96%), *p*-Tol (91%), *o*-Tol (83%), *m*-$MeOC_6H_4$ (99%), *p*-FC_6H_4 (77%)

Since the reaction can proceed at ambient temperature, some electrophilic functional groups can be present in the alkyl electrophiles, as exemplified by products **1–3**. Further optimizations studies on this catalyst led to the development of *N*-heterocyclic carbene (NHC)-derived palladium catalysts for arylations of alkyl chlorides. In this system, it is assumed that the initial reaction of a precatalyst $[Pd(IMes)(NQ)]_2$ **4** and an aryl Grignard reagent should generate a coordinatively unsaturated palladium–carbene complex, which is the actual palladium-catalysis complex with high catalytic activity [9].

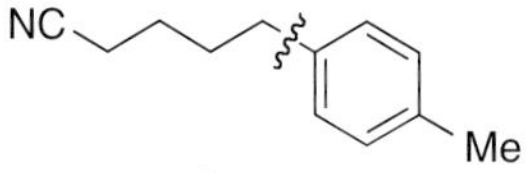

1 (99%)

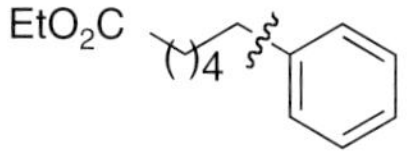

2 (58% with Pd/PCy_3) (93% with **4**)

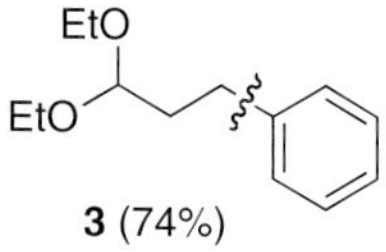

3 (74%)

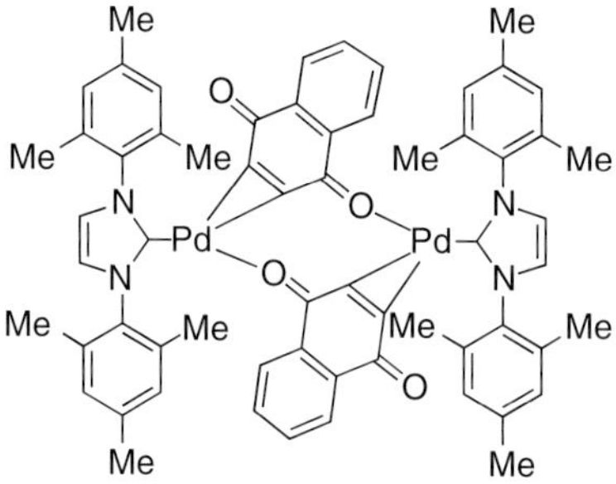

4 $[Pd(IMes)(NQ)]_2$

Palladium-catalyzed cross-coupling reactions of alkyl bromides with arylboron reagents represent another breakthrough for the development of a general method for arylation of alkyl halides. Fu found that the palladium-complexes with a bulky trialkylphosphine ligand P(*t*-Bu)$_2$Me could promote arylations of a variety of primary alkyl bromides with various arylboronic acids (Equation 5.6) [10]. An air-stable commercially available preligand, [HP(*t*-Bu)$_2$Me]BF_4 **5**, also worked well with a variety of arylboronic acids and gave the same products in comparable yields, as shown in the parentheses in Equation 5.6.

Oct–Br + $(HO)_2B$–Ph (1.5 equiv) → Oct–Ph (5.6)

$Pd(OAc)_2$ (5 mol%), $P(t\text{-Bu})_2Me$ (10 mol%), KO*t*-Bu (3.0 equiv); *t*-amyl alcohol, rt; 87% (90% with **5**)

6 68% (66%) **7** 85% (84%) **8** 63% (67%)

9 71% (76%) **10** 97% (93%) **11** 89% (91%)

The efficient synthesis of the tetrasubstituted benzene **11** from a sterically demanding mesitylboronic acid represents a good demonstration of the outstanding catalytic ability of Fu's system. It should be noted that other bulky trialkyl phosphines are much less effective than $P(t\text{-Bu})_2Me$. For example, the ligands PCy_3, PCy_2Et, $P(t\text{-Bu})_2Et$ and $P(t\text{-Bu})_3$ afforded octylbenzene in 63%, 39%, 4% and less than 2% yield, respectively, under otherwise identical reaction conditions of Equation 5.6.

The catalytic system consisting of $[Pd(OAc)_2]$ and $P(t\text{-Bu})_2Me$ was applied to the arylation of a primary alkyl tosylate with 9-phenyl-9-BBN (Equation 5.7) [11]. An isotope-labeling study revealed that the substitution reaction proceeded with an inversion of configuration at the stereogenic carbon center as the net result of an inversion during the oxidative addition step and a retention during the subsequent reductive elimination step (Equation 5.8).

$Pd(OAc)_2$ (4 mol%), $P(t\text{-Bu})_2Me$ (16 mol%), NaOH (1.2 equiv); dioxane, 50 °C; 63% (5.7)

9-phenyl-9-BBN (1.2 equiv)

$Pd(OAc)_2$ (4 mol%), $P(t\text{-Bu})_2Me$ (16 mol%), NaOH (1.2 equiv); dioxane, 70 °C; 54% (5.8)

9-phenyl-9-BBN (1.2 equiv)

inversion major
(ca. 6:1)

Similar palladium-catalyzed arylation reactions using arylsilane and arylzinc reagents are also possible by the judicious choice of bases and additives. As shown in Equation 5.9, the addition of a fluoride donor Bu_4NF promoted efficient cross-couplings of primary alkyl bromides and iodides with aryltrimethoxysilanes [12]. While the reaction showed good functional group compatibility, electron-deficient arylsilanes, such as 4-fluorophenyltrimethoxysilane, showed a rather poor reactivity.

n-$C_{12}H_{25}$–Br + $(MeO)_3$Si–Ph (1.2 equiv) → [$PdBr_2$ (4 mol%), P(t-Bu)$_2$Me (10 mol%), Bu_4NF (2.4 equiv); THF, rt; 75% (88% with **5**)] → n-$C_{12}H_{25}$–Ph (5.9)

X = OMe: 66%
F: 55%

X = OMe: 69%
F: 36%

X = OMe: 82%
F: 50%

The catalytic system [$Pd_2(dba)_3$] and $PCyp_3$, rather than the system consisting of [$PdBr_2$] with P(t-Bu)$_2$Me, turned out to be suitable for the conversion of arylzinc reagents. Representative examples of the arylations of primary alkyl bromides and iodides under the [$Pd_2(dba)_2$]-PCy_3 system are shown in Equations 5.10 and 5.11 [13]. It should be pointed out here that the choice of N-methylimidazole (NMI) as an additive and N-methylpyrrolidinone (NMP) as a cosolvent was crucial to achieve high chemical yields.

EtO(O)C–$(CH_2)_4$–CH_2Br + BrZn–C_6H_4–OMe (1.6 equiv) → [$Pd_2(dba)_3$ (2 mol%), $PCyp_3$ (8 mol%), NMI (1.2 equiv); THF/NMP, 80 °C; 74% (69% with [$HPCyp_3$]BF_4)] → EtO(O)C–$(CH_2)_4$–CH_2–C_6H_4–OMe

(5.10)

Dec–I + BrZn–Ph (1.6 equiv) → [$Pd_2(dba)_3$ (2 mol%), $PCyp_3$ (8 mol%), NMI (1.2 equiv); THF/NMP, 80 °C; 65% (62% with [$HPCyp_3$]BF_4)] → Dec–Ph (5.11)

Terao and Kambe reported a rather unconventional use of 1,3-butadiene as an additive, which affects palladium- and nickel-catalyzed cross-couplings of alkyl bromides and tosylates [14]. In the presence of 50 mol% of 1,3-butadiene and a catalytic amount of $[Pd(acac)_2]$, the cross-coupling of heptyl tosylate with 4-chlorophenylmagnesium bromide was achieved to give the corresponding arylation product in excellent yield (Equation 5.12).

$$\text{Hep-OTs} + \text{BrMg-C}_6\text{H}_4\text{-Cl}\ (1.5\ \text{equiv}) \xrightarrow[\text{THF, 25 °C, 3 h; 96\%}]{\text{Pd(acac)}_2\ (3\ \text{mol\%}),\ \text{1,3-butadiene (50 mol\%)}} \text{Hep-C}_6\text{H}_4\text{-Cl} \tag{5.12}$$

Arylations of primary alkyl bromides with aryl and heteroarylstannane reagents have also been achieved using palladium catalysts. Fu reported that aryltributylstannanes could be used for the arylation of certain functionalized alkyl bromides and iodides (Equation 5.13). An electron-rich phosphine ligand PCy(1-pyrrolidinyl)$_2$ showed a pronounced accelerating effect in this cross-coupling reaction. Notably, not only aryl-substituted stannanes but also a pyridinylstannane, was shown to participate in the functionalization reaction [15].

$$\text{Dec-Br} + \text{Bu}_3\text{Sn-Ph}\ (1.6\ \text{equiv}) \xrightarrow[\text{MeO}t\text{-Bu, MS 3A, RT; 72\%}]{[(?\text{-allyl})\text{PdCl}]_2\ (2.5\ \text{mol\%}),\ \text{PCy(1-pyrrolidinyl)}_2\ (10\ \text{mol\%}),\ \text{Me}_4\text{NF (2.4 equiv)}} \text{Dec-Ph} \tag{5.13}$$

EtO(O)C(CH$_2$)$_4$–C$_6$H$_4$–X: X = Me: 63%; OMe: 71%; CF$_3$: 57%
NC(CH$_2$)$_5$–C$_6$H$_4$–NMe$_2$: 68%
THPO(CH$_2$)$_n$–C$_6$H$_4$–F: 64%
butenyl–pyridine: 53%

As summarized in Table 5.1, a variety of aryl nucleophiles are applicable for the catalytic arylations of alkyl halides in the presence of palladium complexes. Hence, new catalytic systems have been developed to overcome the intrinsic low propensity of alkyl halide substrates to undergo oxidative addition. As seen in this section, one solution to the problem was the unique application of bulky phosphine ligands or diene ligands, which have proven highly effective for these modern cross-couplings of alkyl halides. Despite such remarkable progress, however, new catalysts systems are still required that will further expand the

Table 5.1 Palladium-catalyzed cross-coupling reactions between unactivated alkyl halides and aryl metal reagents.

$$\text{Alkyl–X} + \text{Ar–M} \xrightarrow[\text{solvent}]{\text{palladium catalyst}} \text{Alkyl–Ar}$$

Entry[a]	X	M	Palladium catalyst (or precatalyst)	Solvent	Examples[b]	Yield (%)[c]	Reference
1	I	MgBr	Pd(dppf)	THF	2	53, 88	[4]
2	I	MgBr	Pd(dppf)	THF	1	24	[5]
3	I	9-BBN	$Pd(PPh_3)_4$	dioxane	1	55	[6]
4	Cl	MgBr	$Pd(OAc)_2/PCy_3$	NMP/THF	11	43–99	[8]
5	Br	$B(OH)_2$	$Pd(OAc)_2/P(t\text{-Bu})_2Me$	*t*-amyl alcohol	9	63–97	[10]
6	Br	$B(OH)_2$	$Pd(OAc)_2/[HP(t\text{-Bu})_2Me]BF_4$	*t*-amyl alcohol	9	62–93	[16–19]
7	OTs	9-BBN	$Pd(OAc)_2/P(t\text{-Bu})_2Me$	dioxane	1	63	[11]
8	Br	$Si(OMe)_3$	$PdBr_2/P(t\text{-Bu})_2Me$	THF	20	36–84	[12]
9	Br	$Si(OMe)_3$	$PdBr_2/[HP(t\text{-Bu})_2Me]BF_4$	THF	9	42–88	[16–19]
10	Br, I	ZnBr	$Pd_2(dba)_3/PCyp_3$	NMP/THF	2	65, 74	[13]
11	Br, I	ZnBr	$Pd_2(dba)_3/[HPCyp_3]BF_4$	NMP/THF	2	62, 69	[16–19]
12	Cl	MgBr	$[Pd(IMes)(NQ)]_2$	NMP/THF	13	33–99	[9]
13	OTs	MgBr	$Pd(acac)_2$/1,3-butadiene	THF	1	86	[14]
14	Br, I	$SnBu_3$	$[(\pi\text{-allyl})PdCl]_2/PCy(1\text{-pyrrolidinyl})_2$	THF	10	53–76	[15]

a In the order of received date of the manuscripts.
b Based on the numbers of compounds prepared by the cross-coupling between unactivated alkyl halides and aryl metals.
c Reported yields.

substrate scope of organic electrophiles, more specifically to include bulky alkyl halides and rather inert alkyl chlorides.

5.3 Nickel-Catalyzed Arylations of Alkyl (Pseudo)Halides

Nickel catalysts show a high reactivity and a broad substrate scope in the arylation of alkyl halides. While only alkyl halides without β-hydrogens were used in the early examples of nickel-catalyzed arylations of alkyl halides, highly efficient arylations at the α-position of a *gem*-dimethyl-substituted quaternary center highlighted its synthetic potentials (Equation 5.14) [16, 17].

$$\text{Ph-CH}_2\text{C(CH}_3)_2\text{CH}_2\text{-I} + \text{BrMg-Ph or Ph}_2\text{Zn} \xrightarrow[\text{Et}_2\text{O, reflux}]{\text{NiCl}_2\text{(dppf) (10 mol\%)}} \text{Ph-CH}_2\text{C(CH}_3)_2\text{CH}_2\text{-Ph} \tag{5.14}$$

71% (with PhMgBr)
81% (with Ph_2Zn)

In 1998, Knochel reported that, in the presence of 4-trifluoromethylstyrene, $[Ni(acac)_2]$ efficiently catalyzed cross-couplings between polyfunctional arylzinc derivatives and alkyl halides possessing β-hydrogens (Equation 5.15). While the alkyl halides were limited to primary alkyl iodides, the scope of nickel catalysis was significantly expanded. The role of the electron-deficient olefin, 4-trifluoromethylstyrene, was proposed to accelerate the reductive elimination step by decreasing the electron density at the nickel center of an (alkyl)(aryl)nickel intermediate [18].

$$\text{FG-(CH}_2)_n\text{-I} + \text{BrZn-C}_6\text{H}_4\text{-FG'} \xrightarrow[\text{THF/NMP, }-15\text{ °C}]{\text{Ni(acac)}_2\text{ (10 mol\%), 4-CF}_3\text{C}_6\text{H}_4\text{CH=CH}_2\text{ (1 equiv)}} \text{FG-(CH}_2)_n\text{-C}_6\text{H}_4\text{-FG'} \tag{5.15}$$

Products: $BuCO(CH_2)_3$-C_6H_4-4-OMe, 72%; $EtO_2C(CH_2)_2$-C_6H_4-4-CN, 75%; 2-methyl-1,3-dithiolan-2-yl-$(CH_2)_2$-C_6H_4-4-CN, 80%; $PhS(CH_2)_3$-C_6H_4-3-CO_2Et, 75%

Kambe found that the selectivity could be improved by the use of conjugated diene additives in the nickel-catalyzed cross-couplings of primary alkyl halides or tosylates with various Grignard reagents [19]. Phenylmagnesium bromide can be used for arylations of primary alkyl bromides with $[NiCl_2]$ as catalyst and

1,3-butadiene as ligand, as shown in Equation 5.16. However, in the case of cross-couplings between phenylmagnesium bromide and ethyl tosylate to give ethylbenzene in 56% yield, a stoichiometric amount of 1,3-butadiene was necessary. The diene-based nickel catalyst is effective even for the coupling of alkyl fluorides with alkyl Grignard reagents. However, the reaction of alkyl fluorides with aryl metal species using the same catalytic system was not described in Kambe's report, and is yet to be reported [20].

$$\text{Oct–Br} + \underset{(1.3\ \text{equiv})}{\text{BrMg–Ph}} \xrightarrow[\text{THF, 25 °C, 3 h; 90\%}]{NiCl_2\ (3\ \text{mol\%}),\ \text{1,3-butadiene (30 mol\%)}} \text{Oct–Ph} \tag{5.16}$$

Further investigations into the nature of unsaturated hydrocarbon additives led Kambe and Terao to develop a tetraene-type ligand **12**, which could promote the cross-coupling between primary alkyl bromides or tosylates, and various diorganozinc reagents in the presence of a catalytic amount of [$NiCl_2$], as shown in Equation 5.17 [21]. It is noteworthy that the reaction required the addition of an excess $MgBr_2$ together with the nickel catalyst.

$$\text{Pr–Br} + \underset{(1.3\ \text{equiv})}{Ph_2Zn} \xrightarrow[MgBr_2\ (3.0\ \text{equiv}),\ \text{THF/NMP, 25 °C, 1 h; 86\%}]{NiCl_2\ (3\ \text{mol\%}),\ \mathbf{12}\ (9\ \text{mol\%})} \text{Pr–Ph} \tag{5.17}$$

Kambe and Terao proposed the following catalytic cycle shown in Scheme 5.2. First, a bis(η^3-allyl)nickel intermediate of type **A** is formed from an *in situ*-generated nickel(0) species and a conjugated diene or tetraene ligand. Further reaction with the carbanionic nucleophile then gives rise to formation of the nickelate **B**, which will undergo the subsequent facile oxidative addition and reductive elimination. Notably, bis(η^3-allyl)nickel and palladium complexes were reported to be good catalyst precursors for cross-couplings of alkyl electrophiles with alkyl or aryl Grignard reagents [22]. Although the detailed mechanism is still unclear, an involvement of an intermediate radical species has been experimentally ruled out based on a radical-clock study using cyclopropylmethyl bromide as an electrophile [19, 20].

Recently, bipyridines and amino alcohols, which were rather unique ligands for conventional nickel-catalyzed cross-coupling reactions, have been shown to be effective for nickel-catalyzed arylations of secondary alkyl halides. Fu reported that a combination of [Ni(cod)$_2$] and bathophenanthroline **13** catalyzed the cross-

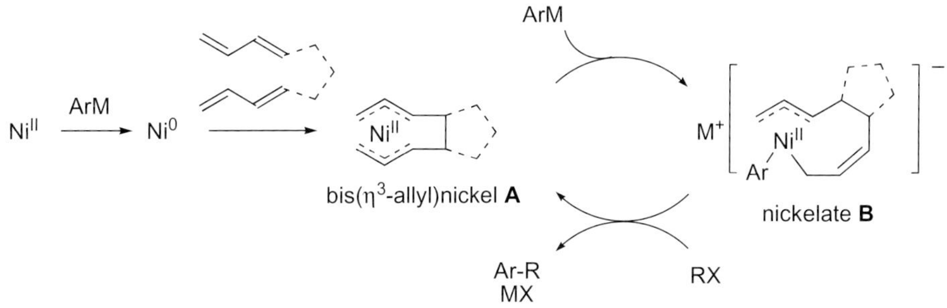

Scheme 5.2 A plausible reaction mechanism for the nickel/1,3-butadiene-catalyzed cross-coupling reaction.

couplings of secondary alkyl bromides and iodides with arylboronic acids. The halides were cleanly displaced by the boronic acid-derived aromatic moieties to provide a variety of tertiary alkyl compounds (Equation 5.18) [23]. The formation of byproducts stemming from isomerization or β-hydride elimination was not reported. The broad synthetic scope of the reaction is illustrated in Equation 5.18.

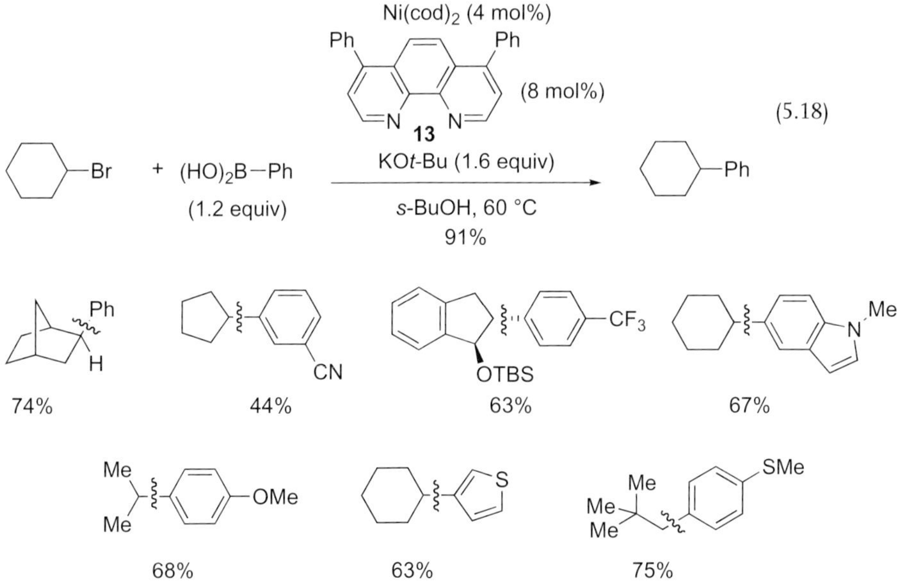

(5.18)

The bathophenanthroline-based nickel catalyst was also found to be applicable to cross-couplings between alkyl halides and arylsilanes. Thus, in the presence of excess CsF, the combination of [$NiBr_2$·diglyme] and bathophenanthroline **13** promoted arylations of secondary alkyl bromides and iodides with aryltrifluorosilanes (Equation 5.19) [24].

4-Bromocyclohexanone + F_3Si–(2-MeC$_6$H$_4$) (1.5 equiv) → [$NiBr_2$·diglyme (6.5 mol%), **13** (7.5 mol%), CsF (3.8 equiv); DMSO, 60 °C] 4-(2-methylphenyl)cyclohexanone, 82%

(5.19)

Fu also reported that the combination of $NiCl_2$ and 2,2′-bipyridine, instead of bathophenanthroline **13**, generates an effective catalyst for cross-couplings between secondary alkyl bromides and organostannanes. For example, a mixture of 2-adamantyl bromide and phenyltrichlorostannane was treated with $NiCl_2$ and 2,2′-bipyridine in the presence of KO*t*-Bu to give 2-phenyladamantane in 74% yield (Equation 5.20). When this catalyst system was applied to other structurally diverse secondary alkyl halides, ranging from acyclic to monocyclic and steroidal substrates, the corresponding products were obtained in high yields [25].

2-Adamantyl bromide + Cl_3Sn–Ph (1.2 equiv) → [$NiCl_2$ (10 mol%), 2,2'-bipyridyne (15 mol%), KO*t*-Bu (7.0 equiv); *t*-BuOH/*i*-BuOH, 60 °C] 2-phenyladamantane, 74%

(5.20)

Me(Hep)CH–Ph 72%; cyclohexyl–(2-MeC$_6$H$_4$) 47%; cycloheptyl–(4-FC$_6$H$_4$) 67%; cholesteryl–(4-MeOC$_6$H$_4$) (*i*-Bu, Me, Me, H, H, H, Me, OMe) 55%

In order to accomplish catalytic arylations of inert secondary alkyl chlorides, Fu and coworkers have developed highly active nickel catalysts, consisting of a nickel halide and a β-aminoalcohol. The cross-coupling between cyclohexyl chloride and phenylboronic acid, thus, takes place smoothly in the presence of a catalytic amount of [$NiCl_2$·dme] and (*S*)-(+)-prolinol, along with 2 equiv. of potassium hexamethyldisilazide (KHMDS), to provide cyclohexylbenzene in 80% yield (Equation 5.21) [26].

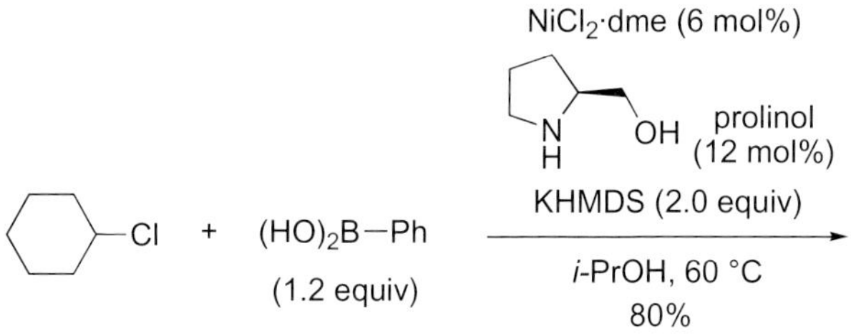

(5.21)

In the presence of [$NiCl_2$], norephedrine, and lithium hexamethyldisilazide (LHMDS), arylsilanes participate in the catalytic arylation of primary and secondary alkyl bromides (Equation 5.22). Interestingly, the addition of a catalytic amount of water was found critical here to obtain good chemical yields [27].

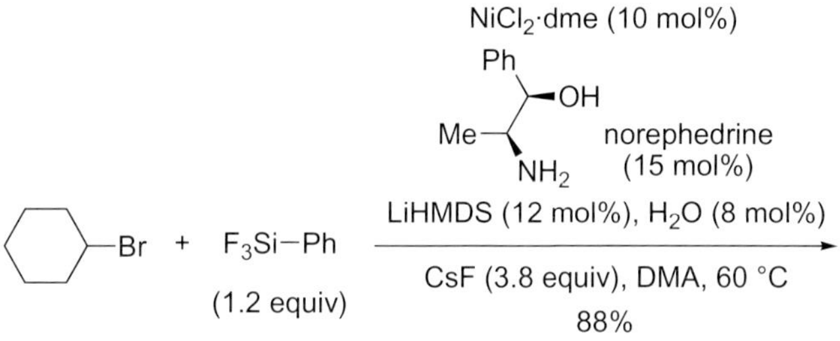

(5.22)

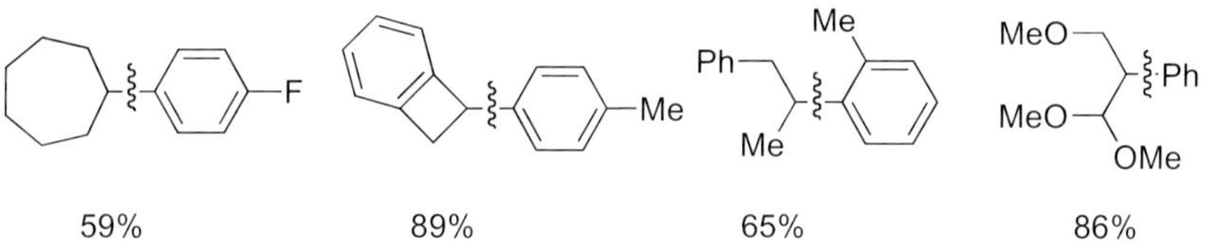

A new phosphine ligand, **14**, bearing a pentamethylcyclopentadienyl moiety, was recently prepared by Oshima and Yorimitsu for cross-couplings of unactivated alkyl bromides and iodides with aryl Grignard reagents (Equation 5.23). Primary alkyl bromides and iodides can be arylated in good yields, while secondary alkyl bromides (e.g. cyclohexyl bromide) were not converted to the corresponding alkylbenzenes in satisfactory yields. It should be noted that a radical clock study suggested the generation of a radical intermediate under these reaction conditions [28].

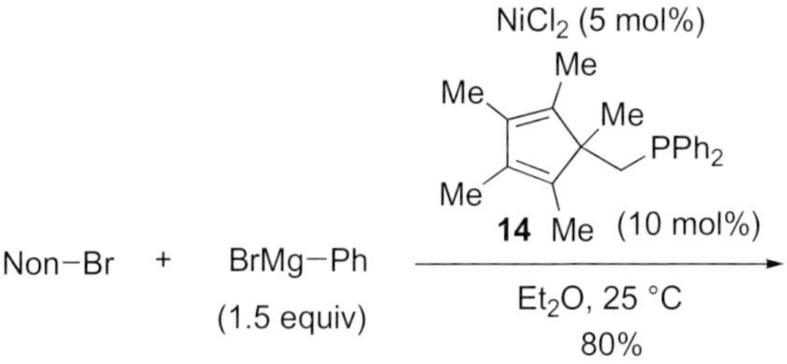

(5.23)

Some recent examples of cross-couplings of alkyl halides and pseudohalides with a variety of arylmetal reagents by the action of various nickel catalysts are summarized in Table 5.2. A quick glance at the table highlights the one limitation that is associated with the currently established nickel catalysts systems, namely that less-nucleophilic electron-deficient aryl groups can be introduced, but only in modest yields. In order to make arylation reactions using nickel catalysis more general, the current limitation of the substrate scope must be overcome, and therefore the development of new catalytic systems – and particularly new ligands, additives and combinations thereof – constitutes a future challenge.

5.4 Iron-Catalyzed Arylations of Alkyl (Pseudo)Halides

Despite the long history of iron-catalysis in cross-coupling reactions [29], its synthetic potential has not yet been fully investigated when compared to nickel and palladium catalyses. Studies aimed at controlling the selectivity and reactivity of iron catalysts have been in progress during the past two decades. Cahiez found polar additives, such as *N*-methylpyrrolidinone (NMP), to improve the reactivity and also the 'reliability' of iron-catalyzed cross-couplings of alkenyl halides with organomanganese and magnesium reagents [30, 31]. It should be noted here that the use of organomagnesium reagents was originally reported by Kochi and coworkers during the early days of metal-catalyzed cross-coupling reactions. Fürstner made another important observation with regards to the efficacy of NMP, namely that, in the presence of NMP, iron can catalyze the coupling of alkyl Grignard reagents with rather inert aryl chlorides as well as aryl sulfonates [32].

Significant advancements in iron-catalyzed cross-coupling reactions were made in 2004, when several research groups (including that of the present authors) reported independently that certain iron salts or complexes could catalyze cross-couplings between aryl Grignard reagents and alkyl halides possessing β-hydrogens. Because of the apparent practical advantages associated with the use of iron catalysts, a number of cross-coupling reactions using new iron catalysts have been developed recently. In this section we will summarize the recent progress in iron-catalyzed cross-coupling reactions which, it is hoped, will be helpful in the development of new methodologies for the arylation of unactivated alkyl halides.

In sharp contrast to the palladium- and nickel-catalyzed cross-couplings described above, iron catalysts show a high catalytic activity towards secondary alkyl halides, as reported by Nakamura and coworkers. Thus, cyclohexyl chloride, bromide or iodide can be arylated by a phenyl Grignard reagent in almost quantitative yield in the presence of $FeCl_3$ as catalyst, along with an excess amount of a diamine additive *N,N,N′,N′*-tetramethylethylenediamine (TMEDA). Primary alkyl halides, such as octyl chloride, bromide and iodide, can also be arylated by the action of the same iron catalyst. However, the yields are somehow decreased, especially when primary alkyl chlorides are used as electrophiles (Equations 5.24 and 5.25) [33].

Table 5.2 Nickel-catalyzed cross-coupling reactions between unactivated alkyl halides and aryl metal reagents.

$$\text{Alkyl–X} + \text{Ar–M} \xrightarrow[\text{solvent}]{\text{nickel catalyst}} \text{Alkyl–Ar}$$

Entry[a]	X	M	Nickel catalyst (or precatalyst)	Solvent	Examples[b]	Yield (%)[c]	Reference
1	I	MgBr	$NiCl_2$(dppf)	Et_2O	9	57–94	[16]
2	I	ZnAr	$NiCl_2$(dppf)	Et_2O	8	72–88	[17]
3	I	ZnBr	$Ni(acac)_2$/ *p*-trifluoromethylstyrene	THF/NMP	12	71–80	[18]
4	Br, OTs	MgBr	$NiCl_2$/ 1,3-butadiene	THF	2	56, 90	[19]
5	Br, I	$B(OH)_2$	$Ni(cod)_2$/ bathophenathroline	*s*-BuOH	11	44–90	[23]
6	Br	ZnAr	$NiCl_2$/ 1,3,8,10-tetraene	THF/NMP	1	86	[21]
7	Br, I	SiF_3	$NiBr_2 \cdot$diglyme/ bathophenathroline	DMSO	16	60–82	[24]
8	Br, I	$SnCl_3$	$NiCl_2$/2,2′-bipyridine	*t*-BuOH/ *s*-BuOH	7	47–83	[25]
9	Br, I	$B(OH)_2$	NiI_2/β-aminoalcohol	*i*-PrOH	18	66–97	[26]
10	Cl	$B(OH)_2$	$NiCl_2 \cdot$dme/ β-aminoalcohol	*i*-PrOH	7	65–87	[26]
11	Br, I	MgBr	$NiCl_2/Cp^*CH_2PPh_2$	Et_2O	12	36–84	[28]
12	OTs	MgBr	$Ni(\eta^3\text{-allyl})_2$	THF	1	87	[22]
13	Br, I	SiF_3	$NiCl_2 \cdot$dme/ β-aminoalcohol	DMA	9	59–94	[27]

a In the order of received date of the manuscripts.
b Based on the numbers of compounds prepared by the cross-coupling between unactivated alkyl halides and aryl metals.
c Reported yields.

Cyclohexyl–X + BrMg–Ph → Cyclohexyl–Ph (5.24)

X = Cl, Br, I; BrMg–Ph (1.2–1.5 equiv), *slow addition*

$FeCl_3$ (5 mol%), TMEDA (1.2–1.5 equiv), THF, 0–25 °C

99% (X = Cl)
99% (X = Br)
99% (X = I)

Oct–X + BrMg–Ph → Oct–Ph (5.25)

X = Cl, Br, I; BrMg–Ph (1.2–1.5 equiv), *slow addition*

$FeCl_3$ (5 mol%), TMEDA (1.2–1.5 equiv), THF, 0–40 °C

45% (X = Cl)
91% (X = Br)
97% (X = I)

The use of an excess amount of TMEDA can be avoided employing [Fe(acac)$_3$] as catalyst and an ether as a solvent, at elevated temperatures. Hayashi reported that a variety of aryl groups, including the bulky mesityl group, can be introduced to primary and secondary alkyl halides. In this case, the yields of the coupling products are slightly diminished due to the formation of the byproducts alkene and alkane (Equation 5.26). Notably, chemoselective arylation was possible under these reaction conditions, which meant that the primary alkyl halide had reacted while the triflate group of the aromatic ring had remained intact. This product can serve as an electrophile for further iron-catalyzed cross-coupling with alkyl Grignard reagents in the presence of NMP [34].

Hex–CH(Me)–Br + BrMg–C_6H_4–Me (2.0 equiv) → Hex–CH(Me)–C_6H_4–Me (5.26)

$Fe(acac)_3$ (5 mol%), Et_2O, reflux, 73%

Oct–C_6H_4–OMe: 73%
Oct–C_6H_4–F: 60%
Oct–$C_6H_2Me_3$: 60%
TfO–C_6H_4–$CH_2CH_2CH_2$–C_6H_4–Me: 69%

Fürstner reported that a low-valent iron(–II) complex catalyzed the cross-couplings of primary and secondary alkyl halides with various aryl Grignard reagents under such mild reaction conditions that chemoselective arylations would be feasible in the presence of potentially reactive functional groups, such as carbonyl, cyano and isocyanato groups. While alkyl bromides and iodides showed excellent reactivities with this iron catalyst system, alkyl chlorides were found to be inert under the reported reaction conditions (Equation 5.27) [35].

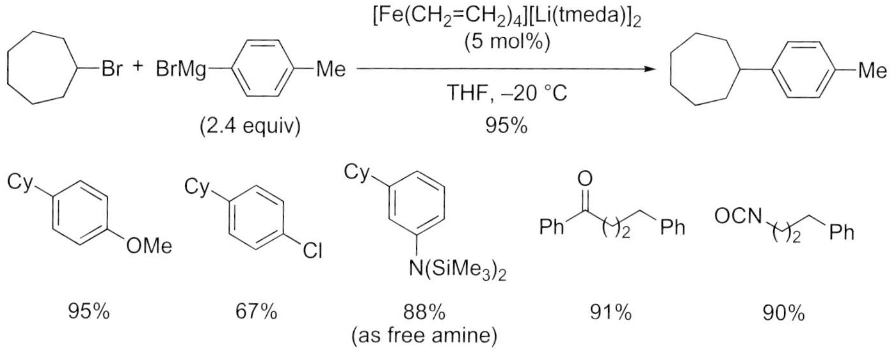

(5.27)

An iron–salen complex can also catalyze the arylation of alkyl halides with organomagnesium compounds. Bedford and coworkers reported that the cross-coupling of primary and secondary alkyl halides with aryl Grignard reagents takes place in the presence of the iron complex **15** to give the corresponding alkyl-substituted arenes in good yields (Equation 5.28) [36].

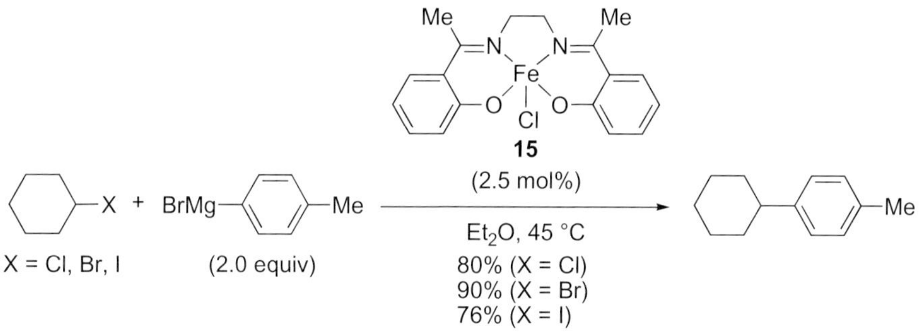

(5.28)

Bedford also found that a catalytic amount of $FeCl_3$ and simple amines, such as Et_3N or DABCO, can affect the arylation of alkyl halides in diethyl ether at elevated temperature (Equation 5.29). Treatment of bromomethylcyclopropane with phenylmagnesium bromide in the presence of $FeCl_3$ and DABCO provided the rearranged acyclic product, suggesting the involvement of an alkyl radical intermediate (Equation 5.30) [37].

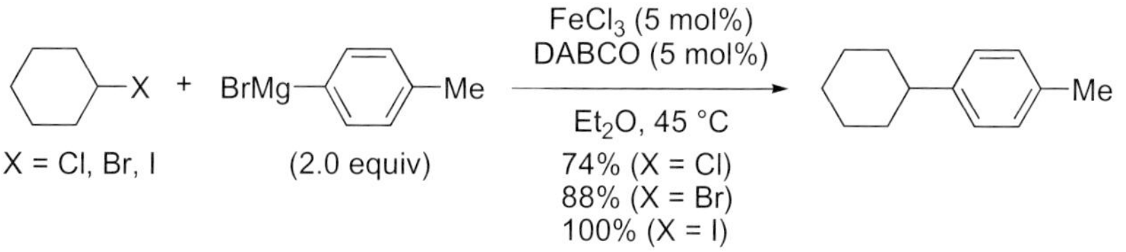

(5.29)

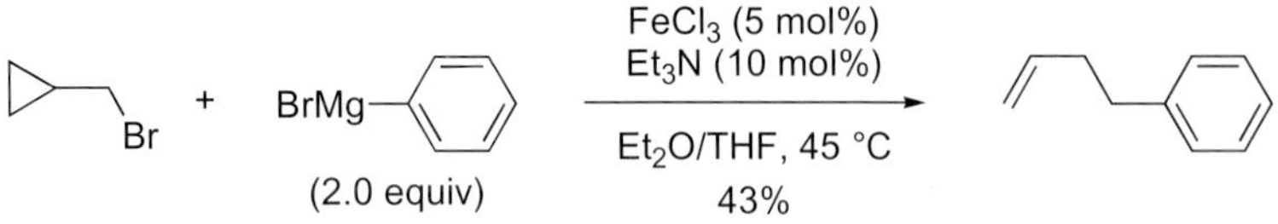

(5.30)

A wide range of ligands, including monodentate or bidentate phosphines, phosphites, arsines and NHCs, have been examined by Bedford in iron-catalyzed cross-couplings between alkyl halides and aryl Grignard reagents. Notably, various of these ligands exerted a beneficial effect on selectivity and chemical yield (Equation 5.31) [38].

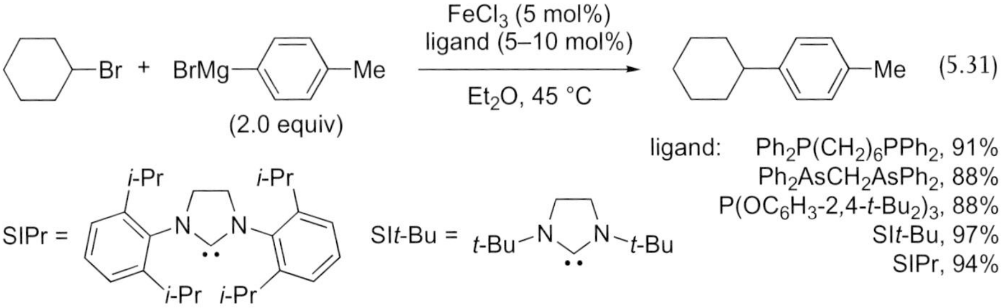

(5.31)

Gaertner reported an interesting recyclable iron catalyst system using imidazolium-derived ionic liquid **16** as the reaction medium. This liquid possesses a ferrate counteranion, which serves as a catalyst for cross-couplings between alkyl bromides and arylmagnesium bromides. As shown in Equation 5.32, the arylation of dodecyl bromide can be performed in this recyclable catalyst system, without losing significant activity even at a fifth cycle [39].

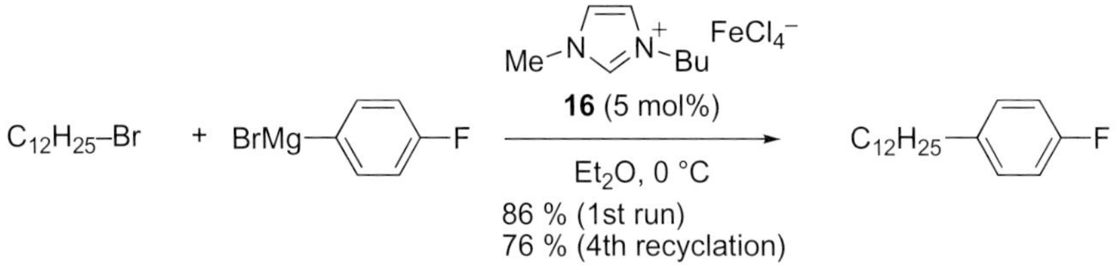

(5.32)

It should be noted here that iron can catalyze the cross-coupling of arylmagnesium reagents with alkyl chloride, bromides and iodides even in the absence of supporting ligands. Thus, Bedford revealed that 'iron nanoparticles', produced by the reduction of $FeCl_3$ with organomagnesium compounds in the presence of polyethylene glycol (PEG 14000), can catalyze the arylation of cyclohexyl chloride in 77% yield (Equation 5.33) [40].

'iron nanoparticles' (5 mol%); Et_2O, 45 °C; 77%

Cl + BrMg—C6H4—Me (2.0 equiv) → C6H11—C6H4—Me

(5.33)

Very recently, Cahiez reported a combination of catalytic amounts of TMEDA and hexamethylenetetramine (**17**), instead of an excess TMEDA alone, for selective cross-couplings between primary as well as secondary alkyl halides and aryl Grignard reagents (Equation 5.34). Furthermore, $[(FeCl_3)_2(tmeda)_3]$ was found to be an effective precatalyst for these transformations (Equation 5.35) [41].

$Fe(acac)_3$ (5 mol%); **17** (5 mol%); TMEDA (10 mol%); THF, 0 °C; 88%

Me(Et)CH—Br + BrMg—Ph → Me(Et)CH—Ph (5.34)

$[(FeCl_3)_2(TMEDA)_3]$ (1.5 mol%); THF, 20 °C; 78%

Me(Et)CH—Br + BrMg—Ph → Me(Et)CH—Ph (5.35)

Iron-catalyzed cross-coupling reactions between alkyl halides and aryl Grignard reagents are summarized in Table 5.3. In contrast to palladium- or nickel-catalyzed arylation transformations, iron-catalyzed reactions currently suffer from a rather limited scope of organometallic nucleophiles that can be used. In most cases, magnesium-based metal reagents are used as nucleophiles, and hence, the functional group compatibility is still limited. Arylations of alkyl halides using organozinc reagents were reported by Nakamura and coworkers. Arylzinc compounds, possessing some functional groups, can be coupled with alkyl halides in the presence of an excess TMEDA (Equations 5.36 and 5.37) or catalytic amounts of diphenylphosphinobenzene.

$FeCl_3$ (5 mol%); TMEDA (2.0 equiv); MgX_2 (X = Cl, Br; 2.0 equiv); THF, 50 °C; 90%

AcO, AcO, AcO, OMe, I, O + Zn(benzodioxolyl)$_2$ (2.0 equiv) → AcO, AcO, AcO, OMe, O

(5.36)

Table 5.3 Iron-catalyzed cross-coupling reactions between unactivated alkyl halides and aryl magnesium reagents.

$$\text{Alkyl–X} + \text{Ar–MgBr} \xrightarrow[\text{solvent}]{\text{iron catalyst}} \text{Alkyl–Ar}$$

Entry[a]	X	Iron catalyst (or precatalyst)	Solvent	Examples[b]	Yield (%)[c]	Reference
1	Cl, Br, I, OTs	$FeCl_3$/TMEDA (excess)	THF	14	45–99	[33], [42][d]
2	Cl, Br, I	$Fe(acac)_3$	Et_2O	9	32–73	[34]
3	Br, I	$[Fe(CH_2{=}CH_2)_4][Li(tmeda)]_2$	THF	17	61–98	[35]
4	Cl, Br, I	FeCl(salen)	THF	7	56–90	[36]
5	Cl, Br, I	$FeCl_3$/DABCO	Et_2O	7	40–100	[37]
6	Cl, Br	$FeCl_3$/NHC	Et_2O	3	71–89	[38]
7	Cl, Br	$FeCl_3/PCy_3$ or DPPHex	Et_2O	7	31–85	[38]
8	Cl, Br	$FeCl_3/[P(OC_6H_3\text{–}2,4\text{-}t\text{-}Bu_2)_3]$	Et_2O	7	47–88	[38]
9	Cl, Br	$FeCl_3$/arsine	Et_2O	1	82	[38]
10	Cl, Br, I	bmim-$FeCl_4$	Et_2O	8	20–89	[39]
11	Cl, Br, I	Iron nanoparticle	Et_2O	11	30–94	[40]
12	Br, I	$Fe(acac)_3$/TMEDA/HMTA	THF	20	39–94	[41]
13	Br	$(FeCl_3)_2(tmeda)_3$	THF	5	75–92	[41]

a In the order of received date of the manuscripts.
b Based on the numbers of compounds prepared by the cross-coupling between unactivated alkyl halides and aryl metals.
c Reported yields.
d Grignard reagents were transmetallated to zinc reagents. Arylzinc halides were also used.

PivO-cyclohexyl-Br + $TMSCH_2Zn$–C_6H_4–CO_2Et (2.0 equiv) $\xrightarrow[\text{THF, 30 °C; 78\%}]{FeCl_3\text{ (5 mol\%)}, \text{ TMEDA (2.0 equiv)}, MgX_2 \text{ (X = Cl, Br; 2.0 equiv)}}$ PivO-cyclohexyl–C_6H_4–CO_2Et (5.37)

It should be noted that addition of Lewis acidic salts, such as $MgBr_2$, is critical in order to achieve an effective catalytic transformation when using arylzinc compounds. This observation indicates that the difficult step of the catalytic cycle is the transmetallation of the aryl group from the zinc reagent to the catalytically active iron complex [42]. While the involvement of an intermediate radical species or a single electron-transfer process is suspected, mechanistic details of these iron-catalyzed cross-coupling reactions remain unclear.

5.5 Copper- and Cobalt-Catalyzed Arylations of Alkyl (Pseudo)Halides

The first catalytic arylation of an alkyl sulfonate was reported by Schlosser and coworkers during the early 1970s, when they showed that a primary alkyl tosylate could be coupled with an alkyl or an aryl Grignard reagent in the presence of catalytic amounts of [Li_2CuCl_4] [43, 44]. The scope of the reaction was considerably expanded by Burns and coworkers in 1997, who reported that the use of a catalytic system consisting of CuBr, LiSPh and LiBr allowed efficient cross-couplings of primary and secondary alkyl tosylates with Grignard reagents. For example, 1,3-bis(tosyloxy)propane was arylated with phenylmagnesium bromide to give 1,3-diphenylpropane in good yield (Equation 5.38) [45, 46].

$$\text{TsO}\diagup\!\!\diagdown\!\!\diagup\text{OTs} + \underset{(2.0\ \text{equiv})}{\text{BrMg–Ph}} \xrightarrow[\text{THF/HMPA, 25 °C; 73\%–76\%}]{\text{CuBr·SMe}_2\ (6\ \text{mol\%}),\ \text{LiBr}\ (6\ \text{mol\%}),\ \text{LiSPh}\ (6\ \text{mol\%})} \text{Ph}\diagup\!\!\diagdown\!\!\diagup\text{Ph} \quad (5.38)$$

Cahiez reported that primary alkyl bromides could be phenylated using phenylmanganese chloride or phenylmagnesium chloride in the presence of a catalytic amount of [Li_2CuCl_4]. It is interesting to note that the polar cosolvent NMP exerted an adverse effect on the *arylation*, giving the arylated products in diminished chemical yields (Equations 5.39 and 5.40) [47, 48].

$$\text{Non–Br} + \underset{(1.2\ \text{equiv})}{\text{ClMn–Ph}} \xrightarrow[\text{THF, 20 °C; 86\% (with NMP: 33\%)}]{\text{Li}_2\text{CuCl}_4\ (3\ \text{mol\%})} \text{Non–Ph} \quad (5.39)$$

$$\text{Oct–Br} + \underset{(1.4\ \text{equiv})}{\text{ClMg–Ph}} \xrightarrow[\text{THF, 20 °C; 70\% (with NMP: 5\%)}]{\text{Li}_2\text{CuCl}_4\ (3\ \text{mol\%})} \text{Oct–Ph} \quad (5.40)$$

In 2003, Kambe reported nickel- and copper-catalyzed cross-couplings between primary alkyl fluorides and organomagnesium compounds. While the addition of a 1,3-butadiene ligand improved the reactions with various alkyl Grignard reagents, cross-couplings with phenyl Grignard reagents proceeded well in the absence of the diene ligand [20]. Hence, phenylation of octyl fluoride was accomplished in the presence of 2 mol% of [$CuCl_2$] to give octylbenzene in 99% yield (Equation 5.41). The reactivity of alkyl halides was found to decrease in the order of fluoride > bromide >> chloride. The origin of the high reactivity of alkyl fluorides and the mechanistic details are still unclear.

$$\text{Oct-F} + \underset{(1.5\ \text{equiv})}{\text{BrMg-Ph}} \xrightarrow[\text{THF, 67°C}\ 99\%]{\text{CuCl}_2\ (2\ \text{mol\%})} \text{Oct-Ph} \quad (5.41)$$

Copper-catalyzed arylations of alkyl chlorides have been achieved using 1-phenylpropyne as an additive. Terao and Kambe recently found that cross-couplings of nonyl chloride, fluoride and mesylate with phenyl Grignard reagent proceeded in THF, when using catalytic amounts of [$CuCl_2$] and 1-phenylpropine, to give nonylbenzene in excellent yields at elevated temperatures. A radial-clock study as well as a reaction with a D-labeled stereochemical probe revealed that the copper-catalyzed reaction proceeded via an S_N2-type mechanism (Equations 5.42 and 5.43). This method also proved applicable to arylations of a primary alkyl mesylate [49].

(5.42): cyclopropylmethyl–Cl + BrMg–Ph (1.5 equiv) → [$CuCl_2$ (2 mol%), Ph–≡–Me (10 mol%); THF, 68°C; 98%] cyclopropylmethyl–Ph

(5.43): adamantyl–CH(D)–CH(D)–Cl (D H / H D) → [BrMg–**Ph**, *cat.* [Cu]; THF, 68°C; 17%] **inversion (major)** + retention; *ca.* 10:1

Although *cobalt catalysis* is somewhat less developed compared to other transition metal catalysis for cross-couplings, but has shown rather unconventional, novel reactivity profiles [50]. For example, Oshima reported a cobalt-catalyzed tandem radical cyclization/cross-coupling reaction between an aryl Grignard reagent and an alkyl halide bearing an ω-alkenyl group (Equation 5.44) [51].

(5.44): BuO–CH(CH₂Br)–O–C(Me)(Me)–CH=CH₂ + BrMg–Ph (2.2 equiv) → [CoCl₂(dppe) (10 mol%); THF, 0 °C; 84%] BuO-substituted tetrahydrofuran (Me Me) bearing CH₂Ph

p-Tol–N pyrrolidine–CH₂Ph: 81% (from iodide)

cyclopentyl–CH₂–C₆H₄–OMe: 59% (from iodide)

Pent / BuO tetrahydrofuran–CH₂–thienyl (S): 63%

Pent / BuO tetrahydrofuran–CH₂–C₆H₄–CF₃: 65%

The phosphine ligand-based cobalt catalyst can be also applied to cross-couplings of allyl Grignard reagents with various unactivated alkyl halides. Tertiary alkyl bromides, possessing multiple β-hydrogens, underwent allylation reactions in the presence of catalytic amounts of $CoCl_2$ and dppp to give the products in fair to good yields [52]. This catalytic system has been proven effective for benzylation, as well as for tandem cyclization/coupling reactions [53, 54]. Yorimitsu and Oshima recently reported the cross-couplings of primary and secondary alkyl bromides and iodides with aryl Grignard reagents, proceeding smoothly at ambient temperature in the presence of catalytic amounts of [$CoCl_2$] and *trans*-1,2-bis(dim ethylamino)cyclohexane (Equation 5.45).

(X = Br, I) + BrMg–Ph (3.0 equiv) → [$CoCl_2$ (5 mol%), NMe_2/NMe_2 (6 mol%), THF, 25 °C; 95% (X = Br), 95% (X = I)] → –Ph (5.45)

Highly stereoselective arylation reactions were reported, as shown in Equation 5.46. With an optically active diamine ligand, a modest asymmetric induction was observed (Equation 5.47). An asymmetric synthesis of a synthetic prostaglandin AH13205 was accomplished using the diastereoselective cobalt-catalyzed cyclization/arylation sequence as key step [55].

+ BrMg–Ar → [$CoCl_2$ (5 mol%), NMe_2/NMe_2 (12 mol%), THF, 25 °C] (5.46)

87% yield, 98% *anti* (Ar = Ph)
82% yield, 98% *anti* (Ar = 1-Naphthyl)
91% yield, 98% *anti* (Ar = 2-Naphthyl)
84% yield, 98% *anti* (Ar = 1-Pyridyl)

rac. –Br + BrMg–Ph → [same as in eq 45; 40% (22% ee)] → –Ph (5.47)

Table 5.4 Copper-catalyzed cross-coupling reactions between unactivated alkyl halides and aryl metal reagents.

$$\text{Alkyl–X} + \text{Ar–M} \xrightarrow[\text{solvent}]{\text{copper catalyst}} \text{Alkyl–Ar}$$

Entry[a]	X	M	Copper catalyst (or precatalyst)	Solvent	Examples[b]	Yield (%)[c]	Reference
1	OTs	MgBr	Li_2CuCl_4	THF/Et_2O	1	90	[43]
2	Br	MnCl	Li_2CuCl_4	THF	1	86	[47]
3	OTs	MgBr	$CuBr \cdot SMe_2/LiBr/LiSPh$	THF/HMPA	4	70–94	[45]
4	Br	MgBr	Li_2CuCl_4	THF	1	70	[48]
5	F, Cl, Br	MgBr	$CuCl_2$	THF	3	42–99	[20]
6	F, Cl, OMs	MgBr	$CuCl_2$/1-phenylpropyne	THF	5	82–98	[49]

a In the order of received date of the manuscripts.
b Based on the numbers of compounds prepared by the cross-coupling between unactivated alkyl halides and aryl metals.
c Reported yields.

Table 5.5 Cobalt-catalyzed cross-coupling (and tandem cyclization/arylation) reactions between unactivated alkyl halides and aryl metal reagents.

$$\text{Alkyl–X} + \text{Ar–M} \xrightarrow[\text{solvent}]{\text{cobalt catalyst}} \text{Alkyl–Ar}$$

Entry[a]	X	M	Cobalt catalyst (or precatalyst)	Solvent	Examples[b]	Yield (%)[c]	Reference
1	Br, I	MgBr	$CoCl_2$(dppe)	THF	13[d]	22–84	[51]
2	Br, I	MgBr	$CoCl_2$(dppe) or $CoCl_2$/DPPP	THF	9	<1–67	[54]
3	Cl, Br, I	MgBr	$CoCl_2$/diamine	THF	28[e]	10–99	[55]

a In the order of received date of the manuscripts.
b Based on the numbers of compounds prepared by the cross-coupling between unactivated alkyl halides and aryl metals.
c Reported yields.
d The arylation reaction is associated with a cyclization of the alkyl halide substrates bearing an ω-alkenyl moiety.
e Including four examples of the tandem cyclization/arylation reactions.

Copper- and cobalt-catalyzed arylations of unactivated alkyl halides and pseudohalides are summarized in Tables 5.4 and 5.5, respectively. Despite the long history and popular applications of organocopper compounds in the substitution reactions of alkyl electrophiles, its full synthetic potential has not yet been tapped, as exemplified by the development of novel alkyne ligands. Additional investigations into the use of ancillary ligands may facilitate further developments of copper-catalyzed coupling reactions. Cobalt catalysts also show interesting reactivity profiles, the reactions proceeding through organic radical intermediates, which were nicely applied to radical cyclization-coupling cascade reactions.

Abbreviations

acac	acetylacetonato
BBN	borabicyclo[3.3.1]nonyl
Bu	butyl
cod	1,5-cyclooctadiene
Cy	cyclohexyl
Cyp	cyclopentyl
dppe	1,2-bis(diphenylphosphino)ethane
dppf	1,1′-bis(diphenylphosphino)ferrocene
DABCO	1,4-diazabicyclo[2.2.2]octane
Dec	decyl
DME	1,2-dimethoxyethane
DMF	*N*,*N*-dimethylformamide
Et	ethyl
Hep	heptyl
Hex	hexyl
HMDS	1,1,1,3,3,3-hexamethyldisilazide
HMPA	hexamethylphosphoric triamide
I*t*-Bu	*N*,*N*′-di(*t*-butyl)imidazol-2-ylidene
IMes	*N*,*N*′-dimesitylimidazol-2-ylidene
IPr	*N*,*N*′-bis(2,6-diisopropylphenyl)imidazol-2-ylidene
Me	methyl
NMI	*N*-methylimidazole
NMP	*N*-methyl-2-pyrrolidone
Non	nonyl
Oct	octyl
PEG	polyethylene glycol
Ph	phenyl
Pr	propyl
TMEDA	*N*,*N*,*N*′,*N*′-tetramethylethylenediamine
TMS	trimethylsilyl
Tol	methylphenyl
Ts	*p*-toluenesulfonyl

References

1 (a) de Meijere, A. and Diederich, F. (eds) (2004) *Metal-Catalyzed Cross-Coupling Reactions*, Wiley-VCH Verlag GmbH, Weinheim;
(b) Tamao, K. (2002) *J. Organomet. Chem.*, **653**, 23–6.

2 (a) Lipshutz, B. (2002) *Organometallics in Synthesis: A Manual*, 2nd edn (ed. M. Schlosser), John Wiley & Sons, Ltd, Chichester;
(b) Lipshutz, B.H. and Sengupta, S. (1992) Organocopper reagents: substitution, conjugate addition, carbo/metallocupration, and other reactions, in *Organic Reactions*, Vol. **41**, John Wiley & Sons, Inc., New York, pp. 135–631.

3 Frisch, A.C. and Beller, M. (2005) *Angew. Chem. Int. Ed. Engl.*, **44**, 674–88.

4 Castle, P.L. and Widdowson, D.A. (1986) *Tetrahedron Lett.*, **27**, 6013–16.

5 Yuan, K. and Scott, W.J. (1989) *Tetrahedron Lett.*, **30**, 4779–82.
6 Ishiyama, T., Abe, S., Miyaura, N. and Suzuki, A. (1992) *Chem. Lett.*, 691–4.
7 Charette, A.B. and Giroux, A. (1996) *J. Org. Chem.*, **61**, 8718–9.
8 Frisch, A.C., Shaikh, N., Zapf, A. and Beller, M. (2002) *Angew. Chem. Int. Ed. Engl.*, **41**, 4056–9.
9 Frisch, A.C., Rataboul, F., Zapf, A. and Beller, M. (2003) *J. Organomet. Chem.*, **687**, 403–9.
10 Kirchhoff, J.H., Netherton, M.R., Hills, I.D. and Fu, G.C. (2002) *J. Am. Chem. Soc.*, **124**, 13662–3.
11 Netherton, M.R. and Fu, G.C. (2002) *Angew. Chem. Int. Ed. Engl.*, **41**, 3910–12.
12 Lee, J.-Y. and Fu, G.C. (2003) *J. Am. Chem. Soc.*, **125**, 5616–7.
13 Zhou, J. and Fu, G.C. (2003) *J. Am. Chem. Soc.*, **125**, 12527–30.
14 Terao, J., Naitoh, Y., Kuniyasu, H. and Kambe, N. (2003) *Chem. Lett.*, **32**, 890–1.
15 Tang, H., Menzel, K. and Fu, G.C. (2003) *Angew. Chem. Int. Ed. Engl.*, **42**, 5079–82.
16 Yuan, K. and Scott, W.J. (1991) *Tetrahedron Lett.*, **32**, 189–92.
17 Park, K., Yuan, K. and Scott, W.J. (1993) *J. Org. Chem.*, **58**, 4866–70.
18 Giovannini, R. and Knochel, P. (1998) *J. Am. Chem. Soc.*, **120**, 11186–7.
19 Terao, J., Watanabe, H., Ikumi, A., Kuniyasu, H. and Kambe, N. (2002) *J. Am. Chem. Soc.*, **124**, 4222–3.
20 Terao, J., Ikumi, A., Kuniyasu, H. and Kambe, N. (2003) *J. Am. Chem. Soc.*, **125**, 5646–7.
21 Terao, J., Todo, H., Watanabe, H., Ikumi, A., Kuniyasu, H. and Kambe, N. (2004) *Angew. Chem. Int. Ed. Engl.*, **43**, 6180–2.
22 Terao, J., Naitoh, Y., Kuniyasu, H. and Kambe, N. (2007) *Chem. Commun.*, 825–7.
23 Zhou, J. and Fu, G.C. (2004) *J. Am. Chem. Soc.*, **126**, 1340–1.
24 Powell, D.A. and Fu, G.C. (2004) *J. Am. Chem. Soc.*, **126**, 7788–9.
25 Powell, D.A. and Fu, G.C. (2005) *J. Am. Chem. Soc.*, **127**, 510–1.
26 González-Bobes, F. and Fu, G.C. (2006) *J. Am. Chem. Soc.*, **128**, 5360–1.
27 Strotman, N.A., Sommer, S. and Fu, G.C. (2007) *Angew. Chem. Int. Ed. Engl.*, **46**, 3556–8.
28 Uemura, M., Yorimitsu, H. and Oshima, K. (2006) *Chem. Commun.*, 4726–7.
29 (a) Tamura, M. and Kochi, J. (1971) *J. Am. Chem. Soc.*, **93**, 1487–9;
(b) Kochi, J.K. (2002) *J. Organomet. Chem.*, **653**, 11–9;
(c) Fürstner, A. and Martin, R. (2005) *Chem. Lett.*, **34**, 624–9.
30 Cahiez, G. and Marquais, S. (1996) *Tetrahedron Lett.*, **37**, 1773–6.
31 Cahiez, G. and Avedissian, H. (1998) *Synthesis*, 1199–205.
32 (a) Fürstner, A. and Leitner, A. (2002) *Angew. Chem. Int. Ed. Engl.*, **41**, 609–12;
(b) Fürstner, A., Leitner, A., Méndez, M. and Krause, H. (2002) *J. Am. Chem. Soc.*, **124**, 13856–63.
33 Nakamura, M., Matsuo, K., Ito, S. and Nakamura, E. (2004) *J. Am. Chem. Soc.*, **126**, 3686–7.
34 Nagano, T. and Hayashi, T. (2004) *Org. Lett.*, **6**, 1297–9.
35 Martin, R. and Fürstner, A. (2004) *Angew. Chem. Int. Ed. Engl.*, **43**, 3955–7.
36 Bedford, R.B., Bruce, D.W., Frost, R.M., Goodby, J.W. and Hird, M. (2004) *Chem. Commun.*, 2822–3.
37 Bedford, R.B., Bruce, D.W., Frost, R.M. and Hird, M. (2005) *Chem. Commun.*, 4161–3.
38 Bedford, R.B., Betham, M., Bruce, D.W., Danopoulos, A.A., Frost, R.M. and Hird, M. (2006) *J. Org. Chem.*, **71**, 1104–10.
39 Bica, K. and Gaertner, P. (2006) *Org. Lett.*, **8**, 733–5.
40 Bedford, R.B., Betham, M., Bruce, D.W., Davis, S.A., Frost, R.M. and Hird, M. (2006) *Chem. Commun.*, 1398–400.
41 Cahiez, G., Habiak, V., Duplais, C. and Moyeux, A. (2007) *Angew. Chem. Int. Ed. Engl.*, **46**, 4364–6.
42 Nakamura, M., Ito, S., Matsuo, K. and Nakamura, E. (2005) *Synlett*, 1794–8.
43 Fouquet, G. and Schlosser, M. (1974) *Angew. Chem. Int. Ed. Engl.*, **13**, 82–3.
44 Seki, M. and Mori, K.This method was applied for the syntheses of some natural product, see: (a) (2001) *Eur. J. Org. Chem.*, 3797–809;
(b) Organ, M.G. and Wang, J. (2003) *J. Org. Chem.*, **68**, 5568–74;

(c) Baradi, P.T., Zarbin, P.H.G., Vieira, P.C. and Corrêa, A.G. (2002) *Tetrahedron: Asymmetry*, **13**, 621–4;
(d) Moreira, J.A. and Corrêa, A.G. (2003) *Tetrahedron: Asymmetry*, **14**, 3787–95.

45 Burns, D.H., Miller, J.D., Chan, H.-K. and Delaney, M.O. (1997) *J. Am. Chem. Soc.*, **119**, 2125–33.

46 Tamagawa, H., Takikawa, H. and Mori, K. This methodology has been used in the syntheses of natural products, see: (1999) *Eur. J. Org. Chem.*, 973–8.

47 Cahiez, G. and Marquais, S. (1993) *Synlett*, 45–7.

48 Cahiez, G., Chaboche, C. and Jézéquel, M. (2000) *Tetrahedron*, **56**, 2733–7.

49 Terao, J., Todo, H., Begum, S.A., Kuniyasu, H. and Kambe, N. (2007) *Angew. Chem. Int. Ed. Engl.*, **46**, 2086–9.

50 (a) Avedissian, H. and Cahiez, G. (1998) *Tetrahedron Lett.*, **39**, 6159–62;
(b) Avedissian, H., Bérillon, L., Cahiez, G. and Knochel, P. (1998) *Tetrahedron Lett.*, **39**, 6163–6.

51 Wakabayashi, K., Yorimtsu, H. and Oshima, K. (2001) *J. Am. Chem. Soc.*, **123**, 5374–5.

52 Tsuji, T., Yorimitsu, H. and Oshima, K. (2002) *Angew. Chem. Int. Ed. Engl.*, **41**, 4137–9.

53 Ohmiya, H., Tsuji, T., Yorimitsu, H. and Oshima, K. (2004) *Chem. Eur. J.*, **10**, 5640–8.

54 Ohmiya, H., Wakabayashi, K., Yorimitsu, H. and Oshima, K. (2006) *Tetrahedron*, **62**, 2207–13.

55 Ohmiya, H., Yorimitsu, H. and Oshima, K. (2006) *J. Am. Chem. Soc.*, **128**, 1886–9.

6
Arylation Reactions of Alkynes: The Sonogashira Reaction

Mihai S. Viciu and Steven P. Nolan

6.1
Introduction

Over the past three decades, the prevalence of cross-coupling reactions in both academia and industrial applications has been second to none. It is quite easy for a newcomer in this area of research to have difficulties in understanding the history of the various named, as well as unnamed, reactions. These reactions can simply be systematically organized by the type of C—C bond being formed. The Sonogashira reaction is defined as being an C(sp^2)—C(sp) bond-forming reaction between an aryl (pseudo)halide and an alkyne, which might occur in the presence of additives, such as copper salts.

In 1975, three different protocols were available in the literature, each describing the synthesis of internal alkynes. Cassar described palladium- or nickel-mediated reactions between aryl or vinyl halides and alkynes complexes with phosphine as ligands in the presence of NaOMe [1]. As a second protocol, Heck published a variation of the Mizoroki–Heck couplings, in which the olefins were replaced by alkynes and coupled with (hetero)aryl, as well as alkenyl bromides or iodides at 100 °C in the presence of a basic amine [2]. More than a decade earlier, Stephens and Castro had described the details of a palladium-free coupling of aryl iodides with cuprous acetylides in refluxing pyridine [3].

The third publication disclosed by Sonogashira and Hagihara described a protocol that can be viewed as a significant improvement of the previously reported methodologies through their elegant combination (Scheme 6.1). In fact, small quantities of CuI added to palladium-mediated couplings of alkynes with aryl-halides increased the yield, and allowed reactions to be conducted at ambient temperature [4].

The convenience of this methodology promoted it as being the most popular protocol for the synthesis of internal alkynes, which is itself (as well as in different variations) closely associated with Sonogashira's name. For instance, couplings involving silver instead of copper salts as additives, as well as copper-free protocols, are also considered Sonogashira reactions. Further, the formation of C(sp^3)—C(sp)

Modern Arylation Methods. Edited by Lutz Ackermann

ISBN: 978-3-527-31937-4

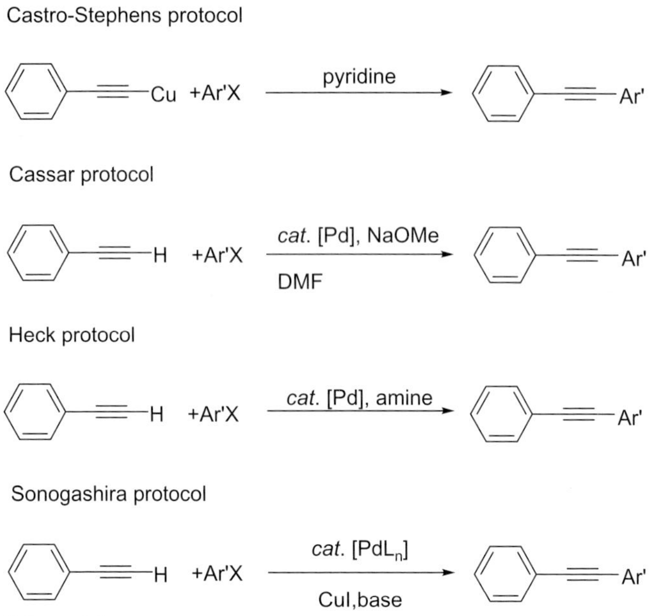

Castro-Stephens protocol

Cu +Ar'X — pyridine → Ar'

Cassar protocol

H +Ar'X — *cat.* [Pd], NaOMe / DMF → Ar'

Heck protocol

H +Ar'X — *cat.* [Pd], amine → Ar'

Sonogashira protocol

H +Ar'X — *cat.* $[PdL_n]$ / CuI,base → Ar'

Scheme 6.1

bonds starting from alkyl halides is described as the Sonogashira reaction. However, many variations in the nature of the metal of the alkynyl nucleophiles are considered as different named reactions, such as the use of alkynylzinc- [5] (Negishi reaction), alkynylboron- [6] (Suzuki–Miyaura reaction) or alkynyltin-derivatives [7] (Stille reaction).

The expanded use of the term 'Sonogashira' reflects that it is not only important to develop a new reaction, but that it at least equally relevant to transform it into a reliable and convenient methodology for organic synthesis. Since its inception, many variations of this transformation have been examined. Thus, various ligands, metal complexes, electrophiles, alkynes and bases were among others examined. In this context, it is somewhat counterintuitive to realize that the mechanism of Sonogashira reactions has received little attention until recently.

The aim of this chapter is to present the main venues of interest in modern Sonogashira reactions, rather than focus on the exceptionally large number of applications. It is beyond doubt that Sonogashira coupling is the dominant methodology used in the synthesis of alkyne derivatives; indeed, the number of synthesized products using this technology is large and is continuing to grow. Consequently, a wide variety of structured materials, electronic polymers, natural products and biologically active intermediates have been prepared. Moreover, a large number of reviews and books are available on the subject, providing a specific overview on targeted synthesis [8].

Recently, this reaction has been used to promote science in early education, with the Tour group at Rice University having applied the Sonogashira coupling to the synthesis of anthropomorphic molecules dubbed 'nanokids'. These 'products' now form the basis of an educational program to promote science at kindergarten through year 12 schooling levels via a dedicated web page (http://cohesion.rice.edu/naturalsciences/nanokids/cast.cfm).

In this chapter, we have considered that it might be more important to provide a perspective of the general factors affecting reactivity in the Sonogashira reaction, rather than to focus on specific applications. Thus, a significant portion of the text is dedicated to the advances made in understanding the mechanism of this powerful reaction. It is hoped that the reader can adapt such knowledge, gained both experimentally and theoretically, to the particularities of his/her research project. Many of the references provides are recent contributions, and were selected to represent the latest advances in our understanding of the Sonogashira reaction.

6.2 Palladium-Catalyzed Reactions: Ligands and Reaction Protocols

6.2.1 Phosphine-Based Ligands

The use of tertiary phosphines as ancillary ligands for palladium- or palladium/copper-mediated cross-coupling of terminal acetylenes and aryl (pseudo)halides is as old as the reaction itself. The reliability of the classical palladium(II) salts, along with PPh_3, amines and CuI, for the coupling of aryl iodides or some activated aryl bromides make them the ligands of first choice, and the benchmark system to evaluate newly developed catalysts. Among palladium salts, $[Pd(OAc)_2]$, $[(CH_3CN)_2PdCl_2]$, $[PdCl_2]$ or M_2PdCl_4 (M = Na or K) are the most common precursors. Usually, the addition of phosphines and/or amines is performed to reduce the palladium(II) complexes *in situ* to the catalytically active complexes. Preformed catalysts of the general formula $[L_2PdCl_2]$ (with L = PPh_3 being the most commonly employed) are believed to be reduced to $[L_2Pd(0)]$ through the reductive elimination of di-acetylene complexes, which in turn are generated by double transmetallation from copper acetylide [9], when copper is present. Palladium(0) complexes, such as $[Pd(PPh_3)_4]$ or $[Pd_2dba_3]$, are other frequently used catalyst precursors. Although $[Pd(PPh_3)_4]$ generates the active species $[L_2Pd(0)]$ by endergonic loss of PPh_3, it is advisable to add excess phosphine to prevent the aggregation and formation of inactive palladium black. The main drawback associated with the use of $[Pd(PPh_3)_4]$ is the rather low concentration of the active species in solution, which is due to an equilibrium between $[Pd(PPh_3)_4]$ and $[Pd(PPh_3)_2]$. $[Pd_2dba_3]$ is much more resilient to oxidation than is $[Pd(PPh_3)_4]$, and phosphines can displace dba even at stoichiometric level. Bidentate ligands, such as dppe, dppf or dppp, are used extensively as ligands for cross-coupling reactions, and their complexes are expected to react without dissociation of the ligand. When generated *in situ* by mixing monodentate phosphines and a palladium source, the active complexes should not be considered *de facto* as 14-electron $[PdL_2]$ species. Stoichiometric studies suggest that, during the catalytic cycle, palladium can be stabilized by only one ligand if the tertiary phosphine is sterically demanding, as was shown for P(*o*-Tol)$_3$ [10].

$$L{-}Pd{-}L \xrightarrow{ArBr} [(Ar)(L)Pd(\mu\text{-}Br)_2Pd(L)(Ar)] \tag{6.1}$$

The stabilization induced by the bridging halogens is probably less significant than that provided by an additional ligand and, thus, leads to a faster initiation that should be borne in mind when testing new reactions.

The main driving forces behind the development of new tertiary phosphine palladium complexes for C(sp^2)–C(sp) couplings have been: (i) a reduction or elimination of side reactions, such as Glaser-type homocouplings; (ii) the development of environmentally friendly reaction protocols, such as copper-free reactions in benign solvents; (iii) the improvement of catalyst stability and activity [higher turnover number (TON) and turnover frequency (TOF)]; and (iv) a cost reduction by using less-expensive aryl bromides, or even aryl chlorides under mild reaction conditions, for example, at ambient temperature.

6.2.1.1 Copper-Free Catalytic Systems

Substantial efforts have been directed toward the elimination of copper salts or their replacement with less-toxic transmetallating reagents. The first reports on copper-free systems mostly made use of the standard reaction conditions simply by increasing the amount of palladium complexes. Thus, 5 mol% [Pd(PPh$_3$)$_4$], along with cyclic amines piperidine or pyrrolidine, can efficiently mediate the coupling of activated electrophiles, such as aryl iodides or triflates, with alkynes even at ambient temperature in high yields [11]. Related protocols, which however use preformed [(PPh$_3$)$_2$PdCl$_2$] at 70 °C in piperidine, can extend the scope of this approach to aryl bromides as electrophiles [12]. Furthermore, changing the base to Et$_3$N and using [Pd(OAc)$_2$] with PPh$_3$ as catalytic system allowed the coupling of conjugated alkenyl tosylates at ambient temperature [13].

A convenient catalyst system, consisting of [Pd(OAc)$_2$] and PPh$_3$ in dimethylsulfoxide (DMSO) as solvent and K$_3$PO$_4$ as base was recently reported [14], and found to be effective for the coupling of aryl halides with terminal alkynes. Among various ligands tested, an iminophosphine ligand gave (in some cases) better conversions than when PPh$_3$ was employed.

Hermann reported copper-free Sonogashira couplings of various aryl bromides with an *in situ*-generated catalyst employing [Pd$_2$dba$_3$] and P(*t*-Bu)$_3$ in tetrahydrofuran (THF) as solvent and with Et$_3$N as base. Worthy of notice here was the low catalyst loading of 0.5 mol% and very mild reaction temperatures. The catalyst choice was based on screening a library of 58 potential ligands using a color assay [15]. A similar system, which used MeCN as solvent and piperidine or DABCO as base was reported by Soheili [16] at Merck, and allowed the quantitative conversion of aryl bromides to internal acetylene; however, the process required

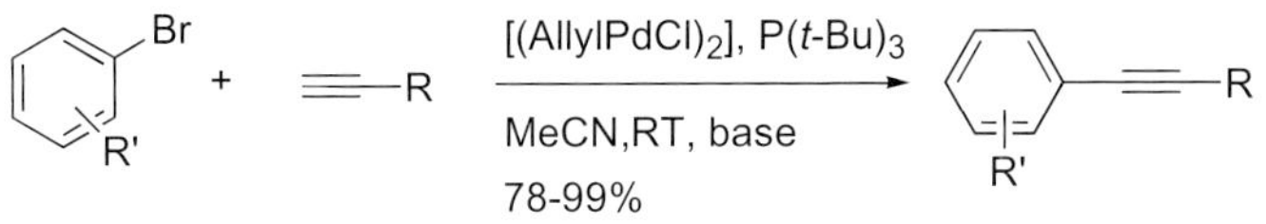

Br
+ R
[(AllylPdCl)$_2$], P(t-Bu)$_3$
MeCN,RT, base
78-99%
R'
R

R = aryl, alkyl
Scheme 6.2

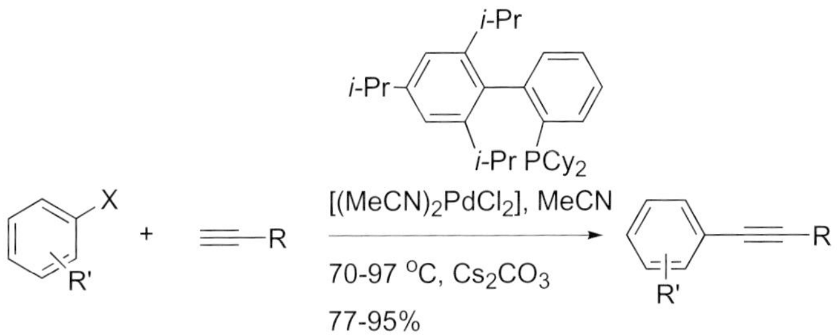

i-Pr
i-Pr
i-Pr PCy$_2$
X
+ R
[(MeCN)$_2$PdCl$_2$], MeCN
70-97 °C, Cs_2CO_3
77-95%
R'
R

R = aryl, alkyl
X = Cl, OTs
Scheme 6.3

a fivefold greater quantity of palladium when compared to Hermann's system (Scheme 6.2).

The excellent performance of P(t-Bu)$_3$ as ligand in cross-coupling reactions is attributed to its electron-rich nature and impressive steric demand [17]. Thus, the oxidative addition step in cross-coupling reactions is facilitated by its electronic-donating properties, and it is therefore understandable why challenging substrates can be activated at mild reaction temperatures using P(t-Bu)$_3$.

Concurrently, Buchwald and coworkers have developed a class of very active, bulky *ortho*-biphenylphosphine ligands (Scheme 6.3), which are able to activate very difficult substrates, such as aryl chlorides and, for the first time, also aryl tosylates [18].

The reactions are carried out at 70–97 °C in MeCN as solvent, with the palladium loading being varied from 1 to 5 mol%. A very interesting observation here is the detrimental role of CuI as additive, and hence the conversions were almost inversely dependent on the amount of copper additive. It can be speculated that the activity of coordinatively unsaturated palladium complexes may be inhibited by coordination of the acetylenic substrate, or by slow transmetallation.

Aryl chlorides substituted with either electron-withdrawing or electron-donating groups can be coupled using a PCy$_3$-based palladium complex at relatively high temperatures in DMSO with Cs_2CO_3 as base [19].

Other bulky phosphines have been employed in cross-coupling reactions, involving numerous substrates with various efficiency, such as BINAP [20], ferrocenyl phosphines (with a maximum TON of 250 000) [21], tetradentate phosphines (0.01 mol% catalyst for the conversion of aryl chlorides) [22], aminophosphines

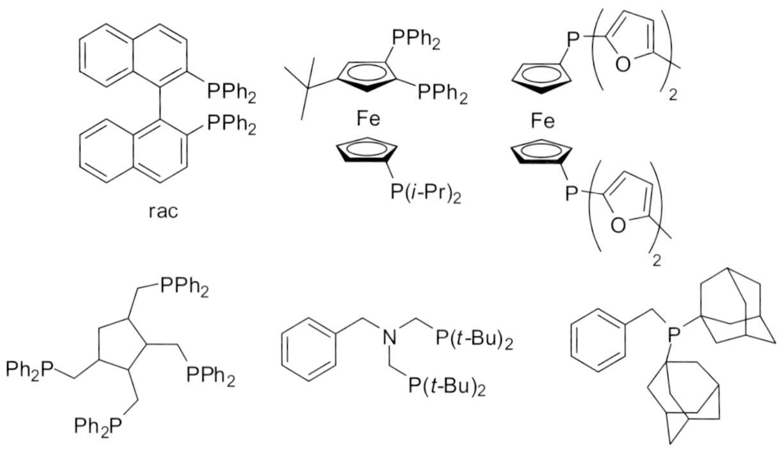

Scheme 6.4

(aryl chlorides at ambient temperature) [23], adamantanyl-substituted phosphines (aryl chlorides and bromides) [24] (Scheme 6.4).

6.2.1.2 **Hemilabile Ligands**

Mixed hemilabile ligands have been suggested as effective for increasing the activity of a catalytic system. In theory, the substrate molecule can easily displace the weakly bound donor ligand moiety. Furthermore, the transition state can be stabilized with a net energetic gain for the overall reaction. In reality, however, it is much more complex to predict the activity of such ligands as they can stabilize both the ground state and the transition state.

P,O-based ligands have been tested in the coupling of iodobenzene with CuI at ambient temperature, and showed moderate catalytic activity (Scheme 6.5) [25].

P,N-ligands, such as iminophosphines, are ligands for active catalysts at loadings as low as 0.1 mol% when aryl iodides or bromides were used as electrophiles in *N*-methylpyrollidone (NMP) at 110 °C (Scheme 6.6) [26].

The variation in phosphine ligand is quite impressive and difficult to cover in the space available here. Rather, the aim of this chapter is to offer a perspective of Sonogashira couplings, rather than to provide a full description of all published methodologies.

Interesting approaches toward the elimination of copper additives are focused on replacing copper with more environmental friendly metal salts, or replacing the terminal alkynes by alkynylsilanes. The first approach was successful when silver halides or silver oxide were used as additives. Whilst the use of silver can be considered as an impediment for future developments of the method due to its high price, it is noteworthy that in some applications elaborate alkynes may exceed the price of silver by orders of magnitude. For instance, an estrogenic malonate compound was synthesized using Sonogashira coupling in the presence

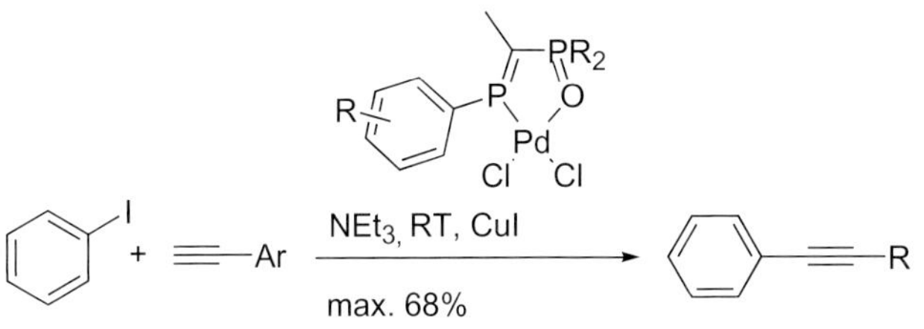

NEt3, RT, CuI

max. 68%

Scheme 6.5

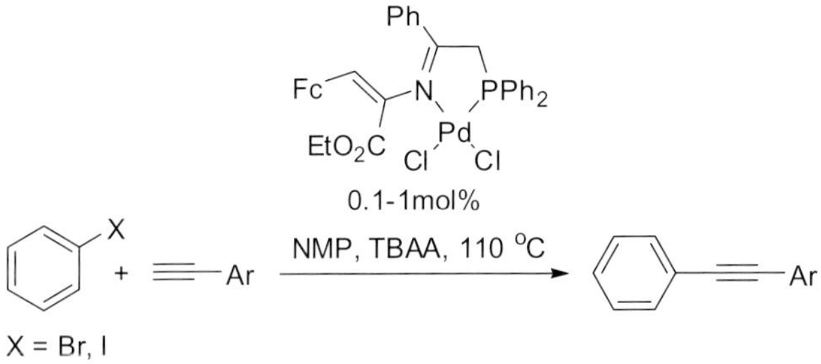

0.1-1mol%

NMP, TBAA, 110 °C

X = Br, I

Scheme 6.6

of [$Pd(PPh_3)_2Cl_2$], along with stoichiometric amounts of Ag_2O, in THF at 60 °C with a yield of 61% [27].

Good conversions of aryl iodides and silyl-substituted alkynes to internal alkynes were obtained with a combination of [$Pd(PPh_3)_4$] and AgX in MeOH as solvent. The efficiency of the catalytic system is based on the formation of silver acetylide as transmetallating agent [28]. The presence of fluoride anion can facilitate the reaction by generating hypervalent silicon-containing intermediates, which transmetallate faster in the presence of silver. This protocol, which involves three different acetylides, may appear cumbersome but was applied in the cross-coupling of alkenyl triflates with trimethylsilyl acetylenes using tetrabutylammonium fluoride (TBAF) as the fluoride source [29]. A silver-free version uses [$Pd(OAc)_2$] and $P(o\text{-Tol})_3$ in DMF as solvent and tetrabutylammonium chloride (TBAC) as additive for the coupling of aryl iodides or bromides at 100 °C, under microwave irradiation [30]. The role of the tetrabutylammonium halides may extend beyond that of a simple base needed to neutralize HX. In a recent example [31], a wide range of aryl halides ArX, with X = I, Br or Cl, could be coupled only in the presence of TBAF and a palladium source. Here, TBAF acted as base, and also probably stabilized the palladium nanoparticles that may form during the course of the reaction [32].

Another method of increasing the efficiency of Sonogashira couplings is to use microwave irradiation, which benefits from the use of polar solvents (by far the most popular solvents for Sonogashira reactions). This technology drastically affects the required reaction time for quantitative conversion, and occasionally the purity of the products may even be higher than that of those obtained when using conventional heating. Aryl iodides and bromides (and on occasion also aryl chlorides) can be coupled in the presence of CuI, $[PdCl_2(PPh_3)_2]$ and secondary amines in dimethylformamide (DMF) as solvent within minutes [33], as well as in the presence of catalysts supported either on polystyrene resin [34] or alumina [35].

During recent years, many important problems were tackled that were related to the environmental impact of cross-coupling protocols with regards to the solvents employed or the recyclability of the catalyst.

6.2.1.3 Ionic Liquids as Reaction Media

The low volatility of ionic liquids and the easy separation of catalysts (which usually remain in these polar media) have made ionic liquids an interesting alternative to typically used organic solvents. Rather unsatisfactory results have been obtained in both copper-mediated [36] and copper-free [37] Sonogashira reactions, with aryl iodides being the only aromatic electrophiles coupled at reaction temperatures between 60 and 80 °C. It should further be noted that imidazolium-based ionic liquids are not necessarily innocent solvents, but can be deprotonated in the presence of bases to generate *N*-heterocyclic carbenes (NHCs).

6.2.1.4 Reactions in Aqueous Media

The use of water as reaction medium is considered a desirable industrial goal, and efforts in this direction are mainly driven by economic and environmental reasons. However, the required substrates and catalysts for cross-coupling reactions are rather nonpolar, to a degree where their limited solubility in pure water may limit the efficiency of the reaction. Thus, most reported protocols for catalytic Sonogashira reactions have used water in combination with an organic solvent. These strategies can be divided mainly into: (i) the design of new soluble ligands or more soluble variants of known ligands; and (ii) the use of additives or cosolvents, along with conventional ligands in water or water-containing solvent mixtures.

6.2.1.4.1 **Hydrophilic Catalytic Systems** The most commonly used procedure to increase the solubility of tertiary phosphines in water is to attach highly polar groups to the organic moiety of the phosphine. Monosulfonated triphenyl phosphine (TPPMS) palladium complexes can mediate the coupling of aryl iodides or bromides in a solvent mixture containing MeCN and water at ambient temperature, but require CuI as additive (Scheme 6.7) [38].

This protocol requires large quantities of catalyst, and in the coupling of propargylamines up to 20 mol% were employed. However, variations, such as the use of secondary amines as bases and additives, can permit a decrease of the palladium loading to 2.5 mol% [39]. The sulfonated version of Buchwald's catalyst retains its high efficiency in solvent mixtures of MeCN and water, when using 2.5 mol% $[PdCl_2(MeCN)_2]$ for couplings of aryl bromides and chlorides, even in the absence of CuI [40]. Carboxylated PPh_3 ligands are very efficient for the coupling of aryl iodides at only 1 mol% palladium loading at 60 °C in solvent mixtures containing water and MeCN [41].

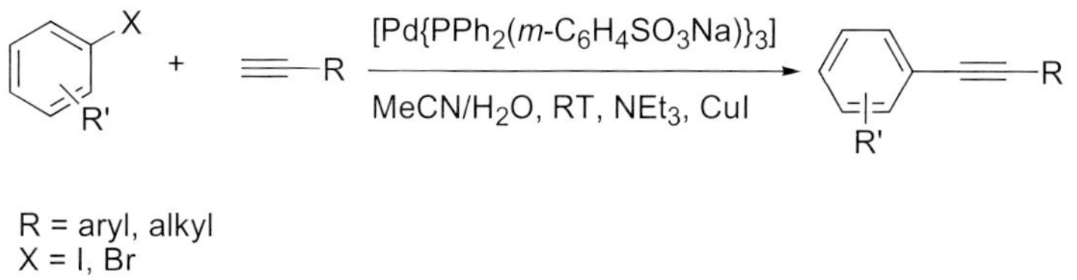

[Pd{PPh_2(m-$C_6H_4SO_3Na$)}$_3$]
MeCN/H_2O, RT, NEt_3, CuI

R = aryl, alkyl
X = I, Br

Scheme 6.7

Other methods for this derivatization approach include the use of phosphinous acids, such as (t-Bu)$_2$P(OH) for couplings of aryl chlorides [42] in the presence of tetrabutylammonium bromide (TBAB). Further, ammonium salts of hindered phosphines, such as 2-(di-*tert*-butylphosphino)ethyltrimethylammonium chloride (t-Bu-Amphos) or 4-(di-*tert*-butylphosphino)-*N,N*-dimethylpiperidinium chloride (t-Bu-Pip-phos), were employed for efficient couplings of unactivated aryl bromides at mild temperatures [43].

6.2.1.4.2 **Conventional Catalytic Systems in Water** More conventionally, non-functionalized ligands or palladium complexes have been employed for the coupling of aryl iodides at ambient temperature in aqueous K_2CO_3 [44], or for couplings of both aryl iodides and bromides in aqueous ammonia at ambient temperature [45].

6.2.1.5 Recyclable Phosphine-Based Catalytic Systems

Recovery of the palladium catalysts remains a serious problem for the large-scale application of cross-coupling reactions. Many methods to immobilize the catalytically active species have been designed in order to simplify catalyst separation and recyclability, the most popular strategies being filtration, centrifugation and biphasic extraction.

Perfluoro-tagged triphenyl phosphine and its palladium complexes have been shown to interact preferentially with a fluoro-modified surface of silica gel. Palladium complexes of these phosphines serve as an active catalyst for electronically activated aryl bromides with 2 mol% catalyst loading, and the supported catalyst can be recycled up to three times [46]. Similar fluorinated Teflon-based supports were used by Wang to immobilize phosphine-based palladium species, with the catalysts being recycled up to four times and showing good conversions at a much lower catalyst loading of 0.5 mol%, when using 2-iodothiophene as electrophile [47].

Three-dimensional aminomethyl-polystyrene can act as an insoluble solid support for an aminomethylphosphine palladium complex, and this enabled the Sonogashira coupling of aryl iodides with a catalyst loading of 1.6 mol%. Moreover, the catalyst could be recycled four times without any loss of activity [48].

Metal complexes immobilized either in or on dendritic systems have the unique advantage of being soluble under the reaction conditions. However, they can be easily precipitated when the polarity of the solvent is changed. Astruc applied this concept [49] to the covalent binding of an electron-rich phosphine on an amine-derived polydendritic support. The Sonogashira reactions were performed in Et_3N under copper-free reaction conditions, and gave good results for aryl bromides and iodides, but performed less satisfactorily for aryl chlorides as coupling partners.

The ring-opening metathesis polymerization (ROMP) of a norbornene-substituted phosphine led to a soluble polymeric ligand which, upon coordination to palladium, was used for the coupling of activated aryl iodides in Et_3N at elevated temperature. The recovery of the catalyst was achieved by simple filtration, although its activity was decreased after each cycle [50].

Salts of sterically demanding phosphines were anchored on styrenic polymers [51] or MeO-PEG [52], and generated the active phosphine by the simple addition of amines. In biphasic systems these polymers are active catalysts for the coupling of unactivated aryl bromides.

6.2.2
N-Heterocyclic Carbene Ligands for Sonogashira Coupling

N-heterocyclic carbenes were first predicted as stable singlet carbenes during in the 1960s, with bulky substituents as well as electronic effects being proposed as responsible for their stability. In the most common type of NHC, the nitrogen atoms adjacent to the carbenic center stabilize the electron pair via a synergic combination of inductive and mesomeric effects. As a result, the energetic gap between the occupied and unoccupied molecular orbitals increases to almost 80 kcal. Such a steric (and thus kinetic) stabilization is widely accepted in the academic community, despite the fact that some isolated imidazole-based carbenes displayed substituents as small as the methyl group [53]. Even the requirement for two electronegative groups for electronic stabilization of the carbenic center was challenged by the synthesis of stable carbenes, bearing only one adjacent nitrogen in the α-position [54].

Spectroscopic–and also especially calorimetric–studies confirmed dominant σ-donating properties, with little to insignificant back-donation [55]. The similarities with tertiary phosphine ligands prompted an extensive evaluation of their role as ligands in a variety of cross-couplings, including the Sonogashira reaction. The first report of NHC-supported palladium catalysts in alkynylation reactions was made by Herrmann in 1998 [56], where the well-defined catalyst displayed a chelating *N,N′*-dimethyl-substituted bis-carbene ligand in a *cis* conformation. In addition to excellent yields in Heck-type reactions of activated aryl chlorides and bromides, the system showed good catalytic activity for couplings of 4-bromoacetophenone or 4-bromofluorobenzene with phenylacetylene in the presence of NEt_3 (Scheme 6.8). The yields were in excess of 70% at a temperature of 90 °C and with a 48 h reaction time. The results of Herrmann's pioneering studies proved that NHCs are not exotic species, but rather efficient ligands that benefit from a high electron density at the carbenic center and a high thermal stability of their metal complexes. At the same time, the difficulties encountered with these complexes in the

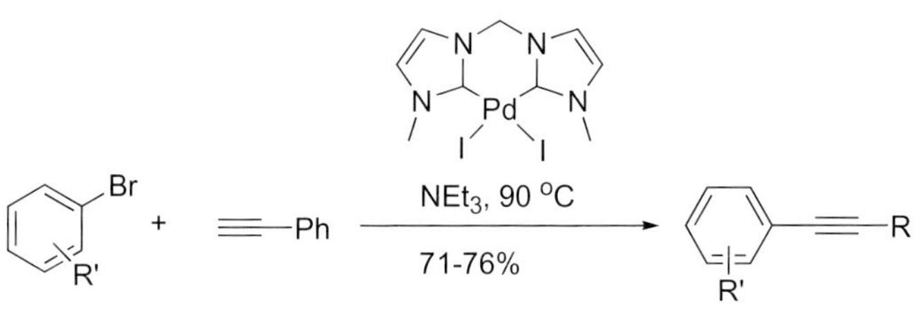

Scheme 6.8

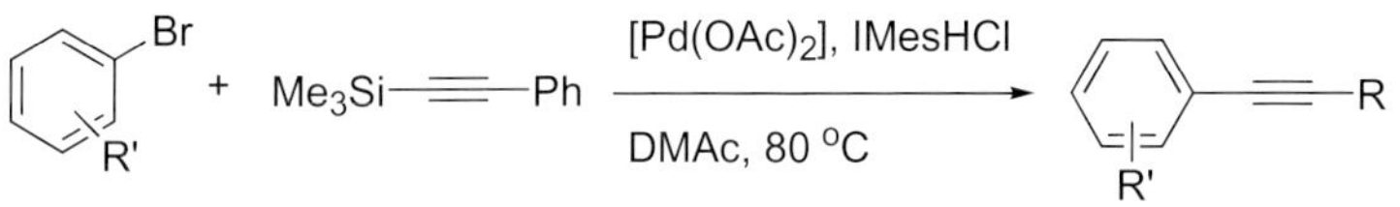

Scheme 6.9

Sonogashira reaction appeared to be a recurring theme for all systems involving palladium–carbene complexes.

Indubitably, the success of NHC-modified catalysts in the Sonogashira reaction still does not match that of phosphines in terms of substrate scope and activity. Nonetheless, the ability of electron-rich NHC-stabilized palladium complexes to activate difficult bonds, such as C—Br or C—Cl bonds in aryl halides, suggests that they could function as efficient ancillary ligands for the coupling of aryl halides with various acetylenes. In a study conducted by Nolan [57], inactivated 4-bromoanisole and phenylacetylene were used as standard substrates. Subsequent optimization studies revealed a detrimental role for NEt_3 (one of the most widely used bases), which led to decomposition of the catalyst, and the formation of a large amount of dimerized phenylacetylene was also observed. In order to overcome these problems, the alkyne was replaced with 1-phenyl-2-(trimethylsilyl)acetylene, a silyl-protected alkyne. Optimized reaction conditions for its conversion involved the use of 3 mol% [$Pd(OAc)_2$], 6 mol% 1,3-bis(2,4,6-trimethylphenyl)imidazolium chloride (IMesHCl), Cs_2CO_3 as base and DMAc as solvent at 80 °C. The methodology proved to be quite general in scope, with high yields of coupling products being obtained with various aryl bromides as coupling partners (Scheme 6.9).

Interestingly, some reactions can be performed in the absence of CuI with minimal loss of catalytic activity. Electron-withdrawing groups (such as carbonyls) present on the arene moiety of the electrophile are well tolerated and lead to a faster conversion to the product. This observation is consistent with the oxidative addition being rate-determining, although to date the available data are limited and hence this assumption is at present speculative. The addition of CuI accelerates the reactions (notably for deactivated aryl bromides), thus highlighting the important role of the transmetallating step on the rate of the reaction. Sterically encumbered mono- or di-*ortho*-substituted substrates perform well under these reaction conditions, and the system shows moderate activity when aryl bromides are replaced with unactivated aryl chlorides and CuI is used as an additive. The reactions were not fully analyzed in terms of TON and TOF, although TOFs higher than 100 h^{-1} can be achieved for selected substrates. The protocol involves the *in situ* generation of the catalytically active species, which is presumed to be either [$(NHC)_2Pd$] or highly unsaturated [(NHC)Pd]. The presence of Cs_2CO_3 as base assists in the deprotonation of the imidazolium salts and generation of the free NHC. Similar conditions were used to generate [$(NHC)_2Pd$] complexes and, while the possibility of forming C-4-carbene species is limited, it cannot yet be ruled out [58].

Variation in the size rather than the structure of the NHC ligand was shown to have a dramatic effect on product yield. An increase in the steric hindrance of nitrogen substituents, when replacing mesityl with 2,6-diisopropylphenyl-substituents or with extended aromatic phenanthrenyl groups, led to efficient Sonogashira couplings under mild temperatures within relatively short times [59]. The reaction conditions involved KO*t*-Bu as base and 18-crown-6 as additive in THF. The common bases used for the Sonogashira reaction, such as K_2CO_3, Cs_2CO_3, KF or NEt_3, either failed or gave unsatisfactory yields in the coupling of PhBr with phenylacetylene. Palladium–NHC complexes with 9-phenanthryl or 2,9-dicyclohexyl-10-phenanthryl substituents on nitrogen showed pronounced differences in reactivity, with the latter complex providing 77% yield of the desired product in 2 h at 65 °C. As expected, aryl iodides (and in some cases also aryl bromides) could be coupled at lower or even ambient temperature, despite the copper-free reaction conditions. Aryl chlorides gave only traces of the product, and the scope of these ligands seemed to be limited to more-activated electrophiles. Although the nature of the catalytically active species is not clear at present, it is speculated to be a monomeric species of unknown metal to ligand stoichiometry. Interestingly, this protocol relies on the use of a phosphine-based palladium complex [$Pd(PPh_3)_2Cl_2$] as catalyst precursor. Phosphine-free palladium precursors display lower reactivity (Scheme 6.10).

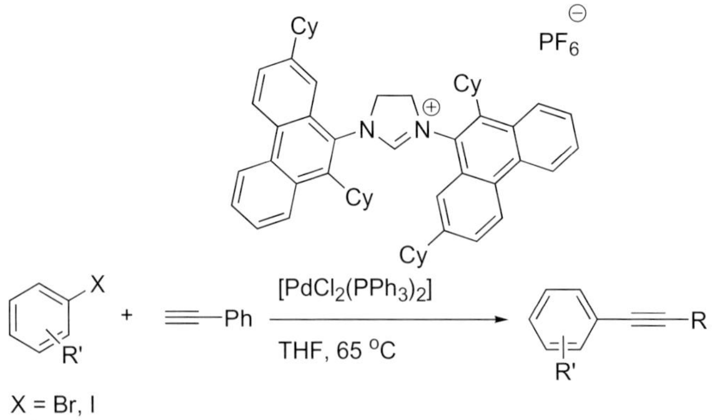

Cy
PF6
Cy
N N
Cy
Cy
X
R'
+ ≡—Ph
[PdCl2(PPh3)2]
THF, 65 °C
≡—R
R'
X = Br, I

Scheme 6.10

Substituted NHCs have been used as ligands under various conditions and with widely variable success. For example, Batey used NHC-precursors with pendant carbonyl groups that could induce a stronger acidity of the C—H bond in position C-2, and allow for a faster generation of the free carbene under basic reaction conditions [60]. Under neutral reaction conditions, however, a rather unusual complex was isolated and used, along with free PPh_3, for the coupling of aryl bromides and iodides with both aryl and alkyl alkynes (Scheme 6.11).

It is interesting to note that the activity of the catalyst decreased by half when the tertiary phosphine PPh_3 was omitted from the reaction. For the coupling of unactivated aryl bromides, the optimized conditions involved Cs_2CO_3 as base, in DMF as solvent at 80 °C, although the corresponding iodides were converted more efficiently in the presence of NEt_3 at ambient temperature. A palladium : PPh_3 ratio of 1 : 1 was shown to be critical for the catalytic system to function.

The findings of an extensive study on NHC-based ligands with semilabile pendant functionalities were reported by Cavell [61]. Here, better conversions were achieved in Sonogashira reactions if the complex took advantage of the donor capability of one NHC ligand, and an additional stabilization was induced by the presence of an N-donor pyridine. Despite long reaction times, a maximum TON of 540 was achieved (Scheme 6.12).

Bis-chelating NHCs with the same pyridine side-arm gave a TON of only 195 under similar reaction conditions.

Tridentate pincer palladium–carbene complexes [62] were used for the coupling of aryl iodides and bromides with phenylacetylene in pyrrolidine as solvent, both

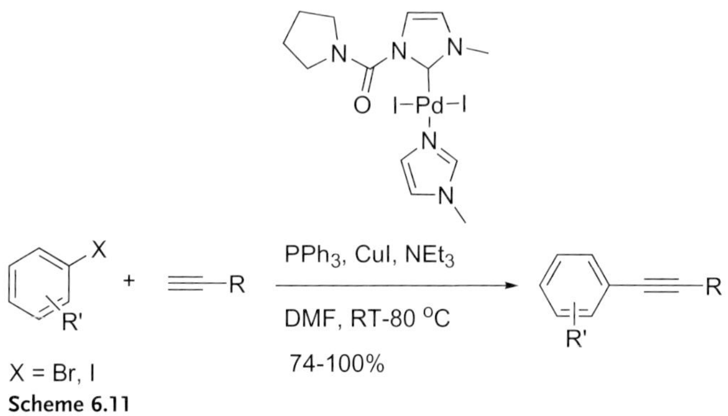

PPh_3, CuI, NEt_3; DMF, RT-80 °C; 74-100%

X = Br, I

Scheme 6.11

NEt_3, 90 °C; TON = 540

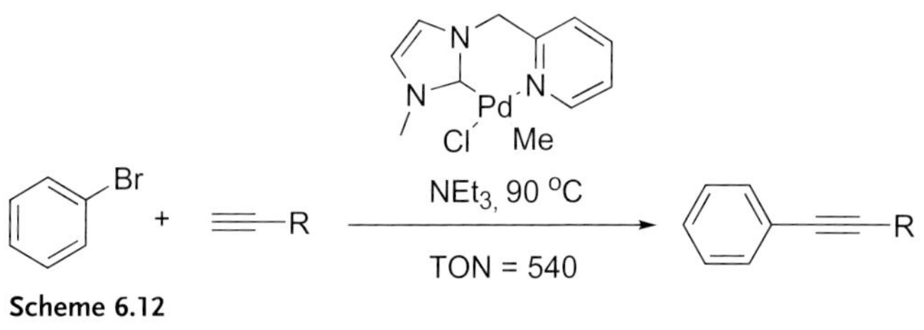

Scheme 6.12

in the presence and absence of CuI. In the presence of CuI a full conversion to the desired products was achieved within 45 min, whereas in its absence the yield of the catalytic process fell to 38%.

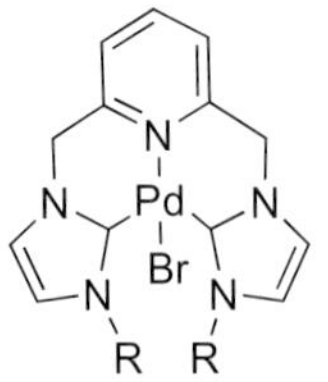

R = Me, *n*-Bu, Bn

The long-term stability of the complex was tested by its immobilization on three types of porous solid, namely montmorillonite K10, bentonite A and bentonite B. The heterogenized systems promise great stability and reusability, despite the fact that they were tested over only two successive catalytic cycles.

In addition to the well-examined cyclic carbenes, acyclic carbenes derived from formamidinium salts are also of interest, especially as their carbenic centers are presumably more basic entities. The lack of any conformational restriction usually imposed by the rigidity of NHCs leads to larger $N{-}C_{carbenic}{-}N$ bond angles in acyclic carbenes, and thus provides more steric pressure on the metal center. In the Sonogashira reaction, the bis(diisopropylamino)carbene ligand [63] was generated *in situ* by deprotonation of the corresponding formamidinium salt with lithium diisopropylamide (LDA), and trapped by $[Pd(allyl)Cl]_2$. The catalytic system is copper-free and gave excellent yields for the coupling of unactivated aryl bromides at ambient temperature, using Cs_2CO_3 as base. Steric hindrance of the substrates and heterocyclic moieties did not affect the catalytic efficacy (Scheme 6.13).

An interesting application of NHC-ligated palladium complexes in the Sonogashira reaction, among other cross-couplings, was achieved by an immobilization of the catalytic species on polystyrene beads, or magnetic nanoparticles, after which recycling of the catalyst for up to five times could be accomplished [64]. This combination takes advantage of the best of the homogeneous and heterogeneous catalyses – that is, a single-site catalyst, as well as recyclability and easy separation,

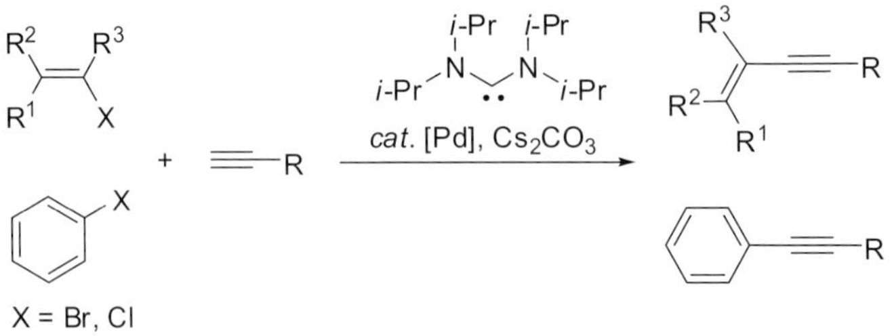

cat. [Pd], Cs_2CO_3

X = Br, Cl

Scheme 6.13

respectively. The reaction rates were, somewhat surprisingly, higher in the case of magnetic nanobeads than for the polystyrene-based catalysts, although the authors ruled out any intrinsic catalytic activity of the bare γ-Fe_2O_3 support. The size of the coated solid support (ca. 11 nm) allowed partial solubility of the supported catalyst in organic solvents and, at the same time, permitted an easy separation without the aggregation that was observed when magnetic separation was used. Consequently, the separated nanoparticles could be washed and reused for the same, or even a different, reaction.

6.2.3 Palladacycles as Catalysts in Sonogashira Reactions

The industrial application of cross-coupling protocols requires active catalysts not only of broad scope but also with chemical and thermal stability, and palladacycles can, in many instances, fulfill all of these requirements. In fact, useful palladacycles are stabilized by the presence of additional donor ligands, such as tertiary phosphines, NHCs or nitrogen-based ligands. The catalytic activity of palladacycles must be initiated by reduction of the palladium(II) center, and in this sense palladacycles can be regarded not only as catalyst precursors but also probably as 'stabilized resting states'. The presence of donating ligands assures the efficiency of the oxidative addition step and provides well-defined metal to ligand ratios. Further, recent studies have shown that the formation of active, palladium colloids, with high surface areas, should be considered another very likely source of palladacycles activity [65]. The seminal studies of Herrmann, in which the catalytic activity of palladacycles in C—C bond formations (particularly in Heck coupling [66]) was examined, prompted a search for active catalysts in other cross-coupling reactions, including the Sonogashira reaction.

Phosphapalladacycles stabilized by highly hindered phosphines showed very good catalytic activity in the copper-free cross-coupling of aryl bromides with phenylacetylenes in NEt_3 with a maximum TON of 8000.

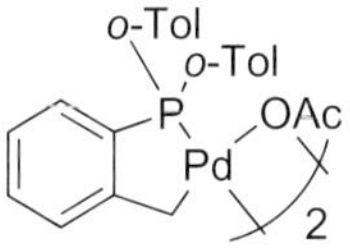

o-Tol
o-Tol
P
OAc
Pd
2

The addition of other bases, including amines, may have a detrimental effect on catalytic activity, with a sensitive association–dissociation equilibrium at the catalytically active organopalladium species being the most likely source of this effect. Despite high TONs in the case of aryl bromides, the reaction failed when aryl chlorides were used as electrophiles. When a closely related palladacycle bearing a less-sterically crowded phosphine ligand was used as polymer-supported catalyst [67], a full conversion and good recovery of up to four times of the catalyst were reported, although catalyst leaching or deactivation led to the need for a high initial catalyst loading of 5 mol%. As a consequence, sulfinimine-based palladacycles afforded cross-couplings of aryl iodides in NEt_3 with a maximum TON of a few hundreds [68].

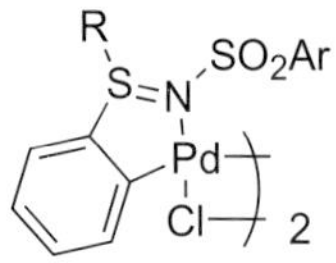

Oxime-based palladacycles have been studied extensively by Nájera, and may be the most active palladacycles for the Sonogashira transformation (Scheme 6.14) [69].

The initial reactions were conducted in the presence of copper cocatalysts in pyrrolidine as both base and solvent, with the choice of base being dictated by the presence of large amounts of diyne formed when NEt_3 or piperidine were used. At 0.5 mol% catalyst loading, the maximum TON observed was 150, which compares poorly with the value of 10^5 observed for Heck reactions. Improved protocols allowed the coupling of aryl iodides or bromides with phenylacetylene under copper-free conditions. The reaction conditions involved TBAOAc as base and NMP as solvent and, at an operating temperature of 110 °C, the catalyst loading could be decreased to 0.001 mol%. Interestingly, when alkynylsilanes were the coupling partners for aryl iodides or bromides with copper as additive and pyrrolidine as solvent, a double arylation was seen to take place in part; however, when the base was changed to TBAB the main product was the corresponding silylated alkyne. Oxime-derived palladacycles anchored by polyethylene glycol (PEG) can be used for copper-free Sonogashira couplings of 4-bromoacetophenone with acetylenes in the presence of CsOAc, albeit at a reaction temperature of 150 °C [70].

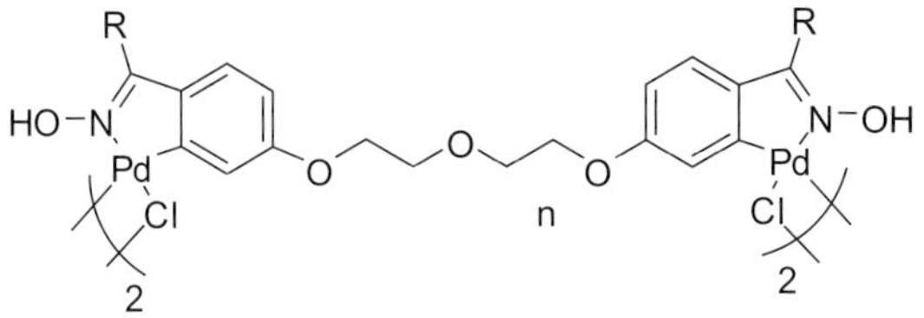

The authors acknowledged decomposition of the catalyst to nanoparticles stabilized by PEG. Whilst the catalytic efficiency was unaffected when the PEG gel was

Scheme 6.14

reused, it proved difficult to differentiate between the reactivity derived from the nanoparticles and that of the palladacycle itself.

Highly active mono- and binuclear cyclometallated palladium complexes containing bridging N,O- and terminal N-imidate ligands have been reported by Fairlamb [71]. Imidato ligands, such as maleimidates, phthalimidates and succinimidates, have the ability to act as mono or bidentate ligands that might confer stabilization of the intermediates of the catalytic cycle.

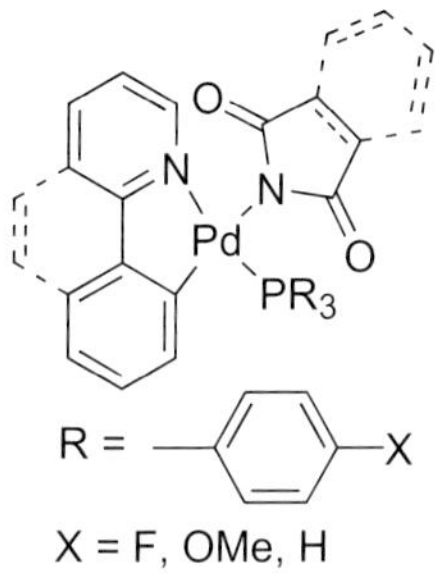

R = —C$_6$H$_4$—X

X = F, OMe, H

It had been noted that whilst a higher catalyst loading increased the rates, this was at the expense of chemoselectivity. Yet, lowering the catalyst loading led to only a marginal decrease in conversion. This behavior translated into higher TOFs, which were almost inversely proportional to the palladium loading. This somewhat counterintuitive observation was explained by the formation of large amounts of metal aggregates which have a low catalytic activity but may act as reservoirs for smaller aggregates. A maximum TON of 37 000 was obtained at 0.001 mol% catalyst loading. Furthermore, the addition of CuI favored higher reaction rates, although the reaction was then less chemoselective. A synergistic interaction between electron-rich phosphines and the copper salt was also noted.

There is no doubt that palladacycles have great potential as robust catalysts in the cross-coupling between alkynes and aryl halides, and their activities (as reflected by TONs or TOFs) may be far from the impressive results observed for Heck couplings. However, the peculiarities of the Sonogashira reaction, in terms of electronic and coordination properties of the substrates, may be an inherent problem for long-lived catalysts in this chemical transformation.

6.2.4 Nitrogen-Coordinating Ligands

Although phosphines had been established as ligands of choice in the Sonogashira coupling, non-phosphorus-based ligands had also provided some promising results. It is difficult, however, to differentiate between amines acting solely as a base, or both as base and ligand, during the catalytic reaction, and consequently the role of amines is very complex (*vide infra*). Even the concept of 'ligands' can be challenged if one considers the ligand as a species that is permanently or transiently coordinated to the metal center. Research in the area of amine ligands has been driven mainly by the drawbacks associated with the use of phosphines and the formation of phosphine-based palladium intermediates, which require in many cases an inert atmosphere due to the low oxidation potential of the phosphorus ligand. The large-scale application of coupling reactions still require, with respect to oxidation, robust catalytic systems, although much of the recent literature regarding the use of amine as ancillary ligands for Sonogashira reactions has gravitated around aromatic amines.

Monoligated oxazolines derived from commercially available 2-ethyl-2-oxazoline can form *trans*-square–planar palladium complexes by simple treatment of this ligand with inorganic palladium salts in methanolic solutions (Scheme 6.15) [72]. The catalysis, when performed in pyrrolidine, gave a yield of 32%, which corresponded to a TON of 190. The Sonogashira reaction was carried out under aerobic conditions with no precautions to exclude moisture from the solvent or substrates.

Pyridine-based ligands, such as 2,2-dipyridylmethylamine, can be easily derived from di-2-pyridyl ketone or its oxime, and can be further anchored to solid supports via the amino group. The corresponding palladium complexes are synthesized by reacting [H_2PdCl_4] with the chelating ligand (Scheme 6.16) [73].

Aryl bromides and iodides can be coupled with terminal aryl- or alkyl-substituted alkynes even in the absence of copper additives. The reactions were performed in NMP as solvent TBAOAc as base at 110 °C. The robustness of the catalyst was tested in water with pyrrolidine as base and, despite the small amounts of diyne or enyne formed during the reactions of up to 20%, TONs of up to 970 were observed. By using a structurally related catalyst, however, the reactions could be

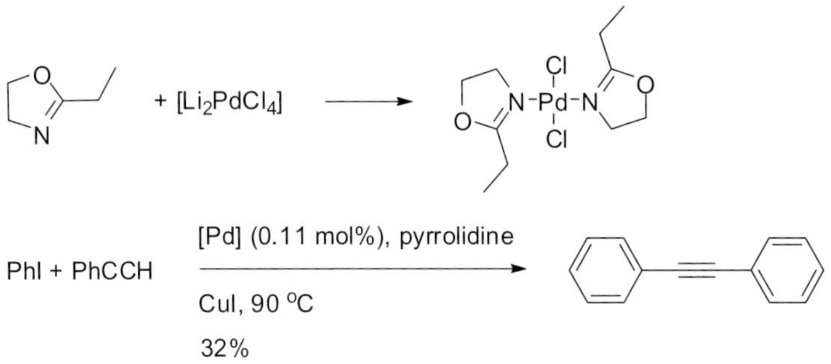

+ [Li_2PdCl_4] →

PhI + PhCCH —[Pd] (0.11 mol%), pyrrolidine, CuI, 90 °C, 32%→

Scheme 6.15

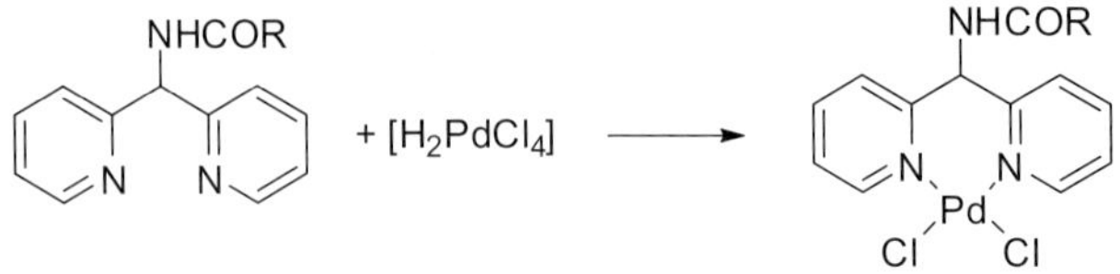

+ [H_2PdCl_4] →

Scheme 6.16

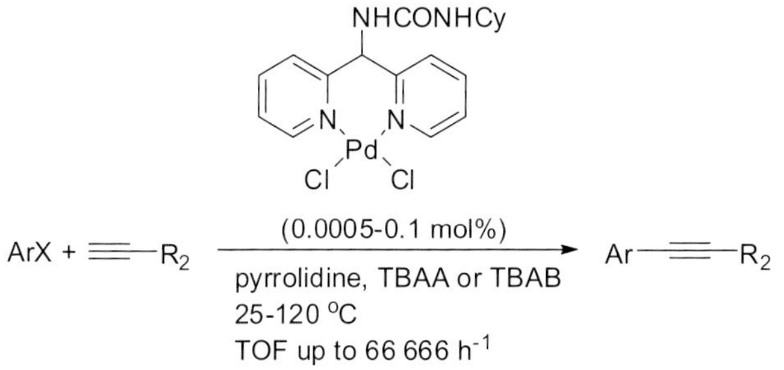

NHCONHCy

N N Pd Cl Cl

ArX + ≡—R_2 → Ar—≡—R_2

(0.0005-0.1 mol%)

pyrrolidine, TBAA or TBAB

25-120 °C

TOF up to 66 666 h^{-1}

Scheme 6.17

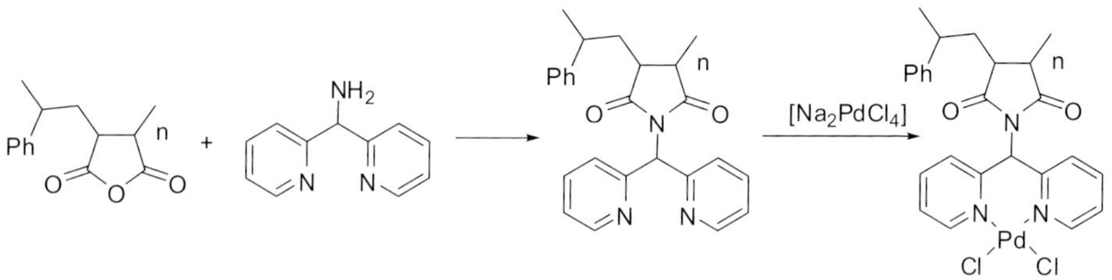

Ph n O O O + NH_2 N N → Ph n O N O N N → $[Na_2PdCl_4]$ Ph n O N O N N Pd Cl Cl

Scheme 6.18

carried out at 0.01 mol% in water or 0.0005 mol% in NMP, which corresponds to TOFs of 6666 h^{-1} and 66 666 h^{-1}, respectively (Scheme 6.17) [74].

The polymer-supported version of this catalyst can be synthesized by reacting a poly(styrene/maleic anhydride) resin with di-(2-pyridyl)methylamine (Scheme 6.18). Here, the catalyst loading can be as low as 0.001 mol% in the case of activated aryl iodides with a corresponding TOF of 3225 h^{-1}. Interestingly, when the same reaction was carried out with the corresponding monomeric palladium catalyst, the TOF did not exceed 730 h^{-1}. Under microwave irradiation, the yields of coupled products were modest to good, and up to five reactions could be carried out with retention of the catalytic activity. The authors also pointed out that no palladium leaching was observed during the synthesis of the complexes. The proven stability of this catalyst, coupled with its high TOF, may represent a successful approach for future industrial catalysts.

Simple bis-imidazole-based palladium complexes were used in ionic liquids for the coupling of aryl iodides with alkynes [75]. The reactions did not require copper salts and were performed in $[bmim]PF_6$ at temperatures varying from 60 to 120 °C. Organic amines performed significantly better than inorganic bases, such as K_2CO_3, Cs_2CO_3. When the reactions had been completed the products were extracted and the ionic liquid layer was washed with water in order to remove the amine salts. In this way the catalyst could be recycled up to four times (Scheme 6.19).

On a similar note, palladium complexes bearing bis-pyrazolyl-derived ligands with pendant imidazolyl-substituents have been used in ionic liquids, and could be recycled up to six times at an initial catalyst loading of 1 mol% [76].

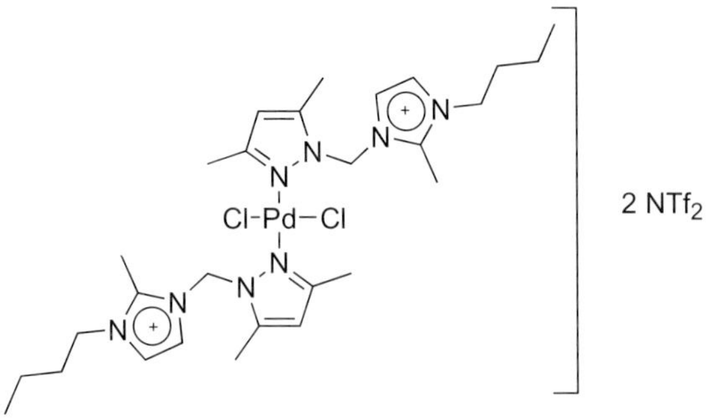

Cl-Pd-Cl

2 NTf_2

Li and coworkers [47] reported ligands of the pyrimidine class allowing for the coupling of aryl iodides with alkynes at ambient temperature under copper-free reaction conditions, albeit at high loadings of palladium. Among the pyrimidines used, 2-aminopyrimidine-4,6-diol was found to be the most efficient. The authors did not speculate on the structure of the active species or mode of ligand coordination, but an increase from 38% to 80% yield was observed upon addition of the ligand when compared to the reaction in the absence of this ligand. Changing the ratio of metal to ligand from 1:1 to 1:2 was found responsible for a further increase in yield to 92%. Optimization studies showed the beneficial role of inorganic bases Cs_2CO_3 and polar solvents (Scheme 6.20) [47].

In the same study the authors also described the use of aryl bromides as electrophiles, although in this case a higher reaction temperature of 60 °C and a longer reaction time were needed.

Previously, Buchmeiser and coworkers developed catalysts which were based on bis-pyrimidine building blocks [77] and proved to be competent in cross-coupling reactions involving aryl iodides, aryl bromides and, with some limitations, also

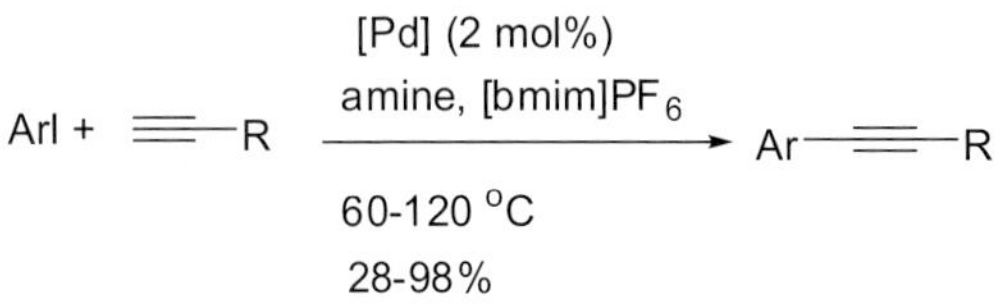

ArI + ≡—R → Ar—≡—R

[Pd] (2 mol%), amine, [bmim]PF_6, 60-120 °C, 28-98%

Scheme 6.19

ArX + ≡—R → Ar—≡—R

X = I, Br

$[Pd(OAc)_2]$, Cs_2CO_3, MeCN, 20-60 °C, 60-100%

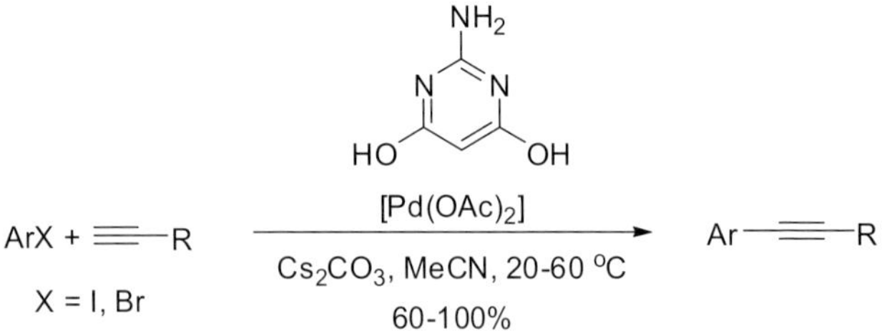

Scheme 6.20

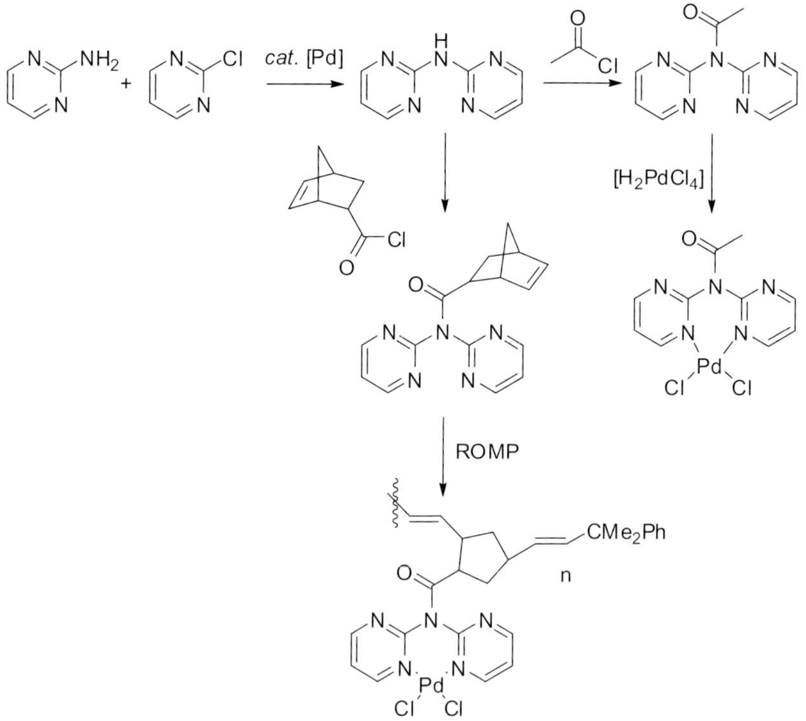

Scheme 6.21

aryl chlorides [78]. A long-term stability under air was also demonstrated. The synthesis of the ligand precursor involved a C—N bond-forming reaction mediated by a palladium–phosphine complex (Scheme 6.21).

The supported version of the catalyst was obtained via ROMP using the molybdenum-based Schrock catalyst. An inductively coupled plasma-optical emission spectroscopy (ICP-OES) analysis of the polymer revealed a palladium to bis(pyrimidine) ratio of 2:1. The authors speculated that this could be explained by coordination of palladium by alternate and repetitive bis(pyrimidine) units of the polymer chains.

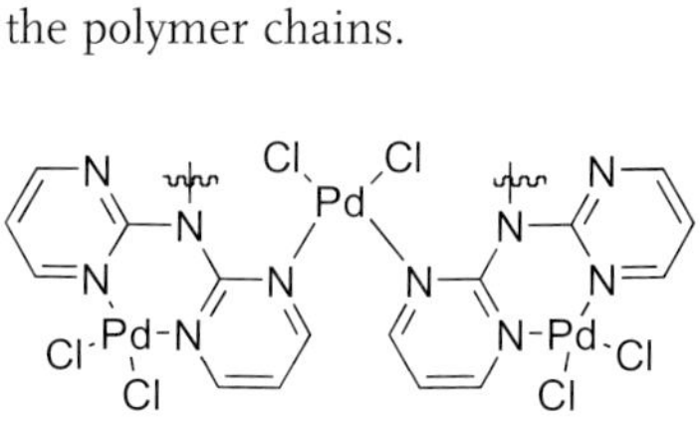

The activity of the supported catalyst was probed in the coupling of phenylacetylene with iodo-, bromo- and chlorobenzene in the absence of additives. Although the required reaction time was relatively long, impressive activities were obtained,

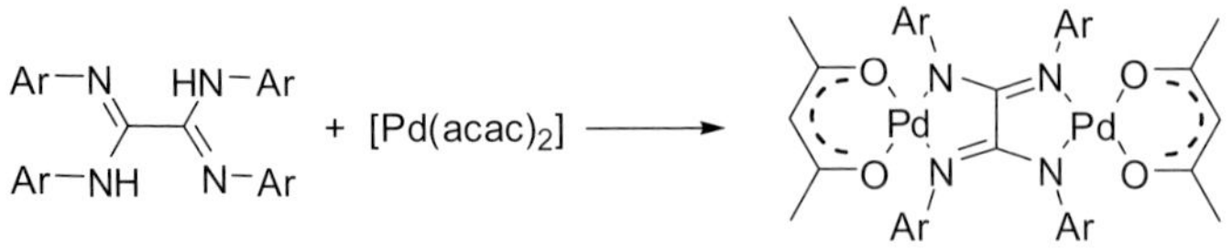

Scheme 6.22

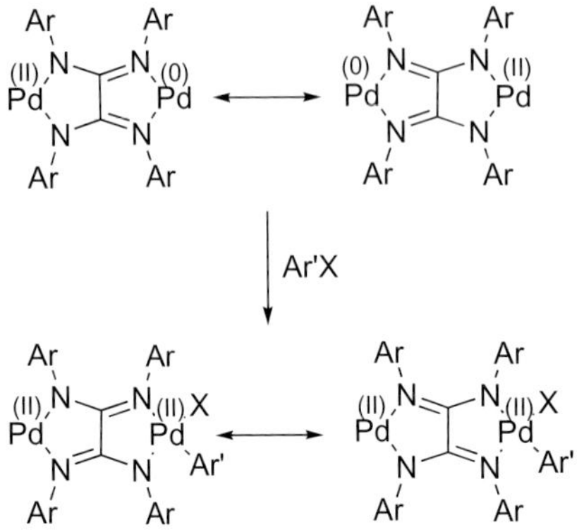

Scheme 6.23

with TONs of 22000 and 15000 for aryl iodides, and aryl chlorides, respectively, at catalyst loadings varying from 0.004 to 0.007 mol%.

Both binuclear and trinuclear oxalamidinate palladium complexes were employed by Rau, Walther and coworkers [79] in copper-free Sonogashira reactions. The peculiarity of the system was seen to reside in the presence of chemical bridges that allowed electronic communication between metal centers (Scheme 6.22). The anionic structure of the ligands was considered to induce poor π-accepting properties, and thus poor coordination to palladium(0) species (Scheme 6.23). It was speculated that the catalytic activity was a result of the reduction of one palladium(II) center under the reaction conditions. This palladium center was stabilized as a mixed-valence complex, and entered the catalytic cycle via oxidative addition of the aryl halide. It was also shown that heterotrimetallic complexes, bearing additional bridging Zn atoms, gave poor conversions and led to the formation of palladium black. A maximum TON of 490 was observed in the reaction of 4-bromoacetophenone with phenylacetylene at 0.2 mol% catalyst loading in NEt_3 at 90 °C. An important observation here was that the addition of CuI did not improve the catalytic activity; rather, the use of this additive caused an overall decrease in the chemical yield.

Macquarrie [80] reported the activity of *N,N*-chelated palladium complexes immobilized on mesoporous silica (Scheme 6.24). The catalytic system showed a remarkable selectivity switch depending on the presence or absence of CuI as additive (Scheme 6.25). In the reaction of aryl iodides, the ratio of cross-coupled versus homo-coupled products varied from more than 50 : 1 to 3.5 : 1 for 3-FC_6H_4I and C_6F_5I as electrophiles, respectively, while for the conversion of 3-FC_6H_4I the

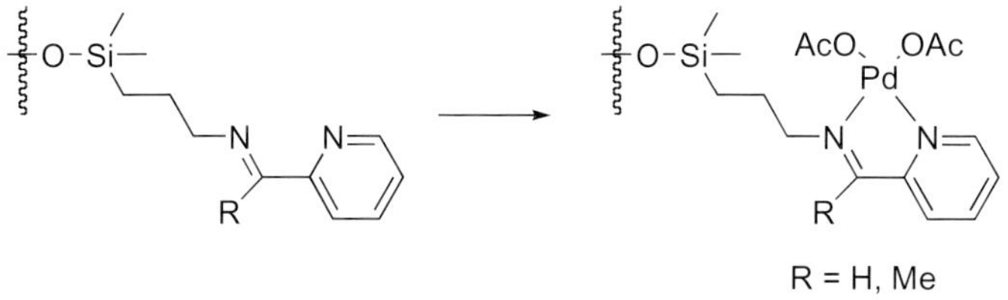

R = H, Me

Scheme 6.24

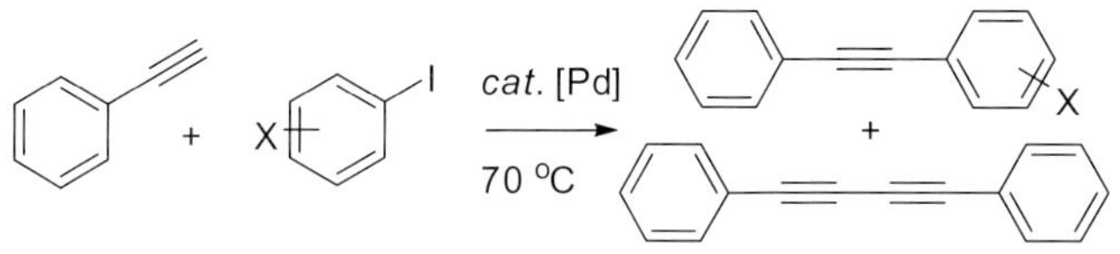

cat. [Pd]

70 °C

Scheme 6.25

addition of CuI led to a decrease in selectivity from 50:1 to 12:1. When iodophenols were used as the coupling partners, an inversion of chemoselectivity was observed from 1.2:1 to 1:50, favoring the homo-coupled product. This makes the system an efficient protocol for the synthesis of 1,3-diynes. Despite the lack of selectivity for a wide range of substrates, the value of the methodology was illustrated by the fact that the reactions used only 0.1 mol% catalyst, yet gave satisfactory yields of up to 72% of the cross-coupled products.

Diamine functionalized porous silica gel was used by Wang [81] as support for palladium complexes, which could be recycled up to 30 times. The copper-free reactions of aryl iodides and some aryl bromides were performed in EtOH as protic solvent with K_2CO_3 as base. Although these reactions could be conducted at ambient reaction temperature by immobilizing palladium on MCM-41 zeolites, they required the presence of Cu(I) additives and piperidine as the base of choice.

The use of amines as supporting ligands in cross-coupling reactions will most likely be extended in the coming years, due mainly to their low cost, low toxicity and wide availability. Many Sonogashira reactions can be performed under copper-free conditions with moderate to very good yields. Moreover, the stability of these systems, together with their 'non-leaching' nature offers a good incentive for the further exploration of alternative ligands compared to the standard phosphine-based catalytic systems.

6.3 Alternative Metal Catalysts

6.3.1 Nickel-Catalyzed Sonogashira Reaction

Nickel complexes have been used as alternatives for palladium catalysts in almost all cross-coupling reactions. However, a potential coordination of the nickel

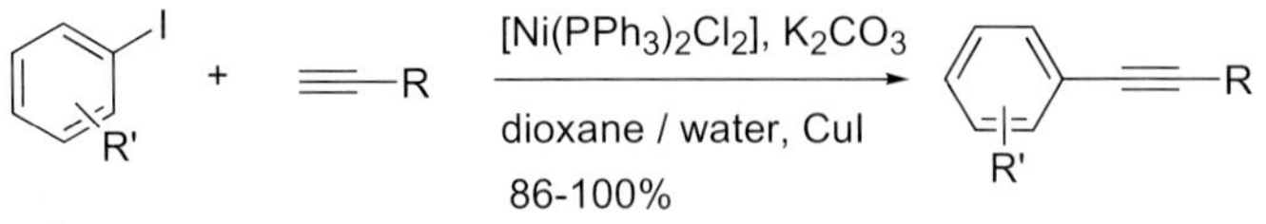

[Ni(PPh$_3$)$_2$Cl$_2$], K$_2$CO$_3$
dioxane / water, CuI
86-100%

Scheme 6.26

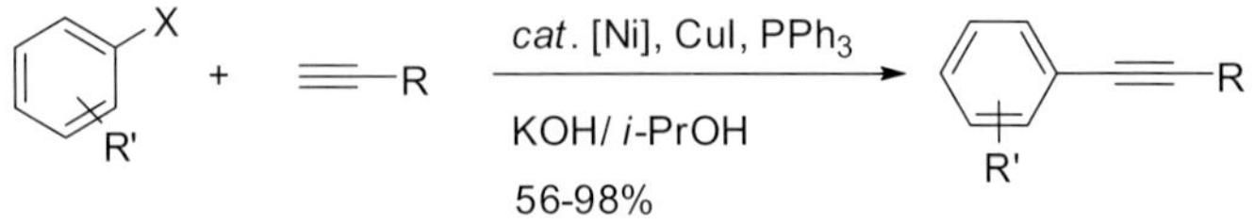

cat. [Ni], CuI, PPh$_3$
KOH/ i-PrOH
56-98%

X = I, Br
Scheme 6.27

complex by the alkyne may have led to the fact that the first report of a Sonogashira reaction catalyzed by a well-defined nickel complex appeared only few years ago [82].

The reaction is highly dependent on the nature of the solvent, such that a solvent mixture of 1,4-dioxane and water led to a yield of 78% 4-methoxytolane at 100 °C. The optimization of the nickel catalyst led to some surprising results, with the only active complex for this transformation identified as [Ni(PPh$_3$)$_2$Cl$_2$], while any minor modification of the ligand structure led to a significant decrease in reactivity. Furthermore, addition of free phosphine resulted in a retardation of the reaction. Unfortunately, simple nickel salts did not display any reactivity in the presence of added PPh$_3$; rather, the system afforded very good yields of coupling products for aryl iodides, but produced only sluggish conversions of aryl bromides (Scheme 6.26).

A simplified procedure uses nickel nanoparticles as catalyst with CuI and PPh$_3$ [83], where the size of the particles had an important effect on the effectiveness of the system. Hence, nickel(0) powder with a 100 nm diameter provided a high catalytic activity and quantitative conversion of 4-MeC$_6$H$_4$I, whereas no conversion was observed when 100-mesh powder was used, while reducing the size further to submicron dimensions also decreased the conversion to 57%. Polar solvents, such as *i*-PrOH or DMF, with KOH or K$_2$CO$_3$ as bases, gave consistently excellent yields. The reaction was also seen to tolerate (hetero)aryl halides and alkenyl iodides, as well as alkyl or aryl alkynes (Scheme 6.27).

The heterogeneous nature of the catalyst permits an easy separation and reuse up to seven times, without any significant loss of reactivity.

6.3.2
Ruthenium-Based Catalytic Systems

Chang and coworkers [84] expanded the scope of previously explored ruthenium-catalyzed Suzuki–Miyaura and Heck-type reactions [85] to the Sonogashira couplings of activated aryl iodides with terminal acetylenes. In spite of the high reaction temperatures of 90 °C, long reaction times and the use of $CuCl_2$ as an additive rather than the more commonly employed CuI, there was little restriction in terms of substrate scope. As with many other examples of solid-supported catalyst, the nature of the porous material affected the efficiency; under identical reaction conditions, the yield of product could vary from 7% with MgO to 97% when using alumina as the support. In the latter case, the absence of a copper cocatalyst decreased the yield to 70%.

6.3.3 Indium-Based Catalytic Systems

A rather strange, yet notable, synthesis of internal alkynes, starting from commonly used substrates for Sonogashira couplings, was reported by Prajapati and coworkers [86]. The peculiarity of this transformation was the nature of the metal center, namely indium. Aryl halides, including aryl iodides, bromides, chlorides and even fluorides, were converted to the corresponding products within 4 h. Regardless of the exact reaction mechanism, it is clear that activation of the aryl halides is not the rate-determining step (Scheme 6.28).

No additional phosphines or cooper salts were required, and $InCl_3$ itself catalyzed this transformation. Additional studies are required in order to carefully evaluate the mechanism of this transformation, which undoubtedly is different from that of the palladium-mediated Sonogashira reactions. Nonetheless, the simplicity and activity of the catalytic system, as well as its generality with respect to the nature of the leaving group, renders this a promising alternative for the synthesis of internal alkynes.

6.3.4 Copper-Based Catalytic Systems

Copper-mediated alkynylations of alkenyl- or aryl-substituted compounds should be considered as variations of the original Castro–Stephens reaction under catalytic reaction conditions. Copper(I) salts stabilized by aryl-substituted phosphines can mediate the coupling of aryl iodides in polar aprotic solvents using either conventional heating [87] or microwave irradiation [88]. More recently, well-defined 1,10-phenanthroline-derived copper complexes were shown to be effective catalysts for the coupling of activated and deactivated aryl iodides at 10 mol% catalyst loading in an apolar solvent (Scheme 6.29) [89].

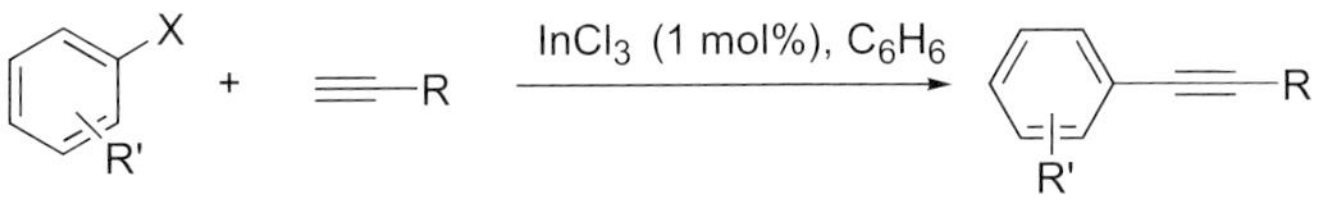

X R' + ≡—R —$InCl_3$ (1 mol%), C_6H_6→ R'-aryl—≡—R

X = I, Br, Cl, F

Scheme 6.28

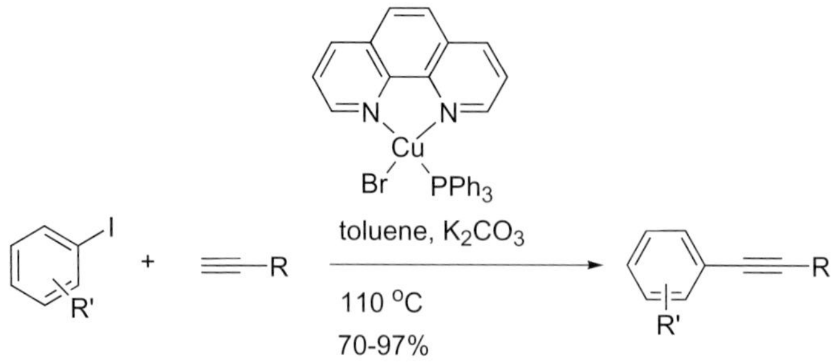

Br PPh$_3$ toluene, K_2CO_3 110 °C 70-97%

Scheme 6.29

Copper nanoclusters are active in Sonogashira reactions involving aryl iodides and some bromides when TBAA is used as base [90].

Generally speaking, palladium-free catalysts for Sonogashira reactions must be regarded with extreme caution, since it is known that cross-coupling reactions can be performed when only ppb levels of palladium are present in some inorganic bases [91].

6.4 Mechanism of the Sonogashira Reaction

Whilst the mechanism of the Sonogashira reaction remains a subject of debate, most investigators agree that the catalytic cycle consists of elementary steps that are closely related to other palladium-catalyzed reactions, such as the Heck coupling of olefins with aryl halides. While oxidative addition, transmetallation and reductive elimination are generally accepted, the detailed nature of these individual steps is not fully understood. The use of copper salts, as reported by Sonogashira in 1975, greatly enhances the reactivity of the catalytic system, and complicates its working mode through an additional copper-based catalytic cycle, which involves the transfer of acetylide from copper to palladium (Scheme 6.30). Therefore, an understanding of the Sonogashira reaction mechanism requires the elucidation of two different scenarios: (i) reactions performed in the presence of copper additives; and (ii) transformations conducted in their absence. Thus, it is highly likely that there is not a singular mechanism operating in all catalytic Sonogashira reactions.

Reaction mechanisms, particularly in complex catalytic sequences, are most often obscured by the formation of transient species or intermediates in very low concentrations, which are often not observable by standard means of analysis, such as chromatography or nuclear magnetic resonance (NMR) and infrared (IR) spectroscopies. As a consequence, a mechanism should be considered as the best model involving all available data at a given time, and thus may naturally be subject to further refinements.

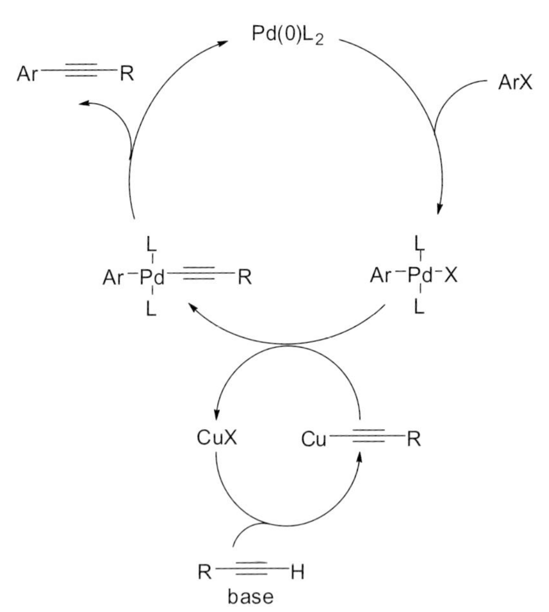

Scheme 6.30

6.4.1 Palladium- and Copper-Based Catalytic Systems

It is widely believed that the Sonogashira coupling performed in the presence of both palladium and copper salts proceeds via the transmetallation of alkynyl substituents from copper to palladium(II) complexes, which in turn is generated by the oxidative addition of aryl halides to an electron-rich palladium(0) center. This assumption is based on the fact that many *in situ*-generated, as well as isolated, alkynylcopper reagents react easily with transition metals from Group V to Group X to form σ-alkynyl complexes [92]. The earliest reports on alkynyl ligand transfer from copper or silver to arylpalladium intermediates were made by Yamamoto [93] and Espinet [94]. In the latter study, very stable complexes [$Pd(C_6F_5)(C{\equiv}CPh)L_2$], where L = PPh_3, showed no sign of reductive elimination from either its *cis*- or *trans*-isomers. Thus, it could be speculated that the electron-withdrawing character of the perfluorinated phenyl-substituent would stabilize this complex and prevents elimination.

In the former case, the reaction of [PdAr(I)Ln], where L is electron-rich PEt3, was examined (Scheme 6.31). Here, the reaction between the palladium complex, with R = Me, and 1 equiv. of [$Cu(CCPh)(PPh_3)]_4$ led to 74% of the Sonogashira product along with unreacted starting material. When the palladium : copper ratio was decreased to 1 : 2, full conversion to the coupling product was observed. The intermediate of this reaction, an aryl(alkynyl)palladium complex, was formed in 65% as its *trans*-isomer, when the reaction temperature was lowered to −30 °C. The

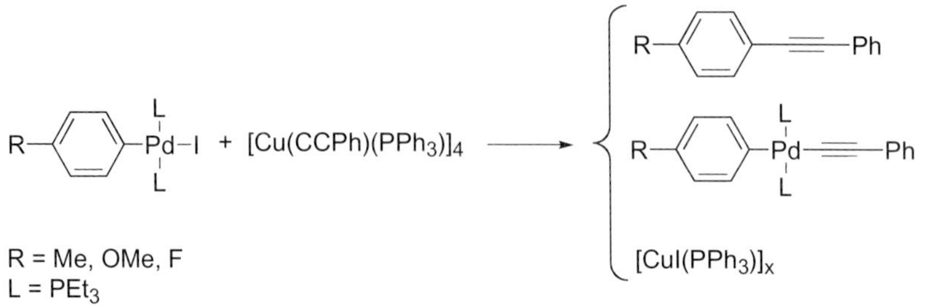

R–C₆H₄–Pd(L)₂–I + [Cu(CCPh)(PPh3)]4 → R–C₆H₄–C≡C–Ph; R–C₆H₄–Pd(L)₂–C≡C–Ph; [CuI(PPh3)]x

R = Me, OMe, F
L = PEt3

Scheme 6.31

authors proposed a rapid reductive elimination of the internal alkyne from the transient *cis*-aryl(alkynyl)palladium, while its *trans*-isomer was thought to undergo reductive elimination via a dissociative mechanism. The required phosphine dissociation of this dissociative mechanism is inhibited by an excess of ligand present in the reaction mixture. Variable-temperature NMR experiments showed ratios of aryl(alkynyl)palladium and the unreacted starting complex of 36 : 64 and 49 : 51 at −30 °C and −10 °C reaction temperatures, respectively. Only at 25 °C, was the formation of a coupling product observed. The alkynyl group migration from copper to palladium is reversible, as was proven by the reaction of aryl(alkynyl)palladium with CuI in the presence of added PPh_3. This reaction led to the originally used aryl(iodo)palladium and copper acetylide. However, an irreversible process was observed when PPh_3 was replaced by PEt_3, which was attributed to its much more basic character. Two plausible mechanisms for the reductive elimination were proposed in these systems: (i) the CuI-assisted removal of a phosphine ligand, leading to a three-coordinate palladium species, and subsequent reductive elimination of the product; and (ii) a reversible alkynyl transfer between metal centers, corresponding to an overall a *cis*/*trans*-isomerization of *trans*-$Pt(C{\equiv}CPh)_2(PPh_2Me)_2$ (Scheme 6.32) [95].

The selective and reversible alkynyl-transfer from copper to palladium is kinetically favored over an aryl-transfer, and thereby prevents formation of the biaryls products [96].

6.4.2
Copper-Free Catalytic Systems

Oxidative addition of aryl halides to an electron-rich palladium(0) species depends substantially on the nature of the $C(sp^2)$–X bond. Relatively weak bonds, as in aryl iodides or triflates, lead to a faster oxidative addition, but the activations of stronger bonds, as in aryl bromides – and especially chlorides – require very good donor ligands. This dependence of oxidative addition rates on the nature of the electrophile in multistep reactions may lead to different rate-determining steps. Apart from the nature of the aromatic substrate, the possible interaction of the terminal alkyne starting material and the internal alkyne product with the metal center of the catalyst can alter the kinetics of oxidative addition. The kinetics of addition of

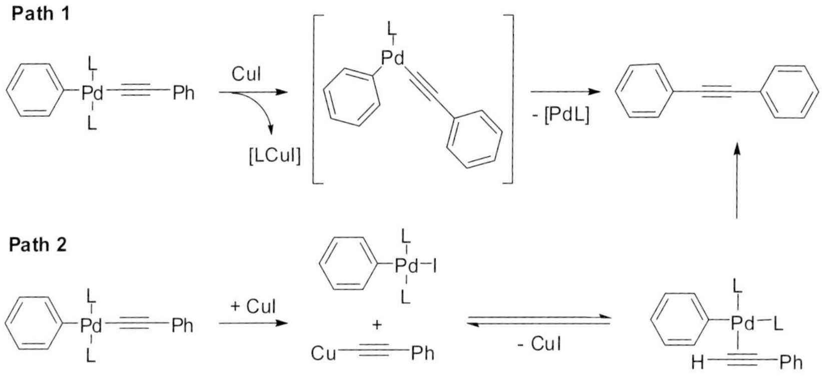

Path 1

Path 2

Scheme 6.32

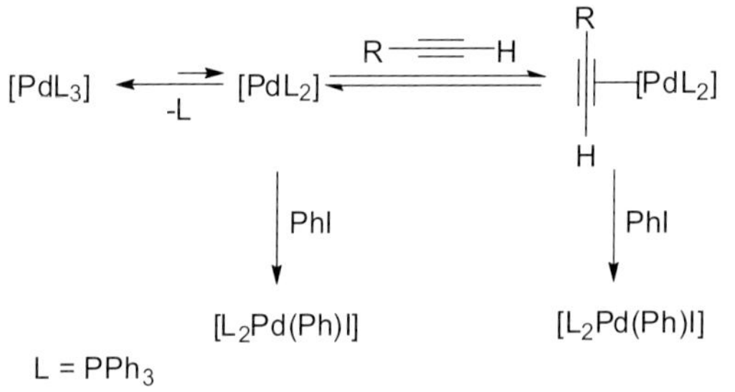

L = PPh_3

Scheme 6.33

PhI to $[Pd(PPh_3)_4]$ in the presence of terminal alkynes was investigated in detail by Amatore and Jutand [97], using NMR and amperometric techniques. In the absence of PhI, the palladium(0) complex was found to interact with phenylacetylene. The formation of palladium hydride species by oxidative addition of the terminal alkyne substrate was ruled out, *inter alia*, by the lack of characteristic hydride NMR resonances. The authors postulated the formation of a η^2-alkyne–palladium coordination complex with only two phosphine ligands. Considering that 14-electron complexes $[PdL_2]$ are typically active species in oxidative additions, it is understandable that a coordinated alkyne will decrease the rate of oxidative addition of PhI to a metal center. Electron-poor ethyl propiolate as substrate is involved in a similar equilibrium; however, the rate of oxidative addition of PhI to the metal complex showed here a saturation behavior at high concentrations of the alkyne. At low concentrations of alkyne, the dominant species for oxidative addition remained $[Pd(PPh_3)_2]$. In contrast, at high concentrations the equilibrium was completely shifted towards the alkyne-coordinated palladium complex, which then acted as the active catalyst (Scheme 6.33).

The equilibrium constant for the displacement of one PPh_3 ligand from $[Pd(PPh_3)_3]$ by ethyl propiolate and phenylacetylene was found to be 29 and 0.012, respectively. As expected, electron-poor alkynes are much better ligands for an electron-rich palladium center when compared to olefins. Similar coordination of olefins [98] leads to poorly active catalysts in Heck or Stille reactions, such as styrene ($K_0 = 0.0048$). By analogy, electron-rich internal alkynes affect the rates in a less significant manner. It can be speculated that the products of the Sonogashira coupling have, thus, rather a limited influence on the reaction rates, even at higher conversions when the concentration of the internal alkyne product is much higher than that of the terminal alkyne starting material. This decelerating effect of the alkyne may be beneficial for the overall catalysis through a leveling effect. Thereby, the rate of the slow transmetallation step will be closer to the rate of oxidative addition, in the case of more reactive substrates, such as aryl iodides [99].

The complexity of the Sonogashira catalytic cycle is evident when one considers that amines are often used in large excess [100], or even as solvents [101], in order to deprotonate the alkyne coordinated to the metal center. The coordination of the internal alkyne to palladium or copper has been speculated, based on the fact that amines by themselves are not basic enough to deprotonate free terminal alkynes. However, amines in large excess can interact with many of the catalytically competent intermediates. In copper-free Sonogashira couplings, the acetylene should therefore be coordinated not only to the palladium(0) species but also to the product of oxidative addition. In doing so, the acidity of the coordinated acetylene would be sufficient so as to allow its deprotonation by amines as bases. In a catalytic cycle involving $[Pd(PR_3)_2]$ as a potentially active species, this can be translated as the displacement of a phosphine ligand by the acetylenic substrate. This reaction pathway may seem less likely and, indeed, Jutand and coworkers [102] subsequently showed that reactions could be performed in the presence of amines, whereas in the absence of alkynes the reaction might follow a different path, where a phosphine is substituted by an amine. This substitution was reversible, and the amount of amino(phosphino)palladium(II) complex would increase at the expense of $[(PPh_3)_2Pd(II)]$, when amine was added in excess (Scheme 6.34).

However, substitution of the second phosphine has never been observed and, under stoichiometric reaction conditions the substitution equilibrium is strongly shifted toward bis-phosphine complexes. Under catalytic reaction conditions, where an excess of amine is present, a substantial amount of amino(phosphino) palladium(II) complex is present in solution. Under these circumstances, a substitution of an amine, rather than of a phosphine, by the alkyne is of greater importance. The details of a comprehensive mechanistic study on these interactions were reported recently for the coupling of PhI and a terminal alkyne mediated by $[PdL_2]$, with L = PPh_3 or $AsPh_3$, in the presence of secondary amines (Scheme 6.35) [103].

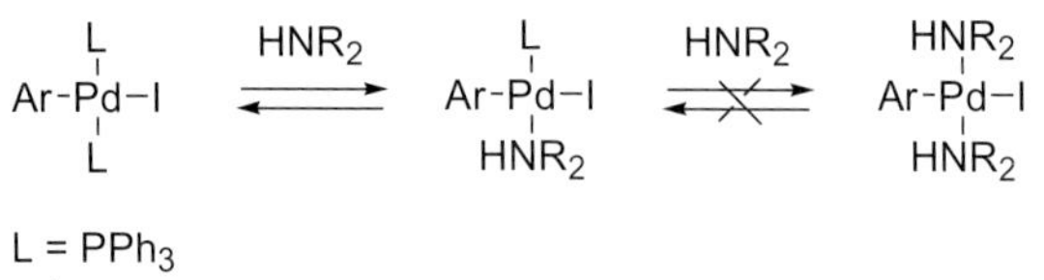

$$\mathrm{Ar{-}Pd(L)_2{-}I} \underset{}{\overset{HNR_2}{\rightleftharpoons}} \mathrm{Ar{-}Pd(L)(HNR_2){-}I} \underset{}{\overset{HNR_2}{\not\rightleftharpoons}} \mathrm{Ar{-}Pd(HNR_2)_2{-}I}$$

L = PPh_3

Scheme 6.34

PhI + HC≡C–CH₂CH₂–OH $\xrightarrow[\text{HNR}_2,\ \text{RT}]{\text{Pd(dba)}_2,\ 2\text{ L}}$ Ph–C≡C–CH₂CH₂–OH

Scheme 6.35

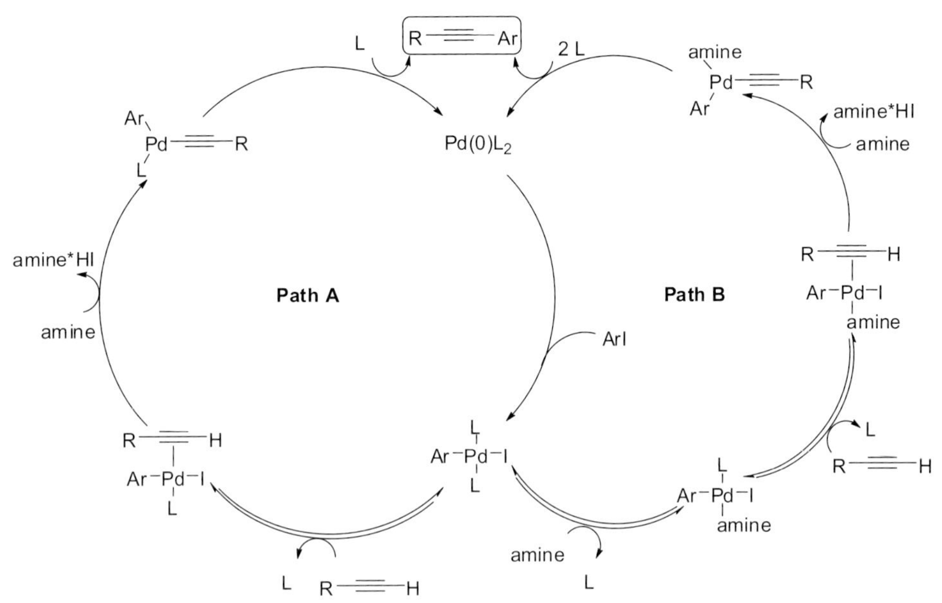

Scheme 6.36

The results showed that [Pd(dba)$_2$] and 2 equiv. of PPh$_3$ generated a much more active catalytic system when compared to [Pd(PPh$_3$)$_4$], which in turn underwent faster oxidative additions with ArI than with the *in situ*-generated complex. This reversal in reactivity was attributed to oxidative addition being rate-determining. The influence of the free phosphine points further to its involvement in the rate-determining step. Similarly, different rates were observed when the nature of the amine was modified, while the reactions also proceeded faster in the presence of the amine. This observation was attributed to a reversible formation of amino(ph osphino)palladium(II) intermediates. Differences in the ability of amines to substitute ligands in the transmetallation step suggest a branched catalytic sequence. In Scheme 6.36, while the alkyne displaces one ligand in a transmetallation step in Path A, the amine substitutes a ligand along Path B. For example, with PPh$_3$ as ligand, its substitution by the amine is not favored, and reaction Path A becomes dominant. In contrast, weaker ligands such as AsPh$_3$ will divert the reactivity towards Path B.

The relative rates of substitution of a given ligand L by either amine or alkyne will influence the reaction pathways. Because these rates depend on the concentration of the reaction partners, a change in the relative concentrations may lead to a switch from Path A to Path B, or vice versa.

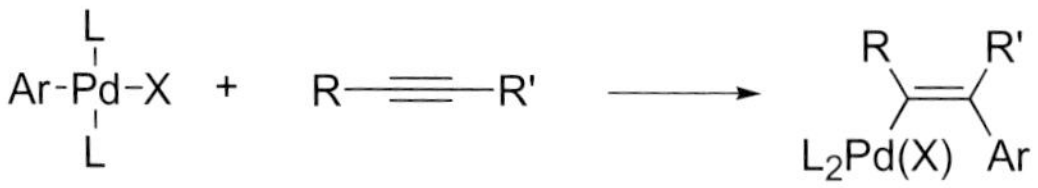

Scheme 6.37

Despite recent mechanistic progress in both copper-assisted couplings and copper-free reactions, numerous questions remain unanswered. For example, the catalytic cycle of Sonogashira reactions employing inorganic bases is less clear. Generally, new reports on metal-catalyzed coupling reactions often challenge our understanding, and invoke new mechanisms or intermediates. In this context, electrochemical studies conducted by Amatore and Jutand are worthy of mention, as these demonstrated a significant contribution of anionic intermediates [104], if excess of halides or bases (such as MO*t*-Bu) were used. Furthermore, the tri-coordinated oxidative addition intermediates of T- or Y- geometry may form if very bulky ligands are used with a ligand : palladium ratio of 1 : 1 [105]. These highly destabilized intermediates may act as substantially more active catalytic species, because the *trans* to *cis* isomerization required for the reductive elimination step in four-coordinated complexes is irrelevant for these [LPd(Ar)X] species.

These mechanistic aspects are not only driven by pure academic interests, but may also find relevance in industrial processes. An in-depth understanding of the reaction steps can lead to better catalysts in terms of, for example, TON, TOF or scope, whereas a limited understanding of the chemistry involved – for the Sonogashira coupling in this case – may lead to the incorrect choice of base, ligand or ligand : metal ratio. Indeed, the incorrect choice of can lead to a fundamental change in the course of the reaction towards single or even multiple insertions of alkynes into palladium–aryl bonds – the so-called carbopalladation reactions (Scheme 6.37) [106]. Carbopalladation may by itself in some cases serve as a relevant step in Sonogashira-type reactions.

On many occasions, poorly designed catalysts can easily decompose to palladium black or to partially soluble palladium nanoclusters, and therefore the assignment as an 'active catalyst' should always be verified by performing background reactions. Recent advances in the chemistry of metallic nanoclusters in catalysis have revealed that particles of approximately 2 nm diameter can be solubilized in polar solvents [107], and then serve as soluble catalysts with kinetics similar to that of monomeric palladium complexes. It should be noted that low concentrations of palladium lead to these soluble species, whereas higher concentrations give rise to larger aggregates with a lower catalytic activity. Consequently, for some poorly active catalysts this aggregation can be considered as a source of catalytic activity.

6.5
Concluding Remarks

The Sonogashira coupling of alkynes with aryl halides has undergone amazing developments during the past three decades, with the numbers of potential ligands, metal precursors, solvents, bases or additives being very large, and those of their combinations staggering. Nonetheless, new ligands continue to be developed, the main aim being to improve catalytic activity towards problematic substrates and to improve system stability. An inverse relationship is observed between price and ease of activation for most aryl halides, with aryl iodides normally being at least an order of magnitude more expensive than their aryl chloride counterparts. Yet, the number of reliable systems available for the coupling of aryl chlorides is small, and for aryl tosylates it is even smaller. In future, the concept of 'designer' ligands will be expanded to develop new classes of ligands, while both steric demand and electron-donating ability will remain dominant themes when creating novel ligands for new reactions. Many attempts at ligand design have been directed towards nonphosphine systems, partly for environmental reasons, and simple amines, ligands that form palladacycles, heterocycles (such as carbenes) and their combinations have led to promising results, notably in terms of catalyst robustness. Catalyst recyclability, separation techniques, heterogenized systems, polymer supports and dendrimers, as well as biphasic and multiphasic conditions, are more frequently encountered when considering industrial applications.

Despite the many advances made in Sonogashira coupling protocols, the results remain rather modest compared to those for related transformations, such as Heck reactions, with typical differences in the TONs of these reactions being at least an order of magnitude. Although much of the catalytic sequence is now understood (mostly in analogy with other palladium-catalyzed reactions), aspects such as the involvement of copper acetylides as transmetallating agents in copper-assisted Sonogashira reactions are still not fully understood. Likewise, copper-free reactions are highly complex and remain the subject of further, detailed investigations. It is believed that an understanding of the complex role of alkynes, bases and ligands may represents the only rational means of improving catalyst activity and selectivity and, while most studies have addressed the need to use copper additives, very few have explored potential alternatives such as silver salts or silylated alkynes.

It is very unlikely that future research in this area will provide the answers to all of the problems associated with the Sonogashira reaction. Whilst for large-scale industrial applications the main considerations are price, simplicity and robustness of the catalytic system, the small-scale syntheses that are associated with pharmaceuticals will seek catalytic systems tailored to minimize adverse side reactions and the loss of precious substrates. Clearly, there is still much to be done, and many surprises lie in wait!

Abbreviations

Amphos	[α-methyl-2-(diphenylphosphino)-benzyl]dimethylamine
BINAP	2,2′-bis(diphenylphosphino)-1,1′-binaphthyl
(bmim)BF4	1-butyl-3-methylimidazolium tetrafluoroborate

dba	dibenzylidene acetone
DMAc	*N,N*-dimethylacetamide
DMF	dimethylformamide
dppe	1,2-bis(diphenylphosphino)ethane
dppf	1,1′-bis(diphenylphosphino)ferrocene
dppp	1,3-bis(diphenylphosphino)propane
Fc	ferrocene
IMes	bis-1,3-(2,4,6-trimethylphenyl)imidazol-2-ylidene
LDA	lithium diisopropylamide
NHC	*N*-heterocyclic carbene
NMP	*N*-methylpyrrolidone
PEG	polyethylene glycol
ROMP	ring-opening metathesis polymerization
TBAA	tetrabutylammonium acetate
TBAB	tetrabutylammonium bromide
TBAC	tetrabutylammonium chloride
TBAF	tetrabutylammonium fluoride
TOF	turnover frequency
TON	turnover number
TPPMS	triphenylphosphine monosulfonate

References

1 Cassar, L. (1975) *J. Organomet. Chem.*, **93**, 253.

2 Diek, H.A. and Heck, F.R. (1975) *J. Organomet. Chem.*, **93**, 259.

3 Stephens, R.D. and Castro, C.E. (1963) *J. Org. Chem.*, **28**, 2163.

4 Sonogashira, K., Tohda, Y. and Hagihara, N. (1975) *Tetrahedron Lett.*, **16**, 4467.

5 (a) King, A.O., Okukado, N. and Negishi, E. (1977) *J. Chem. Soc. Chem. Commun.*, 683;
(b) King, A.O., Negishi, E., Villani, F.J., Jr and Silveira, A., Jr (1978) *J. Org. Chem.*, **43**, 358;
(c) Negishi, E. (1978) *Aspects of Mechanism and Organometallic Chemistry* (ed. J.H. Brewster), Plenum Press, New York, pp. 285–317;
(d) Negishi, E. (1982) *Acc. Chem. Res.*, **15**, 340;
(e) Negishi, E. and Xu, C. (2002) *Handbook of Organopalladium Chemistry for Organic Synthesis* (ed. E. Negishi), John Wiley & Sons, Inc., New York, pp. 531–49.

6 (a) Soderquist, J.A., Matos, K., Rane, A. and Ramos, J. (1995) *Tetrahedron Lett.*, **36**, 2401;
(b) Soderquist, J.A., Rane, A.M., Matos, K. and Ramos, J. (1995) *Tetrahedron Lett.*, **36**, 6847;
(c) Fürstner, A. and Seidel, G. (1995) *Tetrahedron*, **51**, 11165;
(d) Fürstner, A. and Nikolakis, K. (1996) *Liebigs Ann. Chem.*, 2107.

7 (a) Kashin, A.N., Bumagina, I.G., Bumagin, N.A., Beletskaya, I.P. and Reutov, O.A. (1980) *Izv. Akad. Nauk SSSR, Ser. Khim.*, 479;
(b) Bumagin, N.A., Bumagina, I.G. and Beletskaya, I.P. (1983) *Dokl. Akad. Nauk SSSR*, **272**, 1384;
(c) Stille, J.K. (1986) *Angew. Chem. Int. Ed. Engl.*, **25**, 1;
(d) Farina, V. and Krishnan, B. (1991) *J. Am. Chem. Soc.*, **113**, 9585;

(e) Farina, V., Kapadia, S., Krishnan, B., Wang, C.J. and Libeskind, L.S. (1994) *J. Org. Chem.*, **59**, 5905.

8 (a) Sonogashira, K. (2004) *Metal-Catalyzed Cross-Coupling Reactions*, Vol. **1** (eds A. de Meijere and F. Diederich), Wiley-VCH Verlag GmbH, Weinheim, p. 319;
(b) Negishi, E. and Anastasia, L. (2003) *Chem. Rev.*, **103**, 1979;
(c) Sonogashira, K. (1991) *Comprehensive Organic Synthesis*, Vol. **3** (eds B.M. Trost and I. Fleming), Pergamon, Oxford, p. 521;
(d) A recent comprehensive review on Sonogashira methodology includes an updated, extensive section on applications: Chinchilla, R. and Najera, C. (2007) *Chem. Rev.*, **107**, 874.

9 Sonogashira, K. (2002) *J. Organomet. Chem.*, **653**, 46.

10 Paul, F., Patt, J. and Hartwig, J.F. (1994) *J. Am. Chem. Soc.*, **116**, 5969.

11 Alami, M., Ferri, F. and Linstrumelle, G. (1993) *Tetrahedron Lett.*, **34**, 6403.

12 (a) Leadbeater, N.E. and Tominack, B.J. (2003) *Tetrahedron Lett.*, **44**, 8653;
(b) Woon, E.C.Y., Dhami, A., Mahon, M.F. and Threadgill, M.D. (2006) *Tetrahedron*, **62**, 4829;
(c) Pal, M., Parasuraman, K., Gupta, S. and Yeleswarapu, K.R. (2002) *Synlett*, 1976.

13 Fu, X., Zhang, S., Yin, J. and Schumacher, D. (2002) *Tetrahedron Lett.*, **43**, 6673.

14 Shirakawa, E., Kitabata, T., Otsuka, H. and Tsuchimoto, T. (2005) *Tetrahedron*, **61**, 9878.

15 Böhm, V.P.H. and Hermann, W.A. (2000) *Eur. J. Org. Chem.*, 3679.

16 Soheili, A., Albazene-Walker, J., Murry, J.A., Dormer, P.G. and Hughes, D.L. (2003) *Org. Lett.*, **5**, 4191.

17 Barrios-Landeros, F. and Hartwig, J.F. (2005) *J. Am. Chem. Soc.*, **127**, 6944.

18 Gelman, D. and Buchwald, S.L. (2003) *Angew. Chem. Int. Ed. Engl.*, **42**, 5993.

19 Yi, C. and Hua, R. (2006) *J. Org. Chem.*, **71**, 2535.

20 Juo, Y., Gao, H., Li, Y., Huang, W., Lu, W. and Zhang, Z. (2006) *Tetrahedron*, **62**, 2465.

21 (a) Hierso, J.-C., Fihri, A., Amardeil, R., Meunier, P., Doucet, H. and Santelli, M. (2005) *Tetrahedron Lett.*, **61**, 9759;
(b) Hierso, J.-C., Fihri, A., Amardeil, R., Meunier, P., Doucet, H., Santelli, M. and Ivanov, V.V. (2004) *Org. Lett.*, **6**, 3473.

22 Feuerstein, M., Doucet, H. and Santelli, M. (2005) *Tetrahedron Lett.*, **46**, 1717.

23 Méry, D., Heuzé, K. and Astruc, D. (2003) *Chem. Commun.*, 1934.

24 Remmele, H., Köllhofer, A. and Plenio, H. (2003) *Organometallics*, **22**, 4098.

25 Nishide, K., Liang, H., Ito, S. and Yoshifuji, M. (2005) *J. Organomet. Chem.*, **690**, 4809.

26 Arques, A., Auñón, D. and Molina, P. (2004) *Tetrahedron Lett.*, **45**, 4337.

27 Gabano, E., Cassino, C., Bonetti, S., Prandi, C., Colangelo, D., Ghiglia, A.L. and Osella, D. (2005) *Org. Biomol. Chem.*, **3**, 3531.

28 Halbes-Létinois, U. and Pale, P. (2003) *J. Organomet. Chem.*, **687**, 420.

29 (a) Bertus, P., Halbes, U. and Pale, P. (2001) *Eur. J. Org. Chem.*, 4391;
(b) Halbes, U., Bartus, P. and Pale, P. (2001) *Tetrahedron Lett.*, **42**, 8641.

30 Sorensen, U.S. and Pombo-Villar, E. (2005) *Tetrahedron*, **61**, 2697.

31 Liang, Y., Xie, Y.-X. and Li, J.-H. (2006) *J. Org. Chem.*, **71**, 379.

32 (a) Roucoux, A., Schultz, J. and Patin, H. (2002) *Chem. Rev.*, **102**, 3757;
(b) Moreno-Mañas, M. and Pleixats, R. (2003) *Acc. Chem. Res.*, **36**, 638;
(c) Reetz, M.T. and Westermann, E. (2000) *Angew. Chem. Int. Ed. Engl.*, **39**, 165;
(d) Reetz, M.T. and De Vries, J.G. (2004) *Chem. Commun.*, 1559.

33 (a) Erdélyi, M., Langer, V., Karlén, A. and Gogoll, A. (2002) *New J. Chem.*, **26**, 834;
(b) Han, J.W., Castro, J.C. and Burgess, K. (2003) *Tetrahedron Lett.*, **44**, 9359;
(c) Melucci, M., Barbarella, G., Zambianchi, M., Di Pietro, P. and Bongini, A. (2004) *J. Org. Chem.*, **69**, 4821;
(d) Kuang, C., Yang, Q., Senboku, H. and Tokuda, M. (2005) *Tetrahedron*, **61**, 4043;
(e) Zheng, S.-L., Reid, S., Lin, N. and Wang, B. (2006) *Tetrahedron Lett.*, **47**, 2331;

(f) Erdélyi, M. and Gogoll, A. (2001) *J. Org. Chem.*, **66**, 4165.
34 Erdélyi, M. and Gogoll, A. (2003) *J. Org. Chem.*, **68**, 6431.
35 Kabalka, G.W., Wang, L., Namboodiri, V. and Pagni, R.M. (2000) *Tetrahedron Lett.*, **41**, 5151.
36 Kmentova, I., Gotov, B., Gajda, V. and Toma, S. (2003) *Monatsh. Chem.*, **134**, 545.
37 Fukuyama, T., Shinmen, M., Nishitani, S., Sato, M. and Ryu, I. (2002) *Org. Lett.*, **4**, 1691.
38 Casalnuovo, A.L. and Calabrese, J.C. (1990) *J. Am. Chem. Soc.*, **112**, 4324.
39 (a) Amatore, C., Blart, E., Genêt, J.P., Jutand, A., Lemaire-Audoire, S. and Savignac, M. (1995) *J. Org. Chem.*, **60**, 6829;
(b) DeVashner, R.B., Moore, L.R. and Shaughnessy, K.H. (2004) *J. Org. Chem.*, **69**, 7919.
40 Anderson, K.W. and Buchwald, S.L. (2005) *Angew. Chem. Int. Ed. Engl.*, **44**, 6173.
41 Genin, E., Amengual, R., Michelet, V., Savignac, M., Jutand, A., Neuville, L. and Genêt, J.-P. (2004) *Adv. Synth. Catal.*, **346**, 1733.
42 Wolf, C. and Lerebours, R. (2004) *Org. Biomol. Chem.*, **2**, 2161.
43 DeVashner, R.B., Moore, L.R. and Shaughnessy, K.H. (2004) *J. Org. Chem.*, **69**, 7919.
44 Bumagin, N.A., Sukhomlinova, L.I., Luzikova, E.V., Tolstaya, T.P. and Beletskaya, I.P. (1996) *Tetrahedron Lett.*, **37**, 897.
45 Ahmed, M.S.M. and Mori, A. (2004) *Tetrahedron*, **60**, 9977.
46 Tzschucke, C.C., Markert, C., Glatz, H. and Bannwarth, W. (2002) *Angew. Chem. Int. Ed. Engl.*, **41**, 4500.
47 Li, Y., Li, Z., Li, F., Wang, Q. and Tao, F. (2005) *Tetrahedron Lett.*, **46**, 6159.
48 Gonthier, E. and Breinbauer, R. (2003) *Synlett*, 1049.
49 Heuzé, K., Méry, D., Gauss, D. and Astruc, D. (2003) *Chem. Commun.*, 2274.
50 Yang, Y.-C. and Luh, T.-Y. (2003) *J. Org. Chem.*, **68**, 9870.
51 Datta, A. and Plenio, H. (2003) *Chem. Commun.*, 1504.
52 Köllhofer, A. and Plenio, H. (2003) *Chem. Eur. J.*, **9**, 1416.
53 Bourissou, D., Guerret, O., Gabbai, F.P. and Bertrand, G. (2000) *Chem. Rev.*, **100**, 39.
54 Sole, S., Gornitzka, H., Schoeller, W.W., Bourissou, D. and Bertrand, G. (2001) *Science*, **292**, 1901.
55 (a) Huang, J., Stevens, E.D., Nolan, S.P. and Petersen, J.L. (1999) *J. Am. Chem. Soc.*, **121**, 2674;
(b) Huang, J., Schanz, H.J., Stevens, E.D. and Nolan, S.P. (1999) *Organometallics*, **18**, 2370.
56 Herrmann, W.A., Reisinger, C.-P. and Spiegler, M. (1998) *J. Organomet. Chem.*, **557**, 93.
57 Yang, C. and Nolan, S.P. (2002) *Organometallics*, **21**, 1020.
58 Lebel, H., Janes, M.K., Charette, A.B. and Nolan, S.P. (2004) *J. Am. Chem. Soc.*, **126**, 5046.
59 Ma, Y., Song, C., Jiang, W., Wu, Q., Wang, Y., Liu, X. and Andrus, M.B. (2003) *Org. Lett.*, **5**, 3317.
60 Batey, R.A., Shen, M. and Lough, A.J. (2002) *Org. Lett.*, **4**, 1411.
61 McGuinness, D.S. and Cavell, K.J. (2000) *Organometallics*, **19**, 741.
62 Mas-Marza, E., Segarra, A.M., Claver, C., Peris, E. and Fernandez, E. (2003) *Tetrahedron Lett.*, **44**, 6595.
63 Dhudshia, B. and Thadani, A.N. (2006) *Chem. Commun.*, 668.
64 Stevens, P.D., Li, G., Fan, J., Yen, M. and Gao, Y. (2005) *Chem. Commun.*, 4435.
65 (a) Alacid, E., Alonso, D.A., Botella, L., Najera, C. and Pacheco, M.C. (2006) *Chem. Rec.*, **6**, 117;
(b) Bedford, R.B. (2003) *Chem. Commun.*, 1787;
(c) Beletskaya, I.P. and Cheprakov, A.V.K. (2004) *J. Organomet. Chem.*, **689**, 4055.
66 (a) Herrmann, W.A., Bohm, V.P.W. and Reisinger, C.-P. (1999) *J. Organomet. Chem.*, **576**, 23;
(b) Herrmann, W.A., Oefele, K., von Preysing, D. and Schneider, S.K. (2003) *J. Organomet. Chem.*, **687**, 229.
67 Lin, C.-A. and Luo, F.-T. (2003) *Tetrahedron Lett.*, **44**, 7565.

68 Thakur, V.V., Ramesh Kumar, N.S.C. and Sudalai, A. (2004) *Tetrahedron Lett.*, **45**, 2915.

69 (a) Alonso, D.A., Nájera, C. and Pacheco, M.C. (2002) *Tetrahedron Lett.*, **43**, 9365;
(b) Alonso, D.A., Nájera, C. and Pacheco, M.C. (2003) *Adv. Synth. Catal.*, **345**, 1146.

70 Corma, A., García, H. and Leyva, A.J. (2006) *J. Catal.*, **240**, 87.

71 Fairlamb, I.J.S., Kapdi, A.R., Lee, A.F., Sanchez, G., López, G., Serrano, J.J., García, L., Pérez, J. and Pérez, E. (2004) *Dalton Trans.*, 3970.

72 Gossage, R.A., Jenkins, H.A. and Yadav, P.N. (2004) *Tetrahedron Lett.*, **45**, 7689.

73 Nájera, C., Gil-Moltó, J., Karström, S. and Falvello, L.R. (2003) *Org. Lett.*, **5**, 1451.

74 Gil-Moltó, J. and Nájera, C. (2005) *Eur. J. Org. Chem.*, 4073.

75 Park, S.B. and Alper, H. (2004) *Chem. Commun.*, 1306.

76 Wang, R., Piekarski, M.M. and Shreeve, J.M. (2006) *Org. Biomol. Chem.*, **4**, 1878.

77 (a) Sinner, F., Buchmeiser, M.R., Tessadri, R., Mupa, M., Wurst, K. and Bonn, G.K. (1998) *J. Am. Chem. Soc.*, **120**, 2790;
(b) Buchmeiser, M.R. and Wurst, K. (1999) *J. Am. Chem. Soc.*, **121**, 11101.

78 Buchmeiser, M.R., Schareina, T., Kempe, R. and Wurst, K. (2001) *J. Organomet. Chem.*, **634**, 39.

79 Rau, S., Lamm, K., Görls, H., Schöffel, J. and Walther, D. (2004) *J. Organomet. Chem.*, **689**, 3582.

80 Bandini, M., Luque, R., Budarin, V. and Macquarrie, D.J. (2005) *Tetrahedron*, **61**, 9860.

81 Li, P.-H. and Wang, L. (2006) *Adv. Synth. Catal.*, **348**, 681.

82 Beletskaya, I.P., Latyshev, G.V., Tsvetkov, A.V. and Lukashev, N.V. (2003) *Tetrahedron Lett.*, **44**, 5011.

83 (a) Wang, L., Li, P. and Zhang, Y. (2004) *Chem. Commun.*, 514;
(b) Wang, M., Li, P. and Wang, L. (2004) *Synth. Commun.*, **34**, 2803.

84 Park, S., Kim, M., Hyun, D. and Chang, S. (2004) *Adv. Synth. Catal.*, **346**, 1638.

85 Na, Y., Park, S., Han, S.B., Han, H., Ko, S., and Chang, S. (2004) *J. Am. Chem. Soc.*, **126**, 250.

86 Borah, H.N., Prajapati, D. and Boruah, R. C. (2005) *Synlett*, 2823.

87 Okuro, K., Furuune, M., Enna, M., Miura, M. and Nomura, M. (1993) *J. Org. Chem.*, **58**, 4716.

88 Wang, J.-X., Liu, Z., Hu, Y., Wei, B. and Kang, L. (2002) *Synth. Commun.*, **32**, 1937.

89 (a) Gujadhur, R.K., Bates, C.G. and Venkataraman, D. (2001) *Org. Lett.*, **3**, 4315;
(b) Ma, D. and Feng, L. (2004) *Chem. Commun.*, 1934.

90 Thathagar, M.B., Beckers, J. and Rothenberg, G. (2004) *Green Chem.*, **6**, 215.

91 Arvela, R.K., Leadbeater, N.E., Sangi, M.S., Williams, V.A., Granados, P. and Singer, R.D. (2005) *J. Org. Chem.*, **70**, 161.

92 (a) Abu Salah, O.M. and Bruce, M.I. (1976) *Aust. J. Chem.*, **29**, 73;
(b) Sonogashira, K., Fujikura, Y., Yatake, T., Toyoshima, N., Takahashi, S. and Hagihara, N. (1978) *J. Organomet. Chem.*, **145**, 101;
(c) Fujikura, Y., Sonogashira, K. and Hagihara, N. (1975) *Chem. Lett.*, 1067;
(d) Ogawa, H., Joh, T., Takahashi, S. and Sonogashira, K. (1988) *J. Chem. Soc. Chem. Commun.*, 561;
(e) Sonogashira, K., Yatake, T., Tohda, Y., Takahashi, S. and Hagihara, N. (1977) *J. Chem. Soc. Chem. Commun.*, 291;
(f) Ohshiro, N., Takei, F., Onitsuka, K. and Takahashi, S. (1996) *Chem. Lett.*, 871;
(g) Cross, R.J. and Davidson, M.F. (1986) *J. Chem. Soc. Dalton Trans.*, 1987;
(h) Osakada, K., Takizawa, T. and Yamamoto, T. (1995) *Organometallics*, **14**, 3531;
(i) Bruce, M.I., Clark, R., Howard, J. and Woodward, P. (1972) *J. Organomet. Chem.*, **42**, C107;
(j) Abu Salah, O.M., Bruce, M.I. and Redhouse, A.D. (1974) *J. Chem. Soc. Chem. Commun.*, 855;
(k) Abu Salah, O.M. and Bruce, M.I. (1974) *J. Chem. Soc. Dalton Trans.*, 2302;
(l) Bruce, M.I., Abu Salah, O.M., Davis, R.E. and Reghavan, N.V. (1974)

J. Organomet. Chem., **64**, C48;
(m) Abu Salah, O.M. and Bruce, M.I. (1975) *J. Chem. Soc. Dalton Trans.*, 2311;
(n) Abu Salah, O.M., Bruce, M.I., Churchill, M.R. and Bezman, S.A. (1972) *J. Chem. Soc. Chem. Commun.*, 858;
(o) Churchill, M.R. and Bezman, S.A. (1974) *Inorg. Chem.*, **13**, 1418;
(p) Clark, R., Howard, J. and Woodward, P. (1974) *J. Chem. Soc. Dalton Trans.*, 2027;
(q) Yamazaki, S. and Deeming, A.J. (1993) *J. Chem. Soc. Dalton Trans.*, 3051;
(r) Yamazaki, S., Deeming, A.J., Hursthouse, M.B. and Malik, K.M.A. (1995) *Inorg. Chim. Acta*, **235**, 147;
(s) Tanaka, S., Yoshida, T., Adachi, T., Yoshida, T., Onitsuka, K. and Sonogashira, K. (1994) *Chem. Lett.*, 877;
(t) Osakada, K., Sakata, R. and Yamamoto, T. (1997) *J. Chem. Soc. Dalton Trans.*, 1265.

93 Osakada, K., Sakata, R. and Yamamoto, T. (1997) *Organometallics*, **16**, 5354.

94 Espinet, P., Forniés, J., Martinez, F., Sotes, M., Lalinde, E., Moreno, M.T., Ruiz, A. and Welch, A.J. (1991) *J. Organomet. Chem.*, **403**, 253.

95 (a) Cross, R.J. and Gemmill, J. (1984) *J. Chem. Soc. Dalton Trans.*, **199**, 205;
(b) Cross, R.J. and Davidson, M.F. (1985) *Inorg. Chim. Acta*, **97**, L35;
(c) Anderson, G.K. and Cross, R. (1980) *J. Chem. Soc. Rev.*, **9**, 185;
(d) Ozawa, F., Ito, T., Nakamura, Y. and Yamamoto, A. (1981) *Bull. Chem. Soc. Jpn.*, **54**, 1868;
(e) Ozawa, F., Kurihara, K., Yamamoto, T. and Yamamoto, A. (1985) *J. Organomet. Chem.*, **279**, 233;
(f) Koten, G. and Noltes, J.G. (1975) *J. Organomet. Chem.*, **84**, 129.

96 (a) Ozawa, F., Fujimori, M., Yamamoto, T. and Yamamoto, A. (1986) *Organometallics*, **5**, 2144;
(b) Yamamoto, T., Wakabayashi, S. and Osakada, K. (1992) *J. Organomet. Chem.*, **428**, 223;
(c) Osakada, K., Sato, R. and Yamamoto, T. (1994) *Organometallics*, **13**, 4645;
(d) van Koten, G. and Noltes, J.G. (1975) *J. Organomet. Chem.*, **84**, 129.

97 Amatore, C., Bensalem, S., Ghalem, S., Jutand, A. and Medjour, Y. (2004) *Eur. J. Org. Chem.*, 366.

98 (a) Jutand, A., Hii, K.K., Thornton-Pett, M. and Brown, J.M. (1999) *Organometallics*, **18**, 5367;
(b) Amatore, C., Carre, E., Jutand, A. and Medjour, Y. (2002) *Organometallics*, **21**, 4540.

99 Amatore, C., Carre, E. and Jutand, A. (1999) *J. Organomet. Chem.*, **576**, 254.

100 (a) Böhm, V.P.W. and Herrmann, W.A. (2000) *Eur. J. Org. Chem.*, 3679;
(b) Heidenreich, R.G., Kölher, K., Krauter, J.G.E. and Pietsch, J. (2002) *Synlett*, 1118;
(c) Fukuyama, T., Shinmen, M., Nishitani, S., Sato, M. and Ryu, I. (2002) *Org. Lett.*, **4**, 1691;
(d) Leadbeater, N.E. and Tominack, B.J. (2003) *Tetrahedron Lett.*, **44**, 8653.

101 (a) Alami, F. and Ferri, G. (1993) *Tetrahedron Lett.*, **34**, 6403;
(b) Pal, M., Parasuraman, K., Gupta, S. and Yeleswarapu, K.R. (2002) *Synlett*, 1976.

102 Jutand, A., Negri, S. and Principaud, A. (2005) *Eur. J. Inorg. Chem.*, 631.

103 Touguerti, A., Negri, S. and Jutand, A. (2007) *Chem. Eur. J.*, **13**, 666.

104 Amatore, C. and Jutand, A. (2000) *Acc. Chem. Res.*, **33**, 314.

105 Stambuli, J.P., Buhl, M. and Hartwig, J.F. (2002) *J. Am. Chem. Soc.*, **124**, 9346.

106 For reviews, see: (a) Cacchi, S. (1984) *Tetrahedron Lett.*, **25**, 3137;
(b) Cacchi, S. (1999) *J. Organomet. Chem.*, **576**, 42.

107 (a) de Vries, A.H.M., Mulders, J., Mommers, J.H.M., Henderick, H.J.W. and de Vries, J.G. (2003) *Org. Lett.*, **5**, 3285;
(b) Gniewek, A., Trzeciak, A.M., Ziolkoski, J.J., Kepinski, L., Wrzyszcz, J. and Tylus, W. (2005) *J. Catal.*, **229**, 332;
(c) Thathagar, M.B., Kooyman, P.J., Boerleider, R., Jansen, E., Elsevier, C.J. and Rothenberg, G. (2005) *Adv. Synth. Catal.*, **347**, 1965;
(d) Cassol, C.C., Umpierre, A.P., Machado, G., Wolke, S.I. and Dupont, J. (2005) *J. Am. Chem. Soc.*, **127**, 3298;
(e) Burello, E. and Rothenberg, G. (2005) *Adv. Synth. Catal.*, **347**, 1969.

7
Palladium-Catalyzed Arylation Reactions of Alkenes (Mizoroki–Heck Reaction and Related Processes)

Verena T. Trepohl and Martin Oestreich

7.1
Introduction

This chapter describes the basic, as well as current, aspects of three fundamentally important palladium-catalyzed alkene arylation reactions, all of which are major synthetic techniques for C—C bond formation. Although discovered in inverted chronological order, the palladium(0)-catalyzed arylations of alkenes (Sections 7.2 and 7.3) are described first, followed by the related palladium(II)-catalyzed processes (Section 7.4). This arrangement accounts for the impact that these transformations have had to date on organic synthesis. The Mizoroki–Heck reaction, which was reported independently by Mizoroki *et al.* [1] and Heck *et al.* [2], is certainly one of today's key C—C bond-forming reactions. Over the course of three decades, both intermolecular and intramolecular variants, often regio- and stereocontrolled, have been elaborated to meet the highest synthetic standards in terms of yield and selectivity. It is almost impossible to grasp all facets of this reaction in a single chapter, and therefore we will provide here only the essential principles of Mizoroki–Heck chemistry, as well as recent trends highlighting particularly attractive examples (see Section 7.2). A cognate process – the reductive Mizoroki–Heck-type arylation – was reported by Cacchi *et al.* [3] and Larock *et al.* [4] almost two decades later. Due to the limited yet interesting number of examples, an almost comprehensive survey of this chemistry is included (Section 7.3). The Fujiwara–Moritani reaction, which often is termed an oxidative Mizoroki–Heck-type process, actually predated the Mizoroki–Heck reaction [5]. While the latter palladium(0) catalysis has certainly outshone the former palladium(II) catalysis, recent strong interest in C—H instead of C—X bond activation has put the Fujiwara–Moritani reaction back on the 'chemical map'. In the final section we will summarize the currently emerging area of oxidative Mizoroki–Heck-type chemistry, including details of seminal palladium(II)-mediated contributions (Section 7.4).

Modern Arylation Methods. Edited by Lutz Ackermann

ISBN: 978-3-527-31937-4

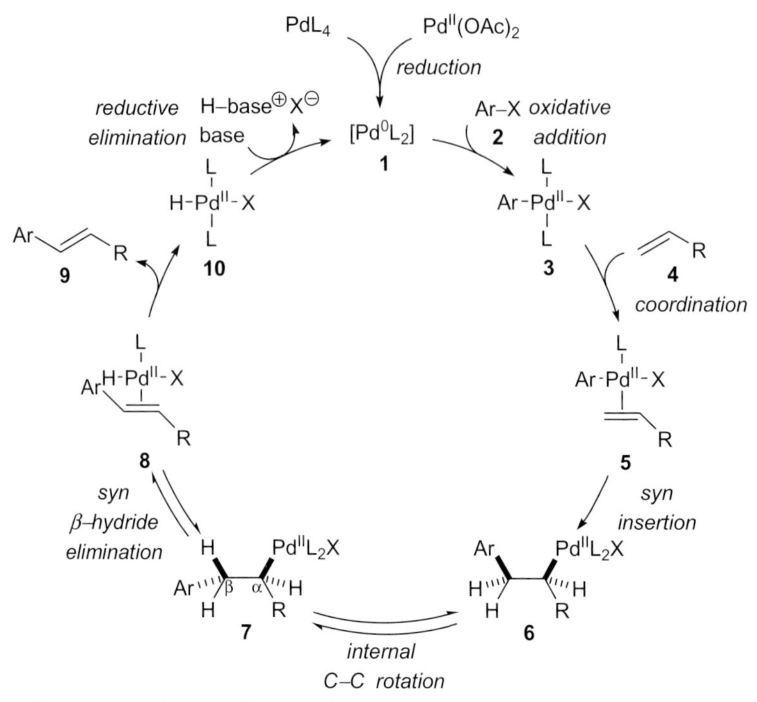

Scheme 7.1 Textbook mechanism of the Mizoroki–Heck reaction.

7.2 Mizoroki–Heck Arylations

7.2.1 Mechanistic Considerations

Although sophisticated mechanistic data were not available during the early days of Mizoroki–Heck chemistry, Dieck and Heck [6] introduced a fundamental mechanism (Scheme 7.1), the basic substeps of which are still generally accepted, albeit refined mechanistically [7–12]. As an entry into the catalytic cycle, the catalytically active 14-electron palladium(0) complex **1** is generated *in situ*, either by ligand dissociation of a Pd(0) precatalyst or by reduction of a Pd(II) precatalyst. The catalysis commences with oxidative addition of the C(sp^2)—X bond of **2** to **1**, thereby forming the *trans*-σ-aryl—palladium(0) complex **3** (**1**→**3**); the leaving group X^- strongly influences the reactivity of **2** in the order $I^- > TfO^- > Br^- \gg Cl^-$ [13]. Displacement of a ligand L at **3** by alkene **4** produces π-complex **5** (**3**→**5**). If the alkene ligand and the aryl substituent are in mutual *cis* disposition, the C=C double bond of alkene **4** might insert into the σ-C(sp^2)—Pd(II) bond, thus forming a C—C bond, as well as a σ-C(sp^3)—Pd(II) bond (**5**→**6**); this migratory insertion is

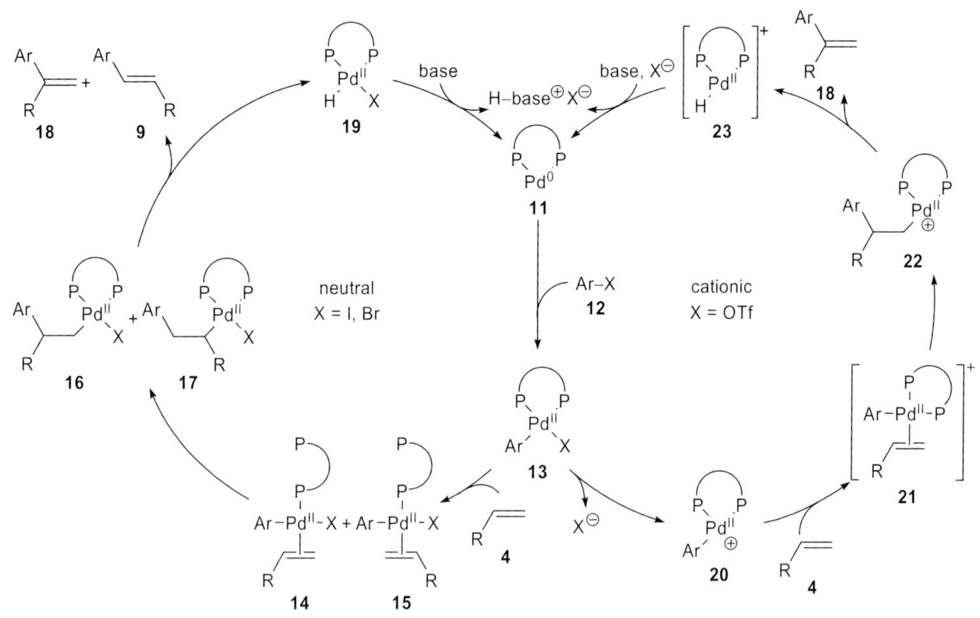

Scheme 7.2 *Neutral* versus *cationic* mechanism (neutral and cationic refers to the formal charge at palladium(II) after oxidative addition/alkene coordination prior to migratory insertion).

also referred to as carbopalladation. Prior to product formation, internal C—C bond rotation in the σ-alkyl—palladium(II) complex **6** is required to bring the $C_\alpha(sp^3)$—Pd(II) and the $C_\beta(sp^3)$—H bonds into a *syn* alignment for the subsequent β-hydride elimination (**6**→**7**). The β-hydride elimination then reinstalls the C=C double bond (**7**→**8**), followed by release of the arylated alkene **9** from the hydridopalladium(II) complex (**8**→**10**). It must be noted that, in principle, this step is reversible (**8**→**7**); readdition with reverse regioselectivity (not shown) might result in alkene isomerization or alkene migration through subsequent β′-hydride elimination. Finally, the active catalyst **1** is regenerated by reductive elimination of HX, which is neutralized by stoichiometric amounts of base (**10**→**1**).

Regiocontrol is a major challenge in intermolecular Mizoroki–Heck reactions, and usually results in a mixture of linear and branched products (cf. Section 7.2.2.3 on directed Mizoroki–Heck chemistry). Both, Cabri *et al.* [14] and Hayashi *et al.* [15] showed that the nature of the ancillary ligand L has a pronounced effect on the regioselectivity when palladium is ligated by bidentate phosphines (Scheme 7.2). The regiochemical outcome of the Mizoroki–Heck reaction is furthermore affected by the (pseudo)halide liberated in the oxidative addition step. Using aryl bromides or iodides, the reaction follows the *neutral* pathway (left cycle, Scheme 7.2). The *trans* effect of the halide anion favors dissociation of one phosphorus donor of the bidentate phosphine ligand in the alkene coordination event (**13→14+15**); the thus-formed neutral π-complexes **14** and **15** contain a Pd(II) fragment which is only weakly coordinated to the π-system of alkene **4**. After migratory insertion (**14/15→16/17**) and β-hydride elimination (**16/17→19**), both the α- and β-arylated products **18** and **9** are expelled from the catalytic cycle. Conversely, using aryl triflates or in the presence of acetate, the reaction proceeds through the *cationic* pathway (right cycle, Scheme 7.2). In this case, dissociation of the weakly coordinating triflate or acetate counterion generates a palladium(II) complex with a formal positive charge (**13→20**). Subsequent alkene capture (**20→21**) induces increased polarization of the alkene moiety, with a lower charge density at the α-carbon than at the β-carbon [14]; consequently, the aryl group will preferentially migrate onto the α-carbon (**21→22**). The branched alkene **18** is then released in the β-hydride elimination (**22→23**).

It might be assumed that the partial dissociation of a chiral bidentate ligand in the neutral pathway renders it less attractive for asymmetric Mizoroki–Heck transformations. A lower degree of organization in the enantioselectivity-determining step (alkene coordination–migratory insertion) and, hence, a diminished level of enantioselection is to be expected. However, Overman *et al.* observed a remarkable exception while investigating asymmetric Mizoroki–Heck cyclizations (cf. Section 7.2.3.2) [16–18]. When employing a (*Z*)-butenanilide iodide as the starting material, a high enantiomeric excess (ee) was obtained (92%); this clearly indicated that significant enantioinduction is possible under neutral conditions. This finding was further corroborated by a control experiment: replacement of bidentate diphosphine (*R*)-BINAP ((*R*)-**74**) by its monodentate analogue led to a poor enantiomeric excess. From these observations, Overman *et al.* concluded that both phosphorus donors of the bidentate ligand remain coordinated to the palladium center during the enantioselectivity-determining step. Instead of the previously discussed mechanistic picture (**25→26→29**), a modified neutral pathway for the asymmetric intramolecular Mizoroki–Heck reaction, which excludes partial phosphine dissociation, was then postulated by these authors (Scheme 7.3). For this, two other scenarios were considered after oxidative addition (**24→25**):

- A cationic mechanism (**25→27**), in which Hal^- leaves the coordination sphere of palladium(II) and the alkene then occupies the free coordination site, was discarded as the other optical antipode (!) is produced under truly cationic conditions (cf. Section 7.2.3.2).
- An associative mechanism for the generation of **27** via pentacoordinated intermediate **28** was proposed instead (**25→28→27**). The direct formation of **29** from **28** cannot be completely ruled out, but is at least unlikely based on literature precedence and quantum chemical calculations [19].

It should be noted here that a recent mechanistic study conducted by Curran *et al.* has shed new light on the stereochemistry-determining step of Overman-type Mizoroki–Heck ring closures (see Schemes 7.24 and 7.25 in Section 7.2.3.2) [20].

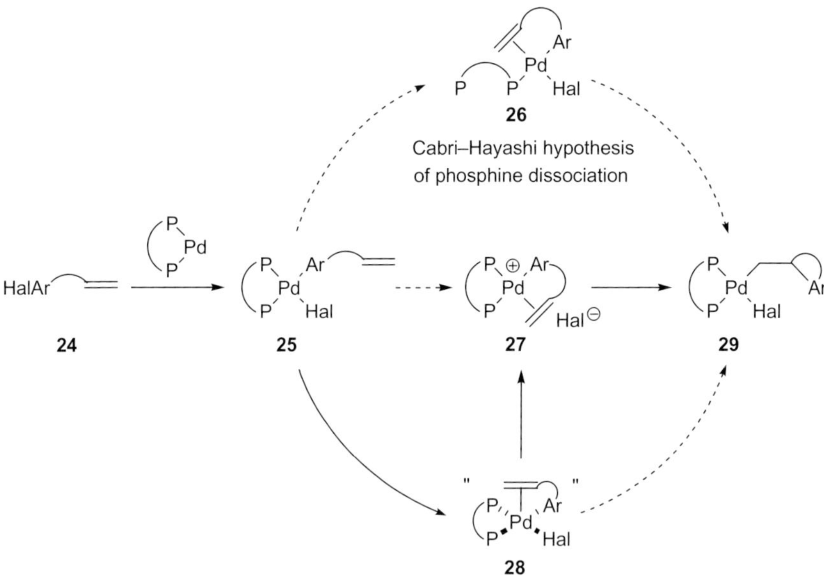

Scheme 7.3 Postulated neutral pathway of asymmetric Mizoroki–Heck cyclizations by Overman *et al.* [16–18].

7.2.2 Intermolecular Mizoroki–Heck Arylations

7.2.2.1 Intermolecular Arylations

Over the course of the past decade, the intermolecular Mizoroki–Heck reaction has witnessed tremendous progress [9, 10]. As a comprehensive survey of this area is certainly beyond the scope of this chapter, the decision was taken to include a few important developments, namely the activation of less-reactive aryl chlorides, waste-minimized processes, and novel catalyst systems.

In a major breakthrough, Beller and coworkers introduced palladacycle **32** as a catalyst for C(sp^2)—Cl activation (Scheme 7.4) [21]. At the time of its discovery, this easy-to-handle palladium complex surpassed all known catalysts in terms of chemical stability and lifetime; **32** is more reactive than conventional catalyst systems and is easily prepared from equimolar amounts of $Pd(OAc)_2$ and $(o\text{-tol})_3P$. The highest reactivity of **32** was found towards 4-chloroacetophenone; in its reaction with *n*-butyl acrylate in the presence of tetra-*n*-butylammonium bromide at extremely low catalyst loadings. Here, turnover numbers (TONs) of 40 000 were achieved that were previously unprecedented in homogeneous catalysis (**30**→**33**).

In a landmark publication, Fu *et al.* reported the first Mizoroki–Heck couplings of electron-poor aryl chlorides that proceed at ambient temperature [22]. The same group also accomplished the C—C bond formation of activated aryl chlorides to form trisubstituted alkenes (**34**→**36**; Scheme 7.5). The combination of $Pd_2(dba)_3$

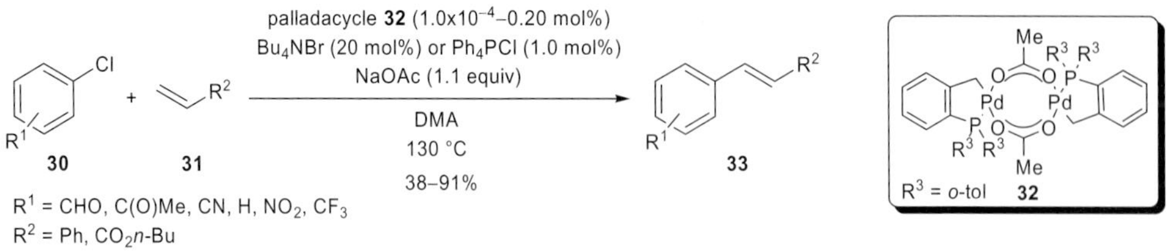

palladacycle **32** (1.0×10^{-4}–0.20 mol%), Bu_4NBr (20 mol%) or Ph_4PCl (1.0 mol%), NaOAc (1.1 equiv), DMA, 130 °C, 38–91%

R^1 = CHO, C(O)Me, CN, H, NO_2, CF_3
R^2 = Ph, CO_2n-Bu
R^3 = *o*-tol

Scheme 7.4 Mizoroki–Heck reaction of chloroarenes using palladacycle **32** as catalyst.

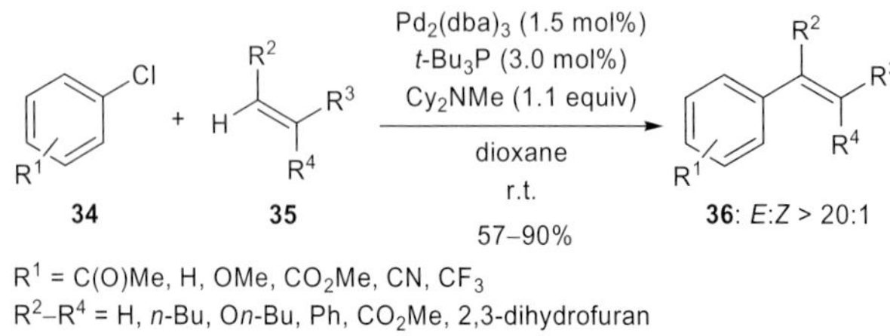

$Pd_2(dba)_3$ (1.5 mol%), *t*-Bu_3P (3.0 mol%), Cy_2NMe (1.1 equiv), dioxane, r.t., 57–90%

36: *E*:*Z* > 20:1

R^1 = C(O)Me, H, OMe, CO_2Me, CN, CF_3
R^2–R^4 = H, *n*-Bu, O*n*-Bu, Ph, CO_2Me, 2,3-dihydrofuran

Scheme 7.5 Mizoroki–Heck arylation of activated aryl chlorides at ambient temperature.

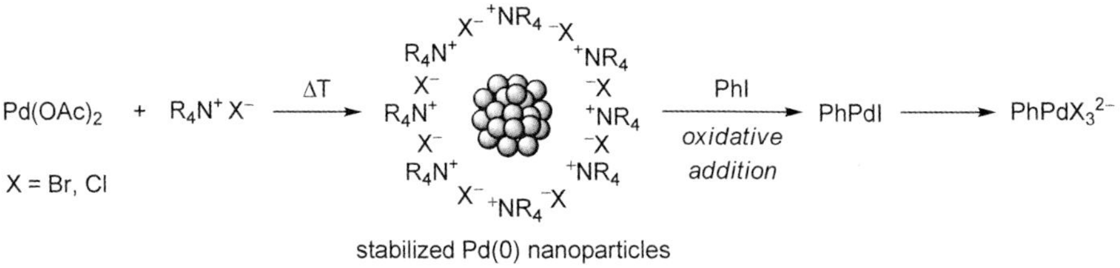

$Pd(OAc)_2$ + $R_4N^+X^-$ →(ΔT) stabilized Pd(0) nanoparticles →(PhI, oxidative addition) PhPdI → $PhPdX_3^{2-}$

X = Br, Cl

Scheme 7.6 Thermal preparation of ammonium salt-stabilized palladium colloids.

and sterically hindered, electron-rich P(*t*-Bu)$_3$ in the presence of Cy_2NMe secured high levels of regioselection, as well as diastereoselection ($E:Z > 20:1$), while displaying excellent functional group tolerance.

Reetz *et al.* reported phosphine-free, palladium-catalyzed Mizoroki–Heck reactions of unactivated aryl halides using palladium(0) nanoparticles [23, 24]. The thermolytic decomposition of $Pd(OAc)_2$ at 130 °C in the presence of ammonium salts resulted in the formation of $R_4N^+X^-$-stabilized palladium(0) nanoparticles (Scheme 7.6). The decisive role of palladium colloids was secured by nuclear magnetic resonance (NMR) studies of the oxidative addition step yielding PhPdI and $PhPdX_3^{2-}$.

At a relatively high reaction temperature (150 °C), Reetz *et al.* also accomplished the Mizoroki–Heck reaction of chlorobenzene with styrene in the presence of $Pd(OAc)_2$ and phosphonium salt Ph_4PCl [23a]. Again, the high catalytic activity was ascribed to the presence of nanosized palladium colloids, which are believed to be stabilized by the phosphonium salt.

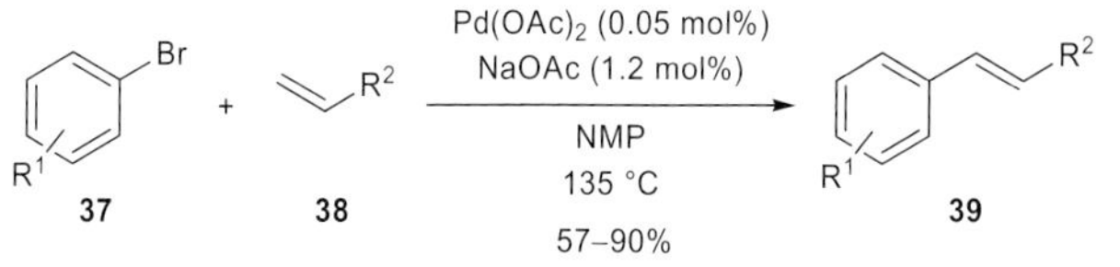

$$\text{37} + \text{38} \xrightarrow[\text{NMP, 135 °C, 57–90\%}]{\text{Pd(OAc)}_2\ (0.05\ \text{mol\%}),\ \text{NaOAc}\ (1.2\ \text{mol\%})} \text{39}$$

R^1 = C(O)Me, H, OMe, CO_2Me, CN, CF_3, NO_2, CHO, Br, COOH
R^2 = Ph, CO_2Bu, OCy, NHC(O)CH_3, CH(OH)CH_3I

Scheme 7.7 Homeopathic, phosphine ligand-free palladium catalyst in Mizoroki–Heck reactions.

Lately, de Vries *et al.* developed a so-called 'homeopathic', phosphine ligand-free Mizoroki–Heck arylation using a wide range of aryl bromides bearing both electron-withdrawing and electron-donating substituents at very low catalyst loadings (**37**→**39**; Scheme 7.7) [25]. Importantly, the palladium catalyst concentration was inversely proportional to the turnover frequency (TOF)! In this process, $Pd(OAc)_2$ is believed to generate soluble Na^+X^--stabilized palladium clusters which not only obviate the formation of inactive palladium black but also serve as a reservoir, which releases highly active, phosphine ligand-free palladium(0) into the catalytic cycle. At lower catalyst concentrations, the equilibrium between the palladium present in these clusters and that participating in the catalysis is shifted towards the catalytically active palladium. However, so far, the 'homeopathic' protocol has not been applicable to Mizoroki–Heck reactions of less-reactive aryl chlorides.

During recent years, the Mizoroki–Heck reaction has increasingly become the focus of industrial applicability, its inherent attractiveness emerging from a high functional group tolerance, as well as good availability and low cost of alkenes as compared with the vinyl metal compounds required for other $C(sp^2)$—$C(sp^2)$ cross-coupling reactions [26]. Nevertheless, Mizoroki–Heck chemistry is still afflicted with several disadvantages, the predominant problem being the need for a leaving group on the arene, $C(sp^2)$—X and not $C(sp^2)$—H (see Section 7.4 for details of Mizoroki–Heck reactions in connection with C—H bond activation). This inevitably produces stoichiometric amounts of strong acids, which are usually trapped by excess base, thus forming equimolar quantities of solid salt waste. From the standpoint of an industrial process, additional factors such as atom economy, toxicity and the recyclability of byproducts must also be considered.

An interesting strategy to achieve a waste-minimized Mizoroki–Heck arylation was published by de Vries *et al.*, who disclosed a process based on cheap aromatic carboxylic anhydrides as aryl sources (Scheme 7.8) [27]. The reaction of carboxylic anhydrides **40** with alkenes **41** in the presence of a phosphine ligand-free palladium catalyst (using a fourfold excess of sodium bromide for its stabilization) afforded **42** in moderate to good yields (**40**→**42**). These authors anticipated that the carboxylic acid formed in the reaction might be reconverted to the anhydride by thermal dehydration. Carbon monoxide produced in the coupling itself, and water formed in the condensation, would then be the sole byproducts. Unfortunately, with the exception of benzoic acid, the dehydration step emerged as troublesome and could not be extended to other carboxylic acids.

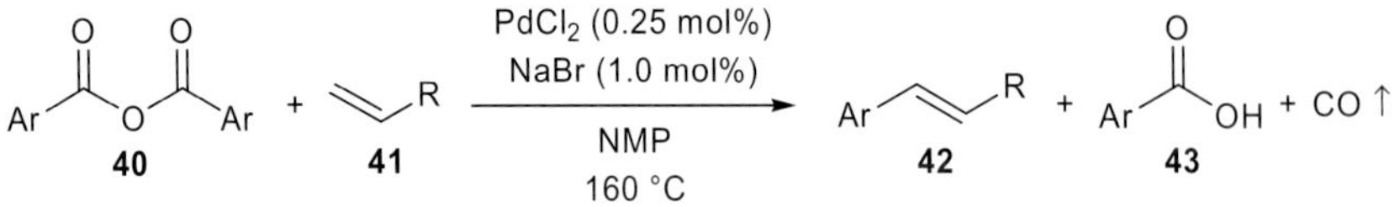

40 + 41 → 42 + 43 + CO ↑

$PdCl_2$ (0.25 mol%)
NaBr (1.0 mol%)
NMP
160 °C
54–77%

Ar = Ph, 2-Furyl, 4-MeO-C_6H_4
R = Ph, CN, CO_2R, CONHR'

Scheme 7.8 Decarbonylative Mizoroki–Heck reaction of carboxylic anhydrides.

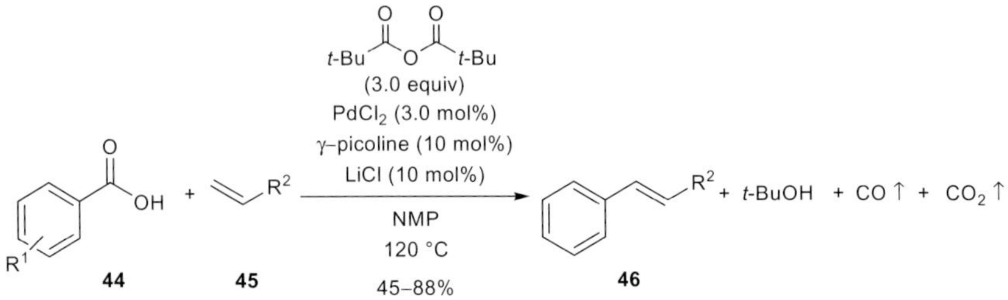

44 + 45 → 46 + t-BuOH + CO ↑ + CO_2 ↑

(t-BuOCO)$_2$O (3.0 equiv)
$PdCl_2$ (3.0 mol%)
γ-picoline (10 mol%)
LiCl (10 mol%)
NMP
120 °C
45–88%

R^1 = H, OMe, aryl, Cl, F, CO_2R
R^2 = Ph,CO_2 *t*-Bu, alkyl

Scheme 7.9 Decarboxylative Mizoroki–Heck arylation of carboxylic acids.

When resuming the studies of de Vries, Gooßen *et al.* reported a simple protocol for a waste-minimized Mizoroki–Heck arylation directly using carboxylic acids (Scheme 7.9) [28]. The treatment of **44** with di-*tert*-butyldicarbamate yielded mixed *tert*-butyloxycarbonyl anhydrides which, in the presence of a palladium catalyst, reacted with several alkenes **45** to give styrenes **46** (**44**→**46**). The Gooßen procedure was certainly a major step forward–formation of gaseous carbon monoxide and carbon dioxide along with *tert*-butanol–yet raising it to larger scale was hampered by the high total weight of volatile byproducts and by the need for costly di-*tert*-butyldicarbamate.

Aiming to overcome the limitations of the previous protocols, Gooßen *et al.* then extended their own studies using easily available carboxylic esters [29]. This indeed paved the way to waste-minimized Mizoroki–Heck reactions, in which any byproduct was efficiently recyclable such that waste formation was limited to carbon monoxide and water (**47**→**48**; Scheme 7.10). Subsequently, this technique has proven to be viable for converting various 4-nitrophenyl esters **47** in the presence of a $PdCl_2$-LiCl-isoquinoline catalyst system into styrenes **48**. Under Lewis acid catalysis, 4-nitrophenol (**49**) cleanly reacts with benzoic acid at the same temperature required for the Mizoroki–Heck arylation (**49**→**47**), thereby regenerating 4-nitrophenyl ester **47**.

A tentative mechanism for the Mizoroki–Heck arylation of carboxylic esters is shown in Scheme 7.11 [29]. The initial step of the catalytic cycle is the oxidative addition of the C(O)—O bond of carboxylic ester **47** to palladium catalyst **1**, forming acylpalladium complex **50** (**1**→**50**). Ligand exchange of alkoxide for halide (**50**→

$PdCl_2$ (3.0 mol%)
LiCl (10 mol%)
isoquinoline (3.0 mol%)
NMP
160 °C
40–95%

47 + 41 → 48 + ArOH + CO↑
49

$PhCO_2H$ (1.0 equiv)
$Sc(OTf)_3$ (3.0 mol%)
neat
160 °C
82%
– H_2O

R^1 = MeO, CN, CH_3CO, F, Me, NO_2, Cl, CF_3
Ar = p-NO_2-C_6H_4
R^2 = Ph, CN, CO_2R, CONHR

Scheme 7.10 Decarbonylative Mizoroki–Heck arylation of aryl esters.

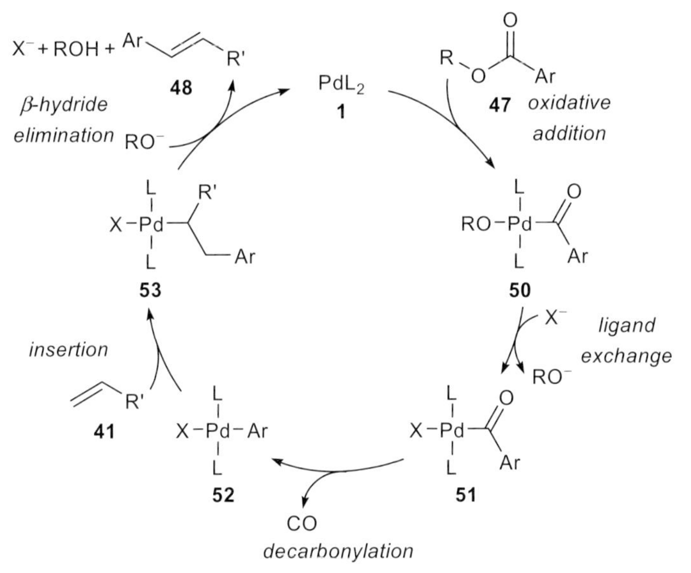

Scheme 7.11 Proposed mechanism for the waste-minimized Mizoroki–Heck reaction of carboxylic esters.

51) is necessary to promote the decarbonylation step (**51**→**52**). The extrusion of carbon monoxide produces the σ-arylpalladium(II) complex **52**, which, after migratory insertion of alkene **41** (**52**→**53**) and β-hydride elimination/reductive elimination (**53**→**1**), liberates the desired arene **48**.

Recently, Gooßen *et al.* further extended the scope of this process to unreactive enol esters **54**, which undergo arylation reactions only yielding the low-molecular-weight byproducts carbon monoxide and acetone (**54**→**56**; Scheme 7.12) [30]. An additional advantage is that enol esters **54** are accessed by waste-free addition of

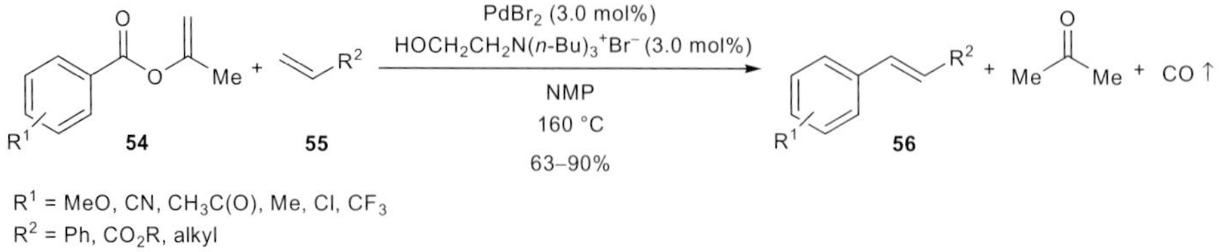

R^1 = MeO, CN, $CH_3C(O)$, Me, Cl, CF_3
R^2 = Ph, CO_2R, alkyl

PdBr$_2$ (3.0 mol%), $HOCH_2CH_2N(n\text{-}Bu)_3^+Br^-$ (3.0 mol%), NMP, 160 °C, 63–90%; 54 + 55 → 56 + Me(CO)Me + CO↑

Scheme 7.12 Salt-free decarbonylative Mizoroki–Heck arylation of enol esters.

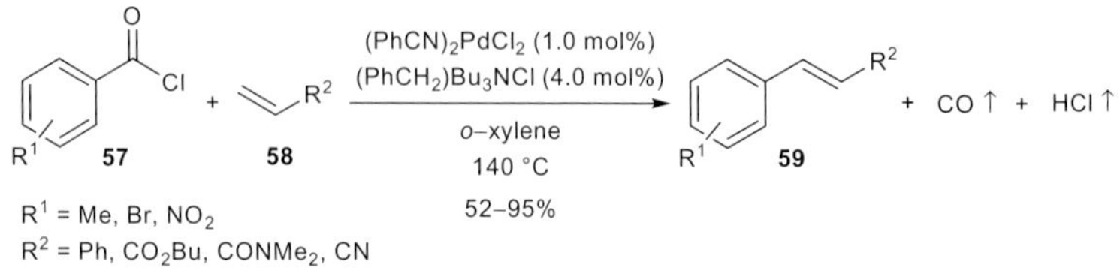

$(PhCN)_2PdCl_2$ (1.0 mol%), $(PhCH_2)Bu_3NCl$ (4.0 mol%), o-xylene, 140 °C, 52–95%; 57 + 58 → 59 + CO↑ + HCl↑

R^1 = Me, Br, NO_2
R^2 = Ph, CO_2Bu, $CONMe_2$, CN

Scheme 7.13 Base-free Mizoroki–Heck arylation using aroyl chlorides.

alcohols to propyne, a side component of the C_3-fraction in steam-cracking. The substrate scope of the reaction was found to be almost similar to that of aryl esters **47** (*vide supra*). A combination of less-common $PdBr_2$ and tri-*n*-butyl(2-hydroxyethyl)ammonium bromide gave the most active catalyst.

Another interesting strategy to achieve a waste-minimized Mizoroki–Heck reaction was recently reported by Miura *et al.* (Scheme 7.13) [31]. Using aroyl chlorides **57**, the Mizoroki–Heck reaction proceeded well under base-free conditions, liberating only hydrochloric acid and carbon monoxide as byproducts (**57**→**59**). However, due to the particularly waste-intensive preparation of acyl chlorides, it remains questionable whether this procedure would be suitable for large-scale alkene arylation.

In the quest for additional active catalysts for the Mizoroki–Heck reaction, the advent of *N*-heterocyclic carbene (NHC) ligands was warmly welcomed [32]. Transition metal–carbene complexes are often thermally more stable than the corresponding transition metal–phosphine complexes, and have therefore attracted considerable interest as competitive alternatives in Mizoroki–Heck chemistry, which requires high reaction temperatures. Since the seminal application of NHC ligands in Mizoroki–Heck arylations by Herrmann *et al.* [33], several research groups have introduced novel palladium catalyst–NHC ligand combinations. These were tested and assessed in standard couplings of simple iodo- or bromoarenes **60** and activated acceptors such as acrylates **61** or styrene (**63**) [32], and a selection of impressive examples is summarized in Scheme 7.14.

The benchmark was set by Herrmann's carbene complex **65** which, under optimized reaction conditions, showed a respectable TON of 13 000. In a similar study by Baker *et al.*, the Mizoroki–Heck arylation of iodobenzene (**60b**) with **63** in the presence of catalyst **66** bearing a chelating NHC ligand with a cyclophane backbone showed an outstanding TON of 7 100 000 [34]. In another study, Buchmeiser *et al.* introduced a new class of NHC based on 1,3-disubstituted tetrahydropyrimi-

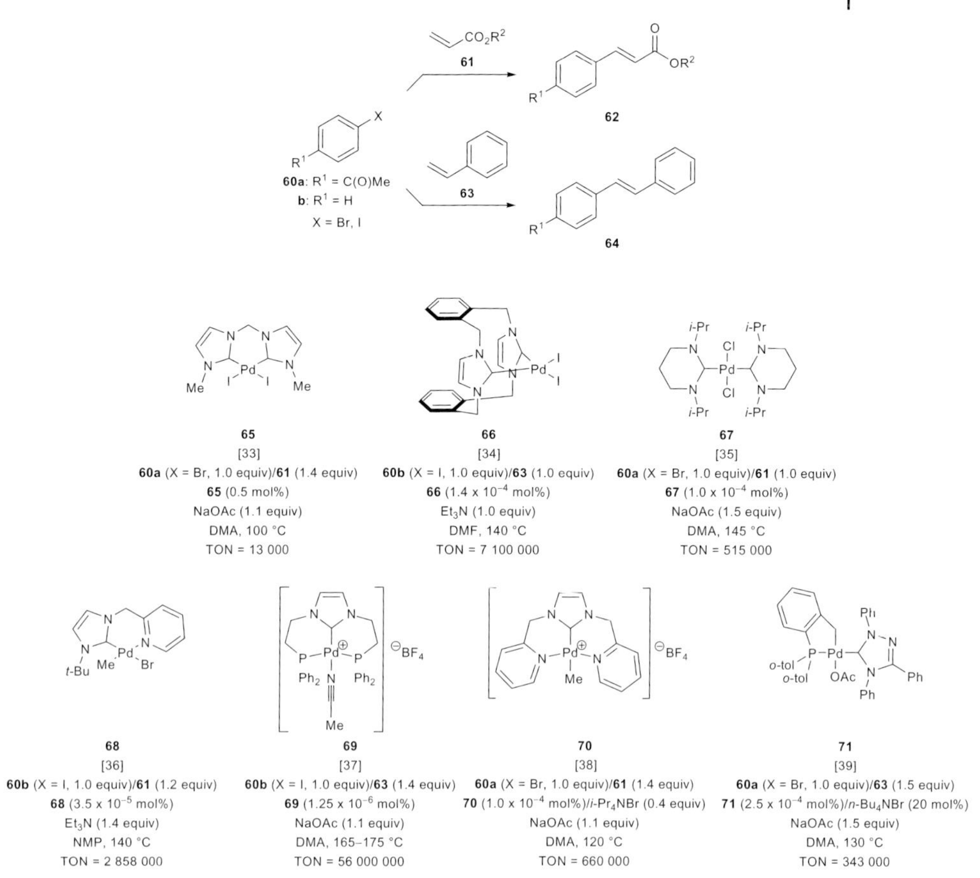

Scheme 7.14 Different protocols for Mizoroki–Heck reactions using NHC-ligands.

din-2-ylidenes [35]. The bis-NHC–palladium(II) complex **67** turned out to be a highly active catalyst (TON = 515 000) for the Mizoroki–Heck coupling of aryl bromides (**60a**→**62a**), and even converted less-reactive but electronically activated aryl chlorides (not shown). Furthermore, Danopoulos and Hursthouse *et al.* obtained a highly active palladium catalyst using hemilabile carbene complexes such as **68**, where the NHC ligand is tethered to a pyridine donor via a methylene bridge [36]. This unusual ligand motif gave an impressive TON of 2 858 000 in the arylation of methyl acrylate with iodobenzene (**60b**→**62b**). Closely related to bidentate **68** are tridentate pincer ligands **69** and **70**. Lee *et al.* presented complex **69** with a $PC^{NHC}P$-based pincer ligand, which showed high reactivity and tremendous catalytic activity [37]. In fact, **69** exhibited a TON of 56 000 000 at catalyst loadings as low as 1.25×10^{-6} mol% (**60b**→**64b**), which represents the highest TON ever measured with a Pd-NHC-catalyst! Further noteworthy catalytic activities were shown by Cavell *et al.* using $NC^{NHC}N$-pincer-coordinated catalyst **70** [38]. Mizoroki–Heck coupling of **60a** with *n*-butyl acrylate **61** in the presence of **70** afforded significant TONs of 660 000. Recently, Herrmann *et al.* presented NHC-substituted phosphapalladacycles (e.g. **71**) as an additional class of catalysts [39]. This new type of palladacycles combines the advantageous stability of the phosphapalladacyclic skeleton with the steric demand, as well as the high σ-donor strength of NHCs. Mizoroki–Heck coupling of **60a** with styrene (**63**) gave a high TON of 343 000. It was also noted here that **71** catalyzed the Mizoroki–Heck reactions of activated aryl chlorides with styrene (with TONs up to 10 800).

7.2.2.2 Asymmetric Intermolecular Arylations

As a C—C bond-forming process, the Mizoroki–Heck reaction is of course synthetically important yet, in principle, no stereogenic carbons are formed. Therefore, asymmetric Mizoroki–Heck reactions will only be possible if either β-hydride elimination, which usually re-establishes the alkene fragment, is steered away from the site of C–C bond formation, or if the place of C–C bond formation is not coinciding with a prochiral center, as in desymmetrizing transformations (cf. Section 7.2.3.3). In order to selectively realize the former – that is, alternative β′-hydride elimination – the β-carbon atom generated in the migratory insertion step must not have any conformationally accessible hydrogens attached to it. This situation is either realized when a quaternary carbon center is formed (cf. Section 7.2.3.2) or is often observed for cyclic (and therefore rigid) skeletons in which a synperiplanar alignment of the C_{β}–H bond and the C_{α}–Pd(II) bond is prevented (as in this chapter).

The development of enantioselective variants of intermolecular Mizoroki–Heck arylations began with the pioneering studies of Ozawa and Hayashi *et al.* (Scheme 7.15) [15, 40]. Whereas, the coupling of dihydrofuran (**72**) with phenyl iodide (**73a**) in the presence of bidentate phosphine ligand (*R*)-BINAP ((*R*)-**74**) was accomplished with no enantiomeric excess and poor yield, the analogous arylation reaction with phenyl triflate (**73b**) afforded the arylated product (*R*)-**75** with 93% ee and in moderate chemical yield. Interestingly, the minor regioisomer, 2-phenyl-2,5-dihydrofuran ((*S*)-**76**), is formed with inverted absolute configuration and significantly diminished enantiomeric purity.

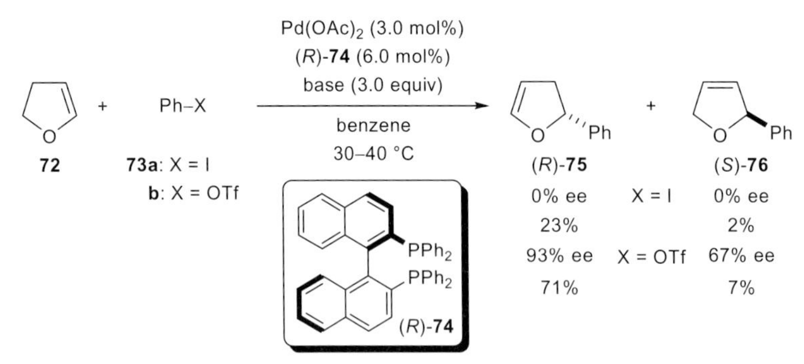

Scheme 7.15 Ligand-induced enantioselective intermolecular Mizoroki–Heck arylation of 2,3-dihydrofuran.

In the case of phenyl iodide (**73a**), the absence of asymmetric induction was attributed to a partial dissociation of the chiral ligand during the alkene coordination–migratory insertion steps, according to the *neutral* pathway. In the case of phenyl triflate (**73b**), high enantioinduction is rationalized assuming the *cationic* pathway to be operative, allowing for dissociation of the triflate counterion, while both donors of the chiral ligand remain coordinated to palladium(II) during the stereochemistry-determining step (cf. Section 7.2.1).

The stereochemical outcome with (*R*)-**75** and (*S*)-**76** having opposite absolute configuration is intriguing. Hayashi *et al.* assumed that a kinetic resolution is operating within the enantioselective Mizoroki–Heck reaction (Scheme 7.16). The catalytic cycle starts with oxidative addition of **73b** to palladium(0) (**77**→**78**), followed by enantiofacial alkene-coordination to form the diastereomeric complexes (*R*)-**79** and (*S*)-**79** (**78**→(*R*)-**79** and **78**→(*S*)-**79**). In both cases, *syn*-selective migratory insertion ((*R*)-**79**→(*R*)-**80** and (*S*)-**79**→(*S*)-**80**) and subsequent stereospecific β-hydride elimination furnishes a pair of diastereomeric alkene–hydridopalladium(II) complexes (*R*)-**81** and (*S*)-**81** ((*R*)-**80**→(*R*)-**81** and (*S*)-**80**→(*S*)-**81**). The π-bound Pd—H fragment in complex (*S*)-**81** dissociates rapidly ((*S*)-**81**→**77**), whereas (*R*)-**81** undergoes hydropalladation but with reversed regioselectivity ((*R*)-**81**→(*R*)-**82**); β-hydride elimination and alkene dissociation then affords (*R*)-**75** ((*R*)-**82**→**77**). The enhanced enantiomeric excess of (*R*)-**75** is therefore due to a kinetic resolution at the stage of the diastereomeric complexes (*R*)-**81** and (*S*)-**81**. According to this, the enantiomeric purity of (*R*)-**75** would increase with increasing formation of (*S*)-**76**.

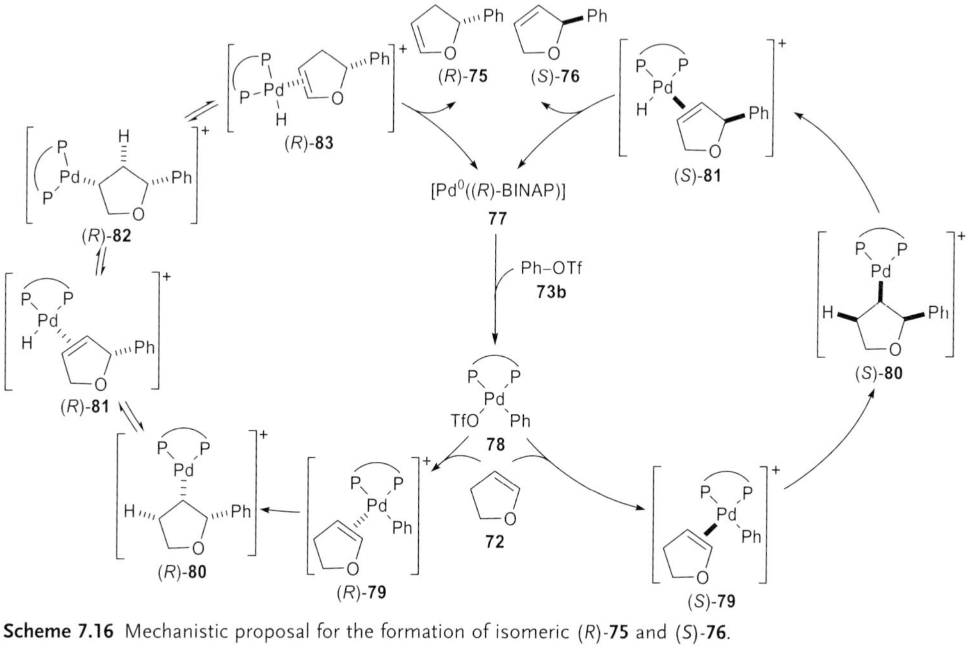

Scheme 7.16 Mechanistic proposal for the formation of isomeric (*R*)-**75** and (*S*)-**76**.

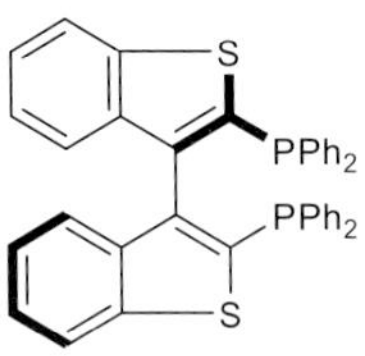

(*R*)-BITIANP (**84**)
[41]
72 (5.0 equiv)/**73b** (1.0 equiv)
$Pd_2(dba)_3 \cdot dba$ (3.0 mol%)
84 (12 mol%)
proton sponge (3.0 equiv)
DMF, 40 °C
(*R*)-**75** : **76**
91% *ee* –
84% nd

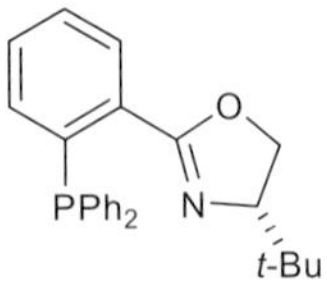

(*S*)-*t*-Bu-PHOX (**85**)
[42]
72 (4.0 equiv)/**73b** (1.0 equiv)
$Pd(dba)_2$ (3.0 mol%)
85 (6.0 mol%)
i-Pr_2NEt (2.0 equiv)
THF, 70 °C
75 : (*R*)-**76**
– 97% *ee*
nd 87%

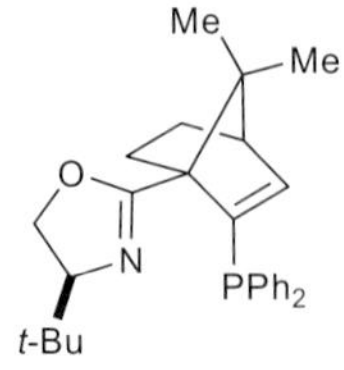

(*S*)-**86**
[43]
72 (5.0 equiv)/**73b** (1.0 equiv)
$Pd_2(dba)_3$ (2.5 mol%)
86 (6.0 mol%)
i-Pr_2NEt (3.0 equiv)
benzene, 70 °C
75 : (*R*)-**76**
– 96% *ee*
nd 100% conv.

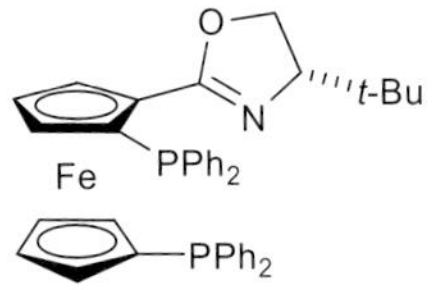

(*S,S*)-ferrocene **87**
[44]
72 (4.0 equiv)/**73b** (1.0 equiv)
$Pd(OAc)_2$ (1.5 mol%)
87 (3.0 mol%)
i-Pr_2NEt (2.0 equiv)
toluene, 60 °C
(*S*)-**75** : (*R*)-**76**
26% ee 97% *ee*
5 : 95

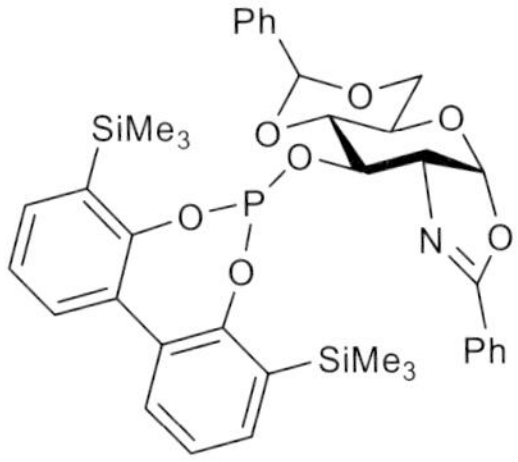

88
[45]
72 (4.0 equiv)/**73b** (1.0 equiv)
$Pd_2(dba)_3 \cdot dba$ (2.5 mol%)
88 (5.6 mol%)
i-Pr_2NEt (2.0 equiv)
THF, 70 °C, microwave
75 : (*R*)-**76**
– 99% *ee*
2 : 98

Figure 7.1 Chiral ligands tested in asymmetric intermolecular Mizoroki–Heck arylations (nd = not determined).

Based on these seminal studies, several research groups have tested novel ligand scaffolds in the asymmetric intermolecular Mizoroki–Heck reaction of **72**, monitoring double bond migration and enantiomeric purity (Figure 7.1) [41–45]. By replacing (*R*)-BINAP ((*R*)-**74**) with (*R*)-BITIANP (**84**), Tietze *et al.* achieved complete control of regioselectivity in favor of (*R*)-**75** with equally high enantiomeric excess and even an improved yield [41]. In contrast to diphosphine-type ligands, Pfaltz *et al.* showed that *P,N*-type ligands, for example (*S*)-*t*-Bu-PHOX (**85**), displayed a low tendency towards C=C double bond migration; (*R*)-**76** was isolated with excellent enantiomeric excess and in high yield, whereas regioisomer **75** was not even detected [42]. This observation was confirmed by Gilbertson *et al.* using the related *P,N*-type ligand (*S*)-**86**, which performed equally well in this transformation [43]. Hou and Dai *et al.* also showed that ferrocene-derived *P,N* ligand **87** gives similar results [44]. Recently, Diéguez *et al.* established sugar-based phosphite-oxazolidine ligands such as **88** as a new ligand class in the intermolecular Mizoroki–Heck reaction, to yield (*R*)-**76** with excellent regioselectivity and superb enantioselectivity [45]. This transformation is also an impressive example of the benefits of microwave irradiation [46], by which the reaction times were dramatically reduced from 15 h to 10 min, without any loss of regio- and enantiocontrol.

7.2.2.3 Directed Intermolecular Arylations

Over the past two decades, directed intermolecular Mizoroki–Heck reactions have been elaborated to remarkably high standards. Recently, Oestreich summarized the major advances in controlling both regio- and diastereoselectivity in these transformations [47]. With regards to the former aspect, the modular preparation of trisubstituted and tetrasubstituted alkenes is certainly a highlight within recent developments in Mizoroki–Heck chemistry [48].

Hallberg *et al.* were the first to deliberately study how the regioselectivity of intermolecular Mizoroki–Heck arylations is influenced by a tethered Lewis basic donor in proximity of the substrate double bond [49]. Within these investigations, the authors discovered a peculiar dependence of the regioisomeric distribution on the nature of the phosphine ligand (Scheme 7.17) [49b]. In the presence of monodentate phosphine Ph_3P, **89** was converted into **91** in excellent regioselectivity (**89**→**91**), whereas using bidentate phosphine dppp gave **90** with complete reversal of the regioselectivity (**89**→**90**). A systematic screening of different bidentate phosphines revealed an interdependence of the ligand bite angle and the regioselectivity; a bite angle of approximately 90° (as in dppp or dppf) ensured optimal α-selectivity.

The possibility of this targeted switch between α- and β-arylation found a noteworthy synthetic application (Scheme 7.18) [50], when Hallberg *et al.* reported the first example of a practical two-step, chelation-controlled triarylation of **89**, which provides β,β-diarylated acetophenone derivatives **93** after hydrolysis (**89**→**93**). In the first step, a highly regiocontrolled α-arylation was secured by dppp (**89**→**90**); in the second step, Ph_3P allowed for selective twofold β-arylation (**90**→**92**); such a

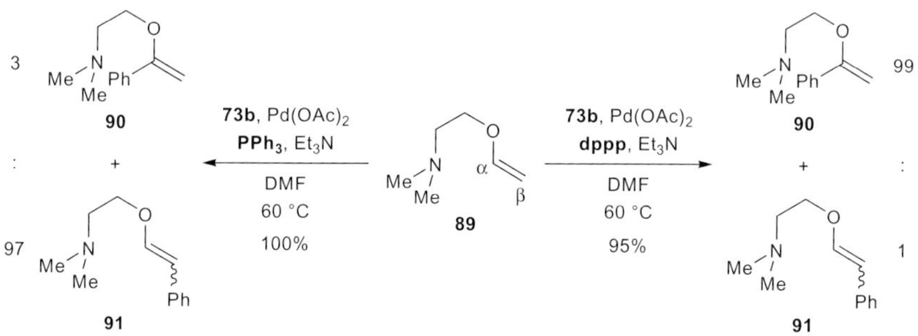

Scheme 7.17 Selective α- and β-arylation by a regiochemical switch.

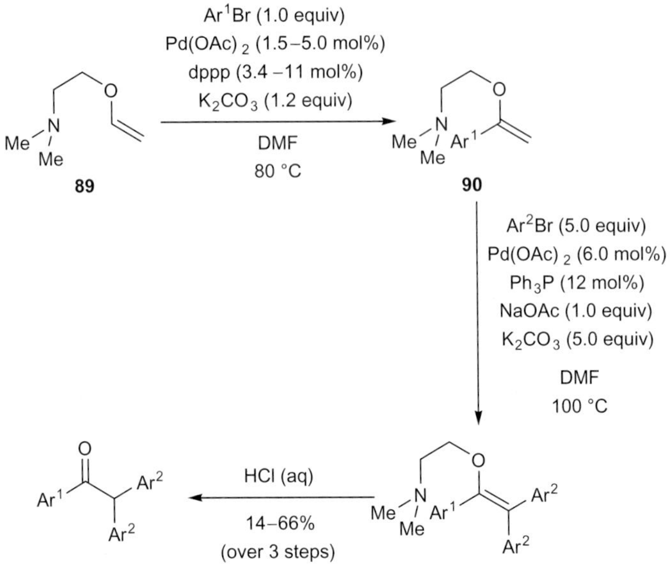

Scheme 7.18 Directed α,β,β-trisubstitution of enol ethers.

one-pot double arylation procedure eludes otherwise troublesome double bond isomerization, which is a general issue in these directed arylations [49]. It should be noted that it was advantageous to perform the α- prior to the β-arylation. This reaction sequence is an instructive example of the advantages of a removable directing group in Mizoroki–Heck chemistry: regiocontrol and, as in intramolecular variants, participation of otherwise unreactive disubstituted and trisubstituted alkenes.

Since Hallberg's first report of the use of nitrogen donor-containing enol ethers as platforms for directed Mizoroki–Heck arylations, several groups have designed related heteroatom-based systems (Figure 7.2) [51–55]. When Badone *et al.* replaced the diamino group (cf. **89**) by the diphenylphosphanyl group (**94**) [51], the phosphorus donor showed the same directing effect to yield β-arylated vinyl ethers in excellent regioselectivity, but with poor control of the double bond geometry. In platform **95**, Carretero *et al.* employed a sulfoxide tether bearing a 2-anilido group as the catalyst-directing donor [52]. C—C Bond formation then occurred selectively in the β-position and – in contrast to Hallberg's and Badone's system – with excellent *E* selectivity of the double bond.

The beauty of this chemistry was then elaborated by Itami and Yoshida *et al.*, who developed 2-pyridyl- and 2-pyrimidyl-directed Mizoroki–Heck multiarylations [53]. The pivotal feature of their platforms **96** and **97** (Figure 7.2) is the tethering bond itself, which serves as a placeholder for an aryl group by unique cross-coupling reactions of the $C(sp^2)$—Si and $C(sp^2)$—S bond, respectively.

An impressive reaction sequence for the modular synthesis of tetraarylated alkenes involving two directed Mizoroki–Heck reactions is outlined in Scheme

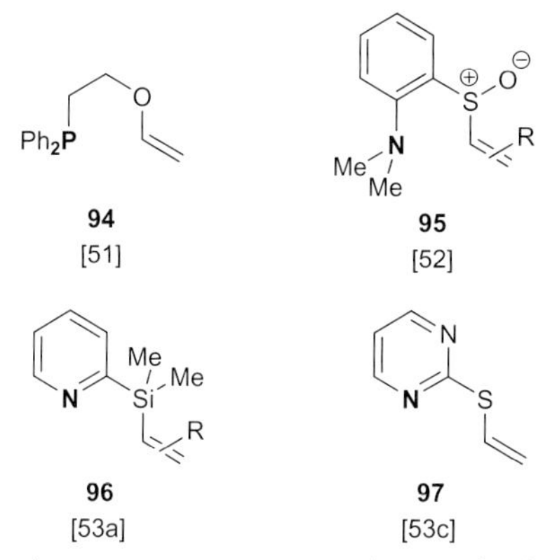

Figure 7.2 Representative substrates for the chelation-controlled Mizoroki–Heck arylation.

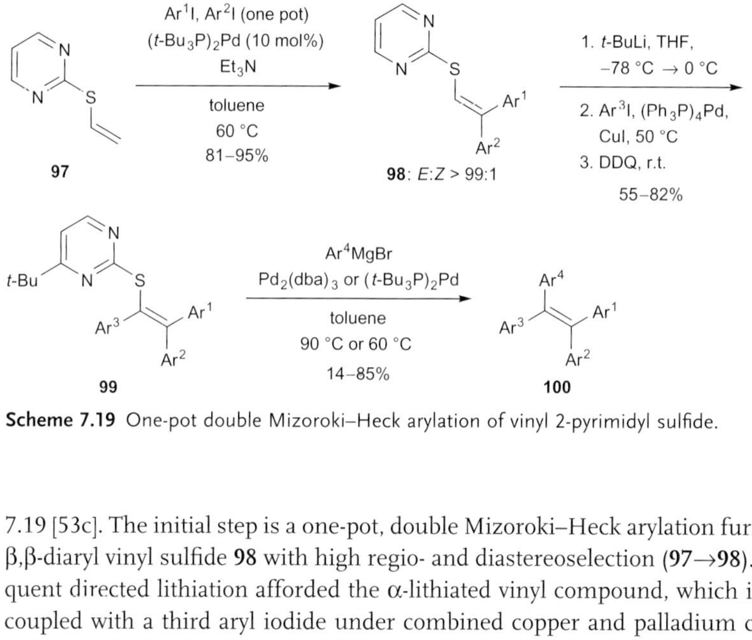

Scheme 7.19 One-pot double Mizoroki–Heck arylation of vinyl 2-pyrimidyl sulfide.

7.19 [53c]. The initial step is a one-pot, double Mizoroki–Heck arylation furnishing β,β-diaryl vinyl sulfide **98** with high regio- and diastereoselection (**97**→**98**). Subsequent directed lithiation afforded the α-lithiated vinyl compound, which is cross-coupled with a third aryl iodide under combined copper and palladium catalysis giving α,β,β-triarylated vinyl sulfide **99** (**98**→**99**). In the final step, the fourth aryl substituent was introduced by a palladium-catalyzed cross-coupling reaction with a Grignard reagent (**99**→**100**).

The first example of an asymmetric directed Mizoroki–Heck reaction was published by Carretero *et al.*, using sulfoxides as chiral auxiliaries (Scheme 7.20) [54a]. By coordination to the nitrogen, the sulfinyl group nicely directs the arylpalladium(II) complex towards one of the diastereotopic faces of the alkene, followed by diastereoselective C—C bond formation [54]. As an example, the reaction of **101** with iodobenzene (**73a**), yielding (2*S*,^{S}*R*)-**102** in very good diastereomeric ratio (dr) of 94 : 6 which, after desulfinylation with zinc, gave (*R*)-**103** in excellent enantiomeric

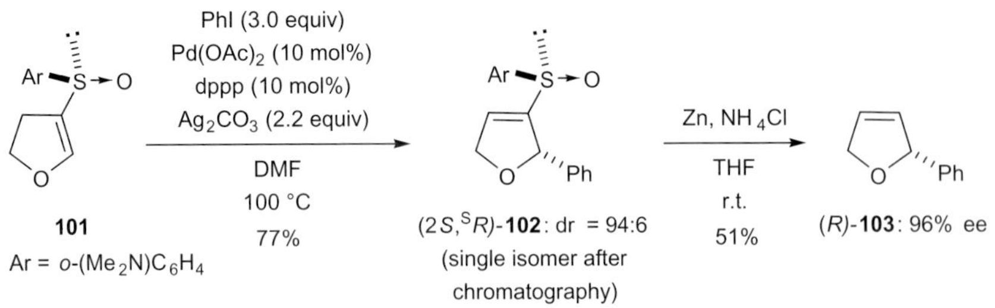

Scheme 7.20 Diastereoselective Mizoroki–Heck arylation directed by a chiral sulfinyl group.

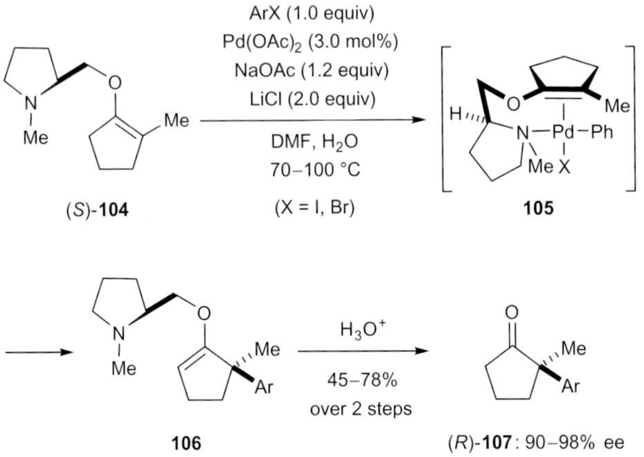

Scheme 7.21 Amino-directed, diastereoselective Mizoroki–Heck reaction.

purity (**101**→(*R*)-**103**). The stereochemical controller might be used for a second diastereoselective arylation prior to its reductive removal. The substrate control was unambiguously secured by a Mizoroki–Heck reaction of the requisite precursor devoid of the *N,N*-dimethylamino donor; arylation under identical reaction conditions clearly favored the other diastereomer (not shown, dr = 23 : 77).

Later, Hallberg *et al.* also reported an asymmetric chelation-controlled Mizoroki–Heck arylation using proline-derived enol ether **104** (Scheme 7.21) [56]. The intermolecular arylation of tetrasubstituted alkene **104** generates a quaternary carbon center with high diastereoselectivity ((*S*)-**104**→**106**) and, after hydrolysis, with excellent enantioselectivity (**106**→(*R*)-**107**). During the reaction, the major diastereomer **106** is likely to be formed through *si*-face insertion via intermediate **105**. By avoiding steric interactions between the methylene unit attached to the oxygen atom and the methyl group at the double bond, complex **105** is expected to adopt a gauche conformation relative to the O–C=C plane. Upon coordination of palladium(II) in a cisoid fashion (with respect to the neighboring stereogenic carbon), the nitrogen atom becomes chiral.

7.2.3
Intramolecular Mizoroki–Heck Arylations

7.2.3.1 Intramolecular Arylations

Until the mid-1980s, the intramolecular version of the Mizoroki–Heck reaction had remained largely unexplored [57]. However, since then the use of this functional group-tolerant methodology for the formation of carbocycles as well as heterocycles – mainly as part of total synthesis endeavors [7] – has flourished. With few exceptions, many of the nonenantioselective protocols [58] are today standard transformations and therefore beyond the scope of this chapter, which is focused on modern arylation techniques (important asymmetric processes are highlighted in the following section). So far, ring sizes ranging from three- to nine-membered (by *exo*-trig cyclization), as well as five- to nine-membered (by *endo*-trig cyclization) have been accessible (including a few macrocycles) [59]. Intramolecular Mizoroki–Heck cascade reactions are also known; deliberate combination of the Mizoroki–Heck reaction with other transformations such as pericyclic reaction [60], for example, 1,3-dipolar cycloadditions [61], or palladium-catalyzed cross-coupling reactions (Suzuki coupling [62] or Stille coupling [63]) have led to a remarkable variety of cascade (or domino) processes [64].

7.2.3.2 Asymmetric Intramolecular Arylations

Almost two decades ago, Shibasaki *et al.* and Overman *et al.* reported independently asymmetric intramolecular Mizoroki–Heck reactions by either indirect (desymmetrization, see Section 7.2.3.3) or direct (congested quaternary carbons) formation of stereogenic carbons [65, 66]. Since that time, the exceptional efficiency of this enantioselective C—C bond-forming process is reflected in the syntheses of numerous structurally intriguing natural products [17, 67].

In particular, the formation of oxindoles having a chiral quaternary center has been studied extensively by the Overman group, from both a synthetic and mechanistic standpoint [16, 18]. In the group's early studies, remarkable observations were made in the syntheses of spiro-oxindoles (*S*)-**109** and (*R*)-**109** starting from common precursor **108** (**108**→**109**; Scheme 7.22) [68]. By using the same optical antipode of the BINAP ligand (**74**), either enantiomer of the cyclic product **109** was obtained by simple variation of the reaction conditions. When using silver phosphate as base, (*S*)-**109** was isolated with good enantiomeric excess and in reasonable yield, whereas with PMP in the absence of a halide scavenger, (*R*)-**109** was also formed with significant enantiomeric enrichment, yet with inverted absolute configuration. Overman *et al.* postulated that two different mechanisms were operating, depending on the presence of a halide scavenger: in the presence of a silver salt, the reaction proceeded through *cationic* palladium(II) intermediates; conversely, when using a tertiary amine base, the reaction was thought to proceed through the *neutral* pathway, involving either a pentacoordinated or a tetracoordinated intermediate, in which the BINAP ligand (**74**) was only coordinated to palladium(II) by one phosphorus donor (cf. Section 7.2.1, Scheme 7.2).

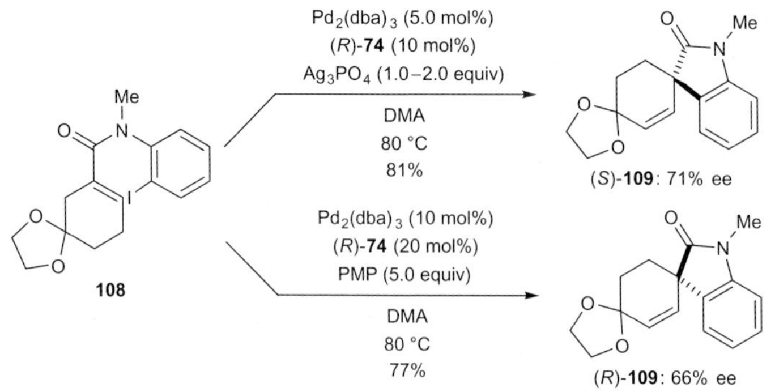

Scheme 7.22 Enantioselective synthesis of either enantiomer of spirooxindoles.

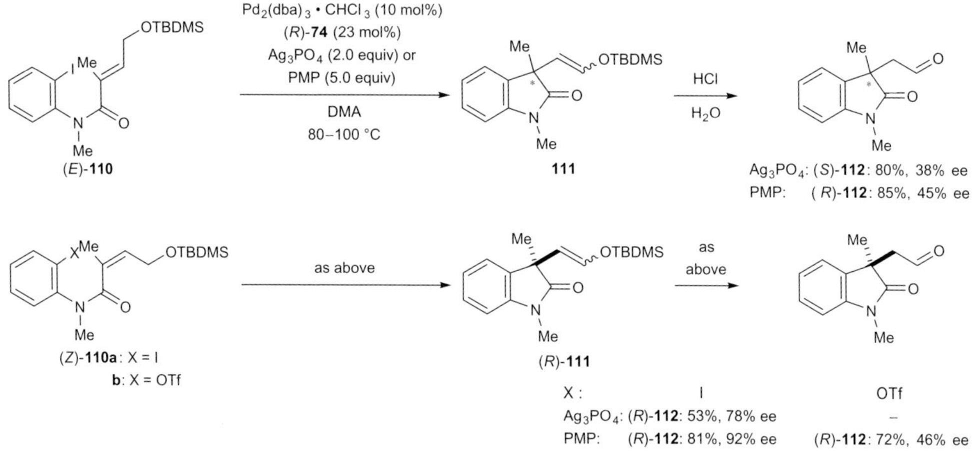

Scheme 7.23 Enantioselective Mizoroki–Heck cyclization of (*E*)- and (*Z*)-α,β-unsaturated haloanilides and analogues.

Similar results were realized in the Mizoroki–Heck cyclizations of an analogous (*E*)-α,β-unsaturated 2-iodoanilide (*E*)-**110** ((*E*)-**110**→**112**; Scheme 7.23) [69]. Again, the sense of asymmetric induction in the ring closure of the (*E*)-**110** stereoisomer was determined by the halide scavenger employed. Although the overall levels of enantioselection were low, cyclization in the presence of Ag_3PO_4 gave, after hydrolysis of intermediate **111**, the *S*-enantiomer of **112**, while (*R*)-**112** was produced when using PMP as base. Interestingly, identical cyclizations of isomeric (*Z*)-**110a** (X = I) proceeded with good to excellent enantioselection, with both affording the *R*-enantiomer ((*Z*)-**110**→(*R*)-**112**; Scheme 7.23); here, the neutral pathway was superior. These results suggest that chiral induction is governed not only by a 'base/additive effect' but also by a 'geometry effect' (*E* versus *Z*). Cyclization of the cognate triflate (*Z*)-**110b** (X = OTf) resulted in a substantial decrease in enantiomeric excess (46% ee) as compared to ring closure of the corresponding iodide (*Z*)-**110a** (X = I) under 'cationic conditions' (78% ee) [16].

Detailed mechanistic studies by Overman *et al.* provided further insight into the effect of halide scavenger and alkene geometry on the yield and enantioselectivity of this transformation [16, 70]. A comparison of BINAP (**74**) and its monophosphine analogue, which was designed to mimic a partially dissociated BINAP chelate, supported the conclusion that BINAP remains chelated during the enantioselectivity-determining step of the *neutral* pathway (cf. **25**→**28**→**27**; see Section 7.2.1, Scheme 7.3). Since substitution at square–planar palladium(II) follows an associative process, coordination of the alkene would therefore initially generate the pentacoordinate intermediate **28**. Overman *et al.* proposed that the stereochemistry is set during displacement of the iodide by the tethered alkene (**25**→**28**→**27**) [16], although several aspects of this process are still not fully understood. Furthermore, this chemistry was beautifully applied to the enantioselective total synthesis of monomeric pyrrolidinoindoline alkaloids [69, 70b] (−)-physostigmine and (−)-physovenine, as well as to higher members of this natural product family (*vide infra*) [71].

Recently, Curran *et al.* presented a low-temperature Mizoroki–Heck reaction of axially chiral, configurationally stable *o*-iodoanilides (*M*)-**113**. Without a chiral palladium catalyst, these ring closures proceeded with efficient chirality transfer from the chiral axis of (*M*)-**113** to the new stereogenic carbonatom in (*R*)-**114** ((*M*)-**113**→(*R*)-**114**; Scheme 7.24) [20].

This fundamental experiment has strong implications on related catalyst-controlled Mizoroki–Heck cyclizations of precursors of this type. As axial chirality in **113** sets the stereochemistry in **114**, enantioinduction was rationalized to arise from a dynamic kinetic resolution of (at elevated temperature) rapidly interconverting enantiomers of **113** in the oxidative addition step, rather than in the alkene coordination–migratory insertion event. Such a dynamic kinetic resolution process has been previously proposed by Stephenson *et al.* within their mechanistic study regarding the conformations of helically chiral 2-iodoanilides in intramolecular asymmetric Mizoroki–Heck reactions [72].

A catalytic cycle of such an enantiospecific Mizoroki–Heck reaction is outlined in Scheme 7.25. The catalytic cycle starts with an oxidative addition of the C—I bond of (*M*)-**113** to palladium(0), yielding intermediate **115** with retention of axial chirality due to the hindered bond rotation of the N—Ar bond (**1**→**115**). Even at this stage, **115** can still provide either enantiomer of **114** because the alkene moiety will have to rotate for the insertion into the C(sp^2)—Pd(II) bond to occur. However, in case of a *si*-face attack, alkene complexation cannot occur because the palladium

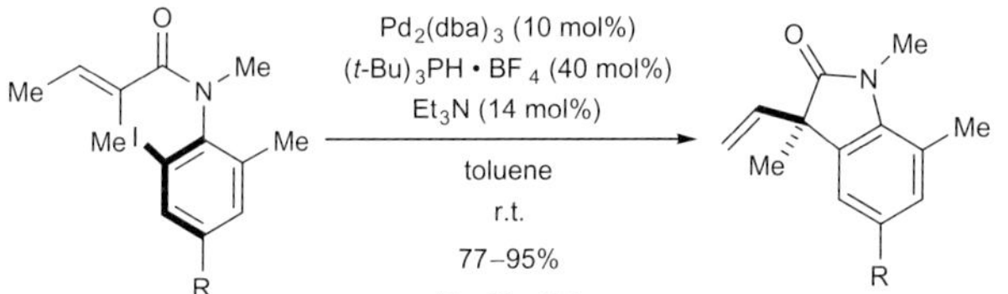

$Pd_2(dba)_3$ (10 mol%)
$(t\text{-}Bu)_3PH \cdot BF_4$ (40 mol%)
Et_3N (14 mol%)
toluene
r.t.
77–95%
(R = Me, Br)

(*M*)-**113**: 99% ee → (*R*)-**114**: 71–78% ee

Scheme 7.24 Enantiospecific Mizoroki–Heck cyclization of axially chiral anilides.

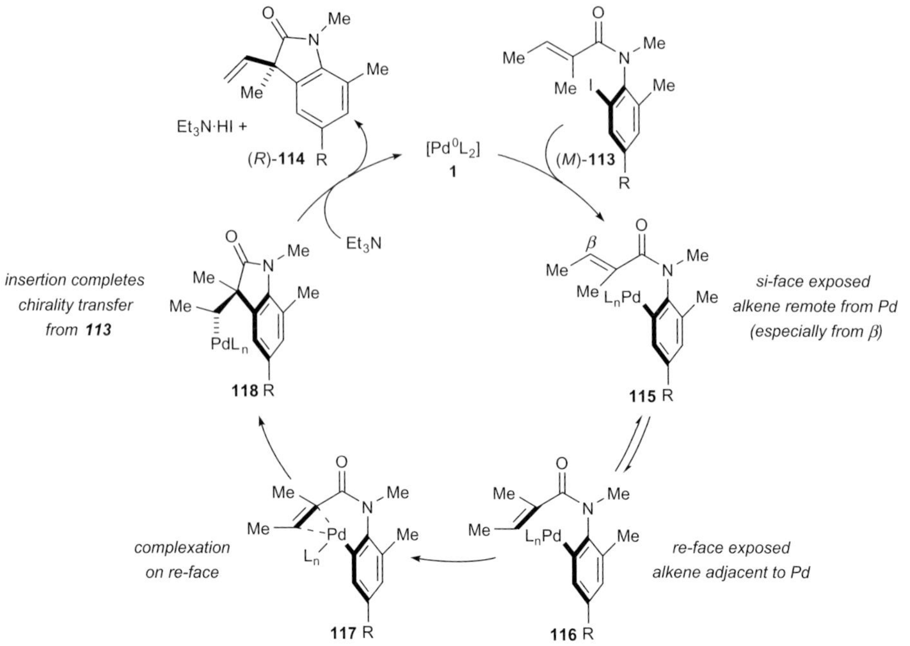

Scheme 7.25 Assumed mechanism for the transfer of axial chirality in low-temperature Mizoroki–Heck cyclization.

atom is not positioned over the alkene π-system in **115**. Conversely, rotation of the C(O)—C bond yields **116**, in which palladium can easily coordinate to the alkene in a *re*-face fashion (**116**→**117**). Subsequent migratory insertion (**117**→**118**) and β-hydride elimination provides (*R*)-**114** with good chirality transfer (**118**→**1**).

Within studies towards the enantioselective total synthesis of a tetrameric member of the pyrrolidinoindoline alkaloid family [71], quadrigemine C (**121**; see Scheme 7.27), Overman *et al.* were, for the first time, able to isolate a stable σ-alkylpalladium(II) intermediate of an asymmetric Mizoroki–Heck reaction that has β-hydrogen atoms residing on a freely rotating β-carbon atom (**119**→**120**; Scheme 7.26) [73]. Compound **120** was prepared from (*Z*)-butenanilide **119** in the presence of stoichiometric amounts of palladium(II) acetate and (*R*)-BINAP ((*R*)-**74**), yielding **120** in high dr, indicating high enantioselection in formation of the oxindole quaternary carbon center. Such palladacyclic structures, especially when palladium is attached to a stereogenic carbon [74], are rare [75].

Its striking total synthesis was however accomplished by Overman *et al.* by following a different route (Scheme 7.27) [76]. *meso*-Chimonantine (**123**), a natural product by itself, was chosen as a starting point as it already contains two of the pyrrolidinoindoline units of target compound **121**. Starting from commercially available oxindole and isatin, *meso*-chimonantine had been synthesized previously in the Overman laboratories in a stereocontrolled 13-step procedure [77]. Now, four additional steps were necessary to access the central intermediate for the

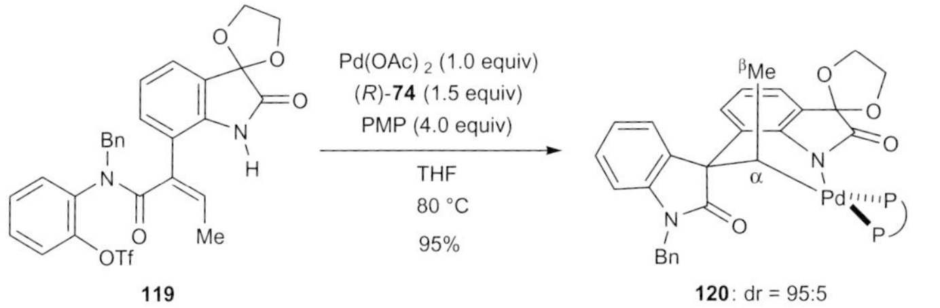

Scheme 7.26 Isolation of σ-alkylpalladium(II) intermediate **120** in an asymmetric intramolecular Mizoroki–Heck reaction.

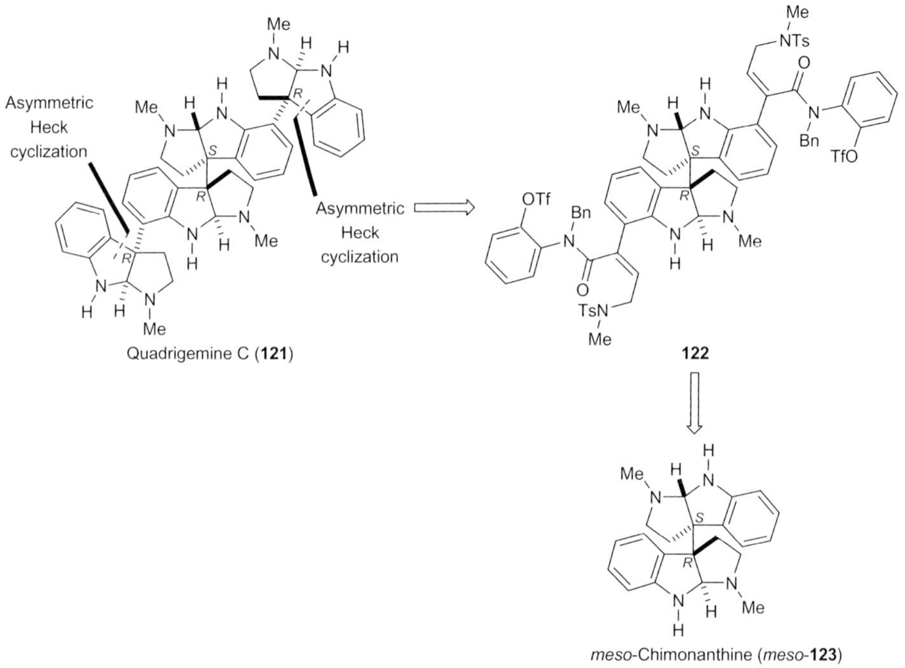

Scheme 7.27 Retrosynthetic analysis of quadrigemine C.

projected key step, a catalyst-controlled double Mizoroki–Heck cyclization of **122**. This reaction was then employed to desymmetrize the *meso*-core of **122**, simultaneously creating the two peripheral biaryl quaternary stereocenters in **121**. By using $Pd(OAc)_2$, (*R*)-tol-BINAP and PMP as base, the C_1-symmetrical cyclization product (not shown) is obtained in 62% yield and 90% ee! Removal of the four protecting groups and deoxygenation of the two oxindoles resulted in the formation of the two final rings of quadrigemine C (**121**). Overall, this complex natural product was accessed in phenomenal 19 steps and 2% overall yield.

In the previous synthesis, *meso*-**123** was used as building block for a more complex natural product. Overman *et al.* also presented an elegant Mizoroki–Heck-based strategy for the diastereoselective construction of the vicinal

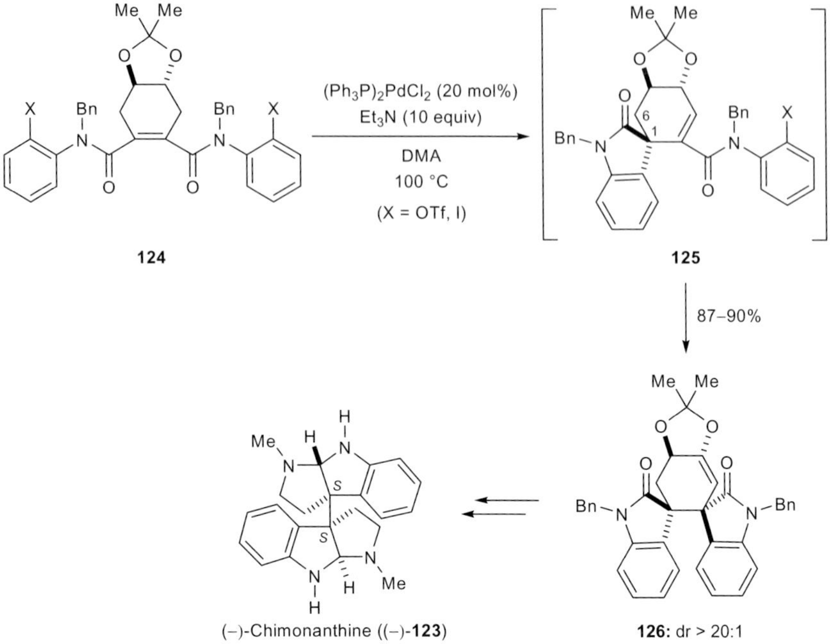

Scheme 7.28 Diastereoselective intramolecular Mizoroki–Heck cyclization as key step in the total synthesis of (–)-chimonanthine.

quaternary centers in (–)-chimonanthine (**123**; Scheme 7.28) [78]. Upon treatment with $Pd(PPh_3)_2Cl_2$ under typical Mizoroki–Heck conditions, diamide **124** delivered bis-oxindole **126** with superb diastereoselectivity (**124**→**126**). Remarkably, the stereochemical outcome of this double Mizoroki–Heck cyclization is controlled by the choice of the diol protecting groups (acetonide or silyl ethers). In an attempt to elucidate the origin of stereoselection, the two ring-closing events were studied separately [79]. High stereoselection (dr > 20 : 1) was only seen in the first intramolecular Mizoroki–Heck cyclization when the substrate contained both a *trans*-acetonide and a tertiary amide substituent at C-2 (**124**→**125**). Consequently, two subtle factors appear to be involved:

- The avoidance of eclipsing interactions between: (i) the forming C—C bond and the pseudoaxial hydrogen atom at C-6; and (ii) the carbonyl carbon of the forming spiro-oxindole and the pseudoequatorial hydrogen atom at C-6 leads to moderate preference for insertion from the alkene face proximate to the pseudoequatorial hydrogen atom at C-6.
- The vinylic amide substituent preferentially adopts a perpendicular conformation, thus placing the sterically bulky NR_2 over the alkene π-bond. For that reason, *syn*-pentane-like interactions between this substituent and the C-3 of cyclohexene are avoided in the favored insertion topography.

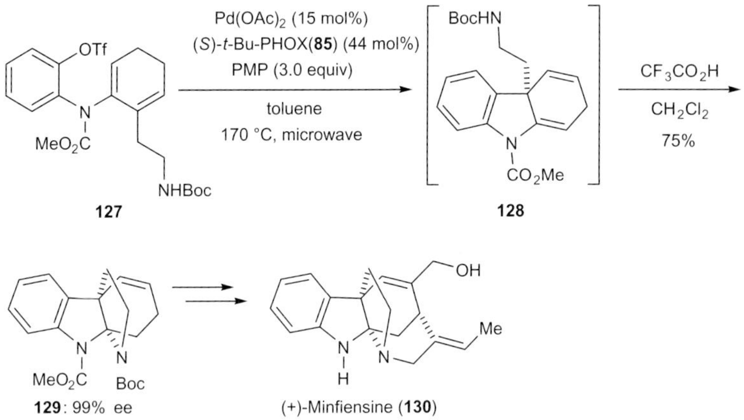

$Pd(OAc)_2$ (15 mol%)
(*S*)-*t*-Bu-PHOX(**85**) (44 mol%)
PMP (3.0 equiv)
toluene
170 °C, microwave

CF_3CO_2H
CH_2Cl_2
75%

127 **128**

129: 99% ee (+)-Minfiensine (**130**)

Scheme 7.29 Total synthesis of (+)-minfiensine.

The combination of these two effects accounts for the high diastereoselection in the first Mizoroki–Heck ring closure; diastereoselectivity of the second cyclization is simply determined by steric interactions with the previously formed spiro-oxindole.

In another beautiful application, Overman *et al.* developed the total synthesis of the natural product (+)-minfiensine (**130**), having a 1,2,3,4,-tetrahydro-9a,4a-(iminoethano)9*H*-carbazol core (Scheme 7.29) [80]. The key step in their synthesis is a tandem process consisting of a catalytic asymmetric Mizoroki–Heck ring closure and an *N*-acyliminium ion cyclization of dienyl carbamate triflate **127** (**127**→**128**). The reaction proceeded smoothly using the Pfaltz ligand (*S*)-*t*-Bu-PHOX (**85**), providing (dihydroiminoethano)carbazole **129** after the addition of excess trifluoroacetic acid, with 75% overall yield and 99% enantiomeric purity (**127**→**129**).

7.2.3.3 Desymmetrizing Intramolecular Arylations

As mentioned above, the enantioselective desymmetrization of prochiral, as well as *meso* compounds, by using a Mizoroki–Heck coupling represents another possibility to effect asymmetric transformations [81–84]. Therefore, the plane of symmetry in these precursors must be broken by differentiation of enantiotopic alkene-containing groups with a chiral palladium complex.

Based on the pioneering studies of Shibasaki *et al.* [65], Feringa *et al.* accomplished the desymmetrizing ring closure of prochiral cyclohexadienones **131** using TADDOL-based monodentate phosphoramidite **132** (**131**→**133**; Scheme 7.30) [81]. The fact that a monodentate ligand induced high enantioselectivity is remarkable in its own right. Moreover, the success of this transformation was highly dependent on the leaving group: aryl iodide **131a** cyclized in high enantiomeric excess (**131a**→**133a**), whereas aryl bromide **131b** was almost inert (**131b**→**133b**, <5% conversion); the reaction of aryl triflate **131c** was sluggish, and **133c** was isolated in moderate enantiomeric excess. A complete understanding of this set of data

$Pd(OAc)_2$ (10 mol%)
132 (20 mol%)
Cy_2MeN (3.0 equiv)
$CHCl_3$
80 °C

131a (X = I)
131b (X = Br)
131c (X = OTf)

133a: 100% conv 96% ee
133b: < 5% conv –
133c: 25% conv 75% ee

Scheme 7.30 Desymmetrizing Mizoroki–Heck arylation using a monodentate ligand.

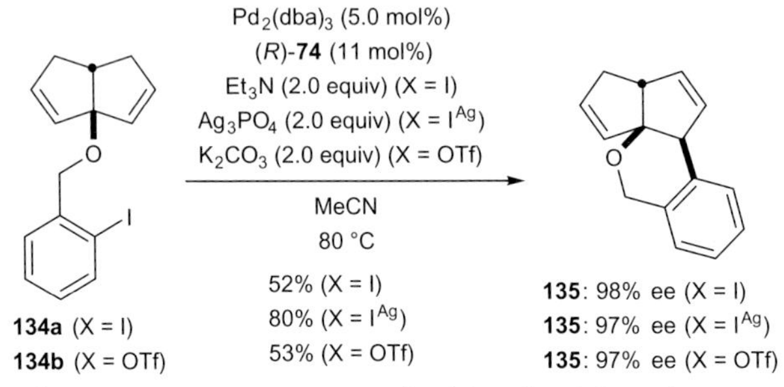

Scheme 7.31 Desymmetrization of a *cis*-bicyclo[3.3.0]octadiene.

remains elusive. However, Feringa *et al.* hypothesized that the cationic complex formed from the triflate is unreactive; instead, a reactive neutral complex is generated from the aryl iodide. It should be noted here that the traditional mechanistic picture of neutral and cationic pathways is developed for bidentate ligands (cf. Section 7.2.1), and might not apply to asymmetric Mizoroki–Heck processes involving monodentate ligands.

In the aforementioned reaction, the stereogenic carbon is not created at the site of C—C bond formation. In an example reported recently by Lautens *et al.*, a stereocenter is also formed at the former $C(sp^2)$-hybridized carbon. Tetracyclic **135** containing three contiguous stereogenic carbons was enantioselectively accessed from *cis*-bicyclo[3.3.0]octadiene precursors (**134**→**135**; Scheme 7.31) [82]. In fact, this desymmetrizing reaction proceeds with superb enantioselectivity, independent of the leaving group. By replacing (*R*)-**74** in this system, (*S*)-*p*-tol-BINAP performed best (87% yield, 99% ee), whereas the *P,N* ligand (*S*)-*t*-Bu-PHOX (**85**) gave only a low chemical yield and enantioselectivity (18%, 29% ee). An extension to a *cis*-bicyclo[4.4.0]decadiene skeleton failed (not shown).

Recently, Bräse *et al.* reported a desymmetrizing enantioselective Mizoroki–Heck reaction producing tetracycle **138**, again with three adjacent stereogenic centers (**136**→**138**; Scheme 7.32) [83]. Interestingly, a Mizoroki–Heck reaction of **136** using conventional chiral ligands such as (*R*)-BINAP ((*R*)-**74**) or (*S*)-*i*-Pr-PHOX resulted in very low enantioselection and/or conversion. Conversely, (*S*,*R*)-

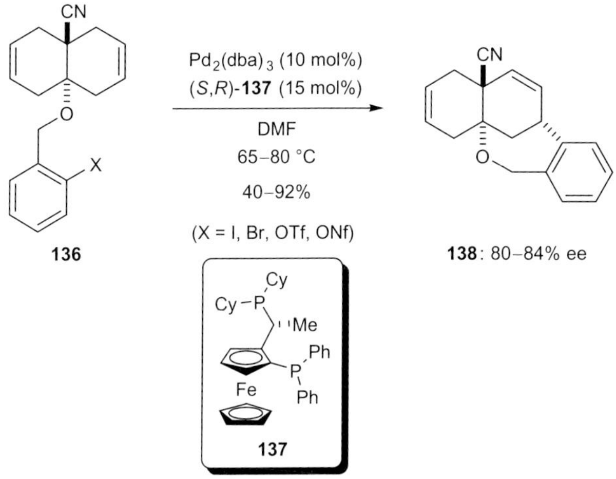

Scheme 7.32 Desymmetrization of *trans*-bicyclo[4.4.0]decadienes **136**.

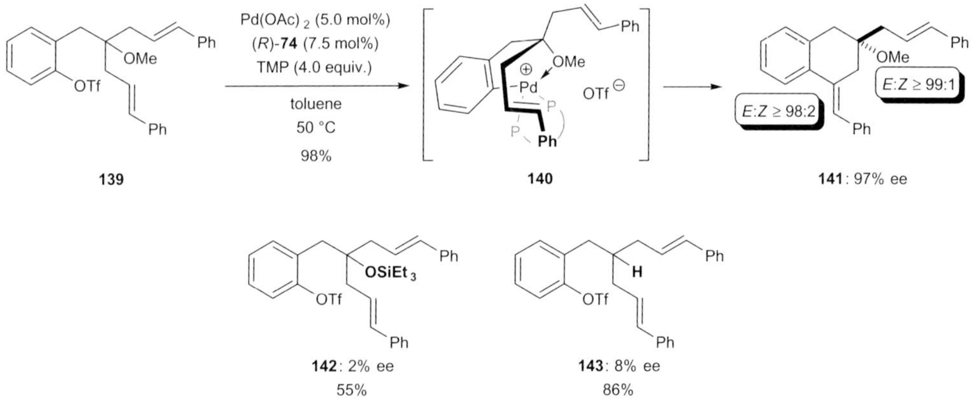

Scheme 7.33 Oxygen-directed, desymmetrizing Mizoroki–Heck cyclization.

JOSIPHOS (**137**) gave **138** in 92% yield and 84% ee. As in the preceding case, the leaving group (iodide, bromide, triflate or nonaflate) had almost no effect on stereoinduction, but the yields were lower.

Until a recent discovery by Oestreich *et al.*, the substrate scope of desymmetrizing Mizoroki–Heck reactions was restricted to cyclic dienes [84], although it had long been suspected that, for efficient differentiation of the enantiotopic unsaturated branches, a rigid framework would be necessary. An intramolecular Mizoroki–Heck reaction of **139** resulted in the formation of **141** with excellent enantiomeric excess and immaculate control of the alkene geometry (Scheme 7.33). Further investigations revealed that the oxygen donor in **139** is critical to success; the *O*-silylated precursor **142**, as well as the deoxygenated substrate **143**, underwent the ring closure to generate the corresponding products in almost racemic form. Straub and Oestreich *et al.* later showed that the Lewis basic oxygen donor reversibly interacts with the weakly Lewis acidic chiral palladium(II) catalyst during the catalysis, thereby controlling the level of enantioselection [84c]. Based on experimental observations, and supported by quantum chemical calculations, Straub and Oestreich *et al.* proposed a mechanism, in which the oxygen donor mediates the equilibration of the diastereomeric alkene–palladium complexes via a six-membered chelate **140** (Scheme 7.33) [84c].

Furthermore, this oxygen-directed, intramolecular Mizoroki–Heck methodology found a first application in the synthesis of an AB synthon of anthracyclines [84b].

7.3
Reductive Mizoroki–Heck-Type Arylations [85]

In conventional Mizoroki–Heck chemistry, catalytic turnover is ensured by a generally accepted sequence of fundamental steps (cf. Section 7.2.1). Within this catalytic process, both migratory insertion – and importantly β-hydride elimination – are diastereospecific, the latter usually necessitating a synperiplanar arrangement of a C_{β}—H bond and the C_{α}—Pd^{II} bond. The absence of conformationally accessible β-hydrogens in the σ-alkyl palladium(II) intermediate will therefore not only thwart β-hydride elimination but also intercept the catalytic cycle. In order to provoke this situation, particular structural features of the substrate are required, which are satisfied by alkenes having rigid bicyclo[2.2.1]heptane or bicyclo[2.2.2]octane skeletons. Instead of suffering β-hydride elimination, the C—Pd^{II} bond in these systems might then enter different reaction channels.

In a seminal communication, Chiusoli *et al.* reported an example of an *intercepted* Mizoroki–Heck reaction for the first time [86]. A palladium-catalyzed reaction of bicyclo[2.2.1]heptene (**144**) with **145** in the presence of a reducing agent, ammonium formate, produced a mixture of **146** and **147** in a reaction temperature-dependent ratio (Scheme 7.34). The minor product **147** was formed by standard intermolecular Mizoroki–Heck alkenylation and subsequent reductive cleavage of the intermediate $C(sp^3)$—Pd^{II} bond; formation of the major product **146** involved an additional intramolecular alkene insertion prior to the reduction step.

These findings clearly foreshadowed that the $C(sp^3)$—Pd^{II} bond might enable synthetically useful C—C, as well as C—H bond-forming reactions, yet almost a decade passed until the latter process was systematically investigated (cf. Section

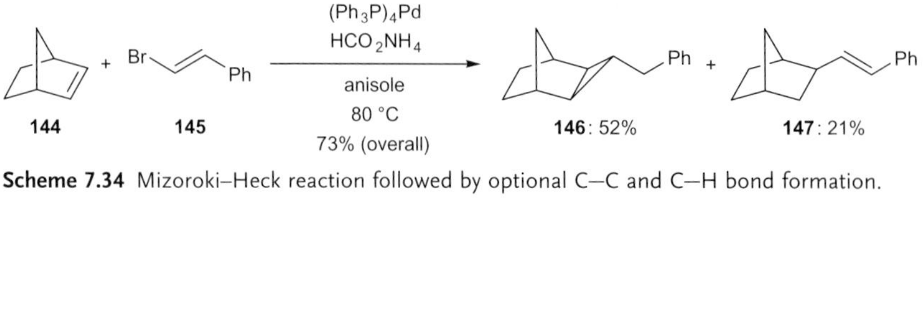

144 + **145** (Br, Ph) → $(Ph_3P)_4Pd$, HCO_2NH_4, anisole, 80 °C, 73% (overall) → **146**: 52% + **147**: 21%

Scheme 7.34 Mizoroki–Heck reaction followed by optional C—C and C—H bond formation.

7.3.2). Apart from standard subsequent transformations, it was later discovered by Catellani *et al.* that the C(sp^3)—Pd^{II} bond in norbornene-derived palladium(II) alkyls activates proximal C(sp^2)—H bonds, thereby rendering this chemistry a general methodology [87].

The present section summarizes reductive Mizoroki–Heck-type arylations – that is, palladium-catalyzed hydroarylation reactions of alkenes – which are essentially limited to (hetero-)norbornenes (Section 7.3.2). It should be noted however, that only a handful of remarkable examples are known that are not based on the bicyclo[2.2.1]heptane framework (see Section 7.3.3).

7.3.1
Mechanistic Considerations

The mechanism of the reductive Mizoroki–Heck process (Scheme 7.35) is related to the mechanistic picture of the Mizoroki–Heck reaction itself (cf. Section 7.2.1). The pivotal difference is the presence of excess formic acid, at least equimolar to added base, as the reducing agent. The catalysis commences with an oxidative addition of the C(sp^2)—X bond of **148** to a palladium(0) catalyst (**148**→**149**). The thus-formed electrophilic C(sp^2)—Pd(II) complex **149** will then capture the alkene **150**, followed by *syn* selective migratory insertion (**149**→**151**). C(sp^3)—Pd(II) intermediates are normally prone to facile β-hydride elimination, yet no conformationally accessible C—H bond, neither at C_β nor at $C_{\beta'}$, is available in **151** (*vide supra*). Instead, salt elimination with formate (**151**→**152**), followed by the extrusion of carbon dioxide, occurs (**152**→**153**). This detour establishes an alternative route to the σ-alkyl palladium(II) hydride intermediate **153**, which could otherwise emerge from **151** if conventional β-hydride elimination were possible. Reductive

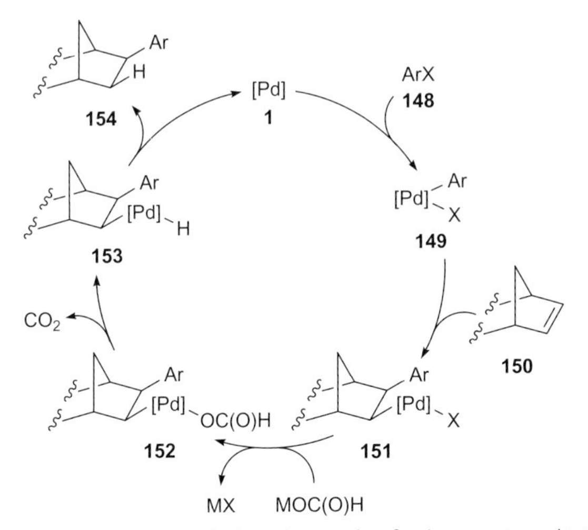

Scheme 7.35 Simplified catalytic cycle of reductive Mizoroki–Heck-type arylations.

elimination, α-elimination at palladium(II), at **153** liberates the hydroarylation product **154** (**153**→**154**) and regenerates catalytically active palladium(0).

7.3.2
Intermolecular Arylations: The Bicyclo[2.2.1]heptane Case

In light of the initial contribution by Chiusoli *et al.*, it is somewhat surprising that efficient procedures for the palladium-catalyzed hydroarylation of norbornene (**144**) were reported almost a decade later by Cacchi *et al.* [3] and, shortly after, by Larock *et al.* [4] (**144**→**155**; Scheme 7.36). These protocols resemble that of Chiusoli *et al.* [86], as well as those for the conjugate hydroarylation of α,β-unsaturated carbonyl compounds (see Chapter 8) [88]. In recent years, catalysts such as palladacycles [89–91] and palladium–carbene complexes [92] were also shown to catalyze the hydroarylation of **144**.

As the desymmetrization of norbornene (**144**) results in the formation of three contiguous stereogenic carbons, several groups have attempted asymmetric variants of the reductive Mizoroki–Heck-type reaction (Figure 7.3) [93–99]. Brunner *et al.* were first to realize this asymmetric hydroarylation using (*R,R*)-norphos (**156**) as the chiral ligand with moderate levels of enantioselection [93]. Replacement of the *P,P* ligand **156** with a *P,N* ligand, valphos-type ligand **157**, and using phenyl triflate instead of the corresponding iodide, substantially improved the enantiomeric excess [94]. These findings by Achiwa *et al.* were further refined by Kaufmann *et al.*, using phenyl nonaflate [95]. The latter examination also notes that reductive dehalogenation/desulfonation of ArX (X = I, OTf and ONf) lowers the chemical yields. This is rationalized by competing reduction of the intermediate $C(sp^2)-Pd^{II}$ complex **149** (Scheme 7.35) prior to alkene insertion. Zhou *et al.* also showed that *N,N* ligands such as **158** perform equally well in this transformation [96–98]. (*R,R*)-duphos-type ligand **159** yielded no further improvement [99].

Extension of the asymmetric hydroarylation to oxanorbornenes, as well as azanorbornenes, was the obvious step forward. Fiaud *et al.* made several remarkable observations while examining the reductive Mizoroki–Heck-type arylation of the annulated 7-oxanorbornene **160** (Scheme 7.37) [100]. First, intermediate σ-alkyl

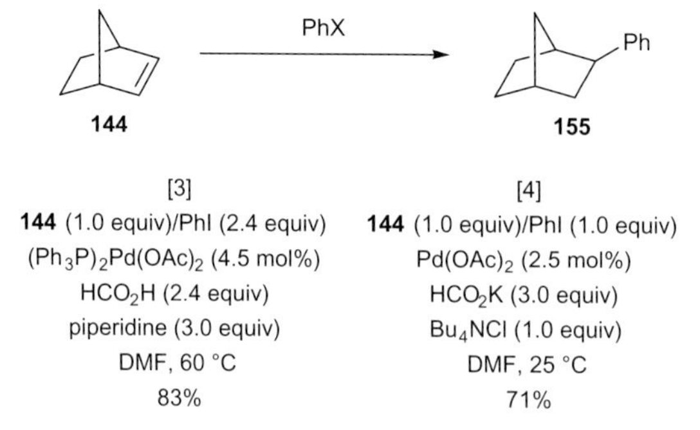

Scheme 7.36 Palladium-catalyzed hydroarylation of norbornene.

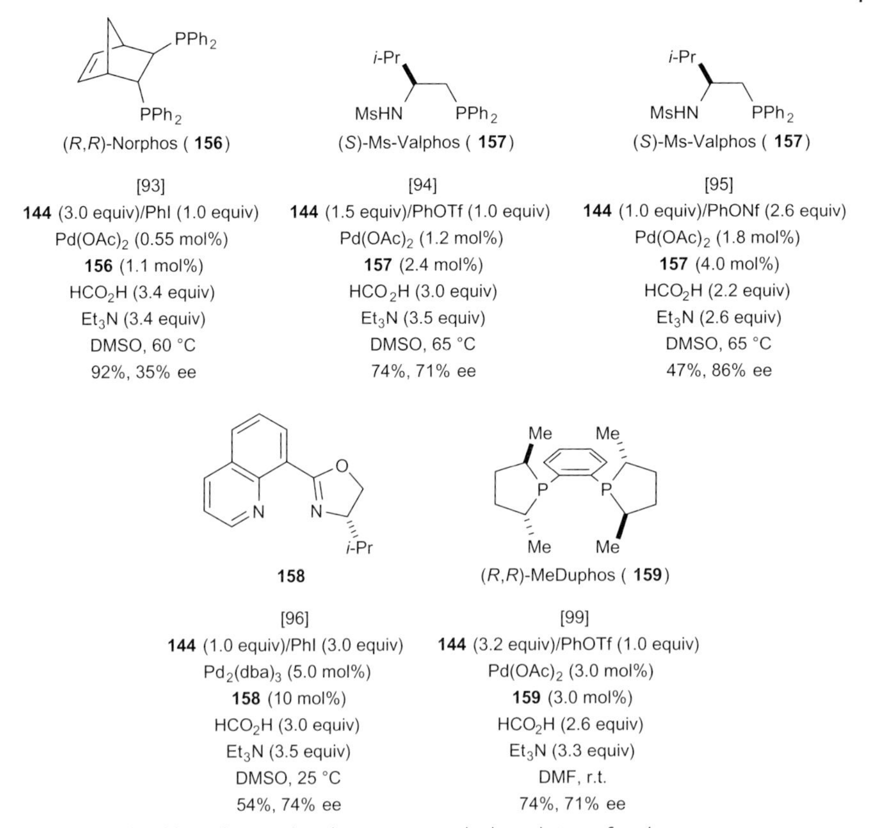

Figure 7.3 Chiral ligands tested in the asymmetric hydroarylation of norbornene.

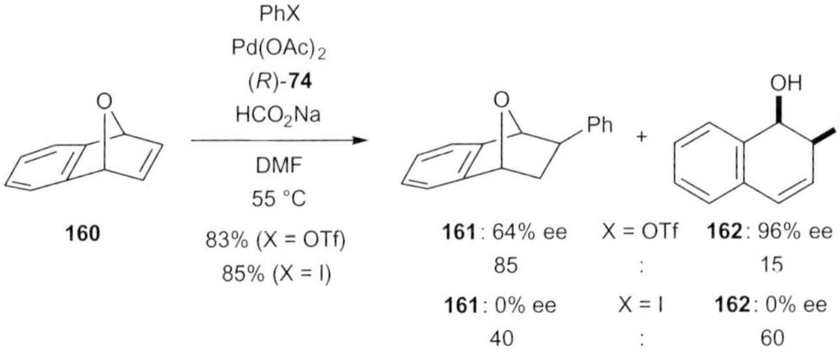

PhX, $Pd(OAc)_2$, (*R*)-**74**, HCO_2Na, DMF, 55 °C

160 → 83% (X = OTf), 85% (X = I)

161: 64% ee : **162**: 96% ee (X = OTf) — 85 : 15

161: 0% ee : **162**: 0% ee (X = I) — 40 : 60

Scheme 7.37 Highly enantioselective hydroarylation of an annulated 7-oxanorbornene (exact reaction conditions not reported).

palladium(II) complexes (**151** or **152**; Scheme 7.35) suffer partial β-alkoxy elimination affording ring-opened **162** [101] along with desired **161**. Second, enantiomeric excesses of **161** (64% ee) and **162** (96% ee) are not identical. Unfortunately, absolute configurations were not determined, although it might be reasonable to assume that one of the two diastereomeric σ-alkyl palladium(II) complexes

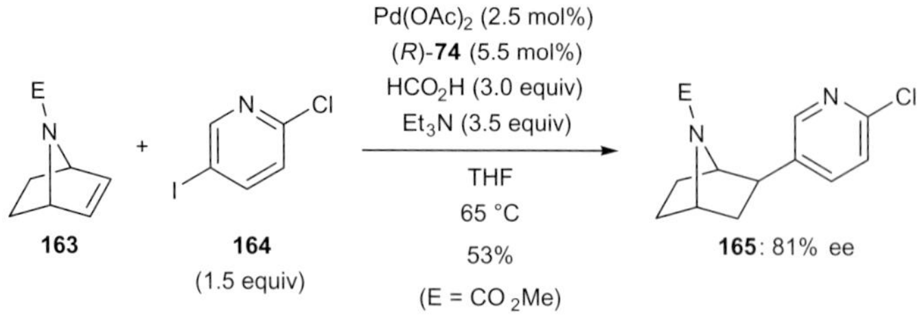

$Pd(OAc)_2$ (2.5 mol%)
(*R*)-**74** (5.5 mol%)
HCO_2H (3.0 equiv)
Et_3N (3.5 equiv)
THF
65 °C
53%
(E = CO_2Me)

163 + **164** (1.5 equiv) → **165**: 81% ee

Scheme 7.38 Enantioselective hydroarylation: synthesis of *N*-protected epibatidine.

preferentially undergoes β-alkoxy elimination. This, in turn, would also account for the exceptionally high enantiomeric excess seen for **162**. Third, enantioselection is strongly influenced by the nature of the leaving group, that is weakly coordinating triflate (96% ee) or coordinating iodide (0% ee). This finds precedence in asymmetric Mizoroki–Heck chemistry (cf. Section 7.2.2.2) [15, 40], which indicates that the possibility of cationic and neutral pathways also applies to reductive Mizoroki–Heck processes.

Reductive Mizoroki–Heck couplings of azanorbornenes would allow for short syntheses of the alkaloid epibatidine [102–104] and analogues thereof [104, 105]. Kaufmann *et al.* elaborated a straightforward one-step enantioselective access to *N*-protected epibatidine **165** (Scheme 7.38) [102, 103]. Hydroarylation of 7-azanorbornene **163** with functionalized aryl iodide **164** produced **165** in moderate yield with good enantiomeric excess [103]. Zhou *et al.* later tested a ligand similar to **158** (Figure 7.3) in the same reaction, and with a number of other heteronorbornenes (49–85% ee) [104].

7.3.3
Reductive Mizoroki–Heck-Type Arylation in Action

The palladium-catalyzed hydroarylation is heavily dependent on the structural constraints imposed by competing β-hydride elimination (*vide supra*). Aside from the standard workhorses – namely bicyclo[2.2.1]heptanes – a few more demanding systems have been subjected to reductive Mizoroki–Heck couplings so far, as was recently summarized by Mitchell *et al.* [106].

In a comprehensive investigation, Diaz *et al.* targeted the asymmetric intramolecular hydroarylation (Scheme 7.39) [107]. In intramolecular reactions, the length of the tether often determines the regiochemistry by kinetically favoring either isomer, for example 5-*exo*-trig over 6-*endo*-trig. As for **166** (Scheme 7.39), regioselective cyclization onto the 1,1-disubstituted alkene generates a $C(sp^3)$—Pd^{II} bond adjacent to a congested quaternary carbon center. In this well-chosen structural motif lacking β-hydrogens, β-hydride elimination cannot compete with the reductive pathway [108]. Under (cationic [14]) conditions using aryl iodide **166** in connection with a silver(I)-exchanged zeolite, the ring closure afforded **167** in

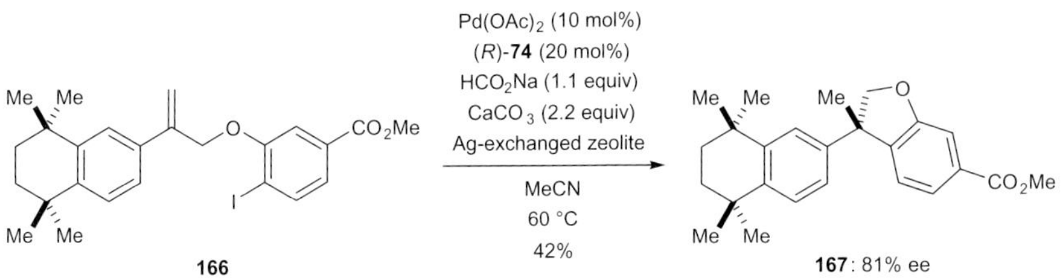

166 167: 81% ee

Scheme 7.39 Catalytic asymmetric intramolecular hydroarylation.

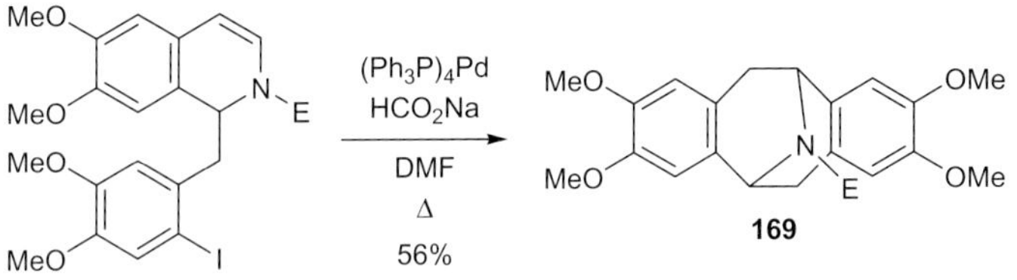

168 (E = CO_2Et)

Scheme 7.40 Intramolecular hydroarylation *en route* to pavine alkaloids.

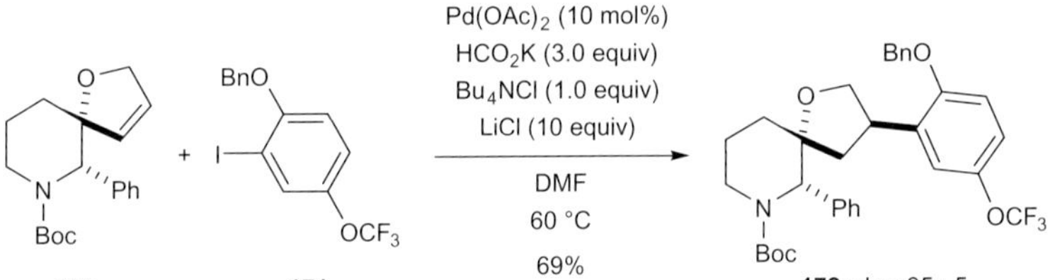

170 171 (3.0 equiv) 172: dr = 95 : 5

Scheme 7.41 Diastereoselective intermolecular hydroarylation.

respectable enantiomeric excess (**166**→**167**; Scheme 7.39). It should be noted here that this has remained the sole example of an enantioselective palladium-catalyzed hydroarylation.

A reductive Mizoroki–Heck-type cyclization, in which again conformational rigidity thwarts β-hydride elimination, was reported by Ruchirawat *et al.* (**168**→**169**; Scheme 7.40) [109], who compared the palladium-catalyzed hydroarylation with the alternative radical cyclization, which emerged as less effective.

In fact, intermolecular hydroarylation of alkenes other than norbornenes was also realized, with Kulagowski *et al.* exploiting the diastereospecificity of both alkene insertion and β-hydride elimination in the sterically driven regioselective coupling of **170** and **171** (**170**→**172**; Scheme 7.41) [110]. However, it was emphasized that an excess of chloride anions was required to prevent dehalogenation [95].

7.4
Oxidative Mizoroki–Heck-Type Arylations [111, 112]

7.4.1
Mechanistic Considerations

In the Mizoroki–Heck reaction, the catalysis begins with the oxidative addition of a C(sp^2)—X bond to a palladium(0) complex to give a C(sp^2)—Pd(II) complex common to almost all palladium(0)-catalyzed cross-coupling reactions (cf. Section 7.2.1). There are, however, alternative ways to generate the central σ-aryl palladium(II) intermediate and to effect a Mizoroki–Heck-type process (Scheme 7.42).

First, the interaction of a Lewis acidic palladium(II) complex with a C(sp^2)—H bond of an electron-neutral or electron-rich (het)arene might result in its electrophilic palladation (**173**→**175**; Scheme 7.42), a fundamental process later referred to as electrophilic C—H bond activation. In contrast, the oxidative addition of a C(sp^2)—X bond to palladium(0) is favored for electron-poor (het)arenes, thus providing an orthogonal entry into Mizoroki–Heck chemistry. After C—H activation, conventional alkene insertion (**175**→**176**) and β-hydride elimination (**176**→**177**) eventually lead to the formation of catalytically inactive palladium(0). In order to achieve catalytic turnover, however, palladium(0) must be reoxidized to palladium(II).

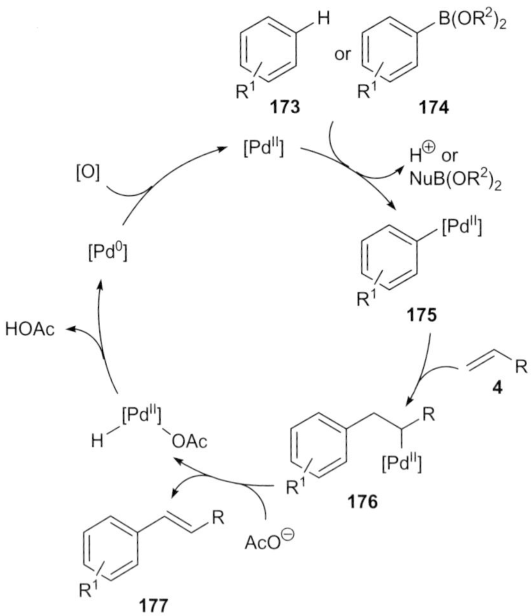

H
or
$B(OR^2)_2$
R^1
R^1
173
174
$[Pd^{II}]$
$H^{\oplus}$ or
$NuB(OR^2)_2$
[O]
$[Pd^{II}]$
$[Pd^0]$
R^1
175
HOAc
R
4
$[Pd^{II}]$
R
H
OAc
$[Pd^{II}]$
R^1
176
R
$AcO^{\ominus}$
R^1
177

Scheme 7.42 Electrophilic C—H bond activation or transmetallation as an entry into oxidative Mizoroki–Heck-type arylations.

This transformation, often termed the oxidative Mizoroki–Heck reaction, was actually reported by Moritani and Fujiwara [5] prior to publication of the initial descriptions of the Mizoroki–Heck reaction [1, 2]. During the early years, the oxidation of palladium(0) by molecular oxygen required copper(II) as a redox mediator, although in recent years modified protocols were developed (see Section 7.4.2.1) [113].

Second, the catalysis might also be initiated by a transmetallation step (**174**→**175**; Scheme 7.42). Several C(sp^2)—E bonds (E = element prone to nucleophilic attack) might serve as precursors for the nucleophilic coupling partner; here, aryl boron compounds **174**, boronic acids (R^2 = H) and boronic esters (R^2 = alkyl) have emerged as particularly effective reagents to date [114]. Again, copper(II) was initially used as a mediator [114a] before procedures involving the direct oxidation of palladium(0) by dioxygen were described [114b–d].

7.4.2 Intermolecular C—C Bond Formation

7.4.2.1 Arenes as Nucleophiles

The synthetic potential of the Fujiwara–Moritani reaction was certainly underappreciated at the time of its discovery. However, more than three decades later, Fujiwara and colleagues themselves resumed their own chemistry and reported several practical procedures for the oxidative (het)arylation of alkenes involving C–H bond activation (**179**→**180**; Scheme 7.43) [113]. A major advance was recently accomplished by Ishii *et al.*, who used heteropolyacids as mediators for the often critical reoxidation step; careful optimization of the reaction conditions was again necessary to obtain good results (Scheme 7.43) [115]. Within a systematic investigation, Jacobs *et al.* found that manganese(III) acetate or benzoic acid (0.20 equiv.) are particularly effective promoters, although their mechanism of action is not yet understood [116].

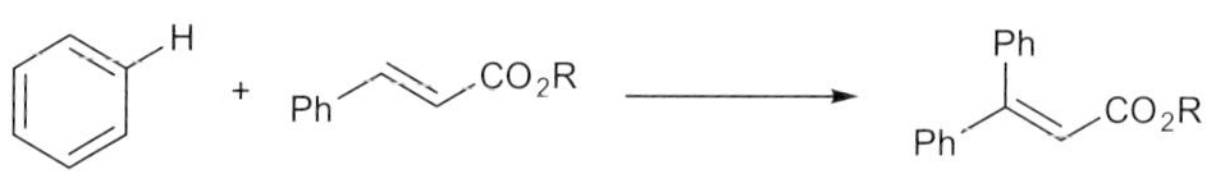

178 + **179** (Ph–CH=CH–CO_2R) → **180** (Ph_2C=CH–CO_2R)

[113]: R = Et	[115]: R = Me	[116]: R = Et
$Pd(OAc)_2$ (1.0 mol%)	$Pd(OAc)_2$ (6.7 mol%)	$Pd(OAc)_2$ (1.0 mol%)
benzoquinone (10 mol%)	acac (6.7 mol%)	$Mn(OAc)_3$ (4.0 mol%)
t-BuOOH (1.3 equiv)	NaOAc (5 mol%)	O_2 (0.8 Mpa)
178 (solvent)	$H_7PMo_8V_4O_{40}$ (1.3 mol%)	**178** (solvent)
AcOH:Ac_2O = 3:1	O_2 (1 atm)	90 °C
90 °C	propionic acid	84%
72%	90 °C	
	84%	

Scheme 7.43 Different protocols of Fujiwara–Moritani arylations.

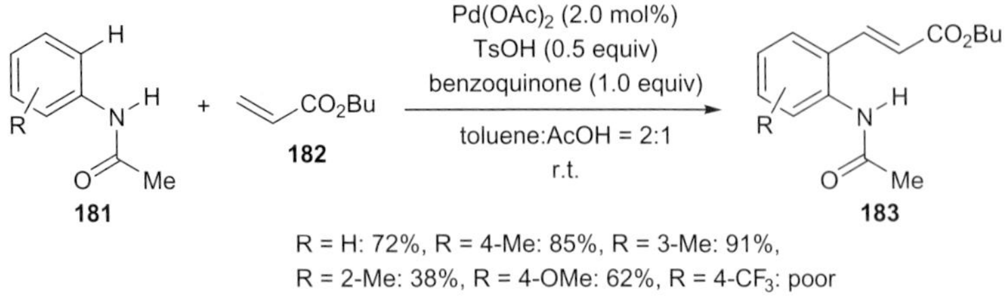

$Pd(OAc)_2$ (2.0 mol%)
TsOH (0.5 equiv)
benzoquinone (1.0 equiv)
toluene:AcOH = 2:1
r.t.

181 + **182** → **183**

R = H: 72%, R = 4-Me: 85%, R = 3-Me: 91%,
R = 2-Me: 38%, R = 4-OMe: 62%, R = 4-CF_3: poor

Scheme 7.44 Directed palladium(II)-catalyzed C—H bond activation.

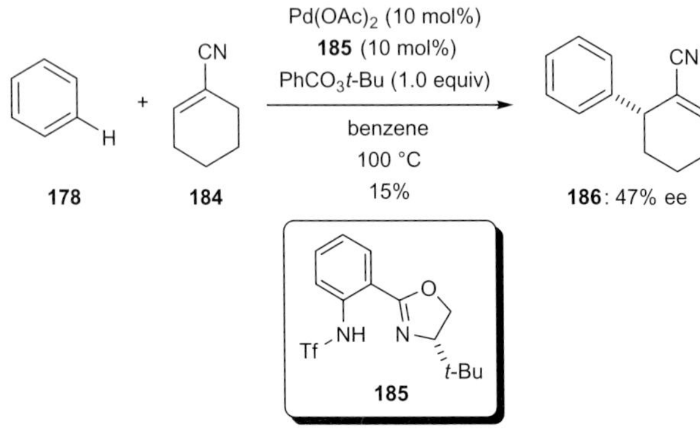

$Pd(OAc)_2$ (10 mol%)
185 (10 mol%)
$PhCO_3t$-Bu (1.0 equiv)
benzene
100 °C
15%

178 + **184** → **186**: 47% ee

185

Scheme 7.45 Catalytic asymmetric intermolecular Fujiwara–Moritani reaction.

The directed C—H bond activation of arenes is an established technique for the regioselective functionalization of (het)arenes (see Chapters 9 and 10), and was also applied to the regioselective oxidative Mizoroki–Heck arylation of anilides (**181**→**183**; Scheme 7.44) [117].

Currently, performing Fujiwara–Moritani reactions in an asymmetric sense appears to be elusive, and to date only two isolated contributions by Mikami *et al.* have been devoted to this challenging task [118]. For example, oxidative Mizoroki–Heck reaction of benzene (**178**) with acceptor **184** in the presence of chiral ligand **185** furnished **186** in moderate enantiomeric excess, yet in poor chemical yield (**184**→**186**; Scheme 7.45) [118a]. Later, these authors reported a comparable level of enantioselection (46% ee) at higher yield (73%) using (*S,S*)-chiraphos, a chiral phosphine ligand [118b]. Surprisingly, enantiomeric excesses of less than 50% still are the benchmark for enantioselective intermolecular oxidative Mizoroki–Heck processes; intramolecular systems have not yet even been reported!

7.4.2.2 **Hetarenes as Nucleophiles**

Although Fujiwara *et al.* had already examined the aerobic palladium(II)-catalyzed hetarylation of alkenes for a limited number of substrates [113a], it is particularly due to the recent studies of Gaunt *et al.* that the C—H bond alkenylation of nitrogen-containing hetarenes has become a synthetically useful methodology [119, 120]. Gaunt's group presented some clever strategies for C-2/C-3-switchable C—C

182: CO_2Bu (butyl acrylate)
Pd(OAc)$_2$ (10 mol%)
t-BuOOBz (1.0 equiv) or air
dioxane:AcOH:DMSO = 9:3:1
35 °C

187 → 188: 75% or 71%, PG = Boc, **electronic control**

187 → 189: 81% or 71%, PG = Si *i*-Pr$_3$, **steric control**

Scheme 7.46 The protective group at nitrogen as a regiochemical switch in intermolecular pyrrole alkenylations.

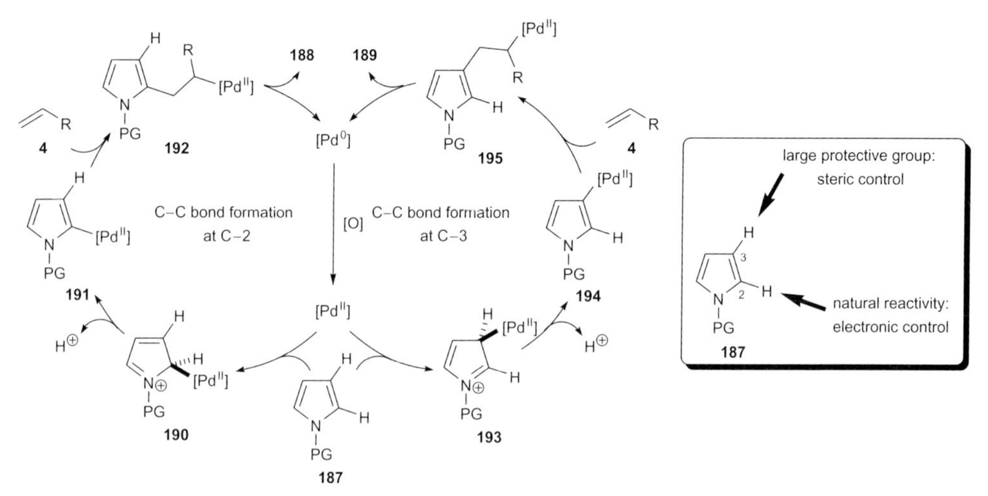

Scheme 7.47 Proposed mechanisms for electronic and steric control in pyrrole alkenylations.

bond formation at both pyrroles [119] (Schemes 7.46 and 7.47) and indoles [120] (Schemes 7.48 and 7.49).

During the course of their studies towards Fujiwara–Moritani couplings of pyrroles **187**, Gaunt *et al.* discovered that the regioselectivity is governed by the steric bulk of the protective group (PG) at nitrogen (Scheme 7.46) [119]. With Boc at nitrogen, electrophilic C—H bond activation occurred selectively at C-2 under electronic control (**187**→**188**), whereas a sterically demanding silyl group at nitrogen completely steered the palladation towards C-3 (**187**→**189**). Whether employing either peroxide *t*-BuOOBz or molecular oxygen, the chemical yields were generally high under mild reaction conditions.

The simplified mechanisms for **187**→**188** (C—C bond formation at C-2) and for **187**→**189** (C—C bond formation at C-3) only differ in the opening steps of the catalytic cycles (**187**→**190**→**191** and **187**→**193**→**194**; Scheme 7.47). According to Gaunt *et al.*, electronically favored C-2 palladation is blocked by large substituents at the adjacent nitrogen. The migratory insertion of **4** (**191**→**192** and **194**→**195**)

CO_2Bu **182**
$Pd(OAc)_2$ (10 mol%)
t-BuOOBz (0.9 equiv)
dioxane:AcOH:DMSO = 15:5:2
70 °C
66%

CO_2Bu **182**
$Pd(OAc)_2$ (20 mol%)
t-BuOOBz (0.9 equiv)
dioxane:AcOH = 3:1
70 °C
51%

197 **196** **198**

Scheme 7.48 Solvent-controlled regioselective indole alkenylations.

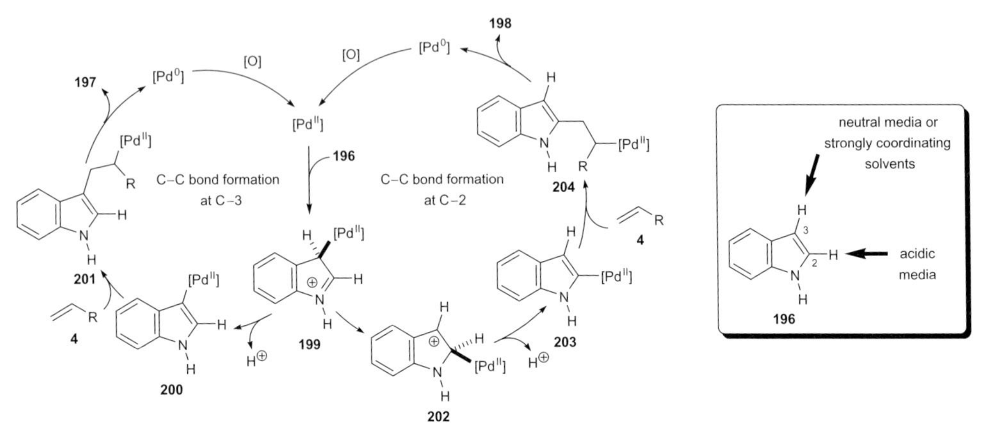

Scheme 7.49 Proposed mechanisms for solvent control in indole alkenylation.

and β-hydride elimination (**192**→**188** and **195**→**189**) conform to the general catalytic cycle outlined in Scheme 7.42.

Controlling the regiochemistry in indole alkenylations [120, 121] (as well as arylations [122]) also attracted considerable interest. Again, it was Gaunt *et al.* who disclosed facile oxidative coupling procedures, which allow for selectively accessing of C-3 (**196**→**197**) and C-2 (**196**→**198**) of the (free) indole core, without the aid of a directing group (Scheme 7.48) [120]. It should be noted here that, at the same time, Brown, Ricci *et al.* reported the directed electrophilic C–H bond activation of C-2 in indoles carrying a 2-pyridylmethyl substituent at nitrogen (not shown) [121].

The method developed by Gaunt *et al.* is remarkable in that the nature of a cosolvent is capable of fully reversing the regiochemical outcome under otherwise identical reaction conditions [120]. In the presence of strongly coordinating solvents, for example dimethylsulfoxide (DMSO), conventional alkenylation at C-3 [113a] was observed (**196**→**197**), whereas in the absence of such an additive the alkenylation was seen at C-2 (**196**→**198**).

Although conclusive evidence is still pending, based on the available experimental data, the authors proposed the unified mechanistic picture depicted in Scheme 7.49 [120]. Both catalyses are considered to originate from the common σ-alkyl palladium(II) intermediate **199**, which arises from regioselective electrophilic palladation at C-3 (**196**→**199**). It is hypothesized that the kinetics of the deprotonation of **199** might then determine its fate: in neutral media, as well as in acidic media but in the presence of a strongly coordinating solvent, the abstraction of a proton (by acetate) occurs rapidly (**199**→**200**). In contrast, this deprotonation might be substantially retarded in purely acidic media, thus opening another reaction channel, namely the 1,2-migration of palladium(II) from C-3 to C-2 (**199**→**202**), followed by rearomatization (**202**→**203**). The next successive steps then lead to the C—C coupling product **197** and **198**, respectively.

Uracil derivatives also participate in catalytic intermolecular oxidative Mizoroki–Heck-type reactions [123].

7.4.3
Intramolecular C—C Bond Formation

7.4.3.1 Arenes as Nucleophiles

In light of the significant impact of the intramolecular Mizoroki–Heck reaction on synthetic organic chemistry (*vide supra*) [17], it is astonishing to see that intramolecular Fujiwara–Moritani-type processes had remained relatively[1)] unexplored until a recent report by Stoltz *et al.* (Scheme 7.50) [128]. These authors basically transferred their experiences gained from the related oxidative cyclization of indoles (cf. Section 7.4.3.2) [129] to the palladium(II)-catalyzed ring closure of electron-rich arenes tethered to an alkene (**205**→**206**; Scheme 7.50) [128]. An electronically tuned pyridine ligand – ethyl nicotinate – is needed in order to achieve sufficient catalytic activity of palladium(II) [128, 129]; this beneficial ligand–catalyst combination was initially described for palladium(II)-catalyzed dioxygen-oxidations [130]. By using dioxygen or benzoquinone, several functionalized benzofurans having a benzylic quaternary carbon were accessible.

In general, such cyclizations catalyzed or mediated by a cationic metal might proceed through two distinct pathways, either electrophilic alkene or C—H bond

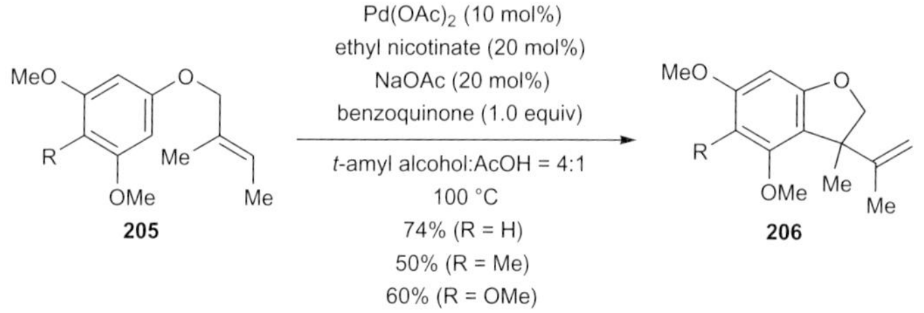

MeO, O, R, Me, OMe, Me — **205**

$Pd(OAc)_2$ (10 mol%)
ethyl nicotinate (20 mol%)
NaOAc (20 mol%)
benzoquinone (1.0 equiv)
t-amyl alcohol:AcOH = 4:1
100 °C
74% (R = H)
50% (R = Me)
60% (R = OMe)

MeO, O, R, OMe, Me, Me — **206**

Scheme 7.50 Intramolecular oxidative Mizoroki–Heck reaction of electron-rich arenes.

1) In fact, Norman *et al.* disclosed the first example of an intramolecular Fujiwara–Moritani reaction back in 1970 [124]; it was, however, understood as a Friedel–Crafts-type process involving alkene activation by palladium(II). Aside from this seminal report, a few oxidative cyclizations of 1,4-quinones using stoichiometric amounts of $Pd(OAc)_2$ were reported [125]. Importantly, Knölker *et al.* developed a reaction protocol for the same substrate class, which required only catalytic amounts of $Pd(OAc)_2$ using $Cu(OAc)_2$ as the stoichiometric oxidant [126]; Åkermark *et al.* later showed that both *t*-BuOOH [127a] and dioxygen [127b] are effective.

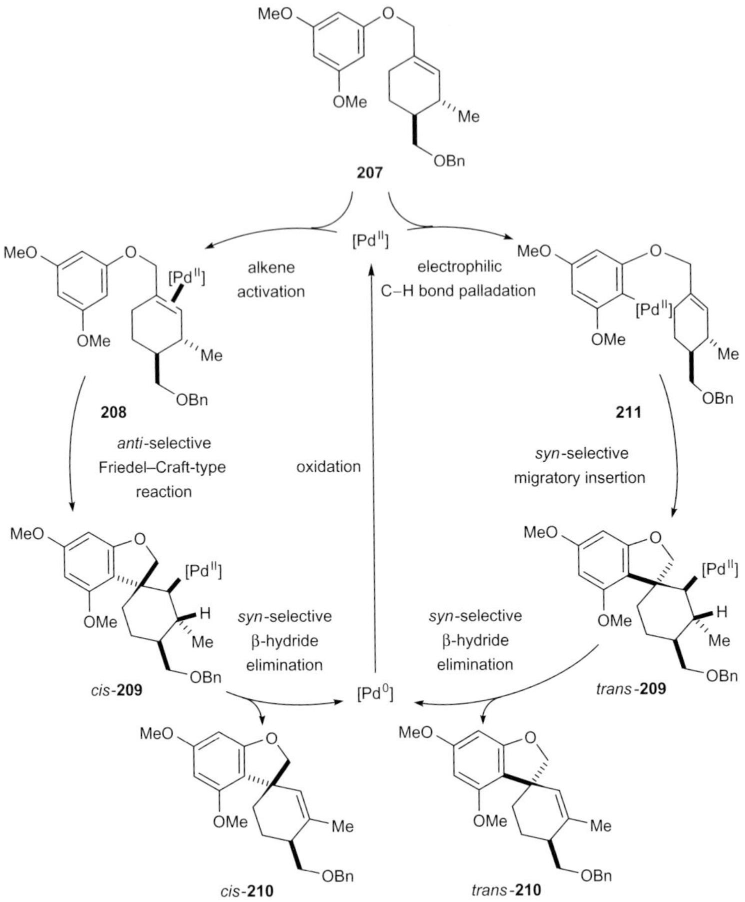

Scheme 7.51 Diastereospecificity as a mechanistic probe: electrophilic alkene or C—H bond activation.

activation. Stoltz *et al.* devised a simple system for rigorously discriminating between these mechanisms (**207**; Scheme 7.51) [128]. The diastereoselective cyclization (**207**→**210**) allows for clear identification of the mode of activation, each of which is characterized by its stereochemical course.

The diastereofacial coordination of palladium(II) to the alkene from the least-hindered face would produce an alkene–palladium(II) complex (**207**→**208**), which would then undergo *anti*-selective electrophilic aromatic substitution – that is, a Friedel–Crafts alkylation (**208**→*cis*-**209**). The alkene unit would subsequently be re-established in the *syn*-selective β-hydride elimination (*cis*-**209**→*cis*-**210**), thereby liberating the cyclization product with *cis* relative configuration. Conversely, electrophilic C—H bond activation (**207**→**211**) would be followed by *syn*-selective alkene insertion (**211**→*trans*-**209**). Again, *syn*-selective β-hydride elimination would finish the catalytic cycle with concomitant product formation (*trans*-**209**→*trans*-

210). The diastereospecificity of both stereoselective steps would unambiguously set the *trans* relative stereochemistry.

The isolated product is *trans* configured, which proves C—H bond activation and, in turn, precludes the electrophilic alkene activation pathway. It should be noted, however, that when using platinum(II) instead of palladium(II) in related cyclizations, the ring closure indeed proceeds through the latter pathway [131]!

7.4.3.2 Hetarenes as Nucleophiles

Intramolecular oxidative Mizoroki–Heck-type reactions of indoles predated the previously discussed example. In fact, Trost *et al.* had already reported oxidative ring closures of an indole onto an azabicyclo[2.2.2]octene during the course of synthetic endeavors (**212** and **213**; Figure 7.4) [132]. Although still stoichiometric in palladium(II) and terminated by reductive cleavage of the stable σ-alkyl palladium(II) intermediate (cf. Section 7.3.2), this seminal contribution not only demonstrated the potential of palladium(II)-mediated cyclizations but also provided conclusive evidence for a C—H bond activation pathway [132]. Later, Williams *et al.* applied the oxidative coupling at C-2 of an indole to the synthesis of paraherquamides B [133a,b] (**214**; Figure 7.4) and A [133c].

Corey *et al.* recently revived this useful transformation in the syntheses of several tryptophan-derived natural products [134]. For example, an unusual palladium(II)-catalyzed ring closure under an atmosphere of dioxygen followed by ring enlargement (cf. Scheme 7.54) is the key step in the total synthesis of okaramine N (**217**) (**215**→**216**; Scheme 7.52). This transformation is particularly remarkable because C—H bond activation is chemoselective; the unprotected indole unit reacted whereas the N-protected indole remained untouched during the catalysis.

Shortly afterwards, Stoltz *et al.* (5- and 6-*exo*-trig cyclizations) [129] and Beccalli *et al.* (6-*exo*-trig cyclization) [135] simultaneously reported palladium(II)-catalyzed oxidative annulations of indoles. The observation that an electronically modified pyridine ligand increases chemical yields emerged from the systematic survey by Stoltz *et al.* (*vide supra*) [129]. Under an atmosphere of dioxygen, a number of indoles cyclized in moderate to good yields (**218**→**221** and **219**→**222**; Scheme 7.53). As a quaternary carbon is formed upon cyclization of the C—Pd^{II} onto C-1

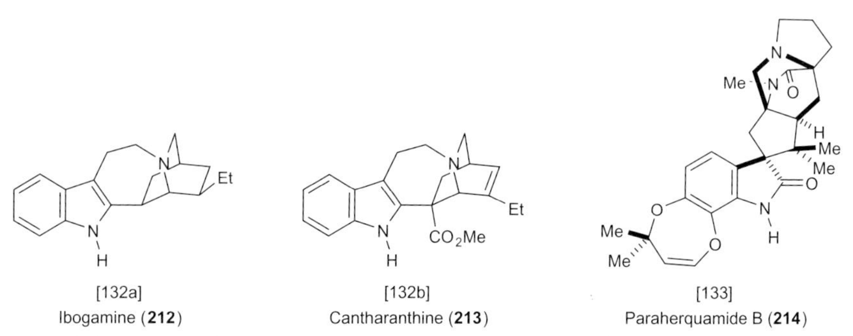

Figure 7.4 Early examples of oxidative indole couplings in total synthesis.

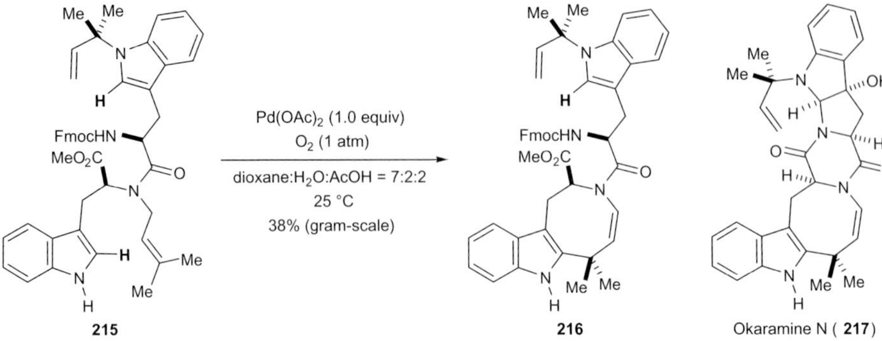

$Pd(OAc)_2$ (1.0 equiv)
O_2 (1 atm)
dioxane:H_2O:AcOH = 7:2:2
25 °C
38% (gram-scale)

215 **216** Okaramine N (**217**)

Scheme 7.52 Demanding palladium(II)-catalyzed ring closure *en route* to okaramine N.

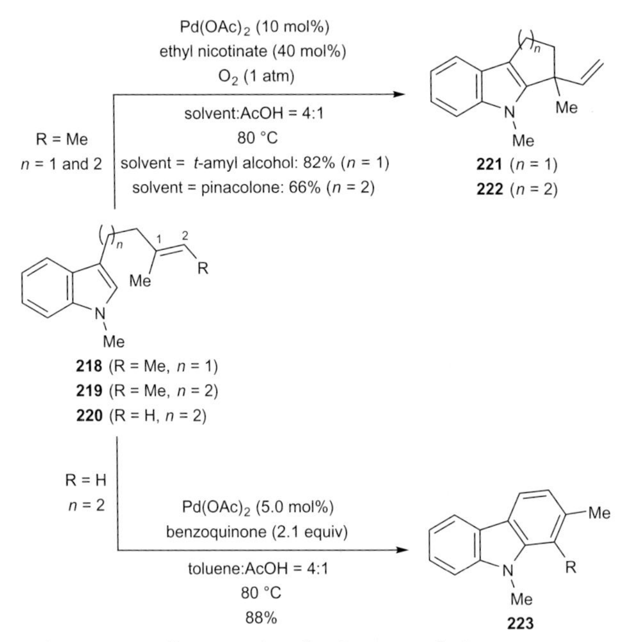

Scheme 7.53 Different modes of cyclization: palladium(II)-catalyzed oxidative Mizoroki–Heck reactions of indoles.

of the alkene, precursors **218** and **219** were modified at C-2 with a C—H bond-containing substituent (R = Me) in order to enable β′-hydride elimination.

As reported recently by Lu *et al.* [136], the cyclization of **220** devoid of such a substituent (R = H) produced a six- rather than a five-membered ring, which subsequently aromatized under the oxidative reaction conditions (**220**→**223**; Scheme 7.53).

Lu *et al.* rationalized this regiochemical outcome by a 6-*endo-trig* ring closure (**224**→**225**) followed by β-hydride elimination (**225**→**226**; Scheme 7.54) [136].

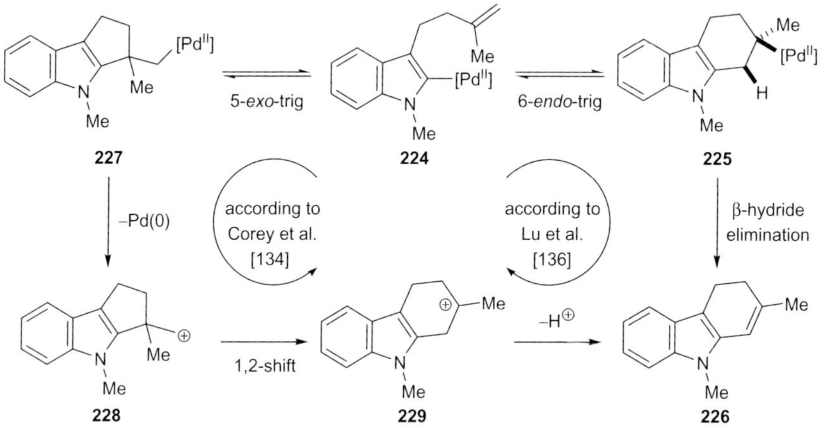

Scheme 7.54 Mechanistic rationale for six-membered ring formation.

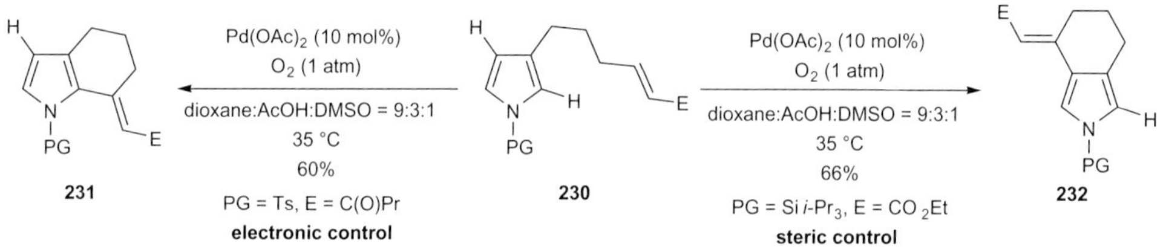

Scheme 7.55 The protective group at nitrogen as a regiochemical switch in intramolecular pyrrole alkenylations.

These authors also postulated that **225** is in equilibrium with dead-end intermediate **227**, which emerges from competing 5-*exo*-trig cyclization (**224**→**227**; Scheme 7.54) [136]. As *retro* alkene insertions are highly uncommon in Mizoroki–Heck chemistry [87], a different mechanism might apply. Based on literature precedence, and according to Baran and Corey [134a], heterolysis of the C–Pd^{II} bond in aqueous acidic media might form a cationic intermediate (**227**→**228**), which would then undergo ring enlargement by 1,2-migration of the electron-rich indolyl group (**228**→**229**) followed by elimination of a proton (**229**→**226**). Therefore, it is proposed that this process might be a tandem 5-*exo* cyclization–cationic rearrangement sequence rather than a 6-*endo* ring closure.

Within their studies on the intermolecular C–C bond formation of pyrroles, exploiting the steric demand of the protective group at nitrogen (cf. Section 7.4.2.2), Gaunt *et al.* also included related examples of 6-*exo-trig* cyclizations (**230**→**231** and **230**→**232**; Scheme 7.55) [119].

Recently, Stoltz *et al.* disclosed an oxidative ring closure – stoichiometric in palladium(II) – of acyl pyrrole **233** to form the central skeleton **234** of dragmacidin F (**235**) (Scheme 7.56) [137]. Notably, conventional Mizoroki–Heck cyclization of the cognate 3-bromosubstituted acyl pyrrole yielded a mixture of isomeric products arising from *exo*- and *endo*-ring closing events, respectively (not shown).

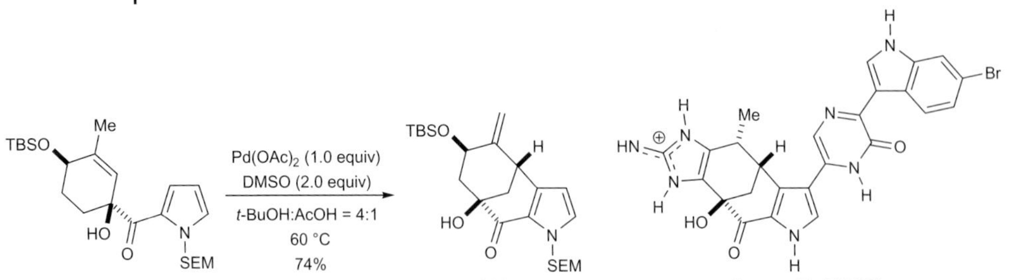

TBSO Me HO O N SEM **233**

$Pd(OAc)_2$ (1.0 equiv)
DMSO (2.0 equiv)
t-BuOH:AcOH = 4:1
60 °C
74%

TBSO H HO O N SEM **234**

Dragmacidin F (**235**)

Scheme 7.56 Oxidative alkenylation of a pyrrole *en route* to dragmacidin F.

Abbreviations

Bn	benzyl
Boc	*tert*-butoxycarbonyl
Bz	benzoyl
dba	*E,E*-dibenzylideneacetone
DMA	*N,N*-dimethylacetamide
DMF	*N,N*-dimethylformamide
DMSO	dimethylsulfoxide
dppf	1,1-bis(diphenylphosphanyl)ferrocene
dppp	1,3-bis(diphenylphosphanyl)propane
Nf	nonafluorobutanesulfonyl
NMP	*N*-methylpyrrolidinone
PMP	1,2,2,6,6-pentamethylpiperidine
TBDMS	*tert*-butyldimethylsilyl
Tf	trifluoromethanesulfonyl
TMP	2,2,6,6-tetramethylpiperidine
Ts	4-toluenesulfonyl

References

1 Mizoroki, T., Mori, K. and Ozaki, A. (1971) *Bull. Chem. Soc. Jpn.*, **44**, 581.

2 Nolley, J.P. and Heck, R.F. Jr (1972) *J. Org. Chem.*, **37**, 2320–2.

3 Arcadi, A., Marinelli, F., Bernocchi, E., Cacchi, S. and Ortar, G. (1989) *J. Organomet. Chem.*, **368**, 249–56.

4 Larock, R.C. and Johnson, P.L. (1989) *J. Chem. Soc., Chem. Commun.*, 1368–70.

5 (a) Moritani, I. and Fujiwara, Y. (1967) *Tetrahedron Lett.*, 1119–22; (b) Fujiwara, Y., Moritani, I., Danno, S., Asano, R. and Teranishi, S. (1969) *J. Am. Chem. Soc.*, **91**, 7166–9.

6 Dieck, H.A. and Heck, R.F. (1974) *J. Am. Chem. Soc.*, **96**, 1133–6.

7 Negishi, E.-i. (2002) *Organopalladium Chemistry for Organic Synthesis*, Vol. **1**, John Wiley & Sons, Inc., New York.

8 de Meijere, A. and Meyer, F.E. (1994) *Angew. Chem. Int. Ed. Engl.*, **33**, 2379–411.

9 Beletskaya, I.P. and Cheprakow, A.V. (2000) *Chem. Rev.*, **100**, 3009–66.

10 Jeffery, T. (1996) *Advances in Metal-Organic Chemistry* (ed. L.S.

Liebeskind), JAI Press, Greenwich, pp. 153–260.

11 Bräse, S. and de Meijere, A. (2004) *Metal-Catalyzed Cross-Coupling Reactions* (eds. A. de Meijere and F. Diederich), Wiley-VCH Verlag GmbH, Weinheim, pp. 217–315.

12 Crisp, G.T. (1998) *Chem. Soc. Rev.*, **27**, 427–36.

13 Amatore, C., Azzabi, M. and Jutand, A. (1991) *J. Am. Chem. Soc.*, **113**, 8375–84.

14 (a) Cabri, W., Candiani, I., Bedeschi, A., Penco, S. and Santi, R. (1992) *J. Org. Chem.*, **57**, 1481–6;
(b) Cabri, W. and Candiani, I. (1995) *Acc. Chem. Res.*, **28**, 2–7.

15 Ozawa, F., Kubo, A. and Hayashi, T. (1991) *J. Am. Chem. Soc.*, **113**, 1417–9.

16 (a) Overman, L.E. and Poon, D.J. (1997) *Angew. Chem. Int. Ed. Engl.*, **36**, 518–21;
(b) Ashimori, A., Bachand, B., Calter, M.A., Govek, S.P., Overman, L.E. and Poon, D.J. (1998) *J. Am. Chem. Soc.*, **120**, 6488–99.

17 Dounay, A.B. and Overman, L.E. (2003) *Chem. Rev.*, **103**, 2945–63.

18 Donde, Y. and Overman, L.E. (2000) *Catalytic Asymmetric Synthesis* (ed. I. Ojima), Wiley-VCH Verlag GmbH, New York, pp. 675–97.

19 (a) Thorn, D.L. and Hoffmann, R. (1978) *J. Am. Chem. Soc.*, **100**, 2079–90;
(b) Samsel, E.G. and Norton, J.R. (1984) *J. Am. Chem. Soc.*, **106**, 5505–12;
(c) Gooßen, L.J., Koley, D., Hermann, H.L. and Thiel, W. (2005) *J. Am. Chem. Soc.*, **127**, 11102–14.

20 Lapierre, A.J.B., Geib, S.J. and Curran, D.P. (2007) *J. Am. Chem. Soc.*, **129**, 494–5.

21 (a) Herrmann, W.A., Brossmer, C., Öfele, K., Reisinger, C.-P., Priermeier, T., Beller, M. and Fischer, H. (1995) *Angew. Chem. Int. Ed. Engl.*, **34**, 1844–8;
(b) Herrmann, W.A., Böhm, V.P.W. and Reisinger, C.-P. (1999) *J. Organomet. Chem.*, **576**, 23–41.

22 (a) Littke, A.F. and Fu, G.C. (1999) *J. Org. Chem.*, **64**, 10–11;
(b) Littke, A.F. and Fu, G.C. (2001) *J. Am. Chem. Soc.*, **113**, 6989–7000.

23 (a) Reetz, M.T., Lohmer, G. and Schwickardi, R. (1998) *Angew. Chem. Int. Ed.*, **37**, 481–3;
(b) Reetz, M.T. and Westermann, E. (2000) *Angew. Chem. Int. Ed.*, **39**, 165–8.

24 Reetz, M.T. and de Vries, J.G. (2004) *Chem. Commun.*, 1559–63.

25 de Vries, A.H.M., Mulders, J.M.C.A., Mommers, J.H.M., Henderickx, H.J.W. and de Vries, J.G. (2003) *Org. Lett.*, **5**, 3285–8.

26 (a) Riermeier, T.H., Zapf, A. and Beller, M. (1997) *Top. Catal.*, **4**, 301–9;
(b) Zapf, A. and Beller, M. (2002) *Top. Catal.*, **19**, 101–9.

27 Stephan, M.S., Teunissen, A.J.J.M., Verzijl, G.K.M. and de Vries, J.G. (1998) *Angew. Chem. Int. Ed.*, **37**, 662–4.

28 Gooßen, L.J., Paetzold, J. and Winkel, L. (2002) *Synlett*, 1721–3.

29 Gooßen, L.J. and Paetzold, J. (2002) *Angew. Chem. Int. Ed.*, **41**, 1237–41.

30 Gooßen, L.J. and Paetzold, J. (2004) *Angew. Chem. Int. Ed.*, **43**, 1095–8.

31 Sugihara, T., Satoh, T. and Miura, M. (2005) *Tetrahedron Lett.*, **46**, 8269–71.

32 Kantchev, E.A.B., O'Brien, C.J. and Organ, M.G. (2007) *Angew. Chem.*, **119**, 2824–70.

33 Herrmann, W.A., Elison, M., Fischer, J., Köcher, C. and Artus, G.R.J. (1995) *Angew. Chem. Int. Ed. Engl.*, **34**, 2371–4.

34 Baker, M.V., Skelton, B.W., White, A.H. and Williams, C.C. (2001) *J. Chem. Soc., Dalton Trans.*, 111–20.

35 Mayr, M., Wurst, K., Ongania, K.-H. and Buchmeiser, M.R. (2004) *Chem. Eur. J.*, **10**, 1256–66.

36 Tulloch, A.A.D., Danopoulos, A.A., Tooze, R.P., Cafferkey, S.M., Kleinhenz, S. and Hursthouse, M.B. (2000) *Chem. Commun.*, 1247–8.

37 Lee, H.M., Zeng, J.Y., Hu, C.-H. and Lee, M.-T. (2004) *Inorg. Chem.*, **43**, 6822–9.

38 Magill, A.M., McGuinness, D.S., Cavell, K.J., Britovsek, G.J.P., Gibson, V.C., White, A.J.P., Williams, D.J., White, A.H. and Skelton, B.W. (2001) *J. Organomet. Chem.*, **617-18**, 546–60.

39 Frey, G.D., Schütz, J., Herdtweck, E. and Herrmann, W.A. (2005) *Organometallics*, **24**, 4416–26.

40 (a) Ozawa, F. and Hayashi, T. (1992) *J. Organomet. Chem.*, **428**, 267–77;
(b) Hayashi, T., Kubo, A. and Ozawa, F. (1992) *Pure Appl. Chem.*, **64**, 421–7;

(c) Ozawa, F., Kubo, A. and Hayashi, T. (1992) *Tetrahedron Lett.*, **33**, 1485–8;
(d) Ozawa, F., Kobatake, Y. and Hayashi, T. (1993) *Tetrahedron Lett.*, **34**, 2505–8;
(e) Ozawa, F., Kubo, A., Matsumoto, Y., Hayashi, T., Nishioka, E., Yanagi, K. and Moriguchi, K.-i. (1993) *Organometallics*, **12**, 4188–96.

41 Tietze, L.F., Thede, K. and Sannicolò, F. (1999) *Chem. Commun.*, 1811–2.

42 (a) Loiseleur, O., Meier, P. and Pfaltz, A. (1996) *Angew. Chem. Int. Ed. Engl.*, **35**, 200–2;
(b) Pfaltz, A. (1996) *Acta Chem. Scand.*, **50**, 189–94;
(c) Loiseleur, O., Hayashi, M., Schmees, N. and Pfaltz, A. (1997) *Synthesis*, 1338–45.

43 Gilbertson, S.R. and Fu, Z. (2001) *Org. Lett.*, **3**, 161–4.

44 Tu, T., Deng, W.-P., Hou, X.-L., Dai, L.-X. and Dong, X.-C. (2003) *Chem. Eur. J.*, **9**, 3073–81.

45 Mata, Y., Pàmies, O. and Diéguez, M. (2007) *Chem. Eur. J.*, **13**, 3296–304.

46 Alonso, F., Beletskaya, I.P. and Yus, M. (2005) *Tetrahedron*, **61**, 11771–835.

47 (a) Oestreich, M. (2005) *Eur. J. Org. Chem.*, 783–92;
(b) Oestreich, M. (2007) *Topics in Organometallic Chemistry*, Vol. **24** (ed. N. Chatani), Springer, Heidelberg, pp. 169–92.

48 (a) Itami, K. and Yoshida, J.-i. (2006) *Synlett*, 157–80;
(b) Itami, K. and Yoshida, J.-i. (2006) *Bull. Soc. Chem. Jpn.*, **79**, 811–24;
(c) Itami, K. and Yoshida, J.-i. (2006) *Chem. Eur. J.*, **12**, 3966–74.

49 (a) Andersson, C.-M., Larsson, J. and Hallberg, A. (1990) *J. Org. Chem.*, **55**, 5757–61;
(b) Nilsson, K. and Hallberg, A. (1992) *J. Org. Chem.*, **57**, 4015–7;
(c) Larhed, M., Andersson, C.-M. and Hallberg, A. (1993) *Acta Chem. Scand.*, **47**, 212–7;
(d) Larhed, M., Andersson, C.-M. and Hallberg, A. (1994) *Tetrahedron*, **50**, 285–304.

50 Nilsson, P., Larhed, M. and Hallberg, A. (2001) *J. Am. Chem. Soc.*, **123**, 8217–25.

51 Badone, D. and Guzzi, U. (1993) *Tetrahedron Lett.*, **34**, 3603–6.

52 Alonso, I. and Carretero, J.C. (2001) *J. Org. Chem.*, **66**, 4453–6.

53 (a) Itami, K., Mitsudo, K., Kamei, T., Koike, T., Nokami, T. and Yoshida, J.-i. (2000) *J. Am. Chem. Soc.*, **122**, 12013–4;
(b) Itami, K., Nokami, T., Ishimura, Y., Mitsudo, K., Kamei, T. and Yoshida, J.-i. (2001) *J. Am. Chem. Soc.*, **123**, 11577–85;
(c) Itami, K., Minero, M., Muraoka, N. and Yoshida, J.-i. (2004) *J. Am. Chem. Soc.*, **126**, 11778–9.

54 (a) Buezo, N.D., Alonso, I. and Carretero, J.C. (1998) *J. Am. Chem. Soc.*, **120**, 7129–30;
(b) de la Rosa, J.C., Diaz, N. and Carretero, J.C. (2000) *Tetrahedron Lett.*, **41**, 4107–11;
(c) Buezo, N.D., de la Rosa, J.C., Priego, J., Alonso, I. and Carretero, J.C. (2001) *Chem. Eur. J.*, **7**, 3890–900.

55 Llamas, T., Arrayás, R.G. and Carretero, J.C. (2004) *Adv. Synth. Catal.*, **346**, 1651–4.

56 Nilsson, P., Larhed, M. and Hallberg, A. (2003) *J. Am. Chem. Soc.*, **125**, 3430–1.

57 (a) Bräse, S. and de Meijere, A. (2002) *Organopalladium Chemistry for Organic Synthesis*, Vol. **1** (eds E.-i. Negishi and A. de Meijere), John Wiley & Sons, Inc., New York, pp. 1223–54;
(b) Dyker, G. (2002) *Organopalladium Chemistry for Organic Synthesis* (eds E.-i. Negishi and A. de Meijere) Vol. **1**, John Wiley & Sons, Inc., New York, pp. 1255–82.

58 (a) Gibson, S.E. and Middleton, R.J. (1996) *Contemp. Org. Synth.*, **3**, 447–71;
(b) Link, J.T. (2002) *Organic Reactions*, Vol. **60** (ed. L.E. Overman), John Wiley & Sons, Inc., New York, pp. 157–534;
(c) Zeni, G. and Larock, R.C. (2006) *Chem. Rev.*, **106**, 4644–80.

59 Bräse, S. and de Meijere, A. (1998) *Metal-Catalyzed Cross Coupling Reactions* (eds F. Diederich and P.J. Stang), Wiley-VCH Verlag GmbH, Weinheim, pp. 99–166.

60 Zezschwitz, P.v. and de Meijere A. (2006) *Topics in Organometallic Chemistry*, Vol. **19** (ed. T.J.J. Müller), Springer, Heidelberg, pp. 49–89.

61 Grigg, R., Millington, E.L. and Thornton-Pett, M. (2002) *Tetrahedron Lett.*, **43**, 2605–8.

62 Grigg, R., Sansano, J.M., Santhakumar, V., Sridharan, V., Thangavelanthum, R., Thornton-Pett, M. and Wilson, D. (1997) *Tetrahedron*, **53**, 11803–26.

63 (a) Burns, B., Grigg, R., Ratananukul, P., Sridharan, V., Stevenson, P., Sukirthalingam, S. and Worakun, T. (1988) *Tetrahedron Lett.*, **29**, 5565–8; (b) Fretwell, P., Grigg, R., Sansano, J.M., Sridharan, V., Sukirthalingam, S., Wilson, D. and Redpath, J. (2000) *Tetrahedron*, **56**, 7525–39.

64 Tietze, L.F., Brasche, G. and Gericke, K. M. (2006) *Domino Reactions in Organic Synthesis*, Wiley-VCH Verlag GmbH, Weinheim.

65 Sato, Y., Sodeoka, M. and Shibasaki, M. (1989) *J. Org. Chem.*, **54**, 4738–9.

66 Carpenter, N.E., Kucera, D.J. and Overman, L.E. (1989) *J. Org. Chem.*, **54**, 5846–8.

67 Overman, L.E. (1994) *Pure Appl. Chem.*, **66**, 1423–30.

68 Ashimori, A. and Overman, L.E. (1992) *J. Org. Chem.*, **57**, 4571–2.

69 Ashimori, A., Matsuura, T., Overman, L.E. and Poon, D.J. (1993) *J. Org. Chem.*, **58**, 6949–51.

70 (a) Ashimori, A., Bachand, B., Overman, L.E. and Poon, D.J. (1998) *J. Am. Chem. Soc.*, **120**, 6477–87; (b) Matsuura, T., Overman, L.E. and Poon, D.J. (1998) *J. Am. Chem. Soc.*, **120**, 6500–3; (c) Dounay, A.B., Hatanaka, K., Kodanko, J.J., Oestreich, M., Overman, L.E., Pfeifer, L.A. and Weiss, M.M. (2003) *J. Am. Chem. Soc.*, **125**, 6261–71.

71 Steven, A. and Overman, L.E. (2007) *Angew. Chem. Int. Ed.*, **46**, 5488–508.

72 McDermott, M.C., Stephenson, G.R., Hughes, D.L. and Walkington, A.J. (2006) *Org. Lett.*, **8**, 2917–20.

73 (a) Oestreich, M., Dennison, P.R., Kodanko, J.J. and Overman, L.E. (2001) *Angew. Chem. Int. Ed.*, **40**, 1439–42; (b) Burke, B.J. and Overman, L.E. (2004) *J. Am. Chem. Soc.*, **126**, 16820–33.

74 Clique, B., Fabritius, C.-H., Couturier, C., Monteiro, N. and Balme, G. (2003) *Chem. Commun.*, 272–3.

75 (a) Arnek, R. and Zetterberg, K. (1987) *Organometallics*, **6**, 1230–5; (b) Zhang, L. and Zetterberg, K. (1991) *Organometallics*, **10**, 3806–13.

76 Lebsack, A.D., Link, J.T., Overman, L.E. and Stearns, B.A. (2002) *J. Am. Chem. Soc.*, **124**, 9008–9.

77 Overman, L.E., Larrow, J.F., Stearns, B.A. and Vance, J.M. (2000) *Angew. Chem. Int. Ed.*, **39**, 213–5.

78 Overman, L.E., Paone, D.V. and Stearns, B.A. (1999) *J. Am. Chem. Soc.*, **121**, 7702–3.

79 (a) Overman, L.E. and Watson, D.A. (2006) *J. Org. Chem.*, **71**, 2587–99; (b) Overman, L.E. and Watson, D.A. (2006) *J. Org. Chem.*, **71**, 2600–8.

80 Dounay, A.B., Overman, L.E. and Wrobleski, A.D. (2005) *J. Am. Chem. Soc.*, **127**, 10186–7.

81 (a) Imbos, R., Minnaard, A.J. and Feringa, B.L. (2002) *J. Am. Chem. Soc.*, **124**, 184–5; (b) Imbos, R., Minnaard, A.J. and Feringa, B.L. (2003) *J. Chem. Soc., Dalton Trans.*, 2017–23.

82 Lautens, M. and Zunic, V. (2004) *Can. J. Chem.*, **82**, 399–407.

83 Lormann, M.E.P., Nieger, M. and Bräse, S. (2006) *J. Organomet. Chem.*, **691**, 2159–61.

84 (a) Oestreich, M., Sempere-Culler, F. and Machotta, A.B. (2005) *Angew. Chem. Int. Ed.*, **44**, 149–52; (b) Oestreich, M., Sempere-Culler, F. and Machotta, A.B. (2006) *Synlett*, 2965–8; (c) Machotta, A.B., Straub, B.F. and Oestreich, M. (2007) *J. Am. Chem. Soc.*, **129**, 13455–63.

85 (a) Tietze, L.F., Ila, H. and Bell, H.P. (2004) *Chem. Rev.*, **104**, 3453–516; (b) Bolm, C., Hildebrand, J.P., Muñiz, K. and Hermanns, N. (2001) *Angew. Chem. Int. Ed.*, **40**, 3284–308.

86 Catellani, M., Chiusoli, G.P., Giroldini, W. and Salerno, G. (1980) *J. Organomet. Chem.*, **199**, C21–3.

87 Catellani, M. (2003) *Synlett*, 298–313.

88 Cacchi, S. (1990) *Pure Appl. Chem.*, **62**, 713–22.

89 Brunel, J.M., Heumann, A. and Buono, G. (2000) *Angew. Chem. Int. Ed.*, **39**, 1946–9.

90 Bravo, J., Cativiela, C., Navarro, R. and Urriolabeitia, E.P. (2002) *J. Organomet. Chem.*, **650**, 157–72.

91 Yuan, K., Zhang, T.K. and Hou, X.L. (2005) *J. Org. Chem.*, **70**, 6085–8.

92 Zhong, J., Xie, J.-H., Wang, A.-E., Zhang, W. and Zhou, Q.-L. (2006) *Synlett*, 1193–6.

93 Brunner, H. and Kramler, K. (1991) *Synthesis*, 1121–4.

94 Sakuraba, S., Awano, K. and Achiwa, K. (1994) *Synlett*, 291–2.

95 Namyslo, J.C. and Kaufmann, D.E. (1997) *Chem. Ber. Recueil*, **130**, 1327–31.

96 Wu, X.-Y., Xu, H.-D., Zhou, Q.-L. and Chan, A.S.C. (2000) *Tetrahedron: Asymmetry*, **11**, 1255–7.

97 Wu, X.-Y., Xu, H.-D., Tang, F.-Y. and Zhou, Q.-L. (2001) *Tetrahedron: Asymmetry*, **12**, 2565–9.

98 Dupont, J., Ebeling, G., Delgado, M.R., Consorti, C.S., Burrow, R., Farrar, D.H. and Lough, A.J. (2001) *Inorg. Chem. Commun.*, **4**, 471–4.

99 Drago, D. and Pregosin, P.S. (2002) *Organometallics*, **21**, 1208–15.

100 Moinet, C. and Fiaud, J.-C. (1995) *Tetrahedron Lett.*, **36**, 2051–2.

101 Duan, J.-P. and Cheng, C.-H. (1993) *Tetrahedron Lett.*, **34**, 4019–22.

102 (a) Clayton, S.C. and Regan, A.C. (1993) *Tetrahedron Lett.*, **34**, 7493–6;
(b) Namyslo, J.C. and Kaufmann, D.E. (1999) *Synlett*, 114–6.

103 Namyslo, J.C. and Kaufmann, D.E. (1999) *Synlett*, 804–6.

104 Li, X.-G., Tang, F.-Y., Xu, H.-D., Wu, X.-Y. and Zhou, Q.-L. (2003) *ARKIVOC (Archive for Organic Chemistry)*, 15–20.

105 Kasyan, A., Wagner, C. and Maier, M.E. (1998) *Tetrahedron*, **54**, 8047–54.

106 Mitchell, D. and Yu, H. (2003) *Curr. Opin. Drug Discov. Dev.*, **6**, 876–83.

107 Diaz, P., Gendre, F., Stella, L. and Carpentier, B. (1998) *Tetrahedron*, **54**, 4579–90.

108 Burns, B., Grigg, R., Santhakumar, V., Sridharan, V., Stevenson, P. and Worakun, T. (1992) *Tetrahedron*, **48**, 7297–320.

109 Ruchirawat, S. and Namsa-aid, A. (2001) *Tetrahedron Lett.*, **42**, 1359–61.

110 Kulagowski, J.J., Curtis, N.R., Swain, C.J. and Williams, B.J. (2001) *Org. Lett.*, **3**, 667–70.

111 Leeuwen, P.W.N.M., de Vries, J.G. (2005) *Handbook of C–H Transformations* (ed. G. Dyker), Wiley-VCH Verlag GmbH, Weinheim, pp. 203–12.

112 (a) Jia, C., Kitamura, T. and Fujiwara, Y. (2001) *Acc. Chem. Res.*, **34**, 633–9;
(b) Fujiwara, Y. and Kitamura, T. (2005) *Handbook of C–H Transformations* (ed. G. Dyker), Wiley-VCH Verlag GmbH, Weinheim, pp. 194–202.

113 (a) Jia, C., Lu, W., Kitamura, T. and Fujiwara, Y. (1999) *Org. Lett.*, **1**, 2097–100;
(b) Jia, C., Piao, D., Oyamada, J., Lu, W., Kitamura, T. and Fujiwara, Y. (2000) *Science*, **287**, 1992–5.

114 (a) Du, X., Suguro, M., Hirabayashi, K., Mori, A., Nishikata, T., Hagiwara, N., Kawata, K., Okeda, T., Wang, H.F., Fugami, K. and Kosugi, M. (2001) *Org. Lett.*, **3**, 3313–6;
(b) Jung, Y.C., Mishra, R.K., Yoon, C.H. and Jung, K.W. (2003) *Org. Lett.*, **5**, 2231–4;
(c) Andappan, M.M.S., Nilsson, P., and Larhed, M. (2004) *Chem. Commun.*, 218–9;
(d) Andappan, M.M.S., Nilsson, P., von Schenck, H. and Larhed, M. (2004) *J. Org. Chem.*, **69**, 5212–8.

115 (a) Yokota, T., Tani, M., Sakaguchi, S. and Ishii, Y. (2003) *J. Am. Chem. Soc.*, **125**, 1476–7;
(b) Tani, M., Sakaguchi, S. and Ishii, Y. (2004) *J. Org. Chem.*, **69**, 1221–6.

116 Dams, M., de Vos, D.E., Celen, S. and Jacobs, P.A. (2003) *Angew. Chem. Int. Ed.*, **42**, 3512–5.

117 Boele, M.D.K., van Strijdonck, G.P.F., de Vries, A.H.M., Kamer, P.C.J., de Vries, J. G. and van Leeuwen, P.W.N.M. (2002) *J. Am. Chem. Soc.*, **124**, 1586–7.

118 (a) Mikami, K., Hatano, M. and Terada, M. (1999) *Chem. Lett.*, 55–6;
(b) Akiyama, K., Wakabayashi, K. and Mikami, K. (2005) *Adv. Synth. Catal.*, **347**, 1569–75.

119 Beck, E.M., Grimster, N.P., Hatley, R. and Gaunt, M.J. (2006) *J. Am. Chem. Soc.*, **128**, 2528–9.

120 Grimster, N.P., Gauntlett, C., Godfrey, C.R.A. and Gaunt, M.J. (2005) *Angew. Chem. Int. Ed.*, **44**, 3125–9.

121 Capito, E., Brown, J.M. and Ricci, A. (2005) *Chem. Commun.*, 1854–6.

122 (a) Lane, B.S. and Sames, D. (2004) *Org. Lett.*, **6**, 2897–900;
(b) Lane, B.S., Brown, M.A. and Sames, D. (2005) *J. Am. Chem. Soc.*, **127**, 8050–7.

123 Hirota, K., Isobe, Y., Kitade, Y. and Maki, Y. (1987) *Synthesis*, 495–6.

124 Bingham, A.J., Dyall, L.K., Norman, R.O.C. and Thomas, C.B. (1970) *J. Chem. Soc. C*, 1879–83.

125 (a) Furukawa, H., Yogo, M., Ito, C., Wu, T.-S. and Kuoh, C.-S. (1985) *Chem. Pharm. Bull.*, **33**, 1320–2;
(b) Bittner, S., Krief, P. and Massil, T. (1991) *Synthesis*, 215–6;
(c) Yogo, M., Ito, C. and Furukawa, H. (1991) *Chem. Pharm. Bull.*, **39**, 328–34;
(d) Knölker, H.-J. and O'Sullivan, N. (1994) *Tetrahedron Lett.*, **35**, 1695–8.

126 (a) Knölker, H.-J. and O'Sullivan, N. (1994) *Tetrahedron*, **50**, 10893–908;
(b) Knölker, H.-J. and Fröhner, W. (1998) *J. Chem. Soc., Perkin Trans. 1*, 173–5;
(c) Knölker, H.-J., Reddy, K.R. and Wagner, A. (1998) *Tetrahedron Lett.*, **39**, 8267–70.

127 (a) Oslob, B., Åkermark, J.D. and Heuschert, U. (1995) *Tetrahedron Lett.*, **36**, 1325–6;
(b) Hagelin, H., Oslob, J.D. and Åkermark, B. (1999) *Chem. Eur. J.*, **5**, 2413–6.

128 Zhang, H., Ferreira, E.M. and Stoltz, B.M. (2004) *Angew. Chem. Int. Ed.*, **43**, 6144–8.

129 Ferreira, E.M. and Stoltz, B.M. (2003) *J. Am. Chem. Soc.*, **125**, 9578–9.

130 Nishimura, T., Onoue, T., Ohe, K. and Uemura, S. (1999) *J. Org. Chem.*, **64**, 6750–5.

131 Han, X. and Widenhoefer, R.A. (2006) *Org. Lett.*, **8**, 3801–4.

132 (a) Trost, B.M., Godleski, S.A. and Genêt, J.P. (1978) *J. Am. Chem. Soc.*, **100**, 3930–1;
(b) Trost, B.M., Godleski, S.A., Belletire, J.L. (1979). *J. Org. Chem*, **44**, 2052–4.

133 (a) Cushing, T.D., Sanz-Cervera, J.F. and Williams, R.M. (1993) *J. Am. Chem. Soc.*, **115**, 9323–4;
(b) Cushing, T.D., Sanz-Cervera, J.F. and Williams, R.M. (1996) *J. Am. Chem. Soc.*, **118**, 557–79;
(c) Williams, R.M., Cao, J., Tsujishima, H. and Cox, R.J. (2003) *J. Am. Chem. Soc.*, **125**, 12172–8.

134 (a) Baran, P.S. and Corey, E.J. (2002) *J. Am. Chem. Soc.*, **124**, 7904–5;
(b) Baran, P.S., Guerrero, C.A. and Corey, E.J. (2003) *J. Am. Chem. Soc.*, **125**, 5628–9.

135 Abbiati, G., Beccalli, E.M., Broggini, G. and Zoni, C. (2003) *J. Org. Chem.*, **68**, 7625–8.

136 Kong, A., Han, X. and Lu, X. (2006) *Org. Lett.*, **8**, 1339–42.

137 (a) Garg, N.K., Caspi, D.D. and Stoltz, B.M. (2004) *J. Am. Chem. Soc.*, **126**, 9552–3;
(b) Garg, N.K., Caspi, D.D. and Stoltz, B. M. (2005) *J. Am. Chem. Soc.*, **127**, 5970–8.

8
Modern Arylations of Carbonyl Compounds

Christian Defieber and Erick M. Carreira

8.1
Introduction

The catalytic enantioselective formation of C—C bonds is one of the most fundamental processes in modern organic synthesis for the rapid construction of complex molecules. Arylation methods have their special place in this area due to the importance of aromatic groups in numerous pharmaceuticals (Figure 8.1) [1]. Considerable research endeavors have been undertaken during the past few years in order to optimize existing protocols and widen the scope of molecules that can be accessed. Since the publication by Bolm, in 2001, of a review of catalytic asymmetric arylation reactions, many innovative and practical processes have been developed for the straightforward enantioselective, catalytic introduction of aryl groups [2].

This chapter, which highlights the novel advances made up to June 2007, is structured according to the various classes of electrophiles. Initially, the enantioselective arylation of aldehydes and imines which give rise to pharmaceutically important chiral diarylmethanols and -amines will be discussed [3]. Considerable progress has also been made in the conjugate addition of arylmetals to various types of α,β-unsaturated carbonyl compounds. Recent progress in the coupling of electron-rich aryl moieties via metal-catalyzed Friedel–Crafts alkylation will be complemented by a section on organic catalytic methods.

8.2
Enantioselective Arylation of Aldehydes

8.2.1
Zinc-Mediated Asymmetric Arylation of Aldehydes

Fu reported the first enantioselective arylation protocol of aldehydes **1** involving diphenylzinc as aryl source and chiral azaferrocene-based ligand **3** (Scheme 8.1;

Modern Arylation Methods. Edited by Lutz Ackermann

ISBN: 978-3-527-31937-4

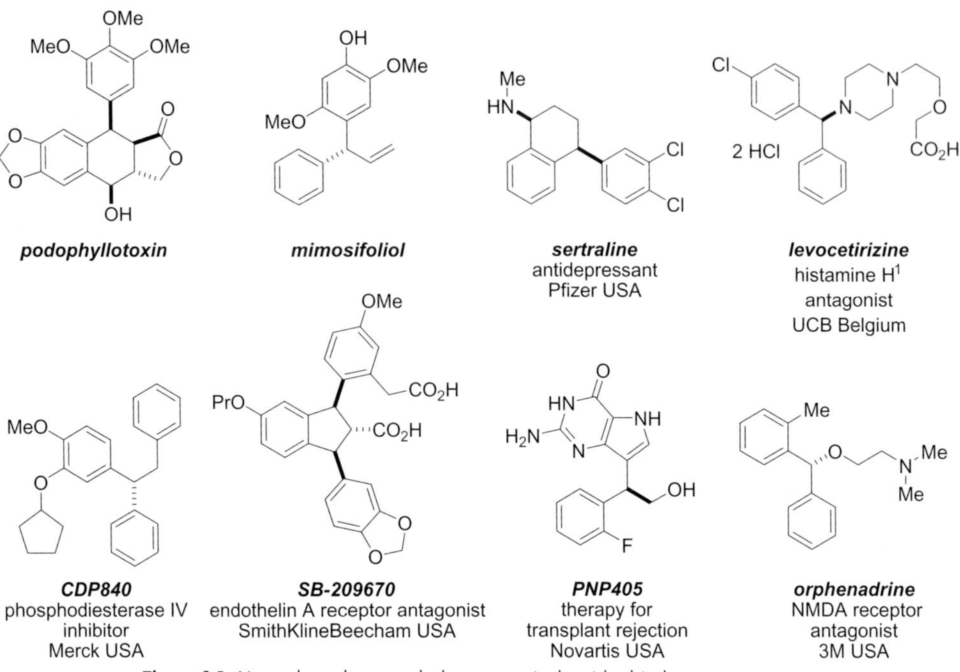

Figure 8.1 Natural products and pharmaceuticals with chiral diarylmethine stereogenic centers.

$$\underset{\mathbf{1}}{\mathrm{ArCHO}} + \mathrm{Ph_2Zn} \xrightarrow{\text{chiral ligand}} \underset{\mathbf{2}}{\mathrm{ArCH(OH)Ph}}$$

Scheme 8.1 Zinc-mediated asymmetric arylation of aldehydes.

Figure 8.2) [4]. The initially obtained enantiomeric excess (ee) in this process was only 57%, and this was attributed to the competing background reaction. Later, Pu described protocols in which certain additives, such as methanol, were used to tame the reactivity of Ph_2Zn [5]. The addition of diethylzinc to the reaction mixture led to the formation of a mixed zinc reagent, PhZnEt, which was found to be less reactive and thus more selective than Ph_2Zn itself [6]. Additionally, Ph_2Zn could be used in substoichiometric amounts (0.65 equiv.), since both its phenyl groups were now available for the reaction.

Most of the catalytic enantioselective protocols are limited to aryl aldehydes as substrates. Pu recently disclosed a process involving chiral bifunctional ligand **6** which is perfectly suited for additions to aliphatic aldehydes [7]. Traditionally used aryl aldehydes were also converted under the reported conditions.

Although significant improvements in the catalytic enantioselective addition of Ph_2Zn to aldehydes have been reported, some distinct limitations remained to be

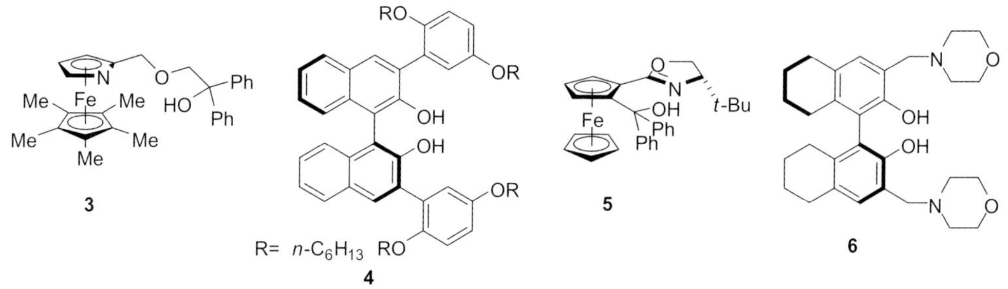

Figure 8.2 Examined ligands.

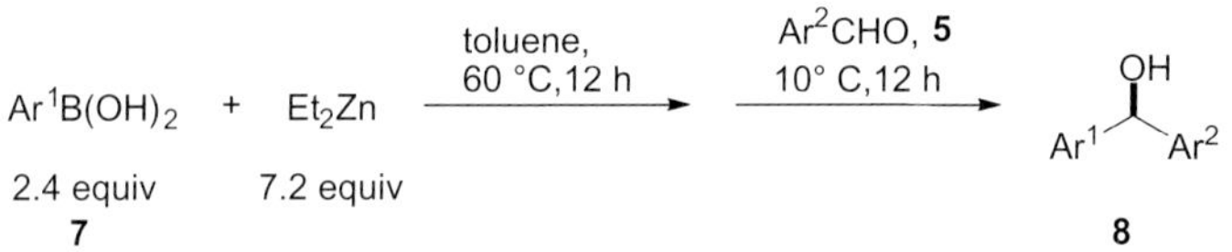

Scheme 8.2 Arylboronic acids **7** as aryl sources.

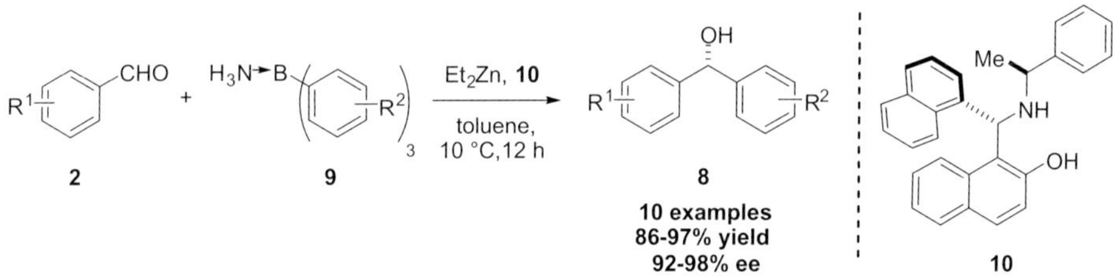

Scheme 8.3 Triarylboranes **9** as aryl sources.

addressed. The scope of the aryl group to be transferred has been limited to the phenyl ring, since only Ph_2Zn is a commercially available aryl zinc reagent. Moreover, Ph_2Zn is a rather expensive and difficult-to-handle reagent due to its air- and moisture sensitivity.

An interesting protocol by Bolm addresses these issues (Scheme 8.2) [8] by taking advantage of the use of boronic acids as the source of nucleophiles. Due to their widespread application in the Suzuki–Miyaura reaction, vast amounts of boronic acids are available commercially. Transmetallation takes place from boron to zinc to form the active arylation reagent ArZnEt, which then selectively undergoes addition to the aldehyde using ligand **5**.

Ph_3B was recently found to be an inexpensive alternative to Ph_2Zn [9]. In order to increase the practicality of the process, Dahmen studied the corresponding readily available air-stable ammonia complexes **9** (Scheme 8.3) [10]. Kinetic investigations of the reaction profile showed that this complex only slowly transmetallates with Et_2Zn and, as a consequence, only small amounts of the arylation reagent relative to the aldehyde are present in the reaction mixture; thus, the

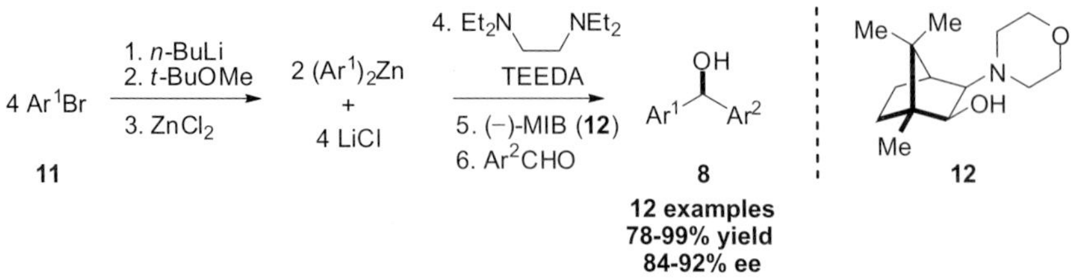

Scheme 8.4 Aryl bromides as aryl sources.

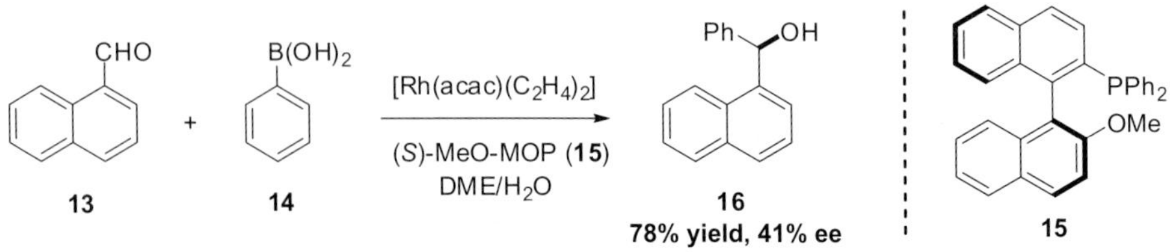

Scheme 8.5 Rhodium-catalyzed enantioselective arylation of aldehyde **13**.

background reaction is efficiently suppressed. It is important to emphasize the ease of preparation of the chiral aminoalcohol **10** employed in this transformation [11].

An attractive one-pot arylation protocol starting from inexpensive and readily available aryl bromides **11** was recently disclosed by Walsh (Scheme 8.4) [12]. Here, the aryl zinc species, which is formed *in situ* by lithiation and transmetallation with Zn, adds selectively, under the influence of chiral aminoalcohol **12**, to a variety of aldehydes. The introduction of a diamine, such as tetraethylethylene diamine (TEEDA), was key to the success of this method, because it prevents the formed LiCl coproduct from promoting the racemic addition reaction. The practicality of the process is further highlighted by the fact that neither filtering away of the salts nor handling of pyrophoric zinc species are necessary.

8.2.2
Rhodium-Catalyzed Asymmetric Arylation of Aldehydes

Due to their widespread application in the Suzuki–Miyaura reaction, arylboronic acids are attractive aryl transfer precursors. However, in the zinc-promoted reaction, an excess of Et_2Zn (up to 7 equiv.) must be added in order for efficient transmetallation to occur. The application of a reactive aryl metal species necessitating only a catalytic amount of metal would thus be advantageous. In recent years, several rhodium-catalyzed enantioselective protocols have been described, as well as processes involving palladium and nickel.

The first rhodium-catalyzed enantioselective arylation was described by Miyaura in 1998 (Scheme 8.5) [13]. The reaction between 1-naphthaldehyde (**13**) and phenylboronic acid (**14**) was catalyzed by a rhodium complex of the monodentate phosphine (*S*)-MeO-MOP (**15**), and resulted in the formation of diarylmethanol

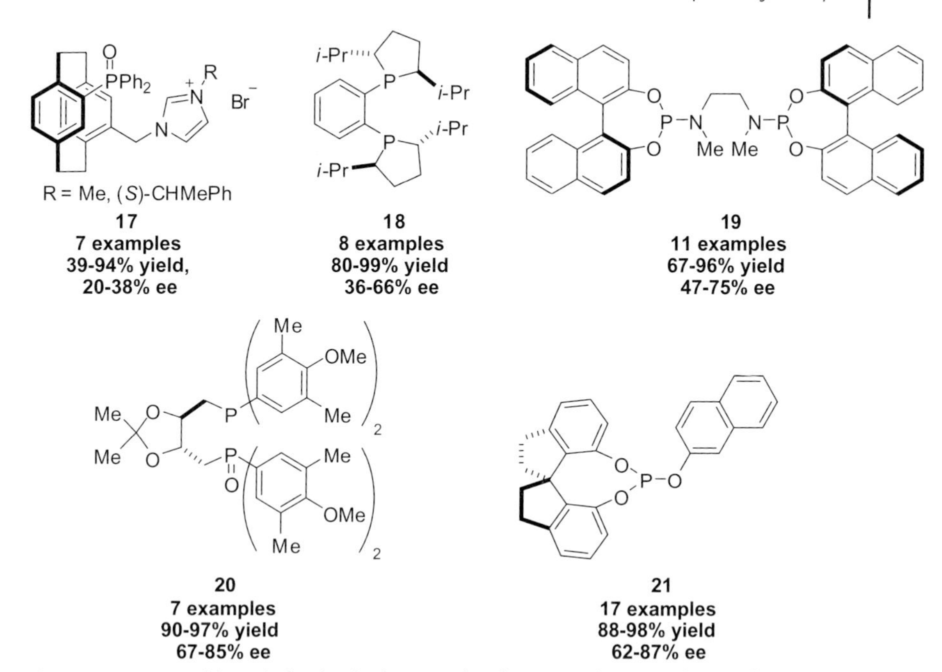

Figure 8.3 Examined ligands for the rhodium-catalyzed enantioselective arylation of aldehydes.

Ar^1CHO (**1**) + $(Ar^2BO)_3$ → [$Ni(cod)_2$], (*R,R*)-Et-DuPhos (**22**), NaO*t*-Bu, DME/H_2O → $Ar^1CH(OH)Ar^2$ (**8**)

10 examples
83-94% yield,
32-78% ee

22

Scheme 8.6 Nickel-catalyzed enantioselective arylation of aldehydes **1**.

16 in 78% yield, but only 41% ee. Such low enantioselectivity was not too surprising in light of the known addition of arylrhodium species to aldehydes [14]. Subsequent investigations aimed at improving the stereochemical outcome of this transformation were thus primarily focused on the screening of various ligand scaffolds (Figure 8.3).

A lower range of ee-values was obtained using chiral carbene ligand **17** [15], *i*Pr-DuPHOS (**18**) [16] and diethyleneamine-bridged phosphoramidite **19** [17]. To date, the best enantioselectivities using this protocol were achieved with Kondo's hemilabile TADDOL-based phosphine–phosphine oxide **20** [18] and Zhou's spirocyclic phosphite **21** (Figure 8.3) [19].

Recently, Kondo and Aoyama reported on a nickel/Et-DuPhos-catalyzed arylation of aldehydes using arylboroxines as aryl source (Scheme 8.6) [20]. Although the reactivity of this protocol is satisfying, enantioselectivities were only found to be in the range of 32 to 78%.

8.3
Enantioselective Arylation of Ketones

8.3.1
Enantioselective Arylation of Aryl-Alkyl-Substituted Ketones

Enantioselective additions of metal aryl species to ketones are more challenging transformations than the use of aldehydes as acceptors. This situation is related not only to the decreased reactivity of ketones as acceptors, but also to the greater difficulty for a Lewis acid to differentiate between the two lone pairs of a ketone carbonyl. An early example of an enantioselective phenyl transfer to aryl-alkyl ketones **23** was reported by Fu in 1998 (Scheme 8.7) [21]. The use of Noyori's dimethylaminoisoborneol (DAIB) ligand **24** provided tertiary alcohols **25** in moderate to high yields and enantioselectivities. The addition of methanol was found to be necessary in order to form a mixed alkoxy phenyl zinc species which is less reactive than Ph_2Zn.

In 2003, both Walsh and Yus reported independently on a titanium-mediated phenyl transfer to alkyl-aryl ketones **23** (Scheme 8.8) [22]. The role of the titanium alkoxide was not only to form the active chiral catalytic species, but also to sequester the tertiary alkoxide generated during the catalytic cycle. Yus has discussed the possibility of using either arylboronic acids or Ph_3B as precursors for the aryl transfer, and in certain cases ee-values greater than 99% were observed [23]. It is worth highlighting at this point that Walsh also used a similar protocol for the aryl transfer to α,β-unsaturated ketones to produce optically enriched tertiary alcohols [24].

Gau demonstrated that triaryl(tetrahydrofuran) aluminum reagents also proved to be competent aryl sources for performing 1,2-additions to aldehydes [25] and ketones (Scheme 8.9) [26]. These highly reactive, easily accessible organometallics

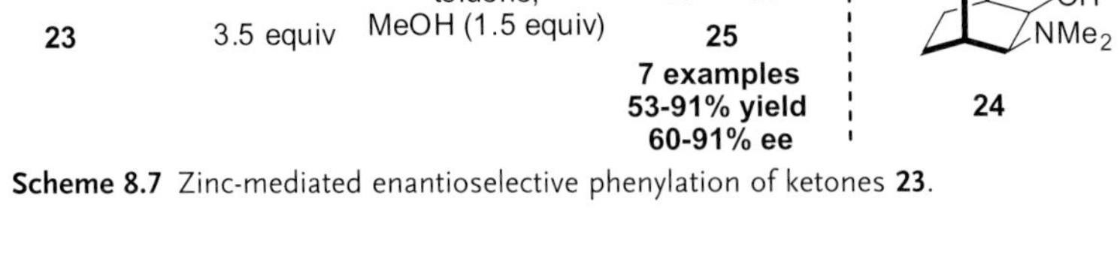

$R^1C(O)R^2$ (**23**) + Ph_2Zn (3.5 equiv) → [(+)-DAIB (**24**), toluene, MeOH (1.5 equiv)] → $R^1R^2C(OH)Ph$ (**25**): **7 examples, 53-91% yield, 60-91% ee**

Scheme 8.7 Zinc-mediated enantioselective phenylation of ketones **23**.

$R^1C(O)R^2$ (**23**) + Ph_2Zn (1.6 equiv) + $Ti(O\text{-}i Pr)_4$ (0.6 equiv) → [**26**, toluene/hexanes] → $R^1R^2C(OH)Ph$ (**25**)

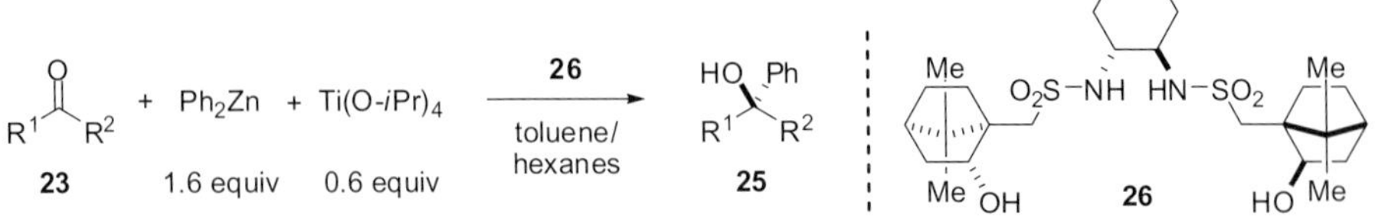

Scheme 8.8 Zinc/titanium-mediated enantioselective phenylation of ketones **23**.

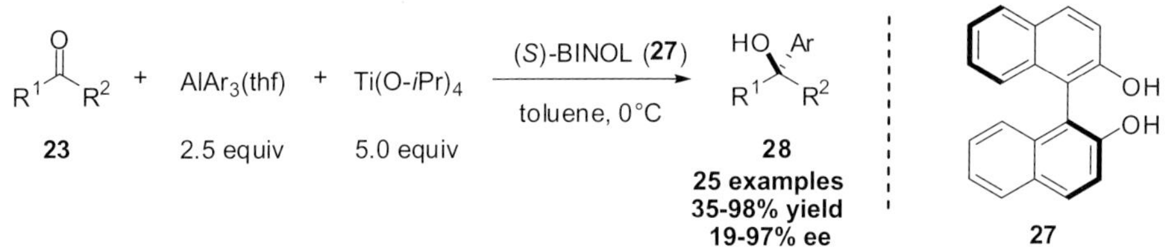

Scheme 8.9 Triarylaluminum as aryl sources.

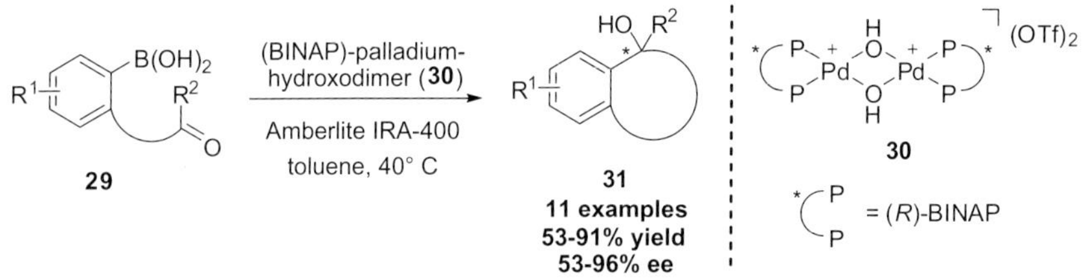

Scheme 8.10 Intramolecular, enantioselective phenylation of ketones.

add to aldehydes in less than 15 min, although the reaction with ketones generally requires a longer reaction time, as well as an excess of Ti(O-*i*Pr)$_4$.

Intramolecular enantioselective arylation reactions of ketones **29** were described by Lu (Scheme 8.10) [27]. The cationic (BINAP)–palladiumhydroxo complex **30** works in combination with a basic anion-exchange resin as an additive to afford optically active cycloalkanols **31** in high yields and enantioselectivities.

8.3.2
Enantioselective Arylation of Isatins

In addition to simple aryl-alkyl-substituted ketones, more reactive acceptors have recently been examined. Due to their widespread occurrence in natural products and pharmaceuticals, isatins **32** have been the subject of intense investigations by the research groups of Hayashi [28] and Feringa [29] (Scheme 8.11). The combination of rhodium and (*R*)-MeO-MOP (**15**) enabled straightforward access to α-arylated oxindoles **33** in high yields and enantioselectivities. Feringa's protocol, which originally was developed as a racemic process, is also based on a rhodium-catalyzed addition of arylboronic acids. The use of readily available phosphoramidite **34** afforded α-phenylated oxindole **35** in 99% yield and 55% ee, which could be upgraded to 94% ee after a single recrystallization.

8.3.3
Enantioselective Arylation of Trifluoromethyl-Substituted Ketones

Trifluoromethyl-substituted ketones **36** are highly reactive electrophiles due to the electron-deficient character of the trifluoromethyl group. As a result of a combinatorial catalyst screening, Feringa identified phosphoramidite **34** as the optimal

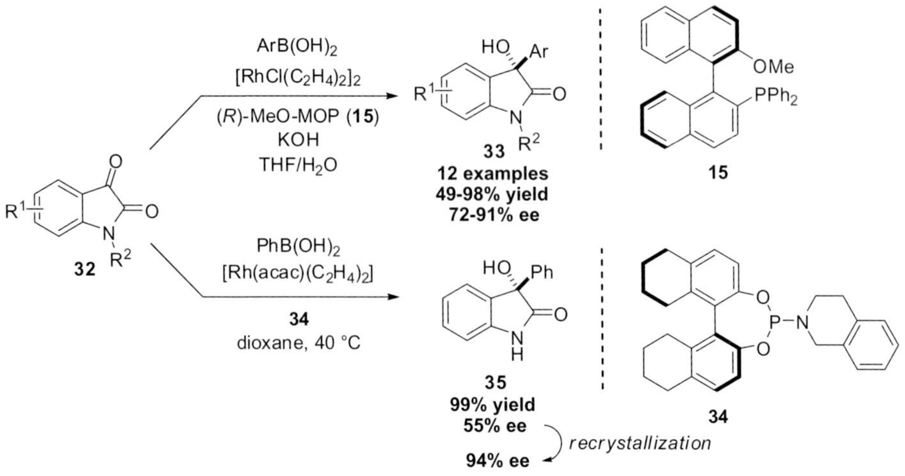

Scheme 8.11 Enantioselective arylation of isatins **32**.

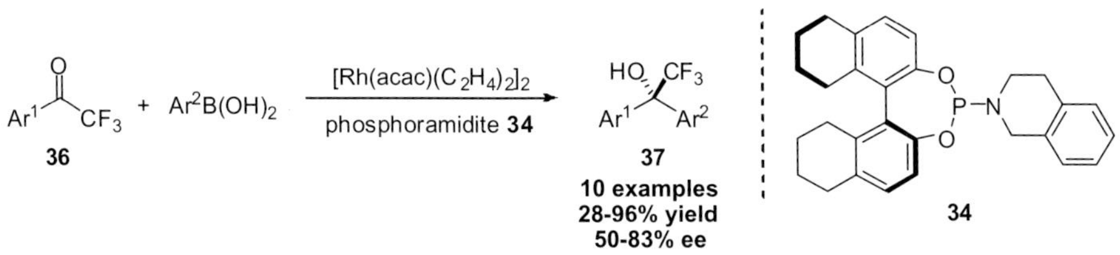

Scheme 8.12 Enantioselective arylation of trifluoromethylketones **36**.

ligand to perform enantioselective additions of arylboronic acids to trifluoromethylketones (Scheme 8.12) [30]. The resulting adducts **37** are of interest for medicinal chemistry applications.

8.4 Enantioselective Arylation of Imines

Pioneering studies of the catalytic, enantioselective arylation of imines date back to 2000, when Hayashi disclosed a rhodium/phosphine-catalyzed addition of arylstannanes to *N*-tosylarylimines [31]. Whereas, the method gave rise to highly enantioenriched diarylmethylamines, five equivalents of the stannane were required to obtain high yields.

8.4.1 Zinc-Mediated Enantioselective Phenylation of Imines

Bolm and Bräse documented a zinc-mediated phenyl transfer to masked formyl imines **38** employing catalytic amounts of paracyclophane-based hydroxyimine

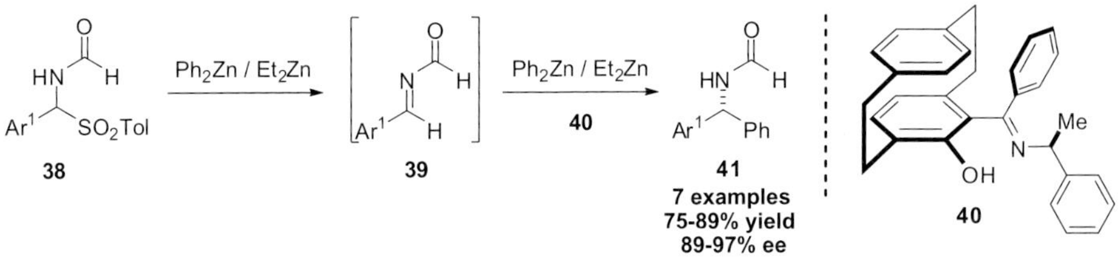

Scheme 8.13 Zinc-mediated enantioselective phenylation of imines **39**.

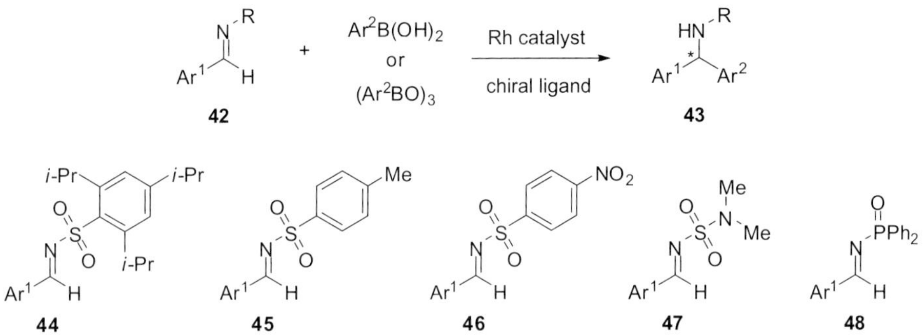

Scheme 8.14 Rhodium-catalyzed enantioselective arylation of imines **42**.

ligand **40** (Scheme 8.13) [32]. The actual substrates **39** are prepared by base-induced elimination of *p*-toluenesulfinic acid from the corresponding masked imines **38**. Furthermore, the straightforward acidic cleavage of the *N*-formyl protecting group in **41** adds synthetic value to this protocol.

8.4.2 Rhodium-Catalyzed Enantioselective Arylation of Imines

Since 2004, considerable progress has been achieved in the rhodium-catalyzed enantioselective addition of arylboronic acids or arylboroxines to various classes of aryl imines **42** (Scheme 8.14).

Tomioka's amidomonophosphane ligands **49** (Figure 8.4) allowed a straightforward entry to optically enriched diarylmethylamines [33]. Hayashi reported on the use of *(S)*-segphos (**50**) for the addition of aryltitanium reagents to *N*-tosylarylimines **45** [34], and chiral dienes proved to be especially valuable in this transformation. Whereas, bicyclo[2.2.2]octadiene **51** [35] and bicyclo[3.3.0]octadiene **53** [36] were efficient ligands for the addition to *N*-tosylarylimines **45**, bicyclo[3.3.1]nonadiene **52** was suited for the addition to *N*-nosylarylamines **46** [37]. The application of a readily cleavable protecting group was also addressed by Feringa [38]. The rhodium-phosphoramidite **54**-catalyzed process involved the addition of arylboronic acids to *N,N*-dimethylsulfamoyl aldimines **47**. Recently, Zhou documented that spirocyclic phosphites **55** serve as efficient ligands not only

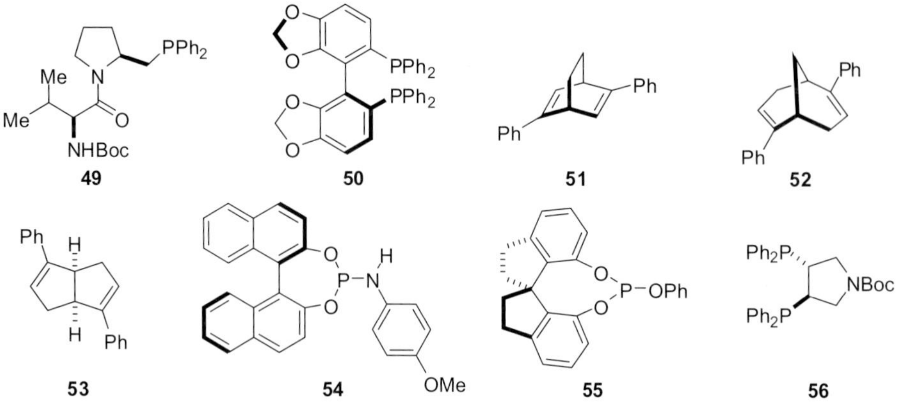

Figure 8.4 Examined ligands in the rhodium-catalyzed arylation of imines.

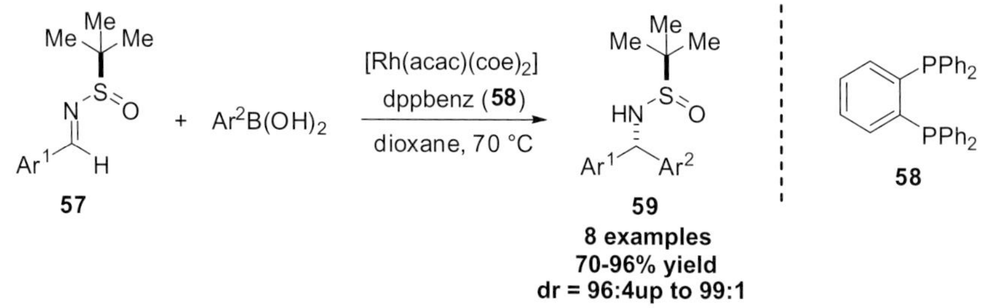

Scheme 8.15 Rhodium-catalyzed diastereoselective arylation of imines **57**.

for the rhodium-catalyzed addition of arylboronic acids to aldehydes but also for the functionalization of imines [39].

8.4.3 Rhodium-Catalyzed Diastereoselective Arylation of Imines

A diastereoselective approach to these chiral building blocks was developed by Ellman (Scheme 8.15) [40]. Thus, *N-tert*-butylsulfinyl aldimines **57** were employed as electrophiles in the rhodium/phosphine **58**-catalyzed addition of arylboronic acids. Whereas Ellman's procedure requires heating, Batey discovered that an amine base allows the reaction to take place at ambient temperature [41]. However, it is worth noting that Ellman's process can also be carried out in an enantioselective way by using *N*-diphenylphosphinoyl aldimines **48** as substrate and Degu-PHOS (**56**) as ligand (87–97% yield, 88–94% ee) (Scheme 8.14; Figure 8.4).

Recently, Ellman expanded the scope to include *N-tert*-butyl imino esters **60** (Scheme 8.16) [42]. The resulting arylglycines **61**, which are unnatural amino acids, are interesting chiral building blocks that might find application in peptide synthesis.

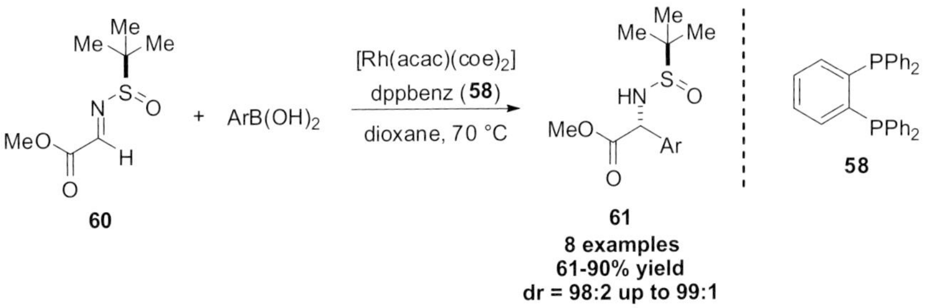

$[Rh(acac)(coe)_2]$
dppbenz (**58**)
dioxane, 70 °C

60 + $ArB(OH)_2$ → **61**

8 examples
61-90% yield
dr = 98:2 up to 99:1

58

Scheme 8.16 Rhodium-catalyzed diastereoselective arylation of imino esters **60**.

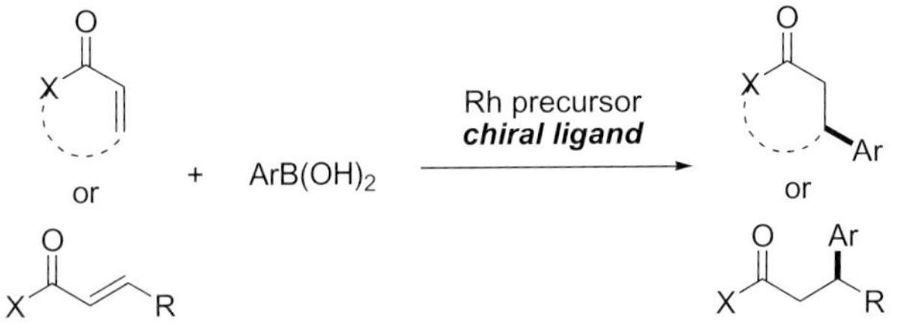

Rh precursor
chiral ligand

Scheme 8.17 Rhodium-catalyzed 1,4-addition of arylboronic acids to α,β-unsaturated carbonyl compounds.

8.5
Conjugate Asymmetric Arylation

The rhodium-catalyzed asymmetric 1,4-addition of arylboronic acids to α,β-unsaturated carbonyl compounds is one of the most versatile and robust methods for the stereoselective introduction of aryl groups (Scheme 8.17). In contrast, copper-catalyzed processes generally transfer alkyl groups with high selectivities, but perform only poorly in aryl transfer [43].

The rhodium-catalyzed addition of arylboronic acids to α,β-unsaturated ketones was first reported by Miyaura in 1997 [44], and only a short time thereafter the first enantioselective addition was documented by Hayashi [45]. Excellent reviews are available that summarize the early developments in this rapidly expanding field of asymmetric catalysis [46]; consequently, only recent advances will be included in this section.

8.5.1
Aryl Sources for the Conjugate Asymmetric Arylation

The most commonly employed aryl sources are arylboronic acids. Due to their widespread application in the Suzuki–Miyaura reaction, organoboronic acids have found wide acceptance, and currently more than 450 different arylboronic acids are available commercially [47]. These compounds unite several beneficial properties such as high air and moisture stability and great functional group

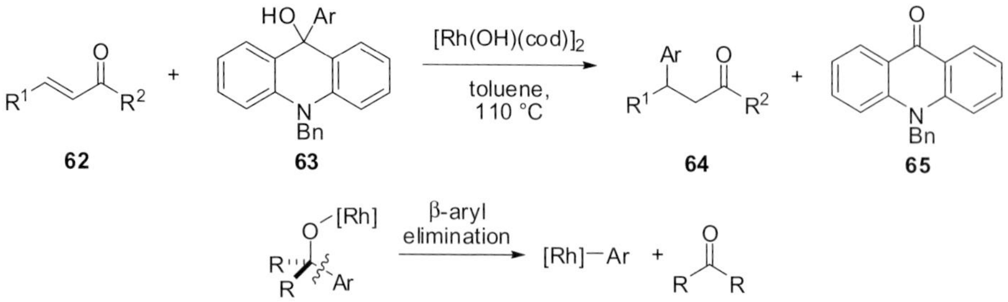

Scheme 8.18 Acridinols **63** as aryl sources.

compatibility. However, the use of boroxines or boronates has also occasionally been documented. Organotrifluoroborates, which are protected forms of boronic acids, also find ready application in the 1,4-addition [48]. In addition to boron, rhodium readily undergoes transmetallation with a wide range of other metals, including titanium [49], silicon [50], zinc [51], zirconium [52], tin [53] and indium [54]. In certain cases, tandem processes are viable using these metals.

An interesting approach which obviates the need for metallated nucleophiles was recently disclosed by Hayashi (Scheme 8.18) [55].

Based on mechanistic investigations conducted by Hartwig [56], β-aryl elimination from trisubstituted aryl methanols **63** derived from acridinone **65** was exploited to generate an aryl rhodium species which then undergoes 1,4-addition to a range of acyclic α,β-unsaturated carbonyl compounds **62**.

8.5.2
Ligand Systems

The combination of a rhodium-based catalyst precursor and BINAP as chiral ligand proved to be a highly efficient catalyst system for performing enantioselective conjugate additions of organoboron reagents to α,β-unsaturated carbonyl compounds [45]. However, equally effective chiral ligands have also been described in the literature (Figure 8.5); among these, amidomonophosphine **66** [57] is noteworthy as well as several carbene-based preligands, such as **67** [58] or **68** [59]. Other bisphosphine structures include *(S)*-P-Phos **69** [60], P-chiral Quinox-P* (**70**) [61], *(S,S)*-Diphonane **71** [62] or cyrhetrene **72** [63]. Phosphoramidites, such as **73**, were also successfully examined for this transformation [64]. This privileged ligand motif deserves special mention, due to its high modular character and ease of preparation. Consequently, it was also exploited in several combinatorial catalysis approaches [65].

Both, Hayashi and Carreira developed, independently, chiral dienes as novel ligands in asymmetric catalysis (Figure 8.6) [66]. Early investigations by Miyaura revealed that the rhodium-catalyzed conjugate addition of arylboronic acids to α,β-unsaturated ketones could be very efficiently catalyzed by a rhodium(I) complex of cyclooctadiene [67]. As a consequence, the development of chiral cod-analogues

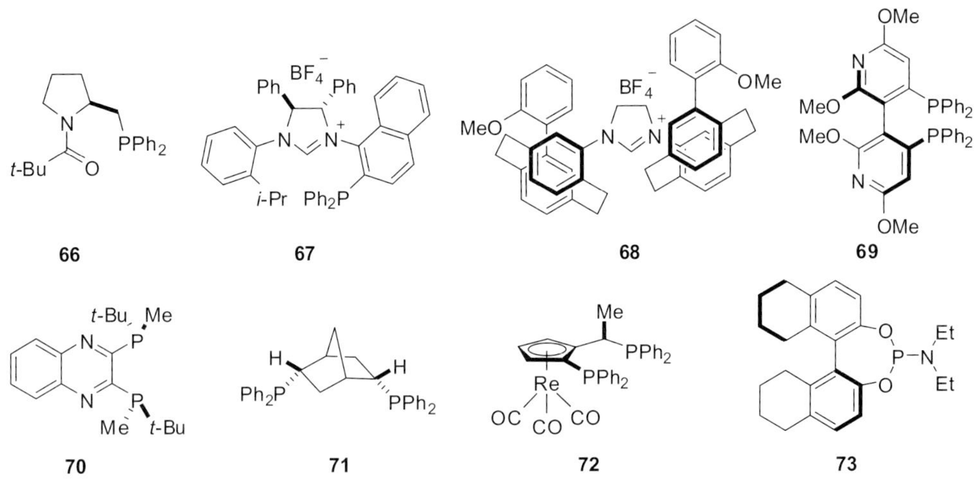

Figure 8.5 Employed ligands for the conjugate asymmetric arylation.

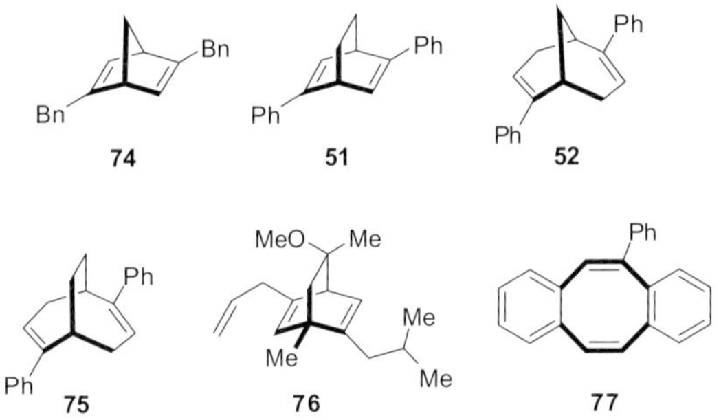

Figure 8.6 Chiral diene ligands for the conjugate arylation.

represented an ideal testing ground for implementation of this novel concept in asymmetric catalysis.

The preferred diene architecture incorporates the coordinating olefin units in a bicyclic scaffold, not only to ensure rigidity of the resulting complex, but also to provide a tight binding to late transition metals. Hayashi's dienes are based on substituted chiral bicyclo[2.2.1]hepta-2,5-dienes **74** [68], bicyclo[2.2.2]octa-2,5-dienes **51** [69], bicyclo[3.3.1]nona-2,6-dienes **52** or bicyclo[3.3.2]deca-2,6-dienes **75** [36b]. The enantiomerically enriched ligands were accessed by resolution via their diastereomers or separation by high-performance liquid chromatography (HPLC) on a chiral phase. In contrast, Carreira's diene **76** is derived from the chiral-pool monoterpene carvone by a short sequence of reactions [70]. Grützmacher prepared a chiral dibenzocyclo-octadiene-based ligand **77** which was resolved upon coordination with a chiral rhodium/diamine complex [71]. Generally, rhodium/diene complexes catalyze the asymmetric 1,4-addition of arylboronic acids to a wide range of cyclic and acyclic ketones and esters, with excellent yields and stereoselectivities

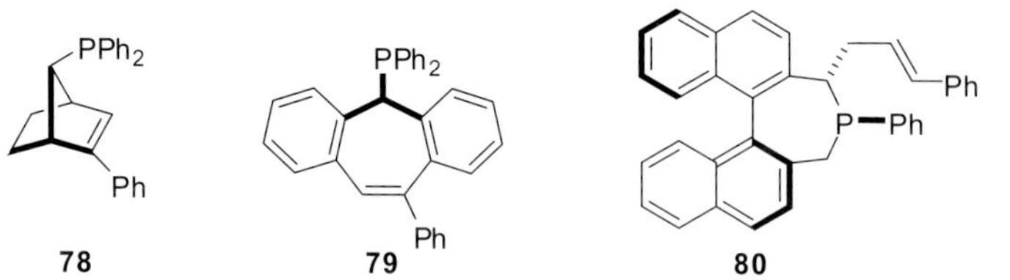

Figure 8.7 Chiral olefin–phosphine hybrid ligands for the conjugate arylation.

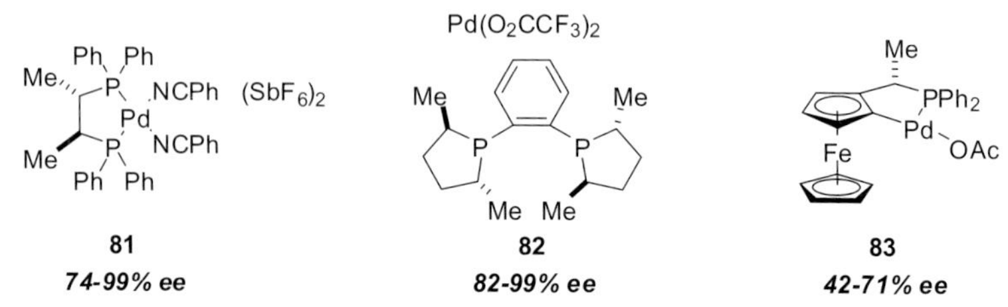

Figure 8.8 Phosphine ligands for the palladium(II) complexes for conjugate arylation.

(typically >90% yield and ee). Importantly, Hayashi also showed that the catalyst loadings could be as low as 0.005 mol%, without any deleterious effects on the stereochemical outcome [turnover frequency (TOF) up to 14 000 h^{-1}] [72].

Phosphine–olefin hybrid ligands developed by Hayashi **78** [73], Grützmacher **79** [74] and Widhalm **80** [75], combined the beneficial aspects of both design concepts. Whereas, phosphine ligands ensure tight binding to the metal, olefins provide better opportunities for chiral discrimination (Figure 8.7).

With these ligands enantioselectivities in excess of 95% were obtained in the rhodium-catalyzed addition of phenylboronic acid to cyclohexenone. Furthermore, the results of kinetic experiments suggested that catalyst turnover occurred much more rapidly with these mixed rhodium complexes of hybrid ligands compared to the corresponding rhodium complexes of BINAP or cod [76].

8.5.3 Conjugate Arylation with Diphosphine–Palladium(II) Complexes

Studies by Miyaura and coworkers revealed that dicationic palladium complexes are also capable of transmetallating with a number of aryl metal species Ph-[M] ([M] = $BiPh_2$, BF_3K, SiF_3) [77]. Enantioselective variants of this protocol were disclosed using chiraphos complex **81** or dipamp as chiral ligand, although the enantioselectivities remained lower than in the rhodium-catalyzed version [78]. However, the use of $Pd(O_2CCF_3)_2$ and *(R,R)*-MeDuPhos (**82**) enabled Feringa to obtain high enantioselectivities (up to 99% ee) for the conjugate addition of arylboronic acids [79] or arylsiloxanes [80] to a wide range of cyclic and acyclic ketones and esters (Figure 8.8).

Ohta reported that palladium(0) complexes coordinated by phosphine ligands catalytically induce 1,2-additions of arylboronic acids to aldehydes [81] as well as

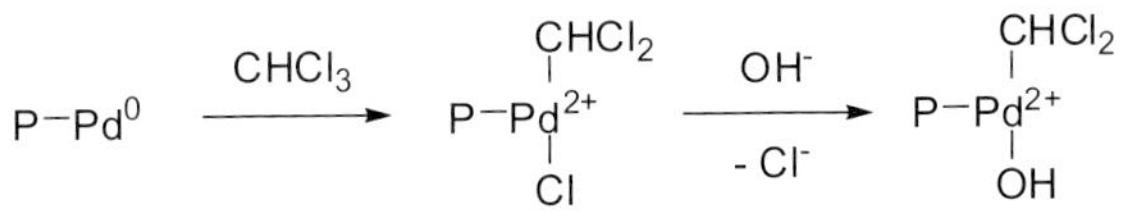

Scheme 8.19 Catalysis by palladium(0) and $CHCl_3$.

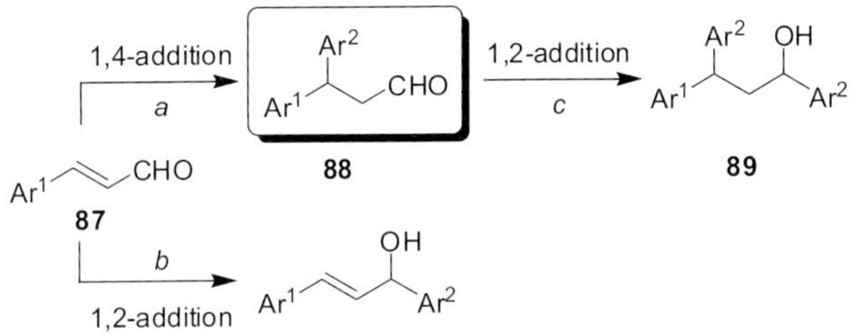

Figure 8.9 Competing reaction pathways in arylations of α,β-unsaturated aldehydes.

1,4-additions to α,β-unsaturated carbonyl compounds [82]. The addition of chloroform was essential in these transformations, and a possible rationale is depicted in Scheme 8.19. Thus, dichloromethylpalladium(II) intermediate **85** is generated by oxidative addition of a C–Cl bond to the phosphine–palladium(0) complex **84**. The active hydroxylpalladium(II) species **86** is then produced from **85** by ligand exchange. Chiral palladacycle **83** (Figure 8.8) was thought to possess similar catalytic activity, but only moderate enantioselectivities (42–71% ee) were achieved following the addition of arylboronic acids to cyclohex-2-enone [83].

8.5.4 Enantioselective Conjugate Arylation of α,β-Unsaturated Aldehydes

In 2005, Carreira and coworkers disclosed a rhodium/diene-catalyzed enantioselective conjugate addition of arylboronic acids to enals **87**, a traditionally challenging class of acceptors (Figure 8.9; path *a*) [84]. Although 1,2-addition of boronic acids to aldehydes has been studied extensively (Figure 8.9; path *b*) [14], a general and reliable enantioselective conjugate addition protocol was lacking in precedence. This is attributed to the fact that any prior effort was thwarted by 1,2-addition either in competition with 1,4-addition or after the formation of **88** to give **89** (Figure 8.9; paths *a* and *c*).

The selectivities that arise when using phosphine ligands for this transformation are depicted in Scheme 8.20. The use of a phosphine-stabilized catalyst for the addition of phenylboronic acid to cinnamaldehyde (**91**) led selectively to the allyl alcohol **93**, whereas use of the diene-modified species process resulted in the formation of the desired 1,4-adduct **92** [13b].

The use of chiral diene **94** in combination with methanol as solvent were optimal reaction parameters for the formation of a wide range of enantiomerically highly

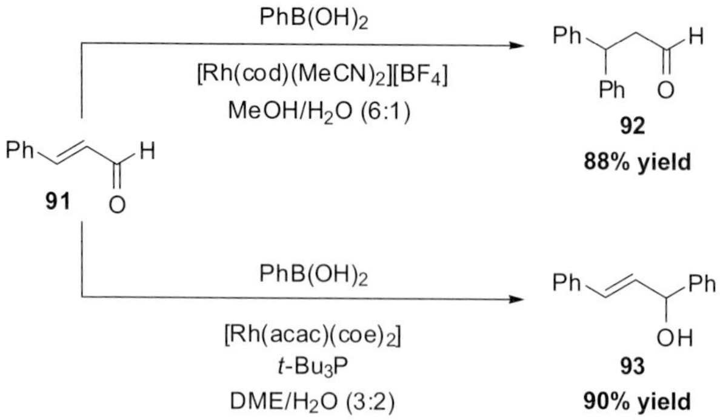

Scheme 8.20 Phosphine versus diene in the rhodium-catalyzed arylation of α,β-unsaturated aldehyde **91**.

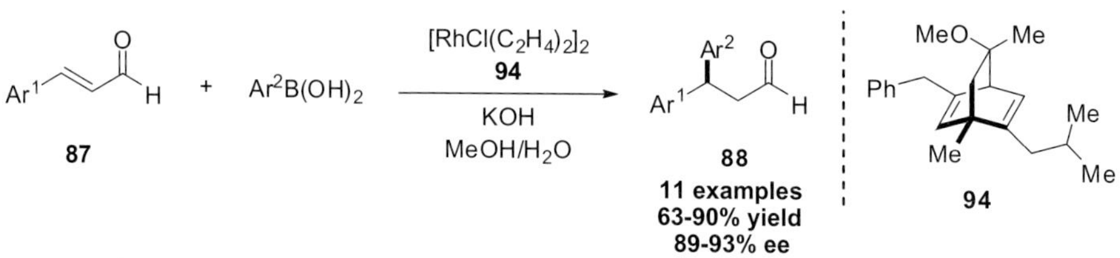

Scheme 8.21 Rhodium/diene-catalyzed conjugate asymmetric arylation of aldehydes.

enriched 3,3-diarylpropanals **88** (Scheme 8.21). In addition to the otherwise challenging diarylmethine stereogenic centers (which are found in many pharmaceuticals and natural products), the process furnishes aldehyde products, thus establishing a convenient handle for further synthetic modification. Inferior results obtained with conventional ligands such as BINAP (33% yield, 89% ee) or a phosphoramidite (19% yield, 56% ee) further illustrate the importance of chiral dienes in this simple reaction setting [85].

A protocol employing bisphosphine ligands such as BINAP is available, but involves the use of the more difficult-to-handle arylzinc reagents [86]. Additionally, Lewis acid activation by trimethylsilane is necessary. Switching the metal to dicationic palladium salts enabled Miyaura to perform this transformation with similar yields and selectivities (Scheme 8.22), and the addition of HBF_4 and/or $AgBF_4$ resulted in significant acceleration in the rate of the process [87].

8.5.5
Enantioselective Conjugate Arylation of Maleimides

Maleimides, such as **96**, also fall into the class of traditionally challenging substrates in conjugate addition chemistry. However, the corresponding 1,4-adducts **97**, α-substituted succinimides, are of certain synthetic interest in relation to their biological activity. As is evident from Scheme 8.23, conventional phosphine-based

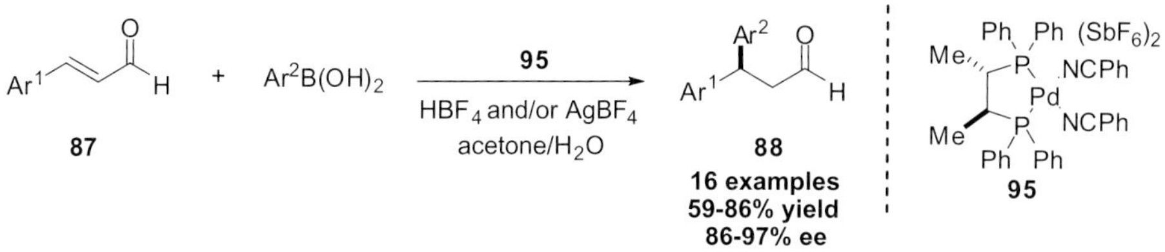

Scheme 8.22 Palladium(II)-catalyzed conjugate asymmetric arylation of aldehydes **87**.

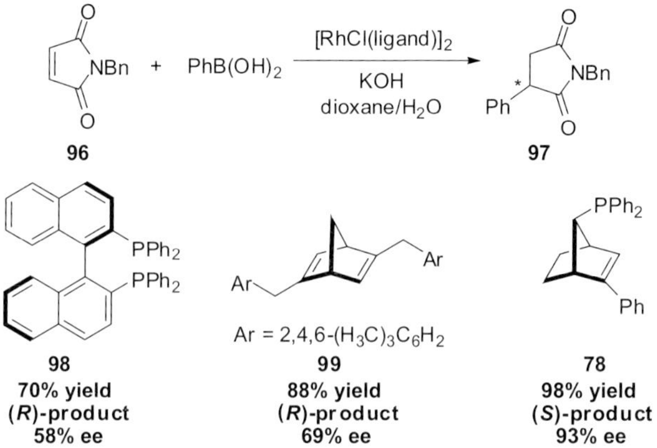

Scheme 8.23 Rhodium/diene-catalyzed enantioselective arylation of maleimide **96**.

ligands such as *(R)*-BINAP (**98**) lead to only moderate enantioselection, while first-generation chiral dienes, such as **99**, showed increased reactivity [88], with enantioselectivity remaining at a moderate level. However, a breakthrough was achieved with use of the phosphine-olefin hybrid ligand **78**, with which excellent yields and enantioselectivities were obtained [73a].

During an examination of the use of substituted maleimides **100**, Hayashi discovered that the regioselectivity in the addition was a function of the ligand employed (Scheme 8.24) [89]. Whereas, rhodium/BINAP-catalyzed processes preferably gave rise to 1,4-adducts with a quaternary stereocenter, rhodium/diene-catalysis led to *cis/trans*-mixtures of **102**.

Axially chiral *N*-arylsuccinimides were prepared by the addition of arylboronic acids to 1-(2-*tert*-butylphenyl)maleimide (**103**) (Scheme 8.25), and excellent diastereo- and enantioselectivities were obtained upon catalysis with the rhodium complex of diene **51** [90].

The resulting chiral C—N axis could then be used as a good template to control the configuration of subsequent transformations as was shown for alkylation and Diels–Alder processes.

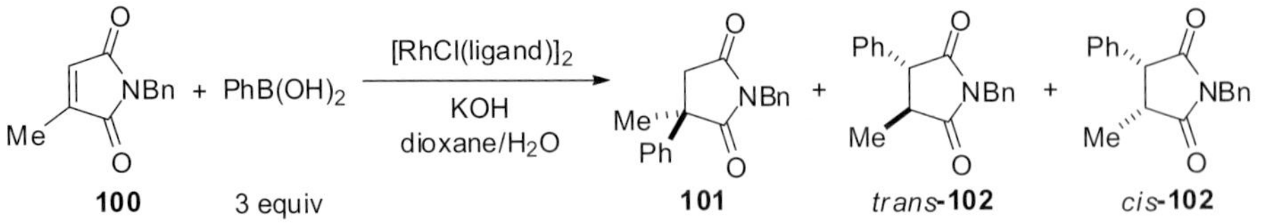

100 + $PhB(OH)_2$ (3 equiv) —[RhCl(ligand)]$_2$, KOH, dioxane/H_2O→ 101 + *trans*-102 + *cis*-102

ligand	yield	**101/102** (*trans/cis*)	ee of **101**	ee of **102**
(*R*)-BINAP **98**	99%	75%/25% (2.1/1)	95%	0%, 96%
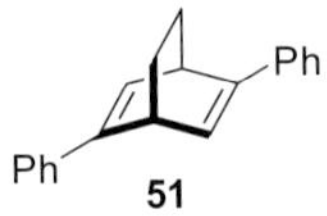 **51**	94%	11%/89% (1/1.4)	93%	79%, 99%
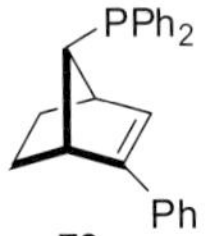 **78**	99%	17%/83% (1/1.2)	77%	95%, >99%

Scheme 8.24 Rhodium/diene-catalyzed enantioselective arylation of substituted maleimide **100**.

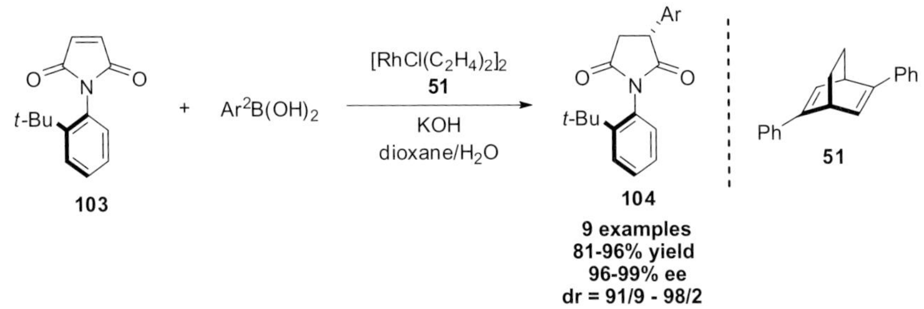

103 + $Ar^2B(OH)_2$ —[RhCl(C_2H_4)$_2$]$_2$, **51**, KOH, dioxane/H_2O→ **104**

9 examples
81-96% yield
96-99% ee
dr = 91/9 - 98/2

Scheme 8.25 Rhodium/diene-catalyzed synthesis of axially chiral *N*-arylsuccinimides **104**.

8.5.6
Additional Acceptors for Rhodium/Diene-Catalyzed Conjugate Arylation

The scope of synthetically interesting acceptors that can be functionalized in the rhodium/diene-catalyzed conjugate addition has been expanded considerably since the initial reports (Figure 8.10). Today, the range comprises not only α,β-unsaturated esters **105**, which are particularly well suited for heterocyclic-substituted substrates [91], but also α,β-unsaturated Weinreb amides **106**, which have found wide application [92]. Both classes of acceptors allow straightforward modification of the resulting adducts. The use of β-silyl-substituted α,β-unsaturated carbonyl compounds **107** as acceptors is of special interest as these compounds can be transformed to β-hydroxyketones via Tamao–Fleming oxidation [93].

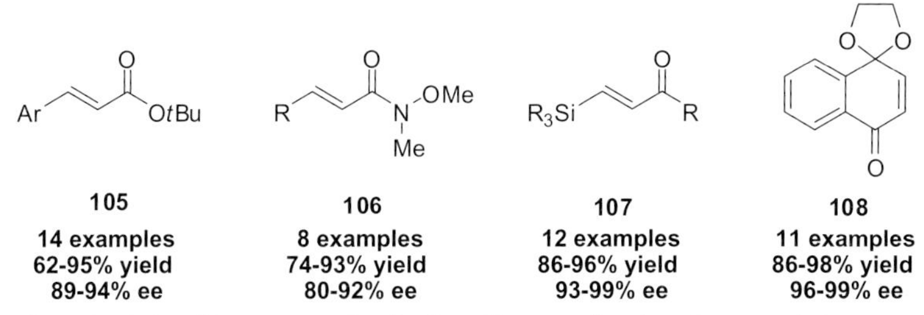

Figure 8.10 Suitable acceptors for rhodium/diene-catalyzed conjugate arylation.

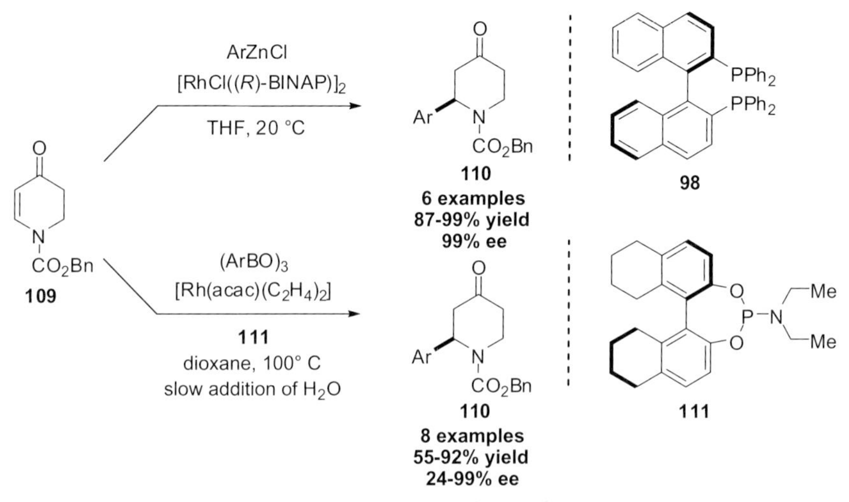

Scheme 8.26 Rhodium-catalyzed enantioselective arylation of 2,3-dihydro-4-pyridone **109**.

Recently, Hayashi reported the details of a rhodium/diene-catalyzed 1,4-addition of organoboron reagents to quinone monoketals **108**, which provides a rapid access to α-arylated tetralones in high yield and stereoselectivity [94].

8.5.7
Enantioselective Conjugate Arylation of 2,3-Dihydro-4-Pyridones

A particularly attractive class of acceptor substrates are 2,3-dihydro-4-pyridones **109** because the resulting adducts, chiral 2-arylpiperidines **110**, form part of a group of notable pharmaceuticals. Hayashi's approach to these building blocks involved the rhodium/BINAP (**98**)-catalyzed addition of arylzinc reagents (Scheme 8.26) [95]. Arylboronic acids were found also to induce high levels of stereoselectivity, although the reactivity was significantly reduced.

Because water is excluded in the handling and use of arylzinc reagents, the possibility of trapping the resulting rhodium enolate with different electrophiles arises. Similar yields and enantioselectivities were obtained when this protocol was employed for the addition to 4-quinolones (72–99% yield, 86–99% ee) [96]. Feringa discovered that arylboroxines are competent nucleophiles for this transformation, as long as water is slowly added during the reaction progress [97]. Phosphoramidite **111** derived from H8-BINOL was used as a highly efficient ligand to obtain optically active adducts in 55–92% yield and 24–99% ee (Scheme 8.26).

8.5.8 Enantioselective Conjugate Arylation of Coumarins

The interest in studying coumarins **112** as α,β-unsaturated acceptor systems was primarily fueled by the fact that the resulting adducts are privileged structural motifs in pharmaceuticals. Hayashi showed that a rhodium/Segphos (**50**)-catalyzed process would provide rapid access to arylated building blocks in excellent enantioselectivities (>99% ee) (Scheme 8.27) [98], and the method was subsequently applied in a straightforward synthesis of the urological drug (*R*)-tolterodine.

8.5.9 Conjugate Arylation of Chiral, Racemic α,β-Unsaturated Carbonyl Compounds

5-Arylcyclohex-2-enones **116** are valuable chiral building blocks in organic synthesis. However, a direct arylation of cyclohexa-2,5-dienone is not a tenable synthetic option. A clever approach to these building blocks was recently disclosed by Tomioka (Scheme 8.28) [99], where a rhodium/phosphane-catalyzed conjugate arylation of racemic 5-(trimethylsilyl)cyclohexenone led to a mixture of **115** which was directly dehydrosilylated by $CuCl_2$ to give **116** in high yields and enantioselec-

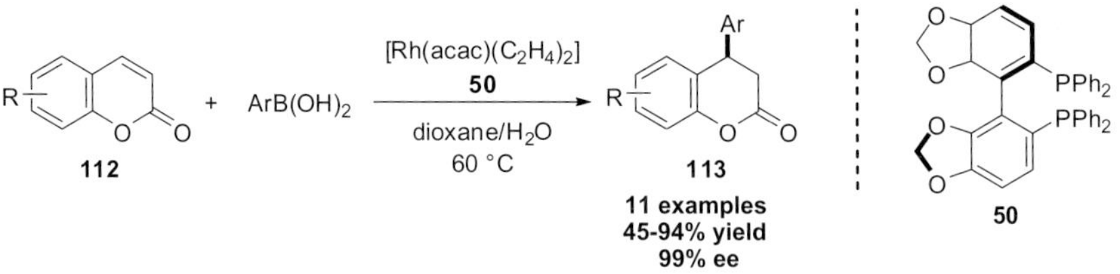

R– (112) + ArB(OH)2 → [Rh(acac)(C2H4)2], 50, dioxane/H2O, 60 °C → 113
11 examples
45-94% yield
99% ee
50: PPh2, PPh2

Scheme 8.27 Rhodium-catalyzed enantioselective arylation of coumarins **112**.

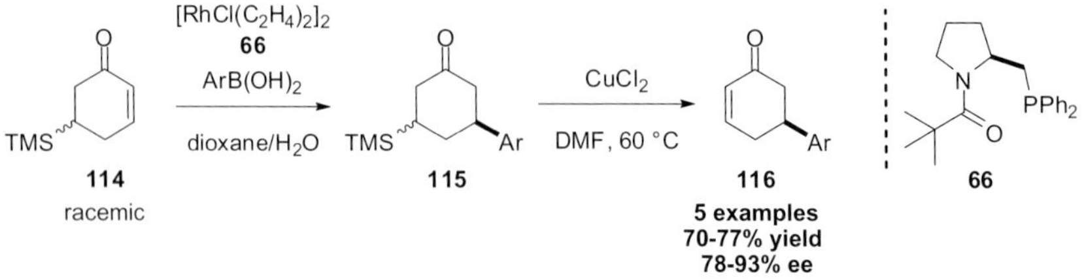

TMS (114, racemic) → [RhCl(C2H4)2]2, 66, ArB(OH)2, dioxane/H2O → TMS, Ar (115) → CuCl2, DMF, 60 °C → Ar (116)
5 examples
70-77% yield
78-93% ee
66: N, PPh2, O

Scheme 8.28 Enantioselective synthesis of 5-arylcyclohex-2-enones **116**.

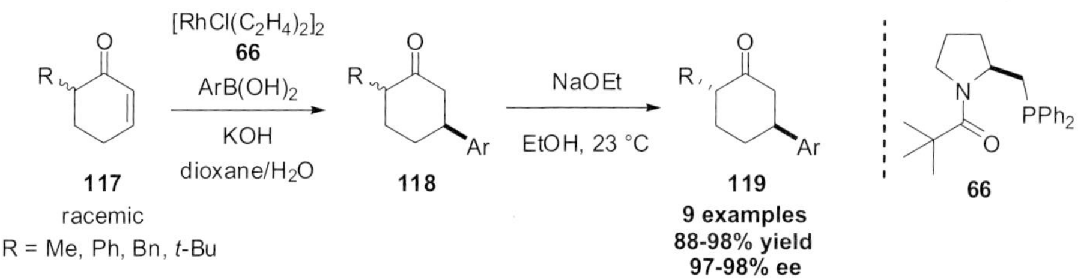

$[RhCl(C_2H_4)_2]_2$
66
$ArB(OH)_2$
KOH
dioxane/H_2O
NaOEt
EtOH, 23 °C
117 racemic
R = Me, Ph, Bn, *t*-Bu
118
119
9 examples
88-98% yield
97-98% ee
66
PPh$_2$

Scheme 8.29 Enantioselective synthesis of *trans*-5-aryl-2-substituted cyclohexanones **119**.

tivity. It is important to point out that the selectivity exerted by the catalyst system overrides the intrinsic substrate bias for *trans*-diastereoselectivity, thus avoiding the occurrence of any kinetic resolution.

Based on initial studies by Krause and Alexakis [100], the highly enantioselective synthesis of *trans*-5-aryl-2-substituted cyclohexanones **119** was accomplished by the conjugate arylation of racemic 6-substituted cyclohexenones **117** (Scheme 8.29) [101]. Although the first step did not proceed in a diastereoselective manner, only the thermodynamically more stable *trans*-disubstituted cyclohexanone **119** was obtained after epimerization with NaOEt.

8.5.10 Conjugate Asymmetric Arylation of 3-Substituted α,β-unsaturated Carbonyl Compounds

The asymmetric synthesis of quaternary centers remains a challenging objective in modern organic synthesis. In terms of conjugate asymmetric arylation methodology, to the best of the present authors' knowledge, only one process has been reported to perform such a difficult task. Hoveyda disclosed a copper/*N*-heterocyclic carbene (NHC)-catalyzed arylation of 3-substituted cycloalkenones **120** (Scheme 8.30) [102], where the resulting adducts **122** were obtained in 88–95% yield and 89–97% ee. Recently, a modification of the catalyst system allowed an expansion of the scope of this reaction [103], such that five- and six-membered γ-keto esters **123** were efficiently phenylated by treatment with Ph_2Zn and catalytic amounts of a NHC–copper complex derived from **124**. It should be noted here that Ph_2Zn, generated *in situ* from the less-expensive commercial grade PhLi, can be employed directly in this transformation.

8.5.11 1,6-Addition of Arylboronic Acids to α,β,γ,δ-Unsaturated Carbonyl Compounds

Recent approaches by Hayashi and others aimed at expanding the arylation chemistry to include also α,β,γ,δ-unsaturated carbonyl compounds as acceptors. In this vein, enynones such as **126** undergo a rhodium–Segphos (**50**)-catalyzed 1,6-addition using lithium aryltitanate-derived reagents (Scheme 8.31), with the

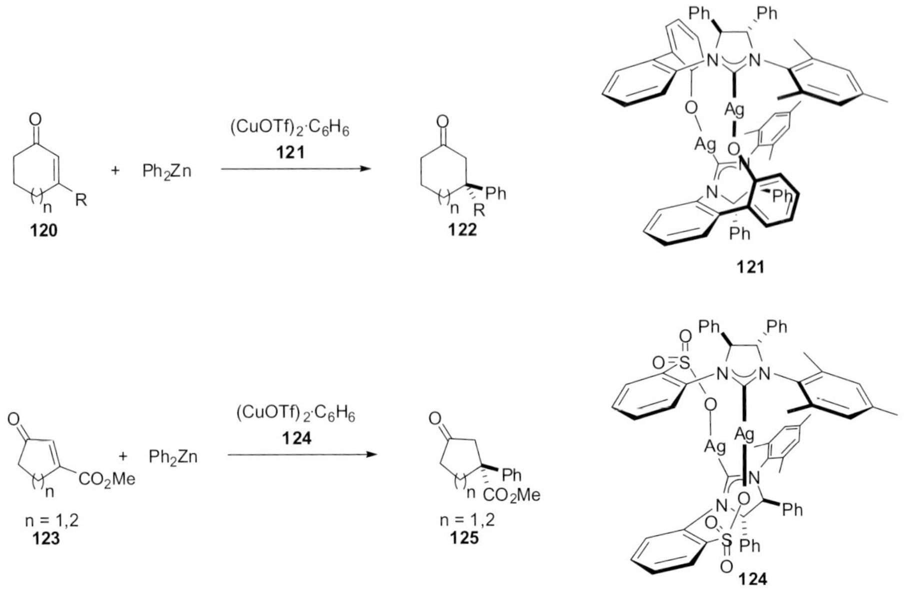

Scheme 8.30 Copper-catalyzed synthesis of quaternary stereocenters.

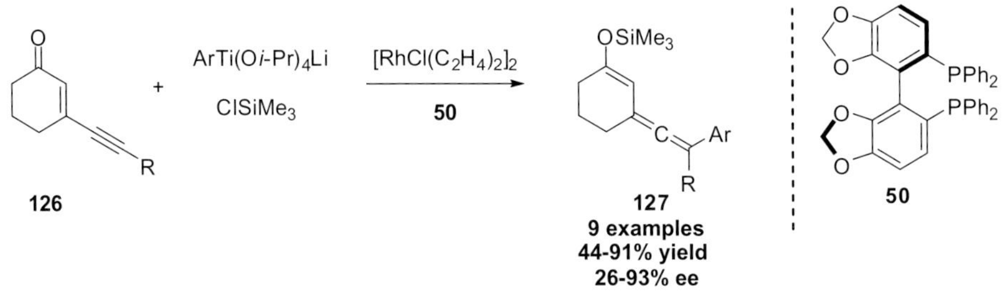

Scheme 8.31 Rhodium-catalyzed enantioselective 1,6-arylation of enynones **126**.

resultant enolates being trapped as their corresponding silyl ethers **127** in up to 93% ee [104].

High regio- and enantioselectivities were obtained in the 1,6-addition of arylzinc reagents to dienones **128** (Scheme 8.32) [105]. Again, the addition of trimethylchlorosilane was found to be necessary to obtain high yields, presumably by Lewis acid activation. The use of arylboronic acids as aryl sources only led to a recovery of starting material in this transformation. However, a metal switch to iridium allowed the use of the more easy-to-handle arylboroxines [106], and high yields of the δ-arylated carbonyl compounds **129** were obtained with excellent 1,6-selectivity.

In a recent report by Csáky, 2,4-dienoate esters **130** were employed as electrophilic partners for a rhodium(I)-catalyzed regioselective 1,6-addition of arylboronic

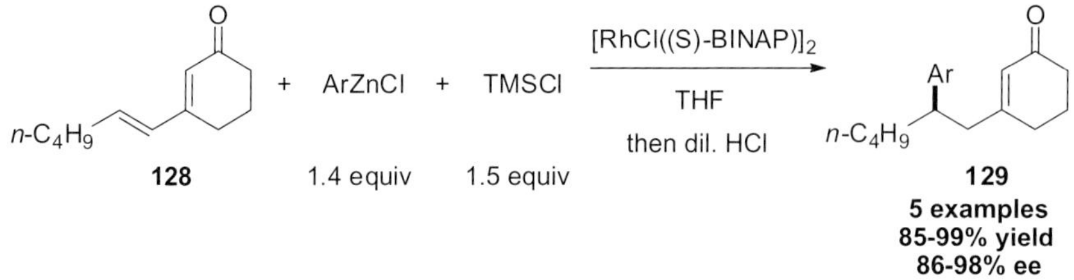

Scheme 8.32 Rhodium-catalyzed enantioselective 1,6-arylation of dienone **128**.

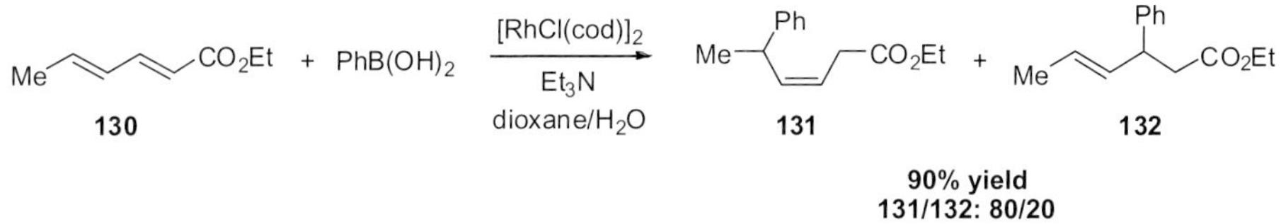

Scheme 8.33 Rhodium-catalyzed 1,6-arylation of dienoate **130**.

acids (Scheme 8.33) [107]. However, the addition of a base such as Et_3N or $Ba(OH)_2$ was crucial to the reactivity of the process.

8.6 Tandem Processes

The conjugate addition of a nucleophile to α,β-unsaturated carbonyl compounds results in formation of the corresponding enolate subsequent to the initial addition step. This reactive intermediate can be trapped by a variety of electrophiles, and thus opens up opportunities for tandem reaction sequences [108].

8.6.1 Rhodium-Catalyzed Enantioselective Conjugate Arylation–Protonation

When the α,β-unsaturated compound contains a β-substituent different from a hydrogen atom, protonation of the oxa-π-allyl metal intermediate will lead to the formation of a new stereocenter. Early investigations by Reetz highlighted the asymmetric addition of arylboronic acids to methyl-2-acetamidoacrylate (**133**) to create protected forms of arylalanines **134** [109]. However, ee-values obtained with ferrocene-based diphosphonite ligands typically were in the range of 50 to 70%. A significant breakthrough was made by Genêt, who employed aryltrifluoroborate salts as aryl sources (Scheme 8.34) [110]. Here, the proton source was crucial to the stereochemical outcome, and phenol derivatives, such as guaiacol, proved to be effective in achieving the desired product in high yields and enantioselectivities [111].

Analogous arylation-protonation chemistry was carried out with a variety of different 1,1′-disubstituted alkenes. For instance, Frost performed arylations using

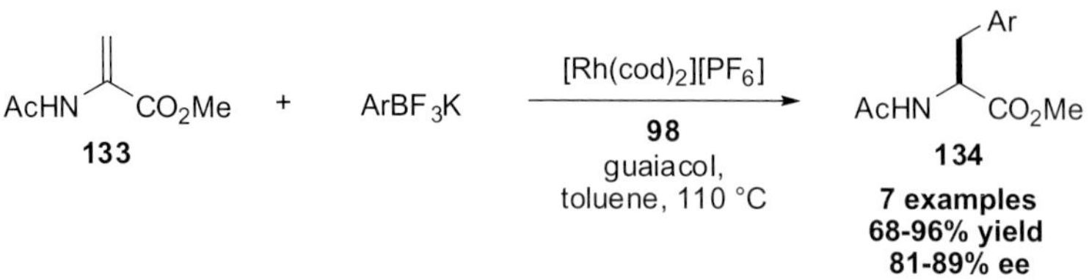

AcHN CO$_2$Me + ArBF$_3$K → [Rh(cod)$_2$][PF$_6$] 98 guaiacol, toluene, 110 °C → AcHN CO$_2$Me (Ar)

133 134

7 examples
68-96% yield
81-89% ee

Scheme 8.34 Rhodium-catalyzed enantioselective arylation of methyl-2-acetamidoacrylate (**133**).

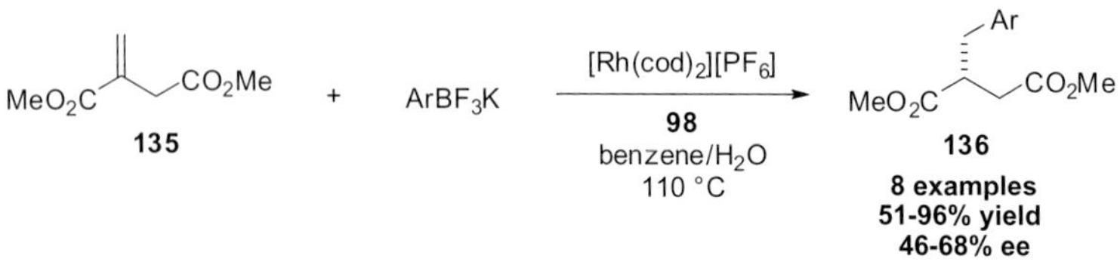

MeO$_2$C CO$_2$Me + ArBF$_3$K → [Rh(cod)$_2$][PF$_6$] 98 benzene/H$_2$O 110 °C → MeO$_2$C CO$_2$Me (Ar)

135 136

8 examples
51-96% yield
46-68% ee

Scheme 8.35 Rhodium-catalyzed enantioselective arylation of dimethyl itaconate (**135**).

138 ← ArZnCl [Rh] ← 137 → ArB(OH)$_2$ or ArSi(OR)$_3$ [Rh] → 139

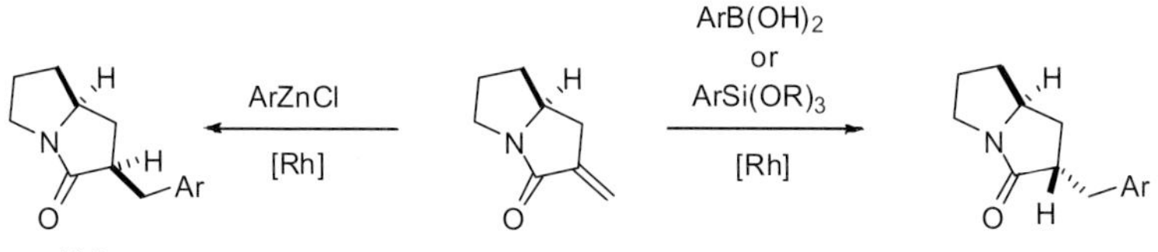

Scheme 8.36 Stereochemical switch as a result of the employed aryl source.

organotrifluoroborates [112] or organosiloxanes [113] to dimethyl itaconate (**135**) (Scheme 8.35). Interestingly, temperatures up to 110 °C were needed to reach high stereoselectivities, whereas lower temperatures led only to racemic mixtures of the product **136**.

Another surprising effect was realized in the context of studying 2-substituted pyrrolizidinones **137** as acceptors (Scheme 8.36) [114] when, depending on the employed aryl source, stereochemical outcome could be tailored to give predominantly one diastereomer.

α-Benzyl acrylates **140** were employed as substrates for a tandem Rh/BINAP-catalyzed arylation–protonation sequence to afford enantiomerically enriched α,α′-dibenzyl esters **141** (Scheme 8.37) [115]. Boric acid proved to be the best proton source for this transformation.

Sibi reported on the enantioselective synthesis of β^2-amino acids by the addition of arylboronic acids to β-amino acrylates **142** (Scheme 8.38) [116]. Phthalimide served as the most efficient proton source, while the use of a catalyst system comprising a difluorphos (**143**) -functionalized rhodium complex led to high yields and enantioselectivities.

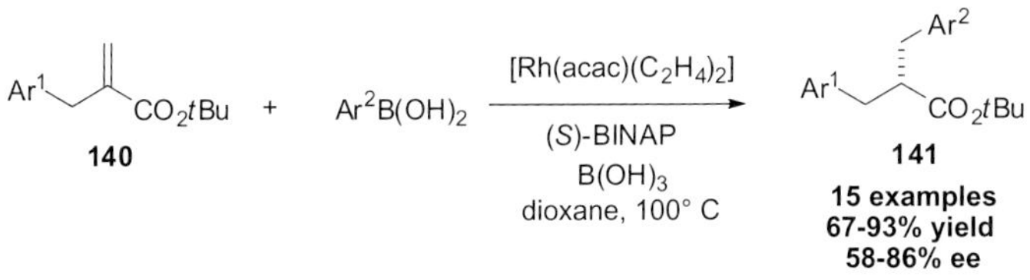

Scheme 8.37 Rhodium-catalyzed enantioselective synthesis of α,α′-dibenzyl esters **141**.

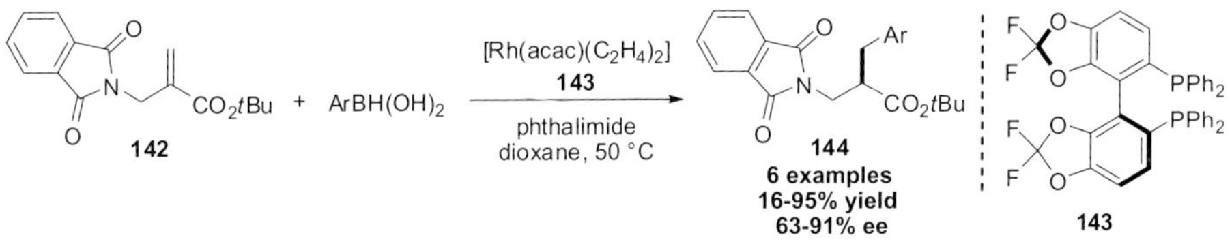

Scheme 8.38 Rhodium-catalyzed enantioselective synthesis of β^2-amino acids.

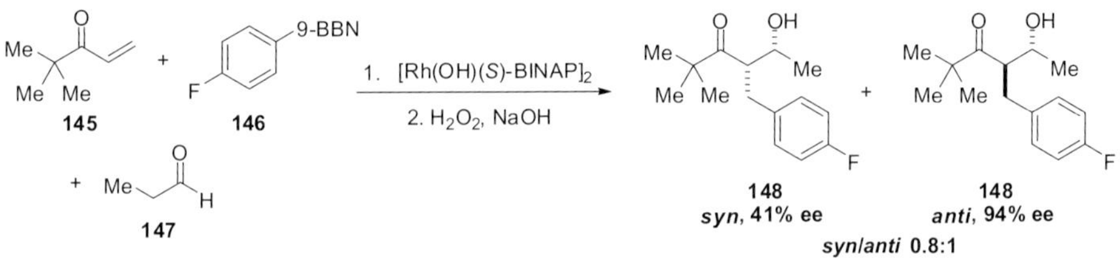

Scheme 8.39 Rhodium-catalyzed conjugate arylation–aldol cyclization.

8.6.2 Rhodium-Catalyzed Conjugate Arylation–Aldol-Addition

Most of the currently applied protocols for rhodium-catalyzed conjugate addition chemistry involve the use of aqueous solvent systems which ensure catalytic turnover by protonation of the intermediate rhodium enolate. Consequently, tandem reaction sequences with electrophiles other than a proton are troublesome. In early investigations, Hayashi reported a rhodium/BINAP-catalyzed conjugate addition–aldol reaction under anhydrous conditions by use of 9-aryl-9-borabicyclo[3.3.1]nonanes (9-Ar-9-BBN) as aryl sources [117]. The reaction between *tert*-butyl vinyl ketone (**145**) with 9-(4-fluorophenyl)-9-BBN (**146**) and propionaldehyde (**147**) led to the formation of a *syn/anti*-mixture of **148** in a 0.8 to 1 ratio (Scheme 8.39).

Krische disclosed an intramolecular tandem 1,4-addition/aldol process in the presence of water (Scheme 8.40) [118]. Here, the rate of the aldol cyclization was faster than the competing hydrolysis of the rhodium enolate, so that **150** was isolated in 88% yield and 88% enantioselectivity. The configuration observed for the aldol reaction could be accounted for by the Zimmerman–Traxler-type transition state of the *(Z)*-enolate.

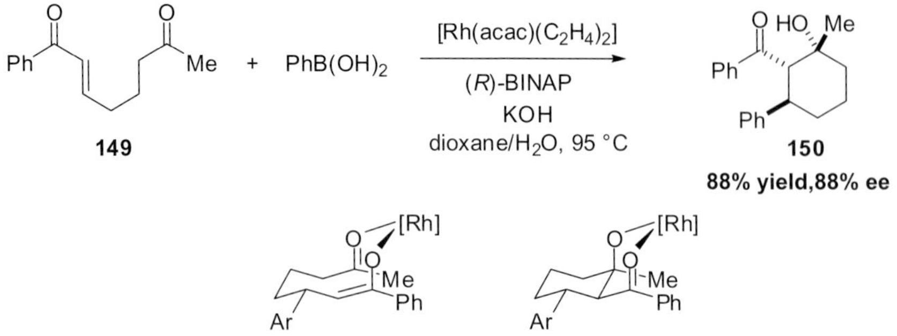

Scheme 8.40 Rhodium-catalyzed intramolecular conjugate arylation–aldol cyclization.

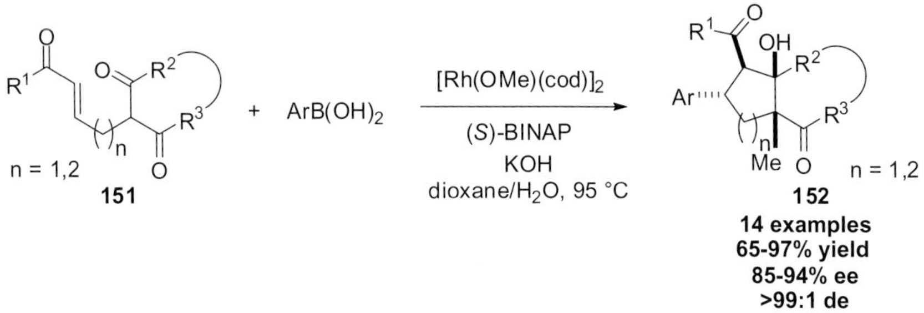

Scheme 8.41 Rhodium-catalyzed desymmetrization of triones **151**.

The methodology was extended to the desymmetrization of triones **151** (Scheme 8.41) [119], and diastereo- and enantioselective conjugate arylation/aldol cyclization led to the formation of five- and six-membered diquinane and hydroquinane structures **152** containing four contiguous stereocenters.

8.6.3
Rhodium-Catalyzed Conjugate Arylation–Allylation

Feringa recently reported a highly enantio- and diastereoselective one-pot conjugate arylation/allylation sequence (Scheme 8.42) [120] where the initial rhodium/phosphoramidite-catalyzed conjugate addition of arylboronic acids in dioxane/water was complemented by a Barbier-type indium-mediated allylation.

8.6.4
Rhodium-Catalyzed Sequential Carbometallation–Addition

Arylrhodium species undergo facile intermolecular 1,2-addition across carbon–carbon triple bonds, with the resulting alkenyl rhodium intermediate providing an entry point for further cascade reactions [121]. As an example, Hayashi applied this concept to the asymmetric synthesis of a number of cycloalkanols **157** starting from ynals **155** and arylboronic acids (Scheme 8.43) [122]. Transmetallation of the

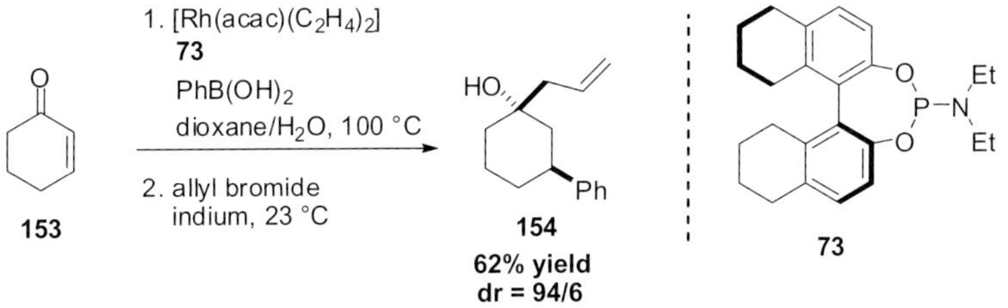

1. $[Rh(acac)(C_2H_4)_2]$ **73**, $PhB(OH)_2$, dioxane/H_2O, 100 °C; 2. allyl bromide, indium, 23 °C. **153** → **154**, 62% yield, dr = 94/6. **73**

Scheme 8.42 Rhodium-catalyzed conjugate arylation–allylation.

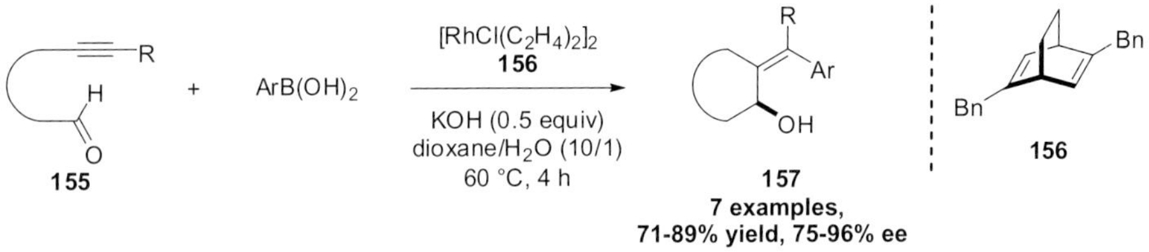

155 + $ArB(OH)_2$ → $[RhCl(C_2H_4)_2]_2$ **156**, KOH (0.5 equiv), dioxane/H_2O (10/1), 60 °C, 4 h → **157**, 7 examples, 71-89% yield, 75-96% ee. **156**

Scheme 8.43 Rhodium/diene-catalyzed arylative cyclization of alkynals **155**.

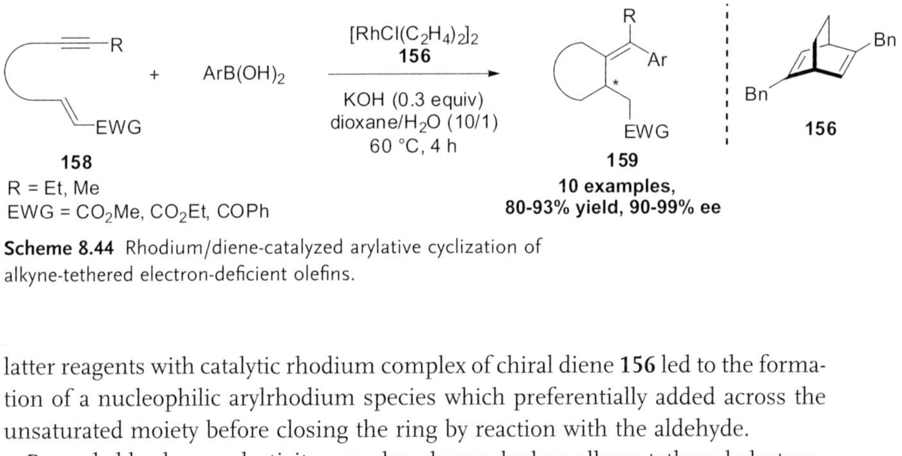

158 + $ArB(OH)_2$ → $[RhCl(C_2H_4)_2]_2$ **156**, KOH (0.3 equiv), dioxane/H_2O (10/1), 60 °C, 4 h → **159**, 10 examples, 80-93% yield, 90-99% ee. **156**

R = Et, Me
EWG = CO_2Me, CO_2Et, COPh

Scheme 8.44 Rhodium/diene-catalyzed arylative cyclization of alkyne-tethered electron-deficient olefins.

latter reagents with catalytic rhodium complex of chiral diene **156** led to the formation of a nucleophilic arylrhodium species which preferentially added across the unsaturated moiety before closing the ring by reaction with the aldehyde.

Remarkable chemoselectivity was also observed when alkyne-tethered electron-deficient olefins **158** were employed as substrates (Scheme 8.44) [123]. Rhodium/diene catalysis led preferentially to carborhodation of the alkyne, followed by 1,4-addition to the α,β-unsaturated moiety to provide adducts **159** in high yield and excellent enantioselectivity. The difference between phosphines and chiral dienes as ligands in this reaction setting was striking. Whereas, rhodium-bisphosphines complexes catalyzed the 1,4-addition to α,β-enoates more effectively than the arylation of alkynes, rhodium/diene catalysts favored the arylation of alkynes over the 1,4-addition.

Another interesting tandem process involving sequential carborhodation and addition was presented by Hayashi in the context of the asymmetric synthesis of 3,3-disubstituted 1-indanones **161** (Scheme 8.45) [124]. The reaction pathway was

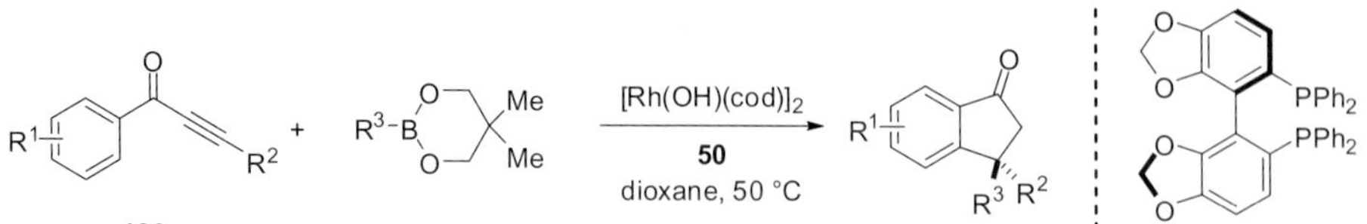

[Rh(OH)(cod)]$_2$
50
dioxane, 50 °C

160 **161** **50**

R^2 = TES, TBS, $GeMe_3$
R^3 = Aryl, Me

16 examples
51-91% yield
87-99% ee

Scheme 8.45 Rhodium-catalyzed asymmetric synthesis of 3,3-disubstituted 1-indanones **161**.

thought to involve an initial insertion of alkyne **160** into the arylrhodium species, after which the generated alkenylrhodium species underwent a 1,4-rhodium migration to form an arylrhodium intermediate [125]. Intramolecular 1,4-addition of this compound then led to the oxa-π-allyl intermediate which, after transmetallation, released the boron alkoxide.

8.7
Enantioselective Friedel–Crafts Arylation

The Friedel–Crafts reaction of electron-rich aromatic and heteroaromatic compounds with carbonyl acceptors is one of the fundamental reactions for forming carbon–carbon bonds. In recent years, several enantioselective approaches have been described utilizing copper and scandium complexes as catalysts. In addition, the use of organocatalytic approaches has been considerably expanded.

8.7.1
Metal-Catalyzed Enantioselective Friedel–Crafts Arylations

Most of the currently reported procedures for carrying out enantioselective metal-catalyzed Friedel–Crafts reactions involve bidentate, chelating substrates that are capable of two-point binding with a suitable Lewis-acidic metal [126]. The ligands of choice for this transformation are C_2-symmetric chiral bis(oxazoline)s [127], while the resulting rigid diastereomeric transition states account for a highly selective attack of the nucleophile. However, in addition to selectivity control, reactivity aspects must also be carefully considered. Most of the substrates that have been used to date comprise only highly electrophilic systems such as glyoxylates [128], pyruvates [129], trifluoropyruvates [130], fluoral [131] or α-imino esters [132]. Lewis acid-catalyzed 1,2-additions to aldehydes and imines **162** are scarce, due to the intrinsic instability of the intermediate benzylic alcohol **163** under acidic reaction conditions. Consequently, double Friedel–Crafts arylations are frequently encountered (Scheme 8.46) [133].

To the best of the present authors' knowledge, only one example of a direct metal-catalyzed 1,2-addition of indoles **165** to tosyl- or nosyl-protected aldimines

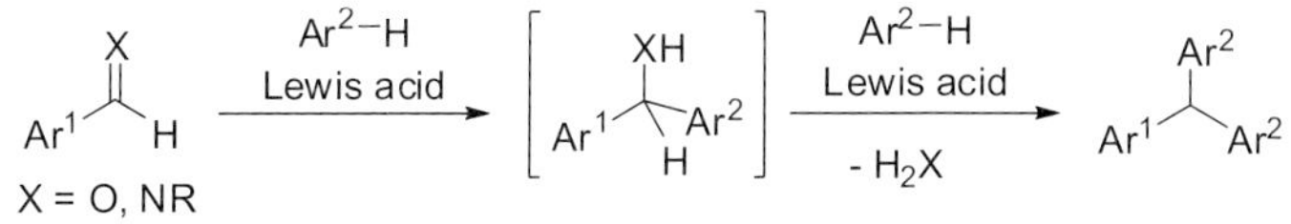

162 **163** **164**

Scheme 8.46 Problematic double Friedel–Crafts arylation.

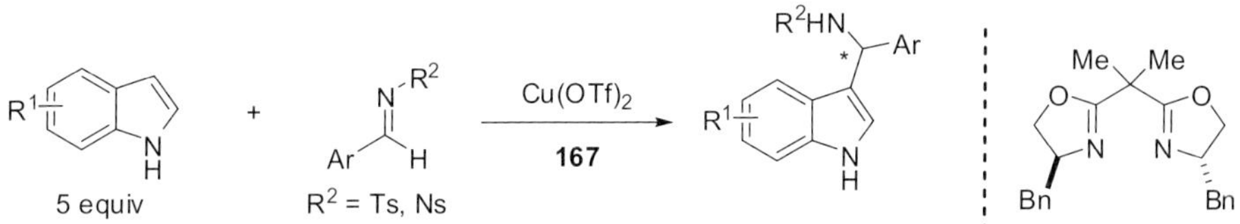

165 **166** **168** **167**

19 examples
47-94% yield
77-96% ee

Scheme 8.47 Copper/box-catalyzed enantioselective Friedel–Crafts arylation of imines **166**.

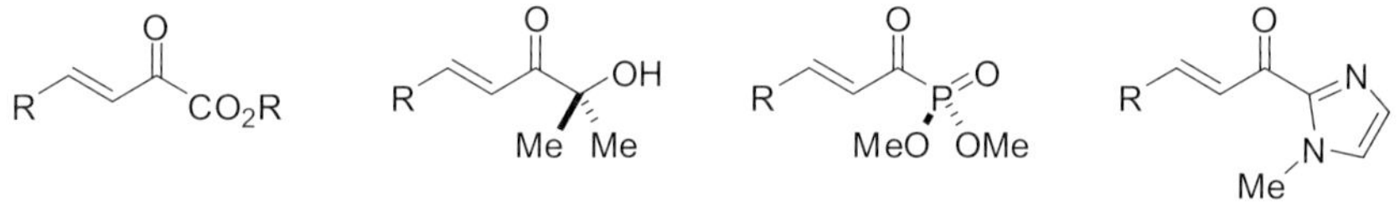

169 **170** **171** **172**

Figure 8.11 α,β-Unsaturated carbonyl compounds investigated for enantioselective metal-catalyzed Friedel–Crafts arylation.

166 has been disclosed (Scheme 8.47) [134]. However, in order to prevent imine hydrolysis, it was necessary to add a large excess of indole (up to 5 equiv.).

Several examples of metal-catalyzed enantioselective Friedel–Crafts arylations to conjugated acceptor systems are known. Similar to the 1,2-addition reaction, most of the substrates are bidentate in nature to ensure a tight-binding, rigid metal–substrate interaction (Figure 8.11).

For instance, an early example by Jørgensen relates to β,γ-unsaturated α-ketoesters **169**, which served as acceptors for indoles, furanes and electron-rich arenes [135]. The same catalyst system, a combination of $Cu(OTf)_2$ and *tert*-butyl-substituted bisoxazoline **173**, was employed by Palomo in the highly enantioselective addition of indoles and pyrroles to α′-hydroxy enones **170** (Figures 8.11 and 12) [136].

Equally effective ester surrogates, acyl phosphonates **171** [137] as well as acyl imidazoles **172** [138], were examined by Evans in transformations which were catalyzed by Sc(III) and pybox-derivative **174** (Figure 8.12). It is important to note that the resulting 1,4-adducts provide ample possibilities for further derivatization to synthetically useful amides, esters, carboxylic acids, ketones and aldehydes by standard organic transformations.

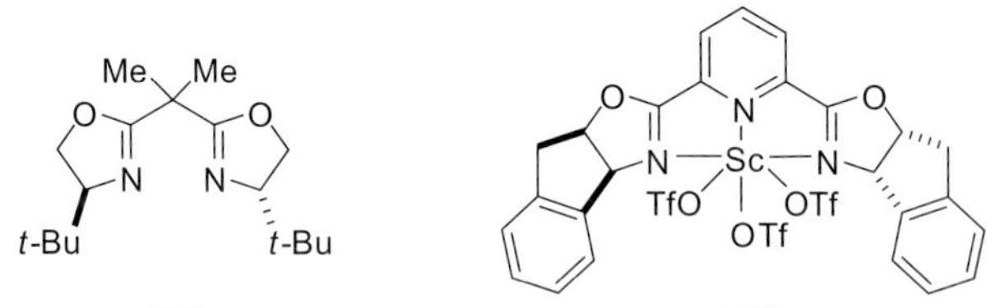

Figure 8.12 Metal-bisoxazoline complexes: catalysts for Friedel–Crafts arylations.

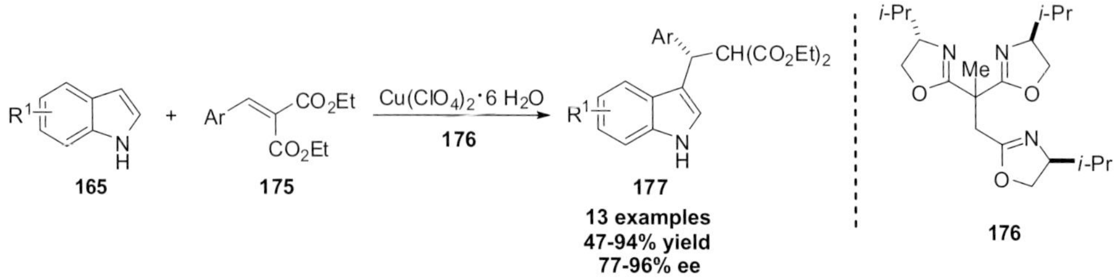

Scheme 8.48 Copper/box-catalyzed enantioselective indole addition to alkylidene malonates **175**.

The use of pseudo-C_3 symmetric tris(oxazoline) **176** proved to be important in the copper-catalyzed indole addition to aryl alkylidene malonates **175** (Scheme 8.48) [139]. A comparison with the traditional *tert*-butyl-substituted bis(oxazoline) **173** resulted in a sharp decrease in enantioselectivity (from 92% to 69% ee). Hence, Tang proposed a sidearm effect exerted by ligand **176** to be essential for this class of acceptors.

8.7.2 Organocatalysis in Friedel–Crafts Arylation

During recent years, asymmetric catalysis by small organic molecules has received much attention [140]. Because these reactions proceed through intermediates that are inherently less reactive, the Friedel–Crafts reactions of electron-rich (hetero)aryls generally seem to be well suited. For instance, Deng described the use of readily accessible cinchona-derived ligand **178** to perform highly enantioselective indole additions to α-ketoesters and even simple aldehydes (Scheme 8.49) [141]. Bisindole adducts, the major side products in many Lewis acid-catalyzed reactions, were formed to only a minor extent.

The additions of indoles to imines were documented in a similar way. In this context, Deng investigated bifunctional cinchona alkaloid **181** for its ability to promote the addition of indoles to a wide range of imines **186** (Figure 8.13; Scheme 8.50) [142], with substituted aryl aldimines being shown to be equally effective acceptors as the traditionally more challenging alkyl imines.

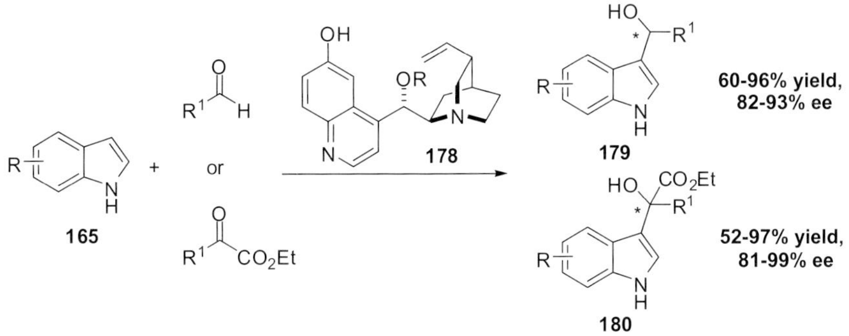

Scheme 8.49 Cinchona-catalyzed enantioselective indole additions to aldehydes and β-ketoesters.

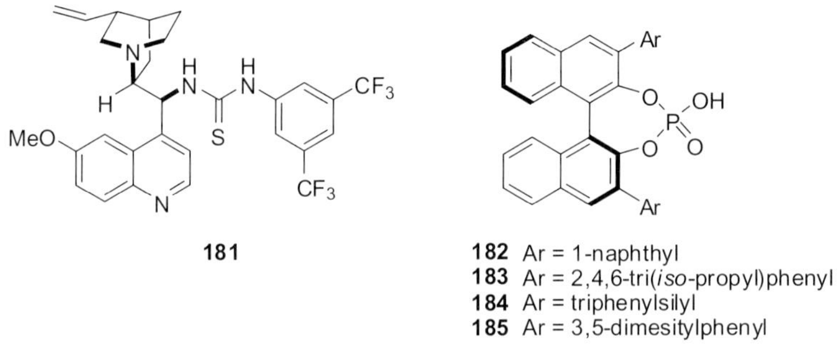

Figure 8.13 Bifunctional cinchona alkaloids and chiral phosphoric acid-derived organocatalysts.

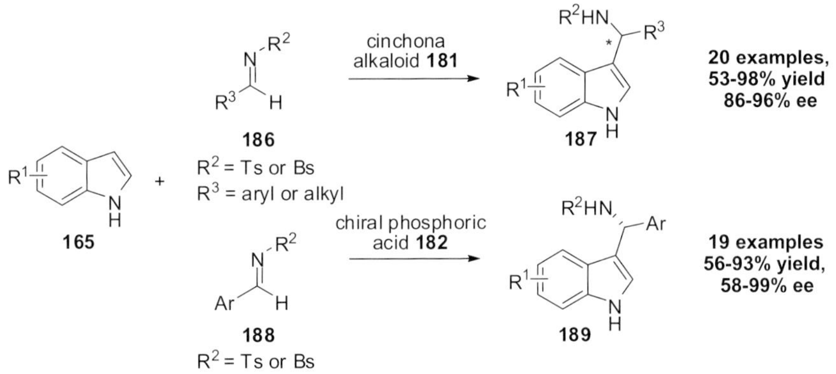

Scheme 8.50 Organocatalytic enantioselective indole additions to imines.

Chiral phosphoric acids, such as **182**, were also found to be suitable catalysts for this transformation [143], although with **182** the substrate scope was limited to *N*-tosyl and *N*-brosyl-substituted aryl imines **188**. In contrast, catalysis by **183** considerably reduced the reaction time (<2 h to reach completion). Antilla reported the use of chiral phosphoric acid **184** in the enantioselective addition of indoles to

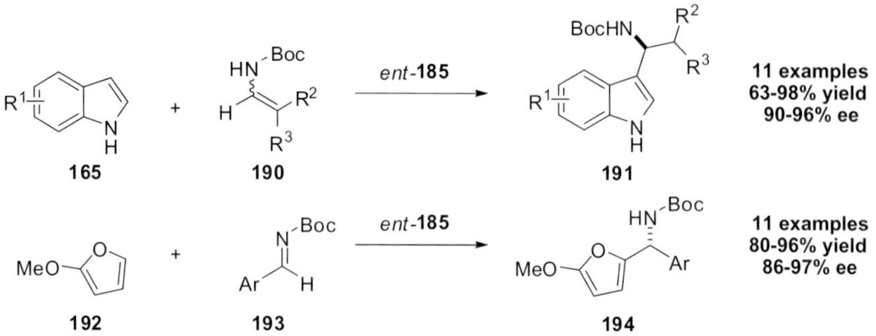

Scheme 8.51 Organocatalytic Friedel–Crafts arylations by chiral phosphoric acids.

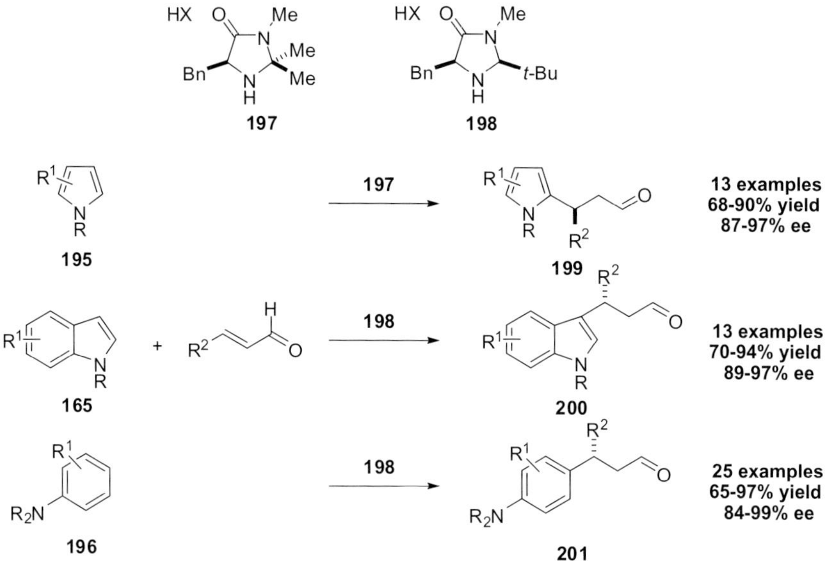

Scheme 8.52 Organocatalytic Friedel–Crafts arylations of α,β-unsaturated aldehydes.

N-acyl imines [144]. A slightly modified chiral Brønsted acid **185** was found to catalytically induce addition of indoles to *N*-Boc-protected enecarbamates **190** in high yields and enantioselectivities (Scheme 8.51) [145]. In a related study, Zhou demonstrated the use of α-aryl enamides to obtain optically enriched tertiary amine products [146].

Terada also reported the enantioselective addition of 2-methoxyfuran (**192**) to a wide range of *N*-Boc-protected aryl aldimines **193** using chiral phosphoric acid **185** [147].

The conjugate addition reaction of substituted pyrroles **195** [148], indoles **165** [149] as well as anilines **196** [150] to α,β-unsaturated aldehydes has been documented by MacMillan to occur in high yield and useful enantioselectivity (Scheme 8.52). By use of chiral imidazolidinones **197** or **198**, iminium activation effectively

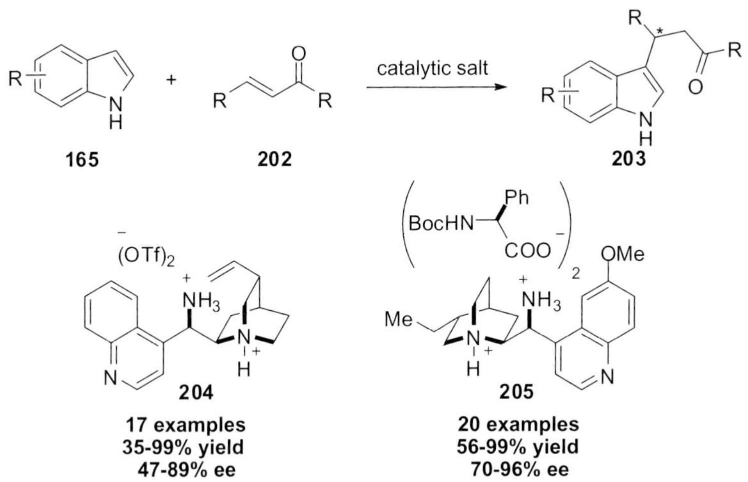

165 + 202 → (catalytic salt) 203

204: $(OTf)_2^-$ — 17 examples, 35-99% yield, 47-89% ee

205: (BocHN–CH(Ph)–COO^-)$_2$ — 20 examples, 56-99% yield, 70-96% ee

Scheme 8.53 Enantioselective Friedel–Crafts arylation of α,β-unsaturated ketones **202**.

lowers the lowest unoccupied molecular orbital (LUMO) of the electrophile in such a way that the corresponding nucleophile can add in a highly regio- and stereoselective manner [151, 152].

α,β-Unsaturated ketones **202** proved to be more difficult substrates for a similar iminium ion activation strategy, although Chen recently suggested that a cinchonine-derived primary amine salt **204** might represent a solution to this problem (Scheme 8.53) [153]. Although the obtained enantioselectivities were good (47–89% ee), an improvement was achieved with catalytic amine salt **205** in which both the cation as well as its counterion were chiral [154].

8.8 Conclusions

In recent years, significant advances have been made in the enantioselective catalytic arylation of a wide range of different carbonyl compounds, including aldehydes, imines and α,β-unsaturated substrates. The application of novel ligand systems, such as chiral dienes, has led to a considerable optimization of existing protocols, rendering them more practical. Moreover, the renaissance in organocatalysis has augmented the 'tool box' of available asymmetric reactions for synthesis. Sophisticated tandem processes avoiding the isolation of intermediates are increasingly examined and becoming available. Due to the widespread occurrence of aryl motifs – not only in a range of natural products but also in a number of pharmaceuticals – the enantioselective arylation of carbonyl compounds deserves continued attention in order to develop efficient, atom-economical and robust protocols which may find application in industry in the future.

Abbreviations

acac	acetylacetonato
BBN	borabicyclo[3.3.1]nonane
BINOL	1,1′-binaphthalene-2,2′-diol
coe	(*Z*)-cyclo-octene
cod	1,5-cyclo-octadiene
(+)-DAIB	(*2R*)-(+)-3-*exo*-(dimethylamino)isoborneol
Dipamp	1,2-bis[(2′-methoxyphenyl)phenylphosphino]ethane
DME	1,2-dimethoxyethane
dppbenz	1,2-(diphenylphosphino)benzene
dr	diastereomeric ratio
ee	enantiomeric excess
(−)-MIB	(*2S*)-(−)-3-*exo*-(morpholino)isoborneol
NHC	*N*-heterocyclic carbene
TADDOL	2,2-dimethyl-$\alpha,\alpha,\alpha',\alpha'$-tetraphenyldioxolane-4,5-dimethanol
TBS	*tert*-butyldimethylsilyl
TEEDA	tetraethylethylene diamine
TES	triethylsilyl

References

1 Farina, V., Reeves, J.T., Senanayake, C. H. and Song, J.J. (2006) *Chem. Rev.*, **106**, 2714–93.

2 Bolm, C., Hildebrand, J.P., Muñiz, K. and Hermanns, N. (2001) *Angew. Chem. Int. Ed.*, **40**, 3284–308.

3 Schmidt, F., Stemmler, R.T., Rudolph, J. and Bolm, C. (2006) *Chem. Soc. Rev.*, **35**, 454–70.

4 (a) Dosa, P.I., Ruble, J.C. and Fu, G.C. (1997) *J. Org. Chem.*, **62**, 444–5;
(b) For an enantioselective arylation with a reagent generated from PhMgBr and $ZnCl_2$, see: Soai, K., Kawase, Y. and Oshio, A. (1991) *J. Chem. Soc. Perkin Trans.*, **1**, 1613–15.

5 Huang, W.-S. and Pu, L. (1999) *J. Org. Chem.*, **64**, 4222–3.

6 Bolm, C., Hermanns, N., Hildebrand, J.P. and Muñiz, K. (2000) *Angew. Chem. Int. Ed.*, **39**, 3465–7.

7 Qin, Y.-C. and Pu, L. (2006) *Angew. Chem. Int. Ed.*, **45**, 273–7.

8 (a) Bolm, C. and Rudolph, J. (2002) *J. Am. Chem. Soc.*, **124**, 14850–1;
(b) Schmidt, F., Rudolph, J. and Bolm, C. (2007) *Adv. Synth. Cat.*, **349**, 703–8.

9 Rudolph, J., Schmidt, F. and Bolm, C. (2004) *Adv. Synth. Cat.*, **356**, 867–72.

10 Dahmen, S. and Lormann, M. (2005) *Org. Lett.*, **7**, 4597–600.

11 Ji, J.-X., Wu, J., Au-Weng, T.T.-L., Yip, C.-W., Haynes, R.K. and Chan, A.S.C. (2005) *J. Org. Chem.*, **70**, 1093–5.

12 Kim, J.G. and Walsh, P.J. (2006) *Angew. Chem. Int. Ed.*, **45**, 4175–8.

13 (a) Sakai, M., Ueda, M. and Miyaura, N. (1998) *Angew. Chem. Int. Ed.*, **37**, 3279–81;
(b) Ueda, M. and Miyaura, N. (2000) *J. Org. Chem.*, **65**, 4450–2.

14 (a) Moreau, C., Hague, C., Weller, A.S. and Frost, C.G. (2001) *Tetrahedron Lett.*, **42**, 6957–60;
(b) Fürstner, A. and Krause, H. (2001) *Adv. Synth. Catal.*, **343**, 343–50;
(c) Imlinger, N., Mayr, M., Wang, D., Wurst, K. and Buchmeiser, M.R. (2004) *Adv. Synth. Cat.*, **346**, 1836–43;
(d) Pucheault, M., Darses, S. and Genêt, J.-P. (2005) *Chem. Commun.*, 4714–16;
(e) Son, S.U., Kim, S.B., Reingold, J.A., Carpenter, G.B. and Sweigart, D.A. (2005) *J. Am. Chem. Soc.*, **127**, 12238–9;

(f) Suzuki, K., Arao, T., Ishii, S., Maeda, Y., Kondo, K. and Aoyama, T. (2006) *Tetrahedron Lett.*, **47**, 5789–92;
(g) Qin, C., Wu, H., Cheng, J., Chen, X., Liu, M., Zhang, W., Su, W. and Ding, J. (2007) *J. Org. Chem.*, **72**, 4102–7.

15 Focken, T., Rudolph, J. and Bolm, C. (2005) *Synthesis*, 429–36.

16 Suzuki, K., Kondo, K. and Aoyama, T. (2006) *Synthesis*, 1360–4.

17 Jagt, R.B.C., Toullec, P.Y., de Vries, J. G., Feringa, B.L. and Minnaard, A.J. (2006) *Org. Biomol. Chem.*, **4**, 773–5.

18 Arao, T., Suzuki, K., Kondo, K. and Aoyama, T. (2006) *Synthesis*, 3809–14.

19 Duan, H.-F., Xie, J.-H., Shi, W.-J., Zhang, Q. and Zhou, Q.-L. (2006) *Org. Lett.*, **8**, 1479–81.

20 (a) Arao, T., Kondo, K. and Aoyama, T. (2007) *Tetrahedron Lett.*, **48**, 4115–17;
(b) Arao, T., Kondo, K. and Aoyama, T. (2007) *Tetrahedron*, **63**, 5261–4.

21 Dosa, P.I. and Fu, G.C. (1998) *J. Am. Chem. Soc.*, **120**, 445–6.

22 (a) Garcia, C. and Walsh, P.J. (2003) *Org. Lett.*, **5**, 3641–4;
(b) Prieto, O., Ramón, D.J. and Yus, M. (2003) *Tetrahedron: Asymmetry*, **14**, 1955–7.

23 Forrat, V.J., Prieto, O., Ramón, D.J. and Yus, M. (2006) *Chem. Eur. J.*, **12**, 4431–45.

24 Li, H., Garcia, C. and Walsh, P.J. (2004) *Proc. Natl Acad. Sci. USA*, **101**, 5425–7.

25 Wu, K.-H. and Gau, H.-M. (2006) *J. Am. Chem. Soc.*, **128**, 14808–9.

26 Chen, C.-A., Wu, K.-H. and Gau, H.-M. (2007) *Angew. Chem. Int. Ed.*, **46**, 5373–6.

27 Liu, G. and Lu, X. (2006) *J. Am. Chem. Soc.*, **128**, 16504–5.

28 Shintani, R., Inoue, M. and Hayashi, T. (2006) *Angew. Chem. Int. Ed.*, **45**, 3353–6.

29 Toullec, P.Y., Jagt, R.B.C., de Vries, J.G., Feringa, B.L. and Minnaard, A.J. (2006) *Org. Lett.*, **8**, 2715–18.

30 Martina, S.L.X., Jagt, R.B.C., de Vries, J.G., Feringa, B.L. and Minnaard, A.J. (2006) *Chem. Commun.*, 4093–5.

31 Hayashi, T. and Ishigedani, M. (2000) *J. Am. Chem. Soc.*, **122**, 976–7.

32 Hermanns, N., Dahmen, S., Bolm, C. and Bräse, S. (2002) *Angew. Chem. Int. Ed.*, **114**, 3692–4.

33 Kuriyama, M., Soeta, T., Hao, X., Chen, Q. and Tomioka, K. (2004) *J. Am. Chem. Soc.*, **126**, 8128–9.

34 Hayashi, T., Kawai, M. and Tokunaga, N. (2004) *Angew. Chem. Int. Ed.*, **43**, 6125–8.

35 Tokunaga, N., Otomaru, Y., Okamoto, K., Ueyama, K., Shintani, R. and Hayashi, T. (2004) *J. Am. Chem. Soc.*, **126**, 13584–5.

36 (a) Otomaru, Y., Tokunaga, N., Shintani, R. and Hayashi, T. (2005) *Org. Lett.*, **7**, 307–10;
(b) Otomaru, Y., Kina, A., Shintani, R. and Hayashi, T. (2005) *Tetrahedron: Asymmetry*, **16**, 1673–9.

37 Wang, Z.-Q., Feng, C.-G., Xu, M.-H. and Lin, G.-Q. (2007) *J. Am. Chem. Soc.*, **129**, 5336–7.

38 Jagt, R.B.C., Toullec, P.Y., Geerdink, D., de Vries, J.G., Feringa, B.L. and Minnaard, A.J. (2006) *Angew. Chem. Int. Ed.*, **45**, 2789–91.

39 Duan, H.-F., Jia, Y.-X., Wang, L.-X. and Zhou, Q.-L. (2006) *Org. Lett.*, **8**, 2567–9.

40 Weix, D.J., Shi, Y. and Ellman, J.A. (2005) *J. Am. Chem. Soc.*, **127**, 1092–3.

41 Bolshan, Y. and Batey, R.A. (2005) *Org. Lett.*, **7**, 1481–4.

42 Beenen, M.A., Weix, D.J. and Ellman, J. A. (2006) *J. Am. Chem. Soc.*, **128**, 6304–5.

43 (a) For general reviews, see: Alexakis, A. and Benhaim, C. (2002) *Eur. J. Org. Chem.*, 3221–36;
(b) López, F., Minnaard, A.J. and Feringa, B.L. (2007) *Acc. Chem. Res.*, **40**, 179–88.

44 Sakai, M., Hayashi, H. and Miyaura, N. (1997) *Organometallics*, **16**, 4229–31.

45 Takaya, Y., Ogasawara, M., Hayashi, T., Sakai, M. and Miyaura, N. (1998) *J. Am. Chem. Soc.*, **120**, 5579–80.

46 (a) For general reviews, see: Fagnou, K. and Lautens, M. (2003) *Chem. Rev.*, **103**, 169–96;
(b) Hayashi, T. and Yamasaki, K. (2003) *Chem. Rev.*, **103**, 2829–44;
(c) Hayashi, T. (2004) *Pure Appl. Chem.*, **76**, 465–75;
(d) Hayashi, T. (2004) *Bull. Chem. Soc. Jpn.*, **77**, 13–21;
(e) Yoshida, K. and Hayashi, T. (2005) *Modern Rhodium-Catalyzed Organic*

Reactions; (ed. P.A. Evans), Wiley-VCH Verlag GmbH, Weinheim, Germany, pp. 55–77;
(f) Yoshida, K. and Hayashi, T. (2005) *Boronic Acids* (ed. D.G. Hall), Wiley-VCH Verlag GmbH, Weinheim, Germany, pp. 171–203;
(g) Yamamoto, Y., Nishikata, T. and Miyaura, N. (2006) *J. Synth. Org. Chem. Jpn.*, **64**, 1112–21;
(h) Christoffers, J., Koripelly, G., Rosiak, A. and Rössle, M. (2007) *Synthesis*, 1279–300.

47 Hall, D.G. (ed.) (2005) *Boronic Acids: Preparation and Applications in Organic Synthesis and Medicine*, Wiley-VCH Verlag GmbH, Weinheim.

48 (a) For general reviews, see: Darses, S. and Genet, J.-P. (2003) *Eur. J. Org. Chem.*, 4313–27;
(b) Molander, G.A. and Ellis, N. (2007) *Acc. Chem. Res.*, **40**, 275–86;
(c) For selected applications, see: Pucheault, M., Darses, S. and Genet, J.-P. (2002) *Eur. J. Org. Chem*, 3552–7;
(d) Duursma, A., Boiteau, J.-G., Lefort, L., Boogers, J.A.F., de Vries, A.H.M., de Vries, J.G., Minnaard, A.J. and Feringa, B.L. (2004) *J. Org. Chem.*, **69**, 8045–52;
(e) Navarre, L., Pucheault, M., Darses, S. and Genet, J.-P. (2005) *Tetrahedron Lett.*, **46**, 4247–50.

49 (a) Hayashi, T., Tokunaga, N., Yoshida, K. and Han, J.W. (2002) *J. Am. Chem. Soc.*, **124**, 12102–3;
(b) Tokunaga, N., Yoshida, K. and Hayashi, T. (2004) *Proc. Natl Acad. Sci. USA*, **101**, 5445–9.

50 (a) Oi, S., Taira, A., Honma, Y. and Inoue, Y. (2003) *Org. Lett.*, **5**, 97–9;
(b) Oi, S., Taira, A., Honma, Y., Sato, T. and Inoue, Y. (2006) *Tetrahedron: Asymmetry*, **17**, 598–602.

51 (a) Shintani, R., Tokunaga, N., Doi, H. and Hayashi, T. (2004) *J. Am. Chem. Soc.*, **126**, 6240–1;
(b) Kina, A., Ueyama, K. and Hayashi, T. (2005) *Org. Lett.*, **7**, 5889–92;
(c) Shintani, R., Yamagumi, T., Kimura, T. and Hayashi, T. (2005) *Org. Lett.*, **7**, 5317–19.

52 (a) Oi, S., Sato, T. and Inoue, Y. (2004) *Tetrahedron Lett.*, **45**, 5051–5;
(b) Kakuuchi, A., Taguchi, T. and Hanzawa, Y. (2004) *Tetrahedron*, **60**, 1293–9.

53 (a) Huang, T., Meng, Y., Venkatraman, S., Wang, D. and Li, C.-J. (2001) *J. Am. Chem. Soc.*, **123**, 7451–2;
(b) Oi, S., Moro, M., Ito, H., Honma, Y., Miyano, S. and Inoue, Y. (2002) *Tetrahedron*, **58**, 91–7.

54 Miura, T. and Murakami, M. (2005) *Chem. Commun.*, 5676–7.

55 Nishimura, T., Katoh, T. and Hayashi, T. (2007) *Angew. Chem. Int. Ed.*, **46**, 4937–9.

56 (a) Zhao, P., Incarvito, C.D. and Hartwig, J.F. (2006) *J. Am. Chem. Soc.*, **128**, 3124–5;
(b) For studies on transmetalation, see: Zhao, P., Incarvito, C.D. and Hartwig, J.F. (2007) *J. Am. Chem. Soc.*, **129**, 1876–7.

57 Kuriyama, M., Nagai, K., Yamada, K.-i., Miwa, Y., Taga, T. and Tomioka, K. (2002) *J. Am. Chem. Soc.*, **124**, 8932–9.

58 Becht, J.-M., Bappert, E. and Helmchen, G. (2005) *Adv. Synth. Cat.*, **347**, 1495–8.

59 Ma, Y., Song, C., Ma, C., Sun, Z., Chai, Q. and Andrus, M.B. (2003) *Angew. Chem. Int. Ed.*, **42**, 5871–4.

60 Shi, Q., Xu, L., Li, X., Jia, X., Wang, R., Au-Yueung, T.T.-L., Chan, A.S.C., Hayashi, T., Cao, R. and Hong, M. (2003) *Tetrahedron Lett.*, **44**, 6505–8.

61 Imamoto, T., Sugita, K. and Yoshida, K. (2005) *J. Am. Chem. Soc.*, **127**, 11934–5.

62 Vandyck, K., Matthys, B., Willen, M., Robeyns, K., van Meervelt, L. and van der Eycken, J. (2006) *Org. Lett.*, **8**, 363–6.

63 Stemmler, R.T. and Bolm, C. (2005) *J. Org. Chem.*, **70**, 9925–31.

64 Martina, S.L.X., Minnaard, A.J., Hessen, B. and Feringa, B.L. (2005) *Tetrahedron Lett.*, **46**, 7159–63.

65 (a) Duursma, A., Peña, D., Minnaard, A.J. and Feringa, B.L. (2005) *Tetrahedron: Asymmetry*, **16**, 1901–4;
(b) Monti, C., Gennari, C. and Piarulli, U. (2007) *Chem. Eur. J.*, **13**, 1547–58.

66 (a) For general reviews, see: Glorius, F. (2004) *Angew. Chem. Int. Ed.*, **43**, 3364–6;
(b) Johnson, J.B. and Rovis, T. (2008) *Angew. Chem. Int. Ed.*, **47**, 840–71;
(c) Defieber, C., Grützmacher, H. and Carreira, E.M. (2008) *Angew. Chem. Int. Ed.*, **47**, 4482–502.

67 Itooka, R., Iguchi, Y. and Miyaura, N. (2001) *Chem. Lett.*, **30**, 722–3.

68 (a) Hayashi, T., Ueyama, K., Tokunaga, N. and Yoshida, K. (2003) *J. Am. Chem. Soc.*, **125**, 11508–9;
(b) Berthon-Gelloz, G. and Hayashi, T. (2006) *J. Org. Chem.*, **71**, 8957–60.

69 Otomaru, Y., Okamoto, K., Shintani, R. and Hayashi, T. (2005) *J. Org. Chem.*, **70**, 2503–8.

70 Defieber, C., Paquin, J.-F., Serna, S. and Carreira, E.M. (2004) *Org. Lett.*, **6**, 3873–6.

71 Läng, F., Breher, F., Stein, D. and Grützmacher, H. (2005) *Organometallics*, **24**, 2997–3007.

72 Chen, F.-X., Kina, A. and Hayashi, T. (2006) *Org. Lett.*, **8**, 341–4.

73 (a) Shintani, R., Duan, W.-L., Nagano, T., Okada, A. and Hayashi, T. (2005) *Angew. Chem. Int. Ed.*, **44**, 4611–14;
(b) Shintani, R., Duan, W.-L., Okamoto, K. and Hayashi, T. (2005) *Tetrahedron: Asymmetry*, **16**, 3400–5.

74 Piras, E., Läng, F., Rüegger, H., Stein, D., Wörle, M. and Grützmacher, H. (2006) *Chem. Eur. J.*, **12**, 5849–58.

75 Kasák, P., Arion, V.B. and Widhalm, M. (2006) *Tetrahedron: Asymmetry*, **17**, 3084–90.

76 Duan, W.-L., Iwamura, H., Shintani, R. and Hayashi, T. (2007) *J. Am. Chem. Soc.*, **129**, 2130–8.

77 (a) Nishikata, T., Yamamoto, Y. and Miyaura, N. (2003) *Angew. Chem. Int. Ed.*, **42**, 2768–70;
(b) Nishikata, T., Yamamoto, Y. and Miyaura, N. (2004) *Organometallics*, **23**, 4317–24.

78 Nishikata, T., Yamamoto, Y., Gridnev, I.D. and Miyaura, N. (2005) *Organometallics*, **24**, 5025–32.

79 Gini, F., Hessen, B. and Minnaard, A.J. (2005) *Org. Lett.*, **7**, 5309–12.

80 Gini, F., Hessen, B., Feringa, B.L. and Minnaard, A.J. (2007) *Chem. Commun.*, 710–12.

81 Yamamoto, T., Ohta, T. and Ito, Y. (2005) *Org. Lett.*, **7**, 4153–5.

82 Yamamoto, T., Iizuka, M., Ohta, T. and Ito, Y. (2006) *Chem. Lett.*, **35**, 198–9.

83 (a) Suzuma, Y., Yamamoto, T., Ohta, T. and Ito, Y. (2007) *Chem. Lett.*, **36**, 470–1;
(b) He, P., Lu, Y., Duo, C.-G. and Hu, Q.-S. (2007) *Org. Lett.*, **9**, 343–6.

84 Paquin, J.-F., Defieber, C., Stephenson, C.R.J. and Carreira, E.M. (2005) *J. Am. Chem. Soc.*, **127**, 10850–1.

85 Hayashi, T., Tokunaga, N., Okamoto, K. and Shintani, R. (2005) *Chem. Lett.*, **34**, 1480–1.

86 Tokunaga, N. and Hayashi, T. (2006) *Tetrahedron: Asymmetry*, **17**, 607–13.

87 Nishikata, T., Yamamoto, Y. and Miyaura, N. (2007) *Tetrahedron Lett.*, **48**, 4007–10.

88 Shintani, R., Ueyama, K., Yamada, I. and Hayashi, T. (2004) *Org. Lett.*, **6**, 3425–7.

89 Shintani, R., Duan, W.-L. and Hayashi, T. (2006) *J. Am. Chem. Soc.*, **128**, 5628–9.

90 Duan, W.-L., Imazaki, Y., Shintani, R. and Hayashi, T. (2007) *Tetrahedron*, **63**, 8529–36.

91 Paquin, J.-F., Stephenson, C.R.J., Defieber, C. and Carreira, E.M. (2005) *Org. Lett.*, **7**, 3821–4.

92 Shintani, R., Kimura, T. and Hayashi, T. (2005) *Chem. Commun.*, 3213–14.

93 Shintani, R., Okamoto, K. and Hayashi, T. (2005) *Org. Lett.*, **7**, 4757–9.

94 Tokunaga, N. and Hayashi, T. (2007) *Adv. Synth. Cat.*, **349**, 513–16.

95 Shintani, R., Tokunaga, N., Doi, H. and Hayashi, T. (2004) *J. Am. Chem. Soc.*, **126**, 6240–1.

96 Shintani, R., Yamagami, T., Kimura, T. and Hayashi, T. (2005) *Org. Lett.*, **7**, 5317–19.

97 Jagt, R.B.C., de Vries, J.G., Feringa, B.L. and Minnaard, A.J. (2005) *Org. Lett.*, **7**, 2433–5.

98 Chen, G., Tokunaga, N. and Hayashi, T. (2005) *Org. Lett.*, **7**, 2285–8.

99 Chen, Q., Kuriyama, M., Soeta, T., Hao, X., Yamada, K.-i. and Tomioka, K. (2005) *Org. Lett.*, **7**, 4439–41.

100 Urbaneja, L.M., Alexakis, A. and Krause, N. (2002) *Tetrahedron Lett.*, **43**, 7887–90.

101 (a) Chen, Q., Soeta, T., Kuriyama, M., Yamada, K.-i. and Tomioka, K. (2006) *Adv. Synth. Cat.*, **348**, 2604–8;
(b) For a related study, see: Urbaneja, L.M. and Krause, N. (2006) *Tetrahedron: Asymmetry*, **17**, 494–6.

102 Lee, K.-S., Brown, M.K., Hird, A.W. and Hoveyda, A.H. (2006) *J. Am. Chem. Soc.*, **128**, 7182–4.

103 Brown, M.K., May, T.L., Baxter, C.A. and Hoveyda, A.H. (2007) *Angew. Chem. Int. Ed.*, **46**, 1097–100.

104 Hayashi, T., Tokunaga, N. and Inoue, K. (2004) *Org. Lett.*, **6**, 305–7.

105 Hayashi, T., Yamamoto, S. and Tokunaga, N. (2005) *Angew. Chem. Int. Ed.*, **44**, 4224–7.

106 Nishimura, T., Yasuhara, Y. and Hayashi, T. (2006) *Angew. Chem. Int. Ed.*, **45**, 5164–6.

107 de la Herrán, G., Murcia, C. and Csáky, A.G. (2005) *Org. Lett.*, **7**, 5629–32.

108 (a) For general reviews, see: Guo, H.-C. and Ma, J.-A. (2006) *Angew. Chem. Int. Ed.*, **45**, 354–66;
(b) Chapman, C.J. and Frost, C.G. (2007) *Synthesis*, 1–21.

109 Reetz, M.T., Moulin, D. and Gosberg, A. (2001) *Org. Lett.*, **3**, 4083–5.

110 Navarre, L., Darses, S. and Genêt, J.-P. (2004) *Eur. J. Org. Chem.*, 69–73.

111 Navarre, L., Darses, S. and Genêt, J.-P. (2004) *Angew. Chem. Int. Ed.*, **43**, 719–23.

112 Moss, R.J., Wadsworth, K.J., Chapman, C.J. and Frost, C.G. (2004) *Chem. Commun.*, 1984–5.

113 Hargrave, J.D., Herbert, J., Bish, G. and Frost, C.G. (2006) *Org. Biomol. Chem.*, **4**, 3235–41.

114 Hargrave, J.D., Bish, G. and Frost, C.G. (2006) *Chem. Commun.*, 4389–91.

115 Frost, C.G., Penrose, S.D., Lambshead, K., Raithby, P.R., Warren, J.E. and Gleave, R. (2007) *Org. Lett.*, **9**, 2119–22.

116 Sibi, M.P., Tatamidani, H. and Patil, K. (2005) *Org. Lett.*, **7**, 2571–3.

117 (a) Yoshida, K., Ogasawara, M. and Hayashi, T. (2002) *J. Am. Chem. Soc.*, **124**, 10984–5;
(b) Yoshida, K., Ogasawara, M. and Hayashi, T. (2003) *J. Org. Chem.*, **68**, 1901–5.

118 Cauble, D.F., Gipson, J.D. and Krische, M.J. (2003) *J. Am. Chem. Soc.*, **125**, 1110–11.

119 Bocknack, B.M., Wang, L.-C. and Krische, M.J. (2004) *Proc. Natl Acad. Sci. USA*, **101**, 5421–4.

120 Källström, S., Jagt, R.B.C., Sillanpää, R., Feringa, B.L., Minnaard, A.J. and Leino, R. (2006) *Eur. J. Org. Chem.*, 3826–33.

121 (a) For a review, see: Miura, T. and Murakami, M. (2007) *Chem. Commun.*, 217–24;
(b) Some notable racemic examples include: Lautens, M. and Marquardt, T. (2004) *J. Org. Chem.*, **69**, 4607–14;
(c) Shintani, R., Okamoto, K. and Hayashi, T. (2005) *Chem. Lett.*, **34**, 1294–5;
(d) Matsuda, T., Makino, M. and Murakami, M. (2005) *Chem. Lett.*, **34**, 1416–17;
(e) Miura, T. and Murakami, M. (2005) *Org. Lett.*, **7**, 3339–41;
(f) Chen, Y. and Lee, C. (2006) *J. Am. Chem. Soc.*, **128**, 15598–9;
(g) Miura, T., Harumashi, T. and Murakami, M. (2007) *Org. Lett.*, **9**, 741–3.

122 Shintani, R., Okamoto, K., Otomaru, Y., Ueyama, K. and Hayashi, T. (2005) *J. Am. Chem. Soc.*, **127**, 54–5.

123 Shintani, R., Tsurusaki, A., Okamoto, K. and Hayashi, T. (2005) *Angew. Chem. Int. Ed.*, **44**, 3909–12.

124 Shintani, R., Takatsu, K. and Hayashi, T. (2007) *Angew. Chem. Int. Ed.*, **46**, 3735–7.

125 Ma, S. and Gu, Z. (2005) *Angew. Chem. Int. Ed.*, **44**, 7512–17.

126 For general reviews, see: Jørgensen, K.A. (2003) *Synthesis*, 1117–25;
(b) Bandini, M., Melloni, A. and Umani-Ronchi, A. (2004) *Angew. Chem. Int. Ed.*, **43**, 550–6.

127 (a) Johnson, J.S. and Evans, D.A. (2000) *Acc. Chem. Res.*, **33**, 325–35;
(b) Desimoni, G., Faita, G. and Jørgensen, K.A. (2006) *Chem. Rev.*, **106**, 3561–651.

128 Gathergood, N., Zhuang, W. and Jørgensen, K.A. (2000) *J. Am. Chem. Soc.*, **122**, 12517–22.

129 Erker, G. and Van der Zeijden, A.A.H. (1990) *Angew. Chem. Int. Ed.*, **29**, 512–14.

130 (a) Zhuang, W., Gathergood, N., Hazell, R.G. and Jørgensen, K.A. (2001) *J. Org. Chem.*, **66**, 1009–13;
(b) Corma, A., García, H., Moussaif, A., Sabater, M.J., Zniber, R. and Redouane, A. (2002) *Chem. Commun.*, 1058–9;
(c) Lyle, M.P.A., Draper, N.D. and Wilson, P.D. (2005) *Org. Lett.*, **7**, 901–4;
(d) Zhao, J.-L., Liu, L., Sui, Y., Liu, Y.-L., Wang, D. and Chen, Y.-J. (2006) *Org. Lett.*, **8**, 6127–30.

131 Ishii, A., Soloshonok, V.A. and Mikami, K. (2000) *J. Org. Chem.*, **65**, 1597–9.

132 (a) Johannsen, M. (1999) *Chem. Commun.*, 2233–4;
(b) Saaby, S., Fang, X., Gathergood, N. and Jørgensen, K.A. (2000) *Angew. Chem. Int. Ed.*, **39**, 4114–16;
(c) Saaby, S., Bayón, P., Aburel, P.S. and Jørgensen, K.A. (2002) *J. Org. Chem.*, **67**, 4352–61;
(d) Shirakawa, S., Berger, R. and Leighton, J.L. (2005) *J. Am. Chem. Soc.*, **127**, 2858–9.

133 (a) Ramesh, C., Banerjee, J., Pal, R. and Das, B. (2003) *Adv. Synth. Catal.*, **345**, 557–9;
(b) For the synthesis of unsymmetrical diaryl amines and triarylmethanes see: Esquivias, J., Arrayás, R.G. and Carretero, J.C. (2006) *Angew. Chem. Int. Ed.*, **45**, 629–33.

134 Jia, Y.-X., Xie, J.-H., Duan, H.-F., Wang, L.-X. and Zhou, Q.-L. (2006) *Org. Lett.*, **8**, 1621–4.

135 Jensen, K.B., Thorhauge, J., Hazell, R. G. and Jørgensen, K.A. (2001) *Angew. Chem. Int. Ed.*, **40**, 160–3.

136 Palomo, C., Oiarbide, M., Kardak, B.G., Garcia, J.M. and Linden, A. (2005) *J. Am. Chem. Soc.*, **127**, 4154–5.

137 Evans, D.A., Scheidt, K.A., Fandrick, K.R., Lam, H.W. and Wu, J. (2003) *J. Am. Chem. Soc.*, **125**, 10780–1.

138 (a) Evans, D.A., Fandrick, K.R. and Song, H.J. (2005) *J. Am. Chem. Soc.*, **127**, 8942–3;
(b) Evans, D.A. and Fandrick, K.R. (2006) *Org. Lett.*, **8**, 2249–52;
(c) Evans, D.A., Fandrick, K.R., Song, H.J., Scheidt, K.A. and Xu, R. (2007) *J. Am. Chem. Soc.*, **129**, 10029–41.

139 Zhou, J. and Tang, Y. (2002) *J. Am. Chem. Soc.*, **124**, 9030–1.

140 (a) For general reviews, see: Berkessel, A. and Gröger, H. (2005) *Asymmetric Organocatalysis*. Wiley-VCH Verlag GmbH, Weinheim;
(b) Dalko, P.I. (2007) *Enantioselective Organocatalysis*, Wiley-VCH Verlag GmbH, Weinheim;
(c) Akiyama, T. (2007) *Chem. Rev.*, **107**, 5744–58.

141 Li, H., Wang, Y.-Q. and Deng, L. (2006) *Org. Lett.*, **8**, 4063–5.

142 Wang, Y.-Q., Song, J., Hong, R. and Deng, H.L.i, L. (2006) *J. Am. Chem. Soc.*, **128**, 8156–7.

143 Kang, Q., Zhao, Z.-A. and You, S.-L. (2007) *J. Am. Chem. Soc.*, **129**, 1484–5.

144 Rowland, G.B., Rowland, E.B., Liang, Y., Perman, J.A. and Antilla, J.C. (2007) *Org. Lett.*, **9**, 2609–11.

145 Terada, M. and Sorimachi, K. (2007) *J. Am. Chem. Soc.*, **129**, 292–3.

146 Jia, Y.-X., Zhang, J., Zhu, S.-F., Zhang, C.-M. and Zhou, Q.-L. (2007) *Angew. Chem. Int. Ed.*, **46**, 5565–7.

147 Uraguchi, D., Sorimachi, K. and Terada, M. (2004) *J. Am. Chem. Soc.*, **126**, 11804–5.

148 Paras, N.A. and MacMillan, D.W.C. (2001) *J. Am. Chem. Soc.*, **123**, 4370–1.

149 Austin, J.F. and MacMillan, D.W.C. (2002) *J. Am. Chem. Soc.*, **124**, 1172–3.

150 Paras, N.A. and MacMillan, D.W.C. (2002) *J. Am. Chem. Soc.*, **124**, 7894–5.

151 Lelais, G. and MacMillan, D.W.C. (2006) *Aldrichim. Acta*, **39**, 79–86.

152 (a) Pederson, R.L., Fellows, I.M., Ung, T.A., Ishihara, H. and Hajela, S.P. (2002) *Adv. Synth. Cat.*, **344**, 728–35;
(b) King, H.D., Meng, Z., Denhart, D., Mattson, R., Kimura, R., Wu, D., Gao, Q. and Macor, J.E. (2005) *Org. Lett.*, **7**, 3437–40;
(c) Kim, S.-G., Kim, J. and Jung, H. (2005) *Tetrahedron Lett.*, **46**, 2437–9.

153 Chen, W., Du, W., Yue, L., Li, R., Wu, Y., Ding, L.-S. and Chen, Y.-C. (2007) *Org. Biomol. Chem.*, **5**, 816–21.

154 Bartoli, G., Bosco, M., Carlone, A., Pesciaioli, F., Sambri, L. and Melchiorre, P. (2007) *Org. Lett.*, **9**, 1403–5.

9
Metal-Catalyzed Direct Arylations (excluding Palladium)

Lutz Ackermann and Rubén Vicente

9.1
Introduction

Biaryls are substructures of various compounds with activities of relevance to organic synthesis, biology or material sciences. Their formation relies heavily on transition metal-catalyzed cross-coupling reactions, which have matured to become one of the most powerful transformations for the formation of C(sp^2)—C(sp^2) bonds [1–3]. Generally, these reactions make use of regioselective couplings between aryl (pseudo)halides as electrophiles and organometallic reagents as nucleophiles (Scheme 9.1a). However, the organometallic reagents – particularly when being functionalized – are often not commercially available or are expensive. Their synthesis from the corresponding arenes requires the use of a number of reactions, during which undesired byproducts are formed. Moreover, the use of organometallic reagents as nucleophilic starting materials in stoichiometric amounts also gives rise to the formation of further byproducts.

Regioselective intermolecular direct arylations represent economically attractive and ecologically benign alternatives to 'traditional' cross-coupling methodologies [4–11]. Here, unfunctionalized (hetero)arenes are directly employed as the starting materials, and functionalized through C—H bond cleavages (Scheme 9.1b). Overall, this strategy is not only advantageous with respect to a reduction of byproduct formation [12], but also allows for a minimization of reaction steps.

Since organic pronucleophiles often display various C—H bonds with comparable dissociation energies, a major challenge constitutes the development of highly regioselective direct arylations for synthetically useful transformations. It is noteworthy, that achieving regioselectivities in intermolecular oxidative arylations through the coupling of two unfunctionalized (hetero)arenes (Scheme 9.1c) constitutes an even more demanding objective.

The regioselectivities of direct arylations can be controlled when the electronic properties of a given arene dominate its reactivity. This holds true for a variety of heteroarenes, and has enabled highly regioselective palladium-catalyzed direct functionalizations of electron-rich, as well as electron-deficient arenes (these are

Modern Arylation Methods. Edited by Lutz Ackermann

ISBN: 978-3-527-31937-4

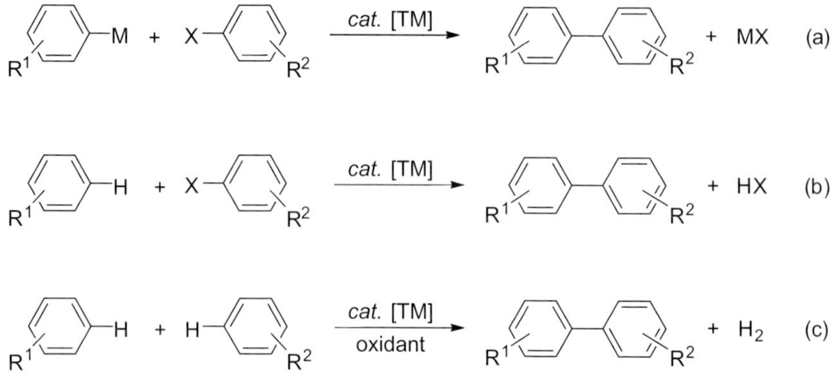

Scheme 9.1 Catalytic strategies for the synthesis of substituted biaryls.

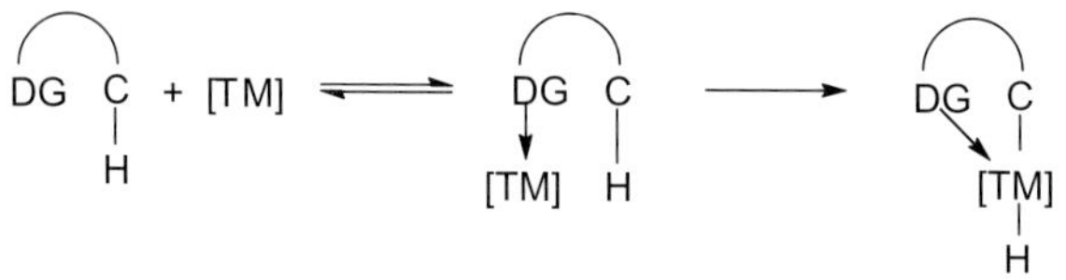

Scheme 9.2 Regioselective intermolecular C—H bond functionalizations through the use of directing groups (DG).

discussed in Chapter 10). However, electronically unbiased arenes often lead to unselective processes, yielding undesired mixtures of regioisomers, which are often difficult to purify. Consequently, strategies have been developed which make use of potentially removable directing groups (DG's). Here, a Lewis-basic functionality coordinates to a transition metal complex, which sets the stage for an intramolecular C—H bond cleavage (Scheme 9.2) [13], such that regioselective direct arylations can be accomplished in an overall intermolecular fashion [7].

Although the direct arylation reactions catalyzed by transition metal complexes (other than palladium) are discussed in this chapter, the corresponding palladium-catalyzed transformations are summarized in Chapter 10. A detailed overview of the mechanistic aspects of metal-catalyzed direct arylations is provided in Chapter 11.

9.2 Rhodium-Catalyzed Direct Arylations

9.2.1 Rhodium-Catalyzed Direct Arylations of Arenes

Rhodium-catalyzed direct arylations of 2-aryl pyridines were efficiently accomplished with arylstannanes through chelation-assistance (Scheme 9.3) [14]. This report constitutes an early example of a metal-catalyzed direct arylation with an

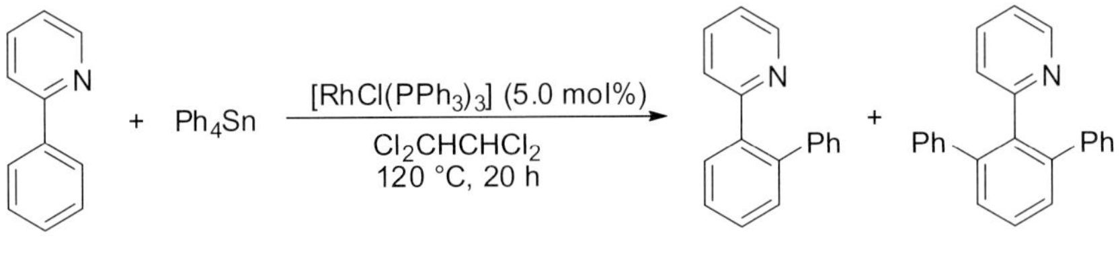

Scheme 9.3 Rhodium-catalyzed direct arylation with stannane **2**.

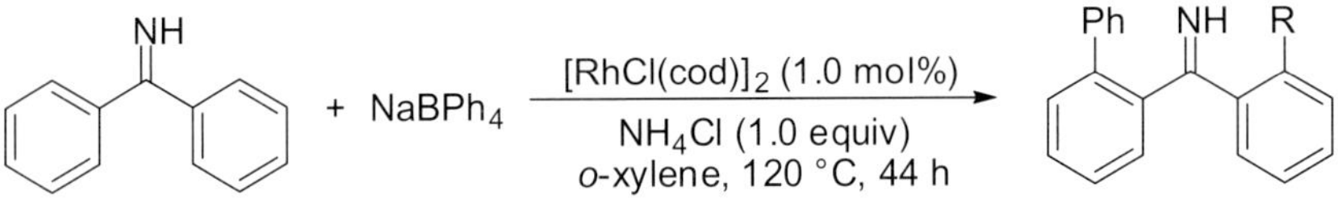

Scheme 9.4 Rhodium-catalyzed direct arylation of imine **5** with boronate **6**.

organometallic reagent. Although detailed information on the working mode of the catalyst was not reported, it is noteworthy that Cl_2CH_2–CH_2Cl_2 was found mandatory for achieving catalytic turnover. Remarkably, during arylations the formation of Cl_2CH=$CHCl_2$ was observed, which suggested that Cl_2CH_2–CH_2Cl_2 acts both as a solvent, as well as an oxidizing reagent [15].

During studies on rhodium-catalyzed Suzuki–Miyaura cross-coupling reactions, Miura and coworkers reported more recently on the use of less-toxic tetraphenylborate **6** for the direct arylation of imines (Scheme 9.4) [16]. Unfortunately, rather low yields of mono- and di-arylated products were obtained, this being due to a reduction of the starting material via a sequence consisting of a rhodium hydride addition and subsequent protonation. The reduction of the imine is mandatory for the regeneration of a rhodium chloride species, and thereby for catalytic turnover.

Highly efficient rhodium-catalyzed direct arylations were accomplished through the use of 2,2′,6,6′-tetramethylpiperidine-*N*-oxyl (TEMPO) as terminal oxidant [17]. Thereby, a variety of pyridine-substituted arenes was regioselectively functionalized with aromatic boronic acids (Scheme 9.5). However, in order for efficient catalysis to proceed, 4 equiv. of TEMPO were required. The use of molecular oxygen as terminal oxidant yielded, unfortunately, only unsatisfactory results under otherwise identical reaction conditions. However, a variety of easily available boronic acids could be employed as arylating reagents.

Additionally, the protocol proved not only applicable to aryl-substituted organometallics, but also enabled the use of an alkenyl boronic acid. Further, aldimine **11** was regioselectively arylated with this catalytic system, yielding – after subsequent hydrolysis – aldehyde **12** (Scheme 9.6) [17].

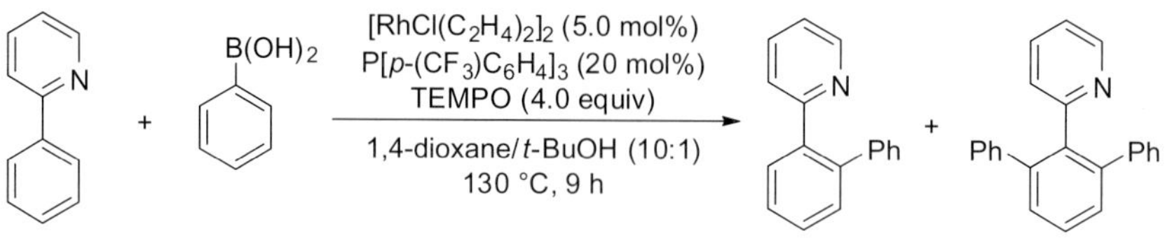

1 + 10 → ($[RhCl(C_2H_4)_2]_2$ (5.0 mol%), $P[p-(CF_3)C_6H_4]_3$ (20 mol%), TEMPO (4.0 equiv); 1,4-dioxane/*t*-BuOH (10:1), 130 °C, 9 h) → 3 (50%) + 4 (18%)

Scheme 9.5 Rhodium-catalyzed direct arylation of pyridine **1** with boronic acid **10**.

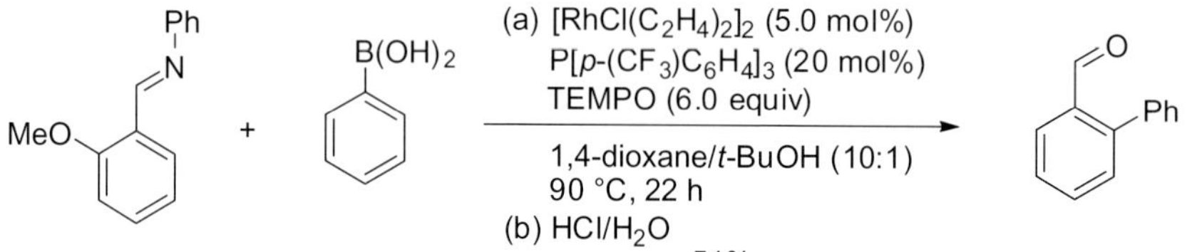

11 + 10 → ((a) $[RhCl(C_2H_4)_2]_2$ (5.0 mol%), $P[p-(CF_3)C_6H_4]_3$ (20 mol%), TEMPO (6.0 equiv), 1,4-dioxane/*t*-BuOH (10:1), 90 °C, 22 h; (b) HCl/H_2O) → 12 (51%)

Scheme 9.6 Rhodium-catalyzed direct arylation of aldimine **11** with boronic acid **10**.

A mechanism proposed for the working mode of this catalyst relies on the conversion of rhodium(I) amino-alkoxide **13** to the corresponding aryl rhodium(I) complex **14** through transmetallation with the boronic acid. Thereafter, 2 equiv. of TEMPO oxidize the generated rhodium(I) species **14**, giving rise to rhodium(III) complex **15**. This sets the stage for a chelation-assisted C—H bond cleavage, yielding regioselectively cyclometallated rhodium(III) complex **17** via intermediate **16**. Finally, reductive elimination from rhodium(III) species **17** liberates the desired arylation product, and regenerates catalytically active rhodium(I) complex **13** (Scheme 9.7).

Rhodium-catalyzed direct arylations with organic halides as electrophiles were accomplished with Wilkinson's catalyst modified with a phosphinite additive. Thereby, Bedford and coworkers developed a chelation-assisted arylation of phenols with aryl bromides (Scheme 9.8) [18]. Notably, the methodology proved capable of tolerating valuable functional groups, including heteroaromatic bromides, as electrophiles. The mechanism of this transformation is based on an *ortho*-metallation of the corresponding phosphinite and its subsequent arylation. Therefore, a trans-esterification of the phosphinite derived from the desired product by the phenol starting material is mandatory for achieving catalytic turnover (for a detailed discussion regarding the mechanism of this reaction, see Chapter 11). As a result, the corresponding phosphinite cocatalyst must be prepared for each individual substrate prior to catalysis in order to avoid the formation of undesired and difficult-to-separate byproducts.

This problem was addressed independently by Oi, Inoue and coworkers [19], as well as by Bedford and colleagues [20], through the use of economical (but toxic) $P(NMe_2)_3$ as cocatalyst. Thus, a catalytic system consisting of $[RhCl(cod)]_2$ and $P(NMe_2)_3$ proved complementary to the previously reported phosphinite-based

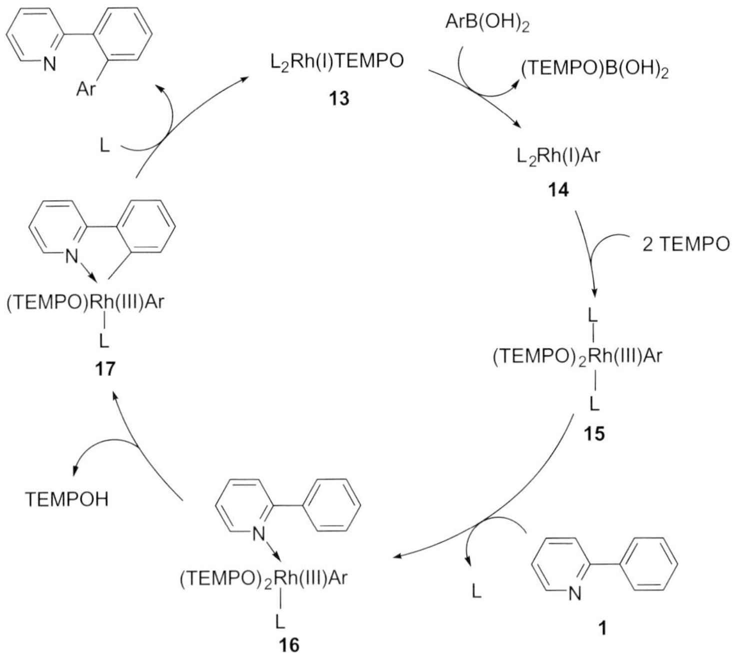

Scheme 9.7 Proposed mechanism for rhodium-catalyzed direct arylations with boronic acids (L=P[*p*-$(CF_3)C_6H_4]_3$).

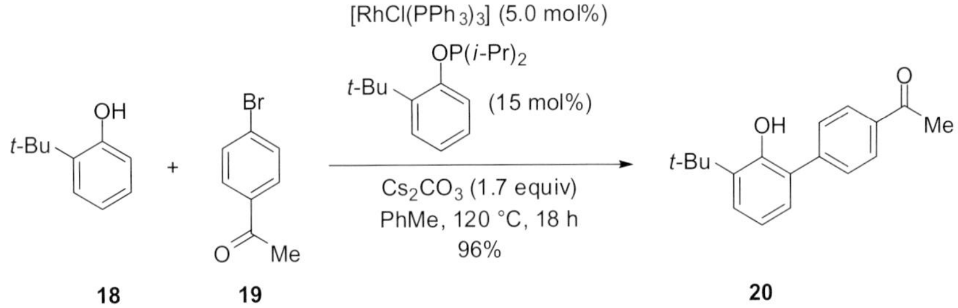

Scheme 9.8 Rhodium-catalyzed direct arylation of phenol **18** with bromide **19**.

catalyst, allowing for efficient functionalizations of phenol substrates displaying no *ortho*-substituents (Scheme 9.9).

As a less-toxic alternative, chlorophosphines [21] were more recently probed as preligands in rhodium-catalyzed direct arylations of phenols [22]. Among a variety of chlorophosphine preligands, *i*-Pr_2PCl (**24**) gave the most efficient catalysis, enabling a significant reduction of catalyst loading (Scheme 9.10). However, satisfactory efficacy was only obtained for *ortho*-substituted phenols, and C–H bond functionalization of phenol itself proceeded with a low yield of isolated product.

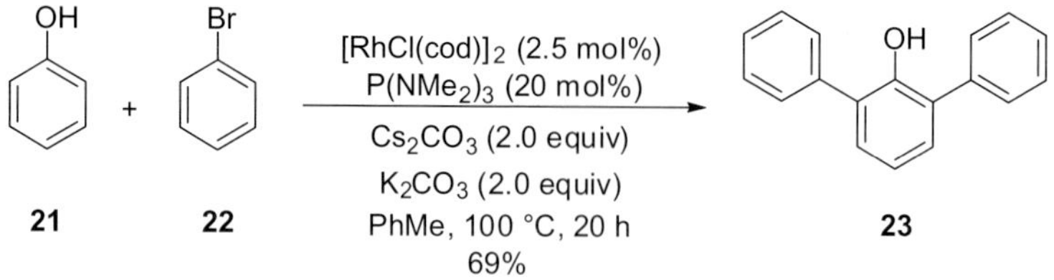

Scheme 9.9 Rhodium-catalyzed direct arylation of phenol (**21**).

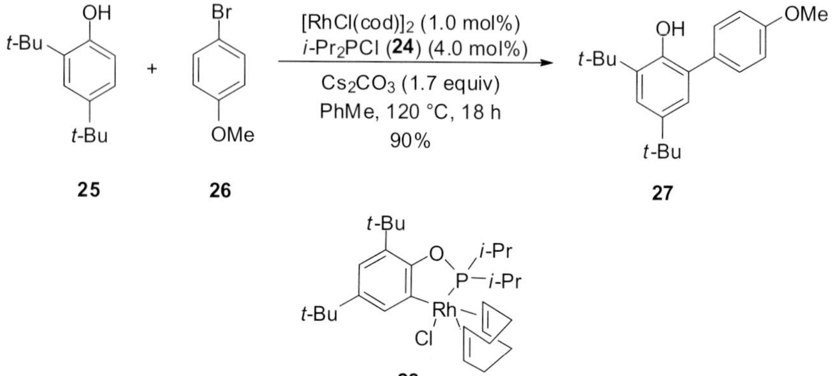

Scheme 9.10 Chlorophosphine **24** as preligand in a rhodium-catalyzed direct arylation.

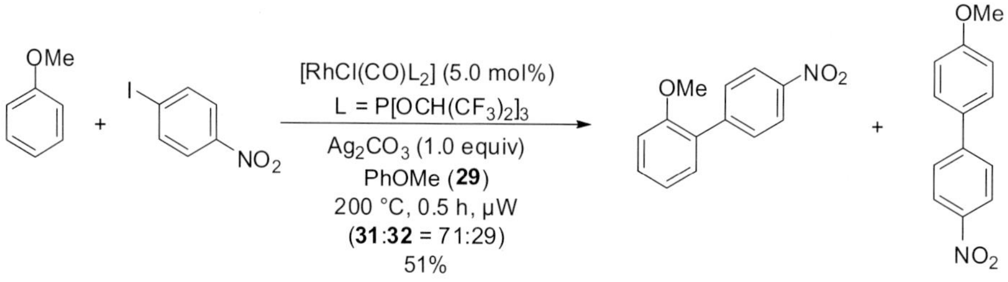

Scheme 9.11 Rhodium-catalyzed direct arylation of anisole (**29**) with aryl iodide **30**.

Notably, an independently synthesized air-stable rhodium complex derived from preligand **24** was shown to be catalytically competent; furthermore, spectroscopic data indicated the formation of complex **28**. Therefore, a mechanism was proposed that proceeded via an *in situ* generation of the corresponding phosphinite, with subsequent cyclometallation.

Rhodium-catalyzed direct arylations of arenes can be achieved in the absence of any directing groups. Thus, a rhodium complex displaying a strong π-accepting ligand $[P(OCH(CF_3)_2]_3$ led to a direct arylation of anisole (**29**) using electron-deficient aryl iodide **30** under microwave irradiation (Scheme 9.11) [23]. As a

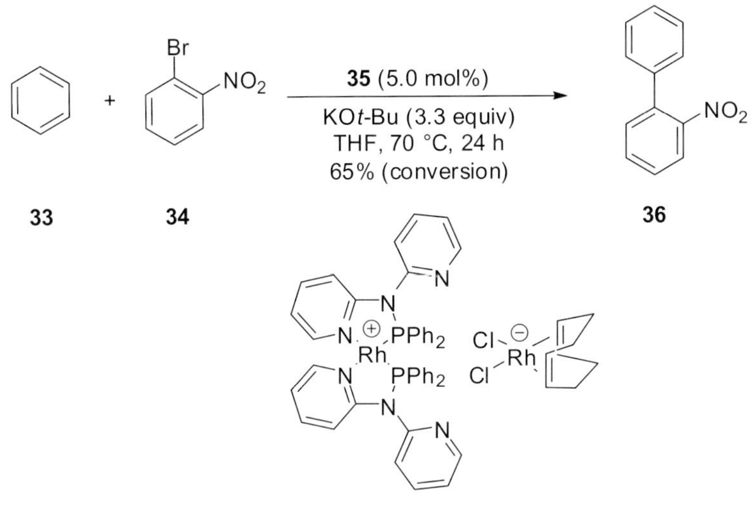

33 + 34 → 36

35 (5.0 mol%)
KO*t*-Bu (3.3 equiv)
THF, 70 °C, 24 h
65% (conversion)

35

Scheme 9.12 Homobimetallic rhodium complex **35** for a catalytic direct arylation.

strongly Lewis-basic directing group was not present in the pronucleophile, a mixture of *ortho*- and *para*-substitution products was obtained. The observed regioselectivity suggested an electrophilic substitution-type mechanism for this rhodium-catalyzed transformation.

A homobimetallic rhodium catalyst derived from a *P,N*-ligand was found to allow for intermolecular direct arylations of unfunctionalized arenes [24]. Interestingly, aryl iodides, bromides–and even chlorides–could be employed as electrophiles, and a variety of valuable functional groups was tolerated by the catalytic system (Scheme 9.12). The C–H bond functionalization of toluene yielded *ortho*-, *meta*- and *para*-substituted regioisomers in a ratio of 71 : 19 : 10. Based on this observation and a Hammett correlation, a mechanism proceeding through radical intermediates was suggested.

9.2.2
Rhodium-Catalyzed Direct Arylations of Heteroarenes

Intermolecular direct arylations of heteroarenes, such as indoles, pyrroles or (benzo)furans, were, thus far, predominantly achieved with palladium catalysts (see Chapter 10). However, rhodium complexes proved also competent for the direct functionalizations of various valuable heteroarenes with comparable or, in some cases, improved catalytic performance. Thus, rhodium-catalyzed C–H bond functionalizations of various *N*-heterocycles, were elegantly developed by Bergman, Ellman and coworkers. Here, the use of a catalytic system comprising $[RhCl(coe)_2]_2$ and PCy_3 led to direct arylations of unprotected benzimidazoles with aryl iodides

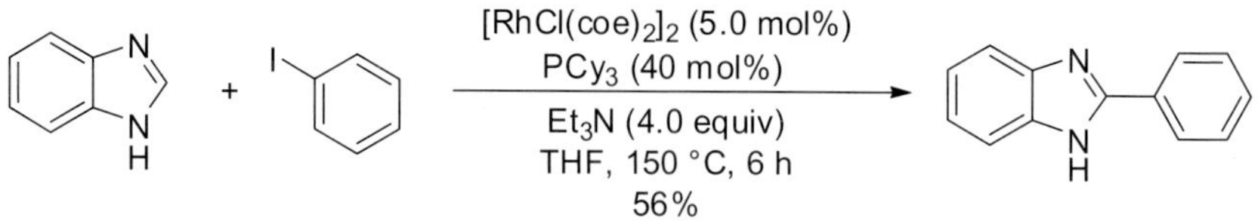

[RhCl(coe)$_2$]$_2$ (5.0 mol%)
PCy$_3$ (40 mol%)
Et$_3$N (4.0 equiv)
THF, 150 °C, 6 h
56%

37 **38** **39**

Scheme 9.13 Rhodium-catalyzed direct arylation of benzimidazole (**37**).

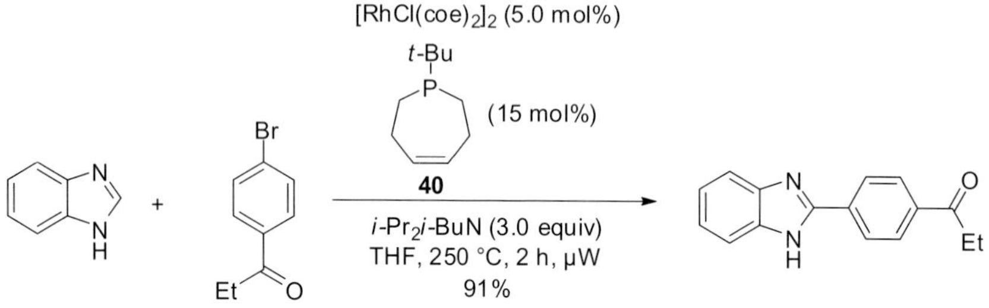

[RhCl(coe)$_2$]$_2$ (5.0 mol%)
t-Bu
(15 mol%)
40
i-Pr$_2$i-BuN (3.0 equiv)
THF, 250 °C, 2 h, μW
91%

37 **41** **42**

Scheme 9.14 Rhodium-catalyzed direct arylation with aryl bromide **41**.

as electrophiles in high yields of isolated products (Scheme 9.13), while aryl bromides or aryl chlorides reacted only sluggishly [25].

Detailed mechanistic studies revealed that this direct arylation reaction occurred through tautomerization of the heteroarene, thereby yielding an *N*-heterocyclic carbene (NHC) intermediate. Unfortunately, significant amounts of hydrodehalogenated byproducts were formed due to C—H bond activation at the organic substituents of coordinated PCy$_3$. This limitation was addressed with phosphine **40**, which gave rise to a highly active, yet thermally stable, rhodium catalyst [26], and this in turn allowed for the use of aryl bromides as electrophiles with significantly reduced loading of ligand **40** [26b]. Additionally, a number of valuable functional groups, such as enolizable ketones, amides or esters, were tolerated by the optimized catalytic system (Scheme 9.14) [26].

Subsequently, it was found that the use of [RhCl(coe)$_2$]$_2$, along with an electron-poor phosphine ligand and CsOPiv as base, allowed for the direct arylation of N—H free pyrroles when employing aryl iodides as electrophiles (Scheme 9.15) [27]. Interestingly, this protocol also proved applicable to the direct arylation of (aza)indoles.

The use of rhodium complexes with strongly π-accepting ligands was not restricted to the arylation of anisole (**29**) (Scheme 9.11), but also enabled an efficient C—H bond functionalization of heteroarenes [23]. Thus, furans or thiophenes were regioselectively arylated with (hetero)aryl iodides as electrophiles when using microwave irradiation. Importantly, the selectivity of this transformation could be controlled through 1,2-dimethoxyethane (DME) as additive, yielding the mono-arylated heteroarene as the sole product. Interestingly, thiophene **46**

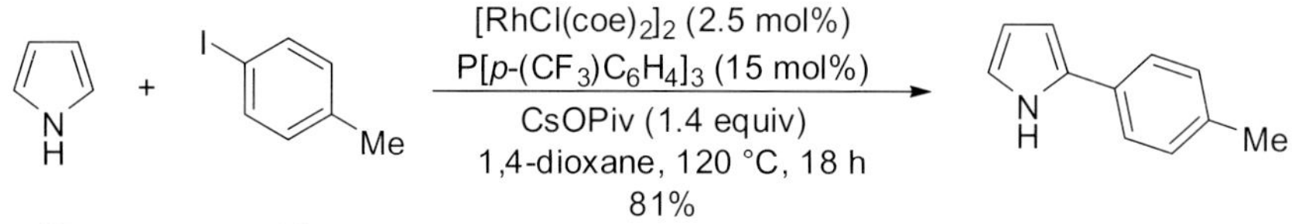

$[RhCl(coe)_2]_2$ (2.5 mol%)
$P[p\text{-}(CF_3)C_6H_4]_3$ (15 mol%)
CsOPiv (1.4 equiv)
1,4-dioxane, 120 °C, 18 h
81%

43 **44** **45**

Scheme 9.15 Rhodium-catalyzed direct arylation of pyrrole (**43**).

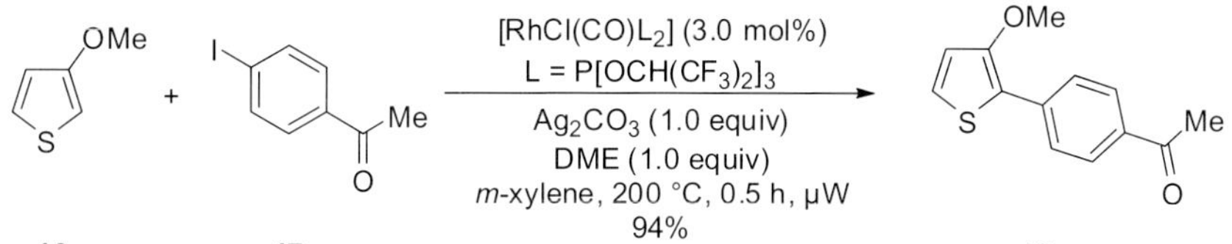

$[RhCl(CO)L_2]$ (3.0 mol%)
L = $P[OCH(CF_3)_2]_3$
Ag_2CO_3 (1.0 equiv)
DME (1.0 equiv)
m-xylene, 200 °C, 0.5 h, μW
94%

46 **47** **48**

Scheme 9.16 Rhodium-catalyzed direct arylation of thiophene **46**.

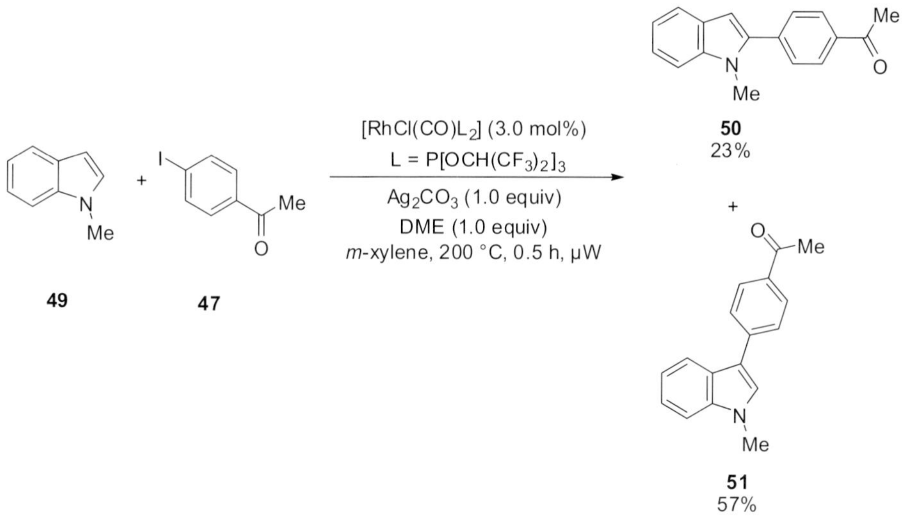

$[RhCl(CO)L_2]$ (3.0 mol%)
L = $P[OCH(CF_3)_2]_3$
Ag_2CO_3 (1.0 equiv)
DME (1.0 equiv)
m-xylene, 200 °C, 0.5 h, μW

49 **47** **50** 23% + **51** 57%

Scheme 9.17 Rhodium-catalyzed direct arylation of indole **49**.

was selectively functionalized at the more sterically encumbered position C-2 (Scheme 9.16).

In contrast, indoles as pronucleophilic substrates gave rather rise to mixtures of regioisomeric products (Scheme 9.17).

Recently, chelation-assisted oxidative rhodium-catalyzed direct arylations of heteroarenes with arylboronic acids were developed [17]. Thus, thiophene derivative **52** was regioselectively arylated at position C-3 with stoichiometric amounts of TEMPO as the terminal oxidant. This methodology proved broadly applicable, allowing the efficient direct arylation with both electron-rich and electron-deficient aryl boronic acids. Notably, more sterically hindered boronic acids led to comparably high yields of isolated products (Scheme 9.18).

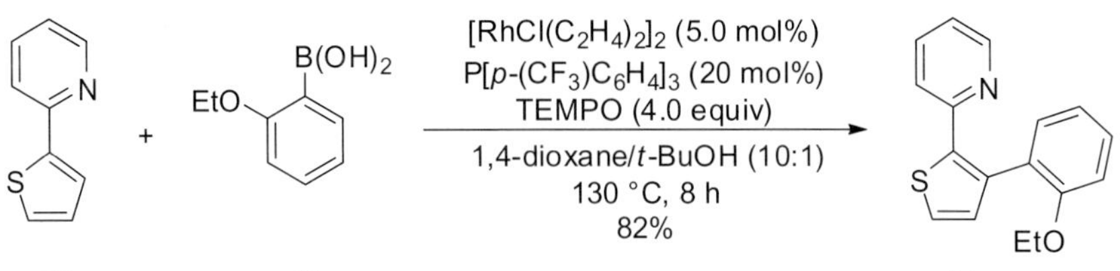

$[RhCl(C_2H_4)_2]_2$ (5.0 mol%)
$P[p\text{-}(CF_3)C_6H_4]_3$ (20 mol%)
TEMPO (4.0 equiv)
1,4-dioxane/*t*-BuOH (10:1)
130 °C, 8 h
82%

52 + **53** → **54**

Scheme 9.18 Chelation-assisted oxidative rhodium-catalyzed direct arylation of thiophene **52**.

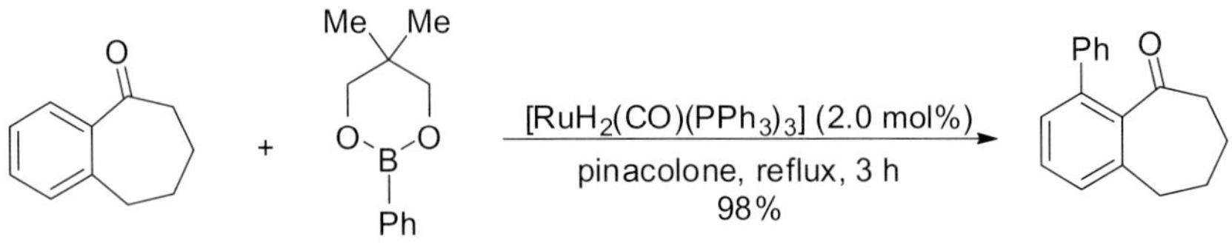

$[RuH_2(CO)(PPh_3)_3]$ (2.0 mol%)
pinacolone, reflux, 3 h
98%

55 + **56** → **57**

Scheme 9.19 Ruthenium-catalyzed arylation of ketone **55** in pinacolone.

9.3 Ruthenium-Catalyzed Direct Arylations

9.3.1 Ruthenium-Catalyzed Direct Arylations with Organometallic Reagents

A ruthenium-catalyzed [28] chelation-assisted approach was developed based on the use of arylboronates as arylating agents [29]. Thereby, a regioselective ruthenium-catalyzed arylation of substrates bearing an oxygen-containing directing group was achieved. A variety of aromatic ketones were efficiently arylated in pinacolone using aryl boronates with both electron-donating, as well as electron-withdrawing substituents (Scheme 9.19).

Mechanistic studies revealed that pinacolone acts here not only as a solvent, but also as an oxidizing agent. Additionally, inter- and intramolecular competition experiments with D-labeled ketones provided evidence for a precoordination of the ruthenium catalyst by the oxygen of the aryl ketone [29b]. Thus, a mechanism was elaborated consisting of: (a) coordination; (b) oxidative addition to yield an *ortho*-metallated ruthenacycle; (c) insertion of pinacolone into the [Ru]—H bond; (d) transmetallation; and finally (e) reductive elimination (Scheme 9.20).

An extension of this reaction to the functionalization of $C(sp^3)$—H bonds was more recently reported. Thus, pyrrolidines and piperidines were efficiently arylated with substituted arylboronates in pinacolone, although the product often contained mixtures of diastereomers (Scheme 9.21) [30].

Jun and coworkers used a related approach for a ruthenium-catalyzed arylation of aldimines [31]. Here, a pyridyl-substituent allowed for the selective arylation employing arylboronates. Methyl vinyl ketone (**61**) as additive led to high isolated yields of the corresponding ketones (Scheme 9.22).

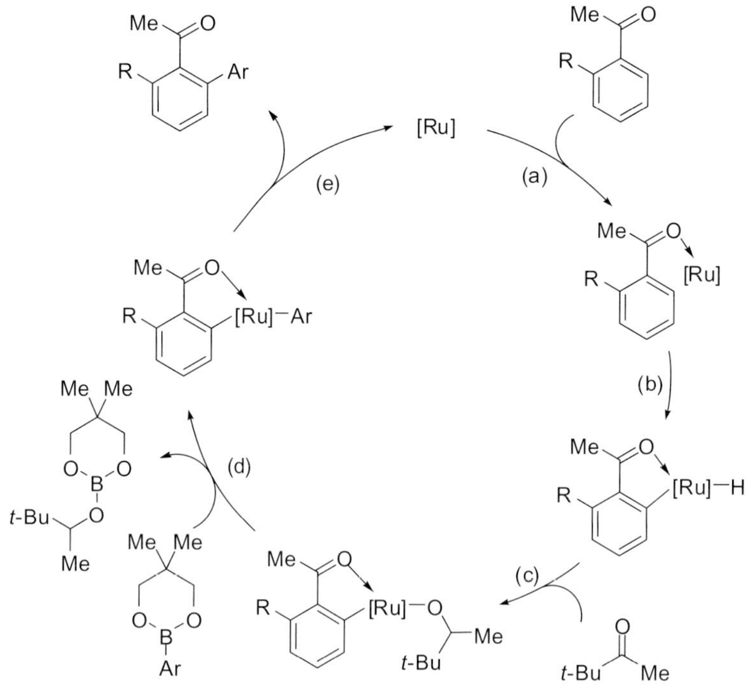

Scheme 9.20 Proposed mechanism for ruthenium-catalyzed arylations of ketones.

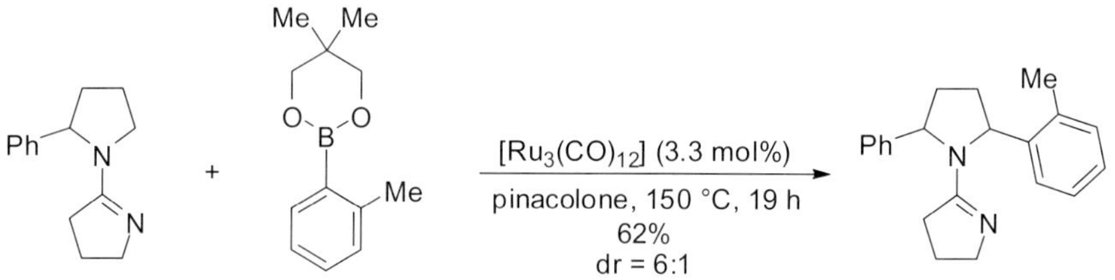

Scheme 9.21 Ruthenium-catalyzed functionalization of a C(sp^3)—H bond in pyrrolidine **58**.

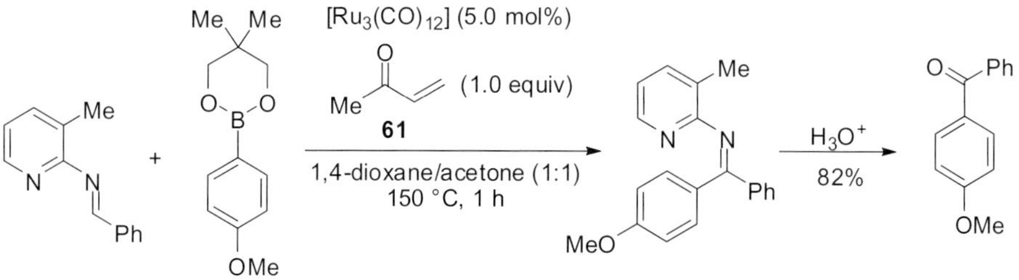

Scheme 9.22 Ruthenium-catalyzed direct arylation of aldimine **62**.

9.3.2
Ruthenium-Catalyzed Direct Arylations with Aryl (Pseudo) Halides

A catalytic system comprising $[RuCl_2(\eta^6\text{-}C_6H_6)]_2$ and PPh_3 was elegantly developed by Oi, Inoue and coworkers for the direct arylation of pyridine derivatives using aryl bromides as electrophiles (Scheme 9.23) [32].

The same protocol proved applicable to directed arylations of imines, imidazolines and oxazolines as pronucleophilic starting materials (Scheme 9.24) [33]. Transformations of the latter substrates should prove useful, since 2-oxazolinyl substituents can easily be converted into a variety of valuable functionalities [34].

Alkenyl C—H bonds were also directly functionalized with aryl bromides using this catalytic system, yielding both regio- and diastereoselectively substituted alkenes (Scheme 9.25) [35].

A phosphine ligand-free ruthenium-catalyzed direct arylation with aryl bromides as electrophiles was recently disclosed. Notably, the use of inexpensive $RuCl_3 \cdot (H_2O)_n$ as catalyst allowed for economically attractive C—H bond functionalizations of

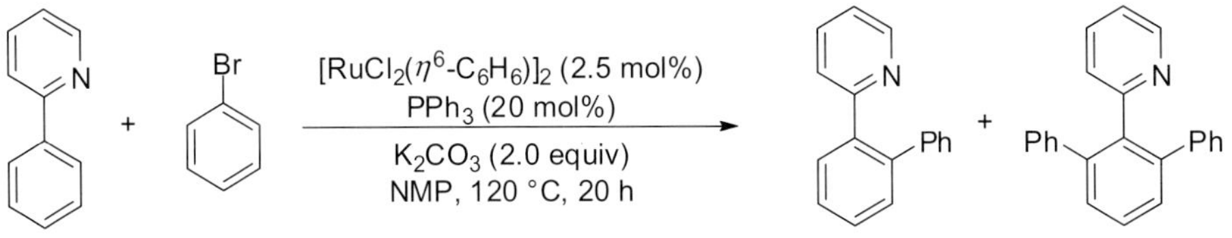

Scheme 9.23 Ruthenium-catalyzed direct arylation of pyridine **1** with bromide **22**.

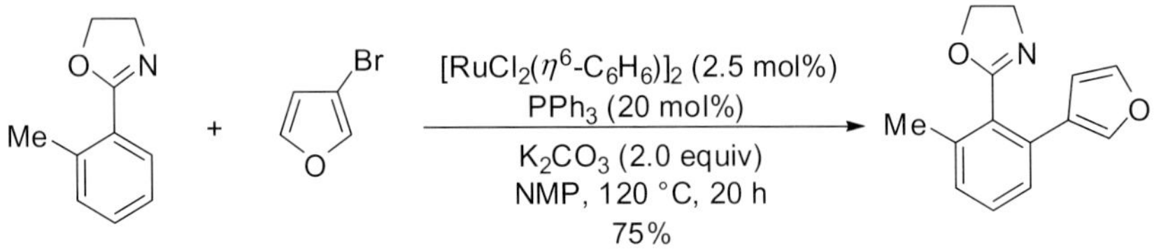

Scheme 9.24 Ruthenium-catalyzed direct arylation with heteroaryl bromide **67**.

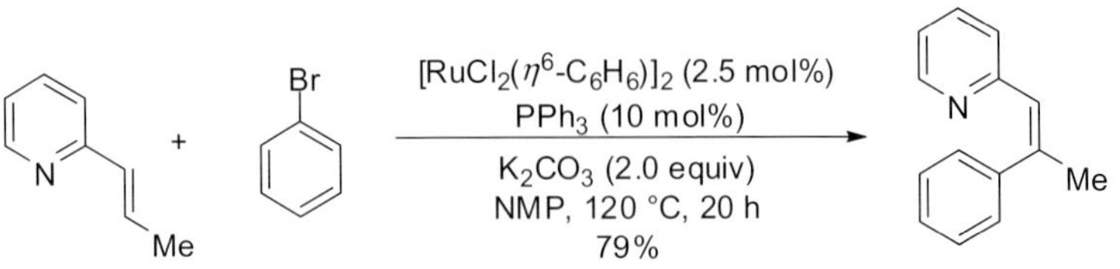

Scheme 9.25 Ruthenium-catalyzed direct arylation of alkene **69**.

pyridine, oxazoline and pyrazole derivatives, also with more sterically hindered *ortho*-substituted aryl bromides as electrophiles (Scheme 9.26) [36, 37].

Among aryl halides, chlorides are arguably the most useful single class of electrophilic substrates, due to their lower cost and wide diversity of (commercially) available compounds. For traditional cross-coupling reactions, the development of stabilizing ligands enabled the use of these inexpensive substrates. However, direct arylations with aryl chlorides were until recently only generally applicable to palladium-catalyzed *intramolecular* transformations (see Chapter 10) [38]. Nonetheless, a broadly applicable intermolecular dircct arylation of various arenes with aryl chlorides was accomplished with a ruthenium complex derived from secondary phosphine oxide (SPO) $(1\text{-Ad})_2P(O)H$ (**73**) as preligand (Scheme 9.27) [39]. Thereby, pyridine and ketimine derivatives were efficiently arylated with functionalized electron-deficient, as well as electron-rich, thus for an oxidative addition electronically deactivated, aryl chlorides.

The use of aryl tosylates as electrophiles is attractive, because they can be prepared from readily available phenols with less-expensive reagents than those required for synthesis of the corresponding triflates. Importantly, tosylates are more stable towards hydrolysis than triflates, yet significantly less reactive as electrophiles. As a result, protocols for traditional cross-coupling reactions were only recently developed (see Chapter 2). In contrast, catalytic direct arylations with aryl tosylates were not reported until recently. Interestingly, a ruthenium complex derived from heteroatom-substituted secondary phosphine oxide (HASPO) preligand **78** [40] allowed for direct arylations with both electron-deficient, as well as electron-rich aryl tosylates [41]. As pronucleophiles, pyridine, oxazoline and pyrazole derivatives could be efficiently functionalized. Selective mono- or diarylation reactions could be accomplished through the judicious choice of the

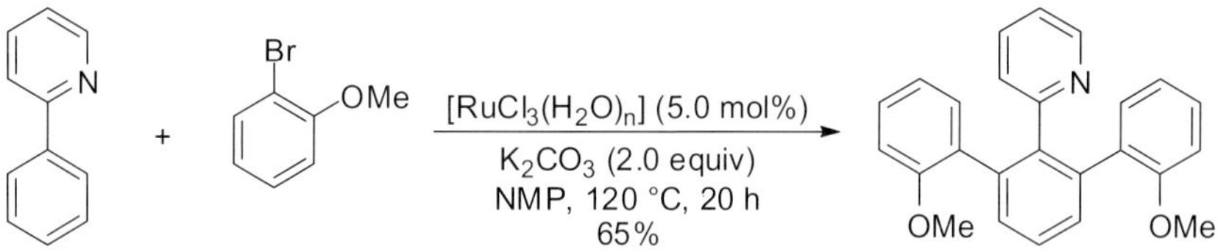

1 + **71** (Br, OMe) → $[RuCl_3(H_2O)_n]$ (5.0 mol%), K_2CO_3 (2.0 equiv), NMP, 120 °C, 20 h, 65% → **72** (OMe, OMe)

Scheme 9.26 Ruthenium-catalyzed phosphine ligand-free direct arylation of pyridine **1**.

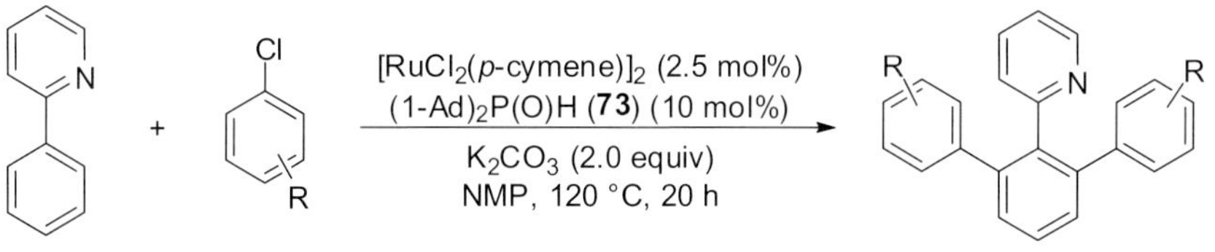

1 + Cl-aryl-R → $[RuCl_2(p\text{-cymene})]_2$ (2.5 mol%), $(1\text{-Ad})_2P(O)H$ (**73**) (10 mol%), K_2CO_3 (2.0 equiv), NMP, 120 °C, 20 h

74 R = 3-CN → R = 3-CN **75** 81%
76 R = 4-OMe → R = 4-OMe **77** 87%

Scheme 9.27 Ruthenium-catalyzed direct arylation with aryl chlorides **74** and **76**.

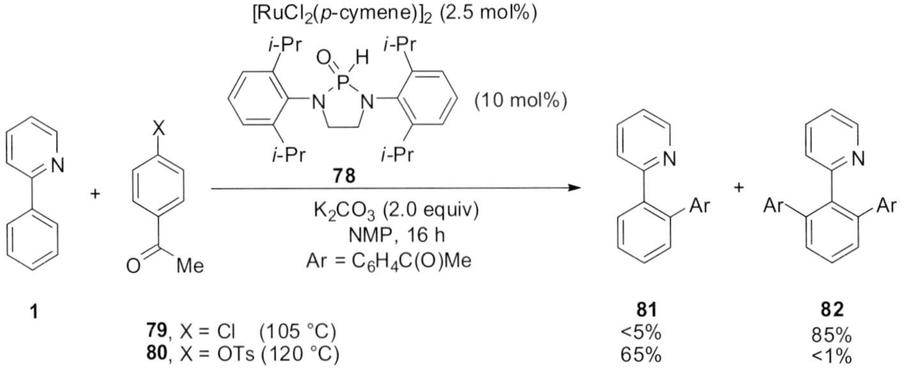

$[RuCl_2(p\text{-cymene})]_2$ (2.5 mol%)
78 (10 mol%)
K_2CO_3 (2.0 equiv)
NMP, 16 h
Ar = $C_6H_4C(O)Me$

1 + 79, X = Cl (105 °C) → 81 <5%, 82 85%
1 + 80, X = OTs (120 °C) → 81 65%, 82 <1%

Scheme 9.28 Selective ruthenium-catalyzed direct arylations through choice of electrophile.

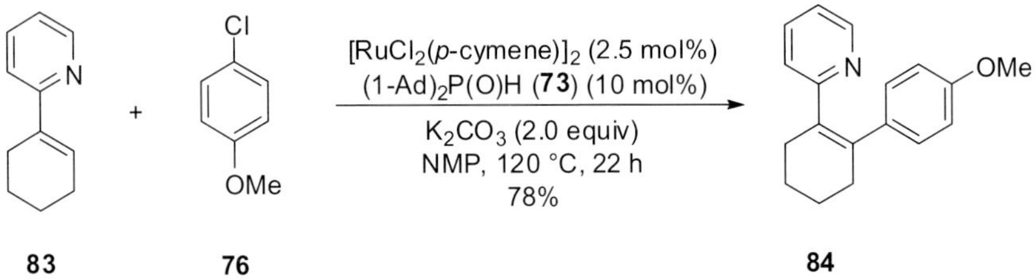

$[RuCl_2(p\text{-cymene})]_2$ (2.5 mol%)
$(1\text{-Ad})_2P(O)H$ (**73**) (10 mol%)
K_2CO_3 (2.0 equiv)
NMP, 120 °C, 22 h
78%

83 + 76 → 84

Scheme 9.29 Ruthenium-catalyzed direct arylation of alkene **83** with chloride **76**.

corresponding electrophile (Scheme 9.28). Thus, while aryl chlorides gave rise to diarylated products, the use of aryl tosylates cleanly afforded the corresponding monoarylated derivatives.

Direct arylations of alkenyl pronucleophiles with inexpensive aryl chlorides proceeded with high efficacy and excellent diastereoselectivity using either ruthenium carbenes or a ruthenium complex derived form air-stable secondary phosphine oxide preligand $(1\text{-Ad})_2P(O)H$ (**73**) as catalyst (Scheme 9.29) [42].

Importantly, the diastereoselectivity of these C—H bond functionalizations proved complementary to both ruthenium-catalyzed cross-metatheses as well as palladium-catalyzed Mizoroki–Heck reactions (Scheme 9.30).

Further, a sequential ruthenium-catalyzed direct arylation/hydrosilylation reaction was developed with this catalytic system (Scheme 9.31).

Unfortunately, experimental studies on the working mode of ruthenium-catalyzed direct arylations with organic (pseudo)halides were thus far not available. However, a beneficial effect of NaOAc in the stoichiometric syntheses of ruthenacycles under mild reaction conditions was previously reported [43], and suggested a cooperative deprotonation/metallation mechanism for the C—H bond activation step. In analogy, a transition-state model **93** could account for the high efficacy observed with (HA)SPO preligands [39, 41, 42] in ruthenium-catalyzed direct arylation reactions (Scheme 9.32).

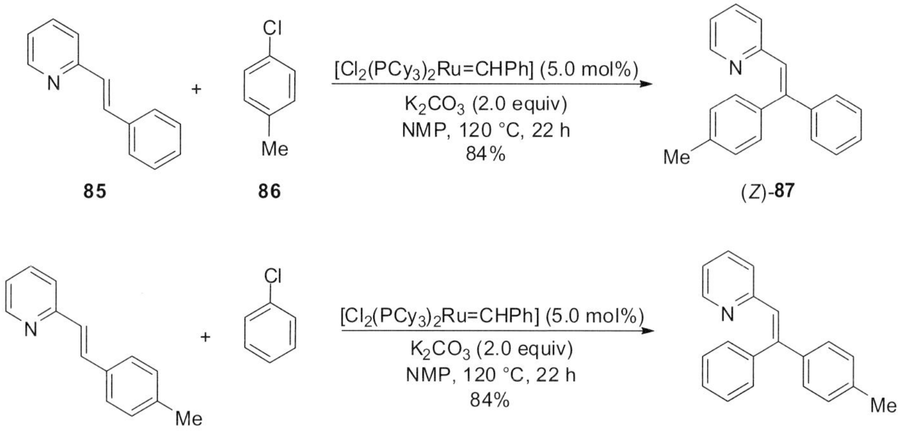

$[Cl_2(PCy_3)_2Ru{=}CHPh]$ (5.0 mol%)
K_2CO_3 (2.0 equiv)
NMP, 120 °C, 22 h
84%

85 + 86 → (*Z*)-87

$[Cl_2(PCy_3)_2Ru{=}CHPh]$ (5.0 mol%)
K_2CO_3 (2.0 equiv)
NMP, 120 °C, 22 h
84%

88 + 89 → (*E*)-87

Scheme 9.30 Ruthenium-catalyzed diastereoselective direct arylation of alkenes **85** and **88**.

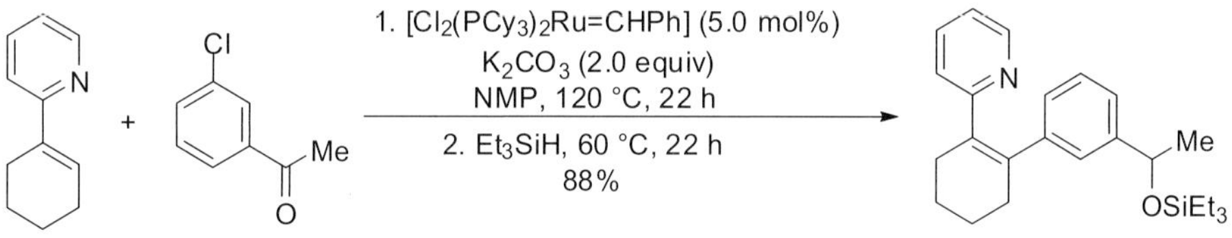

1. $[Cl_2(PCy_3)_2Ru{=}CHPh]$ (5.0 mol%)
K_2CO_3 (2.0 equiv)
NMP, 120 °C, 22 h
2. Et_3SiH, 60 °C, 22 h
88%

83 + 90 → 91

Scheme 9.31 Ruthenium-catalyzed direct arylation/hydrosilylation sequence.

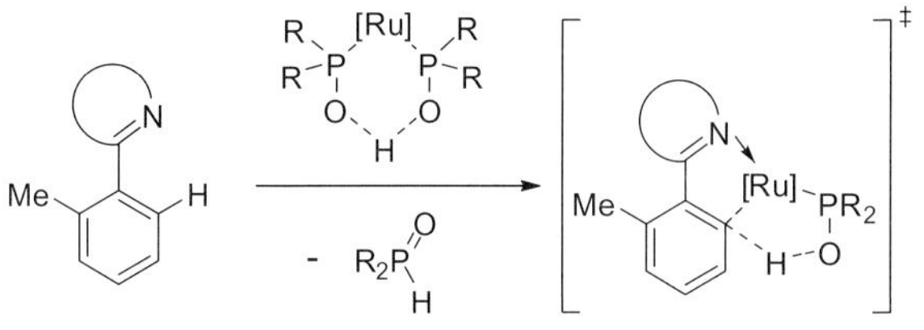

92 → 93 (− $R_2P(O)H$)

Scheme 9.32 Cooperative deprotonation/metallation with (HA)SPO preligands.

Based on this proposal, ruthenium-catalyzed direct arylations with aryl halides in less-coordinating apolar solvents (e.g. toluene) were probed. Interestingly, this catalytic system enabled regioselective C—H bond functionalizations at the aromatic moieties of *N*-aryl-substituted 1,2,3-triazoles (Scheme 9.33) [44]. It is noteworthy, that the regioselectivity of this ruthenium-catalyzed transformation proved complementary to that obtained when applying either palladium- or copper-based catalysts (*vide infra*).

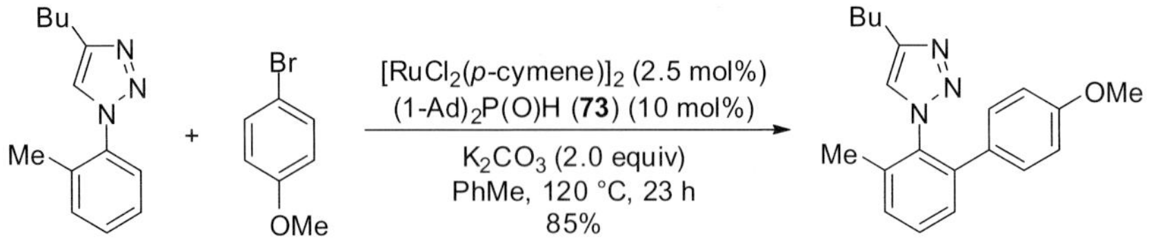

94 + **26** → **95**

$[RuCl_2(p\text{-cymene})]_2$ (2.5 mol%)
$(1\text{-Ad})_2P(O)H$ (**73**) (10 mol%)
K_2CO_3 (2.0 equiv)
PhMe, 120 °C, 23 h
85%

Scheme 9.33 Ruthenium-catalyzed direct arylation of triazole **94**.

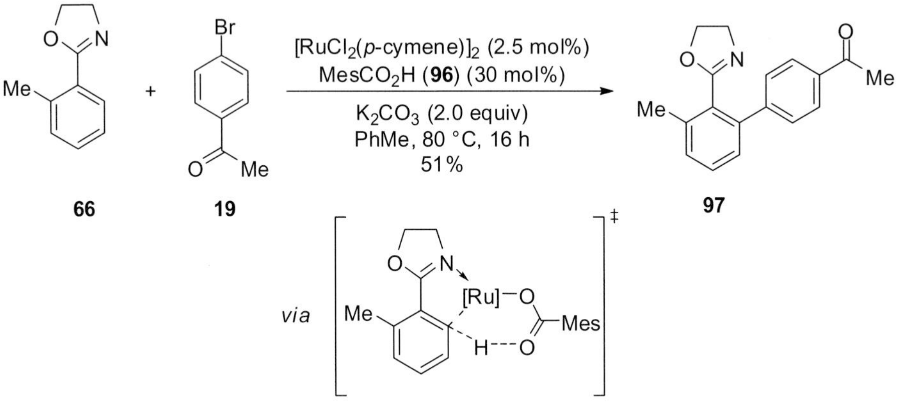

66 + **19** → **97**

$[RuCl_2(p\text{-cymene})]_2$ (2.5 mol%)
$MesCO_2H$ (**96**) (30 mol%)
K_2CO_3 (2.0 equiv)
PhMe, 80 °C, 16 h
51%

via **98**

Scheme 9.34 Catalytic direct arylation with carboxylic acid **96** in toluene as solvent.

As carboxylic acid additives increased the efficiency of palladium catalysts in direct arylations through a cooperative deprotonation/metallation mechanism (see Chapter 11) [45], their application to ruthenium catalysis was tested. Thus, it was found that a ruthenium complex modified with carboxylic acid $MesCO_2H$ (**96**) displayed a broad scope and allowed for the efficient directed arylation of triazoles, pyridines, pyrazoles or oxazolines [44, 46]. With respect to the electrophile, aryl bromides, chlorides and tosylates, including *ortho*-substituted derivatives, were found to be viable substrates. It should be noted here that these direct arylations could be performed at a lower reaction temperatures of 80 °C (Scheme 9.34).

A chelation-assisted ruthenium-catalyzed arylation of aldehyde **99** was accomplished in combination with a palladium complex [47]. This cooperative catalysis [48] proved applicable to organostannanes and aryl iodides as arylating reagents (Scheme 9.35). The direct arylation proceeded most likely through ruthenium-catalyzed C—H bond activation, subsequent transmetallation to palladium, and reductive elimination from a palladium intermediate.

Oi, Inoue and coworkers developed a regioselective ruthenium-catalyzed direct arylation of allyl acetate (**101**) using a catalytic system comprising $[RuCl_2(cod)]_n$ and PPh_3, giving rise to alkenes **102** and **103** (Scheme 9.36) [49].

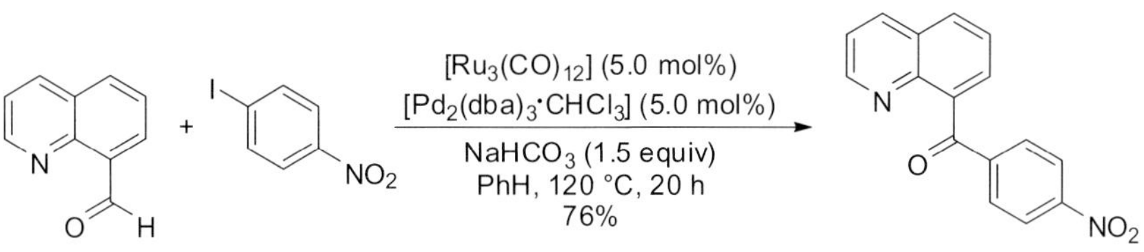

$[Ru_3(CO)_{12}]$ (5.0 mol%)
$[Pd_2(dba)_3{\cdot}CHCl_3]$ (5.0 mol%)
$NaHCO_3$ (1.5 equiv)
PhH, 120 °C, 20 h
76%

99 **30** **100**

Scheme 9.35 Bimetallic catalytic arylation of aldehyde **99**.

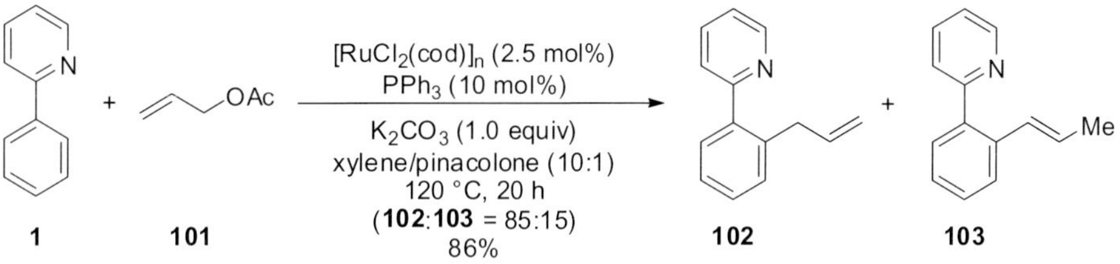

$[RuCl_2(cod)]_n$ (2.5 mol%)
PPh_3 (10 mol%)
K_2CO_3 (1.0 equiv)
xylene/pinacolone (10:1)
120 °C, 20 h
(**102**:**103** = 85:15)
86%

1 **101** **102** **103**

Scheme 9.36 Ruthenium-catalyzed direct arylation of allyl acetate (**101**).

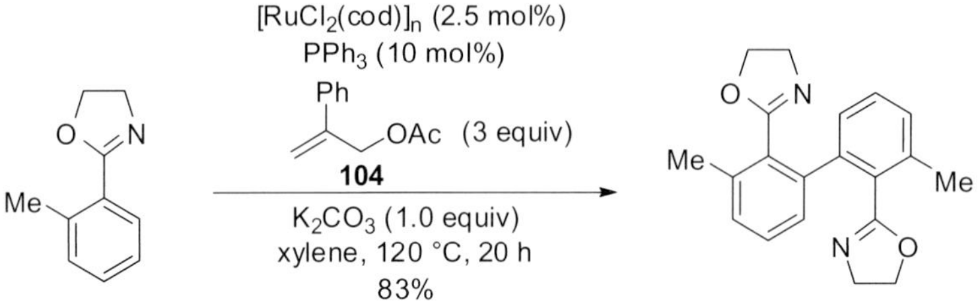

$[RuCl_2(cod)]_n$ (2.5 mol%)
PPh_3 (10 mol%)
104 (3 equiv)
K_2CO_3 (1.0 equiv)
xylene, 120 °C, 20 h
83%

66 **105**

Scheme 9.37 Ruthenium-catalyzed oxidative homo-coupling of oxazoline **66**.

Interestingly, the chemoselectivity of this catalytic system changed dramatically when using aryl oxazolines along with substituted allyl acetate **104**. Thus, oxidative homocoupling reactions occurred under these reaction conditions (Scheme 9.37) [50]. Imidazole, pyrazole or thiazole derivatives could also be efficiently homocoupled.

Finally, a direct arylation of thiophene (**106**) was accomplished with sulfonyl chloride **107** at a relatively high reaction temperature, which most likely proceeded via a radical-based mechanism (Scheme 9.38) [51].

9.4 Iridium-, Copper- and Iron-Catalyzed Direct Arylations

Thus far, palladium-, rhodium- or ruthenium-based complexes have been predominantly employed for catalytic direct arylations. However, promising

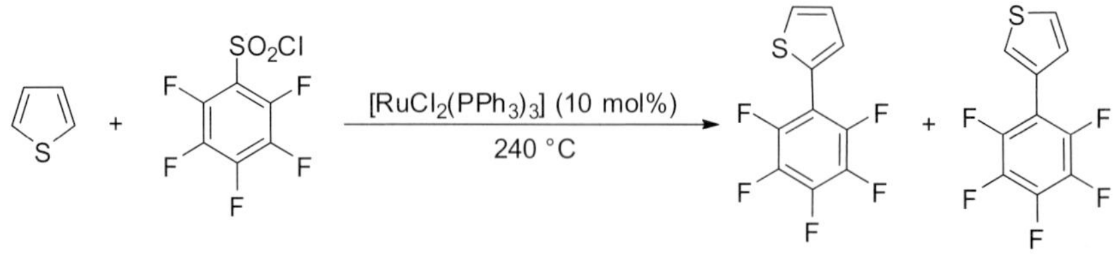

$[RuCl_2(PPh_3)_3]$ (10 mol%)
240 °C

106 **107** **108** 22% **109** 6%

Scheme 9.38 Ruthenium-catalyzed direct arylation of thiophene (**106**).

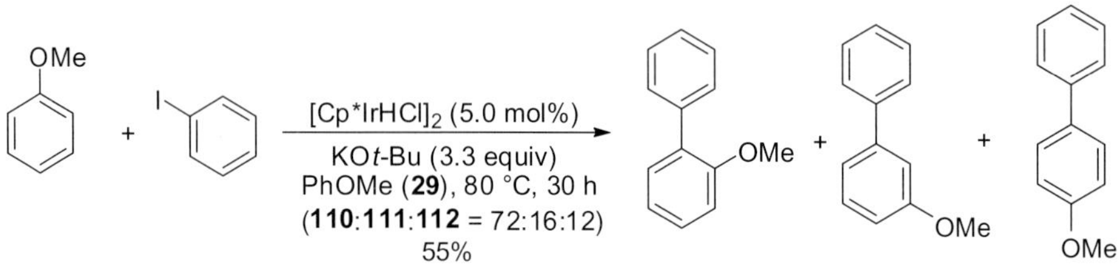

$[Cp^*IrHCl]_2$ (5.0 mol%)
KO*t*-Bu (3.3 equiv)
PhOMe (**29**), 80 °C, 30 h
(**110**:**111**:**112** = 72:16:12)
55%

29 **38** **110** **111** **112**

Scheme 9.39 Iridium-catalyzed direct arylation of anisole (**29**).

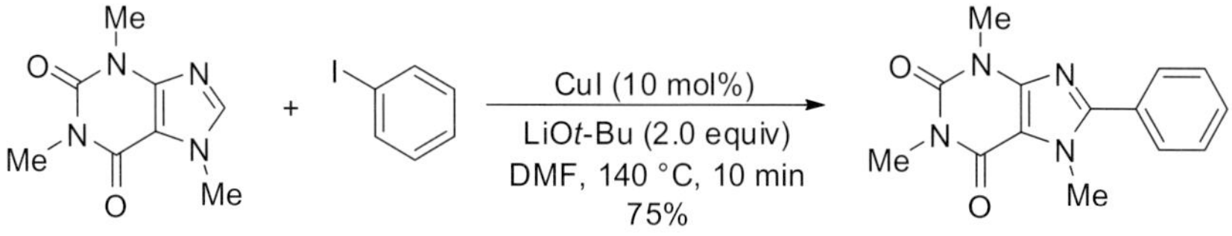

CuI (10 mol%)
LiO*t*-Bu (2.0 equiv)
DMF, 140 °C, 10 min
75%

113 **38** **114**

Scheme 9.40 Copper-catalyzed direct arylation of caffeine (**113**).

methodologies have recently been disclosed, highlighting the potential of additional transition metals in this challenging transformation. Hence, an iridium–hydride complex was found to catalyze direct arylations of arenes [52]. Whereas, this reaction occurred at a relatively low reaction temperature, the protocol proved to be restricted to aryl iodides as electrophiles and to KO*t*-Bu as base. When using anisole (**29**) as a pronucleophile, a mixture of *ortho*-, *meta*- and *para*-substituted regioisomeric products was obtained in a ratio of 72 : 16 : 12, suggesting a radical-based reaction mechanism (Scheme 9.39).

The beneficial effect of copper salts in stoichiometric quantities for palladium-catalyzed direct arylations of *N*-heterocycles was reported by Miura and coworkers in 1998 [53, 54]. However, it was only recently that catalytic amounts of inexpensive CuI were found to enable direct arylations of heteroarenes [55, 56]. Remarkably, a variety of *N*-heterocycles could be arylated in high yields of isolated products with aryl iodides as electrophiles (Scheme 9.40). Unfortunately, this ligand-free catalytic system required the use of a relatively strong base, thereby limiting its functional group tolerance.

Importantly, a copper catalyst modified with *N,N*-bidentate ligand 1,10-phenanthroline (**115**) enabled the direct arylation of electron-deficient fluoroarenes (Scheme 9.41) [57]. The high catalytic efficacy allowed also for the use of aryl bromides as electrophiles, and K_3PO_4 as mild base.

Recently, catalytic amounts of inexpensive CuI were employed for regioselective direct arylations of 1,2,3-triazoles (Scheme 9.42) [58].

As 1,4-disubstituted 1,2,3-triazoles are usually prepared through copper-catalyzed 1,3-dipolar cycloadditions of terminal alkynes with organic azides, the use of a single copper complex for a direct arylation-based sequential catalysis was probed. Thereby, a modular chemo- and regioselective synthesis of fully-substituted 1,2,3-triazoles was achieved (Scheme 9.43). Notably, the overall reaction involved the selective coupling of four components through the formation of one C–C- and three C–N-bonds [58].

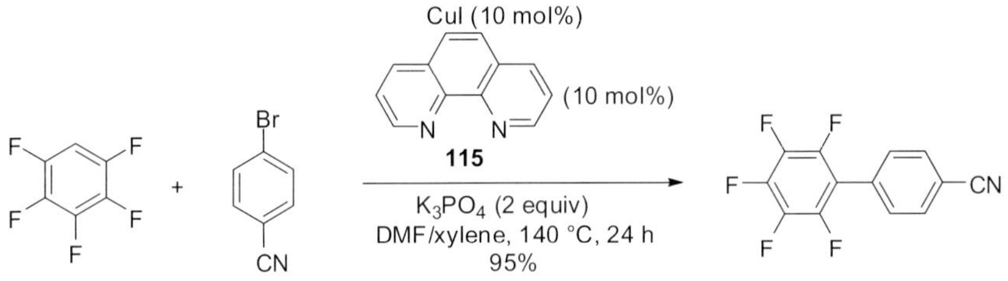

116 + **117** → **118**
CuI (10 mol%), **115** (10 mol%), K_3PO_4 (2 equiv), DMF/xylene, 140 °C, 24 h, 95%

Scheme 9.41 Copper-catalyzed direct arylation of electron-deficient arene **116**.

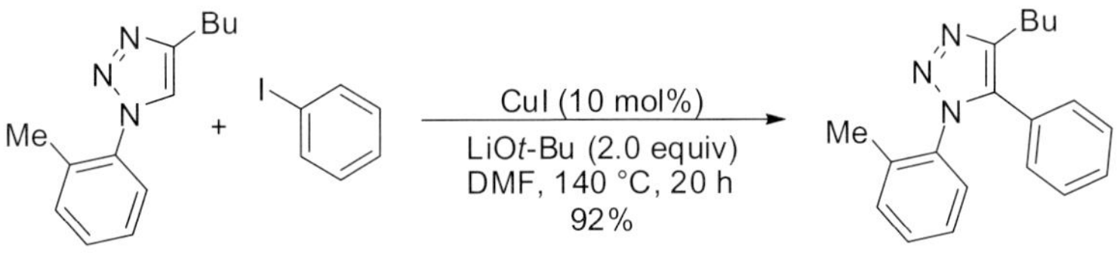

94 + **38** → **119**
CuI (10 mol%), LiO*t*-Bu (2.0 equiv), DMF, 140 °C, 20 h, 92%

Scheme 9.42 Copper-catalyzed direct arylation of 1,2,3-triazole **94**.

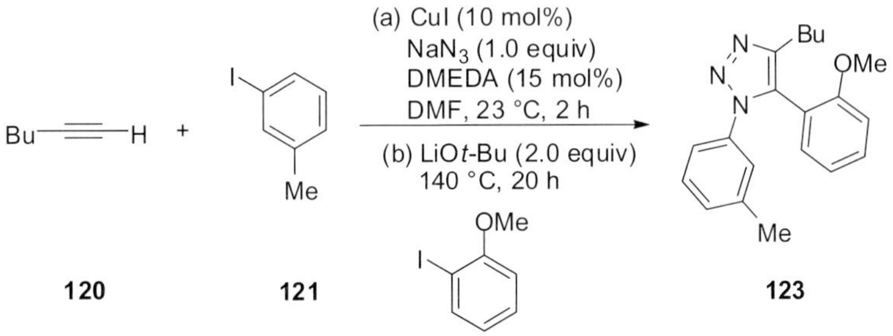

120 + **121** → **123**
(a) CuI (10 mol%), NaN_3 (1.0 equiv), DMEDA (15 mol%), DMF, 23 °C, 2 h
(b) LiO*t*-Bu (2.0 equiv), 140 °C, 20 h, **122**
81%

Scheme 9.43 Copper-catalyzed modular multicomponent synthesis of 1,2,3-triazole **123**.

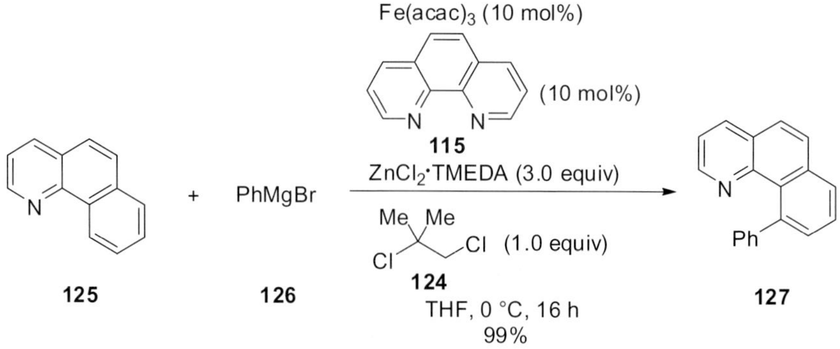

$Fe(acac)_3$ (10 mol%)
115 (10 mol%)
$ZnCl_2 \cdot TMEDA$ (3.0 equiv)
124 (1.0 equiv)
THF, 0 °C, 16 h
99%

125 + PhMgBr 126 → 127

Scheme 9.44 Directed iron-catalyzed direct arylation reaction of quinoline **125**.

Recently, Nakamura and coworkers disclosed an iron-catalyzed direct arylation through chelation assistance [59], whereby pyridine, pyrimidine, pyrazole and quinoline derivatives were efficiently arylated using *in situ*-generated arylzinc reagents (Scheme 9.44). Although low reaction temperatures could be used, an excess of the organometallic reagent, as well as the use of 1,2-dichloroalkane **124**, was found mandatory for achieving catalytic turnover. Notably, the most efficient catalysis took place in the presence of *N,N*-bidentate ligand **115**, as was also observed in previously reported copper-catalyzed transformations.

9.5 Conclusions

Various methodologies for catalytic direct arylations via C—H bond activation employing transition metals other than palladium have been developed in recent years. In particular, rhodium- and ruthenium-based complexes have enabled the development of promising protocols for catalytic direct arylations. Whilst rhodium catalysts were found broadly applicable to the direct arylation of both arenes, as well as heteroarenes, ruthenium-catalyzed chelation-assisted C—H bond functionalizations could be used for the conversion of a variety of attractive organic electrophiles. In addition, inexpensive copper and iron salts have recently been shown as economically attractive alternatives to previously developed more expensive catalysts. Given the economically and environmentally benign features of selective C—H bond functionalizations, the development of further valuable protocols is expected in this rapidly evolving research area.

Abbreviations

acac	acetylacetonate
Ad	adamantyl
cat	catalytic

cod	1,4-cyclooctadiene
coe	cyclooctene
Cp*	pentamethylcyclopentadienyl
Cy	cyclohexyl
dba	dibenzylideneacetone
DG	directing group
DME	1,2-dimethoxyethane
DMEDA	*N,N'*-dimethylethylenediamine
DMF	*N,N*-dimethylformamide
dr	diastereomeric ratio
(HA)SPO	(heteroatom substituted) secondary phosphine oxide
L	ligand
Mes	Mesityl
NMP	1-methyl-2-pyrrolidinone
Piv	Pivalate
TEMPO	2,2′,6,6′-tetramethylpiperidine-*N*-oxyl
THF	tetrahydrofuran
TM	transition metal
TMEDA	*N,N,N',N'*-tetramethylethylenediamine
μW	microwave irradiation

References

1 Corbet, J.P. and Mignani, G. (2006) *Chem. Rev.*, **106**, 2651–710.

2 de Meijere, A. and Diederich, F. (eds) (2004) *Metal-Catalyzed Cross-Coupling Reactions*, Wiley-VCH Verlag GmbH, Weinheim.

3 Beller, M. and Bolm, C. (eds) (2004) *Transition Metals for Organic Synthesis*, Wiley-VCH Verlag GmbH, Weinheim.

4 Alberico, D., Scott, M.E. and Lautens, M. (2007) *Chem. Rev.*, **107**, 174–238.

5 Stuart, D.R. and Fagnou, K. (2007) *Aldrichim. Acta*, **40**, 35–41.

6 Seregin, I.V. and Gevorgyan, V. (2007) *Chem. Soc. Rev.*, **36**, 1173–93.

7 Ackermann, L. (2007) *Top. Organomet. Chem.*, **24**, 35–60.

8 Ackermann, L. (2007) *Synlett*, 507–26.

9 Satoh, T. and Miura, M. (2007) *Chem. Lett.*, **36**, 200–5.

10 Fairlamb, I.J.S. (2007) *Chem. Soc. Rev.*, **36**, 1036–45.

11 (a) Li, B.-J., Yang, S.-D. and Shi, Z.J. (2008) *Synlett*, 949–57;
(b) Schnurch, M., Flasik, R., Khan, A.F., Spina, M., Mihovilovic, M.D. and Stanetty, P. (2006) *Eur. J. Org. Chem.*, 3283–307;
(c) Campeau, L.-C. and Fagnou, K. (2006) *Chem. Commun.*, 1253–64;
(d) Daugulis, O., Zaitsev, V.G., Shabashov, D., Pham, Q.-N. and Lazareva, A. (2006) *Synlett*, 3382–8;
(e) Dyker, G. (ed.) (2005) *Handbook of C–H Transformations*, Wiley-VCH Verlag GmbH, Weinheim;
(f) Kakiuchi, F. and Chatani, N. (2003) *Adv. Synth. Catal.*, **345**, 1077–101;
(g) Kakiuchi, F. and Murai, S. (2002) *Acc. Chem. Res.*, **35**, 826–34;
(h) Miura, M. and Nomura, M. (2002) *Top. Curr. Chem.*, **219**, 211–41;
(i) Ritleng, V., Sirlin, C. and Pfeffer, M. (2002) *Chem. Rev.*, **102**, 1731–70;
(j) Dyker, G. (1999) *Angew. Chem. Int. Ed.*, **38**, 1698–712.

12 (a) Trost, B.M. (1991) *Science*, **254**, 1471–7;
(b) Trost, B.M. (2002) *Acc. Chem. Res.*, **35**, 695–705.

13 Omae, I. (2004) *Coord. Chem. Rev.*, **248**, 995–1023.
14 Oi, S., Fukita, S. and Inoue, Y. (1998) *Chem. Commun.*, 2439–40.
15 Wilkinson's complex was reported to catalyze hydro-dehalogenation reactions. See: Petterson, A.A. and McNeill, K. (2006) *Organometallics*, **25**, 4938–40.
16 (a) Ueura, K., Satoh, T. and Miura, M. (2005) *Org. Lett.*, **7**, 2229–31;
(b) Ueura, K., Miyamura, S., Satoh, T. and Miura, M. (2006) *J. Organomet. Chem.*, **691**, 2821–6.
17 Vogler, T. and Studer, A. (2008) *Org. Lett.*, **10**, 129–31.
18 Bedford, R.B., Coles, S.J., Hursthouse, M.B. and Limmert, M.E. (2003) *Angew. Chem. Int. Ed.*, **42**, 112–14.
19 Oi, S., Watanabe, S., Fukita, S. and Inoue, Y. (2003) *Tetrahedron Lett.*, **44**, 8665–8.
20 Bedford, R.B. and Limmert, M.E. (2003) *J. Org. Chem.*, **68**, 8669–82.
21 (a) For the use of chlorophosphines as preligands in cross-coupling reactions, see: Ackermann, L. and Born, R. (2005) *Angew. Chem. Int. Ed.*, **44**, 2444–7;
(b) Ackermann, L., Spatz, J.H., Gschrei, C.J., Born, R. and Althammer, A. (2006) *Angew. Chem. Int. Ed.*, **45**, 7627–30.
22 Bedford, R.B., Betham, M., Caffyn, A.J.M., Charmant, J.P.H., Lewis-Alleyne, L.C., Long, P.D., Polo-Cerón, D. and Prashar, S. (2008) *Chem. Commun.*, 990–2.
23 Yanagisawa, S., Sudo, T., Noyori, R. and Itami, K. (2006) *J. Am. Chem. Soc.*, **128**, 11748–9.
24 Proch, S. and Kempe, R. (2007) *Angew. Chem. Int. Ed.*, **46**, 3135–8.
25 Lewis, J.C., Wiedemann, S.H., Bergman, R.G. and Ellman, J.A. (2004) *Org. Lett.*, **6**, 35–8.
26 (a) Lewis, J.C., Wu, J.Y., Bergman, R.G. and Ellman, J.A. (2006) *Angew. Chem. Int. Ed.*, **45**, 1589–91;
(b) Lewis, J.C., Berman, A.M., Bergman, R.G. and Ellman, J.A. (2008) *J. Am. Chem. Soc.*, **130**, 2493–500.
27 Wang, X., Lane, B.S. and Sames, D. (2005) *J. Am. Chem. Soc.*, **127**, 4996–7.
28 (a) Kakiuchi, F. and Chatani, N. (2004) *Ruthenium in Organic Synthesis* (eds C. Bruneau and P.H. Dixneuf), Wiley-VCH Verlag GmbH, Weinheim, pp. 219–55;
(b) Kakiuchi, F. and Chatani, N. (2004) *Ruthenium Catalysts and Fine Chemistry*, Vol. **11** (eds C. Bruneau and P.H. Dixneuf), Springer, Berlin-Heidelberg, pp. 45–79.
29 (a) Kakiuchi, F., Kan, S., Igi, K., Chatani, N. and Murai, S. (2003) *J. Am. Chem. Soc.*, **125**, 1698–9;
(b) Kakiuchi, F., Matsuura, Y., Kan, S. and Chatani, N. (2005) *J. Am. Chem. Soc.*, **127**, 5936–45.
30 (a) Pastine, S.J., Gribkov, D.V. and Sames, D. (2006) *J. Am. Chem. Soc.*, **128**, 14220–1;
(b) Comparable products were obtained through ruthenium-catalyzed decarbonylative arylations. Gribkov, D.V., Pastine, S.J., Schnürch, M. and Sames, D. (2007) *J. Am. Chem. Soc.*, **129**, 11750–5.
31 Park, Y.J., Jo, E.-A., and Jun, C.-H. (2005) *Chem. Commun.*, 1185–7.
32 Oi, S., Fukita, S., Hirata, N., Watanuki, N., Miyano, S. and Inoue, Y. (2001) *Org. Lett.*, **3**, 2579–81.
33 (a) Oi, S., Ogino, Y., Fukita, S. and Inoue, Y. (2002) *Org. Lett.*, **4**, 1783–5;
(b) Oi, S., Aizawa, E., Ogino, Y. and Inoue, Y. (2005) *J. Org. Chem.*, **70**, 3113–19;
(c) Oi, S., Funayama, R., Hattori, T. and Inoue, Y. (2008) *Tetrahedron*, **64**, 6051–9.
34 Gant, T.G. and Meyers, A.I. (1994) *Tetrahedron*, **50**, 2297–360.
35 Oi, S., Sakai, K. and Inoue, Y. (2005) *Org. Lett.*, **7**, 4009–11.
36 Ackermann, L., Althammer, A. and Born, R. (2007) *Synlett*, 2833–6.
37 Ackermann, L., Althammer, A. and Born, R. (2008) *Tetrahedron*, **64**, 6115–24.
38 (a) For recent palladium-catalyzed intermolecular direct arylation, see: Daugulis, O. and Chiong, H.A. (2007) *Org. Lett.*, **9**, 1449–51;
(b) Daugulis, O., Chiong, H.A. and Pham, Q.-N. (2008) *J. Am. Chem. Soc.*, **129**, 9879–84.
39 Ackermann, L. (2005) *Org. Lett.*, **7**, 3123–5.
40 Ackermann, L. (2007) *Synlett*, 507–26.
41 Ackermann, L., Althammer, A. and Born, R. (2006) *Angew. Chem. Int. Ed.*, **45**, 2619–22.

42 Ackermann, L., Born, R. and Álvarez-Bercedo, P. (2007) *Angew. Chem. Int. Ed.*, **46**, 6364–7.

43 (a) Davies, D.L., Al-Duaij, O., Fawcett, J., Giardiello, M., Hilton, S.T. and Russel, D.R. (2003) *Dalton Trans.*, 4132–8; (b) Fernández, S., Pfeffer, M., Ritleng, V. and Sirlin, C. (1999) *Organometallics*, **18**, 2390–4.

44 Ackermann, L., Vicente, R. and Althammer, A. (2008) *Org. Lett.*, **10**, 2299–302.

45 (a) Lafrance, M., Lapointe, D. and Fagnou, K. (2008) *Tetrahedron*, **64**, 6015–20; (b) Pascual, S., de Mendoza, P., Braga, A.A.C., Maseras, F. and Echavarren, A. M. (2008) *Tetrahedron*, **64**, 6021–8; and references cited therein.

46 Ruthenium carboxylate complexes are easily obtained under mild reaction conditions. (a) Melchart, M., Habtemariam, A., Parsons, S., Moggach, S.A. and Sadler, P.J. (2006) *Inorg. Chim. Acta*, **359**, 3020–8; (b) A proton abstraction assisted by cesium carbonate was proposed in ruthenium-catalyzed direct arylations with aryl bromides in the presence of imidazolium salts: Özdemir, I., Demir, S., Cetinkaya, B., Gourlaouen, C., Maseras, F., Bruneau, C. and Dixneuf, P. (2008) *J. Am. Chem. Soc.*, **130**, 1156–7.

47 Ko, S., Kang, B. and Chang, S. (2005) *Angew. Chem. Int. Ed.*, **44**, 455–7.

48 Lee, J.M., Na, Y., Han, H. and Chang, S. (2004) *Chem. Soc. Rev.*, **33**, 302–12.

49 Oi, S., Tanaka, Y. and Inoue, Y. (2006) *Organometallics*, **25**, 4773–8.

50 (a) Oi, S., Sato, H., Sugawara, S. and Inoue, Y. (2008) *Org. Lett.*, **10**, 1823–6; (b) Kawashima, T., Takao, T. and Suzuki, H. (2007) *J. Am. Chem. Soc.*, **129**, 11006–7.

51 Kamigata, N., Yoshikawa, M. and Shimizu, T. (1998) *J. Fluorine Chem.*, **87**, 91–5.

52 Fujita, K., Nonogawa, M. and Yamaguchi, R. (2004) *Chem. Commun.*, 1926–7.

53 Pivsa-Art, S., Satoh, T., Kawamura, Y., Miura, M. and Nomura, M. (1998) *Bull. Chem. Soc. Jpn.*, **71**, 467–73.

54 Yoshizumi, T., Tsurugi, H., Satoh, T. and Miura, M. (2008) *Tetrahedron Lett.*, **49**, 1598–600.

55 Do, H.-Q. and Daugulis, O. (2007) *J. Am. Chem. Soc.*, **129**, 12404–5.

56 (a) For selected copper-catalyzed C–H-bond functionalizations, see: Tsang, W.C.P., Zheng, N. and Buchwald, S.L. (2005) *J. Am. Chem. Soc.*, **127**, 14560–1; (b) Chen, X., Hao, X.-S, Goodhue, C.E. and Yu, J.-Q (2006) *J. Am. Chem. Soc.*, **128**, 6790–1; (c) Li, Y., Bohle, D.S. and Li, C.-J. (2006) *Proc. Natl Acad. Sci. USA*, **103**, 8928–33; (d) Brasche, G. and Buchwald, S.L. (2008) *Angew. Chem. Int. Ed.*, **47**, 1932–4.

57 Do, H.-Q. and Daugulis, O. (2008) *J. Am. Chem. Soc.*, **130**, 1128–9.

58 Ackermann, L., Potukuchi, H.K., Landsberg, D. and Vicente, R. (2008) *Org. Lett.*, **10**, 3081–4.

59 Norinder, J., Matsumoto, A., Yoshikai, N. and Nakamura, E. (2008) *J. Am. Chem. Soc.*, **130**, 5858–9.

10
Palladium-Catalyzed Direct Arylation Reactions

Masahiro Miura and Tetsuya Satoh

10.1
Introduction

Today, transition metal-catalyzed cross-coupling is recognized as one of the most useful carbon–carbon bond-formation reactions [1]. The palladium-catalyzed coupling of aryl halides or their synthetic equivalents, such as aryl triflates, with arylmetals (Scheme 10.1a) is very often employed in the synthesis of biaryl molecules, the skeletons of which are found in a wide range of important compounds, including natural products and organic functional materials [1–3].

On the other hand, certain aromatic compounds have been found to undergo arylation directly via C—H bond cleavage on treatment with aryl halides in the presence of transition metal catalysts, including palladium complexes (Scheme 10.1b) [2–11]:

- Suitably functionalized substrates such as phenols and aromatic carbonyl compounds can be arylated regioselectively via precoordination (Scheme 10.2b, i).
- The intramolecular arylative coupling of suitably linked haloaryl-arenes is also generally effective for the construction of polycyclic aromatic compounds (Scheme 10.2b, ii).
- Various five-membered heteroaromatic compounds possessing one or two heteroatoms are known to be so reactive that, even without bearing a functional group, they are capable of undergoing arylation usually at their 2- and/or 5-position(s) on treatment with aryl halides (Scheme 10.2b, iii).

In addition, some related direct arylation reactions using arylmetal reagents are known (Scheme 10.1c).

Apparently, these direct reactions have a significant advantage (that is, the stoichiometric metallation or halogenation of aromatic substrates is not required), and have recently been undergone significant development [2–11]. The representative examples (as well as related examples) that lead to biaryl compounds by means of

Modern Arylation Methods. Edited by Lutz Ackermann

ISBN: 978-3-527-31937-4

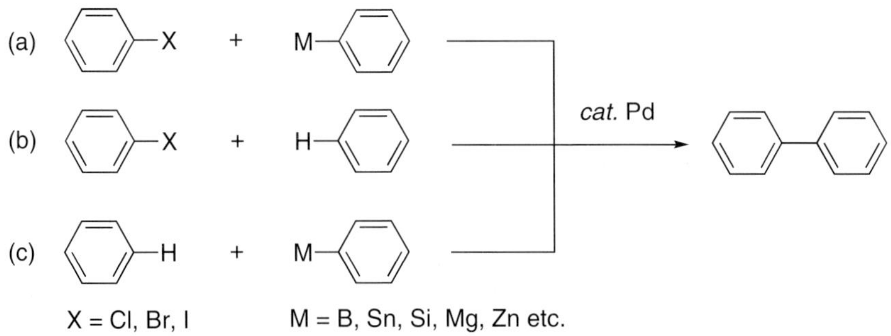

Scheme 10.1 Modes of aryl–aryl coupling by palladium catalysis.

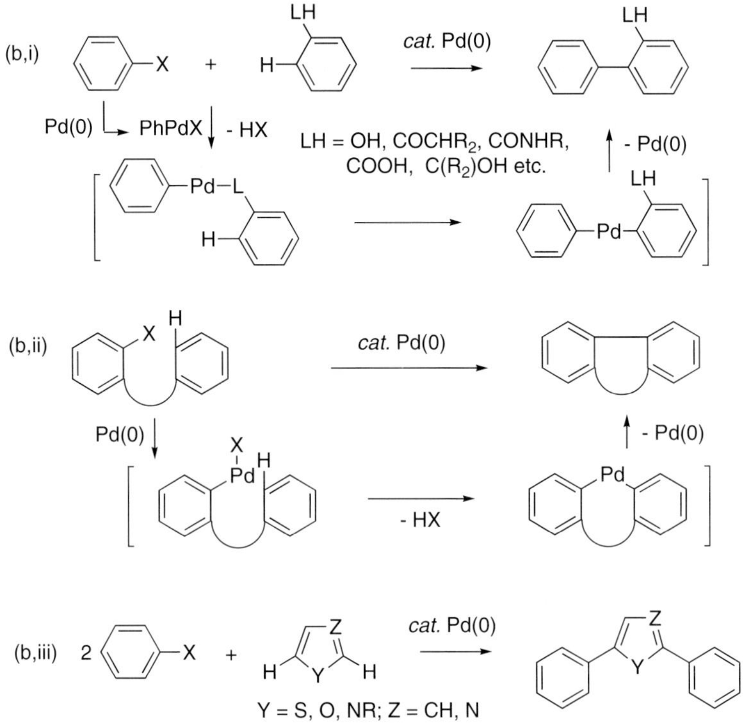

Scheme 10.2 Modes of palladium-catalyzed direct aromatic arylation via C–H bond cleavage.

palladium catalysis are summarized in this chapter, with the information being based on report made up until mid-2007. The reactions with other metal catalysts are described in Chapter 9. It should be noted here that various interesting domino-couplings involving aryl–aryl bond formation through C–H bond cleavage are known to occur in the treatment of aryl halides with unsaturated compounds such as norbornene and diphenylacetylene under palladium catalysis, and several reviews have been prepared describing these reactions [6, 9, 11–13].

10.2
Intermolecular Arylation of Functionalized Arenes

10.2.1
Reaction of Phenols and Benzyl Alcohols

The present authors reported the arylation of 2-phenylphenols with aryl iodides (Scheme 10.3) as one of the first examples to proceed via the sequence shown in Scheme 10.2b, i [14, 15]. The reaction may occur via: (i) oxidative addition of iodobenzene to Pd(0) generated in the reaction medium to give Ph[Pd]I; (ii) coordination of the phenolic oxygen to Pd; (iii) palladation at the 2′-position to form a diarylpalladium intermediate; and (iv) reductive elimination of the product. The use of an inorganic base such as Cs_2CO_3 is important for this coupling. The aromatic arylpalladation via C—H bond cleavage may involve a Friedel–Crafts-type electrophilic substitution (Scheme 10.4i). Other processes, such as σ-bond metathesis, oxidative addition and carbopalladation (Scheme 10.4, ii to iv, respectively) are also possible participants [16, 17]. The fact that the reactions of those substrates which have an electron-donating substituent at the 5′-position proceed smoothly seems consistent with the electrophilic mechanism. However, the strongly electron-withdrawing nitro group does not inhibit the reaction, and therefore other possibilities should also be taken into consideration. Recently, it has been

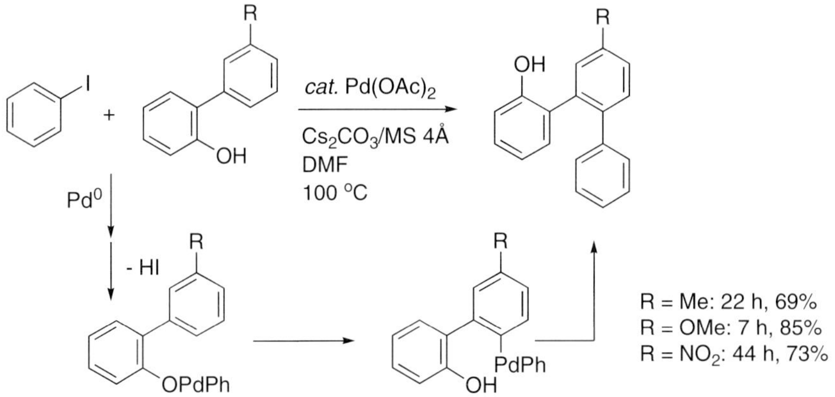

R = Me: 22 h, 69%
R = OMe: 7 h, 85%
R = NO_2: 44 h, 73%

Scheme 10.3 Pd-catalyzed regioselective arylation of 2-phenylphenols.

(i) (ii) (iii) (iv)

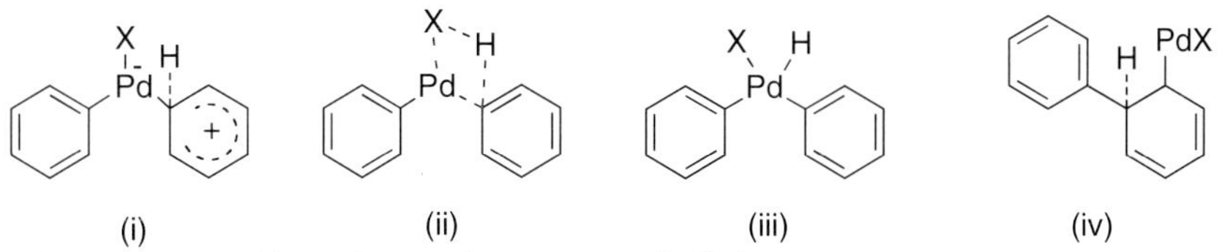

Scheme 10.4 Possible mechanism for aromatic arylpalladation.

suggested that base-assisted C—H bond cleavage processes might play an important role in this type of coupling [7–9, 17]. A detailed discussion of the mechanisms involved is provided in Chapter 11.

Simple 2-phenylphenol with less steric restriction can undergo diarylation to produce a sterically congested 1,2,3-triphenylbenzene derivative (Equation 10.1). 1-Naphthol is arylated at the peri-position selectively (Equation 10.2) [14, 15], while phenol itself can be arylated around the oxygen up to five times by treatment with excess bromobenzene using PPh_3 as ligand (Equation 10.3) [18]. The use of a less-polar solvent such as *o*-xylene is important, since no reaction of phenol occurs in dimethylformamide (DMF). The reaction appears to start with *ortho*-phenylation (by a different mechanism similar to that of α-arylation of ketones; see Chapter 3) [4, 5, 19], followed by the coordination-assisted phenylation. The lack of a hexaphenylated product may be attributed to steric reasons; when the 2- and 6-positions of phenol are masked by *tert*-butyl groups, the para-position is arylated (Equation 10.4) [20]. It is worth noting here that the formation of diphenyl ethers is possible in the reaction of aryl halides with phenols, especially when using bulky phosphines [21].

cat. $PdCl_2$; Cs_2CO_3/DMF; 100 °C, 44 h; 56% (10.1)

cat. $Pd(OAc)_2$; Cs_2CO_3/DMF; 110 °C, 19 h; 73% (10.2)

cat. $Pd(OAc)_2/PPh_3$; Cs_2CO_3/*o*-xylene; 160 °C, 32 h; 58% (10.3)

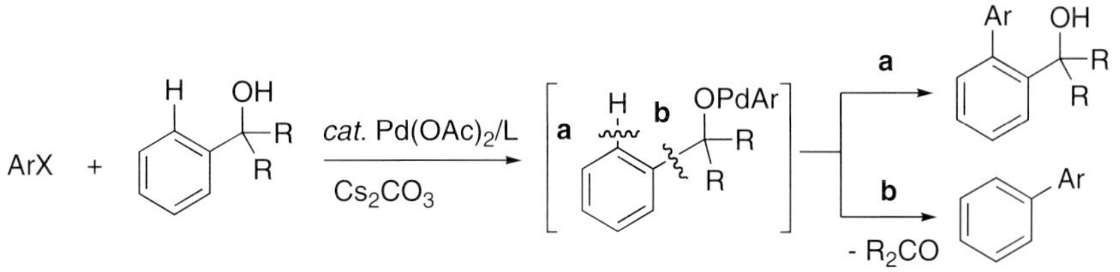

Scheme 10.5 Palladium-catalyzed arylation of *tert*-benzylalcohols via cleavage of C—H and C—C bonds.

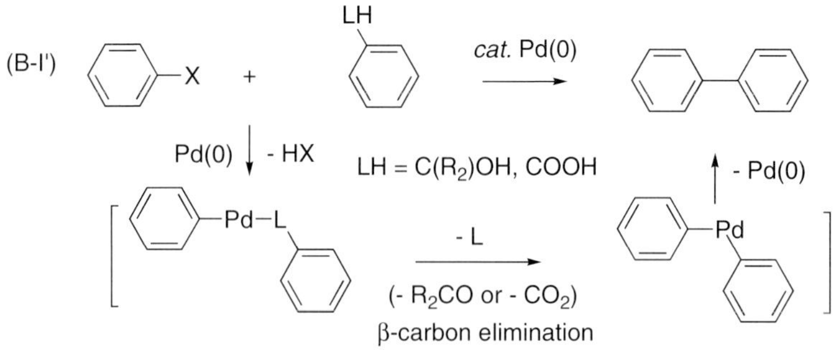

Scheme 10.6 Palladium-catalyzed aryl–aryl coupling via C—C bond cleavage.

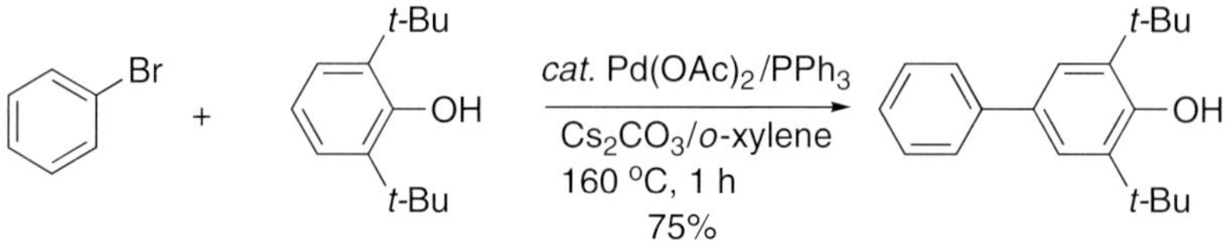

(10.4)

A number of α,α-disubstituted benzylalcohols undergo expected arylation via cleavage of the C(sp^2)—H bond (Scheme 10.5a). Depending on the structure of the substrates and reaction conditions, however, another type of aryl–aryl coupling preferably takes place, accompanied by cleavage of the C(sp^2)–C(sp^3) bond via β-carbon elimination (Scheme 10.5b and Scheme 10.6) [22–24]. As shown in Equation 10.5, the reaction of a simple substrate, 2-phenyl-2-propanol (R=H), with an excess of bromobenzene in the presence of $Pd(OAc)_2/PPh_3$ and Cs_2CO_3 as catalyst and base, respectively, gives mono-, di- and triphenylated products via successive C—H bond cleavage, together with biphenyl and *o*-terphenyl. The latter two products are considered to be formed via the C—C bond cleavage process from the starting alcohol and the monophenylated product, respectively. The reaction with 2-(biphenyl-2-yl)-2-propanol (R=Ph) affords the 1,2,3-triphenylbenzene derivative with an enhanced yield. Interestingly, the reaction of 2-(biphenyl-2-yl)-2-propanol with *ortho*-substituted bromobenzenes proceeds more selectively via C–H bond cleavage. An example is shown in Equation 10.6, where 2-bromotoluene is used as an example, affording a more sterically crowded 1,2,3-triphenylbenzene derivative effectively.

Br + R OH Me Me — *cat.* $Pd(OAc)_2/PPh_3$, Cs_2CO_3/*o*-xylene, 160 °C, 24 h

R = H: 31%, 10%, 15%, 24%, 5%
R = Ph: -, 21%, -, 21%, 45%

(10.5)

Me Br + OH Me Me — *cat.* $Pd(OAc)_2/PPh_3$, Cs_2CO_3/*o*-xylene, 160 °C, 24 h, 76%

(10.6)

In sharp contrast to the reaction in Equation 10.6, that of 2-(2-methylphenyl)-2-propanol with bromobenzene under similar reaction conditions proceeds selectively via C–C bond cleavage to give 2-methylbiphenyl as the single major product (Equation 10.7). Notably, use of PCy_3 in place of PPh_3 as ligand allows the reaction of inexpensive chlorobenzene. The predominant formation of 2-methylbiphenyl indicates that an appropriate *ortho*-substituent on 2-phenyl-2-propanol can selectively induce β-carbon elimination. 2-(1-Naphthyl)-2-propanol, as well as the 9-anthracene and 9-phenanthrene analogues, are also effective substrates that undergo selective coupling via C—C bond cleavage (Equation 10.8). The lack of products via β-carbon elimination with one of the methyl groups is attributable to the fact that such a reaction with an sp^3 carbon is energetically unfavorable.

X + Me OH Me Me — *cat.* $Pd(OAc)_2$/L, Cs_2CO_3/*o*-xylene, 160 °C, 24 h — Me

X = Br, L = PPh_3; 77%
X = Cl, L = PCy_3; 76%

(10.7)

Me Cl Me + Me OH Me — *cat.* $Pd(OAc)_2/PCy_3$, Cs_2CO_3/*o*-xylene, 160 °C, 4 h, quant. → Me Me

(10.8)

10.2.2
Reaction of Aromatic Carbonyl and Pyridyl Compounds

Alkyl aryl ketones are known to be arylated at the α-position of the alkyl group via the corresponding enolates by treatment with aryl halides in the presence of palladium catalysts [4, 5, 19]. The *ortho*-arylation of alkyl aryl ketones is also possible. For example, in the reaction of benzyl phenyl ketone with bromobenzene, the phenylation first occurs at the benzylic position, after which the *ortho*-positions are phenylated via C—H bond cleavage (Equation 10.9) [25]. The *ortho*-phenylation is considered to occur after coordination of the enol oxygen to ArPd(II), which is followed by *ortho*-palladation as in the reaction of 2-phenylphenols, as shown in Scheme 10.3. As expected, the reaction of diphenylmethyl phenyl ketone with 2-bromonaphthalene gives the corresponding diarylated products (Equation 10.10).

Br + CH_2Ph O — *cat.* $Pd(PPh_3)_4$, Cs_2CO_3/*o*-xylene, 160 °C, 20 h, 61% → Ph H O

(10.9)

Br + $CHPh_2$ O — *cat.* $Pd(PPh_3)_4$, Cs_2CO_3/*o*-xylene, 160 °C, 23 h, 43% → $CHPh_2$ O

(10.10)

Benzanilides also undergo *ortho*-arylation (Equation 10.11) [26]. For this reaction, aryl triflates as arylating reagents are more effective than aryl bromides. Here, the reaction is considered to proceed via coordination of amide anion to ArPd(II) as that of benzyl phenyl ketone, since no reaction takes place with secondary amides. *N*-Acylanilines have also been shown to undergo *ortho*-arylation effectively on the aniline moiety on treatment with aryl iodides by using AgOAc as an additive in trifluoroacetic acid (Equation 10.12) [27, 28]. The use of a diphenyliodonium salt as an arylating reagent allows the phenylation of *N*-acylanilines, even of a secondary nature (Equation 10.13) [29]. These reactions are considered to proceed not by the usual Pd(0)–Pd(II), but rather through Pd(II)–Pd(IV) catalytic cycles. On the other hand, the *ortho*-arylation of *N*-acylanilines with aryl(trimethoxy)silanes in place of aryl halides proceeds by a Pd(0)–Pd(II) cycle in the presence of $Cu(OTf)_2$/AgF as reoxidant (Equation 10.14) [30].

OTf + NHPh, O; *cat.* $Pd(OAc)_2/PPh_3$; Cs_2CO_3/toluene, 110 °C, 24 h, 96%; NHPh, O

(10.11)

I + H N *t*-Bu O; *cat.* $Pd(OAc)_2$; $AgOAc/CF_3CO_2H$, 110-130 °C, 91%; NHCO-*t*-Bu

(10.12)

$Ph_2I\ BF_4$ + O N Br; *cat.* $Pd(OAc)_2$; AcOH, 100 °C, 8-24 h, 78%; O N Br

(10.13)

$Si(OMe)_3$ + H N Me O; *cat.* $Pd(OAc)_2$; $Cu(OTf)_2$/AgF, dioxane, 110 °C, 48 h, 74%; NHCOMe

(10.14)

Benzylamines themselves (Equation 10.15) [31] and 2-pyridylbenzenes (Equation 10.16) [29] are directly arylated with aryl iodides or Ph_2IBF_4 by Pd(II)–Pd(IV) catalytic cycles. In the presence of oxone as oxidant, 2-pyridylbenzenes undergo regioselective homocoupling (Equation 10.17) [32]. It is worth noting here that pyridyl nitrogen also acts as an anchor for regioselective aliphatic C—H arylation (Equations 10.18 and 10.19) [33, 34].

I + Br CH_2NH_2 → 1) *cat.* $Pd(OAc)_2$, $AgOAc/CF_3CO_2H$, 130 °C, 2.5 h; 2) $(CF_3CO)_2O$; 68% → Br $CH_2NHCOCF_3$

(10.15)

$Ph_2I\ BF_4$ + Me N → *cat.* $Pd(OAc)_2$, AcOH, 100 °C, 12 h; 88% → Me N

(10.16)

Me N → *cat.* $Pd(OAc)_2$, oxone, *i*-PrOH, 25 °C, 17 h; 86% → Me N N Me

(10.17)

Me I + N → *cat.* $Pd(OAc)_2$, $AgOAc/CH_3CO_2H$, 130 °C, 47 h; 51% → N Me

(10.18)

cat. $Pd(OAc)_2$
AgOAc
110 °C, 5 min
92%

(10.19)

Benzoic acids have been found to undergo *ortho*-arylation with both arylboronates (Equation 10.20) [35] and aryl halides [36]. The reaction has also been applied to regioselective aliphatic C—H arylation (Equation 10.21). It has also been cited that benzoic acids are coupled with aryl halides in the presence of a copper cocatalyst, accompanied by decarboxylation to give the corresponding biaryls. In particular, those having a coordinating group (e.g. a nitro group) react smoothly (Equation 10.22) [37, 38], with the reaction proceeding by the sequence shown in Scheme 10.6. A similar decarboxylative aryl–aryl coupling has been found to occur when using $PdCl_2/AsPh_3$ in dimethylsulfoxide (DMSO) [39].

cat. $Pd(OAc)_2$
benzoquinone
Ag_2CO_3/K_2HPO_4
t-BuOH
120 °C, 3 h
63%

(10.20)

cat. $Pd(OAc)_2$
benzoquinone
Ag_2CO_3/K_2HPO_4
t-BuOH
100 °C, 3 h
38%

(10.21)

cat. $Pd(acac)_2$/CuI/L
K_2CO_3/MS 3Å/NMP
160 °C, 24 h
L = 1,10-phenanthroline
99%

(10.22)

10.2.3
Reaction of Miscellaneous Aromatic Substrates

Benzodioxole has been found to undergo direct arylation at the 3-position selectively in the presence of $Pd(OAc)_2/PMe(t\text{-}Bu)_2$ and $K_2CO_3/AgOTf$ as catalyst and additive, respectively, in *N,N*-dimethylacetamide (DMA). While the mechanism of this reaction is not clear, a cationic ArPd(II) species seems to participate as the reactive intermediate (Equation 10.23) [40]. Interestingly, electron-deficient di- to pentafluorobenzenes couple directly with aryl halides (Equation 10.24) [41, 42]. The use of bulky phosphine ligands is crucial for this reaction; apparently, although the coupling involves no coordination-assisted mechanism, it does occurs efficiently. Furthermore, unactivated benzene itself can be arylated under similar conditions (Equation 10.25) [43]. A mechanism involving base-assisted cooperative deprotonation/metallation has been proposed for the couplings. Cross-coupling between two nonfunctionalized arenes has been reported to occur under acidic oxidative conditions using $K_2S_2O_8/CF_3COOH$, although the turnover numbers are low (Equation 10.26) [44].

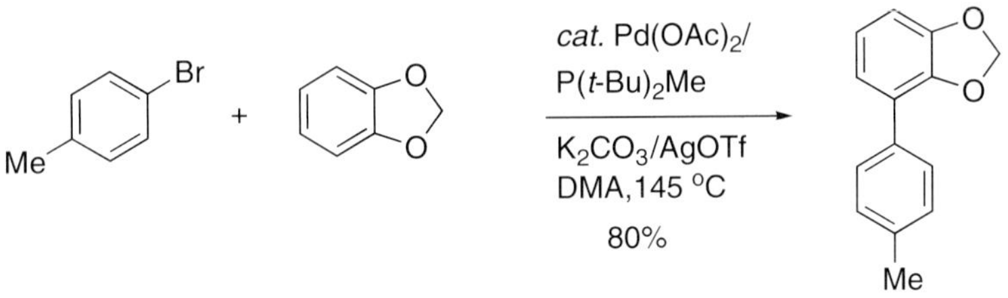

cat. $Pd(OAc)_2$/ $P(t\text{-}Bu)_2Me$, K_2CO_3/AgOTf, DMA, 145 °C, 80%

(10.23)

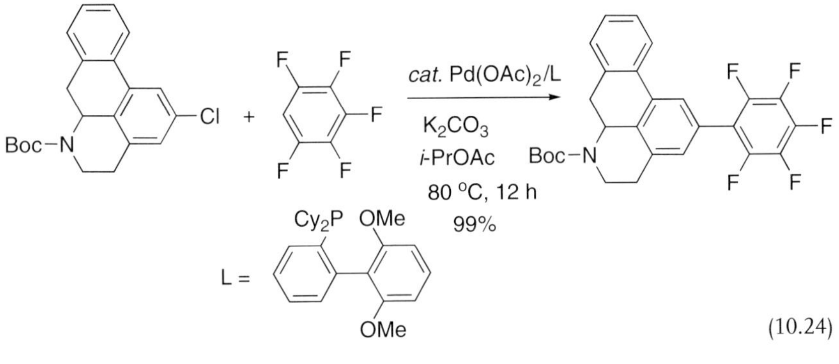

cat. $Pd(OAc)_2$/L, K_2CO_3, *i*-PrOAc, 80 °C, 12 h, 99%

L = Cy_2P, OMe, OMe biphenyl

(10.24)

cat. $Pd(OAc)_2$/L, K_2CO_3/*t*-BuCOOH, DMAc, 120 °C, 82%

L = Cy_2P, NMe_2 biphenyl

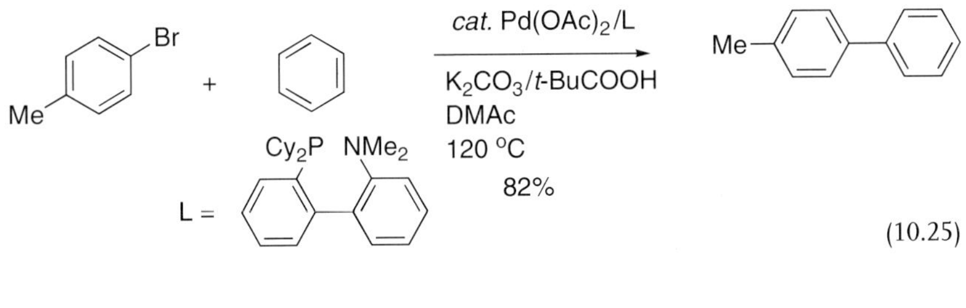

(10.25)

+ *cat.* $Pd(OAc)_2$
$K_2S_2O_8$
CF_3COOH
r. t., 24 h
32%
(10.26)

Cyclopentadienyl anion as a reactive anionic aromatic substrate, which is generated *in situ*, undergoes perarylation on treatment with excess aryl halides to give penta-arylcyclopentadienes (Equation 10.27) [45, 46]. Di(*tert*-butyl)phosphinylferro cene is also perphenylated by chlorobenzene (Equation 10.28) [47], but in this case a coordination-assisted mechanism seems to be operative.

BuO Br + *cat.* $Pd(OAc)_2/P(t\text{-}Bu)_3$
Cs_2CO_3/DMF
140 °C, 24 h
63%
BuO OBu BuO OBu OBu
(10.27)

$P(t\text{-}Bu)_2$ Fe *cat.* $Pd(OAc)_2$
t-BuONa/PhCl
95-110 °C, 12-18 h
60-80%
$P(t\text{-}Bu)_2$ Ph Fe Ph Ph Ph Ph
(10.28)

10.3 Intramolecular Reaction of Haloaryl-Linked Arenes

Ames *et al.* reported the cyclization of bromocinnolines as one of the early important examples of the intramolecular aryl–aryl coupling (Equation 10.29) [48–50]. Today, this type of intramolecular coupling is a standard method for the synthesis of polycyclic compounds [1–9], with the possible cyclization mechanisms perhaps being similar to those for the intermolecular reactions (Scheme 10.4). Among the most recent significant examples is the double cyclization of diiodo compounds by the combination of intramolecular aromatic arylation and *N*-arylation (Equation 10.30) [51]. Interestingly, the reaction of *N*-(2-bromobenzyl)-1-naphthylamines takes place selectively at the peri-position, and not at the 2-position (Equation 10.31) [52]. In this case, coordination of the nitrogen to Pd appears to be a key factor for forming the seven-membered ring.

cat. $Pd(OAc)_2$
NEt_3/MeCN
150 °C, 5 h
55%

(10.29)

cat. $[Pd(dppb)Cl_2]$
KOAc/DMSO
120 °C
98%

(10.30)

cat. $Pd(OAc)_2$/P(*o*-Tol)$_3$
K_2CO_3/DMF
reflux, 2 h
76%

(10.31)

Structurally interesting polycyclic aromatic hydrocarbons can be readily constructed by using a cyclization method, for which the reactions in Equations 10.32–34 are representative examples [53–55]. In the double cyclization of Equation 10.33, the arylation of the $C(sp^3)$–H bond as well as that of the $C(sp^2)$–H bond is involved.

cat. $[Pd(PCy_3)_2Cl_2]$
DBU/DMA
140 °C, 30 h
91%

(10.32)

(10.33)

(10.34)

Fagnou and coworkers have systematically examined the effect of phosphine ligands in a variety of intramolecular aryl–aryl couplings [7, 8], and have shown that suitably bulky phosphines, such as PCy_3 and 2-(dicyclohexylphosphino)-2′-(*N,N*-dimethylamino)biphenyl (Equation 10.35) [56], are often effective.

(10.35)

10.4 Intermolecular Arylation Reactions of Heteroaromatic Compounds

10.4.1 Reaction of Pyrroles, Furans and Thiophenes

The arylation of pyrroles, furans and thiophenes, which are, in general, relatively susceptible to electrophiles, may be considered to proceed through an electrophilic mechanism involving the attack of ArPd(II) species, as judged by the usual substitution pattern (2- and/or 5-arylation) [9, 10]. However, other mechanisms, such as carbopalladation or precoordination of the heteroatoms to Pd, also seem capable of participating. Thus, it is likely that these reactions do not proceed through just one mechanism, but that different reaction pathways occur depending on the substituents on each nucleus and the exact reaction conditions.

One of the earliest significant examples reported is the coupling of 1-substituted indoles with 2-chloro-3,6-dialkylpyridazines (Equation 10.36) [57, 58]. Depending on the 1-substituent, the reaction takes place selectively at either the 2- or 3-position. Substitution at the 3-position is a rare occurrence, and appears to be driven by the electron-withdrawing tosyl group, although the precise mechanism is still not fully understood. As expected, *N*-methylindole (Equation 10.37) [59] and *N*-SEM-protected indole [60] undergo arylation at the 2-position on treatment with aryl iodides.

R = Me or OTs

cat. $Pd(PPh_3)_4$, KOAc/DMA, reflux, 12 h

R = Me, 62%; R = Ts, 4%   +   R = Me, 0%; R = Ts, 54%

(10.36)

cat. $Pd(OAc)_2/PPh_3$, CsOAc, DMA, 125 °C, 24 h, 88%

(10.37)

The arylation of pyrrolyl sodium proceeds selectively at the 2-position in the presence of zinc chloride and an appropriate palladium catalyst system (Equation 10.38) [61]. In this reaction, pyrrolyl zinc chloride, generated *in situ*, may act as a reactive heterocycle. The 2-arylation of indole itself with aryl iodides can be carried out without using any phosphine ligands (Equation 10.39) [62]. The use of either a palladium catalyst with a bulky phosphine ligand (Equation 10.40) [63] or a magnesium base having a bulky nitrogen ligand [59] induces arylation at the 3-position, most likely due to steric interactions. Recently, an interesting oxidative 3-phenylation of *N*-acetylindole with benzene itself has been reported (Equation 10.41) [64]. Indolidines undergo selective arylation on treatment with aryl bromides on the five-membered ring carbon next to the nitrogen (Equation 10.42) [65].

n-Bu, Cl + N Na — *cat.* $Pd(OAc)_2$/ P(*o*-biphenyl)(*t*-Bu)$_2$; $ZnCl_2$/dioxane, 150 °C, 12-15 h, 71% → N H, n-Bu

(10.38)

I + N H — *cat.* $Pd(OAc)_2$; CsOAc, DMA, 125 °C, 24 h, 66% → N H

(10.39)

Br + N H — *cat.* [Pd{P(OH)(*t*-Bu)}$_2$Cl$_2$]; K_2CO_3, dioxane, reflux, 24 h, 72% → N H

(10.40)

H + N Ac — *cat.* $Pd(OCOCF_3)_2$/ 3-nitropyridine; $Cu(OAc)_2$, *t*-BuCOOCs, 140 °C, μW, 87% → N Ac

(10.41)

Br + N — *cat.* [Pd(PPh$_3$)$_2$Cl$_2$]; KOAc/H_2O, NMP, 100 °C, 71% → N

(10.42)

Furan [66, 67] and its 2-formyl derivative (Equation 10.43) [68], as well as 2-alkylfuran [69], are arylated at the 2- and 5- positions. The arylation of 3-ethoxycarbonylfuran as well as its thiophene analogue occurs selectively at the 2-position in the presence of $Pd(PPh_3)_4$ in toluene, whereas the 5-position is attacked preferably using Pd/C in NMP (Equation 10.44) [70]. It has been proposed that carbopalladation and electrophilic attack predominantly participate in the 2- and 5-arylations, respectively.

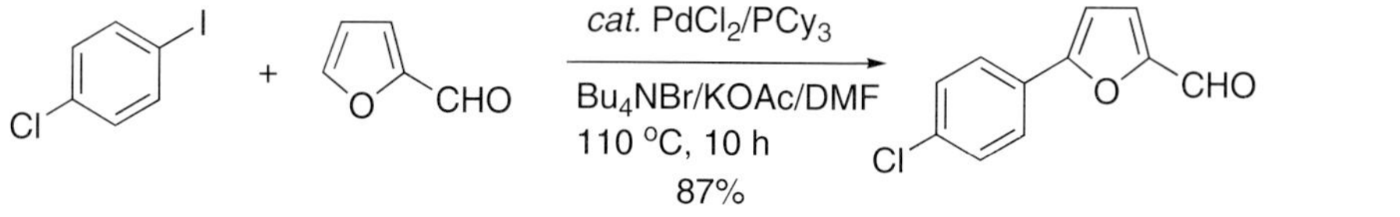

cat. $PdCl_2/PCy_3$
$Bu_4NBr/KOAc/DMF$
110 °C, 10 h
87%

(10.43)

cat. $Pd(PPh_3)_4$, KOAc, toluene, 110 °C	73%	0%
cat. Pd/C, KOAc, NMP, 110 °C	13%	42%

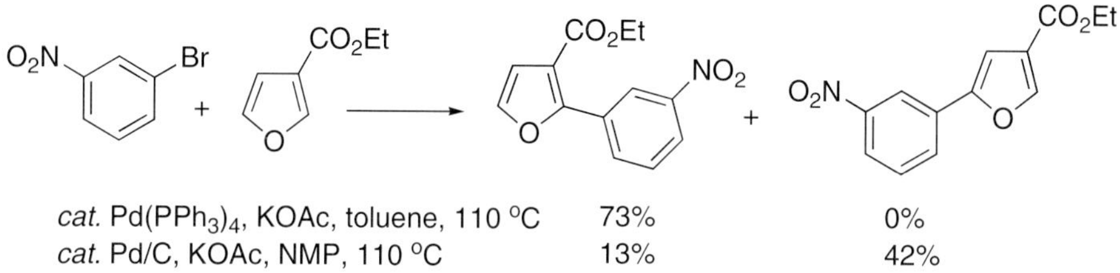

(10.44)

The selective 2-monoarylation of thiophene itself, and furan, can be achieved by the use of an excess of substrate (Equation 10.45) [66]. It has been demonstrated that use of AgF and DMSO as base and solvent, respectively, enables the reaction to proceed at 60 °C (Equation 10.46) [71].

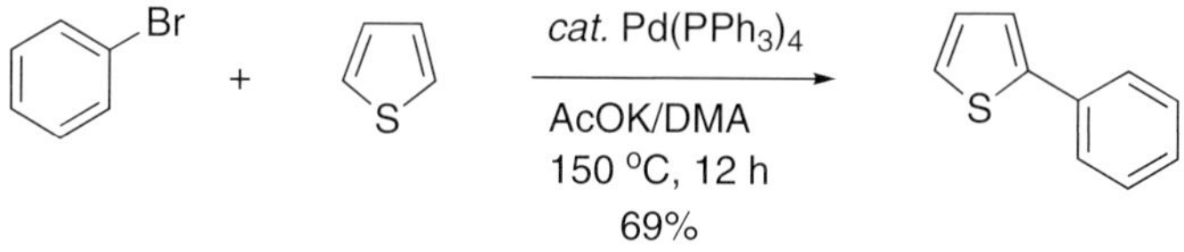

cat. $Pd(PPh_3)_4$
AcOK/DMA
150 °C, 12 h
69%

(10.45)

cat. $[Pd(PPh_3)_2Cl_2]$
AgF or $AgNO_3$-KF
DMSO, 60 °C
60%

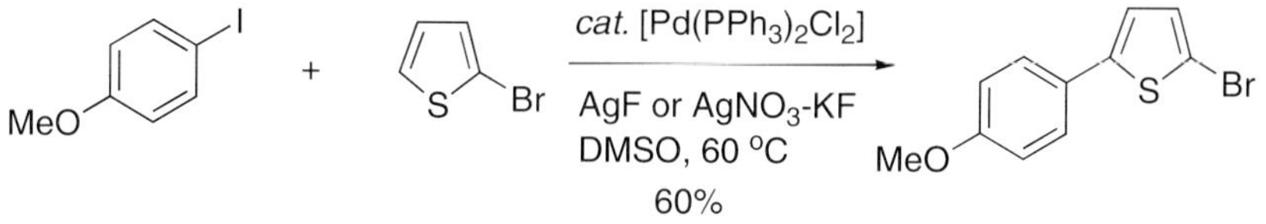

(10.46)

A variety of 2- or 3-substituted thiophenes, as well as benzothiophenes, have been subjected to the catalytic direct arylation [3, 9, 10]. As expected, 2,2′-bithiophene can be diarylated at the 5,5′-positions (Equation 10.47) [72], although the use of a bulky phosphine is of key importance for this reaction. 2,2′-Bithiophene protected by benzophenone at the 5-position reacts with aryl bromides, initially with liberation of the ketone (see Scheme 10.6, to give 5-aryl-2,2′-bithiophene, which is then arylated at the 5′-position (Equation 10.48) [72]. 5-Bromo-2,2′-bithiophenes undergo oxidative homocoupling in the presence of a palladium complex and a silver salt (Equation 10.49) [73].

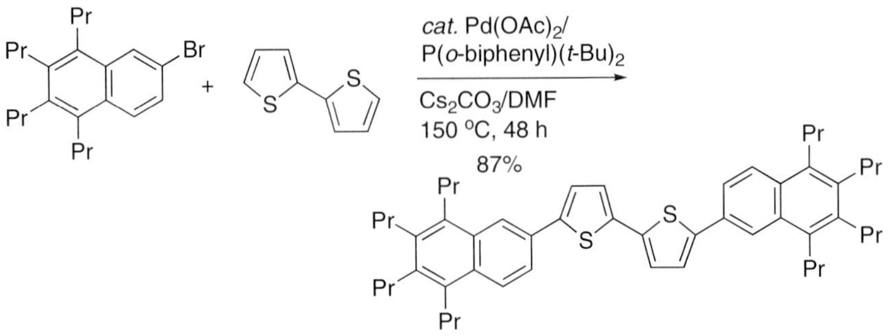

(10.47)

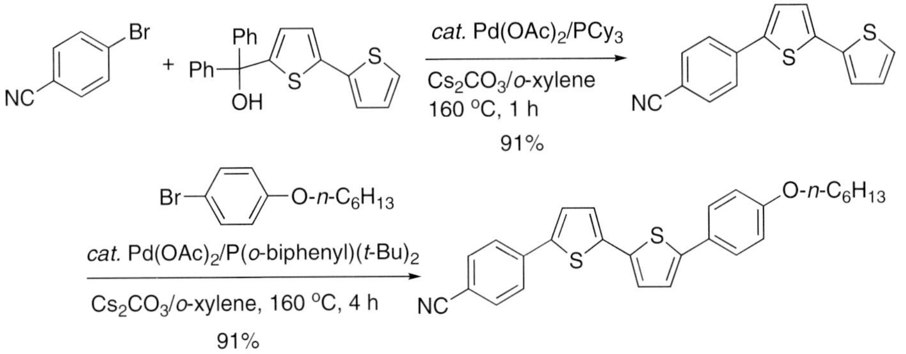

(10.48)

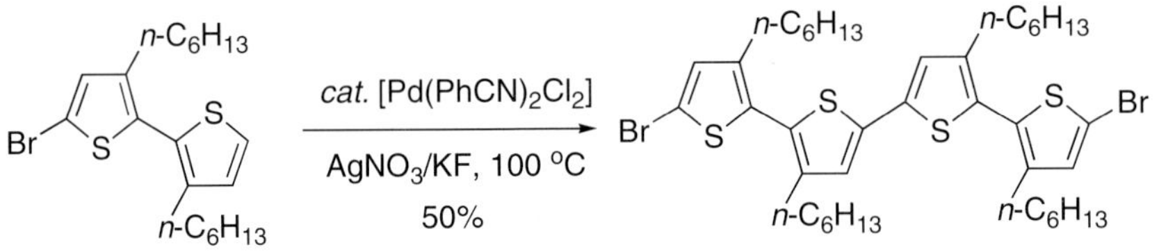

(10.49)

Interestingly, *N*-phenyl-2-thiophenecarboxamides undergo 2,3,5-triarylation accompanied by a formal decarbamoylation upon treatment with excess bromobenzenes (Equation 10.50) [74]. The reaction involves an initial coordination-assisted 3-arylation and successive decarbamoylation which is promoted by a Pd(II) species and the stoichiometric base. A related decarboxylative arylation of 2-thiophenecarboxylic acids at the *ipso*-position has been reported (Equation 10.51) [75]. The introduction of an electron-withdrawing group at the 3-position of thiophene makes the 4-arylation possible, while the reaction at 2- and 5-positions precedes. Thus, the reaction of 3-cyanothiophene affords the corresponding 2,4,5-triarylated products (Equation 10.52) [74]. Whilst the mechanism for the 4-

arylation is not yet fully understood, it has been proposed that the 2-arylation of 3-cyanobenzo[*b*]thiophene proceeds by an insertion mechanism, whereas an electrophilic mechanism predominates in the reaction of 3-methoxybenzo[*b*]thiophene (Equation 10.53) [76].

PhBr + thiophene-2-CONHPh → (cat. $Pd(OAc)_2$/P(*o*-biphenyl)(*t*-Bu)$_2$, Cs_2CO_3/*o*-xylene, 160 °C, 18 h) 2,3,5-triphenylthiophene, 83% (10.50)

PhBr + 3-methylthiophene-2-COOH → (cat. Pd(P*t*-Bu$_3$)$_2$, Bu_4NCl/Cs_2CO_3, DMF, 170 °C (μW)) 3-methyl-2-phenylthiophene, 86% (10.51)

PhBr + 3-cyanothiophene (CN) → (cat. $Pd(OAc)_2$/P(*o*-biphenyl)(*t*-Bu)$_2$, Cs_2CO_3/*o*-xylene, 160 °C, 70 h) 3-cyano-2,4,5-triphenylthiophene (CN), 65% (10.52)

MeO-C$_6$H$_4$-Br + 3-R-benzo[*b*]thiophene → (cat. $Pd(OAc)_2$, Bu_4NCl, K_2CO_3/DMF, 100 °C, 20-28 h) 2-(4-methoxyphenyl)-3-R-benzo[*b*]thiophene (OMe); R = CN, 66%; R = OMe, 61% (10.53)

10.4.2
Reaction of Imidazoles, Oxazoles and Thiazoles

The order of reactivity for the reaction sites of azole compounds in electrophilic reactions is known to be 5 > 4 > 2 [77]. Thus, arylation at the relatively electron-rich 5-position may be considered to be similar to that of pyrroles, furans and thiophenes. In contrast, reaction at the 2-position may be considered to proceed in a different manner; whilst the precise mechanism has still not yet been determined, it may involve base-assisted deprotonative palladation with ArPd(II) species. An insertion of the C—N double bond into the Ar—Pd bond might represent a possible pathway, but nonetheless the 5-arylation usually proceeds faster than the 2-arylation. The order of reactivity can be changed by the use of an additive such Cu(I) species, however (see below).

The reaction of 1-methyl-*1H*-imidazole with 2 equiv. of iodobenzene leads to the production of its 5-phenyl derivative as the major product, along with the 2,5-diphenyl derivative. Interestingly, the addition of 2 equiv. of CuI as promoter produces the 2-phenyl derivative, together with the 2,5-diphenyl derivative, but no 5-phenyl derivative is formed (Equation 10.54) [78]. It has been proposed that the 2-selectivity arises from the participation of a 2-imidazolylcopper species as intermediate [79–81].

I + N, N Me

cat. $Pd(OAc)_2/PPh_3$
$(CuI)/Cs_2CO_3/DMF$
140 °C, 20 h

	N Me (5-phenyl)	N Me (2-phenyl)	N Me (2,5-diphenyl)
- CuI	54%	0%	24%
+CuI	0%	37%	40%

(10.54)

The arylations of condensed imidazoles such as imidazopyrimidines and imidazotriazines have also been reported. The palladium-catalyzed arylation of imidazo[1,2-*a*]pyrimidine takes place regioselectively (Equation 10.55) [82]. This reaction has been employed as a key step in the synthesis of a potential γ-amino-butyric acid (GABA) agonist drug candidate, 2′,4-difluoro-5′-(7-trifluoromethyl-imidazo[1,2-*a*]pyrimidin-3-yl)- biphenyl-2-carbonitrile [83]. The arylation of imidazo[1,2-*b*][1,2,4]triazine has been used for the synthesis of 2′,6-difluoro-5′-[3-(1-hydroxy-1-methylethyl)imidazo[1,2-*b*][1,2,4]triazin-7-yl]biphenyl-2-carbonitrile, which is also a GABA agonist (Equation 10.56) [84, 85]. Caffeine has been found to be a suitable substrate for the direct arylation (Equation 10.57) [86]. The 8-arylation of a purine derivative has also been reported [87]. Recently, 1-substituted 1,2,3-triazoles were shown to undergo selective 5-arylation [88–90].

Br + N N N

cat. $Pd(OAc)_2/PPh_3$
Cs_2CO_3/dioxane
100 °C, 18 h
71%

N N N

(10.55)

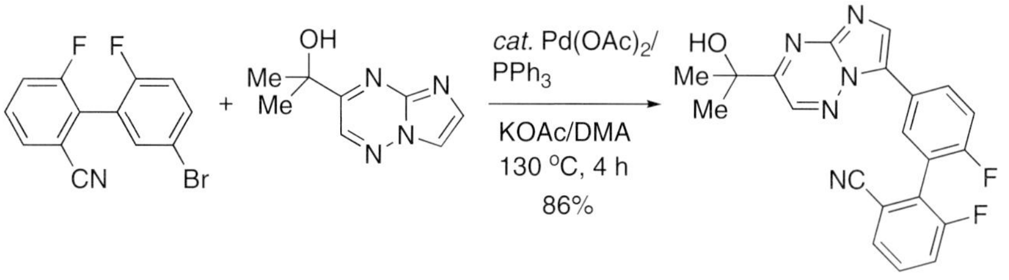

cat. $Pd(OAc)_2$/PPh_3, KOAc/DMA, 130 °C, 4 h, 86%

(10.56)

cat. $Pd(OAc)_2$/$P(n\text{-}Bu)(1\text{-}Ad)_2$, Cs_2CO_3/DMF, 125 °C, 24 h, 86%

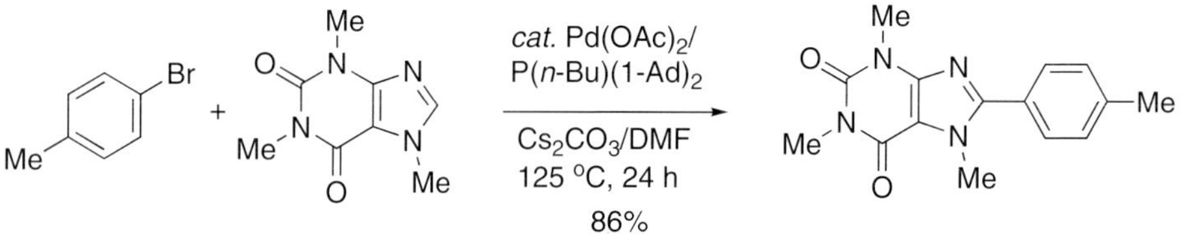

(10.57)

Although, examples of the arylation of oxazole itself are limited, the reaction with 2-chloro-3,6-dialkylpyridazines at the 5-position is known [58]. 2-Phenyloxazole [78] and benzoxazole [58, 78, 91, 92] are good substrates for the direct arylation (Equations 10.58 and 10.59) [78]. The arylation of oxazolo[4,5-*b*]pyridine can be carried out under mild conditions (Equation 10.60) [93]. Treatment of ethyl 4-oxazolecarboxylate with iodobenzene in the presence of a catalyst system consisting of $Pd(OAc)_2$ and IMes affords its 2-phenylated product predominantly (Equation 10.61) [94]. The regioselectivity may be attributable to steric reasons.

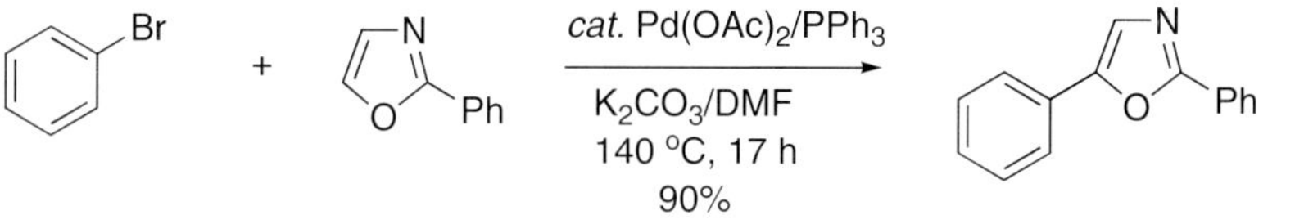

cat. $Pd(OAc)_2$/PPh_3, K_2CO_3/DMF, 140 °C, 17 h, 90%

(10.58)

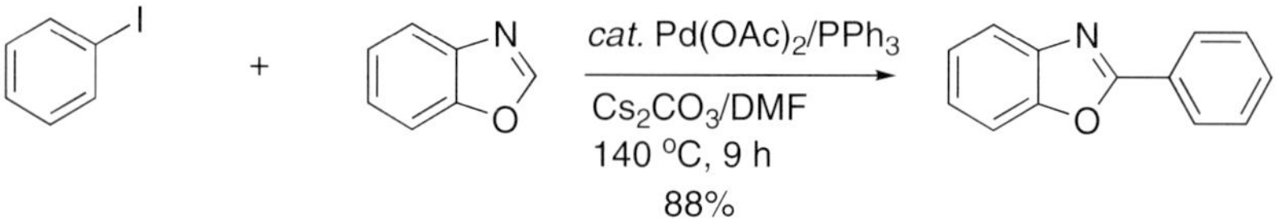

cat. $Pd(OAc)_2$/PPh_3, Cs_2CO_3/DMF, 140 °C, 9 h, 88%

(10.59)

cat. $Pd(OAc)_2$/PPh_3, Cs_2CO_3/acetone, 30 °C, 24 h, 68%

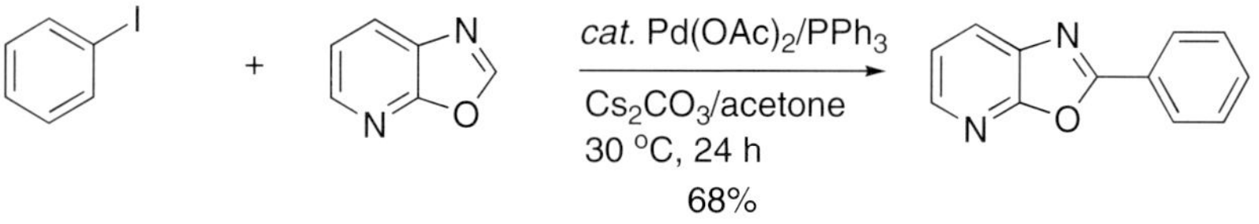

(10.60)

cat. $Pd(OAc)_2$/L; Cs_2CO_3/dioxane, 110 °C, 18 h, 88%

L = Mes–N N–Mes (imidazol-2-ylidene)

(+ 2,5-isomer, 11%)

(10.61)

The 2,5-diarylation of thiazole can be carried out effectively with a bulky phosphine ligand. In this case, no monoarylated product is observed, even in the early stage of the reaction, which suggests that the second arylation proceeds relatively rapidly (Equation 10.62) [95]. The selective 2-arylation of thiazole is accomplished by using CuI and Bu_4NF as cocatalyst and base, respectively (Equation 10.63) [96]. By using $Pd(OH)_2/C$ as catalyst, the 5-position can be arylated selectively (Equation 10.64) [97]. The palladium-catalyzed arylation of thiazole as well as 1-methylpyrrole with a polymer-linked aryl iodide has been reported [98].

$Pd(OAc)_2$/ P(*o*-biphenyl)(*t*-Bu)$_2$; Cs_2CO_3/DMF, 150 °C, 3 h, 69%

(10.62)

cat. $[Pd(PPh_3)_2Cl_2]$/CuI; Bu_4NF/DMSO, 60 °C, 24 h, 82%

(10.63)

cat. $Pd(OH)_2/C$; KOAc/DMA, 140 °C, 12 h, 73%

(10.64)

10.4.3 Reaction of Six-Membered Nitrogen Heterocycles

The direct arylation of electron-deficient heteroaromatics such as pyridine remains a major challenge, as these compounds are much less reactive than the relatively electron-rich, five-membered heteroarenes described above. The phenylation of pyridine at the 2-position has been found to proceed on a heterogeneous palladium catalyst in the presence of zinc and water, albeit with moderate efficiency (Equation 10.65) [99].

Cl + N → (*cat.* Pd/C; Zn/H_2O; 115 °C, 20 h; 52%) N (10.65)

Recently, it was reported that pyridine *N*-oxides and pyridazine *N*-oxide undergo 2-arylation efficiently in the presence of a bulky ligand (Equations 10.66 and 10.67) [100, 101], with both reactions perhaps involving a nucleophilic arylation mechanism.

Me, Br + N^+–O^- → (*cat.* $Pd(OAc)_2/(t\text{-}Bu)_3P{\cdot}HBF_4$; K_2CO_3/toluene; 110 °C, 18 h; 91%) Me, N^+–O^- (10.66)

Me, Br + N, N^+–O^- → (*cat.* $Pd(OAc)_2/(t\text{-}Bu)_3P{\cdot}HBF_4$; K_2CO_3/dioxane; 110 °C, 16 h; 75%) Me, N, N^+–O^- (10.67)

10.5 Concluding Remarks

A variety of methods for the catalytic direct arylation of aromatic and heteroaromatic compounds via the cleavage of C–H bonds has been developed during recent years. As complementary synthetic tools for conventional cross-couplings in the preparation of biaryls and arylated heteroarenes that require neither stoichiometric metallation nor halogenation, these reactions may be both useful and economical in a variety of situations. Consequently, it is highly likely that a major effort will be made in the near future to enhance the catalytic efficiency and regioselectivity of these reactions.

Abbreviations

Ac	acetyl
Ad	adamantyl
Cy	cyclohexyl
DMA	*N,N*-dimethylacetamide
DMF	*N,N*-dimethylformamide
DMSO	dimethylsulfoxide
GABA	γ-aminobutyric acid
IMes	1,3-dimesitylimidazol-2-ylidene
Mes	mesityl
MS	molecular sieves
μW	microwave
NMP	*N*-methylpyrrolidinone
OTf	trifluoromethanesulfonate
SEM	2-(trimethylsilyl)ethoxymethyl
Tol	tolyl

References

1 de Meijere, A. and Diederich, F. (eds) (2004) *Metal-Catalyzed Cross-Coupling Reactions*, 2nd edn Wiley-VCH Verlag GmbH, Weinheim, Germany.

2 Tsuji, J. (2004) *Palladium Reagents and Catalysts*, 2nd edn, John Wiley & Sons, Ltd, Chichester, UK.

3 Hassan, J., Sévignon, M., Gozzi, C., Schulz, E. and Lemaire, M. (2002) *Chem. Rev.*, **102**, 1359–469.

4 Miura, M. and Nomura, M. (2002) *Topics Curr. Chem.*, **219**, 211–41.

5 Miura, M. and Satoh, T. (2005) *Top. Organomet. Chem.*, **14**, 55–83.

6 Dyker, G. (ed.) (2005) *Handbook of C-H Transformations*, Wiley-VCH Verlag GmbH, Weinheim, Germany.

7 Campeau, L.-C. and Fagnou, K. (2006) *Chem. Commun.*, 1253–64.

8 Campeau, L.-C., Stuart, D.R. and Fagnou, K. (2007) *Aldrichim. Acta*, **40**, 35–41.

9 Alberico, D., Scott, M.E. and Lautens, M. (2007) *Chem. Rev.*, **107**, 174–238.

10 Satoh, T. and Miura, M. (2007) *Chem. Lett.*, **36**, 200–5.

11 Catellani, M., Motti, E., Ca', N.D. and Ferraccioli, R. (2007) *Eur. J. Org. Chem*, 4153–65.

12 Catellani, M. (2005) *Top. Organomet. Chem.*, **14**, 21–53.

13 Larock, R.C. (2005) *Top. Organomet. Chem.*, **14**, 147–82.

14 Satoh, T., Kawamura, Y., Miura, M. and Nomura, M. (1997) *Angew. Chem. Int. Ed. Engl*, **36**, 1740–2.

15 Satoh, T., Inoh, J., Kawamura, Y., Kawamura, Y., Miura, M. and Nomura, M. (1998) *Bull. Chem. Soc. Jpn*, **71**, 2239–46.

16 Dyker, G. (1997) *Chem. Ber./Recueil*, **130**, 1567–78.

17 Garcia-Cuadrado, D., de Mendoza, D., Braga, A.A.C., Maseras, F. and Echavarren, A.M. (2007) *J. Am. Chem. Soc.*, **129**, 6880–6.

18 Kawamura, Y., Satoh, T., Miura, M. and Nomura, M. (1999) *Chem. Lett.*, 961–2.

19 Culkin, D.A. and Hartwig, J.F. (2003) *Acc. Chem. Res.*, **36**, 234–45.

20 Kawamura, Y., Satoh, T., Miura, M. and Nomura, M. (1998) *Chem. Lett.*, 931–2.

21 Muci, A.R. and Buchwald, S.L. (2002) *Topics Curr. Chem.*, **219**, 131–209.
22 Satoh, T. and Miura, M. (2005) *Top. Organomet. Chem.*, **14**, 1–20.
23 Terao, Y., Wakui, H., Satoh, T., Miura, M. and Nomura, M. (2001) *J. Org. Chem.*, **123**, 10407–8.
24 Terao, Y., Wakui, H., Nomoto, M., Satoh, T., Miura, M. and Nomura, M. (2003) *J. Org. Chem.*, **68**, 5236–43.
25 Satoh, T., Kametani, Y., Terao, Y., Miura, M. and Nomura, M. (1999) *Tetrahedron Lett.*, **40**, 5345–8.
26 Kametani, Y., Satoh, T., Miura, M. and Nomura, M. (2000) *Tetrahedron Lett.*, **41**, 2655–8.
27 Daugulis, O. and Zaitsev, V.G. (2005) *Angew. Chemie. Int. Ed.*, **44**, 4046–8.
28 Shabashov, D. and Daugulis, O. (2007) *J. Org. Chem.*, **72**, 7720–5.
29 Kalyani, D., Deprez, N.R., Desai, L.V. and Sanford, M.S. (2005) *J. Am. Chem. Soc.*, **127**, 7330–1.
30 Yang, S., Li, B., Wan, X. and Shi, Z. (2007) *J. Am. Chem. Soc.*, **129**, 6066–7.
31 Lazareva, A. and Daugulis, O. (2006) *Org. Lett.*, **8**, 5211–13.
32 Hull, K.L., Lanni, E.L. and Sanford, M.S. (2006) *J. Am. Chem. Soc.*, **128**, 14047–9.
33 Shabashov, D. and Daugulis, O. (2005) *Org. Lett.*, **7**, 3657–9.
34 Zaitsev, V.G., Shabashov, D. and Daugulis, O. (2005) *J. Am. Chem. Soc.*, **127**, 13154–5.
35 Giri, R., Maugel, N., Li, J.-J., Wang, D.-H., Breazzano, S.P., Saunders, L.B. and Yu, J.-Q. (2007) *J. Am. Chem. Soc.*, **129**, 3510–11.
36 Chiong, H.A., Pham, Q.-N. and Daugulis, O. (2007) *J. Am. Chem. Soc.*, **129**, 9879–84.
37 Goossen, L.J., Deng, G. and Levy, L.M. (2006) *Science*, **313**, 662–4.
38 Goossen, L.J., Rodriguez, N., Melzer, B., Linder, C., Deng, G. and Levy, L.M. (2007) *J. Am. Chem. Soc.*, **129**, 4824–33.
39 Becht, J.M., Catala, C., Drian, C.L. and Wagner, A. (2007) *Org. Lett.*, **9**, 1781–3.
40 Champeau, L.-C., Parisien, M., Jean, A. and Fagnou, K. (2006) *J. Am. Chem. Soc.*, **128**, 581–90.
41 Lafrance, M., Shore, D. and Fagnou, K. (2006) *Org. Lett.*, **8**, 5097–100.
42 Lafrance, M., Rowley, C.N., Woo, T.K. and Fagnou, K. (2006) *J. Am. Chem. Soc.*, **128**, 8754–6.
43 Lafrance, M. and Fagnou, K. (2006) *J. Am. Chem. Soc.*, **128**, 16496–7.
44 Li, R., Jiang, L. and Lu, W. (2006) *Organometallics*, **25**, 5973–4.
45 Miura, M., Pivsa-Art, S., Dyker, G., Heiermann, J., Satoh, T. and Nomura, M. (1998) *Chem. Commun.*, 1889–90.
46 Dyker, G., Heiermann, J., Miura, M., Inoh, J., Pivsa-Art, S., Satoh, T. and Nomura, M. (2000) *Chem. Eur. J.*, **6**, 3426–33.
47 Kataoka, N., Shelby, Q., Stambuli, J.P. and Hartwig, J.F. (2002) *J. Org. Chem.*, **67**, 5553–66.
48 Ames, D.E. and Bull, D. (1982) *Tetrahedron*, **38**, 383–7.
49 Ames, D.E. and Opalko, A. (1983) *Synthesis*, 234–5.
50 Ames, D.E. and Opalko, A. (1984) *Tetrahedron*, **40**, 1919–25.
51 Cuny, G., Bois-Choussy, M. and Zhu, J. (2003) *Angew. Chemie. Int. Ed.*, **42**, 4774–7.
52 Harayama, T., Sato, T., Hori, A., Abe, H. and Takeuchi, Y. (2004) *Synthesis*, 1446–56.
53 Wang, L. and Shevlin, P.B. (2000) *Org. Lett.*, **2**, 3703–5.
54 Ren, H., Li, Z. and Knochel, P. (2007) *Chem. Asian J.*, **2**, 416–33.
55 Kim, D., Petersen, J.L. and Wang, K.K. (2006) *Org. Lett.*, **8**, 2313–16.
56 Leblanc, M. and Fagnou, K. (2005) *Org. Lett.*, **7**, 2849–52.
57 Akita, Y., Itagaki, Y., Takizawa, S. and Ohta, A. (1989) *Chem. Pharm. Bull.*, **37**, 1477–80.
58 Aoyagi, Y., Inoue, A., Koizumi, I., Hashimoto, R., Tokunaga, K., Gohma, K., Komatsu, J., Sekine, K., Miyafuji, A., Kunoh, J., Honma, R., Akita, Y. and Ohta, A. (1992) *Heterocycles*, **33**, 257–72.
59 Lane, B.S., Brown, M.A. and Sames, D. (2005) *J. Am. Chem. Soc.*, **127**, 8050–7.

60 Toure, B.B., Lane, B.S. and Sames, D. (2006) *Org. Lett.*, **8**, 1979–82.

61 Rieth, R.D., Mankad, N.P., Calimano, E. and Sadighi, J.P. (2004) *Org. Lett.*, **6**, 3981–3.

62 Wang, X., Girbkov, D.V. and Sames, D. (2007) *J. Org. Chem.*, **72**, 1476–9.

63 Zhang, Z., Hu, Z., Yu, Z., Lei, P., Chi, H., Wang, Y. and He, R. (2007) *Tetrahedron Lett.*, **48**, 2415–19.

64 Stuart, D.R. and Fagnou, K. (2007) *Science*, **316**, 1172–5.

65 Park, C.-H., Ryabova, V., Seregin, I.V., Sromek, A.W. and Gevorgyan, V. (2004) *Org. Lett.*, **6**, 1159–62.

66 Ohta, A., Akita, Y., Ohkuwa, T., Chiba, M., Fukunaga, R., Miyafuji, A., Nakata, T., Tani, N. and Aoyagi, Y. (1990) *Heterocycles*, **31**, 1951–8.

67 Catellani, M., Chiusoli, G.P. and Ricotti, S. (1985) *J. Organomet. Chem.*, **296**, C11–15.

68 McClure, M.S., Glover, B., McSorley, E., Millar, A., Osterhout, M.H. and Roschanger, F. (2001) *Org. Lett.*, **3**, 1677–80.

69 Battace, A., Lemhadri, M., Zair, T., Doucet, H. and Santelli, M. (2007) *Organometallics*, **26**, 472–4.

70 Glover, B., Harvey, K.A., Liu, B., Sharp, M.J. and Tymoschenko, M.F. (2003) *Org. Lett.*, **5**, 301–4.

71 Kobayashi, K., Sugie, A., Takahashi, M., Masui, K. and Mori, M. (2005) *Org. Lett.*, **7**, 5083–5.

72 Yokooji, A., Satoh, T., Miura, M. and Nomura, M. (2004) *Tetrahedron*, **60**, 6757–63.

73 Takahashi, M., Masui, K., Sekiguchi, H., Kobayashi, N., Mori, A., Funahashi, M. and Tamaoki, N. (2006) *J. Am. Chem. Soc.*, **128**, 10930–3.

74 Okazawa, T., Satoh, T., Miura, M. and Nomura, M. (2002) *J. Am. Chem. Soc.*, **124**, 5286–7.

75 Forgione, F., Brochu, M.-C., Onge, M. St-, Thesen, K.H., Bailey, M.D. and Bilodeau, F. (2006) *J. Am. Chem. Soc.*, **128**, 11350–1.

76 Chabert, J.F.D., Joucla, L., David, E. and Lemaire, M. (2004) *Tetrahedron*, **60**, 3221–30.

77 Potts, K.T. (ed.) (1984) *Comprehensive Heterocyclic Chemistry*, Vols 5 and 6, Pergamon Press, Oxford.

78 Pivsa-Art, S., Satoh, T., Kawamura, Y., Miura, M. and Nomura, M. (1998) *Bull. Chem. Soc. Jpn*, **71**, 467–73.

79 Bellina, F., Cauteruccio, S., Mannina, L., Rossi, R. and Viel, S. (2006) *Eur. J. Org. Chem*, 693–703.

80 Bellina, F., Cauteruccio, S., Mannina, L., Rossi, R. and Viel, S. (2005) *J. Org. Chem.*, **70**, 3997–4005.

81 Bellina, F., Cauteruccio, S. and Rossi, R. (2006) *Eur. J. Org. Chem*, 1379–82.

82 Li, W., Nelson, D.P., Jensen, M.S., Hoerrner, R.S., Javadi, G.J., Cai, D. and Larsen, R.D. (2003) *Org. Lett.*, **5**, 4835–7.

83 Cameron, M., Foster, B.S., Lynch, J.E., Shi, Y.-J. and Dolling, U.-H. (2006) *Org. Process Res. Dev.*, **10**, 398–402.

84 Gautier, D.R., Jr, Limanto, J., Devine, P.N., Desmond, R.A. and Szumigala, R. H. (2005) *J. Org. Chem.*, **70**, 5938–45.

85 Jensen, M.S., Hoerrner, R.S., Li, W., Nelson, D.P., Javadi, G.J., Dormer, P.D., Cai, D. and Larsen, R.D. (2005) *J. Org. Chem.*, **70**, 6034–9.

86 Chiong, H.A. and Daugulis, O. (2007) *Org. Lett.*, **9**, 1449–51.

87 Cerna, I., Kleptarova, B. and Hocek, M. (2006) *Org. Lett.*, **8**, 5389–92.

88 Chuprakov, S., Chernyak, N., Dudnik, A.S. and Gevorgyan, V. (2007) *Org. Lett.*, **9**, 2333–6.

89 Iwasaki, M., Yorimitsu, H. and Oshima, K. (2007) *Chem. Asian J.*, **2**, 1430–5.

90 Ackermann, L., Vicente, R. and Born, R. (2008) *Adv. Synth. Catal.*, **350**, 741–8.

91 Alagille, D., Baldwin, R.M. and Tamagnan, G.D. (2005) *Tetrahedron Lett.*, **46**, 1349–51.

92 Sanchez, R.S. and Zhuravlev, F.A. (2007) *J. Am. Chem. Soc.*, **129**, 5824–5.

93 Zhuravlev, F.A. (2006) *Tetrahedron Lett.*, **47**, 2929-32.

94 Hoarau, C., de Kerdaniel, A.D.F., Bracq, N., Grandclaudon, P., Couture, A. and Marsais, F. (2005) *Tetrahedron Lett.*, **46**, 8573–7.

95 Yokooji, A., Okazawa, T., Satoh, T., Miura, M. and Nomura, M. (2003) *Tetrahedron*, **69**, 5685–9.

96 Mori, A., Sekiguchi, A., Masui, K., Shimada, T., Horie, M., Osakada, K., Kawamoto, M. and Ikeda, T. (2003) *J. Am. Chem. Soc.*, **125**, 1700–1.

97 Parisien, M., Valette, D. and Fangou, K. (2005) *J. Org. Chem.*, **70**, 7578–84.

98 Kondo, Y., Komine, T. and Sakamoto, T. (2000) *Org. Lett.*, **2**, 3111–13.

99 Mukhopadhyay, S., Rothenberg, G., Gitis, D., Baidossi, M., Ponde, D.E. and Sasson, Y. (2000) *J. Chem. Soc. Perkin Trans 1*, **2**, 1809–12.

100 Campeau, L.-C., Rousseaux, S. and Fagnou, K. (2005) *J. Am. Chem. Soc.*, **127**, 18020–1.

101 Leclerc, J.-P. and Fagnou, K. (2006) *Angew. Chemie. Int. Ed.*, **45**, 7781–6.

11
Mechanistic Aspects of Transition Metal-Catalyzed Direct Arylation Reactions

Paula de Mendoza and Antonio M. Echavarren

11.1
Introduction

The direct arylation of arenes catalyzed by transition metals offers a more direct alternative for the formation of aryl–aryl bonds [1–7] than methods based on cross-coupling reactions [8, 9]. Although cross-coupling methods – and, in particular, those based on the use of palladium catalysts such as the Suzuki–Miyaura, Stille, Negishi and Sonogashira couplings – still dominate the synthetic arena, direct arylation is today playing a more significant role in the context of the synthesis of complex natural products [4, 10] and other polyarenes [5, 11–23].

The vast majority of methods are based on the use of palladium complexes as catalysts, although copper, ruthenium, rhodium and iridium catalysts have also been used. Progress in the understanding of the mechanisms of these reactions has only been made during the past few years. As comprehensive reviews have been recently published on aryl–aryl bond-formation reactions, covering both mechanistic and synthetic aspects of these reactions [3–7], in this chapter we will summarize only those mechanistic studies on metal-catalyzed arylation reactions that have been carried out in detail.

11.2
Palladium-Catalyzed Intramolecular Direct Arylation

Significant progress has been made recently on the synthetic application of intramolecular direct arylation reactions for the formation of five to seven-membered ring compounds by using different palladium catalysts [3, 24], notably those bearing bulky phosphines [25] or *N*-heterocyclic carbenes (NHC)s as ligands [25e, 26]. The reaction takes place on substrates of general type **1** (where X = Cl, Br, I or a pseudohalide) to form carbo- and heterocycles **2** (Scheme 11.1) [1–5]. The initially formed oxidative addition palladium(II) complex $ArPdXL_2$ could evolve by different mechanisms through intermediates **3–5** (Scheme 11.1). Most authors

Modern Arylation Methods. Edited by Lutz Ackermann

ISBN: 978-3-527-31937-4

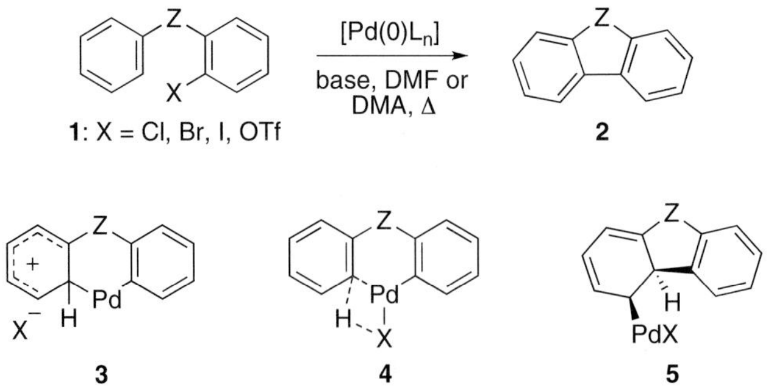

Scheme 11.1

have favored an electrophilic aromatic substitution (S_EAr) via intermediates **3**, considering that the initially formed $ArPdXL_2$ complexes would react as electrophiles with the arene ring [1, 2, 4]. An interesting mechanistic alternative is a sigma bond metathesis via intermediates **4** [22, 27], which seems more likely than processes involving C—H oxidative addition via a Pd(II)/Pd(IV) catalytic cycle. An insertion of the arene to form **5** by a Heck-type process has been proposed. However, the results of recent studies have indicated that this process is rather unlikely [28].

Substituent effects on the intramolecular palladation were studied with alkyl palladium complexes **6a–c** with KOPh as the base to form five-membered ring palladacycles **7a–c** (Scheme 11.2) [29]. This reaction was shown to follow the order X = MeO > H > NO_2 for substituents *meta* to the reacting site. The activation of monodeuterated substrates **6a-d_1** and **8-d_1** with KOPh or Ag_2CO_3 in MeCN at 23 °C led to the palladacycles **7a** and **9**, respectively, which were partially deuterated at C-3. The degree of deuteration was determined by ^{1}H NMR as 48 ± 3%, which showed that there was no isotopic effect for the intramolecular C—H bond activation by the alkylpalladium [29]. The five-membered ring palladacycles **7a–c** were stable complexes at room temperature, and did not suffer any reductive elimination to form strained 2*H*-benzoxete [30].

Similar substituent effects have been determined in the reaction of complexes **10** to form palladacycles **11** (Scheme 11.3) [31]. The opposite process – the intramolecular palladium-catalyzed arylation of alkanes to form dihydrobenzofuranes – has also been examined [32]. For this transformation, a mechanism based on a C—H bond-activation process by the aryl–Pd(II) involving a three-center transition state was found to be more consistent with the experimental kinetic isotope effect (3.6 at 115 °C), as well as with density functional theory (DFT) calculations.

Although the results of these experiments suggest that the palladation proceeds by an electrophilic aromatic substitution, the transformations are probably more complex than the above results suggest. Indeed, the reaction of alkyl palladium complex **6a** with KOPh in MeCN was almost completely inhibited by the addition of 1 equiv. of PPh_3 [29], which indicates that ligand substitution, presumably by an associative mechanism, occurs during the C—H bond-activation process. Biden-

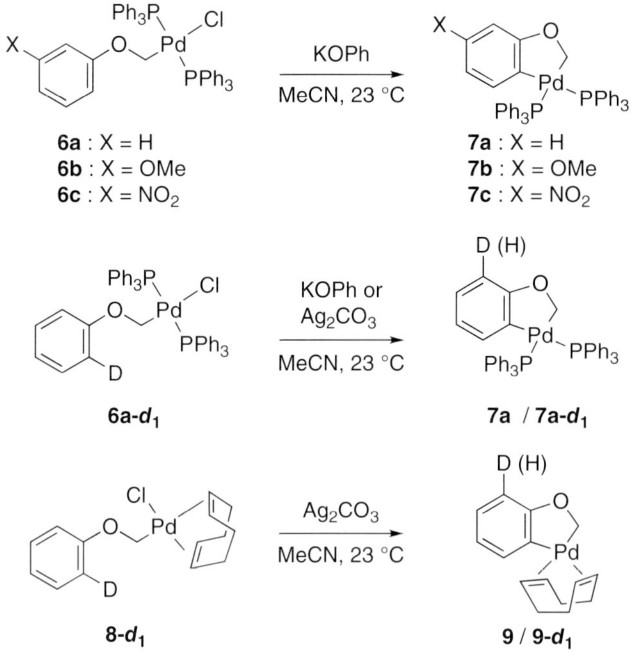

KOPh

MeCN, 23 °C

6a : X = H
6b : X = OMe
6c : X = NO_2

7a : X = H
7b : X = OMe
7c : X = NO_2

KOPh or Ag_2CO_3

MeCN, 23 °C

6a-*d*$_1$

7a / 7a-*d*$_1$

Ag_2CO_3

MeCN, 23 °C

8-*d*$_1$

9 / 9-*d*$_1$

Scheme 11.2

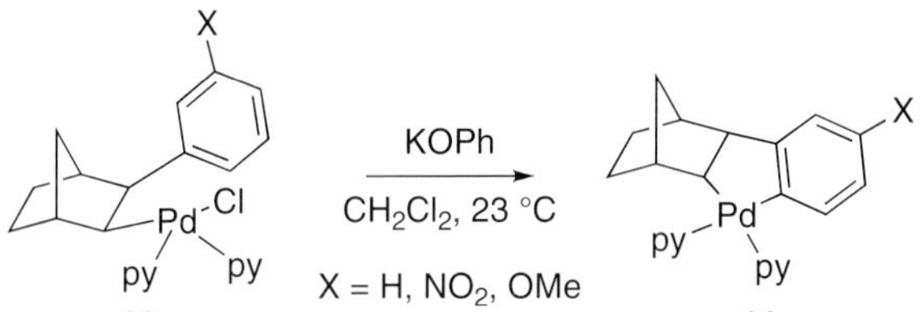

KOPh

CH_2Cl_2, 23 °C

X = H, NO_2, OMe

10

11

Scheme 11.3

tate phosphine ligands also lead to slower palladations. Interestingly, (η^1-arene)alkyl palladium(II) complexes **14** were found to be intermediates in the related intramolecular palladation of complexes **12** to form **13**, a reaction that is promoted by the strong base $NaN(SiMe_3)_2$ (Scheme 11.4) [33].

The isotope effect was also studied in the palladium-catalyzed cyclization of substrates **15** to form oxindoles **16** via C—H bond functionalization (Scheme 11.5) [34]. Whereas, no kinetic isotope effect was observed in the competitive reaction of **15a** and **15a-*d*$_5$**, an intramolecular primary isotope effect of 4 was found in the cyclization of the *ortho*-monodeuterated substrate **15a-*d*$_1$**. The absence of any intermolecular isotope effects suggests that the first step, the oxidative addition, is both slow and rate-determining overall. Although different mechanistic scenarios were considered, the significant intramolecular isotope effect shows that the palladation

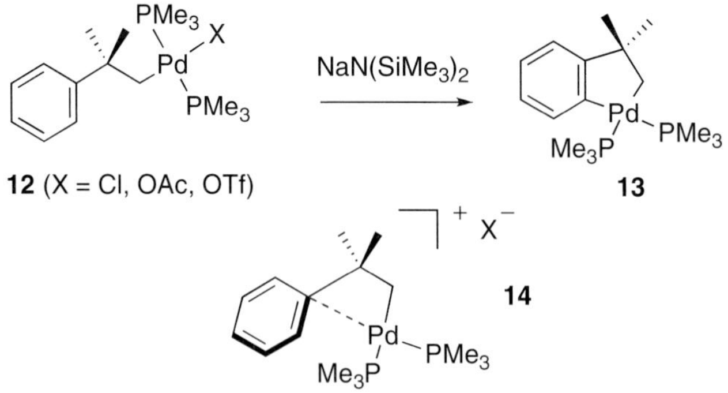

12 (X = Cl, OAc, OTf) $\xrightarrow{NaN(SiMe_3)_2}$ **13**

14

Scheme 11.4

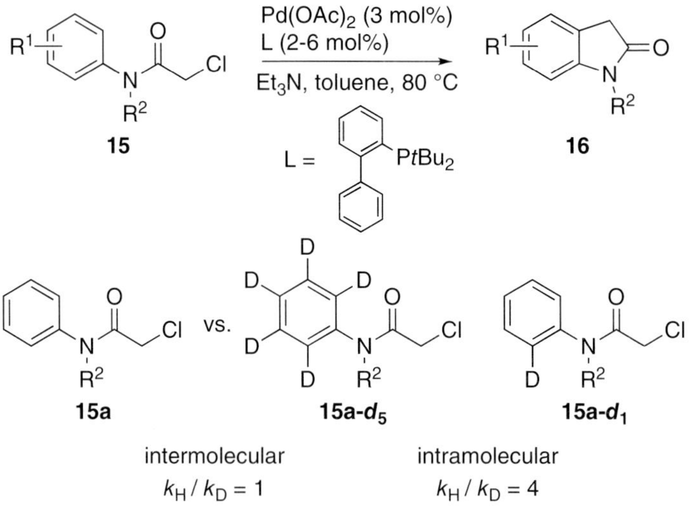

15 $\xrightarrow[\text{Et}_3\text{N, toluene, 80 °C}]{\text{Pd(OAc)}_2\text{ (3 mol\%), L (2-6 mol\%)}}$ **16**

L = $PtBu_2$ biphenyl ligand

15a vs. **15a-d_5**: intermolecular $k_H/k_D = 1$

15a-d_1: intramolecular $k_H/k_D = 4$

Scheme 11.5

process would be reversible and rapid relative to C—H bond cleavage, or that a true C—H bond activation is involved.

Substituent effects have also been examined on several substrates. Thus, the reaction of nitrofluorene **17** affords a 1 : 2 mixture of regioisomers **18a** and **18b**, favoring formation of the C—C bond on the more electron-rich aromatic ring (Scheme 11.6) [35]. However, this slight preference for the formation of **18b** could also be explained by the steric effect exerted by the nitro group. Nitrocarbazole **19**, a less sterically biased substrate, gives an almost equimolar mixture of arylation products under similar conditions, followed by oxidation with MnO_2 to form the lactams **20a** and **20b** [36]. The formation of substantial amounts of arylation products *ortho* or *para* to a strongly electron-withdrawing nitro group is not consistent with an electrophilic aromatic substitution.

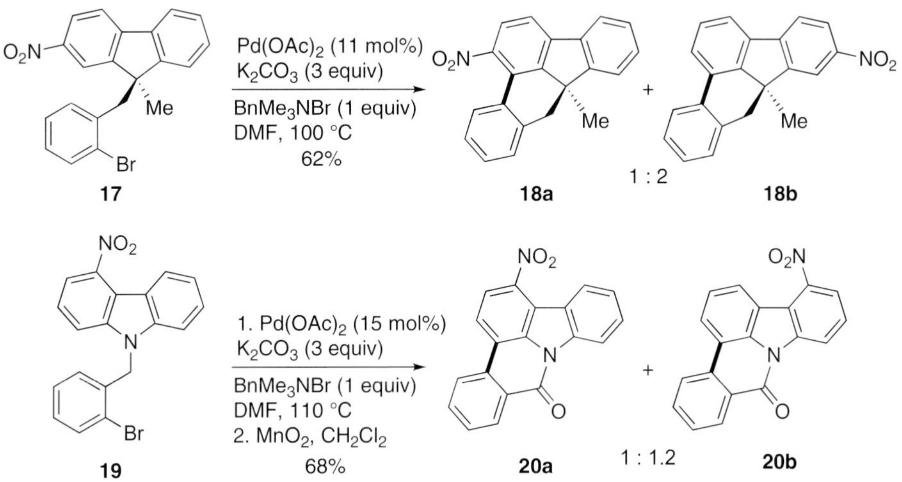

O_2N Me Br **17** → $Pd(OAc)_2$ (11 mol%), K_2CO_3 (3 equiv), $BnMe_3NBr$ (1 equiv), DMF, 100 °C, 62% → **18a** + **18b** (1 : 2)

NO_2 Br **19** → 1. $Pd(OAc)_2$ (15 mol%), K_2CO_3 (3 equiv), $BnMe_3NBr$ (1 equiv), DMF, 110 °C; 2. MnO_2, CH_2Cl_2; 68% → **20a** + **20b** (1 : 1.2)

Scheme 11.6

In the arylation of differently substituted compounds **21**, regioisomers **22a** and **22b** were obtained as the major compounds with good selectivities in most cases (Scheme 11.7) [25e]. Particularly interesting is the arylation of the compound with a fluorine substituent, which leads to isomer **22b** as the major compound. An interesting finding was the observation that aryl iodides reacted poorly under conditions optimized for aryl bromides, which was attributed to catalyst poisoning by an accumulation of the iodide anion [25e].

Low regioselectivities were observed in the reaction of substrates **23** and **25**, which is not characteristic of an electrophilic aromatic substitution mechanism. However, interpretation of the results on substrates such as **25** might be complicated by the competing amide rotation, which may lead to a system far from Curtin–Hammett conditions. Interestingly, the direct arylation of **27** resulted in a primary kinetic isotopic effect of 4.25 (Scheme 11.8) [25e]. These studies revealed a significant C—H bond cleavage step during arylation, which was rationalized by a mechanism proceeding via electrophilic metallation involving either a σ-bond metathesis or a C—H bond functionalization step via an S_E3 process.

Significantly, the palladium-catalyzed arylation reaction is facilitated by electron-withdrawing substituents on the aromatic ring, which is clearly inconsistent with an electrophilic aromatic substitution mechanism. Thus, whereas substrate **29** bearing a single fluorine substituent at C-4 (*meta* to the arylation site) leads to a modest preference for the arylation at the substituted ring (**30a**/**30b** = 1.6 : 1), the selectivity increases dramatically with substrates bearing fluorine substituents *ortho* to the reactive site (Scheme 11.9) [37]. In particular, reaction of substrate **29** with three fluorine substituents occurred almost exclusively at the trifluorophenyl ring to give **30a**. Modest regioisomeric ratios (1.1–2.4 : 1), favoring reaction at the substituted aryl ring, were obtained from substrates of type **29** bearing groups that are either electron-releasing (OMe) or electron-withdrawing (CF_3, Cl) in S_EAr

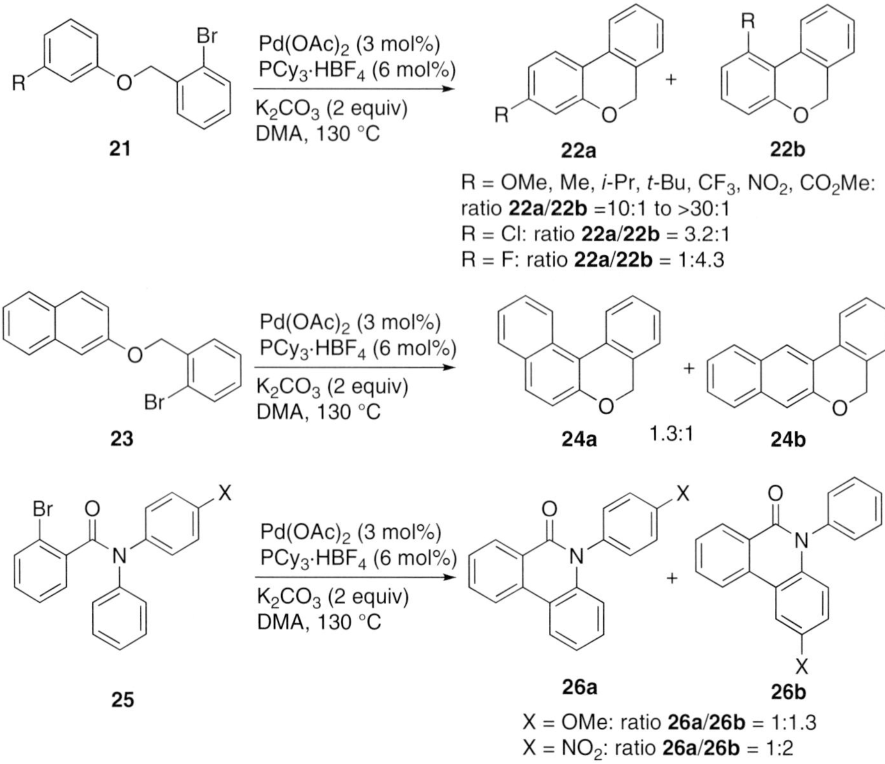

Pd(OAc)$_2$ (3 mol%)
PCy$_3$·HBF$_4$ (6 mol%)
K$_2$CO$_3$ (2 equiv)
DMA, 130 °C

21 → **22a** + **22b**

R = OMe, Me, *i*-Pr, *t*-Bu, CF$_3$, NO$_2$, CO$_2$Me: ratio **22a/22b** =10:1 to >30:1
R = Cl: ratio **22a/22b** = 3.2:1
R = F: ratio **22a/22b** = 1:4.3

23 → **24a** + **24b** (1.3:1)

25 → **26a** + **26b**

X = OMe: ratio **26a/26b** = 1:1.3
X = NO$_2$: ratio **26a/26b** = 1:2

Scheme 11.7

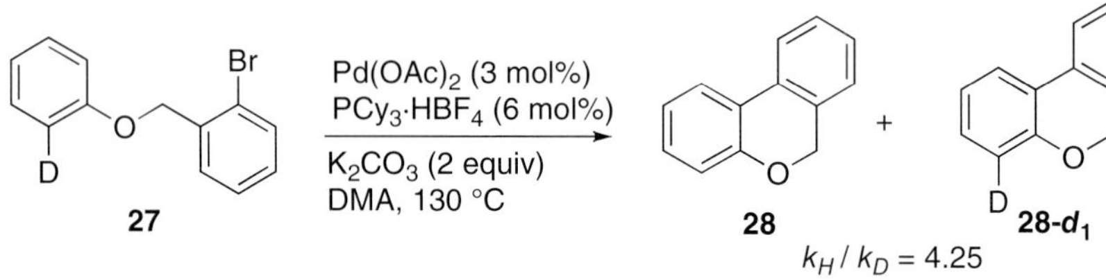

Pd(OAc)$_2$ (3 mol%)
PCy$_3$·HBF$_4$ (6 mol%)
K$_2$CO$_3$ (2 equiv)
DMA, 130 °C

27 → **28** + **28-*d*$_1$**

$k_H / k_D = 4.25$

Scheme 11.8

processes. The arylation of substrates with *t*-Bu and SiMe$_3$ substituents at C-4 took place on the unsubstituted phenyl ring with moderate selectivity. Thus, **31** gave a 1 : 1.5 ratio of **32a** and **32b**, whereas substrate **33** gave 1 : 2.3 ratio of **34a** and **34b**, which is that expected from an approximately additive effects of the fluorine (1.6 : 1) and the *tert*-butyl (1 : 1.5) substituents. In these experiments, aromatization of the crude reaction mixtures with 2,3-dichloro-5,6-dicyano-1,4-benzoquinone (DDQ) led cleanly to the corresponding phenanthrenes, which simplified the determination of regioisomeric ratios by ^{1}H NMR.

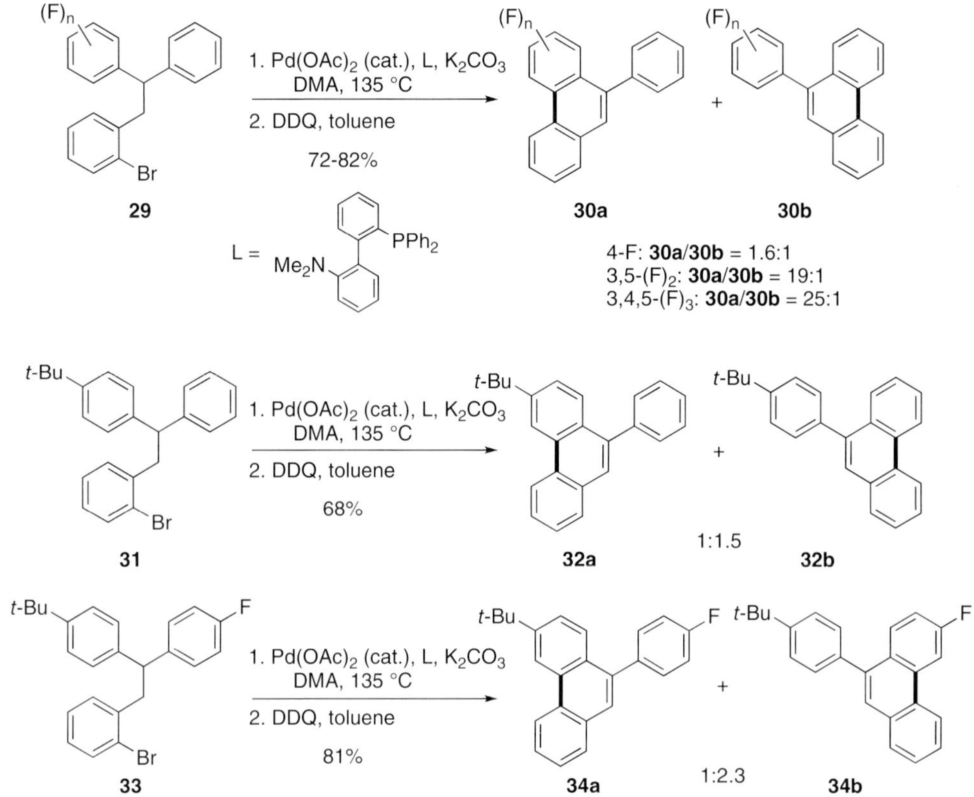

1. $Pd(OAc)_2$ (cat.), L, K_2CO_3 DMA, 135 °C
2. DDQ, toluene
72-82%

29 → 30a + 30b

L = Me_2N ... PPh_2

4-F: **30a/30b** = 1.6:1
3,5-$(F)_2$: **30a/30b** = 19:1
3,4,5-$(F)_3$: **30a/30b** = 25:1

1. $Pd(OAc)_2$ (cat.), L, K_2CO_3 DMA, 135 °C
2. DDQ, toluene
68%

31 → 32a + 32b (1:1.5)

1. $Pd(OAc)_2$ (cat.), L, K_2CO_3 DMA, 135 °C
2. DDQ, toluene
81%

33 → 34a + 34b (1:2.3)

Scheme 11.9

Results on the intramolecular arylation on a 5*H*-indeno[1,2-*b*]pyridine derivative, which proceeded selectively at the pyridine ring, are also inconsistent with an electrophilic aromatic substitution mechanism for this reaction [37].

Isotope effects are also inconsistent with an electrophilic aromatic substitution. Thus, intramolecular competition experiments between phenyl and pentadeuterophenyl groups led to the determination of intramolecular isotope effects k_H/k_D = 5.0 (135 °C) and = 6.7 (100 °C) for this arylation [37]. This led to the proposal of a mechanism based on a proton-abstraction for this reaction [37], which is in line with the findings of other studies [25e]. A related mechanism, in which acetate acts as the basic ligand, has been proposed for the cyclometallation of benzylic amines with $Pd(OAc)_2$ [38].

These results are therefore consistent with a mechanism proceeding by the abstraction of a proton of the aryl ring by the base, in a process in which formation of the metal–carbon bond is concerted with the breaking of the carbon–hydrogen bond. DFT calculations carried out with bicarbonate as a model base for the intramolecular arylation favor two alternatives for the key activation process: (i) a mechanism in which the base coordinates to the Pd(II) center and then acts as an

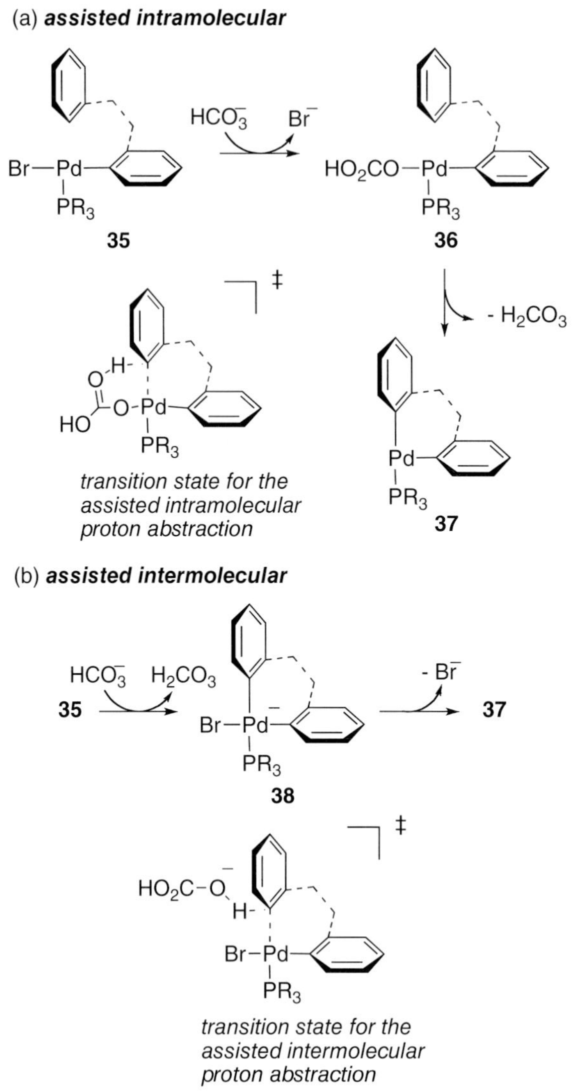

(a) ***assisted intramolecular***

HCO_3^- Br^-

Br–Pd–; PR_3 **35**

HO_2CO–Pd–; PR_3 **36**

‡

- H_2CO_3

HO; O; O-Pd; H; PR_3

transition state for the assisted intramolecular proton abstraction

Pd–; PR_3 **37**

(b) ***assisted intermolecular***

35 HCO_3^- H_2CO_3 Br–Pd$^-$–; PR_3 **38** - Br^- **37**

‡

HO_2C–O^-; H; Br–Pd–; PR_3

transition state for the assisted intermolecular proton abstraction

Scheme 11.10

internal base (*assisted intramolecular*; Scheme 11.10a); or (ii) an intermolecular proton abstraction by an external base (*assisted intermolecular*; Scheme 11.10b) [37]. Models as **35** and **36** are probably reasonably realistic as the phosphines used in these reactions are bulky, and bicarbonate leads experimentally to similar results to those obtained with carbonate as a base in these reactions [37b]. On the other hand, a mechanism in which the bromide ligand acts as an internal base (*non-assisted intramolecular*) appears less likely according to the DFT calculations carried out on models shown in Scheme 11.10 [37].

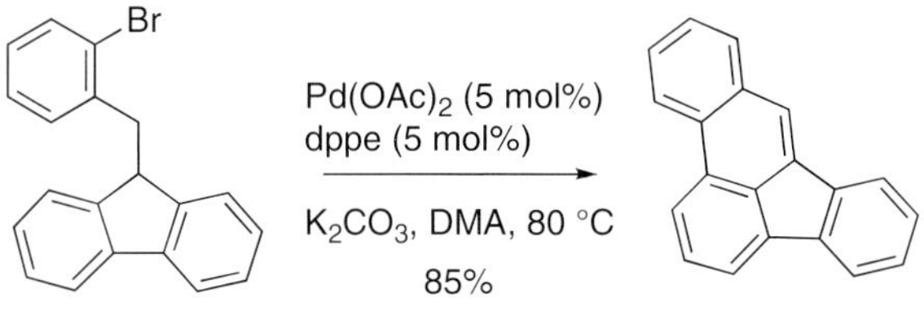

Br

$Pd(OAc)_2$ (5 mol%)
dppe (5 mol%)
K_2CO_3, DMA, 80 °C
85%

39 **40**

Scheme 11.11

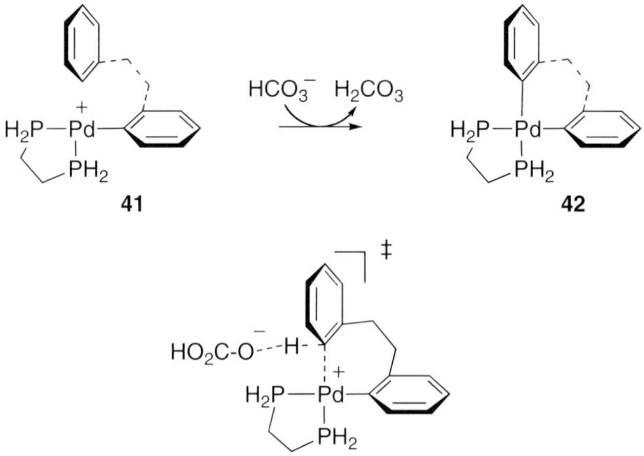

HCO_3^- H_2CO_3

H_2P—Pd—
PH_2

41 **42**

HO_2C-O^----H-

‡

transition state for the assisted intermolecular proton abstraction with bidentate phosphines

Scheme 11.12

The above mechanistic picture pertains to catalytic systems in which palladium bears a bulky phosphine ligand and a base such as carbonate, bicarbonate or a carboxylate is used. Recent studies have shown that bidentate phosphines, such as dppm, dppe, dppf and Xantphos, are also excellent ligands for the palladium-catalyzed arylation, which allows these reactions to be performed under milder conditions. Thus, for example, the arylation of fluorene **39** to give benz[*e*]acephenanthrylene (**40**) can now be carried out in 85% yield with $Pd(OAc)_2$/dppe at 80 °C in 25 h [39], whereas in the absence of the diphosphine a 52% yield was achieved at a higher temperature and over a longer reaction time (130 °C, 48 h) (Scheme 11.11) [35]. An intramolecular isotope effect $k_H/k_D = 5.3$ was determined by using $Pd(OAc)_2$/dppf as the catalyst. It is interesting that bidentate phosphines are also ligands in the intramolecular arylation of alkenyltriflates, which shows an intermolecular isotope effect $k_H/k_D = 5.0$ (80 °C) [40] and other arylations [23].

Arylation reactions with the bidentate phosphines most likely proceed by the intermolecular proton-abstraction mechanism, which is supported by DFT calculations as shown in the transformation of **41** into **42** (Scheme 11.12) [39]. Although the addition of pivalic acid has a beneficial effect in many cases, the fact that the arylation reaction also proceeds satisfactorily with palladium complexes bearing bidentate diphosphines suggests that this carboxylate–like carbonate or bicarbonate–acts only as an external base in the process.

Recently, *o*-alkynyl biaryls have been shown to undergo intramolecular 5-*exo-dig* hydroarylation by a mechanism that proceeds by a C—H bond activation assisted by the alkynyl substituent [41]. This reaction also proceeds with a palladium complex bearing a dibentate phosphine (1,1′-bis(diisopropylphosphinoferrocene)) and shows a significant kinetic isotope effect ($k_H/k_D = 3.5$ for the intramolecular process), which is consistent with a mechanism involving a proton-abstraction mechanism.

11.3 Intermolecular Metal-Catalyzed Direct Arylation of Arenes

The discovery that substrates bearing electron-withdrawing substituents react preferentially in palladium-catalyzed arylations led to the development of an intermolecular version using polyfluoroaromatic compounds to form biaryls **43** (Scheme 11.13) [42]. A kinetic isotope effect of 3.0 was determined for this reaction, which indicates that the C—H bond cleavage occurs in the rate-determining step. The use of the bulky biphenyl phosphine (2-(dicyclohexylphosphino)-2′,6′-dimethoxybiphenyl) (S-Phos), allowed these reactions to be performed in a general way with aryl bromides and chlorides in isopropyl acetate at 80 °C [42b]. A similar reaction has been developed using CuI/phenanthroline as the catalyst with K_3PO_4 as the base [43].

By using pivalate as the base, the intermolecular reaction can even take place with unactivated arenes such as benzene to form biaryls **44** (Scheme 11.14) [44]. The reaction of *p*-bromotoluene with anisole gave a 22:53:25 mixture of *ortho*:*meta*:*para* isomers, which indicates that there is no directing effect by the methoxy substituent, little or no electronic bias, and just a minor steric bias resulting in a small statistical preference for reaction at the *meta* and *para* positions. Both, experimental and theoretical calculations indicate that the pivalate anion is a key component in C—H bond cleaving, lowering the energy barrier of C—H bond cleavage. The use of other carboxylic acids as additives led to poorer results for

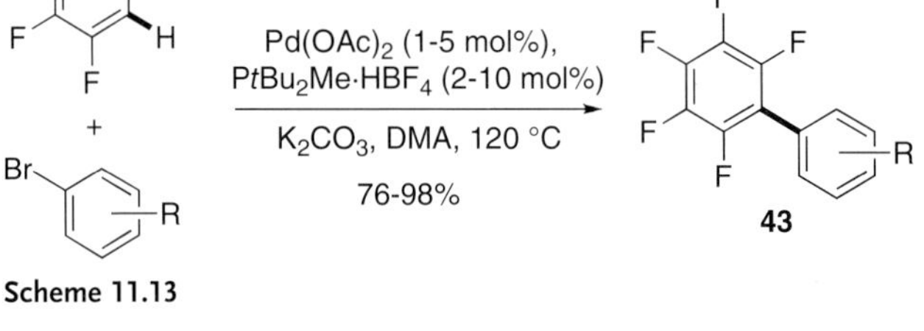

Pd(OAc)$_2$ (1-5 mol%), P*t*Bu$_2$Me·HBF$_4$ (2-10 mol%)

K$_2$CO$_3$, DMA, 120 °C

76-98%

43

Scheme 11.13

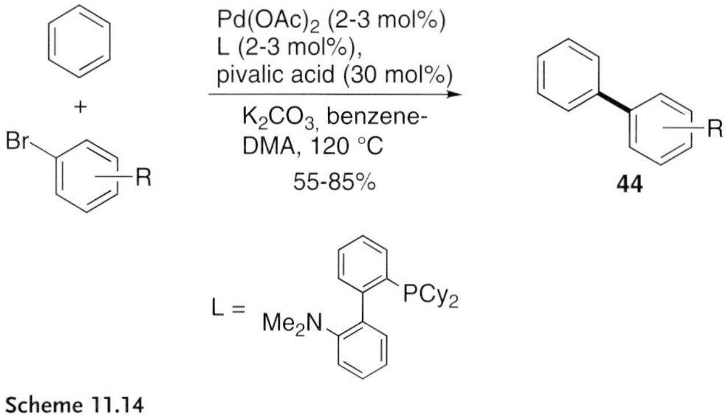

Benzene + Br–C$_6$H$_4$–R → Pd(OAc)$_2$ (2-3 mol%), L (2-3 mol%), pivalic acid (30 mol%); K$_2$CO$_3$, benzene-DMA, 120 °C; 55-85% → **44**

L = Me$_2$N / PCy$_2$ biaryl ligand

Scheme 11.14

C$_6$H$_5$CO$_2$H + Cl–C$_6$H$_4$–R → Pd(OAc)$_2$ (5 mol%), BuAd$_2$P (10 mol%); Cs$_2$CO$_3$, 3Å MS, DMF, 145 °C; 65-91% → **45** (CO$_2$H)

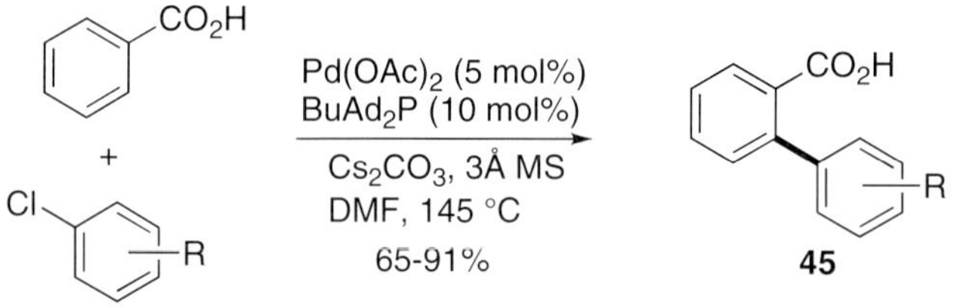

Scheme 11.15

this process. A significant intermolecular isotope effect of 5.5 was determined for this arylation.

The *ortho*-arylation of benzoic acids with aryl chlorides affords biphenylcarboxylic acids **45** in a general way (Scheme 11.15) [45], the best results being obtained when using bulky *n*-butyl-di-1-adamantylphosphine. The corresponding reaction of aryl iodides was carried out with Pd(OAc)$_2$ as catalyst, AgOAc, and acetic acid in the absence of phosphine [45].

In the reaction of Scheme 11.15, identical inter- and intramolecular isotope effects ($k_H/k_D = 4.4$) were observed, which is consistent with a rate-determining step deprotonation of a complex **46** by an external base to form a palladacycle **47**, which undergoes reductive elimination to give biaryl carboxylate **48** (Scheme 11.16) [45].

Interestingly, an isotope effect of 1.8 was determined in the phenylation of palladacycles **49** derived from ferrocenyl oxazolines, which is also consistent with a proton-abstraction by the base in the C–C bond-forming step (Scheme 11.17) [46].

A palladium-catalyzed methylation of 2-arylpyridines with *t*-butylperoxyde, and related peroxides, was recently developed [47]. This transformation also involves a C–H bond-activation step by a Pd(II)R complex, although no isotope effect was observed in an intramolecular competition experiment, which suggests that a radical mechanism might be involved.

The direct arylation of aromatic C–H bonds takes place with [Cp*IrHCl]$_2$ complexes as catalysts in the presence of KO*t*-Bu [48]. The regioselectivity observed in

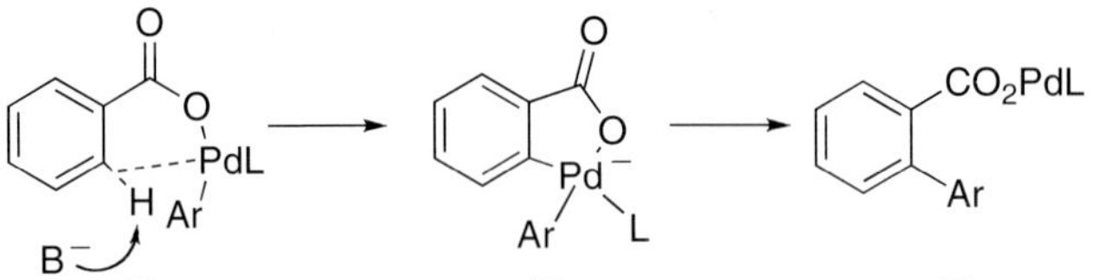

Scheme 11.16

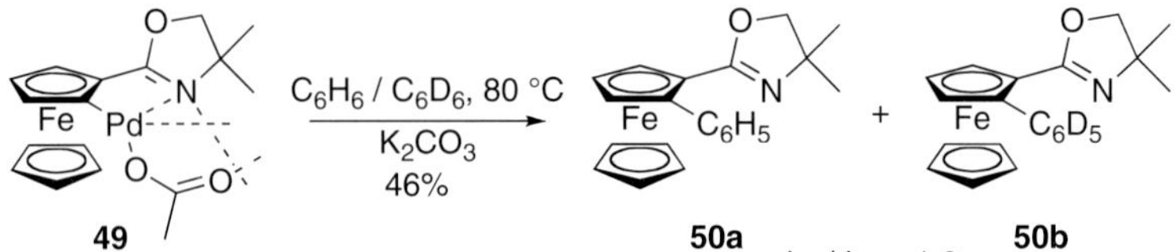

Scheme 11.17

the reactions of aryl iodide with toluene or anisole are similar to those observed for the radical aromatic substitution reactions. The reaction was proposed to proceed via the reduction of a trivalent Cp*Ir(III) complex to Cp*Ir(II), mediated by the base and followed by an electron-transfer process from divalent iridium complex to the aryl iodide to form an aryl iodide radical anion and Cp*Ir(III) catalyst. The elimination of iodide anion from the radical anion then furnishes an aryl radical which can react with benzene to give the biaryl product.

A bimetallic complex, which consists of a $[Rh(cod)Cl_2]^-$ ion and a *P,N*-ligand-stabilized cation $[(PN)_2Rh]^+$, efficiently catalyzes the unactivated arene direct arylation with aryl iodides, aryl bromides and aryl chlorides [49]. As in the case of direct arylation with $[Cp^*IrHCl]_2$, where a reaction mechanism via radical intermediates was proposed [48], the regioselectivity observed in the arylation of toluene with the bimetallic rhodium catalyst is also in accordance with radical processes.

11.4 Metal-Catalyzed Heteroaryl–Aryl and Heteroaryl–Heteroaryl Bond Formation

Many examples exist of synthetically useful intermolecular palladium-catalyzed arylations of π-excessive heterocycles [4, 7, 50–56]. The reaction has also been extended to the direct arylation of pyridine, diazine and azole *N*-oxides with aryl halides [57]. Interestingly, tuning the ligand and base allows selective sp^2 or sp^3 activation of alkyl pyridine or pyrazine *N*-oxides [58]. Palladium-catalyzed sp^3 activation of alkylanines [59] and benzylic arylation of benzoxazoles has also been described [60].

An electrophilic aromatic substitution mechanism has been usually favored for the arylation of electron-rich heterocycles [61, 62]. Thus, for example, no kinetic

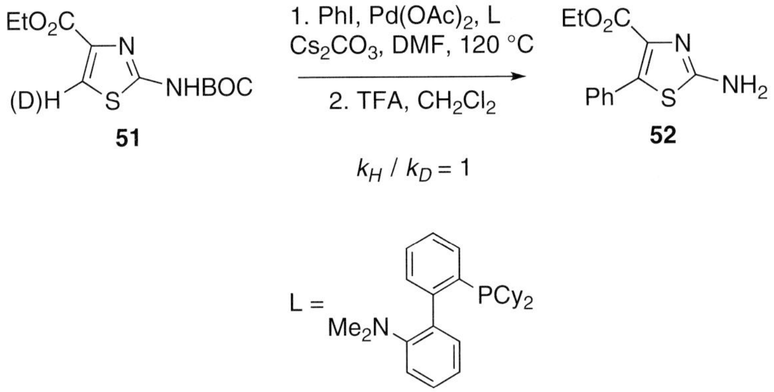

EtO_2C (D)H S N NHBOC **51**

1. PhI, $Pd(OAc)_2$, L, Cs_2CO_3, DMF, 120 °C
2. TFA, CH_2Cl_2

$k_H / k_D = 1$

EtO_2C Ph S N NH_2 **52**

L = Me_2N, PCy_2

Scheme 11.18

isotope effect was found in the reaction of thiazole **51** with iodobenzene to give **52** (Scheme 11.18), which is consistent with an electrophilic aromatic substitution mechanism [63]. However, this result might also indicate that the proton-abstraction step is not rate-determining in this case.

Similar results have been observed for the arylation of indolizidine **53**, which occurs at C-3 to give **54** (Scheme 11.19) [64]. Attempts to trap a possible Heck intermediate in this reaction using indolizidine **55** as the substrate (via **57** and **58**) failed, and led only to a mixture of **56a** and **56b**. For this, and for the palladium-catalyzed alkynylation of related heterocycles with alkynylbromides [65], an electrophilic aromatic substitution was proposed. A Heck process was also excluded in the palladium-catalyzed arylation of imidazolidones, a reaction for which an intramolecular isotope effect of $k_H/k_D = 4.5$ was found [66].

A low isotope effect ($k_H/k_D = 1.3$) was also observed in the palladium-catalyzed phenylation of benzothiazole (Scheme 11.20) [62].

In contrast with arylations of other heterocycles, the palladium-catalyzed arylation of benzoxazoles at C-2 proceeds readily at ambient temperature (Scheme 11.21) [67]. Although, as in other cases no kinetic isotope effect was found for **59a**, a Hammett plot revealed a correlation with σ^- with a positive ρ, which indicates that a phenolate intermediate is formed in this reaction. Therefore, this reaction has been shown to proceed by a totally different mechanism. According to experimental results and DFT calculations, the reactions proceed by the deprotonation of benzoxazoles **59** to form **61**, which is in equilibrium with *o*-phenoxyisocyanide **62**. Coordination of the oxidative addition product $PdI(Ph)L_2$ to the isocyanide then forms **63** which cyclizes to form palladate **64**, from which the 2-phenylbenzoxazoles **60** are formed by reductive elimination.

Recently, the arylation of indoles with halobenzenes has received special attention [54]. An exclusive arylation of indoles at C-2 with aryl iodides has been found with $Pd(OAc)_2$ in the presence of CuI [68], as well as with palladium catalysts with phosphines or NHC ligands and CsOAc as the base [69, 70]. Remarkably, whereas $Pd(OAc)_2$ and CsOAc leads to arylation at C-2 at 125 °C [71], silver(I) carboxylate

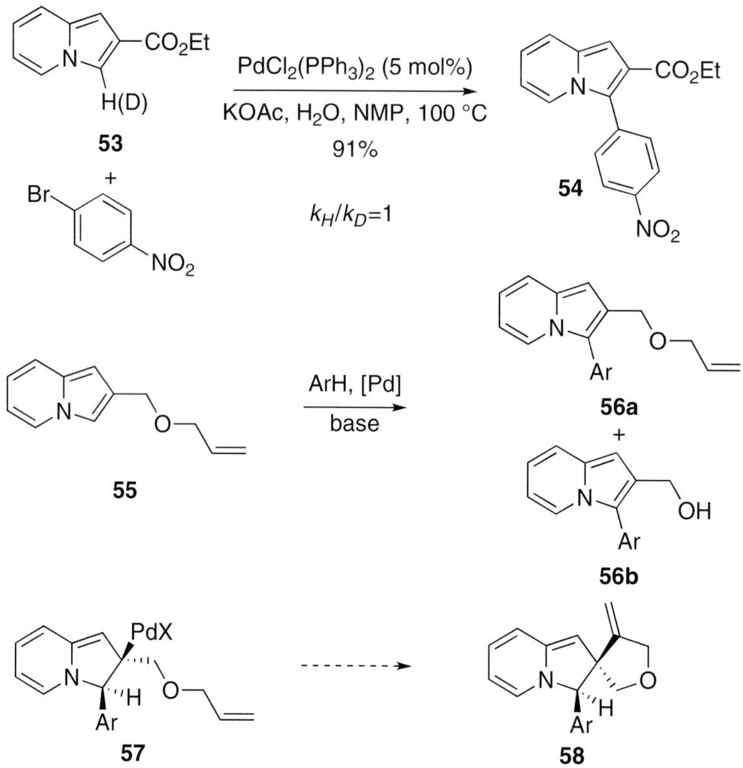

Scheme 11.19

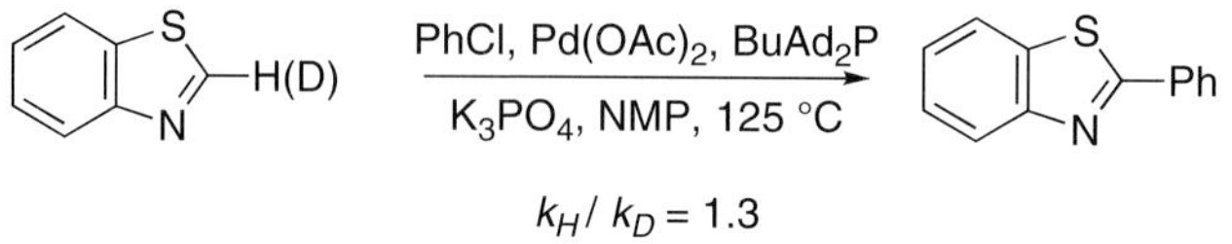

Scheme 11.20

as additive allows this reaction to be performed at ambient temperature [72]. In contrast, the use of palladium–phosphinous acid complexes as catalysts and K_2CO_3, selective arylation at C-3 was observed, which is consistent with an electrophilic aromatic substitution mechanism [73]. A rationale for the C-2 versus C-3 arylation in these reactions has been provided, based on the observation that a substantial secondary isotope effect occurs at C-3 in arylations, leading to C-2 substituted products (Scheme 11.22) [74]. Accordingly, the initial C-3-arylated intermediate **65** would be in equilibrium with **68**, the precursor of the C-2-arylated derivatives **70**. In support of this proposal, the migration of 3-lithio-indole to 2-lithio-indole is known [75]. The proposed electrophilic aromatic substitution reaction is supported by a Hammet plot of the reaction of several C-6-substituted derivatives, while the reactions of several magnesium salts of indole led to selective C-3 arylation [74].

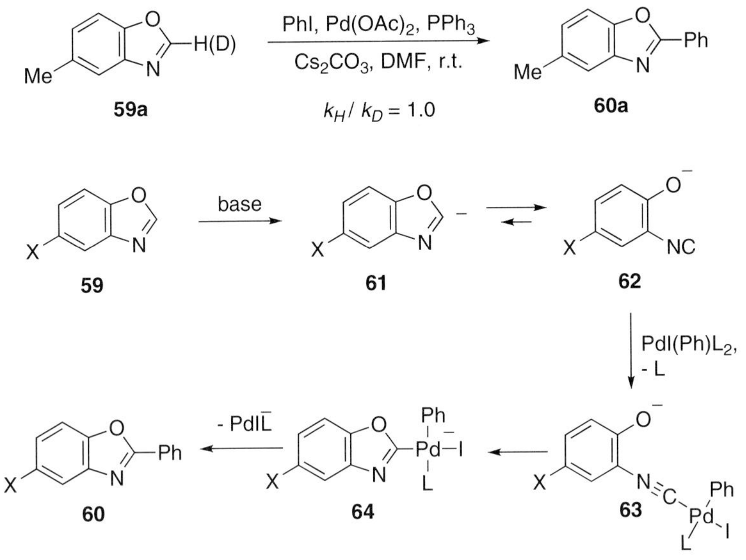

Scheme 11.21

Similarly, the palladium-catalyzed reaction of indoles and other electron-rich arenes with aryl boronic acids was proposed to take place by an electrophilic aromatic substitution, presumably by an initial palladation of the arene or heteroarene followed by a transmetallation process [76]. In this reaction, C-2 arylation was exclusively observed.

At ambient temperature, the palladium-catalyzed arylation of indoles proceeds with reagents $[Ar_2I]BF_4$ to give C-2 arylated derivatives **70** (Scheme 11.23) [77, 78]. This reaction was proposed to proceed by a different mechanism involving Pd(II)/Pd(IV) via complexes **71** and **72**. Importantly, the reaction of $[Ph_2I]OTf$ with Pd(II) and Pt(II) has been reported to give Pd(IV) species by formal transfer of Ph^+ [79].

The copper-catalyzed arylation of benzoxazol, oxazol, thiazol and related heterocycles takes place with iodobenzene, LiO*t*-Bu and /or KO*t*-Bu as the base in dimethylformamide (DMF) at 140 °C [80]. When 4,5-dimethythiazole was allowed to react with C_6D_5Br using KO*t*-Bu as the base, a single hydrogen was introduced in the *ortho* position of the deuterated product (Scheme 11.24). This observation is consistent with a mechanistic pathway proceeding through an aryne intermediate. However, when LiO*t*-Bu was used as the base no hydrogen incorporation was observed. Here, a possible mechanism involves deprotonation of the heterocycle by LiO*t*-Bu, perhaps through copper precoordination by the heterocycle, followed by lithium–copper transmetallation and reaction of the organocopper with aryl iodide.

A very different mechanism occurs in the arylation of benzoxazol, benzimidazol and related heterocycles via rhodium-catalyzed C—H bond functionalization (Scheme 11.25) [81]. Thus, this reaction proceeds by an initial coordination and

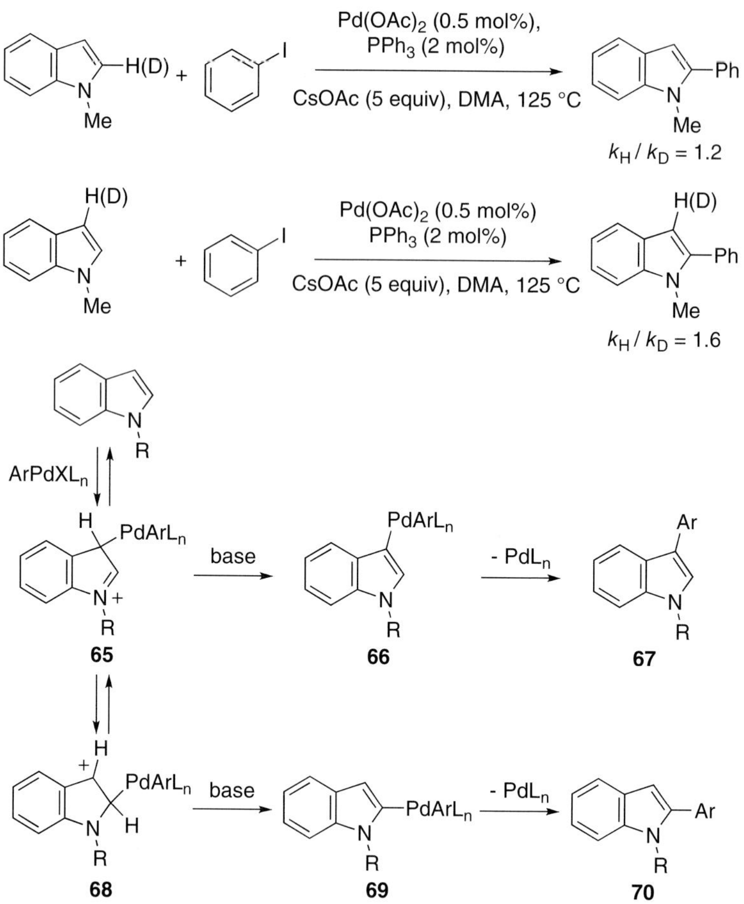

Pd(OAc)$_2$ (0.5 mol%), PPh$_3$ (2 mol%)
CsOAc (5 equiv), DMA, 125 °C
H(D) Me N Ph Me N
$k_H / k_D = 1.2$

Pd(OAc)$_2$ (0.5 mol%) PPh$_3$ (2 mol%)
CsOAc (5 equiv), DMA, 125 °C
H(D) N Me H(D) Ph N Me
$k_H / k_D = 1.6$

N R
ArPdXL$_n$
H PdArL$_n$ N+ R **65** base PdArL$_n$ N R **66** - PdL$_n$ Ar N R **67**

H + PdArL$_n$ H N R **68** base N PdArL$_n$ R **69** - PdL$_n$ Ar N R **70**

Scheme 11.22

N R Pd(OAc)$_2$ OAc Pd N R **71**
[Ar$_2$I]BF$_4$
OAc Pd^{IV} Ar N R **72**
Ar N R **70**

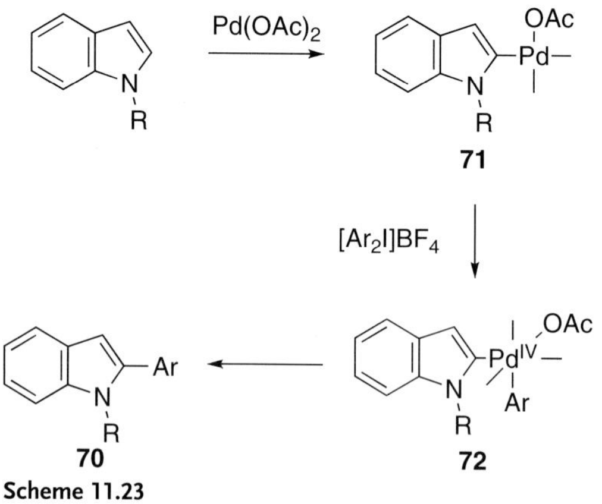

Scheme 11.23

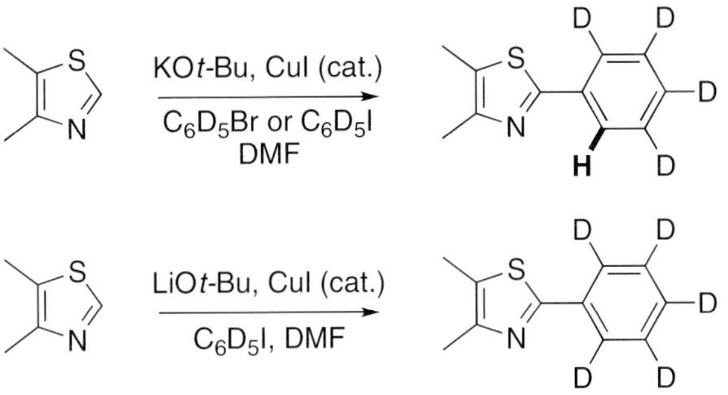

KO*t*-Bu, CuI (cat.)
C_6D_5Br or C_6D_5I
DMF

LiO*t*-Bu, CuI (cat.)
C_6D_5I, DMF

Scheme 11.24

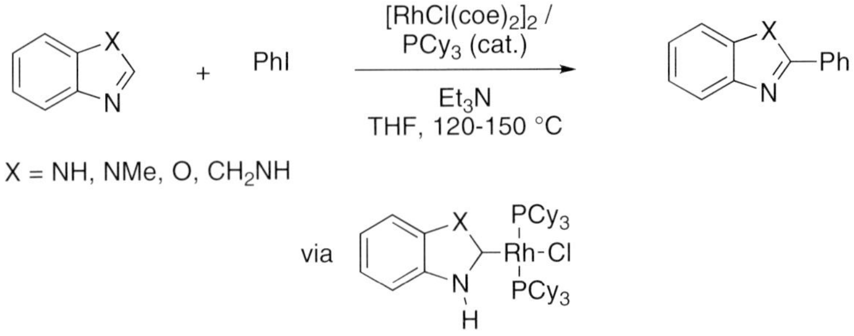

$[RhCl(coe)_2]_2$ / PCy_3 (cat.)
Et_3N
THF, 120-150 °C

X = NH, NMe, O, CH_2NH

via

73

Scheme 11.25

tautomerization of benzimidazole to form NHC complex **73**, followed by phosphine dissociation and oxidative addition of iodobenzene to give the Rh(III) intermediate [81a]. Subsequent iodide dissociation and PCy_3 association leads to an intermediate that suffers reductive elimination to give the arylated compound. Similar NHC–rhodium complexes were proposed as intermediates in the arylation of bromobenzene using 9-cyclohexylbicyclo[4.2.1]-9-phosphonane [81b] or 2,3,6,7-tetrahydrophosphepines as ligands [81c].

The rhodium-catalyzed arylation of (NH)-indoles and pyrroles is also possible using a similar catalytic system consisting of $[RhCl(coe)_2]_2$, $[p\text{-}(CF_3)C_6H_4]_3P$ and CsOPiv (Scheme 11.26) [82]. However, in this case a different mechanism was proposed via complexes **74** to **77** in which the high C-2 selectivity was due to the greater electrophilicity of the Ar–Rh(III) fragment compared to Ar–Pd(II). In addition, pivalate ligand was proposed to assist the C—H bond cleavage as an internal base. The results of kinetic studies of this reaction suggest that the mechanism involves a rate-determining C—H bond activation by the aryl–rhodium complex, followed by a reductive elimination. This proposal was further supported by a significant kinetic isotope effect ($k_H/k_D = 3.0$) at the 2-position of the indole.

The rhodium-catalyzed arylation of heterocycles was also proposed to proceed by an electrophilic aromatic pathway [83]. When using $[RhCl(CO)\{P[OCH(CF_3)_2]_3\}_2$

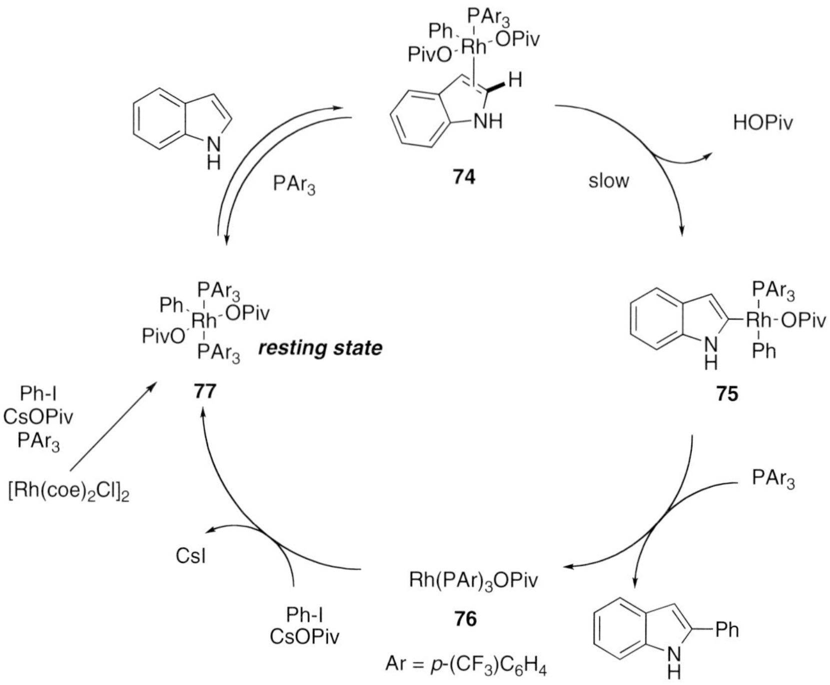

Scheme 11.26

as the catalyst, a regioselective C-2 arylation of thiophene and furanes was observed, whereas *N*-phenylpyrrole reacted at C-3, and *N*-methylindole gave the C-3 arylated derivative as the major compound. When anisole was treated with *p*-nitrophenyl iodide, arylated anisoles were obtained as a mixture of regioisomers (51% yield; $o:m:p = 29:0:71$), which was consistent with an electrophilic aromatic substitution.

11.5 Direct Arylation via Metallacycles

Many arylations are assisted by functional groups that promote *ortho*-metallation. Thus, for example, acetanilides react with arylsilanes at the *ortho* position via palladacycles **78** and **79** to form derivatives **80** (Scheme 11.27) [84]. Mechanistically, this transformation is similar to a cross-coupling reaction, in which the oxidative addition step is replaced by the *ortho*-metallation, although in this case the Pd(0) intermediate must be oxidized *in situ* to generate the reactive Pd(II) species. Unsubstituted benzylamines and *N*-methylbenzylamine are *ortho*-arylated with $Pd(OAc)_2$ in the presence of trifluoroacetic acid (TFA) and silver acetate [85], and a mechanism which differs from the usual Pd(0)/Pd(II) catalytic cycle was suggested for this. *Ortho*-alkylation was also observed in the palladium-catalyzed

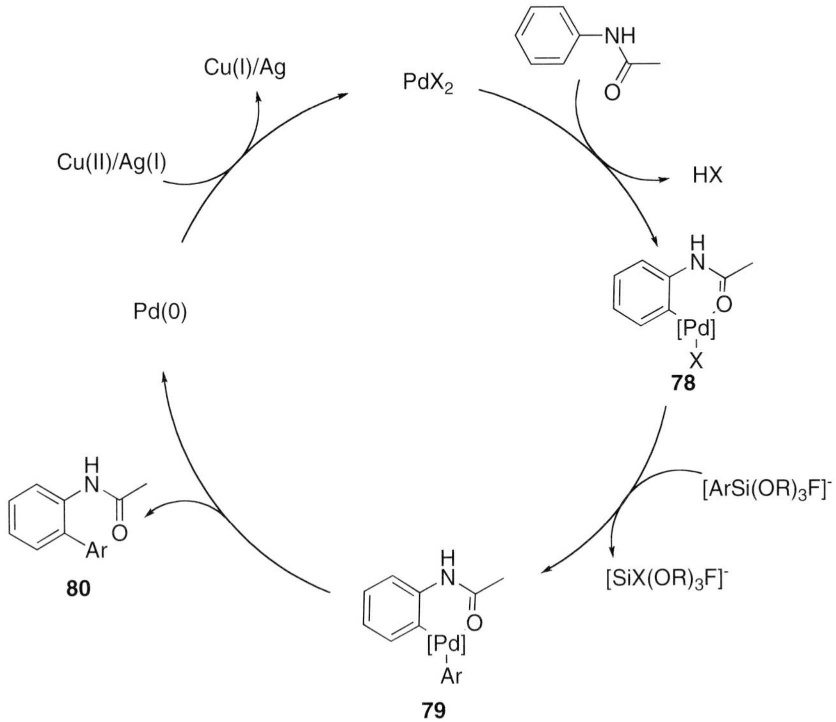

Scheme 11.27

reactions of 2-aryloxazolines with organostannanes [86] and that of 2-arylpyridines with boronic acids or boroxines [87].

The reactions of benzanilides, arylpyridines and quinolines with reagents $[Ar_2I]BF_4$ proceed via palladacycles by a mechanism that involves a Pd(II)/Pd(IV) catalytic cycle [88], as shown above for the corresponding reactions with indoles (Scheme 11.23). No reaction takes place when $[Ph_2I]BF_4$ is replaced with Ph-I or Ph-OTf electrophiles, which undergo rapid oxidative addition to Pd(0). In addition, palladacycle **81** is a competent catalyst for the reaction, and reacts with $[Ph_2I]BF_4$ to give the phenylated derivative **82** (Scheme 11.28) [88]. In contrast, the *ortho*-arylation of anilides occurs readily with aryl iodides using $Pd(OAc)_2$ as catalyst and AgOAc as additive in TFA as solvent [89].

The reaction of norbornene with iodobenzene or bromobenzene in the presence of $[Pd(PPh_3)_4]$ as the catalyst leads to pentacycle **83** (Scheme 11.29) [6, 31, 90–94]. This process is known as the Catellani reaction. The parent transformation involves an insertion of the phenylpalladium(II) complex into the double bond of norbornene to give (η^2-phenyl)norbornyl palladium(II) complex **84**, which then undergoes intramolecular palladation to form **85**. Further reaction of this palladacycle with iodobenzene leads to **83**. Biphenyl, tetracycle **86**, and more complex derivatives **87** and **88**, have also been isolated in this reaction [90, 93c, 94]. Benzocyclobutenes

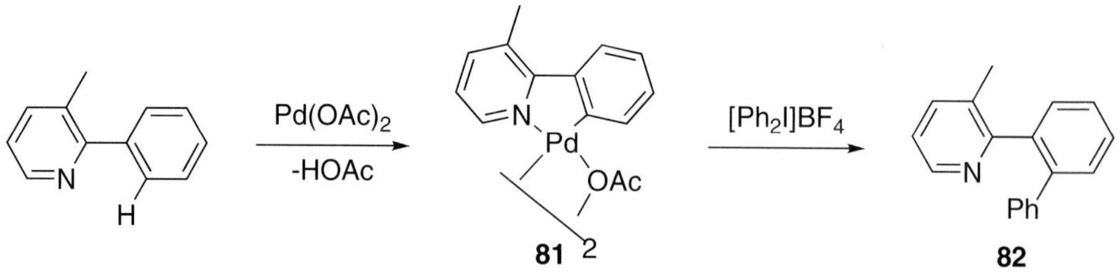

Scheme 11.28

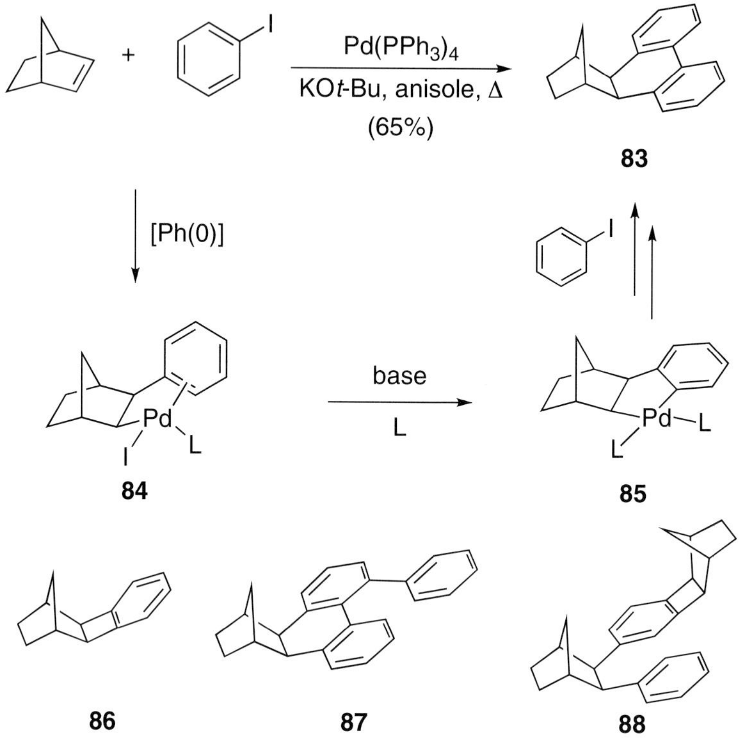

Scheme 11.29

were also observed in the palladium-catalyzed reaction of substituted cyclopentanes with bromoarenes [95]. Recently, new methods have been developed for the simultaneous formation of several C—C bonds based on the insertion of an aryl iodide into the double bond of norbornene, followed by intramolecular C—H bond functionalization [90, 96, 97].

Related reactions leading to polycycles such as **89** have been reported (Scheme 11.30), in which the intermediacy of palladacycles of type **90** was proposed [98]. Complexes of type **90**, which are stabilized versions of palladacycles of type **7** (see Scheme 11.2), have been isolated by using suitable ligands [29].

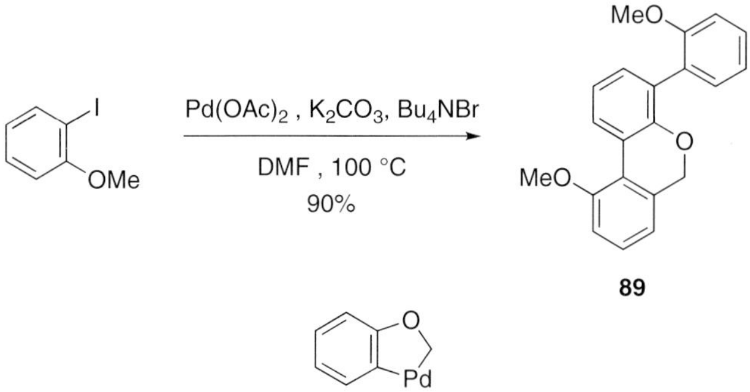

Scheme 11.30

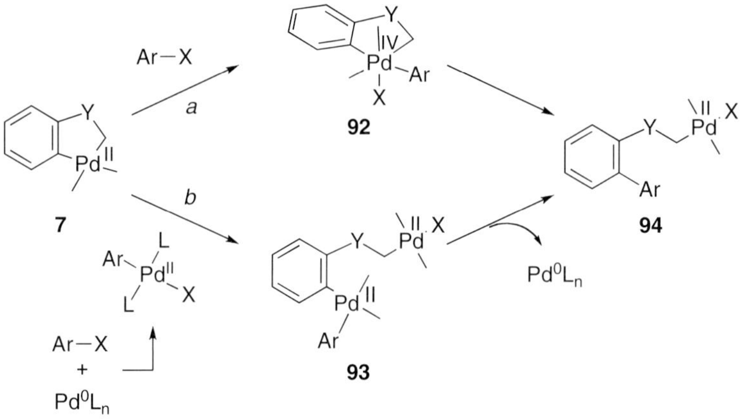

Scheme 11.31

Currently, there is strong evidence for the formation of Pd(IV) intermediates by the oxidative addition of alkyl halides to Pd(II) complexes by S_N2 processes [30, 99–101]. However, although activation of $C(sp^2)$–X electrophiles, such as aryl halides, by oxidative addition has been reported in the case of Ir(I) [102] and Pt(II) [103] complexes, a similar transformation has not been documented for Pd(II) derivatives. Indeed, there is no clear-cut experimental evidence for the oxidative addition of $C(sp^2)$–X electrophiles to Pd(II) complexes [104, 105], although Pd(IV) complexes $[PdArX_3(L\text{-}L)]$ (Ar = C_6F_5; X = Cl, Br) are obtained by oxidative addition of X_2 to [PdArX(L-L)] [106]. Comparison between calculated activation barriers for the oxidative addition of iodobenzene to Pd(0) and Pd(II) precursors show that the former process is more favorable [101].

A recent theoretical study examined the arylation of palladacycles **7** with ArX proceeding by oxidative addition via Pd(IV) intermediate **92** (Scheme 11.31, pathway *a*) or by a transmetallation-type process through **93** (pathway *b*) to afford

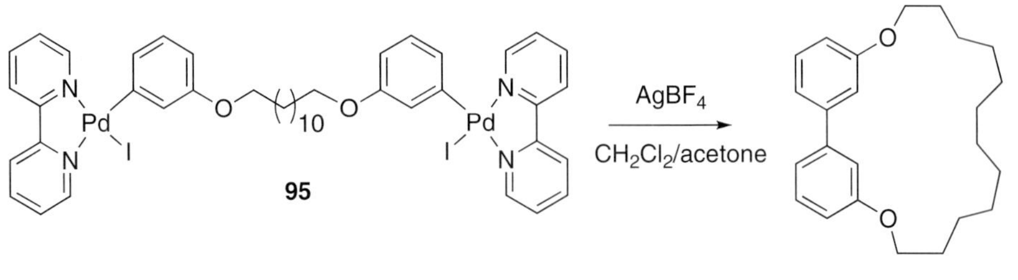

Scheme 11.32

94 [107, 108]. Whereas, the formation of Pd(IV) species in the Catellani reaction involving C(sp^3)–X electrophiles is strongly supported by other experimental studies [99–101], Pd(IV) species are unlikely to be actual intermediates in the formation of C(sp^2)–C(sp^2) bonds in Pd-catalyzed domino reactions involving C(sp^2)–X electrophiles. In these cases, the formation of C(sp^2)–C(sp^2) bonds probably takes place only at the Pd(II) oxidation state. Thus, the easy formation of bridged dinuclear complexes allows a facile intramolecular transmetallation-type process of organic ligands between two Pd centers, leading to diorganopalladium intermediates of type **93** that evolve by reductive elimination. The Pd–Pd transmetallation-type shows lower activation energies than processes occurring via Pd(IV) species, and these computational results are also in agreement with the poorer results obtained in palladium-catalyzed domino reactions in the presence of phosphines as ligands [109]. In addition, experimental evidence also excludes involvement of Pd(II)/Pd(IV) catalytic cycles in Heck and cross-coupling reactions involving palladacycles as precatalysts [110–112].

The exchange of carbon ligands between two Pd(II) centers [113] or between Pd(II) and Pt(II) [114] has been reported; a particularly interesting example is shown in Scheme 11.32. Thus, the reaction of dinuclear arylpalladium complex **95** with $AgBF_4$ gives rise to macrocyclic biaryl **96** via a Pd(II)/Pd(II) transmetallation-like reaction [115].

The formation of phenanthridinones **98** from *o*-bromoamides **97** most likely involves a reaction between two metallacycles (Scheme 11.33) [116, 117]. Accordingly, palladacycles **99** would react by a transmetallation type-process to form **100** [116a], the elimination of which forms a seven-membered ring palladacycle **101**, followed by a reductive elimination to give phenanthridinones **98**. Bimolecular transmetallation-type processes have also been proposed for the formation of other biaryls [69, 118].

Intramolecular arylations are also involved in catalytic processes that are mediated by palladacycles [119–122]. In the case of 4-arylpyridines **102**, a palladium-catalyzed reaction in the presence of biphenylene leads to tetraphenylenes **103** (Scheme 11.34) [123]. Consistent with previous results on the Heck reaction of substrates **102** [120], palladacycles **104** were proposed as reasonable intermediates in the reaction with biphenylene.

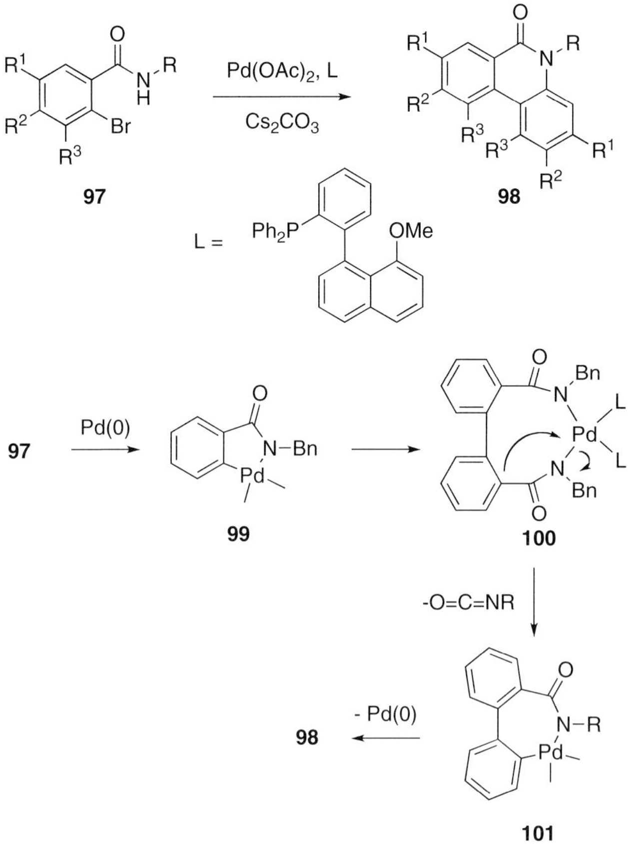

Scheme 11.33

Following studies on the ruthenium-catalyzed C–H bond activation/olefin coupling reaction [124, 125], related ruthenium-catalyzed *ortho*-arylations of a variety of substrates have been developed. The *ortho*-arylation of 2-arylpyridines with organic halides catalyzed by Ru(II) was proposed to proceed by a mechanism which involved arylruthenium (IV) species **105**, generated by the oxidative addition of arylhalide to Ru(II) complex (Scheme 11.35) [126]. Complex **105**, aided by chelation of the pyridyl group, reacts electrophilically with 2-pyridyl benzene to afford the ruthenacycle **106**, followed by reductive elimination to form the arylated product. A similar mechanism was proposed for the ruthenium-catalyzed arylation of aromatic imines [127], 2-alkenylpyridines [128], 2-aryloxazolines and 2-arylimidazolines [129].

Ruthenium(IV)–carbene complexes can be employed for the direct arylation of alkenes with chloroarenes [130]. Phosphine oxides have also been used as preligands in ruthenium-catalyzed direct arylations via C–H bond functionalization using aryl chlorides or aryl tosylates [131–133]. A noteworthy recent finding shows

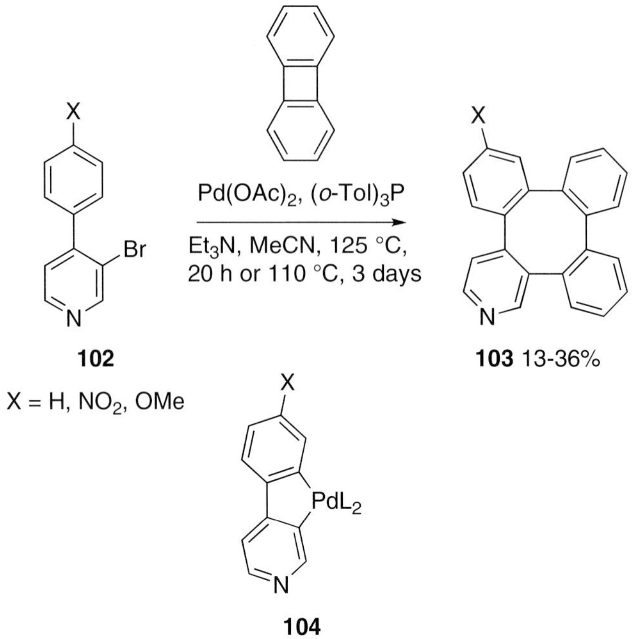

Pd(OAc)$_2$, (*o*-Tol)$_3$P

Et$_3$N, MeCN, 125 °C, 20 h or 110 °C, 3 days

102

X = H, NO$_2$, OMe

103 13-36%

PdL$_2$

104

Scheme 11.34

that the direct arylation reaction of arylpyridines with aryl bromides can be simply catalyzed by [RuCl$_3$(H$_2$O)$_n$] in the absence of any phosphorus-based ligands [134].

The direct functionalization of C—H sp^2 bonds via *ortho*-diarylation of 2-pyridyl benzene with arylbromides using NHC–ruthenium(II) catalysts takes place in the presence of Cs$_2$CO$_3$ (Scheme 11.36) [135]. Interestingly, arylbromides containing electron-donating or electron-withdrawing *para*-substituents lead to similar results. Therefore, in contrast to an electrophilic aromatic substitution process proposed above (see Scheme 11.35), this experimental result – along with DFT calculations – supports a proton-abstraction mechanism by the cooperative action of both the carbonate and the ruthenium(II) catalyst, as shown in **108** (Scheme 11.36). The ruthenium-catalyzed direct arylation proceeds more efficiently and with broad scope in the presence of K$_2$CO$_3$ as the base and substoichiometric amounts of carboxylic acid MesCO$_2$H [136].

The arylation of aromatic ketones with arylboronates proceeds by an *ortho*-ruthenation with RuH$_2$(CO)(PPh$_3$)$_3$ as the catalyst [137]. In this transformation, the final C—C bond is formed in a transmetallation–reductive elimination process. In a different reaction, *ortho*-arylated compounds are obtained from *o*-aminoaryl ketones and arylboronates through substitution of the amino function catalyzed by the same Ru(II) complex [138]. A nitrogen-directed homocoupling of aromatic compounds takes place with Ru(II) catalysts in the presence of allylic chlorides or acetates by a mechanism that presumably involves Ru(IV) intermediates [139].

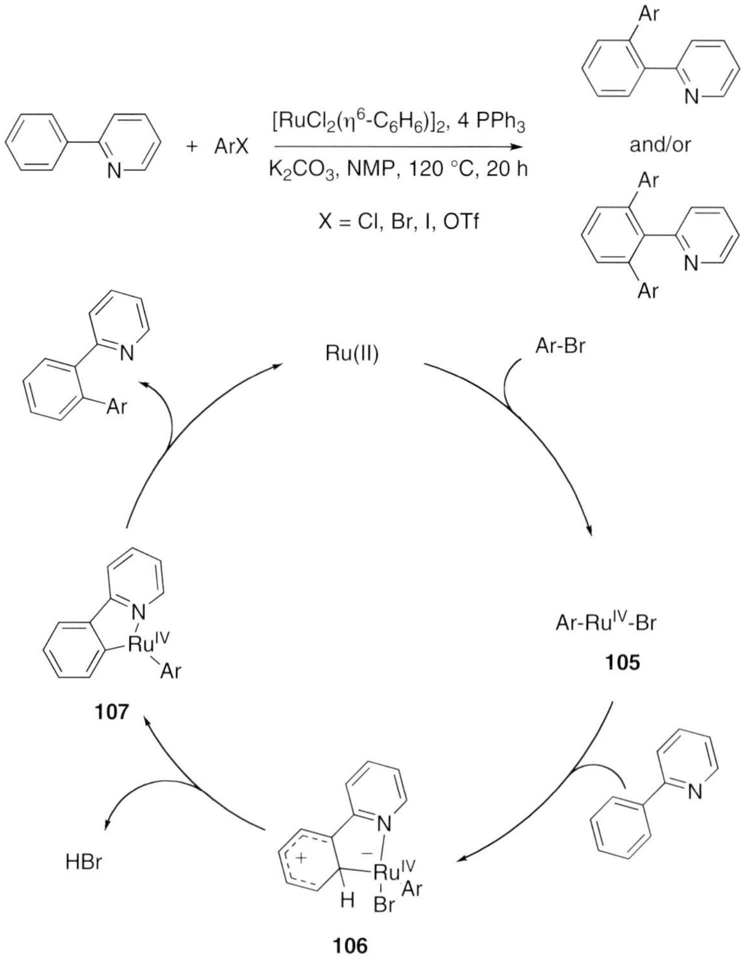

Scheme 11.35

Pyridylbenzenes are directly *ortho*-arylated with tetra-arylstannanes in the presence of a rhodium(I)–phosphine complex as catalyst [140]. A mechanistic pathway was proposed based on the oxidative addition of a rhodium(I) complex to the *ortho* position of the phenyl ring directed by the pyridine nitrogen, followed by arylation by the tetra-arylstannane. A somewhat related reaction of arylboronic acids was achieved with a $[RhCl(C_2H_4)_2]_2$/P[*p*-$(CF_3)C_6H_4]_3$ catalyst system [141]. In this instance, the 2,2,6,6-tetramethylpiperidine-*N*-oxyl (TEMPO) radical was used as a stoichiometric oxidant. Arylboronic acids also arylate benzophenone imines in the presence of Rh(I) catalysts [142].

Phenols can be *ortho*-arylated using Wilkinson catalyst $RhCl(PPh_3)_3$ or other Rh(I) complexes, and $PR_2(OAr)$ [143–145].The proposed mechanism is initiated by oxidative addition of the aryl halide to Rh(I) to form complexes **109**, followed

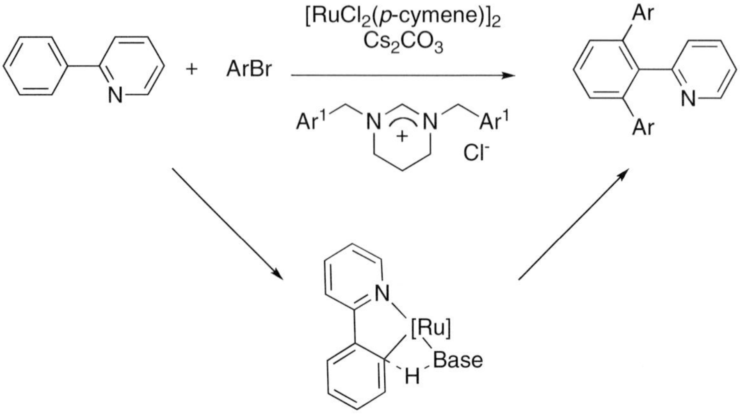

Scheme 11.36

by coordination and *ortho*-metallation of the phosphinite to the rhodium(III) species, affording **110** (Scheme 11.37). A reductive elimination of the ligand and the aryl group leads to regeneration of the active catalyst and to a 2-arylated aryl dialkylphosphinite species **111** that undergoes transesterification with phenol to liberate the regenerated cocatalyst and the 2-arylated phenol. Direct arylation at the *ortho* position of a variety of phenols with aryl bromides using [RhCl(cod)$_2$], hexamethylphosphoramide (HMPT), K_2CO_3 and Cs_2CO_3 in toluene at 100 °C was also shown to proceed by a similar mechanism [146].

11.6
Cross-Dehydrogenative Couplings

The cross-dehydrogenative coupling of *N*-protected acetanilides such as **112** with arenes proceeds with Pd(II) and Cu(II) via palladacycles **113** (Scheme 11.38) [147]. Presumably, palladacycles **113** react with the arenes by a proton-abstraction mechanism via transition state **114** to form Pd(II) complexes **115**, which evolve by reductive elimination to yield the coupled products **116**. A palladacycle was also involved in the related reaction of acetanilides with boronic acids catalyzed by Pd(II) and Cu(II) [148].

Anilides react in a general way as arenes in the presence of Pd(OAc)$_2$/DMSO catalyst and TFA in an atmosphere of O_2 to give products of *ortho*-arylation [149]. In this reaction, arenes with electron-withdrawing substituents, such as fluorobenzene derivatives, gave only poor conversions, which suggests that this reaction does not proceed via a proton-abstraction mechanism.

A palladium-catalyzed cross-dehydrogenative coupling of benzo[*h*]quinoline (**117**) and related substrates with substituted arenes takes place in the presence of benzoquinone [150]. In this reaction, the initially formed palladacycle (as in

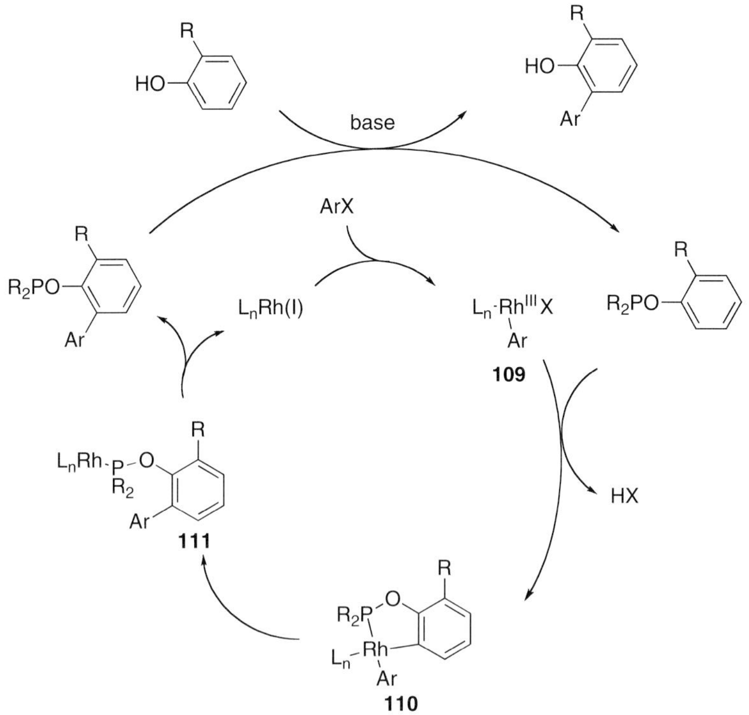

Scheme 11.37

Scheme 11.28) presumably has benzoquinone as a ligand. A mixture of regioisomers was obtained in the reaction of **117** with arenes such as anisole (Scheme 11.39). It is interesting that the reaction is not particularly sensitive to the effect of substituents at the arene. Thus, a competition experiment of **117** with 1,3-dimethyl-2-nitrobenzene and 1,3-dimethyl-2-methoxybenzene gives a 1 : 1.4 ratio of arylation products, slightly favoring reaction with the nitro derivative.

An interesting selectivity was uncovered in the direct cross-dehydrogenative coupling between *N*-protected indoles and arenes (Scheme 11.40) [151]. Thus, whereas 2-arylated indoles **67a** were preferentially obtained from *N*-acetylindole in the presence of $Cu(OAc)_2$, the reaction of *N*-pivalolylindole with AgOAc led to **67b**, with excellent selectivities. The reason for this C-2/C-3 selectivity is most likely due to the formation of higher-order palladium clusters or palladium/copper clusters under the different reaction conditions. A related reaction between arylboronic acids and arenes or heteroarenes also proceeds under oxidative conditions with $Pd(OAc)_2$ as catalyst [76]. A catalytic cycle initiated by an electrophilic attack of Pd(II) on the arene, followed by transmetallation with the aryl boronic acid and reductive elimination, was suggested. In this transformation, $Cu(OAc)_2$ as stoichiometric oxidant could be replaced by O_2, and for indoles, arylation at C-2 was observed.

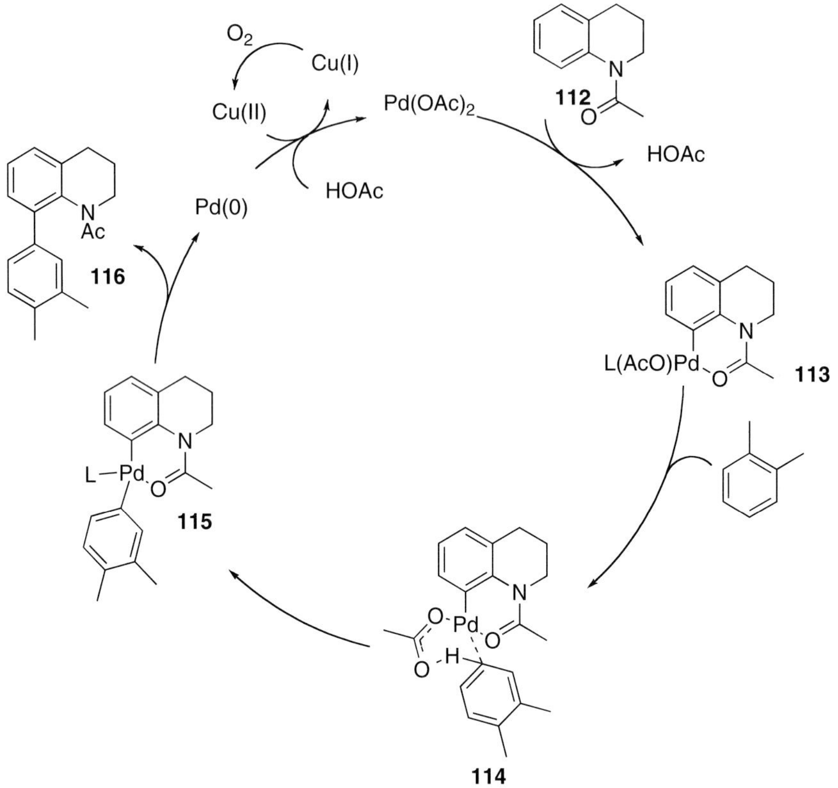

Scheme 11.38

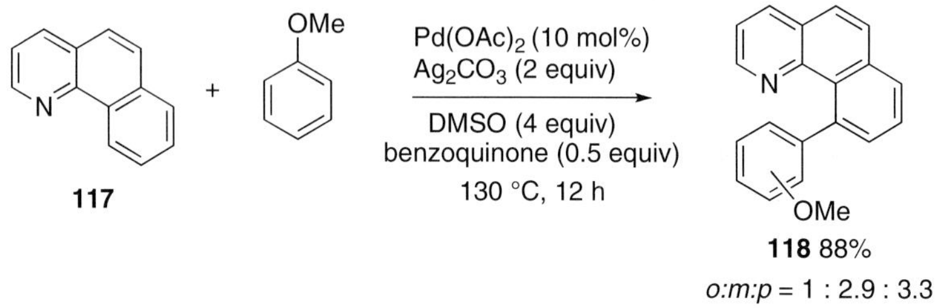

Scheme 11.39

The direct arylation of arenes with stannanes $ArSnCl_3$ is similarly catalyzed by $PdCl_2$ and proceeds with $CuCl_2$ as oxidant [152]. In this instance, a different mechanism was suggested based on the transmetallation between the stannane and a Pd(IV) salt, followed by an electrophilic aromatic substitution and reductive elimination. Arylzinc reagents also arylate 2-arylpyridines in the presence of iron catalysts [153].

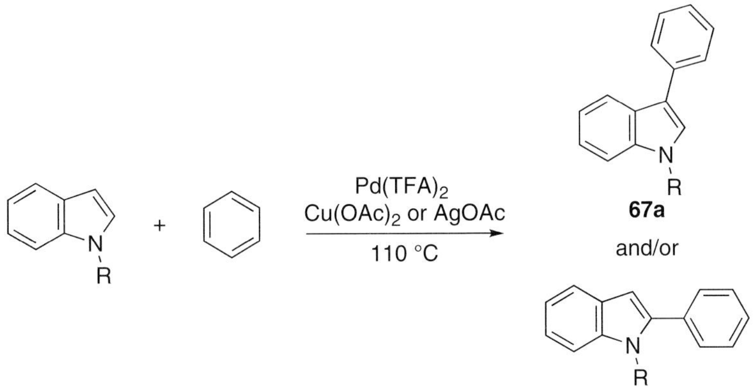

Pd(TFA)$_2$, Cu(OAc)$_2$ or AgOAc, 110 °C; products 67a and/or 67b

Scheme 11.40

11.7 Summary

In this chapter we have presented an outline of our current understanding of the mechanisms of metal-catalyzed direct arylation reactions. Although many investigations are still to be conducted, it is clear that metal-catalyzed direct arylations take place via a broad mechanistic spectrum. In a simplified form, two limiting mechanisms emerge for the most common processes, namely electrophilic aromatic substitution by the organometallic reagent formed *in situ*, and a proton-abstraction assisted by palladium and the base. For the synthetically important palladium-catalyzed direct arylation reactions, proton abstraction by the base is most likely involved as a key C—C bond-forming event. This is similar to the procedure found for palladium-catalyzed direct arylation, and also probably occurs in the ruthenium-catalyzed intermolecular arylation. However, the available data for the direct arylation of many electron-rich heterocycles catalyzed by palladium seem to favor an electrophilic aromatic substitution.

It is important that, whilst mechanisms that occur via the Pd(II)/Pd(IV) catalytic cycle have been excluded from Heck and cross-coupling reactions, this type of catalytic system may well be found in arylation reactions with $[Ar_2I]BF_4$, or in those reactions which occur under oxidative conditions.

Abbreviations

coe	cyclo-octene
cod	1,5-cyclo-octadiene
Cp	cyclopentadienyl
Cp*	pentamethylcyclopentadienyl
DDQ	2,3-dichloro-5,6-dicyano-1,4-benzoquinone

DFT	density functional theory
DMA	dimethylacetamide
DMF	dimethylformamide
DMSO	dimethylsulfoxide
HMPT	hexamethylphosphoramide
MS	molecular sieve
NHC	*N*-heterocyclic carbene
dppe	1,2-diphenylphosphinoethane
NMP	*N*-methylpyrrolidone
Piv	pivalate
TFA	trifluoroacetic acid or trifluoroacetate
Tol	tolyl

References

1 Hassan, J., Sévignon, M., Gozzi, C., Schulz, E. and Lemaire, M. (2002) *Chem. Rev.*, **102**, 1359–470.

2 Bringmann, G., Tasler, S., Pfeifer, R.-M. and Breuning, M. (2002) *J. Organomet. Chem.*, **661**, 49–65.

3 (a) Campeau, L.-C. and Fagnou, K. (2006) *Chem. Commun.*, 1253–64; (b) Campeau, L.-C., Stuart, D.R. and Fagnou, K. (2007) *Aldrichim. Acta*, **40**, 35–41.

4 Alberico, D., Scott, M.E. and Lautens, M. (2007) *Chem. Rev.*, **107**, 174–238.

5 Pascual, S., de Mendoza, P. and Echavarren, A.M. (2007) *Org. Biol. Chem.*, **5**, 2727–34.

6 Catellani, M., Motti, E., Della Ca', N. and Ferraccioli, R. (2007) *Eur. J. Org. Chem.*, 4153–65.

7 Li, B.-J., Yang, S.-D. and Shi, Z.J. (2008) *Synlett*, 949–57.

8 de Meijere, A. and Diederich, F. (eds) (2004) *Metal-Catalyzed Cross-Coupling Reactions*, Wiley-VCH Verlag GmbH, Weinheim.

9 Espinet, P. and Echavarren, A.M. (2004) *Angew. Chem. Int. Ed.*, **43**, 4704–34.

10 Lead references: (a) Abe, H., Takeda, S., Fujita, T., Nishioka, K., Takeuchi, Y. and Harayama, T. (2004) *Tetrahedron Lett.*, **45**, 2327–9; (b) Torres, J.C., Pinto, A.C. and Garden, S.J. (2004) *Tetrahedron*, **60**, 9889–900; (c) Harayama, T., Hori, A., Abe, H. and Takeuchi, Y. (2004) *Tetrahedron*, **60**, 1611–16; (d) Harrowven, D.C., Woodcock, T. and Howes, P.D. (2005) *Angew. Chem. Int. Ed.*, **44**, 3899–901; (e) Ohmori, K., Tamiya, M., Kitamura, M., Kato, H., Oorui, M. and Suzuki, K. (2005) *Angew. Chem. Int. Ed.*, **44**, 3871–4; (f) Abe, H., Nishioka, K., Takeda, S., Arai, M., Takeuchi, Y. and Harayama, T. (2005) *Tetrahedron Lett.*, **46**, 3197–200; (g) Bowie, A.L., Hughes, C.C. and Trauner, D. (2005) *Org. Lett.*, **7**, 5207–9; (h) Shen, D.-M., Liu, C. and Chen, Q.-Y. (2006) *J. Org. Chem.*, **71**, 6508–11; (i) Ackermann, L. and Althammer, A. (2007) *Angew. Chem. Int. Ed.*, **46**, 1627–9; (j) Tamiya, M., Ohmori, K., Kitamura, M., Kato, H., Arai, T., Oorui, M. and Suzuki, K. (2007) *Chem. Eur. J.*, **13**, 9791–823.

11 (a) Tsefrikas, V.M. and Scott, L.T. (2006) *Chem. Rev.*, **106**, 4868–84; (b) Wu, Y.-T. and Siegel, J.S. (2006) *Chem. Rev.*, **106**, 4843–67.

12 Echavarren, A.M., Gómez-Lor, B., González, J.J. and de Frutos, Ó (2003) *Synlett*, 585–97.

13 (a) de Frutos, Ó., Gómez-Lor, B., Granier, T., Monge, M.Á., Gutiérrez-Puebla, E. and Echavarren, A.M. (1999) *Angew. Chem. Int. Ed.*, **38**, 204–7;

(b) Gómez-Lor, B., de Frutos, Ó. and Echavarren, A.M. (1999) *Chem. Commun.*, 2431–2;
(c) Gómez-Lor, B., Koper, C., Fokkens, R.H., Vlietstra, E.J., Cleij, T.J., Jenneskens, L.W., Nibbering, N.M.M. and Echavarren, A.M. (2002) *Chem. Commun.*, 370–1;
(d) Gómez-Lor, B., González-Cantalapiedra, E., Ruiz, M., de Frutos, Ó., Cárdenas, D.J., Santos, A. and Echavarren, A.M. (2004) *Chem. Eur. J.*, **10**, 2601–8.

14 (a) Rice, J.E. and Cai, Z.-W. (1992) *Tetrahedron Lett.*, **33**, 1675–8;
(b) Rice, J.E. and Cai, Z.-W. (1993) *J. Org. Chem.*, **58**, 1415–24;
(c) Rice, J.E., Cai, Z.-W., He, Z.-M. and LaVoie, E.J. (1995) *J. Org. Chem.*, **60**, 8101–4.

15 (a) Reisch, H.A., Bratcher, M.S. and Scott, L.T. (2000) *Org. Lett.*, **2**, 1427–30;
(b) Wegner, H.A., Scott, L.T. and de Meijere, A. (2003) *J. Org. Chem.*, **68**, 883–7;
(c) Wegner, H.A., Reisch, H., Rauch, K., Demeter, A., Zachariasse, K.A., de Meijere, A. and Scott, L.T. (2006) *J. Org. Chem.*, **71**, 9080–7;
(d) Tsefrikas, V.M., Arns, S., Merner, P.M., Warford, C.C., Merner, B.L., Scott, L.T. and Bodwell, G.J. (2006) *Org. Lett.*, **8**, 5195–8;
(e) Jackson, E.A., Steinberg, B.D., Bancu, M., Wakamiya, A. and Scott, L.T. (2007) *J. Am. Chem. Soc.*, **129**, 484–5.

16 Cheng, X.H., Höger, S. and Fenske, D. (2003) *Org. Lett.*, **5**, 2587–9.

17 (a) Wang, L. and Sevlin, P.B. (2000) *Tetrahedron Lett.*, **41**, 285–8;
(b) Wang, L. and Sevlin, P.B. (2000) *Org. Lett.*, **2**, 3703–5 (erratum p. 4115);
(c) Kim, D., Petersen, J.L. and Wang, K.K. (2006) *Org. Lett.*, **8**, 2313–16.

18 Fox, S. and Boyle, R.W. (2004) *Chem. Commun.*, 1322–3.

19 Ohno, H., Iuchi, M., Fujii, N. and Tanaka, T. (2007) *Org. Lett.*, **9**, 4813–15.

20 (a) Bedford, R.B. and Betham, M. (2006) *J. Org. Chem.*, **71**, 9403–10;
(b) Bedford, R.B., Betham, M., Charmant, J.P.H. and Weeks, A.L. (2008) *Tetrahedron*, **64**, 6038–50.

21 Li, C.-W., Wang, C.-I., Liao, H.-Y., Chaudhuri, R. and Liu, R.-S. (2007) *J. Org. Chem.*, **72**, 9203–7.

22 Pinto, A., Neuville, L., Retailleau, P. and Zhu, J. (2006) *Org. Lett.*, **8**, 4927–30.

23 Bernini, R., Cacchi, S., Fabrizi, G. and Sferrazza, A. (2008) *Synthesis*, 729–38.

24 Parisien, M., Valette, D. and Fagnou, K. (2005) *J. Org. Chem.*, **70**, 7578–84.

25 (a) Campeau, L.-C., Parisien, M., Leblanc, M. and Fagnou, K. (2004) *J. Am. Chem. Soc.*, **126**, 9186–7;
(b) Lafrance, M., Blaquière, N. and Fagnou, K. (2004) *Chem. Commun.*, 2874–5;
(c) Leblanc, M. and Fagnou, K. (2005) *Org. Lett.*, **7**, 2849–52;
(d) Leclerc, J.-P., André, M. and Fagnou, K. (2006) *J. Org. Chem.*, **71**, 1711–14;
(e) Campeau, L.-C., Parisien, M., Jean, A. and Fagnou, K. (2006) *J. Am. Chem. Soc.*, **128**, 581–90;
(f) Lafrance, M., Blaquière, N. and Fagnou, K. (2007) *Eur. J. Org. Chem.*, 811–25.

26 Campeau, L.-C., Thansandote, P. and Fagnou, K. (2005) *Org. Lett.*, **7**, 1857–60.

27 Mota, A.J., Dedieu, A., Bour, C. and Suffert, J. (2005) *J. Am. Chem. Soc.*, **127**, 7171–82.

28 Hughes, C.C. and Trauner, D. (2002) *Angew. Chem. Int. Ed.*, **41**, 1569–72.

29 Martín-Matute, B., Mateo, C., Cárdenas, D.J. and Echavarren, A.M. (2001) *Chem. Eur. J.*, **7**, 2341–8.

30 (a) Cárdenas, D.J., Mateo, C. and Echavarren, A.M. (1994) *Angew. Chem. Int. Ed.*, **33**, 2445–7;
(b) Mateo, C., Cárdenas, D.J., Fernández-Rivas, C. and Echavarren, A.M. (1996) *Chem. Eur. J.*, **2**, 1596–606;
(c) Mateo, C., Fernández-Rivas, C., Echavarren, A.M. and Cárdenas, D.J. (1997) *Organometallics*, **16**, 1997–9.

31 Catellani, M. and Chiusoli, G.P. (1992) *J. Organomet. Chem.*, **425**, 151–4.

32 Lafrance, M., Gorelsky, S.I. and Fagnou, K. (2007) *J. Am. Chem. Soc.*, **129**, 14570–1.

33 Cámpora, J., López, J.A., Palma, P., Valerga, P., Spillner, E. and Carmona, E. (1999) *Angew. Chem. Int. Ed.*, **38**, 147–51.

34 Hennessy, E.J. and Buchwald, S.L. (2003) *J. Am. Chem. Soc.*, **125**, 12084–5.

35 González, J.J., García, N., Gómez-Lor, B. and Echavarren, A.M. (1997) *J. Org. Chem.*, **62**, 1286–91.

36 Gómez-Lor, B. and Echavarren, A.M. (2004) *Org. Lett.*, **6**, 2993–6.

37 (a) García-Cuadrado, D., Braga, A.A.C., Maseras, F. and Echavarren, A.M. (2006) *J. Am. Chem. Soc.*, **128**, 1066–7; (b) García-Cuadrado, D., de Mendoza, P., Braga, A.A.C., Maseras, F. and Echavarren, A.M. (2007) *J. Am. Chem. Soc.*, **129**, 6880–6.

38 (a) Davies, D.L., Donald, S.M.A. and Macgregor, S.A. (2005) *J. Am. Chem. Soc.*, **127**, 13754–5; (b) Ryabov, A.D. (1990) *Chem. Rev.*, **90**, 403–24.

39 Pascual, S., de Mendoza, P., Braga, A.A.C., Maseras, F. and Echavarren, A.M. (2008) *Tetrahedron*, **64**, 6021–9.

40 Cruz, A.C.F., Miller, N.D. and Willis, M.C. (2007) *Org. Lett.*, **9**, 4391–3.

41 Chernyak, N. and Gebvorgyan, V. (2008) *J. Am. Chem. Soc.*, **130**, 5636–7.

42 (a) Lafrance, M., Rowley, C.N., Woo, T.K. and Fagnou, K.K. (2006) *J. Am. Chem. Soc.*, **128**, 8754–6; (b) Lafrance, M., Shore, D. and Fagnou, K. (2006) *Org. Lett.*, **8**, 5097–100.

43 Do, H.-Q. and Daugulis, O. (2008) *J. Am. Chem. Soc.*, **130**, 1128–9.

44 Lafrance, M. and Fagnou, K. (2006) *J. Am. Chem. Soc.*, **128**, 16496–7.

45 Chiong, H.A., Pham, Q.-N. and Daugulis, O. (2007) *J. Am. Chem. Soc.*, **129**, 9879–84.

46 Xia, J.-B. and You, S.-L. (2007) *Organometallics*, **26**, 4869–71.

47 Zhang, Y., Feng, J. and Li, C.-J. (2008) *J. Am. Chem. Soc.*, **130**, 2900–1.

48 Fujita, K.-I., Nonogawa, M. and Yamaguchi, R. (2004) *J. Chem. Soc., Chem. Commun.*, 1926–7.

49 Proch, S. and Kempe, R. (2007) *Angew. Chem. Int. Ed.*, **46**, 3135–8.

50 Seregin, I.V. and Gevorgyan, V. (2007) *Chem. Soc. Rev.*, **36**, 1173–93.

51 Leading references on the palladium-catalyzed arylation of furans: (a) McClure, M.S., Glover, B., McSorley, E., Millar, A., Osterhout, M.H. and Roschangar, F. (2001) *Org. Lett.*, **3**, 1677–88; (b) Glover, B., Harvey, K.A., Liu, B., Sharp, M.J. and Tymoschenko, M.F. (2003) *Org. Lett.*, **5**, 301–4.

52 Leading references on the palladium-catalyzed arylation of thiophenes: (a) Gozzi, C., Lavenot, L., Ilg, K., Penalva, V. and Lemaire, M. (1997) *Tetrahedron Lett.*, **38**, 8867–70; (b) Lavenot, L., Gozzi, C., Ilg, K., Orlova, I., Pealva, V. and Lemaire, M. (1998) *J. Organomet. Chem.*, **567**, 49–55; (c) Okazawa, T., Satoh, T., Miura, M. and Nomura, M. (2002) *J. Am. Chem. Soc.*, **124**, 5286–7; (d) Kobayashi, K., Sugie, A., Takahashi, M., Masui, K. and Mori, A. (2005) *Org. Lett.*, **7**, 5083–5. (e) Turner, G.L., Morris, J.A. and Greaney, M.F. (2007) *Angew. Chem. Int. Ed.*, **46**, 7996–8000; (f) Nakano, M., Tsurugi, H., Satoh, T. and Miura, M. (2008) *Org. Lett.*, **10**, 1851–4.

53 Lead reference on the arylation of (a) oxazoles: Besselièvre, F., Mahuteau-Betzer, F., Grierson, D.S. and Piguel, S. (2008) *J. Org. Chem.*, **73**, 3278–80; (b) imidazoles: Bellina, F., Cauteruccio, S. and Rossi, R. (2007) *J. Org. Chem.*, **72**, 8543–6; (c) 1,2,4-Trizazoles: Ackermann, L., Vicente, R. and Born, R. (2008) *Adv. Synth. Catal.*, **350**, 741–8.

54 Leading references on early studies on the palladium-catalyzed arylation of indoles: (a) Akita, Y., Itagaki, Y., Takizawa, S. and Ohta, A. (1989) *Chem. Pharm. Bull.*, **37**, 1477–80; (b) Itahara, T. (1981) *J. Chem. Soc., Chem. Commun.*, 254–5; (c) Akita, Y., Inoue, A., Yamamoto, K. and Ohta, A. (1985) *Heterocycles*, **23**, 2327–33.

55 Cerna, I., Pohl, R., Kelpetarova, A.B. and Hocek, M. (2006) *Org. Lett.*, **8**, 5389–92.

56 Koubachi, J., El Kazzouli, S., Berteina-Raboin, S., Mouaddib, A. and Guillaumet, G. (2006) *Synlett*, 3237–42.

57 (a) Campeau, L.-C., Rousseaux, S. and Fagnou, K. (2005) *J. Am. Chem. Soc.*, **127**, 18020–1;
(b) Leclerc, J.-P. and Fagnou, K. (2006) *Angew. Chem. Int. Ed.*, **45**, 7781–6;
(c) Campeau, L.-C., Bertran-Laperle, M., Leclerc, J.-P., Villemure, E., Gorelsky, S. and Fagnou, K. (2008) *J. Am. Chem. Soc*, **130**, 3276–7.

58 Campeau, L.-C., Schipper, D.J. and Fagnou, K. (2008) *J. Am. Chem. Soc.*, **130**, 3266–7.

59 Watanabe, T., Oishi, S., Fujii, N. and Ohno, H. (2008) *Org. Lett.*, **10**, 1759–62.

60 Salcedo, A., Neuville, L. and Zhu, J. (2008) *J. Org. Chem.*, **73**, 3600–3.

61 Pivsa-Art, S., Satoh, T., Kawamura, Y., Miura, M. and Nomura, M. (1998) *Bull. Chem. Soc. Jpn.*, **71**, 467–73.

62 Chiong, H.A. and Daugulis, O. (2007) *Org. Lett.*, **9**, 1449–51.

63 Priego, J., Gutiérrez, S., Ferritto, R. and Broughton, H.B. (2007) *Synlett*, 2957–60.

64 Park, C.-H., Ryabova, V., Seregin, I.V., Sromek, A.W. and Gevorgyan, V. (2004) *Org. Lett.*, **6**, 1159–62.

65 Seregin, I.V., Ryabova, V. and Gevorgyan, V. (2007) *J. Am. Chem. Soc.*, **129**, 7742–3.

66 Lu, J., Tan, X. and Chen, C. (2007) *J. Am. Chem. Soc.*, **129**, 7768–9.

67 (a) Zhurvalev, F.A. (2006) *Tetrahedron Lett.*, **47**, 2929–32;
(b) Sánchez, R.S. and Zhuralev, F.A. (2007) *J. Am. Chem. Soc.*, **129**, 5824–5.

68 Bellina, F., Calandri, C., Cauteruccio, S. and Rossi, R. (2007) *Tetrahedron*, **63**, 1970–80.

69 Lane, B.S. and Sames, D. (2004) *Org. Lett.*, **6**, 2897–900.

70 Touré, B.B., Lane, B.S. and Sames, D. (2006) *Org. Lett.*, **8**, 1979–82.

71 Wang, X., Gribkov, D.V. and Sames, D. (2007) *J. Org. Chem.*, **72**, 1476–9.

72 Lebrasseur, N. and Larrosa, I. (2008) *J. Am. Chem. Soc.*, **130**, 2926–7.

73 Zhang, Z., Hu, Z., Yu, Z., Lei, P., Chi, H., Wang, Y. and He, R. (2007) *Tetrahedron Lett.*, **48**, 2415–19.

74 Lane, B.S., Brown, M.A. and Sames, D. (2005) *J. Am. Chem. Soc.*, **127**, 8050–7. Correction: *J. Am. Chem. Soc.*, 2007, **129**, 241.

75 Saulnier, M.G. and Gribble, G.W. (1982) *J. Org. Chem.*, **47**, 757–61.

76 Yang, S.-D., Sun, C.-L., Fang, Z., Li, B.-J., Li, Y.-Z. and Shi, Z.-J. (2008) *Angew. Chem. Int. Ed.*, **47**, 1473–6.

77 Deprez, N.R., Kalyani, D., Krause, A. and Sanford, M.S. (2006) *J. Am. Chem. Soc.*, **128**, 4972–3.

78 Deprez, N.R. and Sanford, M.S. (2007) *Inorg. Chem.*, **46**, 1924–35.

79 (a) Bayler, A., Canty, A.J., Ryan, J.H., Skelton, B.W. and White, A.W. (2000) *Inorg. Chem. Commun.*, **3**, 575–8;
(b) Canty, A.J. and Rodemann, T. (2003) *Inorg. Chem. Commun.*, **6**, 1382–4;
(c) Canty, A.J., Patel, J., Rodemann, T., Ryan, J.H., Skelton, B.W. and White, A.H. (2004) *Organometallics*, **23**, 3466–73.

80 Do, H.-Q. and Daugulis, O. (2007) *J. Am. Chem. Soc.*, **129**, 12404–5.

81 (a) Lewis, J.C., Wiedeman, S.H., Bergman, R.G. and Ellman, J.A. (2004) *Org. Lett.*, **6**, 35–8;
(b) Lewis, J.C., Wu, J.Y., Bergman, R.G. and Ellman, J.A. (2006) *Angew. Chem. Int. Ed.*, **45**, 1589–91;
(c) Lewis, J.C., Berman, A.M., Bergman, R.G. and Ellman, J.A. (2008) *J. Am. Chem. Soc.*, **130**, 2493–500.

82 Wang, X., Lane, B.S. and Sames, D. (2005) *J. Am. Chem. Soc.*, **127**, 4996–7.

83 Yanagisawa, S., Sudo, T., Noyori, R. and Itami, K. (2006) *J. Am. Chem. Soc.*, **128**, 11748–9.

84 Yang, S., Li, B., Wan, X. and Shi, Z. (2007) *J. Am. Chem. Soc.*, **129**, 6066–7.

85 Lazareva, A. and Daugulis, O. (2006) *Org. Lett.*, **8**, 5211–13.

86 Chen, X., Li, J.-J., Hao, X.-S., Goodhue, C.E. and Yu, J.-Q. (2006) *J. Am. Chem. Soc.*, **128**, 78–9.

87 Chen, X., Goodhue, C.E. and Yu, J.-Q. (2006) *J. Am. Chem. Soc.*, **128**, 12634–5.

88 Kalyani, D., Deprez, N.R., Desai, L.V. and Stanford, M.S. (2005) *J. Am. Chem. Soc.*, **127**, 7330–1.

89 Daugulis, O. and Zaitsev, V.G. (2005) *Angew. Chem. Int. Ed.*, **44**, 4046–8.

90 Catellani, M., Motti, E., Faccini, F. and Ferraccioli, R. (2005) *Pure Appl. Chem.*, **77**, 1243–8.

91 (a) Catellani, M. and Chiusoli, G.P. (1988) *J. Organomet. Chem.*, **346**, C27–30;
(b) Catellani, M. and Mann, B.E. (1990) *J. Organomet. Chem.*, **390**, 251–5;
(c) Catellani, M., Chiusoli, G.P. and Castagnoli, C.J. (1991) *J. Organomet. Chem.*, **407**, C30–3;
(d) Catellani, M. and Chiusoli, G.P. (1992) *J. Organomet. Chem.*, **437**, 369–73;
(e) Bocelli, G., Catellani, M. and Ghelli, S. (1993) *J. Organomet. Chem.*, **458**, C12–5;
(f) Catellani, M. and Fagnola, M.C. (1994) *Angew. Chem. Int. Ed.*, **33**, 2421–2;
(g) Catellani, M., Marmiroli, B., Fagnola, M.C. and Acquotti, D. (1996) *J. Organomet. Chem.*, **507**, 157–62;
(h) Catellani, M. and Ferioli, L. (1996) *Synthesis*, 769–72;
(i) Catellani, M., Frignani, F. and Rangoni, A. (1997) *Angew. Chem. Int. Ed.*, **36**, 119–22;
(j) Catellani, M., Motti, E. and Minari, M. (2000) *J. Chem. Soc., Chem. Commun.*, 157–8;
(k) Catellani, M., Mealli, C., Motti, E., Paoli, P., Perez-Carreño, E. and Pregosin, P.S. (2002) *J. Am. Chem. Soc.*, **124**, 4336–46;
(l) Motti, E., Ippomei, G., Deledda, S. and Catellani, M. (2003) *Synthesis*, 2671–8;
(m) Catellani, M. (2003) *Synlett*, 298–313;
(n) Ferraccioli, R., Carenzi, D., Rombolà, O. and Catellani, M. (2004) *Org. Lett.*, **6**, 4759–62;
(o) Motti, E., Rossetti, M., Bocelli, G. and Catellani, M. (2004) *J. Organomet. Chem.*, **689**, 3741–9;
(p) Motti, E. and Catellani, M. (2008) *Adv. Synth. Catal.*, **350**, 565–9;
(q) Motti, E., Ca, N.C., Ferraccioli, R. and Catellani, M. (2008) *Synthesis*, 995–7.

92 (a) Reiser, O., Weber, M. and de Meijere, A. (1989) *Angew. Chem. Int. Ed.*, **28**, 1037–8;
(b) Albrecht, K., Reiser, O., Weber, M. and de Meijere, A. (1991) *Tetrahedron*, **50**, 383–401.

93 (a) Li, C.-S., Cheng, C.-H., Liao, F.-L. and Wang, S.-L. (1991) *J. Chem. Soc., Chem. Commun.*, 710–1;
(b) Li, C.-S., Jou, D.-C. and Cheng, C.-H. (1993) *Organometallics*, **12**, 3945–54;
(c) Liu, C.-H., Li, C.-S. and Cheng, C.-H. (1994) *Organometallics*, **13**, 18–20.

94 (a) Markies, B.A., Wijkens, P., Kooijman, H., Spek, A.L., Boersma, J. and van Koten, G. (1992) *J. Chem. Soc., Chem. Commun.*, 1420–3;
(b) Markies, B.A., Canty, A.J., Boersma, J. and van Koten, G. (1994) *Organometallics*, **13**, 2053–8, references cited therein.

95 Bertrand, M.B. and Wolfe, J.P. (2007) *Org. Lett.*, **9**, 3073–5.

96 (a) Lautens, M. and Piguel, S. (2000) *Angew. Chem. Int. Ed.*, **39**, 1045–6;
(b) Lautens, M., Paquin, J.-F., Piguel, S. and Dahlmann, M. (2001) *J. Org. Chem.*, **66**, 8127–34;
(c) Lautens, M., Paquin, J.-F. and Piguel, S. (2002) *J. Org. Chem.*, **67**, 3972–4;
(d) Pache, S. and Lautens, M. (2003) *Org. Lett.*, **5**, 4827–30;
(e) Wilhelm, T. and Lautens, M. (2005) *Org. Lett.*, **7**, 4053–6;
(f) Bressy, C., Alberico, D. and Lautens, M. (2005) *J. Am. Chem. Soc.*, **127**, 13148–9;
(g) Mariampillai, B., Herse, C. and Lautens, M. (2005) *Org. Lett.*, **7**, 4745–7;
(h) Alberico, D., Paquin, J.-C. and Lautens, M. (2005) *Tetrahedron*, **61**, 6283–97;
(i) Blaszykowski, C., Aktoudianakis, E., Alberico, D., Bressy, C., Hulcoop, D.G., Jafarpour, F., Joushagani, A., Laleu, B. and Lautens, M. (2008) *J. Org. Chem.*, **73**, 1888–97;

97 (a) Mauleón, P., Núñez, A.A., Alonso, I. and Carretero, J.C. (2003) *Chem. Eur. J.*, **9**, 1511–20;
(b) Alonso, I., Alcamí, M., Mauleón, P. and Carretero, J.C. (2006) *Chem. Eur. J.*, **12**, 4576–83.

98 (a) Dyker, G. (1992) *Angew. Chem. Int. Ed. Engl.*, **31**, 1023–5;
(b) Dyker, G. (1994) *Chem. Ber.*, **127**, 739–42;
(c) Dyker, G. (1994) *Angew. Chem. Int. Ed. Engl.*, **33**, 103–5;
(d) Dyker, G., Nerenz, F., Siemsen, P., Bubenitschek, P. and Jones, P.G. (1996) *Chem. Ber.*, **129**, 1265–9;
(e) Dyker, G., Siemsen, P., Sostmann, S., Wiegand, A., Dix, I. and Jones, P.G. (1997) *Chem. Ber.*, **130**, 261–5;
(f) Dyker, G., Borowski, S., Henkel, G., Andreas, K., Dix, I. and Jones, P.G. (1999) *Angew. Chem. Int. Ed.*, **38**, 1699–712;
(g) Dyker, G. (2000) *Tetrahedron Lett.*, **41**, 8259–62.

99 (a) Canty, A.J. (1992) *Acc. Chem. Res.*, **25**, 83–90;
(b) Canty, A.J. (1993) *Platinum Metals Rev.* **37**, 2–7.

100 Reviews: (a) Canty, A.J. (2002) *Handbook of Organopalladium Chemistry for Organic Synthesis*, Vol. 1 (ed. E.-I. Negishi), John Wiley & Sons, Inc., New York, Chapter II.4, pp. 189–211;
(b) Kruis, D., Markies, B.A., Canty, A.J., Boersma, J. and van Koten, G. (1997) *J. Organomet. Chem.*, **532**, 235–42;
(c) Catellani, M., Chiusoli, G.P. and Costa, M. (1995) *J. Organomet. Chem.*, **500**, 69–80.

101 Mateo, C., Fernández-Rivas, C. Cárdenas, D.J. and Echavarren, A.M. (1998) *Organometallics*, **17**, 3661–9.

102 Chock, P.B. and Halpern, J. (1966) *J. Am. Chem. Soc.*, **88**, 3511–14.

103 Oxidative addition of aryl halides to Pt(II) complexes only proceeds intramolecularly: Baar, C.R., Hill, G.S., Vittal, J.J. and Puddephatt, R.J. (1998) *Organometallics*, **17**, 32–40, references cited therein.

104 The X-ray structure of a Pd(IV) complex formed by oxidative addition of PhI to a Pd(II) complex was published: (a) Brunel, J.M., Hirlemann, M.-H., Heumann, A. and Buono, G. (2000) *Chem. Commun.*, 1869–70;
(b) Brunel, J.M., Hirlemann, M.-H., Heumann, A. and Buono, G. (2001) *Chem. Commun.*, 1896, 2298. However, this work was retracted (*Chem. Commun.* 16 May 2002).

105 Theoretical work on a hypothetical mechanism for the Heck reaction based on Pd(II)/Pd(IV) indicates that the oxidative addition of PhI to Pd(II) would be the rate-determining step: Sundermann, A., Uzan, O. and Martin, J.M.L. (2001) *Chem. Eur. J.*, **7**, 1703–11.

106 (a) Usón, R., Forniés, J. and Navarro, R. (1975) *J. Organomet. Chem.*, **96**, 307–12;
(b) Forniés, J. and Navarro, R. (1997) *Synth. React. Inorg. Met. Org. Chem.*, **7**, 235–41.

107 Cárdenas, D.J., Martín-Matute, B. and Echavarren, A.M. (2006) *J. Am. Chem. Soc.*, **128**, 5033–40.

108 See also: (a) Mota, A.J. and Dedieu, A. (2006) *Organometallics*, **25**, 3130–42;
(b) Mota, A.J. and Dedieu, A. (2007) *J. Org. Chem.*, **72**, 9669–78.

109 Dyker, G. and Kellner, A. (1998) *J. Organomet. Chem.*, **555**, 141–4.

110 Muñoz, M.P., Martín-Matute, B., Fernández-Rivas, C., Cárdenas, D.J. and Echavarren, A.M. (2001) *Adv. Synth. Catal.*, **343**, 338–42.

111 (a) de Vries, A.H.M., Mulders, J.M.C.A., Mommers, J.H.M., Henderickx, H.J.W. and de Vries, J.G. (2003) *Org. Lett.*, **5**, 3285–8;
(b) de Vries, J.G. (2006) *Dalton Trans.*, 421–9.

112 Thathagar, M.B., ten Elshof, J.E. and Rothenberg, G. (2006) *Angew. Chem. Int. Ed.*, **45**, 2886–90.

113 (a) Ozawa, F., Hidaka, T., Yamamoto, T. and Yamamoto, A. (1987) *J. Organomet. Chem.*, **330**, 253–63;
(b) Ozawa, F., Fujimori, M., Yamamoto, T. and Yamamoto, A. (1986) *Organometallics*, **5**, 2144–9.

114 (a) Suzaki, Y. and Osakada, K. (2004) *Bull. Chem. Soc. Jpn.*, **77**, 139–45;
(b)For aryl ligands bridging two Pt(II) centers, see: Konze, W.V., Scott, B.L. and Kubas, G.J. (2002) *J. Am. Chem. Soc.*, **124**, 12550–6.

115 Suzaki, Y. and Osakada, K. (2003) *Organometallics*, **22**, 2193–5.

116 (a) Furuta, T., Kitamura, Y., Hashimoto, A., Fujii, S., Tanaka, K. and Kan, T. (2007) *Org. Lett.*, **9**, 183–6;
(b) Caddick, S. and Kofie, W. (2002) *Tetrahedron Lett.*, **43**, 9347–50.

117 Ferraccioli, R., Carenzi, D., Motti, E. and Catellani, M. (2006) *J. Am. Chem. Soc.*, **128**, 722–3.

118 (a) Amatore, C., Carre, E. and Jutand, A. (1998) *Acta Chem. Scand.*, **52**, 100–6;
(b) Ozawa, F., Hidaka, T., Yamamoto, T. and Yamamoto, A. (1987) *J. Organomet. Chem.*, **330**, 253–63.

119 Dupont, J., Consorti, C.S. and Spencer, J. (2005) *Chem. Rev.*, **105**, 2527–72.

120 Karig, G., Moon, M.-T., Thasana, N. and Gallagher, T. (2002) *Org. Lett.*, **4**, 3115–8.

121 (a) Campo, M.A. and Larock, R.C. (2002) *J. Am. Chem. Soc*, **124**, 14326–7;
(b) Zhao, J., Campo, M.A. and Larock, R.C. (2005) *Angew. Chem. Int. Ed.*, **44**, 1873–5;
(c) Campo, M.A., Zhang, H., Yao, T., Ibdah, A., McCulla, R.D., Huang, Q., Zhao, J., Jenks, W.S. and Larock, R.C. (2007) *J. Am. Chem. Soc.*, **129**, 6298–307.

122 Singh, A. and Sharp, P.R. (2006) *J. Am. Chem. Soc.*, **128**, 5998–9.

123 Masselot, D., Charmant, J.P.H. and Gallagher, T. (2006) *J. Am. Chem. Soc.*, **128**, 694–5.

124 Kakiuchi, F. and Murai, S. (2002) *Acc. Chem. Res.*, **35**, 826–34.

125 Martinez, R., Chevalier, R., Darses, S. and Genet, J.P. (2006) *Angew. Chem. Int. Ed.*, **45**, 8232–5.

126 Oi, S., Fukita, S., Hirata, N., Watanuki, N., Miyano, S. and Inoue, Y. (2001) *Org. Lett.*, **3**, 2579–81.

127 Oi, S., Ogino, Y., Fukita, S. and Inoue, Y. (2002) *Org. Lett.*, **4**, 1783–5.

128 Oi, S., Sakai, K. and Inoue, Y. (2005) *Org. Lett.*, **7**, 4009–11.

129 Oi, S., Aizawa, E., Ogino, Y. and Inoue, Y. (2005) *J. Org. Chem.*, **70**, 3113–19.

130 Ackermann, L., Born, R. and Álvarez-Bercedo, P. (2007) *Angew. Chem. Int. Ed.*, **46**, 6364–7.

131 Ackermann, L. (2005) *Org. Lett.*, **7**, 3123–5.

132 Ackermann, L., Althammer, A. and Born, R. (2006) *Angew. Chem. Int.*, **45**, 2619–22.

133 Ackermann, L. (2006) *Synthesis*, **10**, 1557–71.

134 Ackermann, L., Althammer, A. and Born, R. (2007) *Synlett*, 2833–6.

135 Özdemir, I., Demir, S., Çetinkaya, B., Gourlaouen, C., Maseras, F., Bruneau, C. and Dixneuf, P.H. (2008) *J. Am. Chem. Soc.*, **130**, 1156–7.

136 Ackermann, L., Vicente, R. and Althammer, A. (2008) *Org. Lett.*, **10**, 2299–302.

137 (a) Kakiuchi, F., Kan, S., Igi, K., Chatani, N. and Murai, S. (2003) *J. Am. Chem. Soc.*, **125**, 1698–9;
(b) Kakiuchi, F., Matsuura, Y., Kan, S. and Chatani, N. (2005) *J. Am. Chem. Soc.*, **127**, 5936–45.

138 Ueno, S., Chatani, N. and Kakiuchi, F. (2007) *J. Am. Chem. Soc.*, **129**, 6098–9.

139 Oi, S., Sato, H., Sugawara, S. and Inoue, Y. (2008) *Org. Lett.*, **10**, 1823–6.

140 Oi, S., Fukita, S. and Inoue, Y. (1998) *Chem. Commun.*, 2439–40.

141 Vogler, T. and Studer, A. (2008) *Org. Lett.*, **10**, 129–31.

142 Ueura, K., Satoh, T. and Miura, M. (2005) *Org. Lett.*, **7**, 2229–31.

143 Bedford, R.B. and Limmert, M.E. (2003) *J. Org. Chem.*, **68**, 8669–82.

144 Bedford, R.B., Coles, S.J., Hursthouse, M.B. and Limmert, M.E. (2003) *Angew. Chem. Int. Ed.*, **42**, 112–4.

145 Bedford, R.B., Betham, M., Caffyn, A.J.M., Charmant, J.P.H., Lewis-Alleyne, L.C., Long, P.D., Polo-Cerón, D. and Prashar, S. (2008) *Chem. Commun.*, 990–2.

146 Oi, S., Watanabe, S.-I., Fukita, S. and Inoue, Y. (2003) *Tetrahedron Lett.*, **44**, 8665–8.

147 (a) Wan, X., Ma, Z., Li, B., Zhang, K., Cao, S., Zhang, S. and Shi, Z. (2006) *J. Am. Chem. Soc.*, **128**, 7416–17;
(b) Li, B., Tian, S.-L., Fang, Z. and Shi, Z. (2008) *Angew. Chem. Int. Ed.*, **47**, 1115–18.

148 Shi, Z., Li, B., Wan, X., Cheng, J., Fang, Z., Cao, B., Qin, C. and Wang, Y. (2007) *Angew. Chem. Int. Ed.*, **46**, 5554–8.

149 Brasche, G., García-Fortanet, J. and Buchwald, S.L. (2008) *Org. Lett.*, **10**, 2207–10.

150 Hull, K.L. and Sanford, M.S. (2007) *J. Am. Chem. Soc.*, **129**, 11904–5.

151 (a) Stuart, D.R. and Fagnou, K. (2007) *Science*, **316**, 1172–5; (b) Stuart, D.R., Villemure, E. and Fagnou, K. (2007) *J. Am. Chem. Soc.*, **129**, 12072–3.

152 Kawai, H., Kobayashi, Y., Oi, S. and Inoue, Y. (2008) *Chem. Commun.*, 1464–6.

153 Norinder, J., Matsumoto, A., Yoshikai, N. and Nakamura, E. (2008) *J. Am. Chem. Soc.*, **130**, 5858–9.

12
Arylation Reactions Involving the Formation of Arynes

Yu Chen and Richard C. Larock

12.1
Introduction

Speculation as to the existence of an aryne first occurred more than one hundred years ago. However, an aryne was only first proposed as an intermediate in 1942 by Wittig [1], and its structure was confirmed by Roberts [2] during the early 1950s using ^{14}C isotope labeling. An aryne is an uncharged transient species derived from an aromatic system by abstraction of two hydrogen atoms. The name benzyne is used for the simplest aryne, bisdehydrobenzene (C_6H_4).

Benzyne is an extremely reactive species due to the nature of its triple bond. In normal acetylenic species, such as ethyne, the unhybridized p orbitals are parallel to each other above and below the molecular axis, and this facilitates maximum orbital overlap. In benzyne, however, the p orbitals are distorted to accommodate the triple bond within the ring system, thus reducing their effective overlap. Benzyne can also be drawn as a diradical, where the triple bond is drawn as a double bond with a single electron on each carbon. Benzyne can exist as either an *ortho*-, a *meta*- or a *para*-benzyne, where the diradical can be a 1,2-, a 1,3- or a 1,4-diradical species, respectively. The 1,4-diradical species has been identified in the Bergman cyclization [3]. In this chapter, we will focus solely on the 1,2-dehydrobenzene (*o*-benzyne species). The term 'aryne' here will be used to refer specifically to 1,2-dehydrobenzenes and derivatives.

Arynes are highly electrophilic and can react with various nucleophiles. Due to their high chemical reactivity, arynes have considerable synthetic potential in synthesis involving arylation. In general, the reactions of arynes can be divided into three categories: electrophilic coupling reactions; pericyclic reactions; and transition metal-catalyzed reactions. Their properties, as well as their chemical reactivities, have been well summarized in several excellent reviews. As previous reviews have incorporated the literature up to 2002 [4], this chapter is an update on recent progress in the application of arynes as synthetic building blocks for the construction of complex aromatic compounds, and covers the period from the year 2000 to mid 2007.

Modern Arylation Methods. Edited by Lutz Ackermann

ISBN: 978-3-527-31937-4

12.2
Generation of Arynes

Many methods have been developed for the generation of benzyne (**1**). In general, these methods can be divided into two categories: (i) the first group involves β-elimination from an aryl anion **2** generated from the corresponding precursor **3** (Equation 12.1); (ii) the second group involves a retro-cycloaddition process of benzo-fused heterocycles, such as **4** (Equation 12.2). Most of the early studies on aryne generation employed the reaction of an aryl monohalide with a strong base to generate the aryne (Equation 12.3). The halogen–metal exchange of 1,2-dihalobenzenes with organolithium or -magnesium compounds has also been employed in a number of protocols for the generation of arynes (Equation 12.4). Generating a benzyne by the decomposition of benzenediazonium 2-carboxylates provides another popular method of aryne generation (Equation 12.5), although the explosive nature of diazonium compounds represents a major drawback of this approach.

X, Y → Y →
3 **2** **1**
(12.1)

N, N, N, NH_2
4 **1**
(12.2)

X base X $-X^-$
X = F, Cl, Br, I, OTf **1**
(12.3)

X RM M $-$ MY
Y Y
X = Cl, Br, I
Y = F, Cl, OTf, OTs
1
(12.4)

NH_2 RONO N_2^+ $-N_2$
CO_2H CO_2^- $-CO_2$
1
(12.5)

In 1983, Kobayashi developed a convenient approach to arynes by the fluoride-induced 1,2-elimination of *o*-(trimethylsilyl)aryl triflates (Equation 12.6) [5]. The mild reaction conditions involved in this protocol are compatible with a variety of reagents, substrates, functionalities and even transition metal catalysts. Today, this approach is the most widely used and the most efficient method for aryne generation.

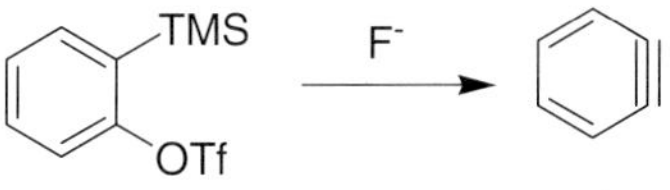

(12.6)

Besides the methods discussed above, other approaches for generating aryne species include Kitamura's and Cadogen's methods. The former involves the fluoride-induced desilylation of phenyl[*o*-(trimethylsilyl)phenyl]iodonium triflate (**5**), which can be prepared from *o*-bis(trimethylsilyl)benzene and $PhI(OAc)_2$ (Scheme 12.1) [6].

Cadogen's method involves the *in situ* generation of an *N*-nitrosoacetanilide (**6**) from acetanilide, followed by rearrangement to the azoacetate **7**, which then dissociates to form the diazonium acetate **8** [7]. Deprotonation by acetate at the β-position and elimination of dinitrogen affords the benzyne (Scheme 12.2).

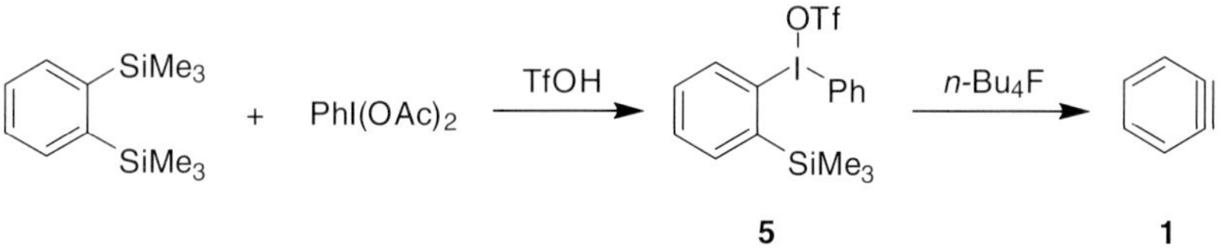

Scheme 12.1

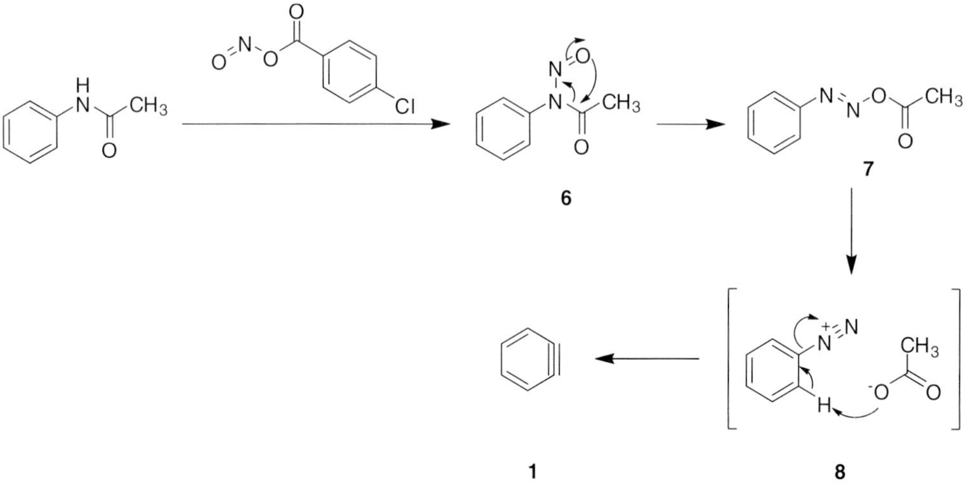

Scheme 12.2

Arynes can also be prepared by the thermal or photolytic fragmentation of benzo-fused compounds. One example reported by Knight is shown in Equation 12.7 [8]. Upon treatment with *N*-iodosuccinimide (NIS), 1-aminobenzotriazole **9** degrades to form aryne **10**.

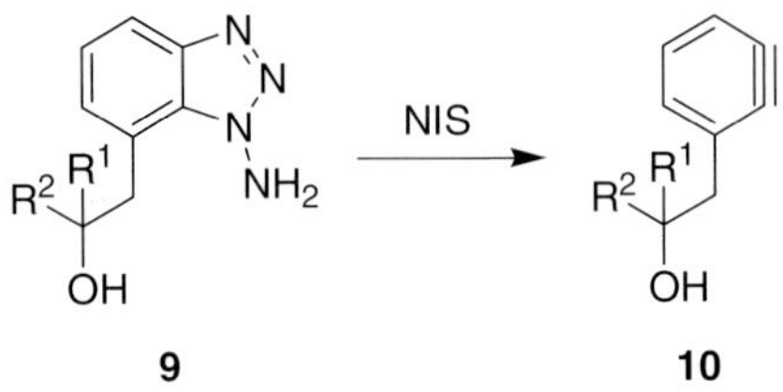

(12.7)

In practice, the selection of a benzyne generation method is often governed by the nature of the reaction conditions employed and the functionality present in the substrates undergoing reaction, all of which must be compatible with each other. The accessibility and cost of the benzyne precursor can also be an important factor.

12.3
Electrophilic Coupling of Arynes

Arynes are highly electrophilic species, and can react with various nucleophiles, resulting in aryl carbanions; the latter can then further react with electrophiles either intermolecularly or intramolecularly. Due to their high electrophilicity, considerable attention has been paid to the electrophilic coupling of arynes. A wide variety of anionic and uncharged nucleophiles are known to add readily to arynes. In general, these coupling reactions proceed through initial nucleophilic attack on the aryne, resulting in the formation of a zwitterion, followed by abstraction of an electrophile by the aryl carbanion generated. These types of reaction can be further divided into three categories depending on the characteristics of the nucleophile and electrophile employed:

- Trapping of the carbanion by the simplest electrophile, a proton, leads to the formation of a monosubstituted arene; this type of reaction will be discussed in Section 12.3.1.
- An aryne insertion reaction occurs when the electrophilic and nucleophilic groups are linked by a σ-bond [9]; these processes will be discussed in Section 12.3.2.
- The electrophile may be trapped after the addition of the nucleophile in a three-component coupling reaction, a process which will be discussed in Section 12.3.3.

12.3.1
Formation of Monosubstituted Arenes by Proton Abstraction

Beller *et al.* have reported a transition metal-free amination reaction of an aryl chloride in the presence of KO*t*-Bu by intermediate aryne formation (Equation 12.8) [10].

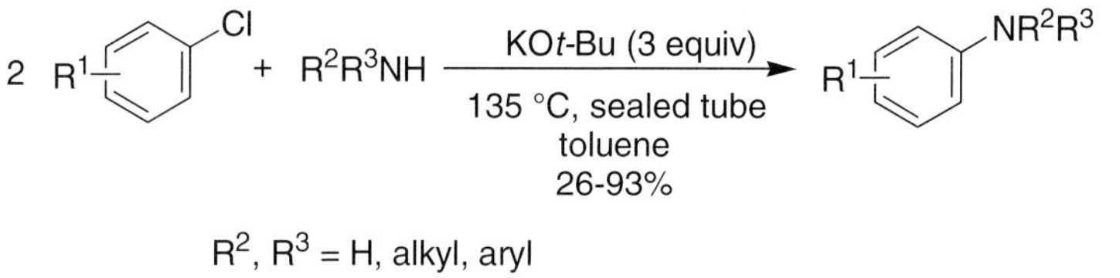

$$2\ R^1\text{-}C_6H_4\text{-}Cl + R^2R^3NH \xrightarrow[\text{135 °C, sealed tube, toluene, 26-93\%}]{\text{KO}t\text{-Bu (3 equiv)}} R^1\text{-}C_6H_4\text{-}NR^2R^3 \quad (12.8)$$

R^2, R^3 = H, alkyl, aryl

The reaction conditions are similar to the palladium-catalyzed Buchwald–Hartwig amination (see Chapter 3) with one distinct advantage; when the base NaO*t*-Bu used in the Buchwald–Hartwig procedure is replaced with KO*t*-Bu, the reaction proceeds without the need for a palladium catalyst. Primary and secondary amines, as well as aromatic and aliphatic amines, undergo the amination smoothly; however, only *meta*-substituted anilines can be obtained.

When a vinyl group is present either *ortho*- or *meta*- to the chlorine atom on the benzene ring, a domino hydroamination/intramolecular aryne amination reaction takes place between the chlorostyrene and an aryl amine, leading to *N*-aryl-2,3-dihydroindoles (Scheme 12.3). The latter products are readily dehydrogenated to their corresponding indole derivatives in the presence of 10 mol% Pd/C and a stoichiometric amount of ammonium formate. The base KO*t*-Bu plays a crucial role in this transformation, as it not only catalyzes hydroamination of the chlorostyrene but also leads to formation of the aryne intermediate. As this reaction requires the use of a sealed tube and high temperatures (135 °C), these harsh reaction conditions are a slight drawback.

By employing *o*-(trimethylsilyl)phenyl triflate derivatives (**11**) as the aryne precursors, Larock and coworkers have developed a facile transition metal-free *N*-arylation method for amines, sulfonamides and carbamates under very mild reaction conditions (Scheme 12.4) [11]. Aromatic and aliphatic, as well as primary and secondary amines, react well, affording good to excellent yields of the desired products. Secondary amines generally react faster than primary amines in

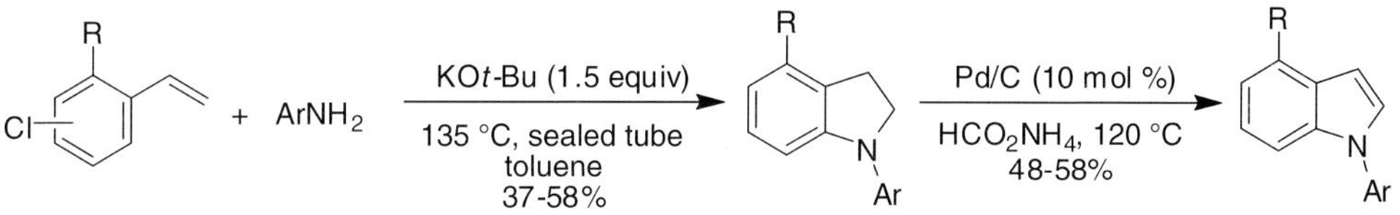

R, Cl, + $ArNH_2$ → KO*t*-Bu (1.5 equiv); 135 °C, sealed tube; toluene; 37-58% → R, N, Ar → Pd/C (10 mol %); HCO_2NH_4, 120 °C; 48-58% → R, N, Ar

Scheme 12.3

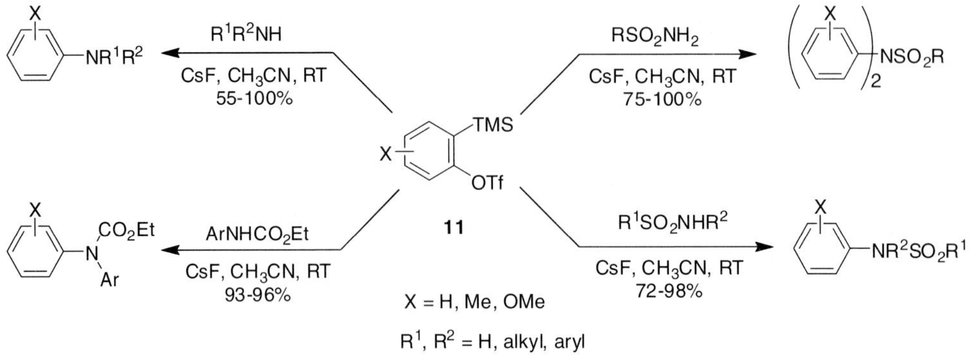

Scheme 12.4

this protocol. In the case of primary amines, one can selectively prepare either secondary or tertiary amines by simply controlling the stoichiometry. The *N*-arylation of synthetically more challenging sulfonamide substrates was also successfully achieved under the reaction conditions. Primary and secondary, as well as alkane- and arenesulfonamides, react efficiently, although the monoarylation of primary sulfonamides was not successful under these reaction conditions. The corresponding diarylation products are generated exclusively in high yield. Although the arylation of simple carboxamides produced only very low yields of *N*-arylamides, the *N*-arylation of phthalimide, as well as *N*-arylcarbamates, afforded the corresponding products in good to excellent yields. A variety of functional groups, including nitro, cyano, methoxy, hydroxyl, halide, ester, ketone, amide, alkene and alkyne functionalities, are readily accommodated in this process, which is a major advantage of this methodology. Shortly after the details of this methodology were published, it was employed by Fagnou in the synthesis of a cytotoxic carbazole [12].

Larock and coworkers have also reported the transition metal-free *O*-arylation of phenols and carboxylic acids by arynes [11b, 13]. Phenols bearing either electron-donating or electron-withdrawing substituents react smoothly, leading to the corresponding diaryl ethers in excellent yields (Equation 12.9).

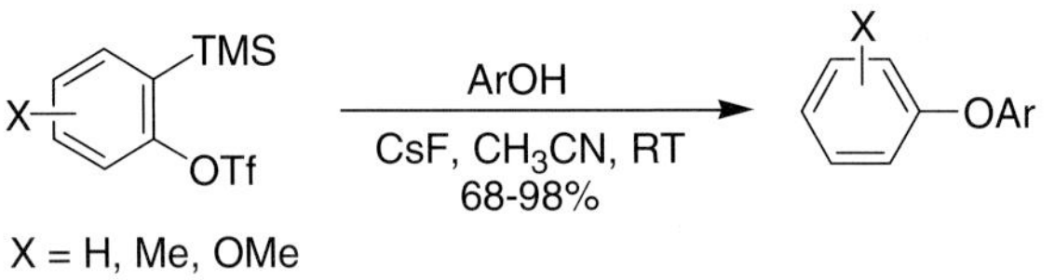

(12.9)

The sterically hindered 2,4,6-trimethylphenol also worked well, but aliphatic alcohols did not react well with arynes under the reaction conditions. On the other hand, arene thiols reacted well with arynes, affording the desired *S*-arylated products in fairly good yields (Equation 12.10).

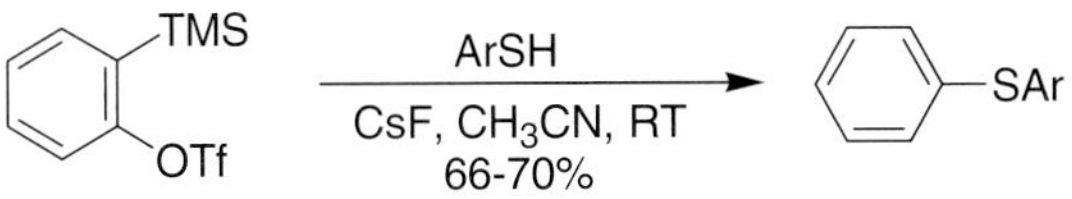

(12.10)

In addition, the *O*-arylation of aromatic carboxylic acids with arynes proceeded smoothly under the reaction conditions, affording the corresponding aryl esters in excellent yields (Equation 12.11). In contrast, aliphatic carboxylic acids afforded only low yields of esters.

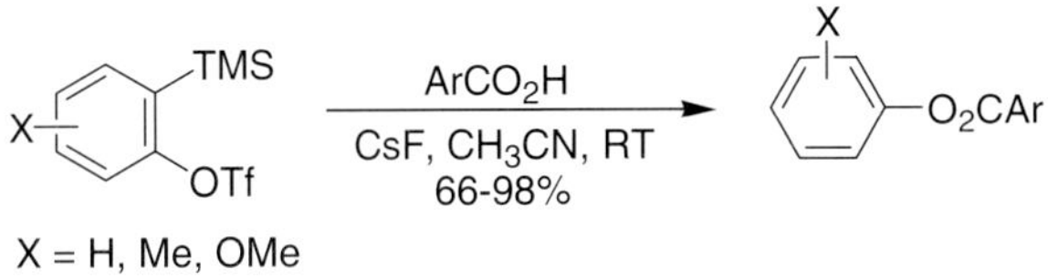

(12.11)

As with the *N*-arylation protocol discussed above, a variety of functional groups, including halide, nitro and methoxy groups, were found to be compatible with the reaction conditions.

Arynes undergo nucleophilic addition by *N*-alkylimidazoles to afford *N*-alkyl-*N'*-arylimidazolium salts after subsequent abstraction of a proton from the solvent (Equation 12.12) [14]. In contrast, if *N*-arylimidazoles are employed, none of the desired product is formed. In the reaction of 3-methoxybenzyne with *N*-methylimidazole, the addition occurred regioselectively at the position *meta* to the methoxy group. Similar regioselectivity has also been observed in nucleophilic additions to 3-methoxybenzyne by both Larock [11, 13] and Beller [10], as well as in some of the examples which follow. Both, steric and electronic effects favor nucleophilic attack *meta* to the methoxy group, which explains the observed regioselectivity (Figure 12.1).

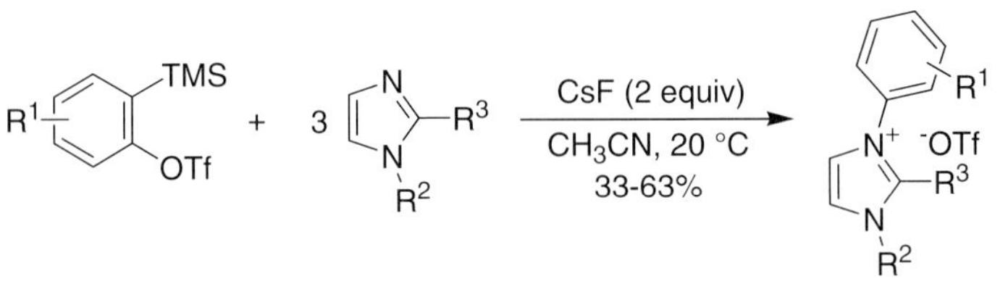

(12.12)

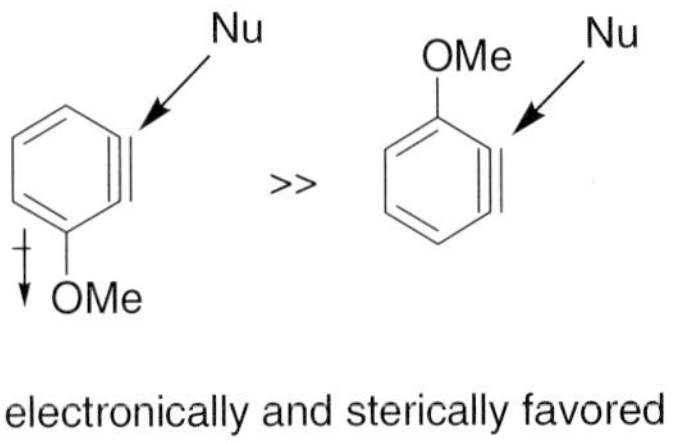

Figure 12.1 Nucleophilic addition to 3-methoxybenzyne.

β-Enamino esters and ketones react with arynes in an interesting manner. Arylation occurs at the α-carbon to form the *C*-arylation product, instead of the expected *N*-arylation product (Equation 12.13) [15]. In addition, a variety of functional groups that are known to react with arynes, such as amino and hydroxyl groups, as well as alkenes and furans, are all well tolerated in this reaction.

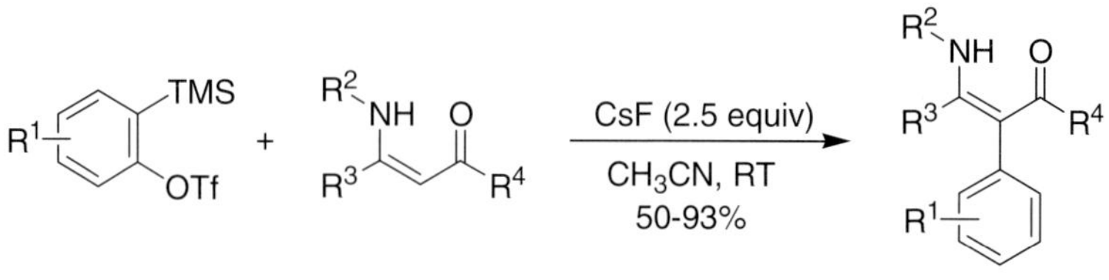

(12.13)

The intramolecular trapping of arynes by phenols has been reported by Knight *et al.* (Scheme 12.5) [16]. Upon treatment of **12** with 20% trifluoroacetic acid, free amine **13** was formed. Without purification, the latter was further treated with NIS, resulting in aryne intermediate **14** which was rapidly trapped in an intramolecular fashion by the phenol moiety. Interestingly, in a final step the aryl anion abstracted an iodo group, instead of a proton, giving rise to the iodoxanthenes **15** in good yields.

Similarly, upon treatment with the oxidant NIS, the aryne intermediates generated from 1-aminobenzotriazoles **16** undergo efficient intramolecular trapping by the pendant hydroxyl groups, leading to the fused iodinated oxygen heterocycles **17** (Scheme 12.6) [17].

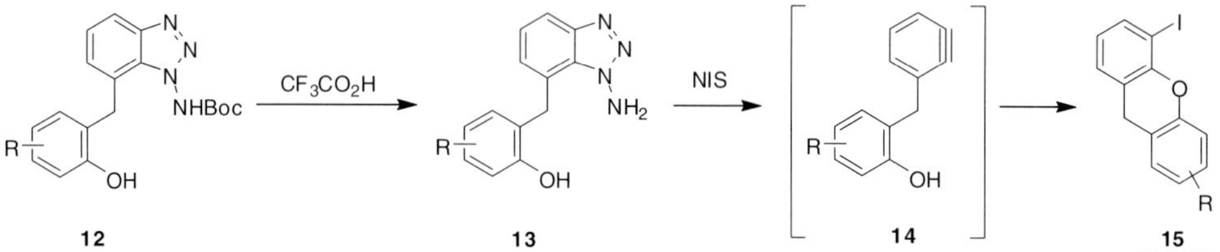

Scheme 12.5

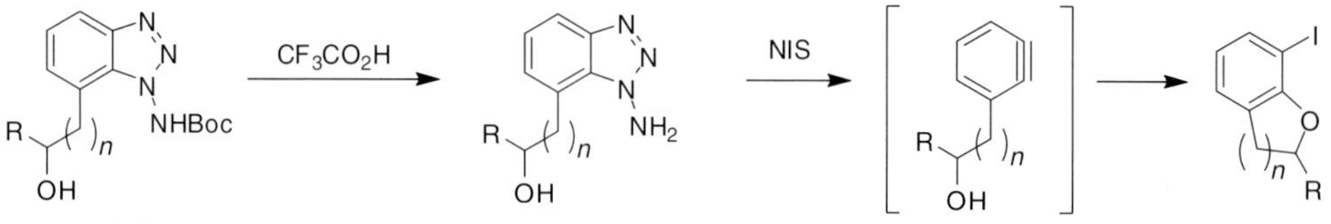

Scheme 12.6

The synthesis of fused heterocyclic systems via cyclization of aryne intermediates has also been achieved on a solid phase (Equation 12.14) [18]. An aryne intermediate generated by dehydrofluorination upon treatment with 2 equiv. of LiO*t*-Bu was subsequently trapped by an intramolecular primary amine moiety, leading to biologically interesting quinoxaline, thiazine and oxazine analogues.

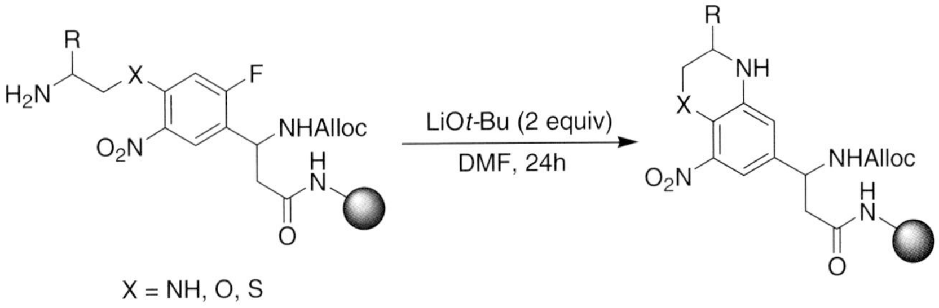

R
H_2N X F
O_2N NHAlloc
H
N
O
X = NH, O, S

LiO*t*-Bu (2 equiv)
DMF, 24h

R
NH
X
O_2N NHAlloc
H
N
O

(12.14)

An intramolecular addition of an ester enolate to a benzyne intermediate generated from the brominated precursor **18** has been reported by Castedo *et al.* as the key step in the total synthesis of (±)-clavizepine (Scheme 12.7) [19].

Similarly, an aryne-mediated cyclization process has been efficiently employed by Couture *et al.* as a key step in the total synthesis of the alkaloids eupolauramine (Scheme 12.8) [20] and nuevamine (Equation 12.15) [21].

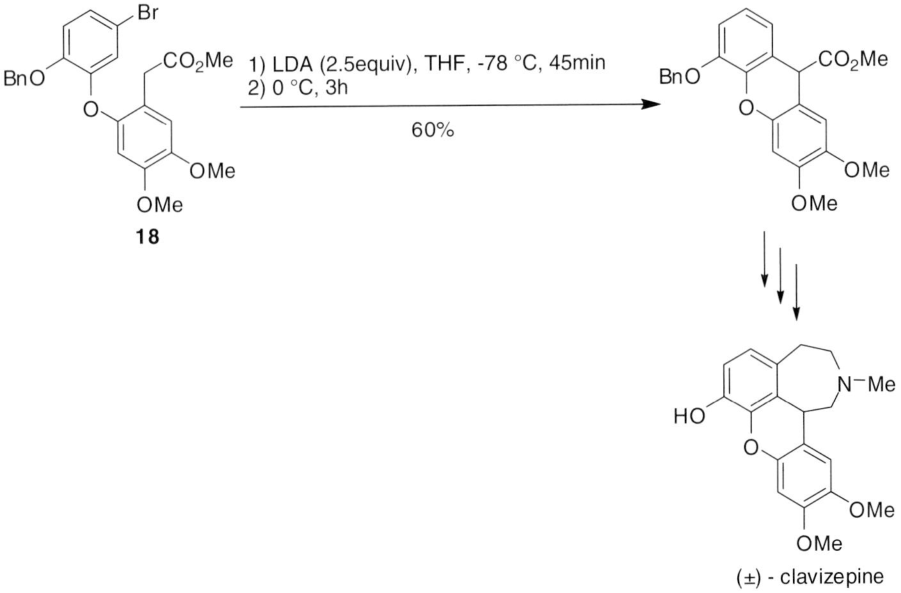

Br
BnO CO_2Me
O
OMe
OMe
18

1) LDA (2.5equiv), THF, -78 °C, 45min
2) 0 °C, 3h
60%

BnO CO_2Me
O
OMe
OMe

HO N−Me
O
OMe
OMe
(±) - clavizepine

Scheme 12.7

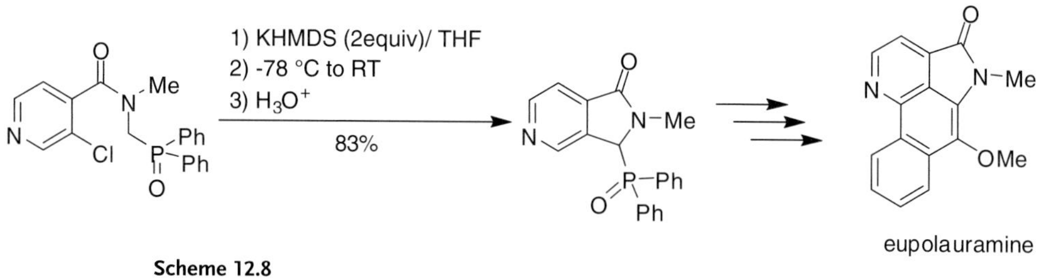

1) KHMDS (2equiv)/ THF
2) -78 °C to RT
3) H_3O^+
83%

eupolauramine

Scheme 12.8

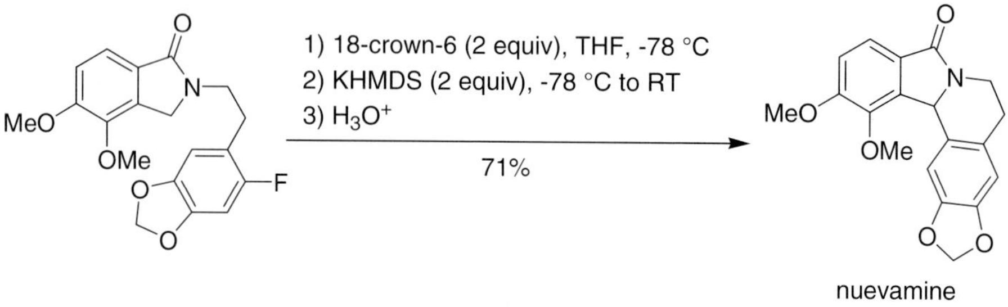

1) 18-crown-6 (2 equiv), THF, -78 °C
2) KHMDS (2 equiv), -78 °C to RT
3) H_3O^+
71%

nuevamine

(12.15)

12.3.2
Aryne Insertion into a Nucleophilic–Electrophilic σ-Bond

During an investigation into the addition reactions of element–element σ-bonds to the C—C triple bonds of arynes, Yoshida *et al.* observed aryne insertion into the S—Sn σ-bond of stannyl sulfides (Equation 12.16) [22]. The reaction is initiated by nucleophilic attack of the sulfur atom of the stannyl sulfide on the aryne triple bond, resulting in the formation of a zwitterion, which undergoes intramolecular nucleophilic substitution at the stannyl moiety. This affords the corresponding 2-(arylthio)arylstannane **19**.

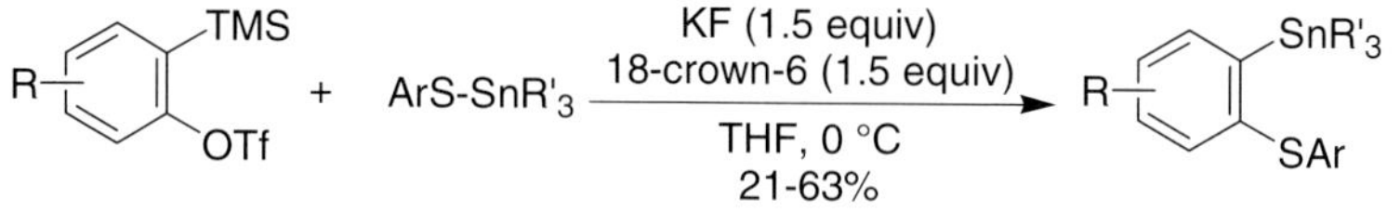

ArS-SnR'3 + KF (1.5 equiv), 18-crown-6 (1.5 equiv), THF, 0 °C, 21-63% → **19** (12.16)

Under similar reaction conditions, arynes can also insert into the N—Si σ-bond of aminosilanes, leading to 2-silylaniline derivatives in a straightforward manner (Equation 12.17) [23]. Both, cyclic and acyclic amine-derived silanes work well in this reaction, which is initiated by a nucleophilic attack of the nitrogen moiety of the aminosilane on the aryne triple bond, forming a zwitterion which subsequently undergoes intramolecular nucleophilic substitution at the silyl moiety to yield the observed product.

$$\text{R-}C_6H_3(\text{TMS})(\text{OTf}) + R'_2N\text{-}SiR''_3 \xrightarrow[\text{THF, 0 °C, 31-72\%}]{\text{KF (1.5 equiv), 18-crown-6 (1.5 equiv)}} \text{R-}C_6H_3(SiR''_3)(NR'_2) \quad (12.17)$$

Arynes can insert into the C—N bond of ureas to afford 2-aminoarenecarboxamides (Equation 12.18) [24]. Both cyclic and acyclic ureas work well, and a high regioselectivity was observed in the reaction of 3-substituted benzynes with urea. The nucleophilic nitrogen atom of the urea attacks the aryne exclusively at the position *meta* to the existing substituent. This regioselectivity is attributed to steric repulsion between the aryne substituent and the incoming nitrogen nucleophile.

$$R_2N\text{-}C(O)\text{-}NR_2 + \text{R'-}C_6H_3(\text{TMS})(\text{OTf}) \xrightarrow[\text{20 °C, 37-89\%}]{\text{CsF}} \text{R'-}C_6H_3(C(O)NR_2)(NR_2) \quad (12.18)$$

The C—N σ-bond of *N*-aryltrifluoroacetamides and the S—N σ-bond of trifluoromethanesulfinamides have both been found to undergo additions across the aryne triple bond to afford 1-[2-(arylamino)aryl]trifluoroethanones and (2-arylamino)aryl trifluoromethyl sulfoxides, respectively (Equations 12.19 and 12.20) [25]. The presence of the CF_3 moiety in the substrates proved to be crucial for these insertion reactions to occur; moreover, the reactions were limited to secondary amides or sulfinamides. No insertion product was detected in the case of the tertiary amide *N*-methyl-*N*-phenyltrifluoroacetamide.

$$\text{R}^1\text{-}C_6H_3(\text{TMS})(\text{OTf}) + \text{R}^2\text{-}C_6H_4\text{-NH-C(O)}CF_3 \xrightarrow[\text{CH}_3\text{CN, RT, 58-88\%}]{\text{CsF (2 equiv)}} \text{R}^1\text{-}C_6H_4\text{-NH-}C_6H_3(\text{C(O)}CF_3)\text{-R}^2 \quad (12.19)$$

$$\text{R}^1\text{-}C_6H_3(\text{TMS})(\text{OTf}) + \text{R}^2\text{-}C_6H_4\text{-NH-S(O)}CF_3 \xrightarrow[\text{THF, RT, 55-91\%}]{\text{TBAF (1.8 equiv)}} \text{R}^1\text{-}C_6H_4\text{-NH-}C_6H_3(\text{S(O)}CF_3)\text{-R}^2 \quad (12.20)$$

In addition to the examples discussed above, arynes have also been found to insert into C–C single bonds. The reaction of either 2- or 3-bromophenol with arylacetonitriles in the presence of a strong base, such as lithium 2,2,6,6-tetramethylpiperidide (LiTMP), led to the same major products, 2-arylmethyl-6-hydroxybenzenecarbonitriles **23** (Scheme 12.9) [26]. In both cases, lithium 2,3-didehydrophenoxide (**21**) was generated. The anion of nitrile **20** then adds solely at the 3-position of **21** to afford adduct **22**, which subsequently leads to the final product **23** after intramolecular rearrangement. When 2-bromo-1-naphthol was used instead of 2-bromophenol, a similar reactivity was observed [27].

In the case of 3-thienylacetonitrile, the reaction did not stop at the α-lithiated thienylmethyl naphthalene carbonitrile **24** (Scheme 12.10) [27]. Rather, this species further cyclized to afford the tetracyclic compound **25**.

In α-lithionitrile additions to benzynes, Durst found that if an aryl iodide, such as iodobenzene or 2,6-dimethoxyiodobenzene, were present in the reaction, the lithiated aryl intermediate **26** undergoes an alternative reaction pathway to form the iodine-ate complex **27**, alongside the cyclization intermediate **28** (Scheme 12.11). Both products **29** and **30** were observed in such reactions [28].

More recently, the coupling of a nitrile with two molar equivalents of an aryne under mild conditions, has been reported by Yoshida *et al.* (Scheme 12.12) [29]. The fluoride anion used in this case acted not only as an aryne generator but also as a base to deprotonate the nitrile. This reaction proceeds in a manner similar to

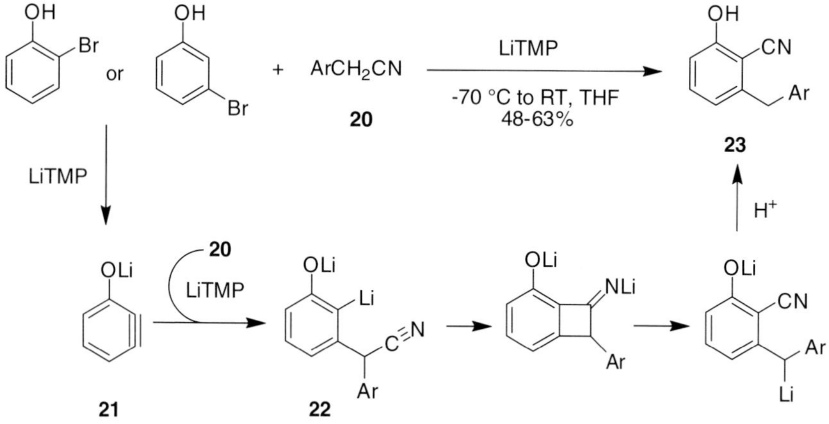

Scheme 12.9

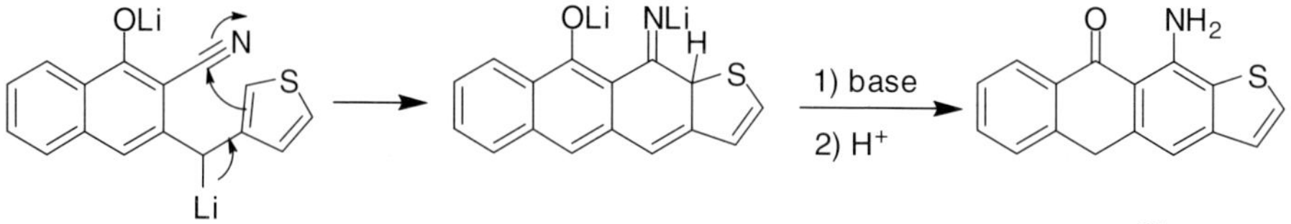

Scheme 12.10

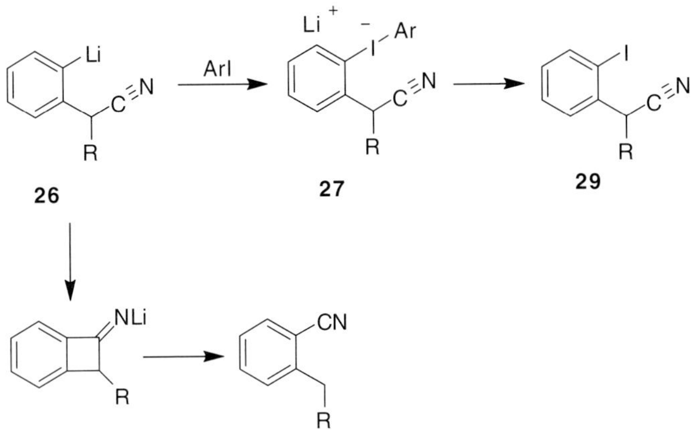

Scheme 12.11

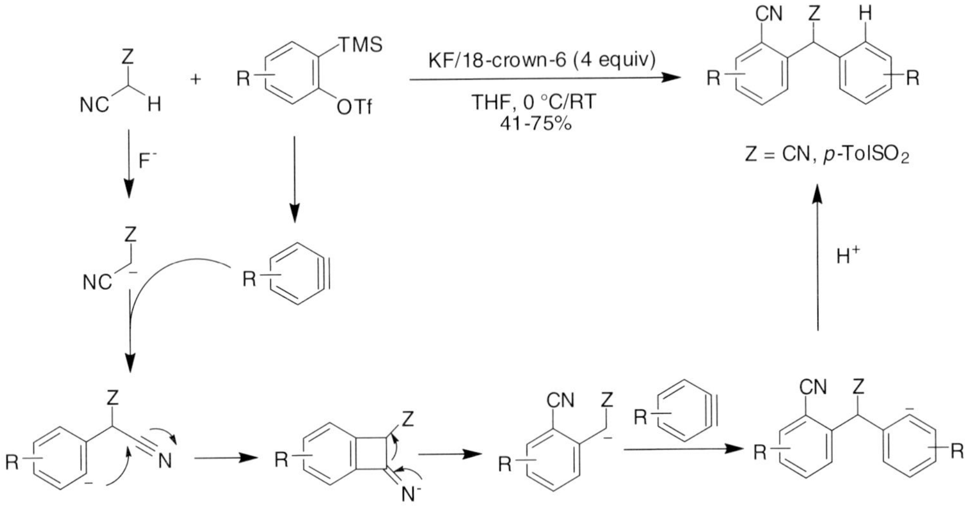

Scheme 12.12

the previous example; however, after an intramolecular rearrangement intermediate **31** added to a second aryne to afford aryl anion **32**.

An early example of aryne insertion into C—C single bonds was described by Guyot *et al.* in the addition of diethyl malonate anion to arynes and the subsequent rearrangement to homophthalic esters, such as **33** (Scheme 12.13) [30]. However, the reported yield was quite low and the desired product accompanied by numerous undesirable side products.

The reaction conditions were later optimized by Danishefsky and coworkers in the condensation of bromoarene **34** with dimethyl malonate anion, where the

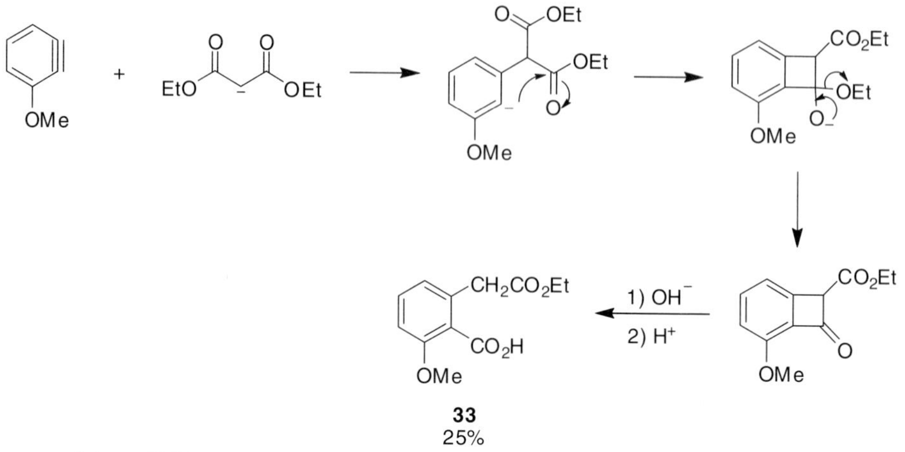

Scheme 12.13

isolated yield of the homophthalic ester **35** was dramatically increased to 71% (Equation 12.21) [31]. This methodology provides an efficient route to homophthalic esters and was soon employed by several research groups.

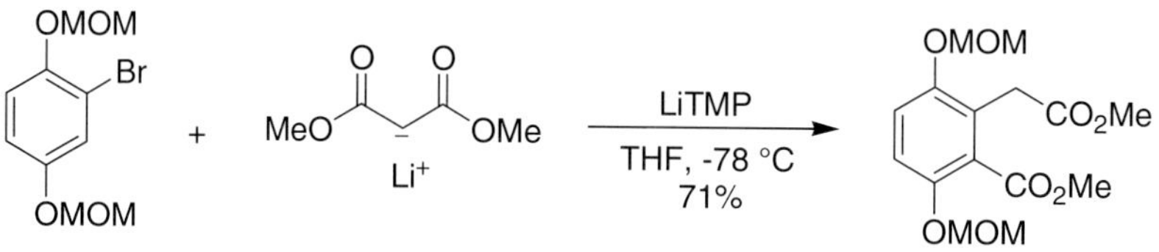

(12.21)

In the total synthesis of fredericamycin A reported by Kita *et al.*, the condensation of dimethyl malonate anion with aryne **36** was used to construct the AB-ring system. The intermediate homophthalic esters **37** and **38** were thus obtained in a 58% combined yield with a modest regioselectivity (2 : 3) (Scheme 12.14) [32].

The same methodology was slightly modified by Bauta *et al.* during the course of the total synthesis of isocoumarin NM-3 [33]. Homophthalic ester **40** was formed as the only regioisomer in the reaction between **39** and sodium dimethyl malonate (Scheme 12.15). The extremely high regioselectivity observed was attributed to *ortho*-stabilization of the aryl anion by the existing methoxy group in the aryne alkylation step [34].

Examples of aryne insertions into the C–C single bonds of β-keto esters have been reported by Tambar and Stoltz to generate *ortho*-disubstituted arenes, presumably via a [2+2] cycloaddition/fragmentation cascade (Equation 12.22) [35]. This methodology is well suited to both acyclic and cyclic β-keto esters. In the latter case, ring expansion of these substrates occurs leading to synthetically challenging medium ring-sized carbocycles (Equation 12.23).

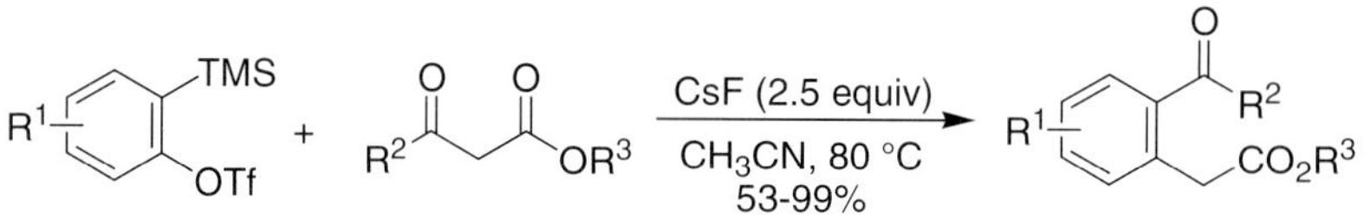

CsF (2.5 equiv), CH_3CN, 80 °C, 53-99%

(12.22)

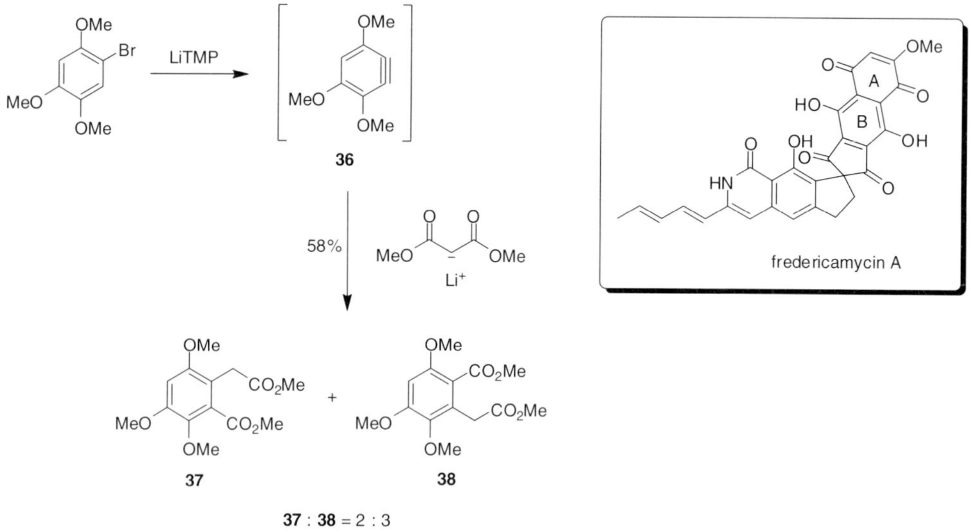

LiTMP

36

58%

Li+

fredericamycin A

37

38

37 : **38** = 2 : 3

Scheme 12.14

LDA (1.5 equiv), NaH (6 equiv)

THF, 0 °C to 5 °C

39

40

1) H_3O^+
2) NaOH, MeOH
3) HCl

44%

isocoumarin NM-3

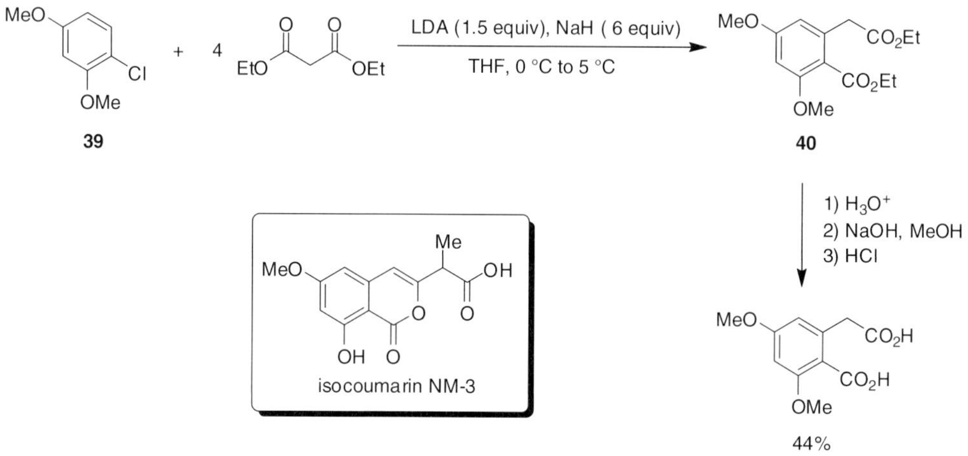

Scheme 12.15

TMS, OTf + MeO, MeO, O, EtO, O — CsF (2.5 equiv) / CH_3CN, 80 °C / 65% → O, OMe, OMe, O, OEt

(12.23)

Yoshida *et al.* extended this reaction to other β-dicarbonyl substrates, including malonates and β-diketones under even milder conditions [36]. Upon treatment of cyclic malonates with benzynes, benzoannulated macrocyclic compounds (with ring sizes up to 19) were readily formed in decent yields (Equation 12.24).

TMS, OTf + O, O, O, O — KF (2 equiv) / 18-crown-6 (2 equiv) / THF, RT, 10 h / 49% → O, O, O, O

(12.24)

The insertion of arynes into α-cyanocarbonyl compounds occurs smoothly under the same mild reaction conditions, and by a similar reaction pathway [37]. In the case of the sterically hindered substrate pivaloylacetonitrile, a single product **41** is formed in excellent yield (Equation 12.25), whereas a mixture of **42** and the double addition product **43** were usually obtained when other α-cyanocarbonyl compounds were used (Equation 12.26).

R, TMS, OTf + CN, *t*-Bu, O — KF/18-crown-6 (2 equiv) / THF, RT / 79-92% → R, CN, O, *t*-Bu

41

(12.25)

TMS, OTf + CN, R, O — KF/18-crown-6 (2 equiv) / THF, RT → CN, O, R + CN, O, R

42 32-68%

43 5-36%

(12.26)

12.3.3 Three-Component Coupling Reactions via Aryl Carbanion Trapping by an External Electrophile

Multicomponent coupling represents a synthetically attractive approach for the assembly of complex and diverse molecules by a cascade of elementary chemical reactions. Multicomponent couplings involving aryne components have recently attracted considerable attention due to the extraordinary reactivity of the aryne species.

Under the usual, mild reaction conditions involved in the electrophilic coupling of arynes, a three-component coupling of arynes, isocyanides and aldehydes takes place, forming benzoannulated iminofurans **45** (Scheme 12.16) [38]. A nucleophilic addition of the carbon atom of the isocyanide group to the aryne then initiates the reaction. The resulting zwitterion **44** is trapped by an aldehyde, followed by subsequent cyclization, to afford the product **45**. Newer coupling reactions have been reported where the aldehydes have been replaced by other electrophiles, such as *N*-tosylaldimines [39] (Equation 12.27), ketones (Equation 12.28) and benzoquinones (Equation 12.29), leading to the corresponding cyclized products **46–48**, respectively [40].

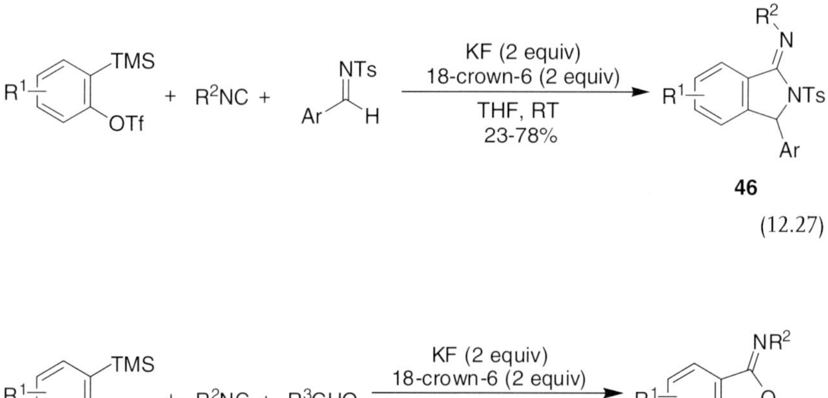

(12.27)

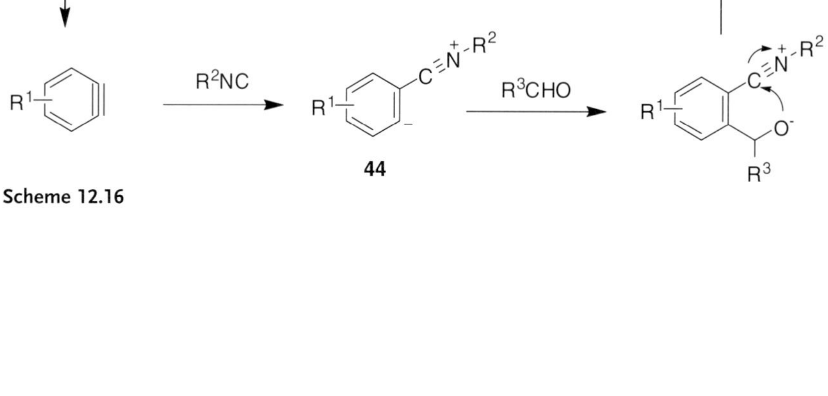

Scheme 12.16

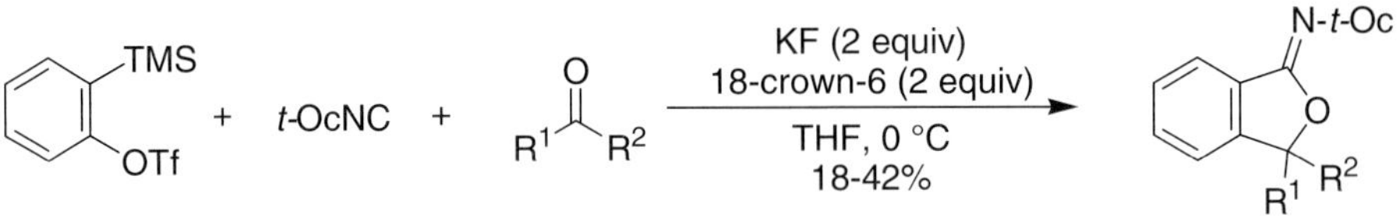

(12.28)

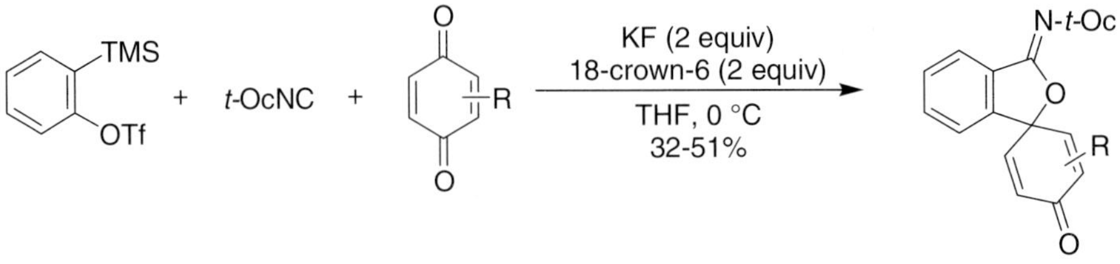

(12.29)

Additionally, CO_2 has been found to take part in a three-component coupling with arynes and imines, leading to six-membered heterocycles such as benzoxazinones **49** (Scheme 12.17) [41]. In contrast to the previous example (Equation 12.27), the imines act as nucleophiles. The use of imines with either steric congestion or decreased nitrogen nucleophilicity significantly slowed the reaction.

Although acetonitrile is a common solvent in aryne chemistry, it is rarely involved as a reactant in the coupling reactions with arynes. Recently, a three-component coupling of arynes, *N*-heteroaromatic compounds, and nitriles was reported by Jeganmohan and Cheng (Scheme 12.18) [42] where the reaction was initiated by a nucleophilic attack of the *N*-heteroaromatic on the aryne, leading to a zwitterionic species **50**. The zwitterion **50** then abstracted a proton from acetonitrile (or other nitrile) to afford intermediate **51**, alongside anion **52**. Anion **52**

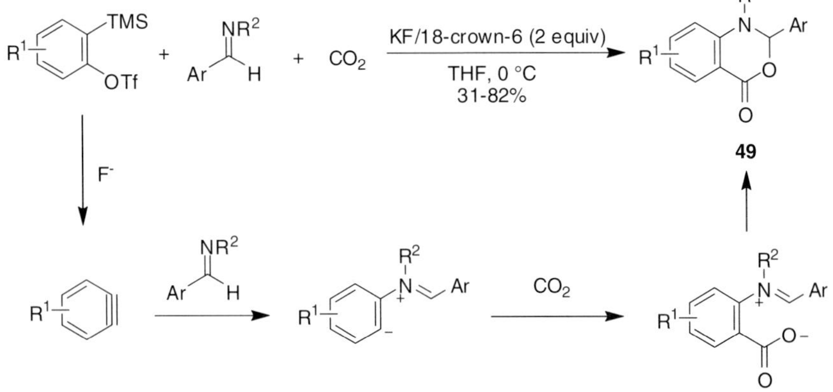

Scheme 12.17

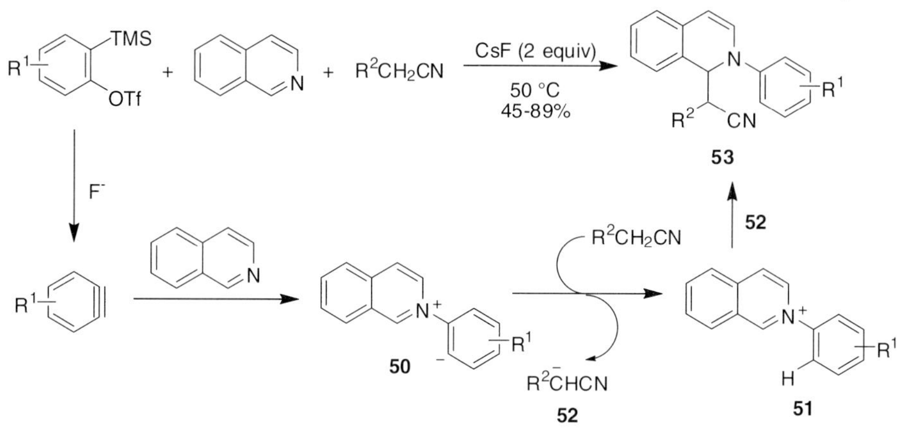

Scheme 12.18

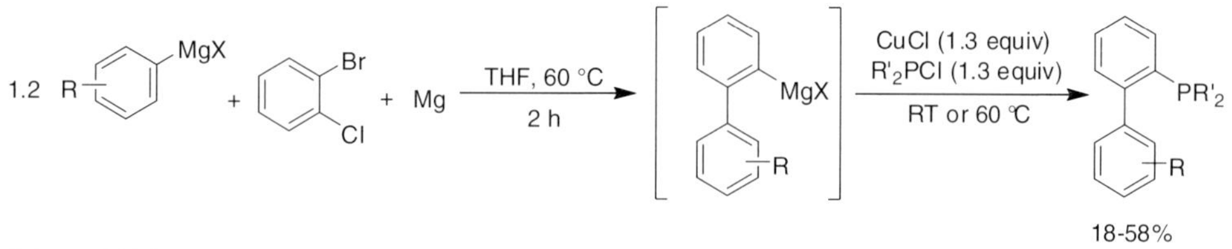

Scheme 12.19

then underwent nucleophilic addition to the C=N double bond of intermediate **51**, giving rise to the coupling product **53**. Isoquinoline, pyridines and quinolines all function well as the *N*-heteroaromatic component, while substituted acetonitriles, such as propionitrile, 2-phenylacetonitrile and 2-(2-thienyl)acetonitrile, also react efficiently.

Buchwald and coworkers have reported a useful synthetic route to functionalized (dialkylphosphino)biphenyl ligands by the addition of arylmagnesium halides to benzyne, followed by further reaction with dialkylchlorophosphines (Scheme 12.19) [43]. This economical one-pot process has been subsequently applied to the synthesis of a number of structurally related biphenyl phosphine ligands, with moderate yields.

Another synthetic protocol to biaryl compounds by an aryne intermediate has been developed by Leroux and Schlosser (Scheme 12.20) [44]. When 1-bromo-2-iodobenzene (**54**) is treated with 0.5 equiv. of *n*-BuLi, the coupling product 2-bromo-2′-iodobiphenyl is produced in 81% yield. The reaction presumably starts by an iodine–lithium exchange between **54** and *n*-BuLi, followed by a partial elimination of LiBr from the organolithium intermediate **55** to generate the transient benzyne species. Subsequent nucleophilic addition of **55** to benzyne generates the biphenyl scaffold. The resulting 2-biaryl lithium species **56** then undergoes lithium–iodine exchange with the starting material **54** to afford 2-bromo-2′-iodobiphenyl, alongside the organolithium species **55**.

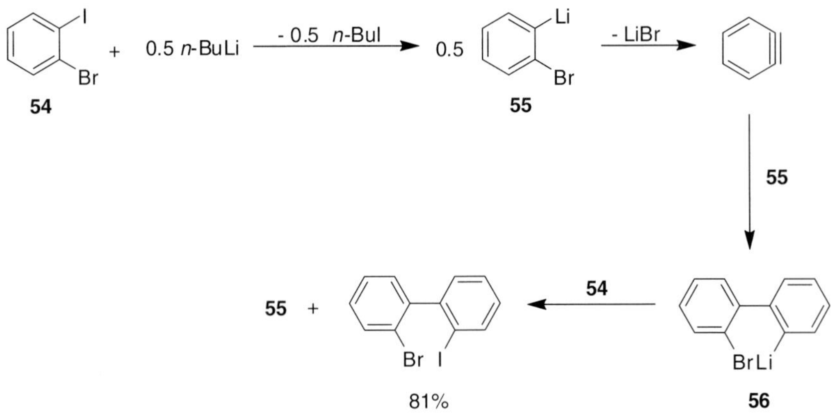

I
Br
54
\+ 0.5 n-BuLi
\- 0.5 n-BuI
0.5
Li
Br
55
\- LiBr
55
55 +
Br I
81%
54
Br Li
56

Scheme 12.20

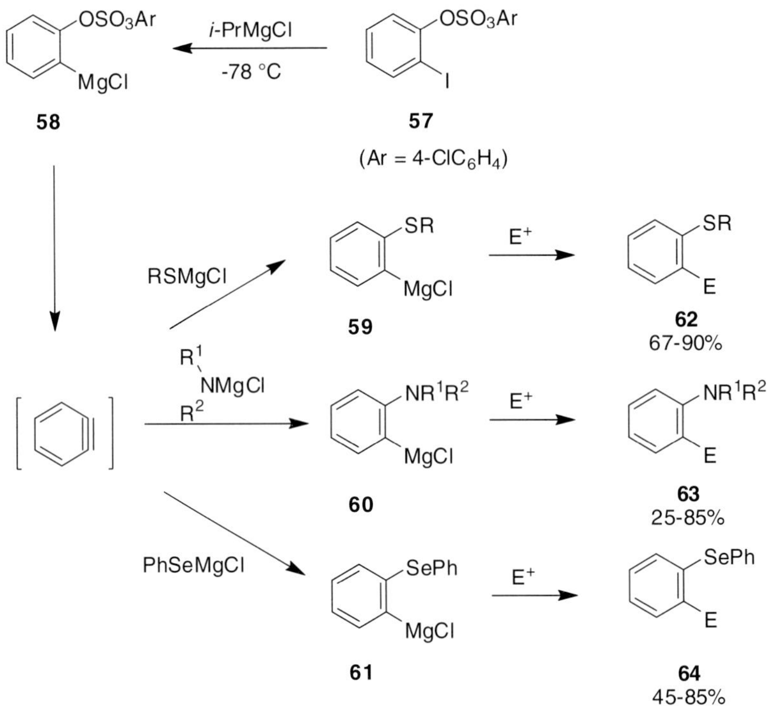

OSO_3Ar
MgCl
58
i-PrMgCl
-78 °C
OSO_3Ar
I
57
(Ar = $4\text{-}ClC_6H_4$)
RSMgCl
SR
MgCl
59
E^+
SR
E
62
67-90%
R^1
NMgCl
R^2
NR^1R^2
MgCl
60
E^+
NR^1R^2
E
63
25-85%
PhSeMgCl
SePh
MgCl
61
E^+
SePh
E
64
45-85%

Scheme 12.21

Arynes prepared from the aryl iodides **57** via 2-magnesiated aryl sulfonates **58** have been reported to undergo nucleophilic addition by magnesium thiolates [45], amides [45] or phenylselenide [46], giving rise to *ortho*-thio, amino, or seleno-substituted arylmagnesium species **59–61**, respectively (Scheme 12.21). These arylmagnesium intermediates can be trapped *in situ* by various electrophiles, such as I_2, dimethylformamide (DMF), acid chlorides and allyl bromide, leading to the *ortho*-disubstituted arenes **62–64**.

A three-component coupling strategy involving benzyne precursor **65**, methallyl Grignard and epoxy-aldehyde **67** has been employed by Barrett and coworkers (Scheme 12.22) [47]. This coupling affords the polysubstituted benzene **68**, a key intermediate in the total synthesis of clavilactone B. Upon treatment with *n*-BuLi, fluorobenzene **65** apparently yields an *o*-fluoroaryl lithium, which fragments at room temperature to a benzyne in the presence of methallyl Grignard, which in turn affords the aryl Grignard species **66**. The 1,2-addition of **66** to epoxy-aldehyde **67** at −78 °C led to **68** in a 65% yield.

The synthesis of polysubstituted naphthols and naphthalenes by a three-component coupling of arynes, β-keto sulfones and Michael acceptors has been described by Huang and Xue (Scheme 12.23) [48]. The coupling is initiated by a nucleophilic attack of the enolate **69** on the aryne to generate aryl anion **70**. The latter then undergoes intramolecular nucleophilic substitution to afford benzyl anion **71**. After subsequent Michael addition of **71** to olefin **72**, intramolecular 1,2-addition to the carbonyl group and elimination of benzenesulfinic acid affords the coupling product, polysubstituted naphthol **73** or naphthalene **74**, in moderate to good yields.

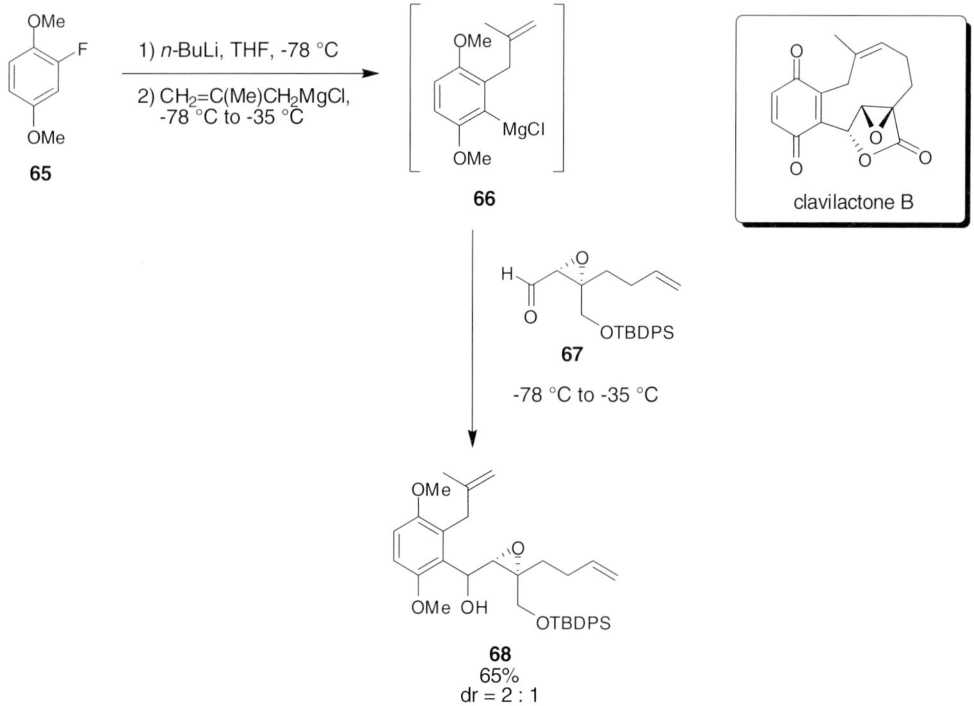

1) *n*-BuLi, THF, -78 °C
2) $CH_2=C(Me)CH_2MgCl$, -78 °C to -35 °C

-78 °C to -35 °C

65 **66** **67** **68** 65% dr = 2 : 1

clavilactone B

Scheme 12.22

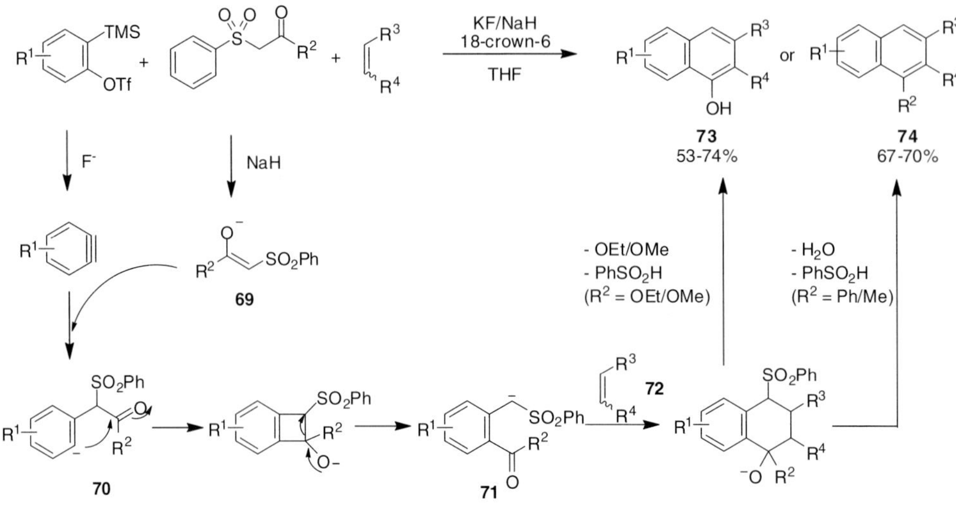

Scheme 12.23

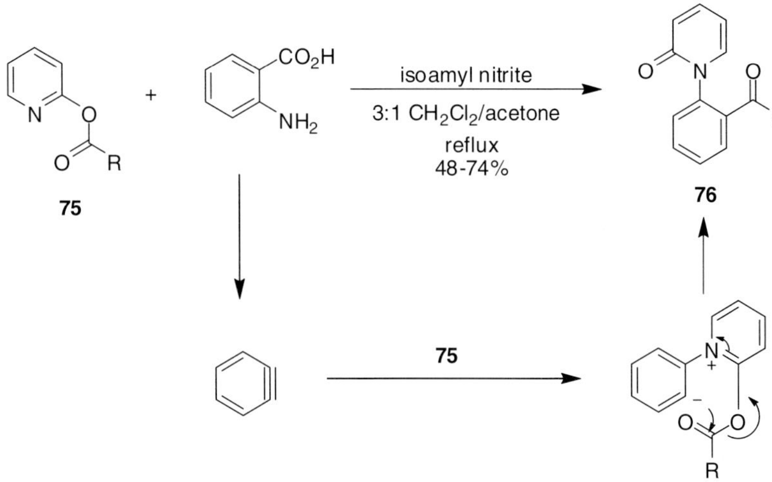

Scheme 12.24

12.3.4
Miscellaneous

2-Pyridyl carboxylates **75** have been found to react with benzyne generated from anthranilic acid, affording 1-(2-acylphenyl)-2-pyridones **76** under mild conditions (Scheme 12.24) [49]. The reaction presumably starts with a nucleophilic addition of the nitrogen atom of the pyridyl group to the benzyne triple bond. The resulting aryl anion then attacks the carbonyl carbon of the carboxylate group, leading to cleavage of the ester bond and subsequent formation of the final product **76**. When

2-pyridyl carboxylates are replaced with 2-pyridyl carbonates, the reactions also take place smoothly and provide the corresponding coupling products in good yields under the same reaction conditions.

A one-pot coupling protocol of arynes with *ortho*-heteroatom-substituted benzoates to prepare xanthones, thioxanthones, and acridones has been reported by Zhao and Larock (Equation 12.30) [50]. The coupling presumably proceeds by a tandem intermolecular electrophilic coupling of the aryne with the *ortho*-heteroatom-substituted benzoates and subsequent intramolecular electrophilic cyclization. Proton abstraction by the aromatic carbanion intermediate is a major competing reaction in this cyclization. This side reaction is, however, largely suppressed by employing tetrahydrofuran (THF) as the solvent.

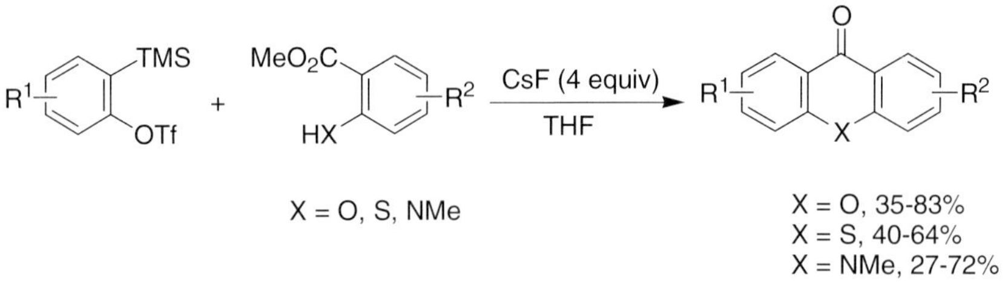

TMS, OTf, R^1 + MeO_2C, HX, R^2 — CsF (4 equiv), THF → O, R^1, R^2, X

X = O, S, NMe

X = O, 35-83%
X = S, 40-64%
X = NMe, 27-72%

(12.30)

Kim and coworkers have reported a synthetic route to diverse heterocyclic compounds through the nucleophilic addition of β-amino carbonyl compounds **79** to 3-halo-4-methoxybenzynes **78** (Scheme 12.25) [51]. The benzyne intermediate **78** was generated from 5-(3-halo-4-methoxyphenyl)thianthrenium perchlorates **77** upon treatment with lithium diisopropylamide (LDA) in THF at reflux. β-Amino carbonyl compounds, such as β-amino ketones, esters, amides and aldehydes, all react smoothly with benzyne **78** under the reported reaction conditions. Moreover, bis(2-aminophenyl) disulfide, 2-aminophenyl benzenesulfonate

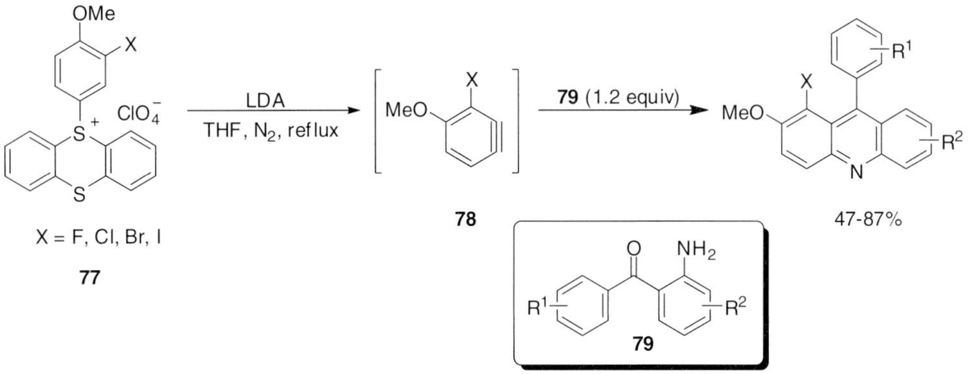

OMe, X, S^+, ClO_4^-, S — LDA, THF, N_2, reflux → [MeO, X] — **79** (1.2 equiv) → MeO, X, R^1, R^2, N

X = F, Cl, Br, I

77

78

47-87%

O, NH_2, R^1, R^2

79

Scheme 12.25

and *N*-(β-amino)-*p*-toluenesulfonate also readily undergo cyclization with benzyne **78**, leading to the corresponding polycyclic heterocycles **80–82** in good yields (Equations 12.31–33).

78 + 1.2 [bis(2-aminophenyl) disulfide] → THF, N_2, reflux, 68% → **80**

(12.31)

78 + 1.2 [2-aminophenyl benzenesulfonate] → THF, N_2, reflux, 76% → **81**

(12.32)

78 + 1.2 [N-(2-amino-1,2-diphenylethyl)-p-toluenesulfonamide] → THF, N_2, reflux, 75% → **82**

(12.33)

In a study directed towards the total synthesis of mumbaistatin, a highly potent glucose-6-phosphate translocase inhibitor, and its structural analogues, an aryne/phthalide annulation reaction was employed by Schmalz *et al.* as a key step in the construction of an anthraquinone ring system (Scheme 12.26) [52]. The anthraquinone building block **85** was thus prepared in a 45% yield by an aryne/phthalide annulation. Anthraquinone **86** was obtained after reductive methylation and selective hydrolysis of the acetal moiety, which was then coupled with a lithiated arene **87**. After subsequent oxidation and deprotection, the mumbaistatin analogue **88** was successfully synthesized in a total of 15 steps.

An anionic [4+2] cycloaddition of furoindolone **89** with 3-pyridyne (**90**) has been demonstrated by Mal *et al.* in the preparation of indoloquinones **91** and **92** (Scheme 12.27), both of which are useful synthetic intermediates for the total synthesis of the biologically important pyrido[4,3-*b*]carbazole alkaloids [53].

The methodology for the synthesis of functionalized carbazoles, dibenzofurans and dibenzothiophenes from *ortho*-fluoroanilines, -phenols or -thiophenols has recently been described by Sanz *et al.* (Scheme 12.28) [54]. The key step in this process involves the generation of a benzyne-tethered aryllithium intermediate **93**,

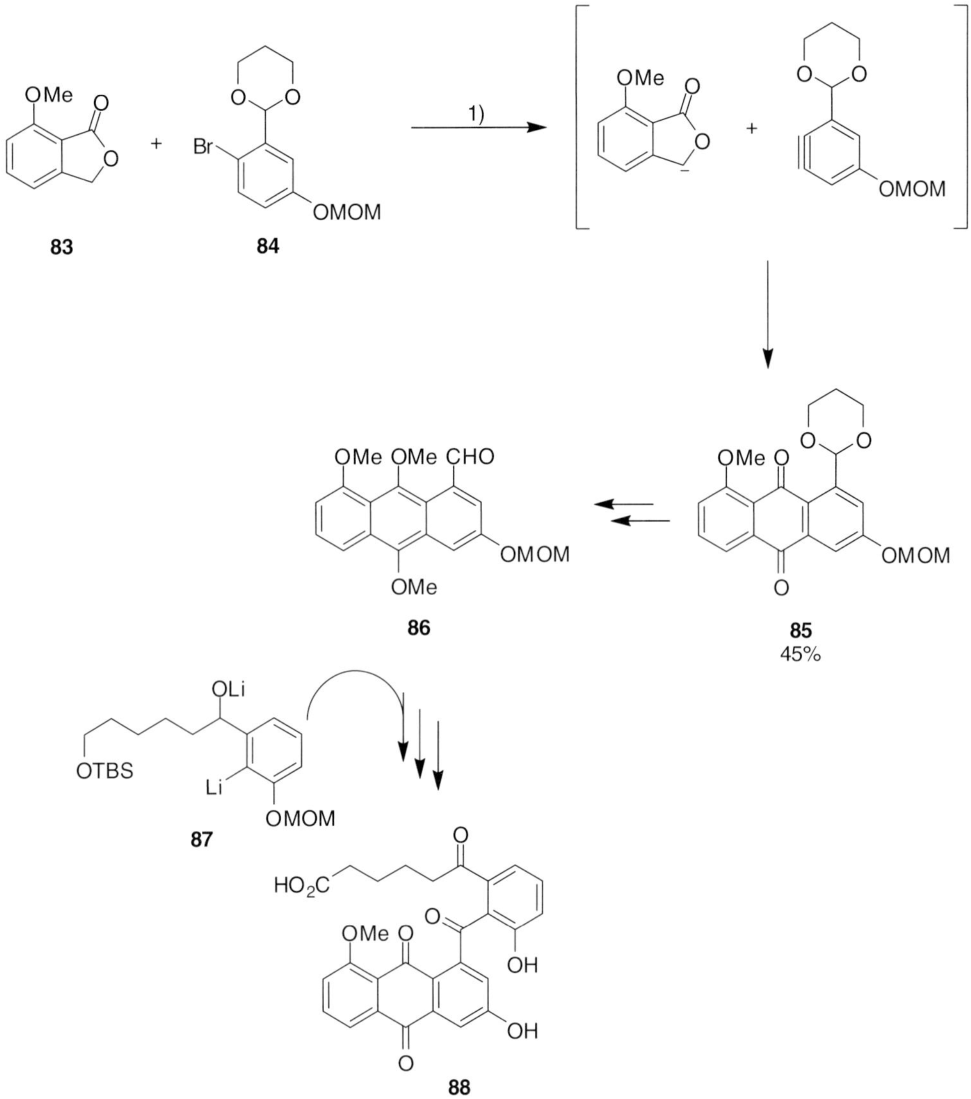

1) **83**, 4 equiv. of LiTMP, THF, -78 °C, then 2 equiv. of **84**, -43 °C to RT, then air

Scheme 12.26

which undergoes an intramolecular anionic cyclization to provide a regiospecifically metallated heterocycle **94**. Further trapping of an electrophile resulted in a dibenzo-fused heterocycle **95**.

Under similar reaction conditions, the synthesis of six-membered ring dibenzo-fused *N*-, *O*- and *S*-heterocycles has also been effectively achieved (Equation 12.34) [55].

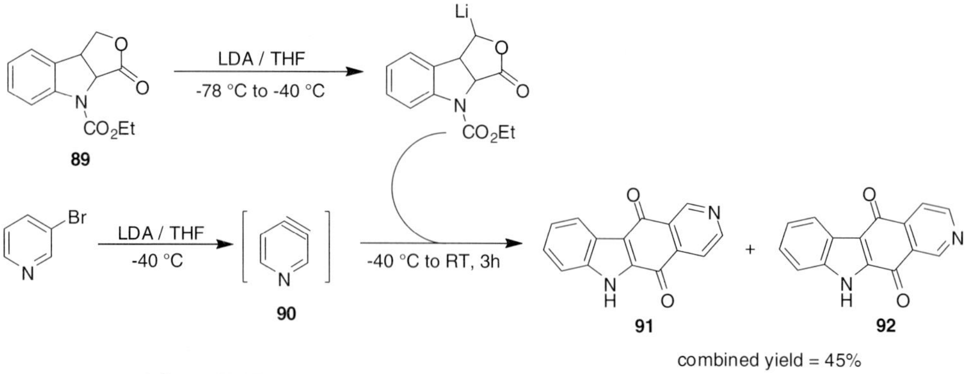

Scheme 12.27

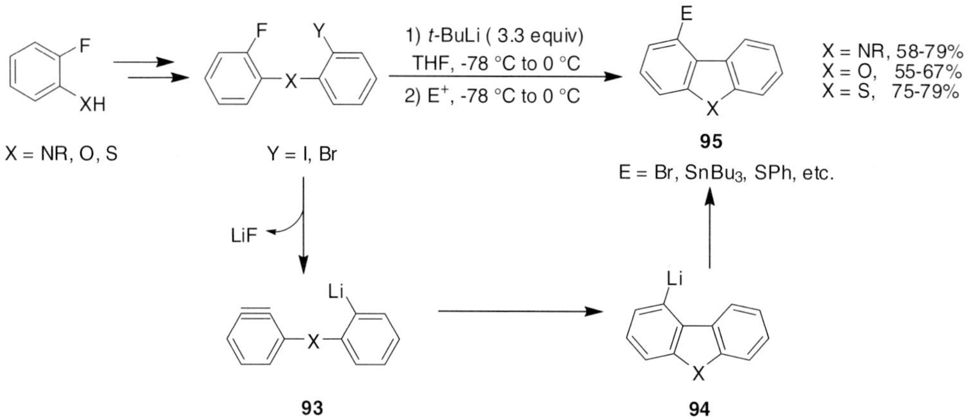

Scheme 12.28

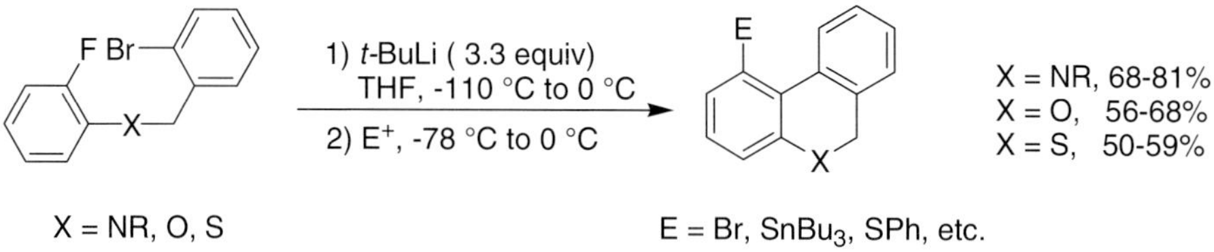

(12.34)

This methodology has been employed in the total synthesis of trisphaeridine and *N*-methylcrinasiadine, providing a concise synthetic route to these phenanthridine alkaloids [56]. The intramolecular anionic benzyne cyclization has also been successfully extended to benzyne-tethered alkenyllithium compounds, but the methodology is limited to the preparation of indole derivatives. Extensions of this

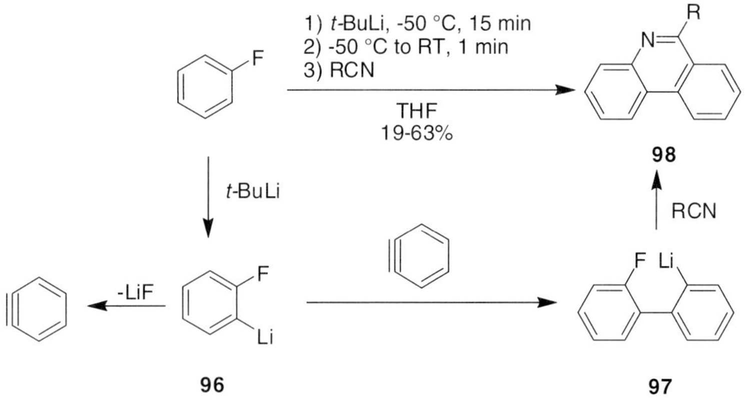

Scheme 12.29

methodology to the synthesis of benzofurans and benzothiophenes have proved inadequate [55].

The 6-substituted phenanthridines **98** have been prepared by a one-pot coupling of fluoroarenes with nitriles (Scheme 12.29) [57]. *Ortho*-lithiation of fluorobenzene with *t*-BuLi led to *o*-fluorophenyllithium (**96**), and benzyne was subsequently generated from **96** by the loss of LiF. A nucleophilic addition of **96** to benzyne led to the formation of biaryl lithium species **97**, while product **98** was generated after nucleophilic addition of **97** to a nitrile and subsequent intramolecular S_NAr reaction.

12.4 Pericyclic Reactions of Arynes

12.4.1 Diels–Alder Reactions

The Diels–Alder reaction is the most prominent reaction of arynes, providing not only a powerful tool to construct various carbocycles and heterocycles of synthetic importance, but also a useful method to detect the generation and existence of transient aryne species. Due to their high electrophilicity, arynes have been shown to react with a wide variety of dienes.

Furan has generally been observed to be a good diene in Diels–Alder reactions with arynes, and has been widely used in trapping aryne intermediates. Upon treatment with one equivalent of lanthanum, 1-halo-2-iodobenzenes effectively generate benzynes, which are trapped with furan to afford the Diels–Alder product **99** (Equation 12.35) [58].

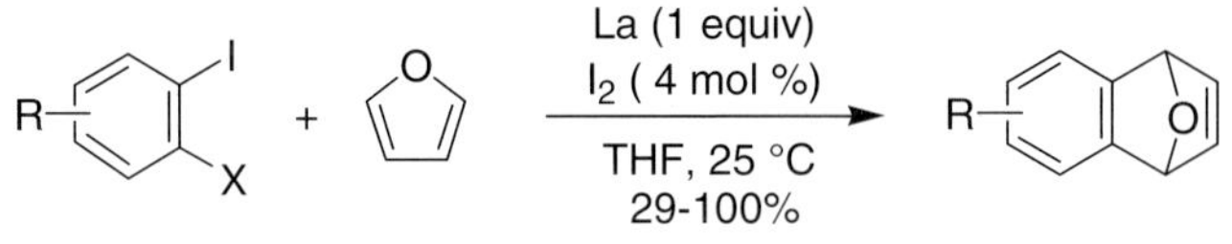

La (1 equiv), I_2 (4 mol %), THF, 25 °C, 29-100%; X = F, Cl, Br, I; **99** (12.35)

Upon treatment with *i*-PrMgCl, arynes are readily generated from 2-iodoaryl sulfonates **100** by the corresponding 2-magnesiated aryl sulfonates **101**. These arynes have been allowed to react with furan *in situ* to afford the Diels–Alder products **102** (Scheme 12.30) [59]. Functional groups, such as an ester, cyano, nitro or ketone group, are all well tolerated under the reaction conditions.

Under very mild reaction conditions, acylbenzynes **104** generated from hypervalent iodine compounds **103** have been trapped by furan, leading to the cycloadducts **105** in good yields (Scheme 12.31) [60].

In their studies on the synthesis of functionalized naphthalenes, Schlosser and coworkers found the Diels–Alder cycloaddition between arynes and furans to be a synthetically useful route to such compounds. Upon treatment with the correct organolithium base, halogenated benzenes **106** bearing substituents, such as trifluoromethyl [61], trifluoromethoxy [62] and trimethylsilyl [63] groups, undergo cycloadditions with suitably substituted furans to afford 1,4-epoxy-1,4-dihydronaphthalenes **107** (Scheme 12.32). The latter can be readily transformed to various substituted naphthalene derivatives by further elaboration, such as deoxygenation, bromination or ring opening.

2,2′-Bifuryl (**109**) undergoes a Diels–Alder reaction with 2 equiv. of the benzyne precursor 2-chloro-1,4-dimethoxybenzene (**108**) to yield the double cycloadduct **110** (Scheme 12.33) [64]. The latter is easily aromatized in the presence of trifluoroacetic acid to provide binaphthyl **111**.

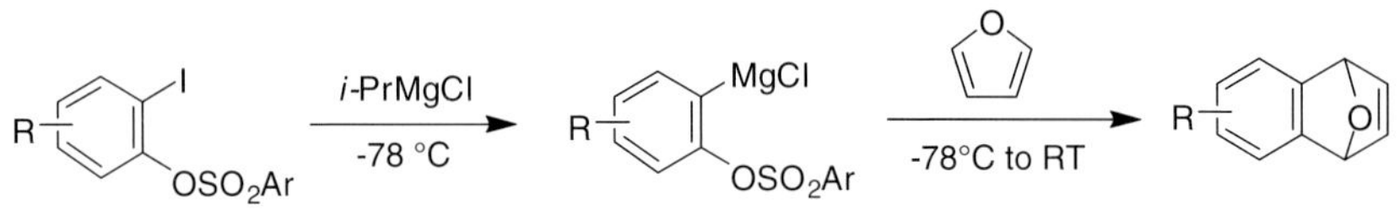

100 (OSO_2Ar, Ar = 4-ClC_6H_4) → *i*-PrMgCl, -78 °C → **101** (MgCl, OSO_2Ar) → furan, -78°C to RT → **102** 68-93%

Scheme 12.30

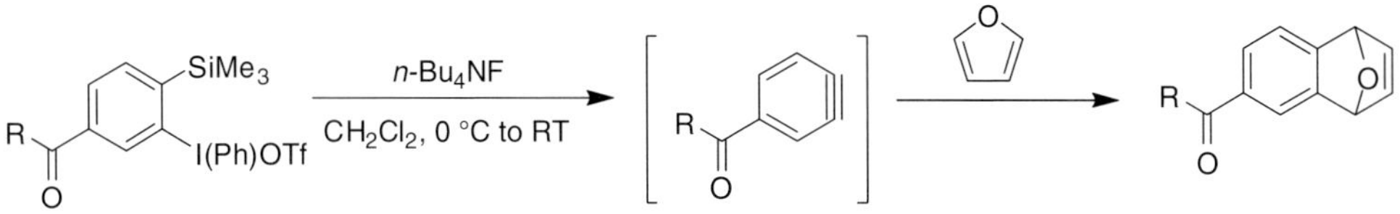

103 ($SiMe_3$, I(Ph)OTf) → *n*-Bu_4NF, CH_2Cl_2, 0 °C to RT → [**104**] → furan → **105** 74-98%

Scheme 12.31

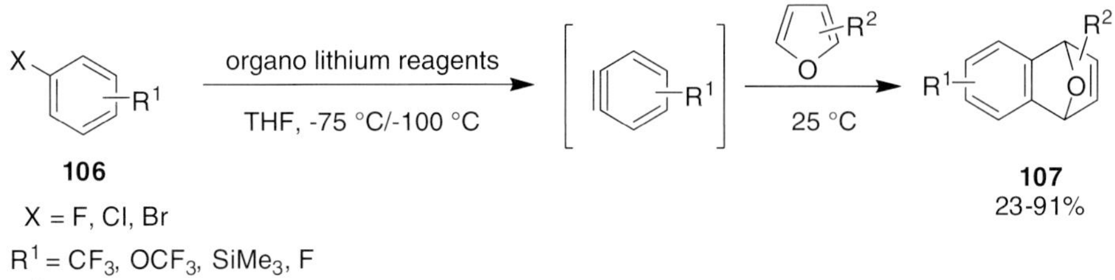

organo lithium reagents, THF, -75 °C/-100 °C; 25 °C

106 → **107** 23-91%

X = F, Cl, Br

$R^1 = CF_3$, OCF_3, $SiMe_3$, F

Scheme 12.32

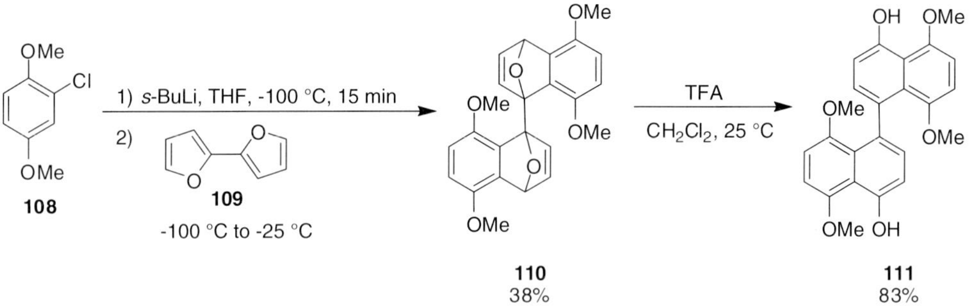

1) *s*-BuLi, THF, -100 °C, 15 min; 2) **109**, -100 °C to -25 °C; TFA, CH_2Cl_2, 25 °C

108 → **110** 38% → **111** 83%

Scheme 12.33

112 → 1), 2), 3) → **113** 9% + **114** 12% + **115** 7% + **116** 5%

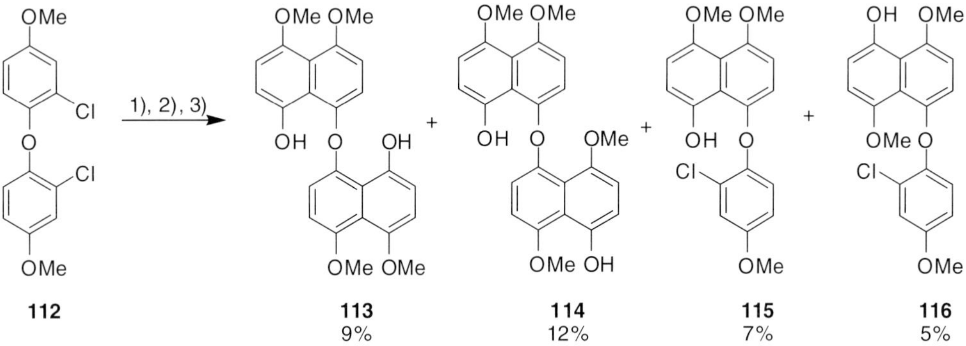

1) *s*-BuLi, THF, -100 °C; 2) 2-methoxyfuran, -100 °C to -25 °C; 3) TFA, CH_2Cl_2, 25 °C

Scheme 12.34

When the dichloro diaryl ether **112** was treated with *s*-BuLi, a dibenzyne was generated and subsequently trapped by 2-methoxyfuran, affording dinaphthyl ethers **113** and **114** after aromatization (Scheme 12.34) [64]. The monocycloadducts **115** and **116** were also present in the reaction mixture. With careful control of the stoichiometry of the butyl lithium, an iterative double benzyne–furan Diels–Alder reaction was achieved between benzyne precursor 1,4-difluoro-2,5-dimethoxybenzene and two different furans.

Polysubstituted anthracenols have also been formed by sequential Diels–Alder cycloadditions and aromatizations (Scheme 12.35) [65]. This iterative double

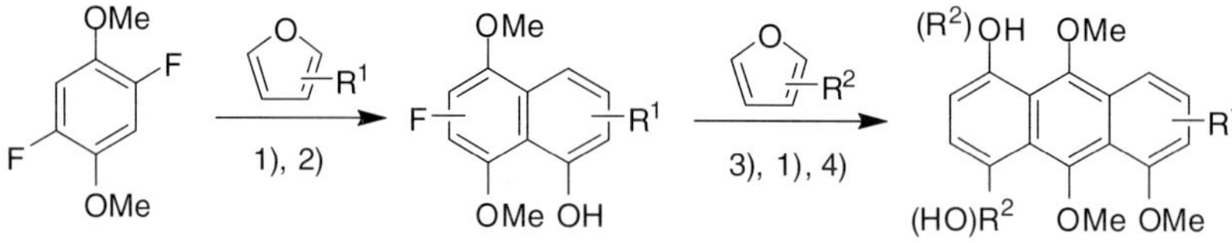

1) n-BuLi, THF, -78 °C to 0 °C; 2) HCl, H_2O, MeOH, heat; 3) MeI, NaH, DMF, 20 °C; 4) HCl, H_2O, THF, heat

Scheme 12.35

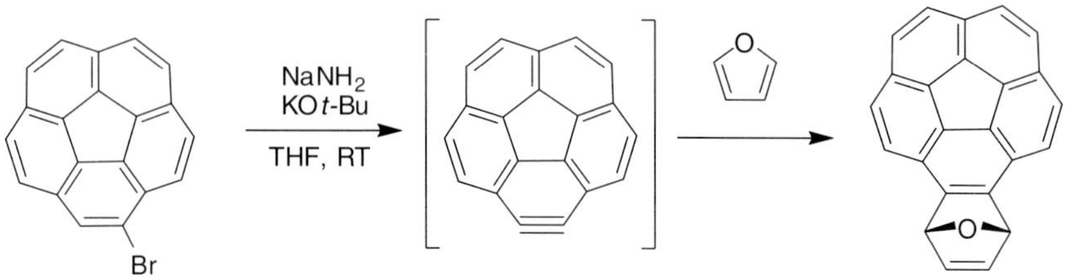

Scheme 12.36

benzyne–furan Diels–Alder reaction was subsequently employed in the synthesis of a model for the total synthesis of the angucycline antibiotic, Sch 47555 [66].

Furan has also been employed by Sygula *et al.* to trap the first buckybowl aryne, corannulyne **118** [67]. Treatment of bromocorannulene **117** with strong bases in the presence of furan led to the formation of cycloadduct **119** in an 80% yield (Scheme 12.36).

Martin and coworkers have examined the Diels–Alder cycloaddition of a substituted benzyne with a glycosyl furan [68], after which the approach was subsequently employed in the total synthesis of the *C*-aryl glycoside antibiotic galtamycinone (Scheme 12.37) [69]. In the 16-step total synthesis of vineomycinone B_2 methyl ester, the domino intramolecular Diels–Alder cycloaddition of a substituted benzyne with a glycosyl furan was again employed as a key step, successfully furnishing a rapid assembly of the glycosyl-substituted aromatic core of *C*-aryl glycosides (Scheme 12.38) [70].

During the course of the total synthesis of rubromycins, Reißig and coworkers have employed an aryne–furan cycloaddition to construct the key building block naphthaldehyde **120** (Scheme 12.39) [71].

While preparing benzonorbornadienes by the Diels–Alder cycloaddition of substituted benzyne intermediates with furan, Caster and coworkers discovered significantly different regiochemistries between bromobenzene **121** and chlorobenzene **124** in the *n*-BuLi induced metallation step (Schemes 12.40 and 12.41) [72]. In the case of bromobenzene **121**, benzyne intermediate **122** was generated via bromine–lithium exchange, followed by elimination of LiF; trapping by furan then afforded 6-fluoro-9-oxabenzonorbornadiene (**123**). In contrast, aryne **125** was produced

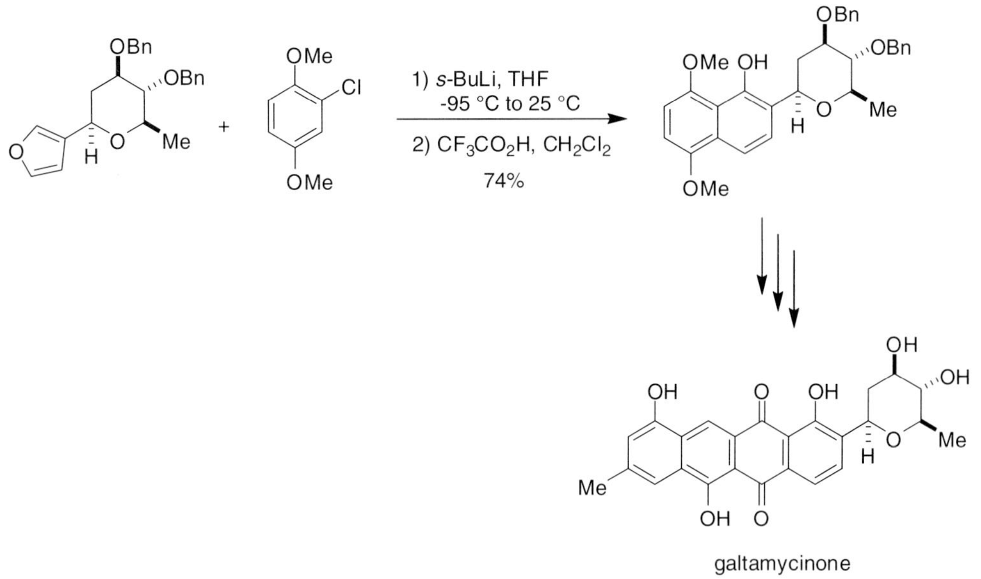

1) s-BuLi, THF -95 °C to 25 °C
2) CF_3CO_2H, CH_2Cl_2
74%

galtamycinone

Scheme 12.37

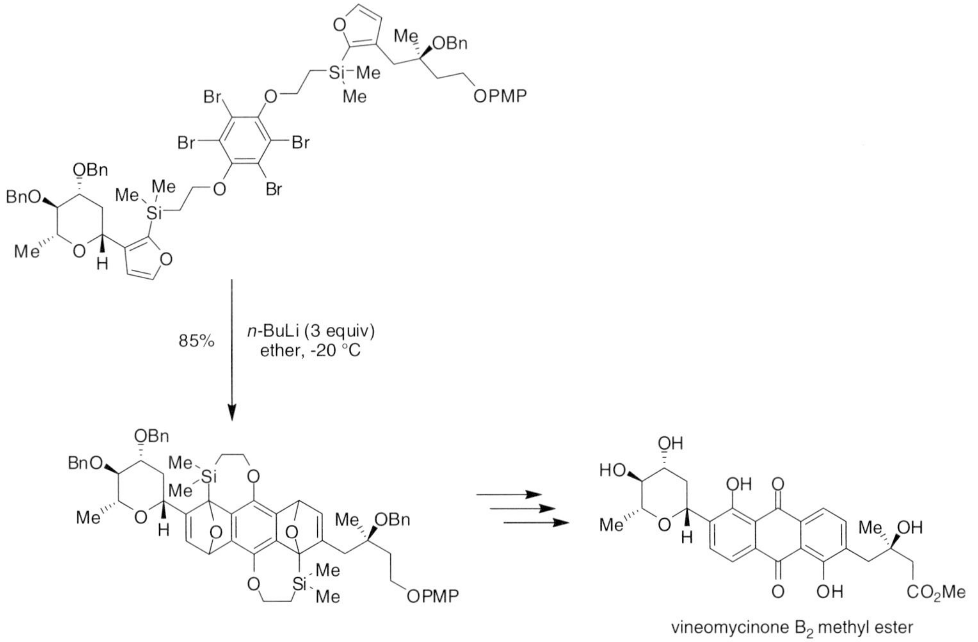

85%
n-BuLi (3 equiv) ether, -20 °C

vineomycinone B_2 methyl ester

Scheme 12.38

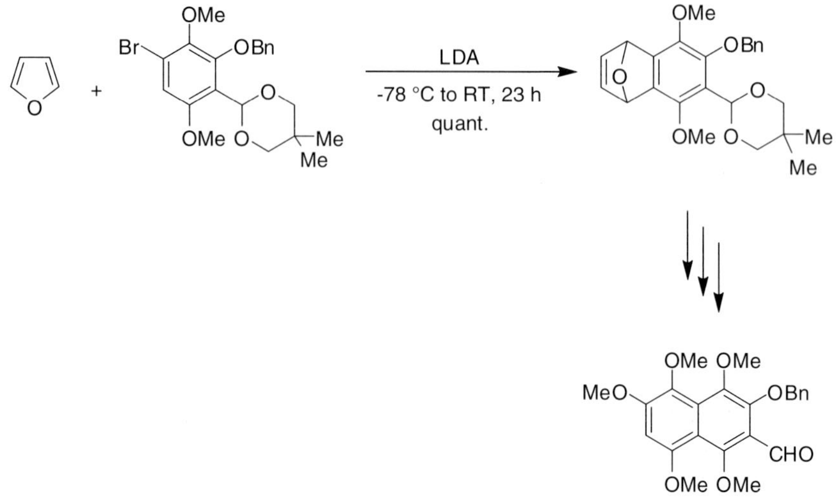

OMe
Br
OBn
+
O
OMe O
Me
Me
LDA
-78 °C to RT, 23 h
quant.
OMe
OBn
O
OMe O
Me
Me

OMe OMe
MeO
OBn
CHO
OMe OMe
120

Scheme 12.39

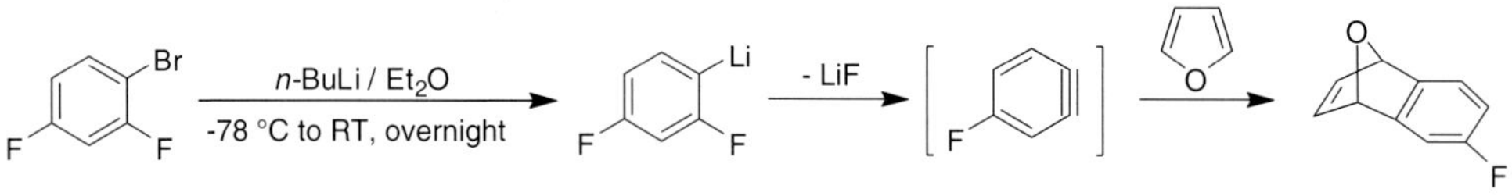

Br
F
F
n-BuLi / Et_2O
-78 °C to RT, overnight
Li
F
F
- LiF
F
O
O
F
121
122
123
94%

Scheme 12.40

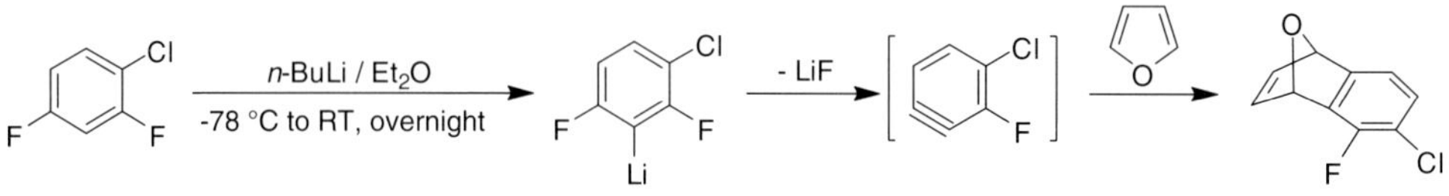

Cl
F
F
n-BuLi / Et_2O
-78 °C to RT, overnight
Cl
F
F
Li
- LiF
Cl
F
O
O
F
Cl
124
125
126
90%

Scheme 12.41

from chlorobenzene **124** by a deprotonation–elimination pathway, leading to 6-chloro-5-fluoro-9-oxabenzonorbornadiene (**126**). In addition, strong solvent effects were observed in these studies. For example, while the cycloaddition reactions took place with high regioselectivities and excellent yields in diethyl ether, a synthetically useless product mixture was produced when THF was used as the reaction solvent.

Aryne intermediates generated from the *ortho*-(trimethylsilyl)triphenylenyl triflate **127** have been found to undergo Diels–Alder cycloaddition with 1,3-diphenylisobenzofuran (**128**) upon treatment with TBAF, affording the polycyclic products **129** (Equation 12.36) [73].

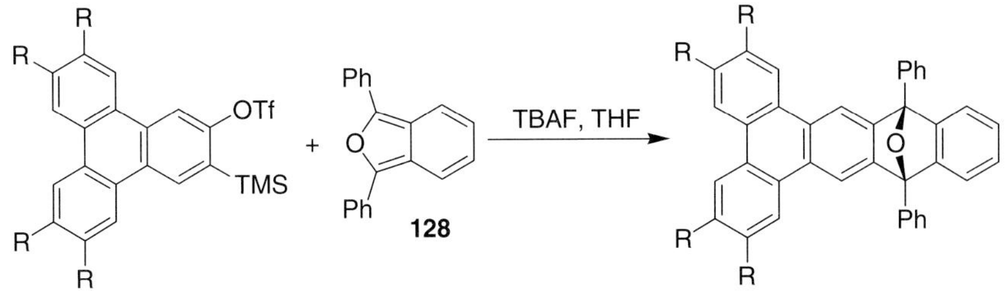

TBAF, THF

127a: R = H
127b: R = OC_6H_{13}

128

129a, 65%
129b, 54%

(12.36)

The Diels–Alder cycloaddition between 5,6-bis(trimethylsilyl)benzo[*c*]furan (**130**) and arynes has been demonstrated by Wong *et al.* in the synthesis of linear polycyclic aromatic compounds, such as **131** and **132** (Scheme 12.42) [74].

Besides furan, the Diels–Alder reaction between other heteroaromatics and arynes is also known. The [4+2] cycloaddition of benzyne with 3,4-dimethylphosphole pentacarbonylmolybdenum complexes **133** has recently been reported by Mathey (Equation 12.37) [75]. The cycloaddition mainly takes place on the less-hindered side of the phosphole ring, affording 2,3-benzo-7-phosphanorbornadiene complexes **134** in moderate yields.

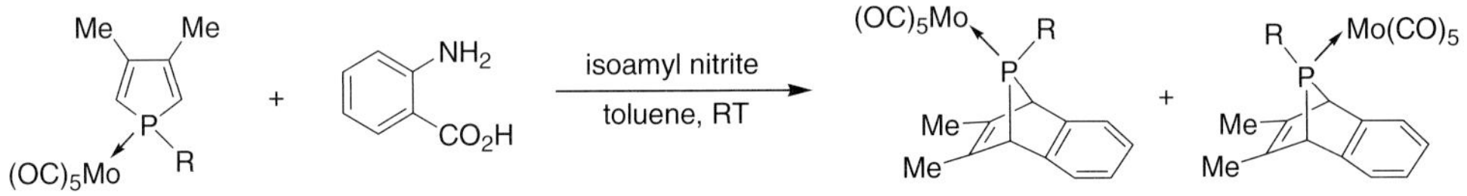

isoamyl nitrite, toluene, RT

133a, R = Ph
133b, R = Me

134a, R = Ph, 19.5 %
134b, R = Me, 8.5 %

trace

(12.37)

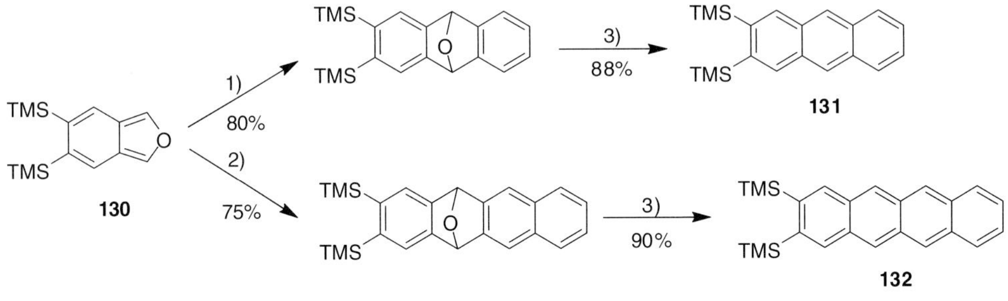

1) phenyl[2-(trimethylsilyl)-1-phenyl]iodonium triflate, TBAF, CH_2Cl_2;
2) phenyl[3-(trimethylsilyl)-2-naphthyl]iodonium triflate, TBAF, CH_2Cl_2;
3) $TiCl_4$, $LiAlH_4$, Et_3N, THF

Scheme 12.42

Piers has described the synthesis of 1-borabenzobarrelene **136** by the cycloaddition of borabenzene **135** and benzyne (Equation 12.38) [76].

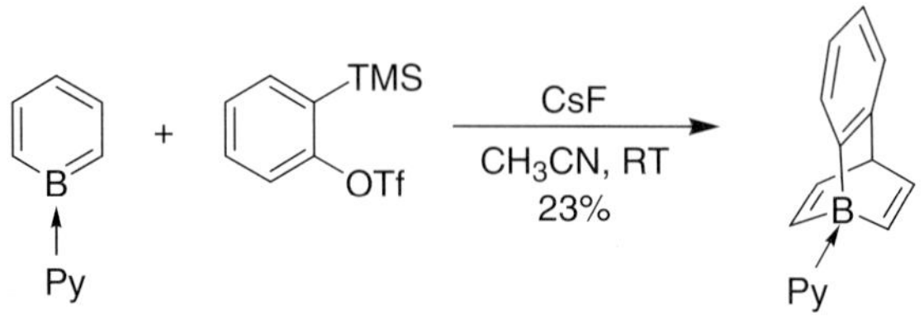

TMS; OTf; CsF; CH_3CN, RT; 23%; B; Py; **135**; **136**

(12.38)

A cycloaddition between phosphabenzene **137** and benzyne has also been reported by Breit and coworkers, giving rise to phosphabarrelene **138** in moderate yields (Scheme 12.43) [77].

In Section 12.3.1, an addition reaction of *N*-alkylimidazoles to arynes was described by Yoshida *et al.* (Equation 12.12) [14], which led to the synthesis of *N*-alkyl-*N'*-arylimidazolium salts. Xie and Zhang, however, have recently reported a completely different result from the reaction between benzyne and *N*-substituted imidazoles (Scheme 12.44) [78].

The molar ratio between benzyne and imidazole proved to have a critical influence on the direction of the reaction: the benzyne/imidazole ratio was 1 : 3 in Yoshida's process, whereas a 1 : 1 ratio was employed by Xie and Zhang. The difference in substrate ratio apparently led to two entirely different reaction pathways. In Zhang's case, benzyne first underwent a Diels–Alder reaction with imidazole to produce the nitrogen-bridged isoquinoline intermediate **139**. A subsequent retro Diels–Alder reaction led to intermediate **140** after the loss of hydrogen cyanide. This was followed by a second Diels–Alder reaction with benzyne to generate intermediate **141**. Intermolecular nucleophilic addition of **138** to a third benzyne then gave rise to the final product arylamine **142**.

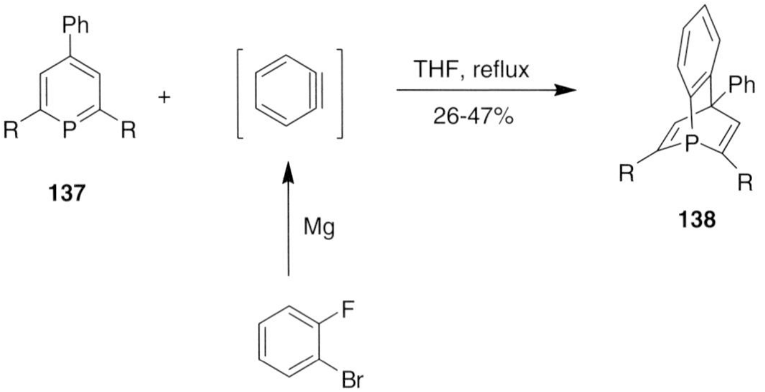

Ph; R; P; R; **137**; THF, reflux; 26-47%; Mg; F; Br; Ph; P; R; R; **138**

Scheme 12.43

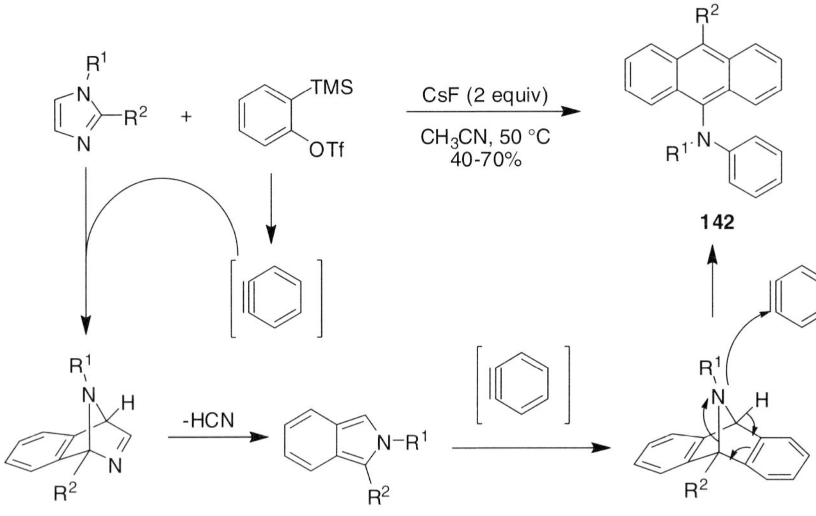

Scheme 12.44

Pascal has reported a synthetic protocol for the preparation of 9,11,20,22-tetraphenyltetrabenzo[*a,c,l,n*]pentacene (**145**), a polycyclic aromatic hydrocarbon, by double Diels–Alder reaction between bisbenzyne precursor **143** and phencyclone **144** (Equation 12.39) [79]. Unfortunately, however, the reaction was found to be extremely low-yielding, with compound **145** being obtained in only 1–2% yield after purification. This problem was most likely a result of the high temperature used to generate the bisbenzyne intermediate.

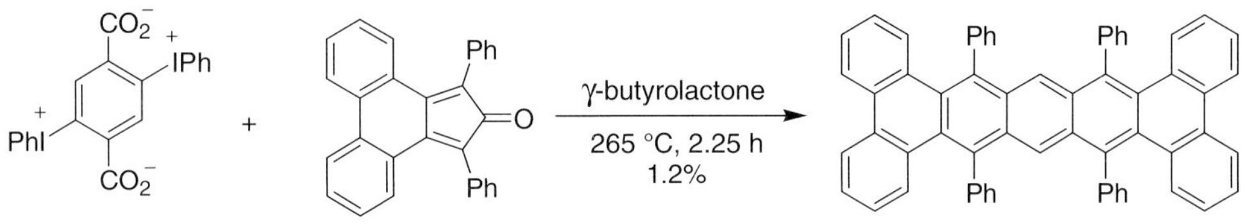

(12.39)

By considering the mild conditions required for the generation of the bisbenzyne species in such cycloaddition reactions, Wudl employed compound **146** as the bisbenzyne precursor in the synthesis of 6,8,15,17-tetraphenyl-1.18,4.5,9.10,13.14-tetrabenzoheptacene (**147**) (Equation 12.40) [80]. In comparison to Pascal's protocol, Wudl's method afforded the cycloaddition product, heptacene **147**, in a much higher (22%) yield.

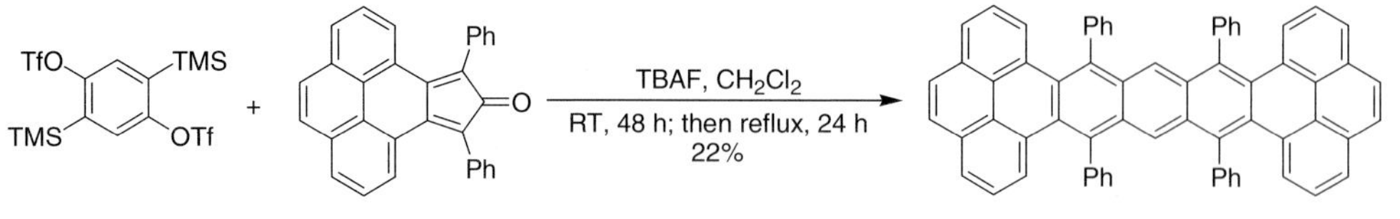

(12.40)

While employing cyclopentadiene **149** as the Diels–Alder trap, a research group at Pfizer studied the effect of solvent in the formation of 3-halobenzynes from the di- and trihalobenzene precursors **148** (Scheme 12.45) [81]. The fluoro adduct **150**, derived by LiCl elimination from **148**, was the predominant product in THF, whereas the chloro adduct **151** was obtained as the major product in toluene by loss of LiF.

Linear and angular benzobisoxadisiloles **152** and **153** have been prepared by Lee as precursors for the stepwise generation of synthetic equivalents of 1,4- and 1,3-benzdiynes [82]. As the example shown in Scheme 12.46 illustrates, two identical or different ring systems can be installed on the benzdiyne equivalent to afford fused benzene rings. Benzynes **155** and **157**, generated from the corresponding iodonium triflate intermediates **154** and **156** respectively, undergo sequential cycloaddition reactions, leading to [4+2] and [3+2] cycloadducts, respectively.

The six-membered ring of η^5-indenyl metal complexes is known to exhibit typical butadiene-like characteristics [83]. The Diels–Alder cycloaddition between η^5-indenyl iron complexes and benzyne has been reported by Wang (Equation 12.41) [84].

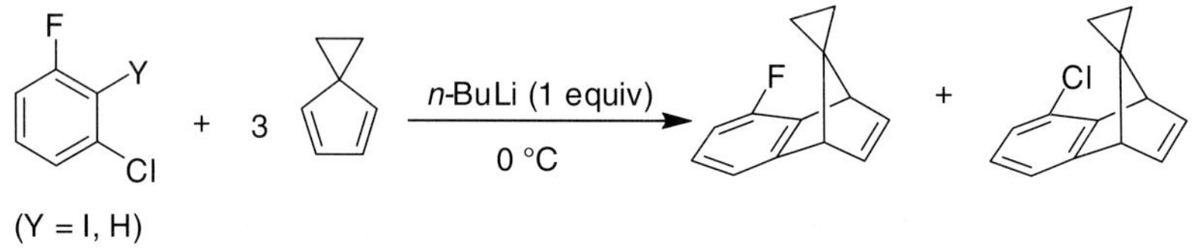

Y	Solvent	**150/151** (GC)	Yield (%)
H	THF	> 97:3	46
H	Toluene	> 3:97	41
I	THF	> 97:3	43
I	Toluene	> 3:97	64

Scheme 12.45

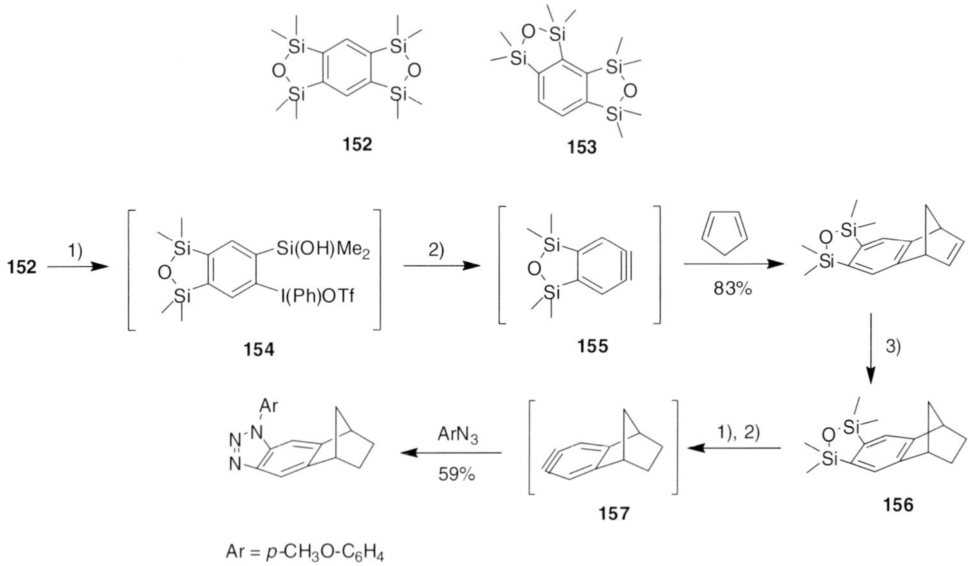

152

153

152 → 1) → [154] → 2) → [155] → cyclopentadiene, 83% → 3) → 156 → 1), 2) → [157] → ArN_3, 59%

Ar = p-CH_3O-C_6H_4

1) $PhI(OAc)_2$, TfOH, CH_2Cl_2, 0 °C to RT; 2) n-Bu_4NF, i-Pr_2NH, THF, RT; 3) H_2, Pd/C, RT

Scheme 12.46

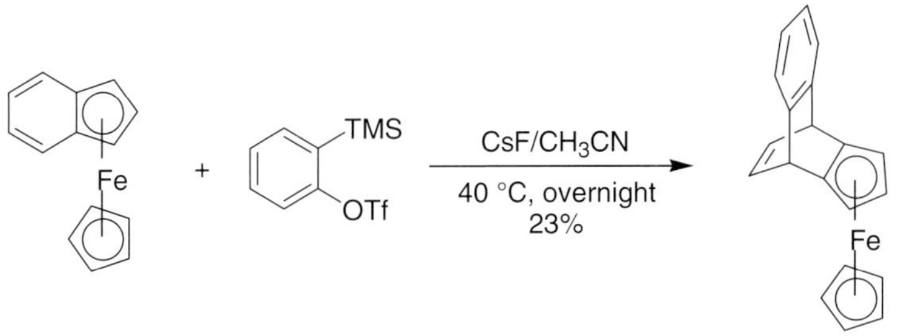

CsF/CH_3CN, 40 °C, overnight, 23% (12.41)

The high reactivity of benzynes as dienophiles in Diels–Alder reactions has also been observed in reactions with aromatic hydrocarbons, a class of compounds usually considered as inert as dienes. In a two-step synthesis of triptycene di- and tetracarboxylic acids, arynes generated *in situ* from anthranilic acids **158** cyclized with anthracenes **159** affording di- or tetramethyltriptycenes **160** in 41–69% yields (Scheme 12.47) [85]. Acids **161** were obtained after subsequent oxidation of the cycloadducts **160** with potassium permanganate.

Benzynes generated by the fluoride-induced elimination of TMS and OTf groups from o-trimethylsilyl [6]- and [7]-helicenol triflates selectively undergo an intramolecular Diels–Alder reaction with one of the benzene rings within the helicenes, rather than an external furan molecule, giving rise to bridged derivatives of coronene (Equation 12.42) [86].

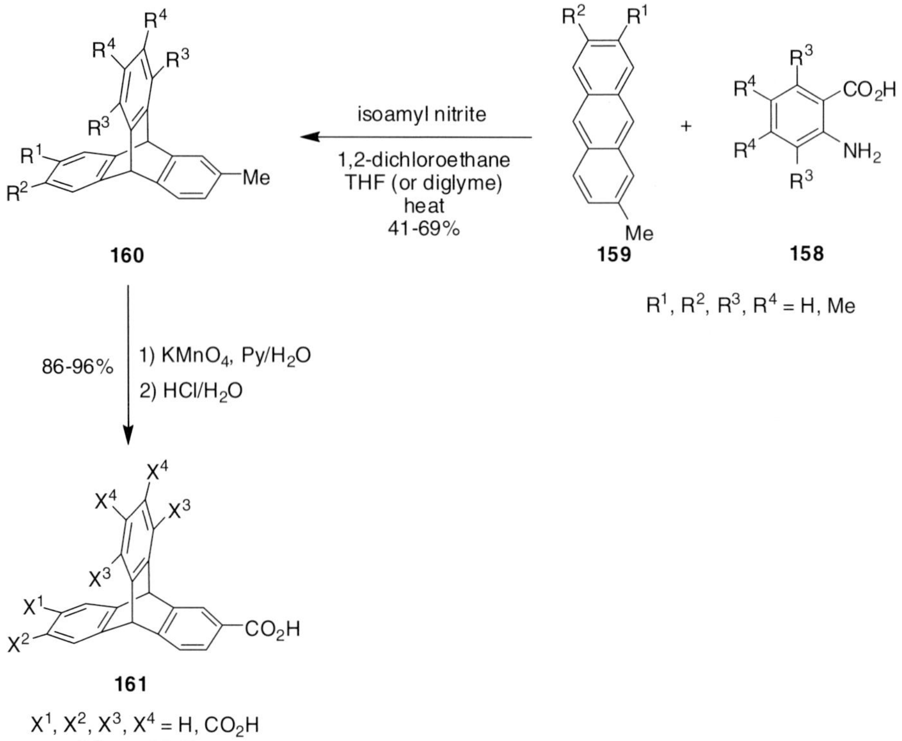

R⁴ R³ R¹ R² Me isoamyl nitrite 1,2-dichloroethane THF (or diglyme) heat 41-69% CO$_2$H NH$_2$

160 159 158

R^1, R^2, R^3, R^4 = H, Me

86-96% 1) KMnO$_4$, Py/H$_2$O 2) HCl/H$_2$O

161

X^1, X^2, X^3, X^4 = H, CO$_2$H

Scheme 12.47

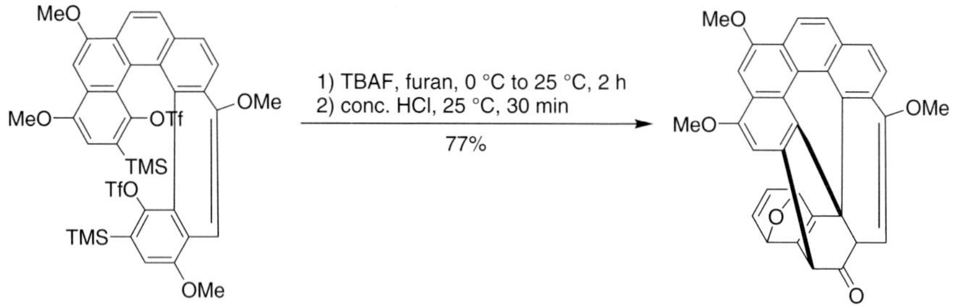

MeO OMe OTf TMS TfO 1) TBAF, furan, 0 °C to 25 °C, 2 h 2) conc. HCl, 25 °C, 30 min 77%

(12.42)

The arynes 4,5-dehydro- and 4,5,15,16-bis(dehydro)octafluoro[2.2]paracyclophanes, **162** and **163** (Figure 12.2), generated from the corresponding aryl iodide [87] or acetamide [88], have been captured in Diels–Alder reactions by Dolbier and coworkers. Aromatics, such as benzene, naphthalene, anthracene and furan

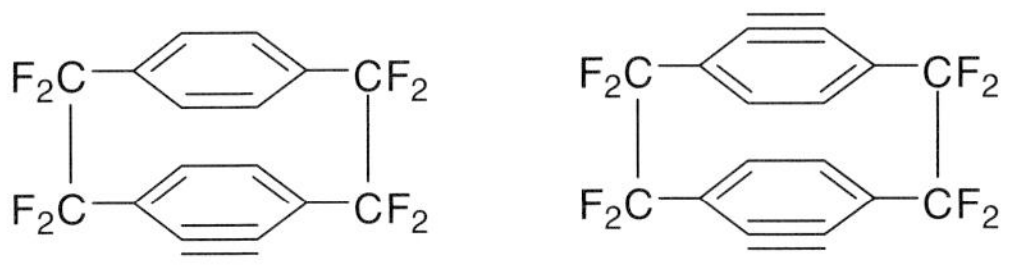

Figure 12.2 4,5-Dehydro-octafluoro[2.2]paracyclophane (**162**) and 4,5,15,16-bis(dehydro)octafluoro[2.2]paracyclophane (**163**).

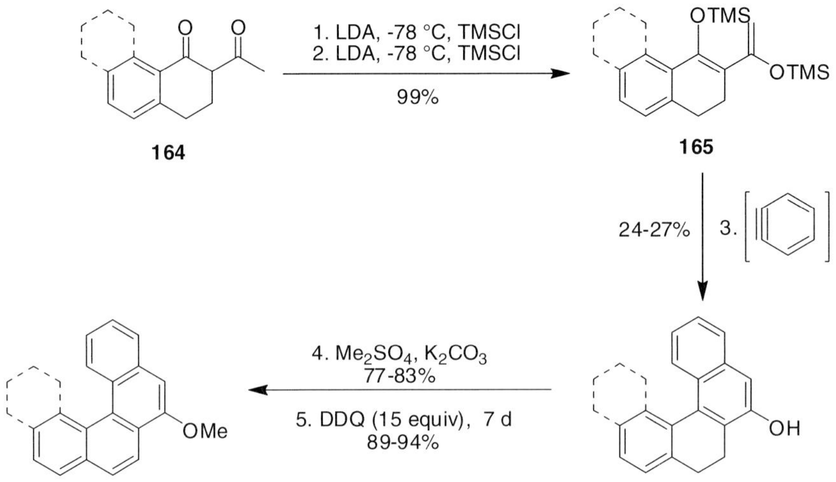

Scheme 12.48

all undergo this Diels–Alder cycloaddition efficiently, leading to the corresponding cycloadducts in yields greater than 80%.

Compared to cyclic dienes, acyclic dienes are less commonly used in aryne Diels–Alder reactions, although one example has been reported by Sarandeses *et al.* in the synthesis of helicenes (Scheme 12.48) [89]. Functionalized [4]/[5]helicenes **166** have been synthesized in five steps starting from **164**, using a synthetic approach based on a Diels–Alder reaction between the 1,3-bis(trimethylsilyloxy)-1,3-dienes **165** and benzyne [90].

Another example of the use of acyclic dienes has been recently reported by Lautens *et al.* in the synthesis of dihydronaphthalenes (Equation 12.43) [91]. The cycloadditions of benzyne generated from benzenediazonium-2-carboxylate with the carbonyl-substituted dienes **167** afford cycloadducts **168**. Excellent diastereoselectivity was observed when the diene was attached to Oppolzer's sultam as a chiral auxiliary. The cycloaddition was employed in the multistep synthesis of racemic sertraline.

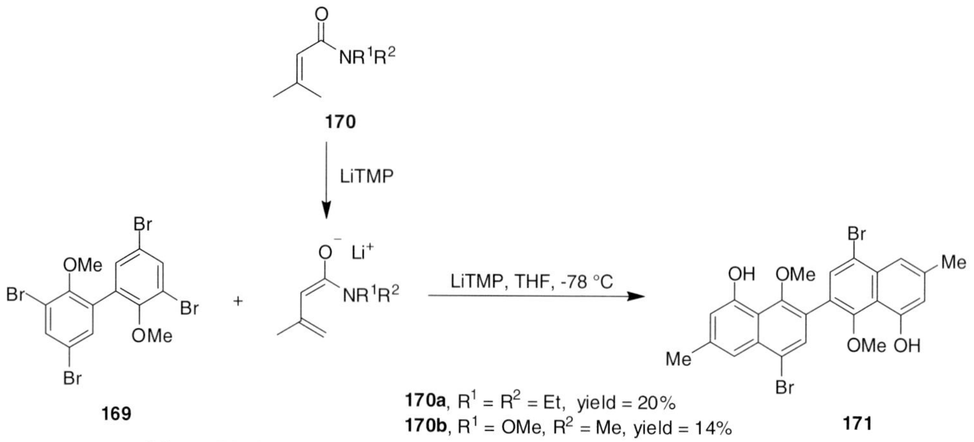

170a, $R^1 = R^2 = Et$, yield = 20%
170b, $R^1 = OMe$, $R^2 = Me$, yield = 14%

Scheme 12.49

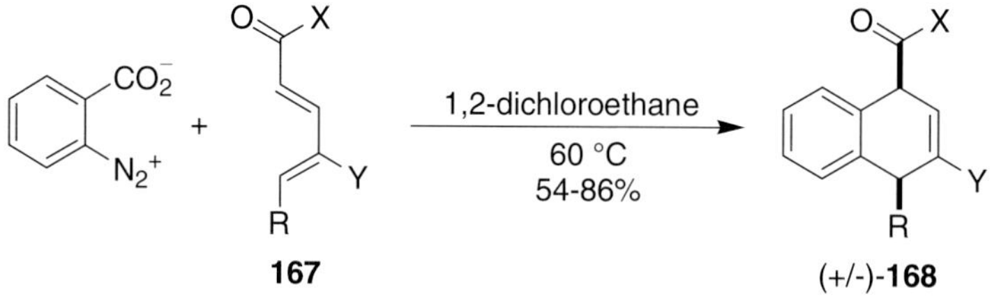

X = OEt, OH, OBn, N(*i*-Pr)$_2$, OMe
Y = H, Me, Br, CO_2Et

(12.43)

An aryne-based Diels–Alder reaction between tetrabromobiphenyl ether **169** and the dienolate anion generated from α,β-unsaturated amide **170** has been reported by Ruchirawat *et al.* in the synthesis of diospyrol derivatives, where the biphenyl ether **169** was converted to binaphthyl ether **171** in the presence of the sterically hindered base LiTMP (Scheme 12.49) [92].

The intramolecular [4+2] cycloaddition of a benzyne with conjugated enynes has been developed by Danheiser, furnishing condensed polycyclic aromatic compounds (Equation 12.44) [93]. Tetra-*n*-butylammonium triphenyldifluorosilicate (TBAT), a nonhygroscopic fluoride source, was specifically employed, instead of the other fluoride sources usually used in benzyne generation from *o*-(trimethylsilyl)aryl triflates. When the enyne double bond was incorporated in an aromatic or heteroaromatic ring, the [4+2] cycloadditions proceed efficiently, affording the corresponding polycyclic compounds in yields greater than 50%. An efficient intramolecular Diels–Alder reaction of a benzyne with an acyclic diene was also achieved under similar reaction conditions.

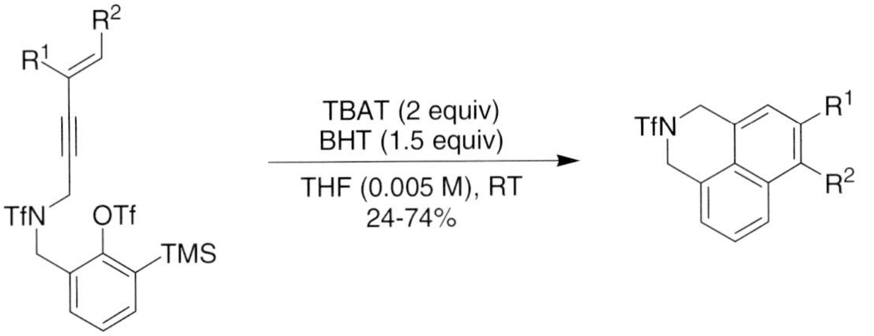

TBAT (2 equiv)
BHT (1.5 equiv)
THF (0.005 M), RT
24-74%

(12.44)

12.4.2
[2+2] Cycloadditions

Arynes have also been employed in a variety of [2+2] cycloadditions. By employing Stevens's and Bisacchi's methodology for the synthesis of benzocyclobutenones [94], Santelli *et al.* have examined the synthesis of benzo-biscyclobutenone **173** by the [2+2] cycloaddition of a dibenzyne with 2-methylene-1,3-dioxepane (Scheme 12.50) [95]. The biscycloadduct **172** was obtained in a 20% yield and smoothly transformed to the corresponding bisketone **173** after hydrolysis. The structure of **172** was confirmed by single-crystal X-ray analysis.

A highly regioselective [2+2] cycloaddition of 3-methoxybenzyne with ketene silyl acetal **175** has been reported by Suzuki *et al.* (Equation 12.45) [96]. The methoxy group on the benzyne precursor **174** was the decisive factor in the high regioselectivity, and the sterically more congested isomer **176** was formed exclusively in the reaction.

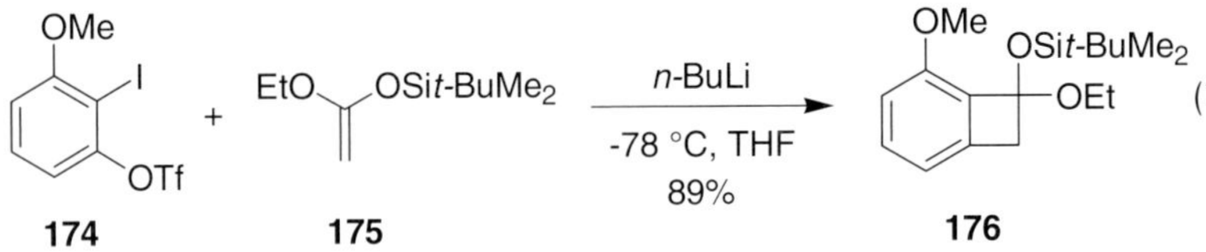

174 + **175** → **176** (n-BuLi, -78 °C, THF, 89%) (12.45)

The same group found that a four-membered ring fused to a benzyne also had a powerful regioselective directing effect in the [2+2] cycloadditions of benzyne, presumably through the severe ring strain [97]. In the reaction of ketene silyl acetal

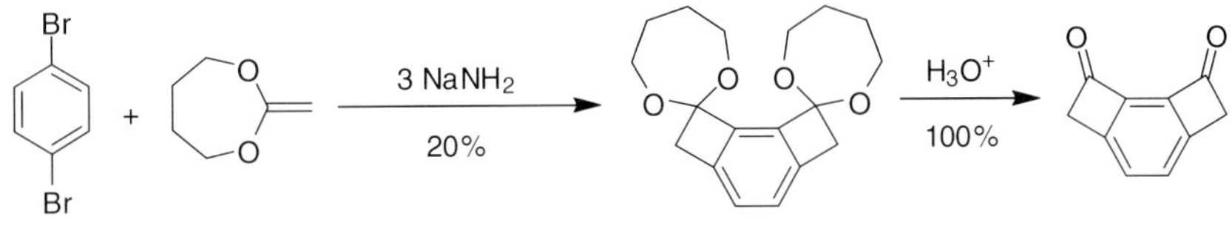

3 $NaNH_2$, 20% → **172**; H_3O^+, 100% → **173**

Scheme 12.50

175 with benzyne precursor 177, cycloadduct 178 containing the greater steric congestion was produced predominantly, alongside a small amount of regioisomer 179 (Equation 12.46).

EtO OSi*t*-BuMe$_2$ + OTf I $\xrightarrow[\text{-78 °C, THF}]{n\text{-BuLi}}$ OSi*t*-BuMe$_2$ OEt + OEt OSi*t*-BuMe$_2$

175 **177** **178** 71% **179** 2.3%

(12.46)

Based on the regioselective [2+2] cycloadditions described above, the theoretically interesting poly-oxygenated tricyclobutabenzenes **180** and **181** were synthesized by iterative cycloadditions between benzynes and ketene silyl acetals (Figure 12.3) [98]. In these multistep syntheses, an interesting question arises as to which is the more influential directing group – an alkoxy group or a four-membered fused ring – if both are present on the benzyne? The answer was provided during the course of the synthesis of hexaoxotricyclobutabenzene **182**, in which the directing ability of a benzyloxy group overrode that of a four-membered fused ring [99].

The Suzuki group subsequently reported a dual benzyne cycloaddition protocol starting from bis(sulfonyloxy)diiodobenzene **183** [100], which can be viewed as a synthetic equivalent of 3-methoxy-1,4-benzdiyne (**184**). When **183** was treated with 1.05 equiv. of *n*-BuLi in the presence of ketene silyl acetal **175** at –95 °C, a benzyne intermediate was generated on the iodo-triflate side of the arene, affording the mono-cycloadduct **185** in 72% yield (Scheme 12.51).

The generation of a benzyne intermediate from the iodo-tosylate side of arene **183** required a higher reaction temperature. When **183** was treated with 2.3 equiv. of *n*-BuLi at an initial temperature of –95 °C, and then warmed up to –78 °C, the bis-cycloadduct **186** was obtained in a 72% yield as the exclusive product. Therefore, by manipulating the stoichiometry of the *n*-BuLi and the reaction tempera-

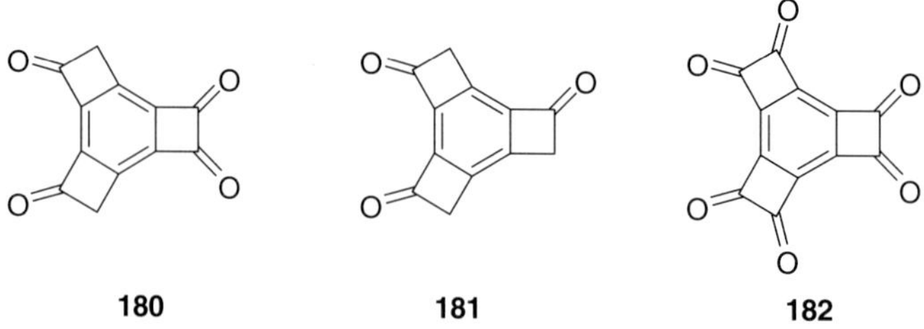

Figure 12.3 Poly-oxygenated tricyclobutabenzenes.

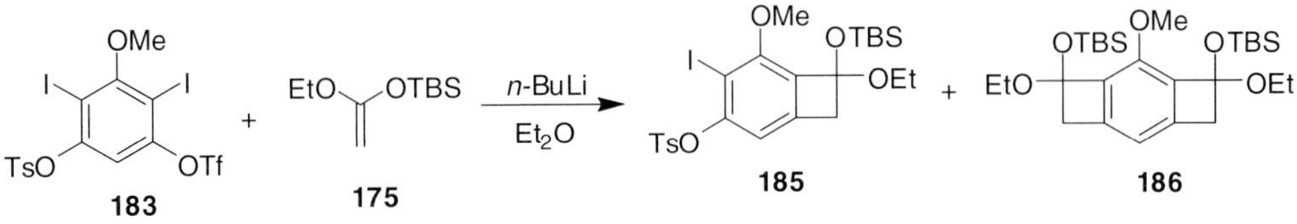

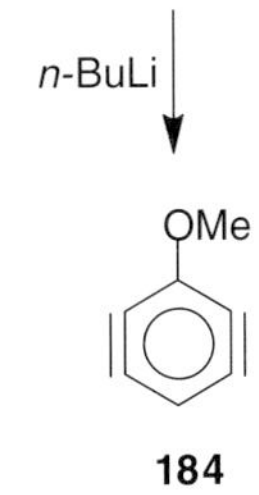

		185	186
-95 °C	n-BuLi (1.05 equiv)	72%	8%
-95 °C to -78 °C	n-BuLi (2.3 equiv)	–	72%

Scheme 12.51

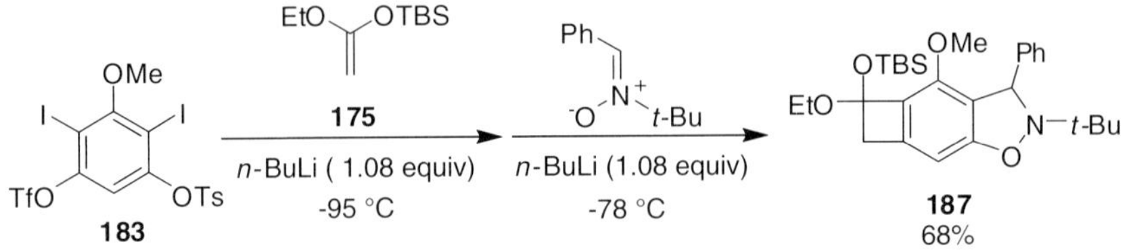

Scheme 12.52

ture, two benzyne species can be generated in a stepwise manner from **183**, enabling tandem cycloadditions with either two identical or two different arynophiles. A representative example is shown in Scheme 12.52, where bis-cycloadduct **187** was generated as a single regioisomer in a one pot reaction by successive [2+2] and [3+2] cycloadditions.

Yoshida *et al.* have reported the coupling of an aromatic aldehyde and 2 equiv. of aryne, affording 9-arylxanthenes (Scheme 12.53) [101]. The reaction presumably starts with a nucleophilic attack of the carbonyl oxygen of **188** on the aryne, leading to zwitterion **189**, which subsequently undergoes cyclization to afford benzoxete **190**. Subsequent isomerization of **190** leads to *o*-quinone methide **191** which then undergoes [4+2] cycloaddition with a second molecule of aryne to afford the xanthene **192**.

12.4.3
[3+2] Cycloadditions

Arynes have also been reported to undergo [3+2] cycloadditions. Indeed, a [3+2] cycloaddition reaction of lithium trimethylsilyldiazomethane [TMSC(Li)N_2] with

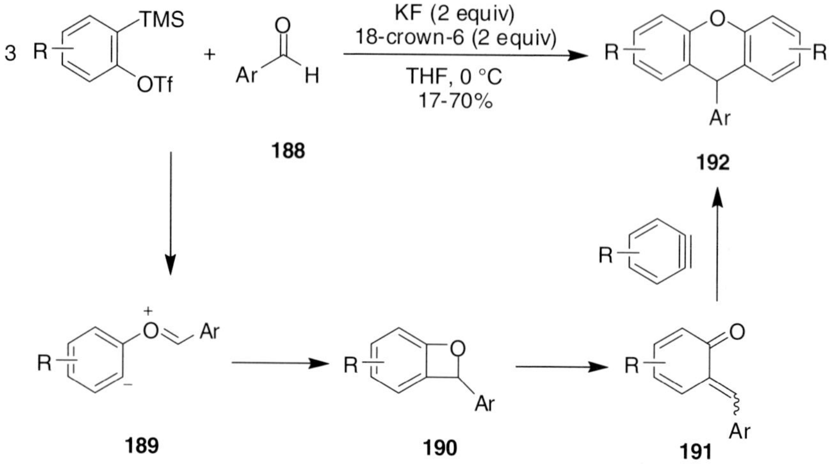

Scheme 12.53

benzyne has been reported by the Aoyama group during their investigation of the utility of TMSC(Li)N_2 as a [C-N-N] azole synthon for the preparation of azoles (Equation 12.47) [102]. The product 3-(trimethylsilyl)indazoles **193** was obtained as a mixture of regioisomers, as usually observed in aryne chemistry. A stepwise mechanism was proposed for the cycloaddition.

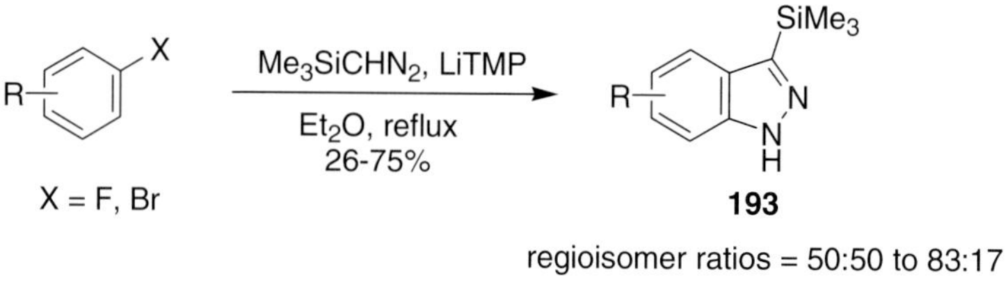

(12.47)

By employing *ortho*-(trimethylsilyl)aryl triflates as benzyne precursors, Jin and Yamamoto have reported the synthesis of indazoles under mild reaction conditions by the [3+2] cycloaddition of benzyne with diazomethane derivatives, followed by a 1,3-hydrogen shift (Scheme 12.54) [103]. Depending on the stoichiometry of the reagents and the reaction conditions, either 1-arylindazoles **194** or 1*H*-indazoles **195** are readily produced in fair to good yields.

A similar process has been developed simultaneously by the Larock group (Scheme 12.55) [104]. Both, 1-arylindazoles **194** and 1*H*-indazoles **195** are readily produced in fair to good yields by simply adjusting the stoichiometry of the reagents and the reaction conditions. When dicarbonyl-containing diazo com-

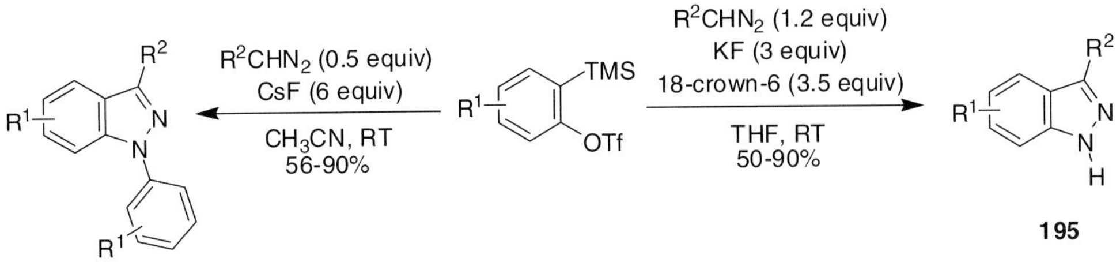

Scheme 12.54

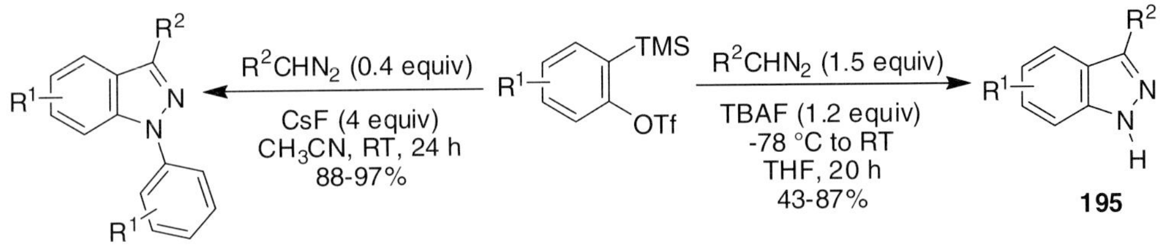

Scheme 12.55

pounds are employed in this process, acyl migration is observed to follow the initial [3+2] cycloaddition, selectively leading to the 1-acyl indazoles **196** (Equation 12.48). Ketone groups are observed to migrate in preference to ester groups.

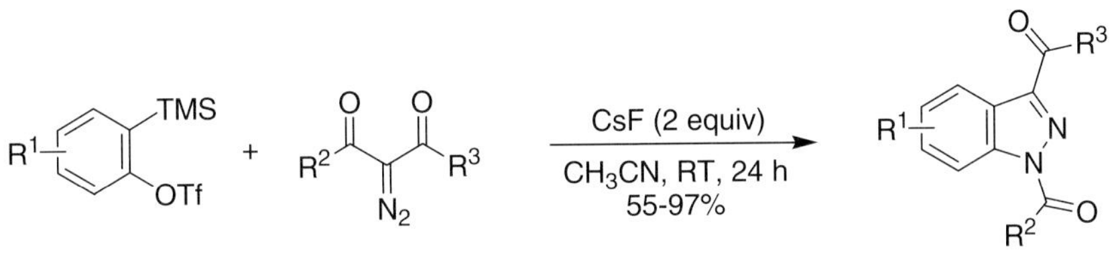

(12.48)

A regioselective synthesis of 3-(2-hydroxyaryl)pyridines **197** initiated by the [3+2] cycloaddition of arynes with pyridine *N*-oxides has been reported by Larock and coworkers (Scheme 12.56) [105]. The initial cycloadduct **198** readily rearranges to form intermediate **199**, which affords the observed 3-arylpyridines rather than

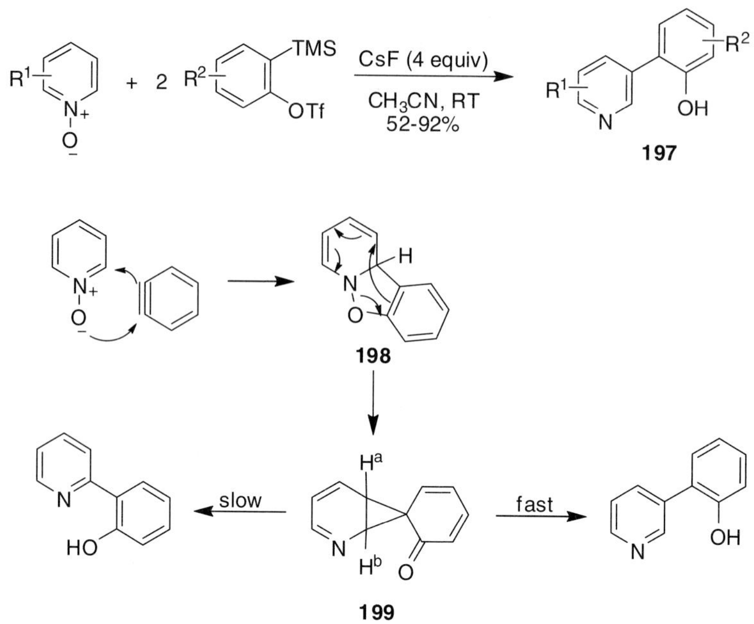

Scheme 12.56

breaking down to form the 2-arylpyridines. Due to the difference in the acidity of the two hydrogens H^a and H^b, the carbonyl group in intermediate **199** preferentially attacks the more acidic H^a to form the 3-substituted pyridines. However, when an electron-withdrawing cyano group is present in the 4 position of the pyridine *N*-oxide, the corresponding 2-substituted pyridine is observed as the sole product.

12.4.4
Ene Reaction

Despite the long history of arynes and their potential as a powerful enophile in the ene reaction (because of their low-lying LUMOs), reports on the ene reaction of arynes remain rare, especially with alkynes. Recently, an ene reaction of arynes with alkynes has been reported by Cheng and coworkers, which readily affords arylallenes (Equation 12.49) [106].

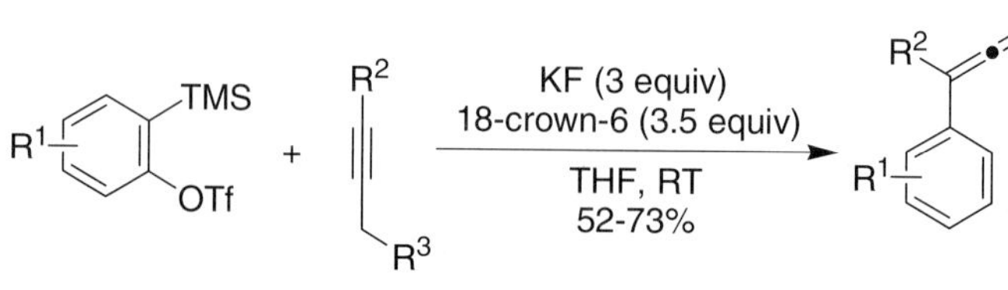

(12.49)

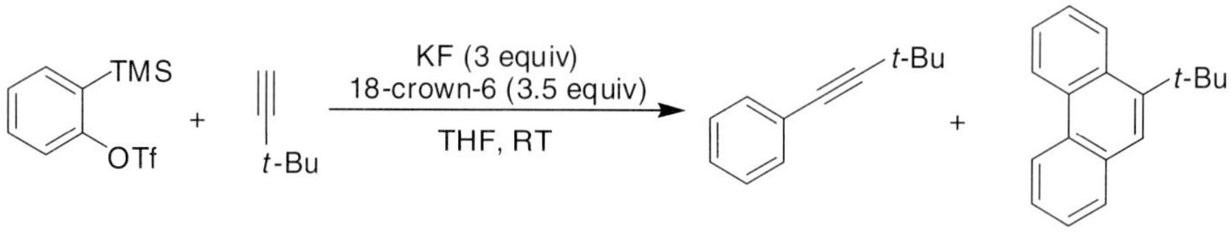

200 : 201	202	203
1 : 4	83%	~5%
2 : 1	trace	92%

Scheme 12.57

During the course of studying the ene reaction, these authors identified an interesting reaction between benzyne and *tert*-butylacetylene, in which two products were observed (Scheme 12.57). When a 1:4 ratio of the substrates **200** and **201** was employed, product **202** from direct addition of the acetylenic C–H bond across the benzyne was obtained as the major product. Only ~5% of another product, **203**, generated from the Diels–Alder reaction between benzyne and product **202** was obtained. On the other hand, if the substrate ratio of **200** to **201** was switched to 2:1, the anthracene **203** was formed predominantly in 92% yield.

Aryne **162** generated from the [2.2]paracyclophane acetanilide **204** using Cadogan's methodology [107] showed high reactivity in ene reactions [88]. Upon treatment of **204** with 4-chlorobenzoyl nitrite (PCBN) in the presence of an excess of 1-octene at 145 °C in dibutyl ether, a high-yielding ene reaction takes place, affording a 2.4:1 ratio of *trans* to *cis* olefinic isomers (Scheme 12.58). Likewise, reactions with cyclopentene and cyclohexene both lead to good yields of the corresponding ene products as mixtures of diastereomers. Interestingly, no ene reaction was observed when the aryne species **162** was generated from the corresponding aryl iodide with potassium *tert*-butoxide, using the Cram [108] methodology.

12.4.5 Miscellaneous

One other miscellaneous cycloaddition reaction has been reported. Arynes generated from anthranilic acids **205** have been found to undergo cyclization with Barton esters **206** to afford polycyclic heterocycles **207** in moderate yields (Scheme 12.59) [109], with the reaction presumably proceeding through a radical pathway.

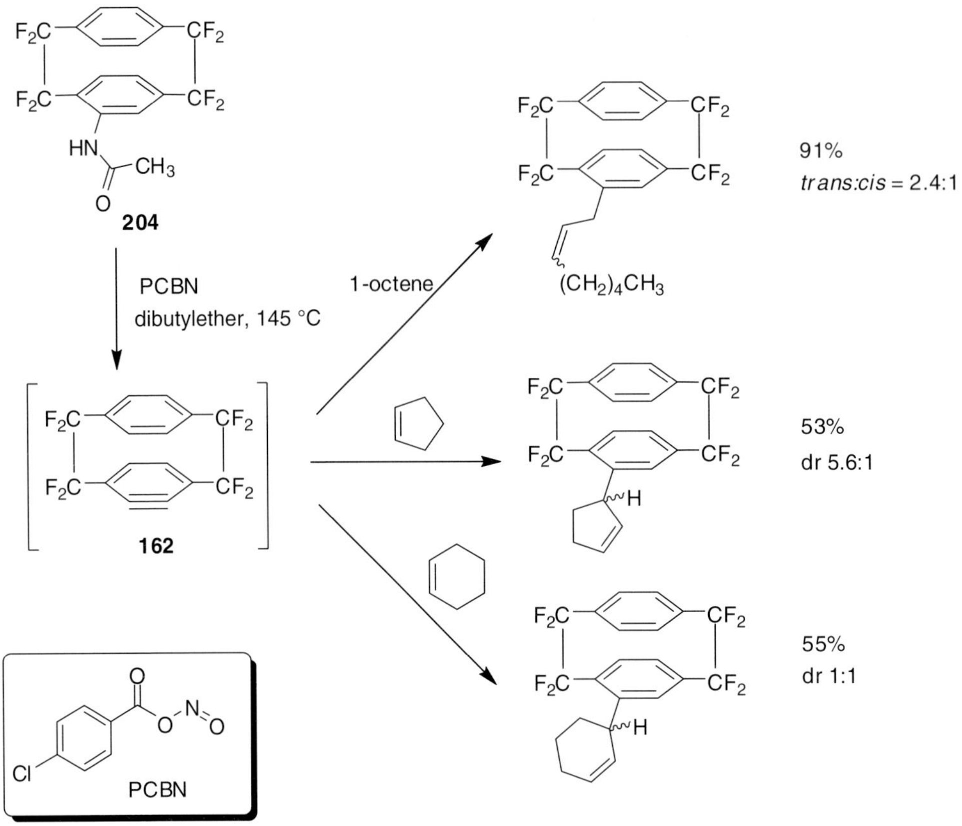

Scheme 12.58

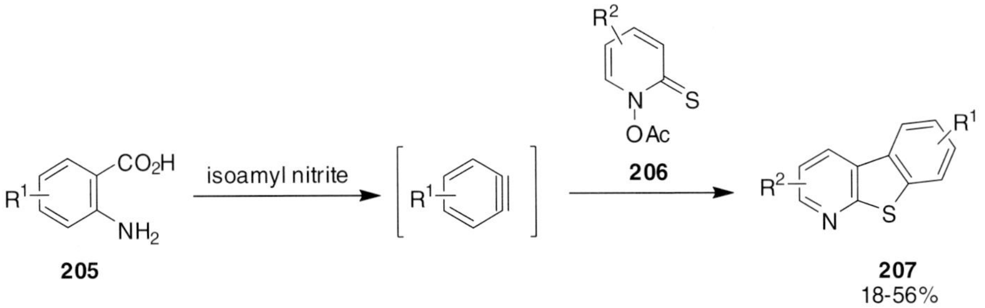

Scheme 12.59

12.5 Transition Metal-Catalyzed Reactions of Arynes

12.5.1 Transition Metal-Catalyzed Cyclizations

Although details of the metal-mediated cycloaddition chemistry of arynes are well known [110], the use of arynes was for a long period of time restricted to stoichiometric reactions. An excellent example of this is the [2+2+2] cycloaddition reaction developed by Bennett and Wenger [111], in which one of the participating triple bonds is a nickel–benzyne complex that reacts with 2 equiv. of an alkyne to generate substituted naphthalenes (Equation 12.50). The development of new methods of generating arynes under mild conditions has led to the exploitation of arynes in metal-catalyzed cycloaddition reactions.

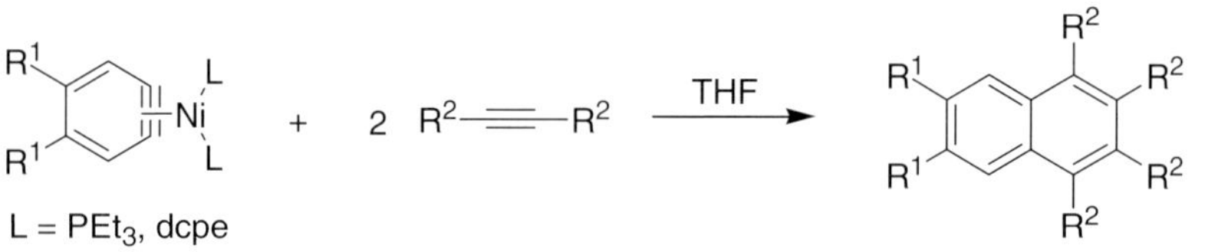

R^1 R^1 L Ni L + 2 R^2—≡—R^2 —THF→ R^1 R^1 R^2 R^2 R^2 R^2

L = PEt_3, dcpe

(12.50)

12.5.1.1 Palladium/Nickel-Catalyzed [2+2+2] Cycloadditions

The first example of a transition metal-catalyzed cyclization of arynes was reported in 1998 [112], when benzyne generated from *o*-(trimethylsilyl)phenyl triflate was shown to undergo cyclotrimerization in the presence of a catalytic amount of $Pd(PPh_3)_4$ to afford triphenylene (**208**) (Scheme 12.60).

Subsequent extensions of this palladium-catalyzed [2+2+2] cycloaddition from benzyne to polycyclic arynes have opened the door to the synthesis of a number

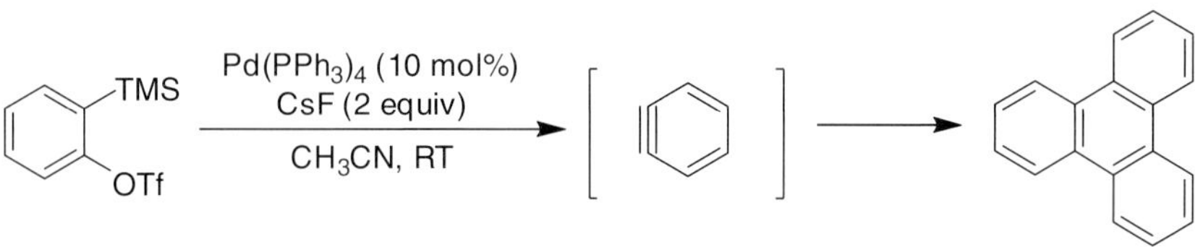

TMS OTf —$Pd(PPh_3)_4$ (10 mol%), CsF (2 equiv), CH_3CN, RT→ [ ] → **208** 83%

Scheme 12.60

of structurally diverse polycyclic aromatic hydrocarbons. When sterically demanding polycyclic arynes are employed, $Pd_2(dba)_3$ becomes the palladium(0) source of choice (Equation 12.51) [113].

TMS
OTf
CsF (2 equiv)
$Pd_2(dba)_3$ (5 mol %)
CH_3CN, RT
+
209
210
209 : **210** = 2.7 : 1
combined yield = 60%

(12.51)

None of the expected trimers is obtained in the presence of 10 mol% of $Pd(PPh_3)_4$ due to the steric bulkiness and strong coordinating ability of PPh_3. On the other hand, in the presence of 10 mol% $Pd(PPh_3)_4$, the triphenylenes **127** undergo cyclotrimerizations, affording the extended triphenylenes **211** in low yields (Equation 12.52) [73].

OTf
TMS
127a: R = H
127b: R = OC_6H_{13}
$Pd(PPh_3)_4$ (10 mol %)
127a: TBAF (1.5 equiv) / THF
127b: CsF (3 equiv) / 3:1 CH_3CN:CH_2Cl_2
211a: 18%
211b: 28%

(12.52)

When different arynes are generated in the same reaction vessel in the presence of a catalytic amount of a palladium(0) catalyst, a mixture of homotrimers (**208** and **211b**) and heterotrimers (**212** and **213**) is obtained (Equation 12.53) [73].

$$4\ \text{(2-(trimethylsilyl)phenyl triflate)} + \mathbf{127b} \xrightarrow[\text{RT, 14 h}]{\text{Pd(PPh}_3)_4\ (50\ \text{mol\%}),\ \text{CsF (15 equiv)},\ \text{CH}_3\text{CN/CH}_2\text{Cl}_2\ (3:1)} \mathbf{212}\ (42\%) + \mathbf{213}\ (9\%) + \mathbf{211b}\ (2\%) \quad (12.53)$$

Didehydrobiphenylenes **215** generated from the corresponding 3-trimethylsilyl-2-biphenylenyl triflates **214** undergo trimerization in the presence of 10 mol% $Pd(PPh_3)_4$, affording the C_3-symmetric trimers **216** in good yields (Scheme 12.61) [114]. Small amounts of 2-fluorobiphenylenes **217** resulting from nucleophilic attack of fluoride on the arynes **215** have been isolated as side products.

Alkynes have been reported to participate in cocyclization with arynes under the same reaction conditions, affording a mixture of phenanthrenes **218**, naphthalenes **219** and triphenylenes (Scheme 12.62) [115]. The chemoselectivity depends on both the palladium catalyst and the alkyne substrate used. For example, when dimethyl acetylenedicarboxylate (DMAD) is employed as the alkyne, a remarkable chemoselectivity is achieved by appropriate selection of the

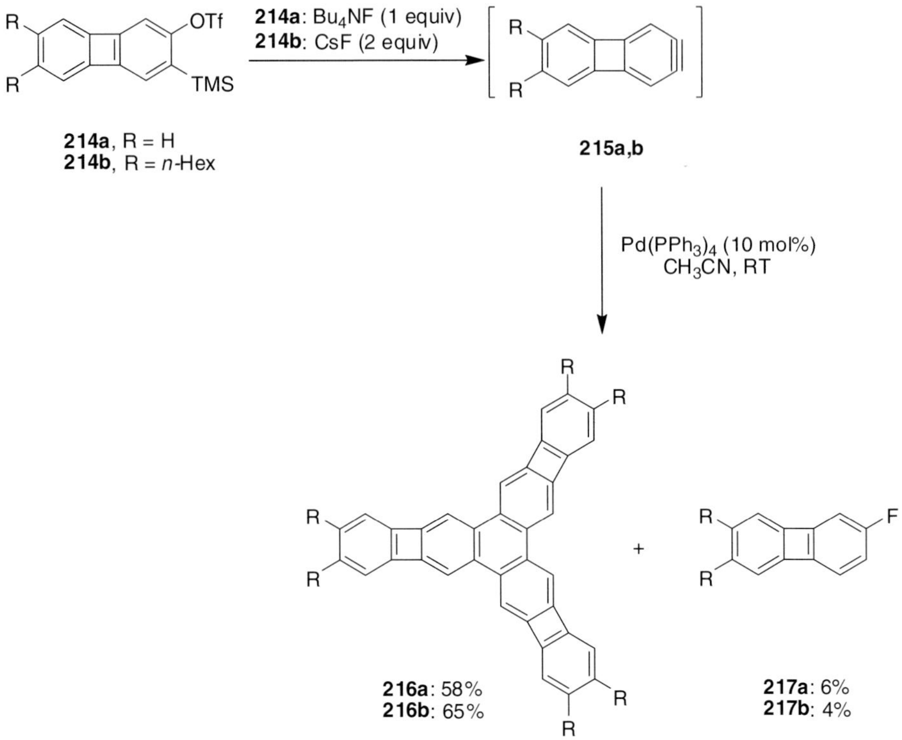

R, R, OTf, TMS; $\mathbf{214a}$: Bu_4NF (1 equiv); $\mathbf{214b}$: CsF (2 equiv); R, R

214a, R = H
214b, R = *n*-Hex

215a,b

$Pd(PPh_3)_4$ (10 mol%)
CH_3CN, RT

216a: 58%
216b: 65%

+ R, R, F

217a: 6%
217b: 4%

Scheme 12.61

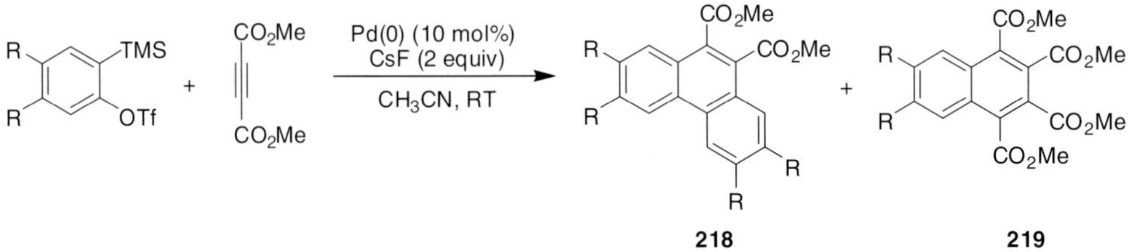

R, R, TMS, OTf + CO_2Me, CO_2Me; Pd(0) (10 mol%), CsF (2 equiv), CH_3CN, RT

CO_2Me, CO_2Me, R, R, R, R — **218**

+ CO_2Me, CO_2Me, CO_2Me, CO_2Me, R, R — **219**

R	Pd(0)	Yield (%) 218	219	Triphenylene (%)
H	$Pd(PPh_3)_4$	84	7	2
H	$Pd_2(dba)_3$	10	83	–
F	$Pd(PPh_3)_4$	64	8	8
F	$Pd_2(dba)_3$	9	54	–

Scheme 12.62

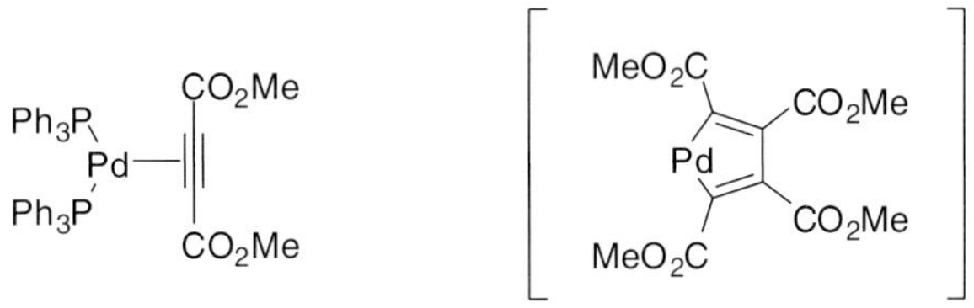

220 221

Figure 12.4 Intermediate complexes in palladium-catalyzed cocyclizations.

palladium catalyst. Phenanthrenes are obtained as the major product when $Pd(PPh_3)_4$ is used, while naphthalenes become the dominant product in the presence of $Pd_2(dba)_3$. In both cases, only a little to trace amounts of the cyclotrimer triphenylenes are observed. Similar results are obtained even when sterically more hindered arynes, such as naphthalyne, phenanthryne [116], triphenylyne [73] and 2,3-didehydrobiphenylene [114] are employed in the cocyclization with DMAD. In general, $Pd(PPh_3)_4$ selectively catalyzes the cocyclization of two molecules of aryne with one molecule of alkyne to afford phenanthrenes, while $Pd_2(dba)_3$ favors the cocyclization of one molecule of aryne with two molecules of alkyne to produce naphthalenes.

The nature of the ligand on palladium in these complexes is thought to be the cause of the observed differences in chemoselectivity. When $Pd(PPh_3)_4$ is used, an intermediate complex **220** is observed and isolated (Figure 12.4); this complex is assumed to be involved in formation of the phenanthrenes **218**. On the other hand, $Pd_2(dba)_3$ is assumed to form a palladacyclopentadiene complex **221** by oxidative coupling of two molecules of DMAD, which leads to formation of the naphthalenes **219** upon reaction with one molecule of aryne [117].

The above reaction conditions are only suitable for electron-deficient alkynes. When electron-rich alkynes are used, the corresponding phenanthrenes are obtained in only low yields (up to 34%). At about the same time, the Yamamoto group independently developed a palladium-catalyzed cotrimerization of benzyne with alkynes [118] where the yields of phenanthrene products generated from electron-rich alkynes could be dramatically increased to >59% in the presence of a $Pd(OAc)_2/P(o\text{-tolyl})_3$ catalyst system. This process was, however, later shown to proceed through a Pd(0) insertion–carbopalladation process, rather than a free benzyne process [119].

The catalyst $Pd_2(dba)_3$ also successfully promotes the [2+2+2] cocyclotrimerization of arynes with suitably functionalized benzodiynes, affording benzo[*b*]fluorenones in moderate yields (Equation 12.54) [120]. The best result was achieved when the electron-deficient diyne **222** was employed, when the corresponding benzo[*b*]fluorenone **223** was obtained in 54% yield.

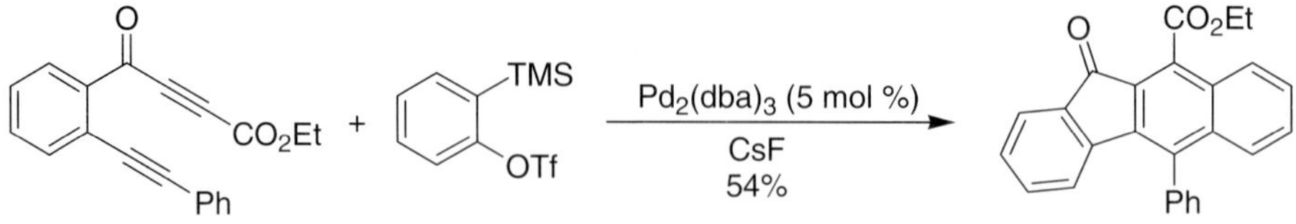
TMS
OTf
CO2Et
Ph
Pd2(dba)3 (5 mol %)
CsF
54%
CO2Et
Ph

222 **223**

(12.54)

A protocol for constructing biaryls by the palladium(0)-catalyzed [2+2+2] cocyclization of arynes with diynes has been developed by Sato and coworkers (Scheme 12.63). Total syntheses of the arylnaphthalene lignans taiwanin C, taiwanin E and dehydrodesoxypodophyllotoxin have been achieved by employing a [2+2+2] cocyclization as a key step [121].

An enantioselective [2+2+2] cycloaddition between two molecules of 7-methoxy-1-trimethylsilyl-2-naphthyl triflate (**224**) and one molecule of DMAD was achieved (Equation 12.55) [122]. This process was carried out in the presence of a chiral palladium–BINAP complex, affording the 9,12-dimethoxypentahelicene **225** in 66% enantiomeric excess (ee), along with the regioisomers **226** and **227** in a 66% overall yield.

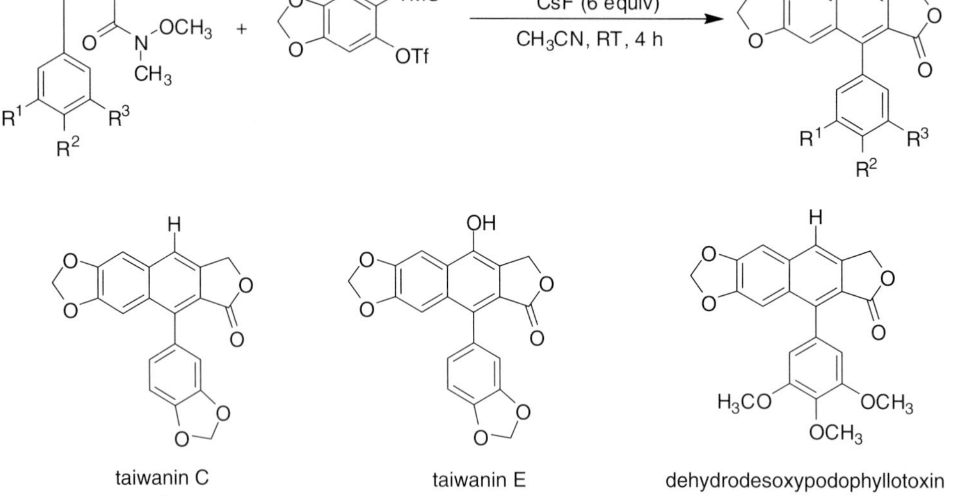
OCH3
CH3
TMS
OTf
Pd2(dba)3 (5 mol %)
P(o-tol)3 (40 mol %)
CsF (6 equiv)
CH3CN, RT, 4 h
CH3
OCH3
R1
R2
R3
H
OH
H3CO
OCH3
OCH3
taiwanin C
taiwanin E
dehydrodesoxypodophyllotoxin

Scheme 12.63

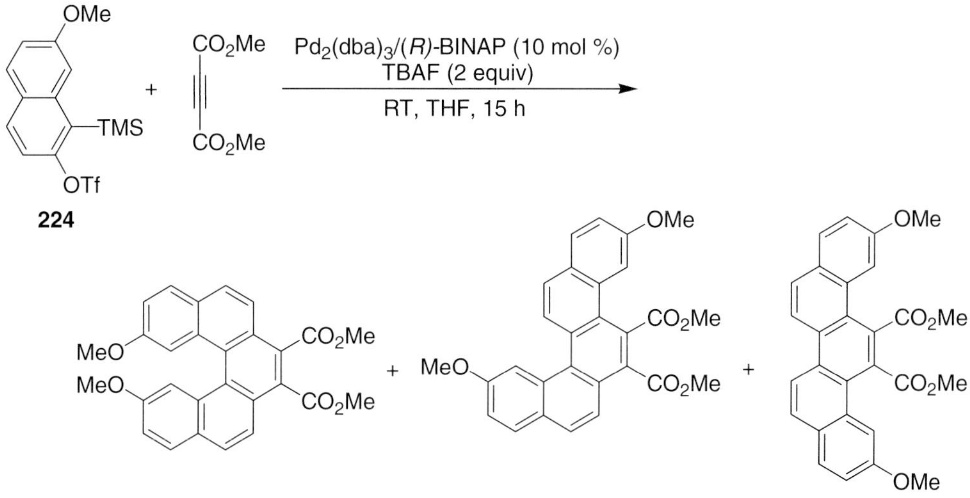

(M)-**225** 16% 66% ee

226 24%

227 26%

(12.55)

The symmetrically-substituted 1,3-diyne **228a** reacts with nickel(0)–benzyne complexes in the presence of an excess of triethylphosphine with regioselective formation of the 2,3-dialkynylnaphthalene **229** (Equation 12.56) [123]. The high regioselectivity observed is attributed to the electronic influence of the 1,3-conjugated diyne. This electronic preference, however, can be overcome when the diyne possesses particularly bulky substituents, such as diyne **228b**, in which case the 1,4-dialkynylnaphthalene **230** is formed exclusively.

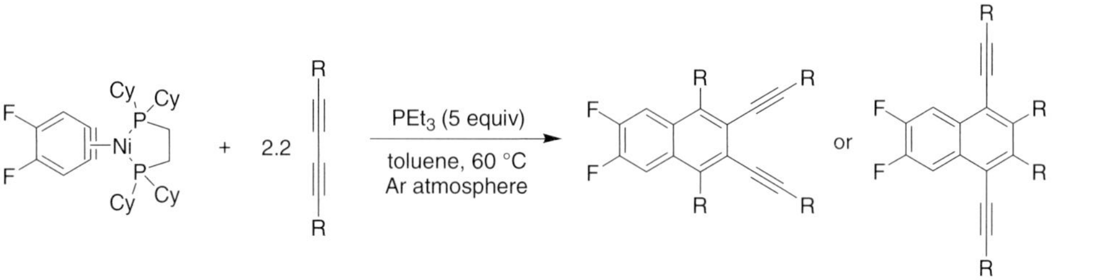

228a, R = Ph
228b, R = $CH(CH_3)O$-*t*-Bu

R = Ph
229
85%

R = $CH(CH_3)O$-*t*-Bu
230
70%

(12.56)

The [2+2+2] cocyclotrimerization of arynes with diynes has also been effectively catalyzed by $NiBr_2$(dppe)–Zn, leading to substituted naphthalene derivatives in moderate to good yields (Equation 12.57) [124]. Depending on the nature and length of the chain X, five- to seven-membered carbo- or heterocycle-fused naphthalenes are generated. This reaction is, however, sensitive to the steric bulkiness of the substituents on both the diyne and the benzyne moieties. No cocyclization product is obtained when both of the distal ends of the triple bonds of the diyne contain substituents.

$NiBr_2$(dppe) (5 mol %)
Zn (15 mol %)
CsF (2 equiv)
CH_3CN, 80 °C
37-89%

X = CH_2; $(CH_2)_2$; $(CH_2)_3$; O; NTs

(12.57)

The $NiBr_2$(dppe)–Zn system has also been effective in the [2+2+2] cocyclotrimerization of arynes with allenes, leading to 10-methylene-9,10-dihydrophenanthrenes (Equation 12.58) [125]. The reaction is highly selective in the case of monosubstituted allenes, and only the internal C=C double bond is involved in the cyclization. With 1,1-disubstituted allenes, on the other hand, both C=C double bonds participate in the reaction.

$NiBr_2$(dppe) (10 mol %)
Zn (2.75 equiv)
CsF (4 equiv)
CH_3CN, 80 °C, 24 h
46-69%

(12.58)

The scope of the [2+2+2] methodology was later extended to the cocyclization of arynes and alkenes. Here, one molecule of bicyclic alkene was found to react with two molecules of aryne in the presence of $PdCl_2(PPh_3)_2$, leading to a norbornane anellated 9,10-dihydrophenanthrene in moderate to good yields and high stereoselectivities with the *exo* isomers as the exclusive products (Equation 12.59) [126].

$PdCl_2(PPh_3)_2$ (5 mol %)
CsF (3 equiv)
CH_3CN, RT, 8 h
54-90%

X = CH_2, O, NCO_2Et

(12.59)

Arynes also undergo transition metal-catalyzed cotrimerization with appropriately functionalized acyclic alkenes. In the presence of a catalytic amount of a palladium–phosphine complex, one molecule of the electrophilic alkenes **231** substituted with either one or two electron-withdrawing groups undergo cotrimerization with two molecules of aryne to form a mixture of dihydrophenanthrene **232** and *ortho*-alkenyl biaryls **233** (Scheme 12.64) [127].

The product ratio of dihydrophenanthrene **232** and *ortho*-olefinated biaryl **233** is highly dependent on the nature of the phosphine ligand employed. Dihydrophenanthrene **232a** is isolated as the major product when PPh_3 is used, yet generation of the biaryl **233a** is favored when $P(o\text{-tol})_3$ is employed. Nickel-based catalysts have been found to be more efficient than their palladium alternatives in the cotrimerization of disubstituted alkenes. The reaction between dimethyl fumarate (**231b**) and *o*-(trimethylsilyl)phenyl triflate affords 61% yield with $Ni(cod)_2/PPh_3$, whereas only 31% yield is obtained in the presence of $Pd_2(dba)_3/PPh_3$ (not shown).

12.5.1.2 Palladium-Catalyzed Cyclization Involving Carbopalladation of Arynes

π-Allylpalladium species can also participate in annulation processes with arynes, yielding phenanthrene derivatives **234** and **235** in a ratio of 58 : 42 (Equation 12.60) [119, 128].

2 (TMS, OTf-substituted benzene) + Me–CH=CH–CH₂Cl → [$[Pd_2(dba)_3]\cdot CHCl_3$ (2.5 mol %), dppf (5 mol %), CsF (4 equiv), 1:1 CH_3CN/THF, 60 °C] → **234** + **235**

234:**235** = 58:42
yield = 66%

(12.60)

The reaction, initiated by carbopalladation of the aryne by stable π-allylpalladium intermediates, is followed by the insertion of a second molecule of aryne into the arylpalladium bond and subsequent intramolecular carbopalladation of the alkene by the newly formed biarylpalladium species. When the second equivalent of aryne is replaced with an internal alkyne, naphthalene **236** is formed in a 47% yield (Equation 12.61).

2 (TMS, OTf-substituted benzene) + CH₂=CH–CH₂Cl + *n*-Pr–≡–*n*-Pr → [$[Pd_2(dba)_3]\cdot CHCl_3$ (5 mol %), dppf (10 mol %), CsF (4 equiv), 1:1 CH_3CN/THF, 60 °C, 47%] → **236** (Me, *n*-Pr, *n*-Pr)

(12.61)

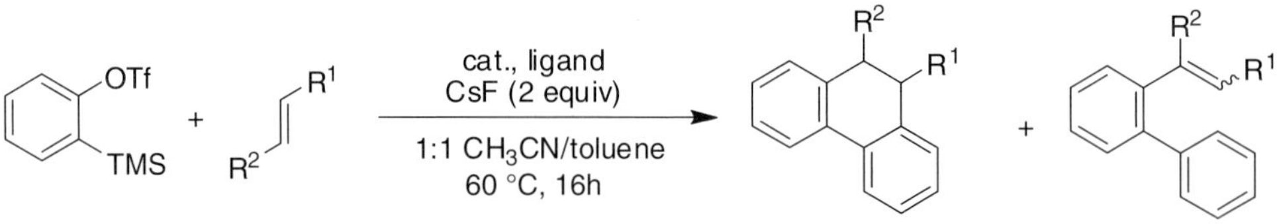

cat., ligand
CsF (2 equiv)
1:1 CH_3CN/toluene
60 °C, 16h

231a,b 232a,b (E/Z)-233a,b

a, R^1 = CO_2Me, R^2 = H
b, R^1 = R^2 = CO_2Me

	Catalyst (mol%)	Ligand (mol%)	Yield (%)	Ratio [**232**:*(E)*-**233**:*(Z)*-**233**]
231a	$Pd_2(dba)_3$ (5)	PPh_3 (20)	95	77:14:9
231a	$Pd_2(dba)_3$ (5)	P(*o*-tol)$_3$ (20)	69	8:68:24
231b	$Ni(cod)_2$ (10)	PPh_3 (20)	61	100:0:0

cod = 1,5-cyclo-octadiene

Scheme 12.64

Due to the instability and high reactivity of aryl and alkenyl palladium species generated by the oxidative addition of aryl and alkenyl halides to palladium(0), and the propensity of arynes to cyclotrimerize in the presence of palladium(0), annulation processes requiring these species to react with one another are more challenging. Despite these difficulties, a palladium-catalyzed carboannulation of arynes by *o*-haloarenecarboxaldehydes **237** to generate fluoren-9-ones has been achieved by Zhang and Larock (Scheme 12.65) [129]. An arylpalladium(IV) intermediate **238** generated from oxidative cyclization of palladium(0) with the aryne and subsequent reaction with the aryl halide **237** may be involved in this catalytic cycle.

This palladium-catalyzed annulation has also been extended to arynes and 2-halobiaryls to synthesize fused polycyclic aromatic hydrocarbons (Equation 12.62) [130].

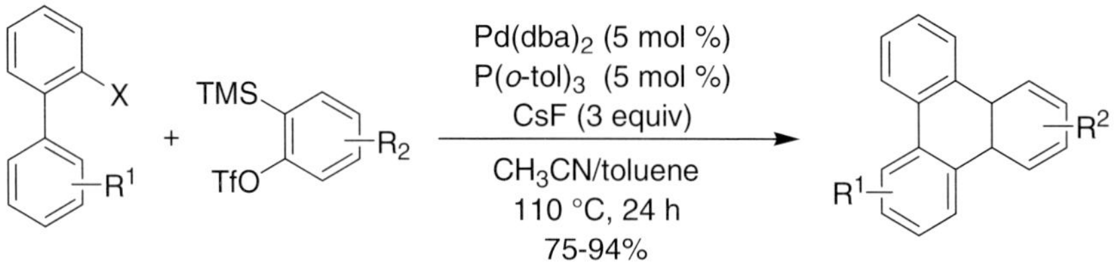

$Pd(dba)_2$ (5 mol %)
P(*o*-tol)$_3$ (5 mol %)
CsF (3 equiv)
CH_3CN/toluene
110 °C, 24 h
75-94%

X = I, Br

(12.62)

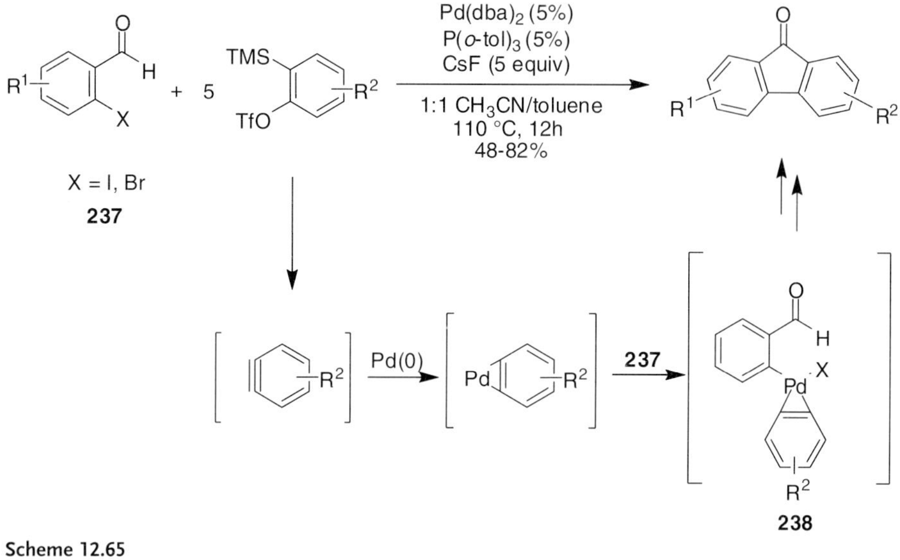

Scheme 12.65

The reaction conditions are compatible with a variety of biaryls bearing substituents with different electronic properties. In addition, a number of heterocyclic biaryls, including an indole, a benzofuran and a chromone, as well as alkenyl halides, have also been used successfully in this process, affording excellent yields of the corresponding polycyclic aromatics. However, alkenyl halides bearing either an alkyl group or a hydrogen atom on the olefin double bond, alkenyl triflates, 2-chlorobiphenyl, as well as sterically demanding substrates such as 2-iodo-2′-methylbiphenyl, have so far been unsuccessful in this annulation chemistry.

Multisubstituted triphenylenes **239** have been successfully synthesized by the palladium-catalyzed double annulation of arynes with simple aryl halides in the presence of the bidentate phosphine ligand dppf (Equation 12.63) [130b]. Aryl halides bearing electron-withdrawing groups afford higher yields than those with electron-donating groups. Interestingly, aryl bromides – which usually are considered to be less reactive than the corresponding iodides – afford better results.

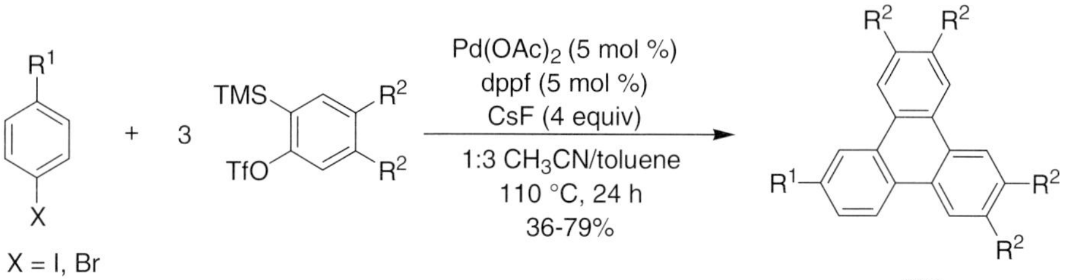

(12.63)

A process similar to this synthesis of substituted triphenylene derivatives has also been reported by Jayanth and Cheng (Equation 12.64) [131]. The palladium-catalyzed carbocyclization of arynes with aryl iodides takes place in the absence of a phosphine ligand. On the other hand, the addition of 1.2 equiv. of TlOAc as an additive was found necessary in order to achieve reliable yields under the reaction conditions. Such conditions, however, are unsuitable for aryl iodides bearing electron-donating groups and arynes substituted with electron-withdrawing groups. These protocols are useful alternatives to the palladium-catalyzed [2+2+2] cocyclotrimerization reactions of arynes, especially for constructing unsymmetrical and functionally substituted triphenylene derivatives.

$Pd(dba)_2$ (5 mol %)
CsF (4 equiv)
TlOAc (1.2 equiv)
CH_3CN, 85 °C, 8 h
72-83%

(12.64)

When one molecule of aryne precursor is replaced with a bicyclic alkene in the previous process, the carbocyclization of an aryl iodide, a bicyclic alkene and an aryne takes place, giving exclusively the *exo* isomer of an annulated 9,10-dihydrophenanthrene **240** (Equation 12.65) [132].

$Pd(dba)_2$ (5 mol %)
$P(2\text{-furyl})_3$ (20 mol %)
CsF (4 equiv)
CH_3CN, RT, 10 h
59-92%

X = CH_2, O

240

(12.65)

This palladium-catalyzed, three-component carbocyclization process is facilitated by addition of the π-acidic monodentate phosphine ligand $P(2\text{-furyl})_3$, with isolated yields of up to 92% being obtained in the presence of $Pd(dba)_2/P(2\text{-furyl})_3$. A wide range of functional groups on the aryl iodides are compatible with the reaction conditions, although aryl iodides bearing electron-withdrawing substituents afford higher yields. The oxabenzonorbornadiene carbocyclization products **241** undergo deoxyaromatization upon treatment with $BF_3 \cdot OEt_2$, yielding the potentially useful electroluminescent and photoluminescent polyaromatic hydrocarbons **242** (Equation 12.66).

(12.66)

Alkynes are also suitable components in this type of cross-coupling reaction. When the bicyclic alkene is replaced with one molecule of an alkyne, a three-component cross-coupling of aryl halides, arynes and alkynes takes place, affording substituted phenanthrenes (Equation 12.67) [133]. Similar to Cheng's results [131], the addition of TlOAc was found crucial for obtaining high yields but, interestingly, the absence of any phosphine ligand afforded a higher yield. The regiochemistry of the coupling product indicates that the reaction proceeds by carbopalladation of the internal alkyne by the arylpalladium species generated from oxidative addition of the aryl iodide to palladium(0), followed by subsequent carbopalladation of the aryne.

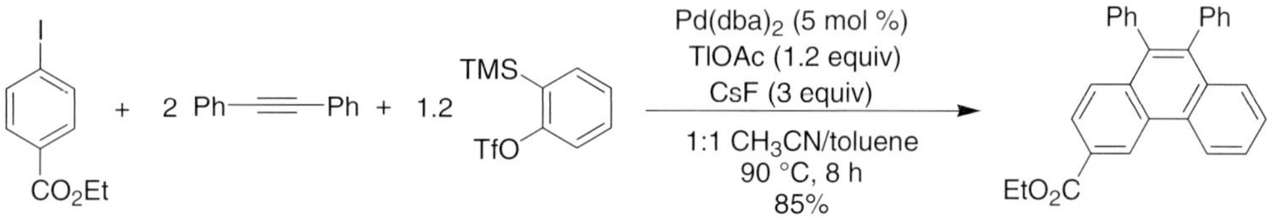

(12.67)

12.5.1.3 Transition Metal-Catalyzed Carbonylations

It has been shown that carbon monoxide can insert into a nickel–aryne bond in a stoichiometric manner [134]. The catalytic carbonylation reaction of arynes induced by transition metals was first reported in 2001 (Equation 12.68) [135], and cobalt carbonyl complexes have since been found to be the catalysts of choice in this transformation. For example, anthraquinone was obtained in 82% yield in the presence of 2 mol% of $Co_4(CO)_{12}$.

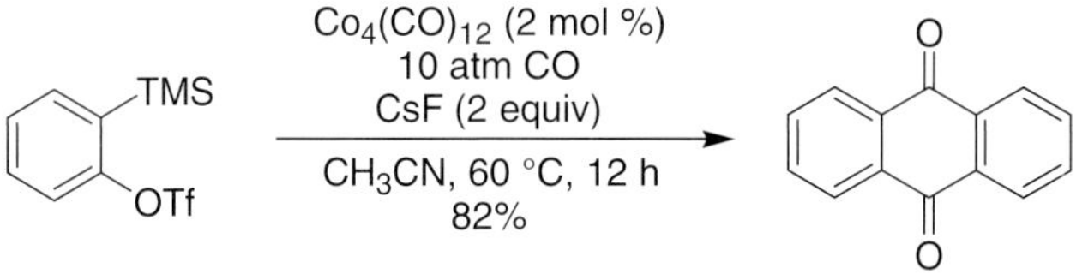

(12.68)

Interestingly, when the complex $[(\pi\text{-}C_3H_5)PdCl]_2$ is used in conjunction with allyl acetates, a different carbonylation process occurs. A three-component reaction of benzyne, allyl acetate and CO leads to the cyclized product 2-methyleneindanone **243** in 80% yield (Equation 12.69) [135].

TMS, OTf + 1.5 OAc; $[(\pi\text{-}C_3H_5)PdCl]_2$ (2.5 mol %), dppe (5 mol %), 1 atm CO, CsF (2 equiv), CH_3CN, 80 °C, 4 h, 80% → **243** (12.69)

12.5.1.4 Miscellaneous

Substituted fluorenes **244** and **245** have been synthesized by a $Pd(OAc)_2$-catalyzed domino reaction of 1-chloro-2-haloarenes or 2-haloaryl tosylates with hindered Grignard reagents (Scheme 12.66) [136]. A palladium(II)-associated aryne species **246** was proposed as the key intermediate, which subsequently undergoes transmetallation with the Grignard reagent, carbopalladation and cross-coupling to afford the substituted fluorenes in a one-pot reaction. The undesired formation of the cross-coupled product **247** involving two molecules of Grignard reagent was suppressed under the reaction conditions described.

A [4+2] benzannulation between acetylenic ketones **248** and a benzenediazonium 2-carboxylate proceeds effectively in the presence of a catalytic amount of AuCl, yielding functionalized anthracenes **250** in good yields (Scheme 12.67) [137]. It is suggested that the reaction involves a reverse electron demand Diels–Alder reaction between benzyne and the benzopyrylium aurate complex **249**.

Wender *et al.* recently reported a rhodium(I)-catalyzed [3+2] cycloaddition of diphenylcyclopropenone with benzyne, affording indenone **251** in 69% yield (Equation 12.70) [138]. The reaction was run in a sealed vial and the benzyne was prepared from the *in situ* reduction of 1,2-diiodobenzene by activated Zn.

O, Ph, Ph + I, I; $[RhCl(CO)_2]_2$ (3 mol%), Zn, EtOH, 120 °C, 69% → O, Ph, Ph **251** (12.70)

12.5.2 Transition Metal-Catalyzed Coupling Reactions

12.5.2.1 Insertion of Arynes into σ-Bonds

The first example of a transition metal-catalyzed carbometallation of an aryne was reported in 2001 [139]. Arynes were found to insert into the C–Sn bond of

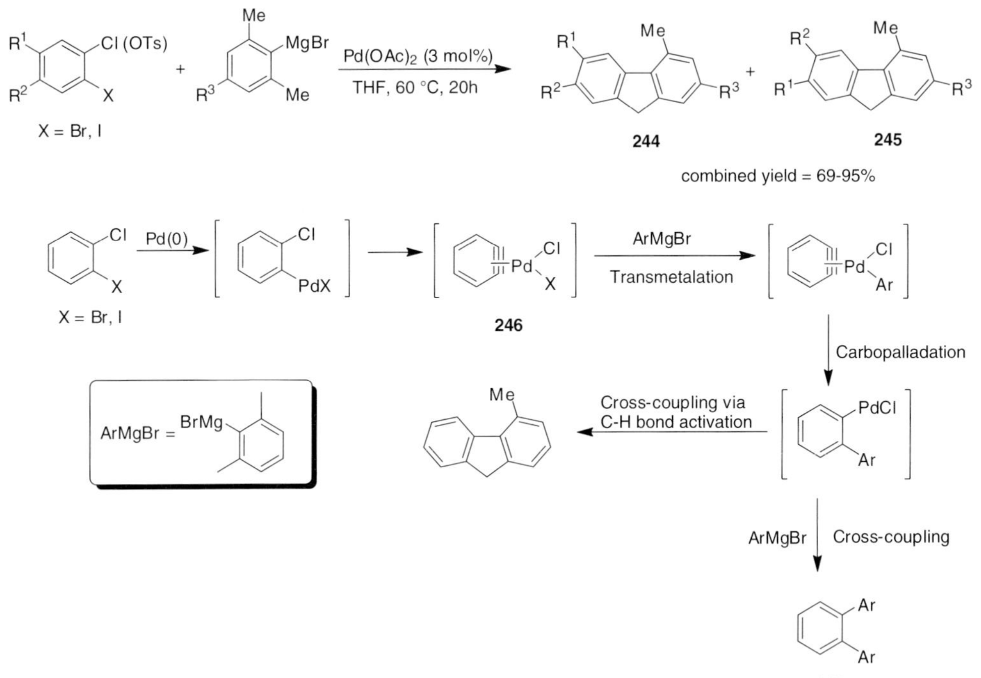

R1, Cl (OTs), R2, X + Me, MgBr, R3, Me — $Pd(OAc)_2$ (3 mol%), THF, 60 °C, 20h → **244** + **245**

X = Br, I

combined yield = 69-95%

Pd(0); **246**; ArMgBr, Transmetalation; Carbopalladation; Cross-coupling via C-H bond activation; ArMgBr, Cross-coupling; **247**

ArMgBr = BrMg–(2,6-dimethylphenyl)

Scheme 12.66

248 + $^{+}N_2$ / $^{-}O_2C$ — AuCl (10 mol %), $ClCH_2CH_2Cl$, 40 °C, 62-87% → **250**

AuCl; **249**

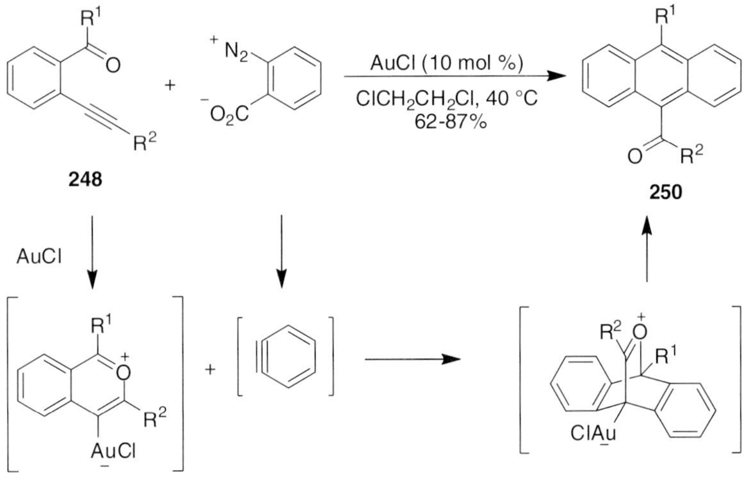

Scheme 12.67

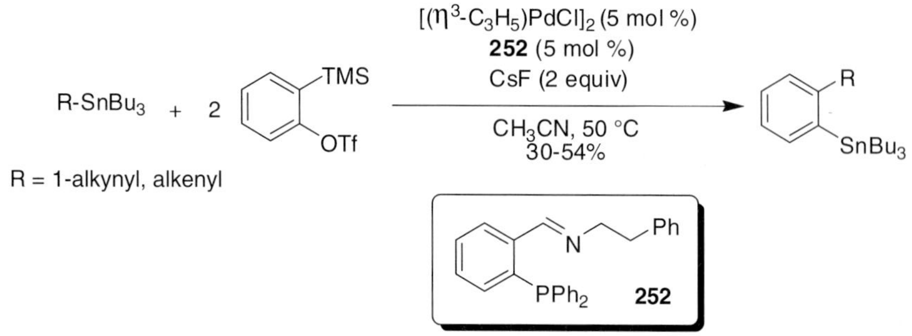

Scheme 12.68

1-alkynyl and alkenyl stannanes in the presence of a catalytic amount of a palladium–iminophosphine complex, leading to *ortho*-alkynyl/alkenyl arylstannanes (Scheme 12.68). These coupling products are excellent substrates for further elaboration to a variety of 1,2-disubstituted arenes using known coupling protocols.

The Si—Si σ-bond of cyclic disilanes is known to add to arynes in the presence of a palladium-1,1,3,3-tetramethylbutyl isocyanide (*t*-OctNC) complex, which is an excellent catalyst for bissilylations of C–C triple and double bonds, yielding benzoannulated disilacarbocycles [140]. Both, five-membered ring (Equation 12.71) and benzo-condensed six-membered ring cyclic disilanes (Equation 12.72) can be employed in this bissilylation, leading to seven- and eight-membered ring disilacarbocycles, respectively. In contrast, none of the desired insertion products was obtained when acyclic disilanes, nonbenzo-condensed six-membered ring cyclic disilanes, or four- or seven-membered ring cyclic disilanes were employed.

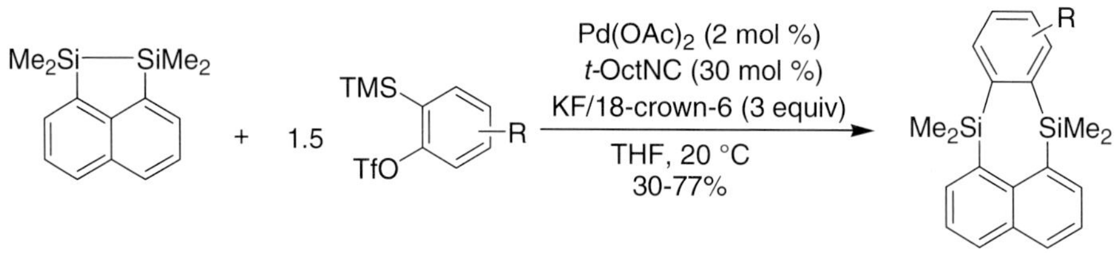

(12.71)

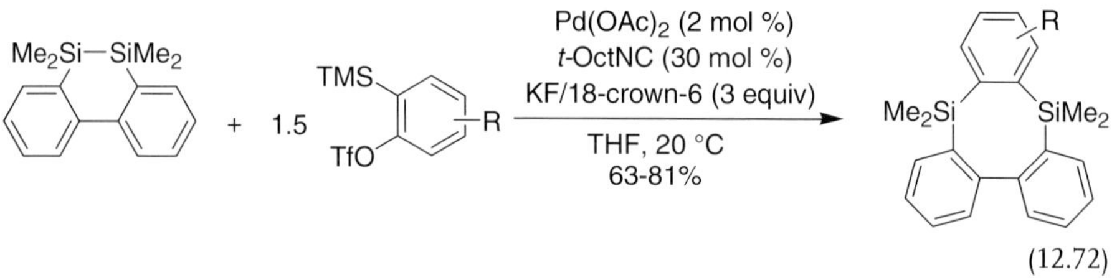

(12.72)

Under the same reaction conditions, the Sn–Sn σ-bond of a distannane also readily adds to the triple bond of arynes, yielding diverse 1,2-distannylarenes which have the potential to act as bis-anion equivalents (Equation 12.73) [141]. Moreover, highly strained cyclohexynes undergo this Sn–Sn σ-bond insertion smoothly under the reaction conditions in yields up to 90%.

R_3Sn-SnR_3 + 1.5 TMS / TfO / R'

$Pd(OAc)_2$ (2 mol %)
t-OctNC (30 mol %)
KF/18-crown-6 (3 equiv)
THF, 20 °C
26-73%

R_3Sn / R_3Sn / R'

(12.73)

A significant ligand effect was observed in the palladium-catalyzed distannylation of arynes (Equation 12.74) [142]. While the reaction with *t*-OctNC yielded **254** as the major product, the insertion of two molar equivalents of aryne into the Sn–Sn σ-bond predominated in the presence of the bicyclic phosphite ETPO (4-ethyl-2,6,7-trioxa-1-phosphabicyclo[2.2.2]octane). 2,2′-Bis(trimethylstannyl)biaryls **253** were thus generated alongside 1,2-bis(trimethylstannyl)benzenes **254**; such 2,2′-distannylbiaryls **253** are synthetically useful compounds and can be readily converted to various substituted biaryls.

3 R / TMS / OTf + Me_3Sn-$SnMe_3$

$Pd(OAc)_2$ (2 mol %)
ETPO (10 mol %)
KF/18-crown-6 (6 equiv)
THF, 20 °C

R / R / $SnMe_3$ / $SnMe_3$ + R / $SnMe_3$ / $SnMe_3$

253 16-62%
254 4-29%

(12.74)

12.5.2.2 Three-Component Coupling of Arynes Involving Carbopalladation

Under mild reaction conditions, arynes react with bis-π-allylpalladium complexes leading to 1,2-diallylated benzene derivatives in moderate to good yields (Equation 12.75) [143].

R^1 / R^2 / TMS / OTf + $SnBu_3$ + Cl

$Pd_2(dba)_3 \cdot CHCl_3$ (2.5 mol %)
dppf (5 mol %)
CsF (2 equiv)
CH_3CN, 40 °C, 12 h
40-81%

R^1 / R^2

(12.75)

The carbopalladation of an aryne with a bis-π-allylpalladium complex was shown to be a key step in the catalytic cycle. This protocol was, however, limited due to a lack of chemoselectivity since, if different allyl groups from allyl chloride and an allylstannane were employed, then a mixture of crossover products was obtained.

If the allylstannane was replaced with other organostannane reagents, such as alkynylstannanes and allenylstannanes, different three-component coupling reactions involving both arynes and allylic chlorides may take place smoothly in one pot, resulting in the formation of two different C–C bonds at the *ortho* positions of the arene ring [144]. Alkynylstannanes afford substituted 1-allyl-2-(1-alkynyl)benzenes in good yields (Equation 12.76), while allenylstannanes produce substituted 1-allyl-2-allenylbenzenes in yields greater than 79% (Equation 12.77) [145]. In both processes, the reactions are initiated by carbopalladation of the aryne by a π-allylpalladium chloride, followed by an intermolecular Stille coupling with the organostannane.

Pd(dba)$_2$ (5 mol %), dppe (5 mol %), CsF (3 equiv); CH$_3$CN, 40 °C, 10 h; 62-93%

(12.76)

Pd(dba)$_2$ (5 mol %), dppe (5 mol %), CsF (3 equiv); CH$_3$CN, RT, 10 h; 79-93%

(12.77)

Based on the above reaction above, it is possible predictably to assemble new multicomponent reactions by simply switching one or two of the reaction components. For example, the Stille coupling step can be replaced with a Suzuki coupling by employing an arylboronic acid. Thus, a three-component coupling of arynes, allylic chlorides and arylboronic acids is possible, leading to *o*-allyl biaryls (Equation 12.78) [146]. However, the reaction conditions need to be slightly modified, including changing the ligand dppe to dppb and increasing the amount of CsF to 4 equiv., in order to achieve optimal results.

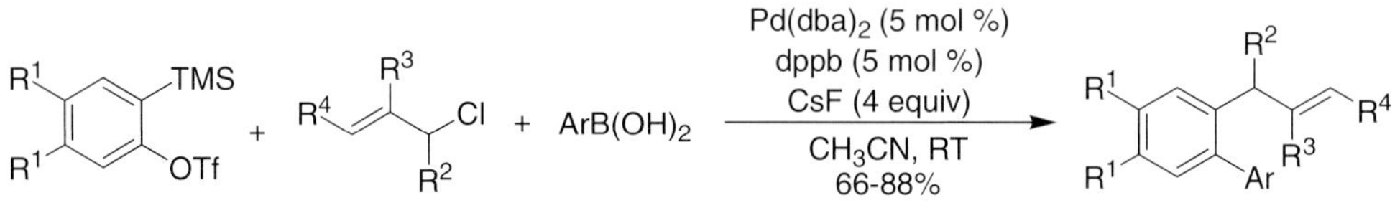

(12.78)

A novel three-component coupling of an aryne, a benzylic bromide, and *tert*-butyl acrylate, based on two successive intermolecular carbopalladation reactions, has recently been reported by Greaney and coworkers (Equation 12.79) [147a]. This methodology provides an alternative to the previous examples by replacing allylic chlorides with benzylic bromides as the initial carbopalladation electrophile, along with the introduction of a Heck reaction into the three-component coupling process. An application of this protocol was demonstrated by a three-step synthesis of the prostenoid EP_3 receptor antagonist **255** (Figure 12.5) in an overall yield of 61%.

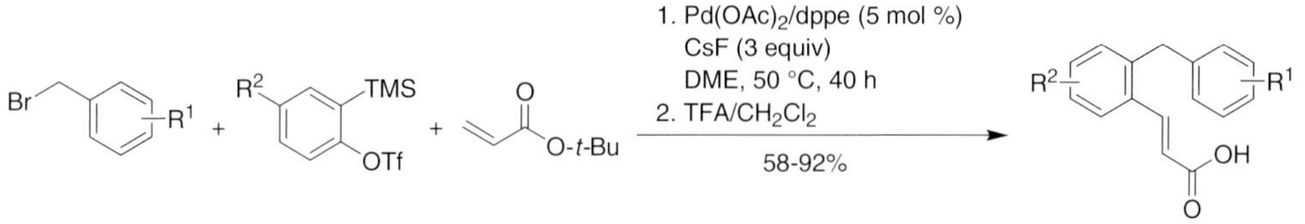

(12.79)

When the catalyst $Pd(OAc)_2$-dppe is replaced by $PdCl_2(dppf)$, methyl bromoacetate can take the place of the benzylic bromide and effectively participate in this coupling reaction (Equation 12.80) [147a].

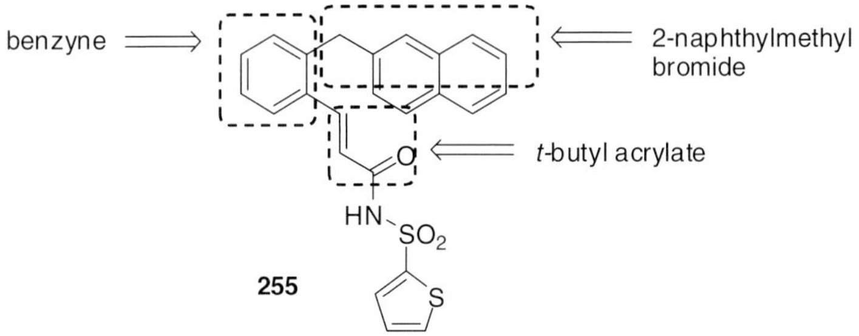

Figure 12.5 Prostenoid EP3 receptor antagonist **255**.

(12.80)

When aryl iodides rather than benzyl bromides are employed as the electrophile in this three-component coupling protocol, a straight Heck coupling between the aryl iodide and the Heck acceptor takes place as the major reaction, presumably because the arylpalladium(II) species formed is far more reactive than the analogous benzylpalladium(II) species and undergoes the two-component Heck coupling more readily. However, this undesired side reaction can be suppressed by using the phosphine ligand $P(o\text{-tol})_3$ and a 1.5 : 1 : 2 molar ratio of the starting materials (Equation 12.81) [147b].

(12.81)

12.6 Summary

In this chapter, we have discussed the exciting new developments in aryne chemistry applied to synthetic organic chemistry during the past seven-and-a-half years. As a highly reactive chemical species and a powerful synthetic tool, arynes have attracted considerable attention from synthetic organic chemists and demonstrated great promise in the construction of a variety of naturally occurring, biologically active, synthetically valuable or theoretically interesting aromatic compounds. The development of new mild reaction conditions for the generation of arynes has significantly increased their chemical compatibility and broadened their application. As a result, aryne chemistry has now been extended to insertion into element–element σ-bonds, multicomponent reactions and transition metal-catalyzed multicomponent reactions, as well as cyclotrimerizations and cocyclotrimerizations.

Abbreviations

BINAP 2,2′-bis(diphenylphosphino)-1,1′-binaphthyl
BHT 2,6-di-*tert*-butyl-4-methylphenol
dcpe 1,2-bis(dicyclohexylphosphino)ethane

DDQ	2,3-dichloro-5,6-dicyano-1,4-benzoquinone
diglyme	bis(2-methoxyethyl)ether
DMAD	dimethyl acetylenedicarboxylate
DMF	*N,N*-dimethylformamide
dppb	1,4-bis(diphenylphosphino)butane
dppe	1,2-bis(diphenylphosphino)ethane
dppf	1,1′-bis(diphenylphosphino)ferrocene
ETPO	4-ethyl-2,6,7-trioxa-1-phosphabicyclo[2.2.2]octane
KHMDS	potassium bis(trimethylsilyl)amide
LDA	lithium diisopropylamide
LiTMP	lithium 2,2,6,6-tetramethylpiperidide
LUMO	lowest unoccupied molecular orbital
NIS	*N*-iodosuccinimide
PCBN	4-chlorobenzoyl nitrite
py	pyridine
RT	room temperature
TBAF	tetrabutylammonium fluoride
TBAT	tetra-*n*-butylammonium triphenyldifluorosilicate
TBDPS	*tert*-butyldiphenylsilyl
TBS	*tert*-butyldimethylsilyl
TFA	trifluoroacetic acid
Tf	trifluoromethanesulfonyl
THF	tetrahydrofuran
TMS	trimethylsilyl

References

1 Wittig, G. (1942) *Naturwissenschaften*, **30**, 696–703.

2 (a) Roberts, J.D., Simmons, H.E., Jr, Carlsmith, L.A. and Vaughan, C.W. (1953) *J. Am. Chem. Soc.*, **75**, 3290–1;
(b) Roberts, J.D., Semenow, D.A., Simmons, H.E., Jr and Carlsmith, L.A. (1956) *J. Am. Chem. Soc.*, **78**, 601–11.

3 Jones, R.R. and Bergman, R.G. (1972) *J. Am. Chem. Soc.*, **94**, 660–1.

4 (a) Pellissier, H. and Santelli, M. (2003) *Tetrahedron*, **59**, 701–30;
(b) Wenk, H.H., Winkler, M. and Sander, W. (2003) *Angew. Chem. Int. Ed.*, **42**, 502–28;
(c) Dyke, A.M., Hester, A.J. and Lloyd-Jones, G.C. (2006) *Synthesis*, 4093–112;
(d) Kessar, S.V. (1991) *Comprehensive Organic Synthesis*, Vol. 4 (eds B.M. Trost and I. Fleming), Pergamon Press, Oxford, chap. 2.3, pp. 483–515.

5 (a) Himeshima, Y., Sonoda, T. and Kobayashi, H. (1983) *Chem. Lett.*, 1211–14;
(b) Peña, D., Cobas, A., Pérez, D. and Guitián, E. (2002) *Synthesis*, 1454–8.

6 (a) Kitamura, T. and Yamane, M. (1995) *J. Chem. Soc. Chem. Commun.*, 983–4;
(b) Kitamura, T., Yamane, M., Inoue, K., Todaka, M., Fukatsu, N., Meng, Z. and Fujiwara, Y. (1999) *J. Am. Chem. Soc.*, **121**, 11674–9.

7 Baigrie, B., Cadogan, J.I.G., Mitchell, J.R., Robertson, A.K. and Sharp, J.T. (1972) *J. Chem. Soc. Perkin Trans. 1*, 2563–7.

8 Birkett, M.A., Knight, D.W. and Mitchell, M.B. (1994) *Synlett*, 253–4.

9 For a review on the insertion of arynes into σ-bonds, see: Peña, D., Pérez, D. and

Guitián, E. (2006) *Angew. Chem. Int. Ed.*, **45**, 3579–81.

10 Beller, M., Breindl, C., Riermeier, T.H. and Tillack, A. (2001) *J. Org. Chem.*, **66**, 1403–12.

11 (a) Liu, Z. and Larock, R.C. (2003) *Org. Lett.*, **5**, 4673–5;
(b) Liu, Z. and Larock, R.C. (2006) *J. Org. Chem.*, **71**, 3198–209.

12 Leclerc, J.-P., André, M. and Fagnou, K. (2006) *J. Org. Chem.*, **71**, 1711–14.

13 Liu, Z. and Larock, R.C. (2004) *Org. Lett.*, **6**, 99–102.

14 Yoshida, H., Sugiura, S. and Kunai, A. (2002) *Org. Lett.*, **4**, 2767–9.

15 Ramtohul, Y.K. and Chartrand, A. (2007) *Org. Lett.*, **9**, 1029–32.

16 Knight, D.W. and Little, P.B. (2001) *J. Chem. Soc. Perkin Trans.*, **1**, 1771–7.

17 Birkett, M.A., Knight, D.W., Little, P.B. and Mitchell, M.B. (2000) *Tetrahedron*, **56**, 1013–23.

18 Dixon, S., Wang, X., Lam, K.S. and Kurth, M.J. (2005) *Tetrahedron Lett.*, **46**, 7443–6.

19 Vázquez, R., de la Fuente, M.C., Castedo, L. and Domínguez, D. (1994) *Synlett*, 433–4.

20 Hoarau, C., Couture, A., Cornet, H., Deniau, E. and Grandclaudon, P. (2001) *J. Org. Chem.*, **66**, 8064–9.

21 Moreau, A., Couture, A., Deniau, E., Grandclaudon, P. and Lebrun, S. (2004) *Tetrahedron*, **60**, 6169–76.

22 Yoshida, H., Terayama, T., Ohshita, J. and Kunai, A. (2004) *Chem. Commun.*, 1980–1.

23 Yoshida, H., Minabe, T., Ohshita, J. and Kunai, A. (2005) *Chem. Commun.*, 3454–6.

24 Yoshida, H., Shirakawa, E., Honda, Y. and Hiyama, T. (2002) *Angew. Chem. Int. Ed.*, **41**, 3247–9.

25 Liu, Z. and Larock, R.C. (2005) *J. Am. Chem. Soc.*, **127**, 13112–13.

26 Tandel, S., Zhang, H., Qadri, N., Ford, G.P. and Biehl, E.R. (2000) *J. Chem. Soc. Perkin Trans.* **1**, 587–9.

27 Tandel, S., Wang, A., Zhang, H., Yousuf, P. and Biehl, E.R. (2000) *J. Chem. Soc. Perkin Trans.* **1**, 3149–53.

28 Tripathy, S., Hussain, H. and Durst, T. (2000) *Tetrahedron Lett.*, **41**, 8401–5.

29 Yoshida, H., Watanabe, M., Morishita, T., Ohshita, J. and Kunai, A. (2007) *Chem. Commun.*, 1505–7.

30 Guyot, M. and Molho, D. (1973) *Tetrahedron Lett.*, **14**, 3433–6.

31 Shair, M.D., Yoon, T.Y., Mosny, K.K., Chou, T.C. and Danishefsky, S.J. (1996) *J. Am. Chem. Soc.*, **118**, 9509–25.

32 Kita, Y., Higuchi, K., Yoshida, Y., Iio, K., Kitagaki, S., Ueda, K., Akai, S. and Fujioka, H. (2001) *J. Am. Chem. Soc.*, **123**, 3214–22.

33 Bauta, W.E., Lovett, D.P., Cantrell, W.R., Jr and Burke, B.D. (2003) *J. Org. Chem.*, **68**, 5967–73.

34 Seconi, G., Taddei, M., Eaborn, C. and Stamper, J.G. (1982) *J. Chem. Soc. Perkin Trans.* **2**, 643–6.

35 Tambar, U.K. and Stoltz, B.M. (2005) *J. Am. Chem. Soc.*, **127**, 5340–1.

36 Yoshida, H., Watanabe, M., Ohshita, J. and Kunai, A. (2005) *Chem. Commun.*, 3292–4.

37 Yoshida, H., Watanabe, M., Ohshita, J. and Kunai, A. (2005) *Tetrahedron Lett.*, **46**, 6729–31.

38 Yoshida, H., Fukushima, H., Ohshita, J. and Kunai, A. (2004) *Angew. Chem. Int. Ed.*, **43**, 3935–8.

39 Yoshida, H., Fukushima, H., Ohshita, J. and Kunai, A. (2004) *Tetrahedron Lett.*, **45**, 8659–62.

40 Yoshida, H., Fukushima, H., Morishita, T., Ohshita, J. and Kunai, A. (2007) *Tetrahedron*, **63**, 4793–805.

41 Yoshida, H., Fukushima, H., Ohshita, J. and Kunai, A. (2006) *J. Am. Chem. Soc.*, **128**, 11040–1.

42 Jeganmohan, M. and Cheng, C.-H. (2006) *Chem. Commun.*, 2454–6.

43 (a) Tomori, H., Fox, J.M. and Buchwald, S.L. (2000) *J. Org. Chem.*, **65**, 5334–41;
(b) Parrish, C. and Buchwald, S.L. (2001) *J. Org. Chem.*, **66**, 3820–7;
(c) Kaye, S., Fox, J.M., Hicks, F.A. and Buchwald, S.L. (2001) *Adv. Synth. Catal.*, **343**, 789–94.

44 Leroux, F. and Schlosser, M. (2002) *Angew. Chem. Int. Ed.*, **41**, 4272–4.

45 Lin, W., Sapountzis, I. and Knochel, P. (2005) *Angew. Chem. Int. Ed.*, **44**, 4258–61.

46 Lin, W., Ilgen, F. and Knochel, P. (2006) *Tetrahedron Lett.*, **47**, 1941–4.

47 Larrosa, I., Da Silva, M.I., Gómez, P.M., Hannen, P., Ko, E., Lenger, S.R., Linke, S.R., White, A.J.P., Wilton, D. and Barrett, A.G.M. (2006) *J. Am. Chem. Soc.*, **128**, 14042–3.

48 Huang, X. and Xue, J. (2007) *J. Org. Chem.*, **72**, 3965–8.

49 Rayabarapu, D.K., Majumdar, K.K., Sambaiah, T. and Cheng, C.-H. (2001) *J. Org. Chem.*, **66**, 3646–9.

50 (a) Zhao, J. and Larock, R.C. (2005) *Org. Lett.*, **7**, 4273–5;
(b) Zhao, J. and Larock, R.C. (2007) *J. Org. Chem.*, **72**, 583–8.

51 Yoon, K., Ha, S.M. and Kim, K. (2005) *J. Org. Chem.*, **70**, 5741–4.

52 (a) Kaiser, F., Schwink, L., Velder, J. and Schmalz, H.-G. (2002) *J. Org. Chem.*, **67**, 9248–56;
(b) Kaiser, F., Schwink, L., Velder, J. and Schmalz, H.-G. (2003) *Tetrahedron*, **59**, 3201–17.

53 Mal, D., Senapati, B.K. and Pahari, P. (2005) *Synlett*, 994–6.

54 Sanz, R., Fernández, Y., Castroviejo, M.P., Pérez, A. and Fañanás, F.J. (2006) *J. Org. Chem*, **71**, 6291–4.

55 Barluenga, J., Fañanás, F.J., Sanz, R. and Fernández, Y. (2002) *Chem. Eur. J*, **8**, 2034–46.

56 Sanz, R., Fernández, Y., Castroviejo, M.P., Pérez, A. and Fañanás, F.J. (2007) *Eur. J. Org. Chem.*, 62–9.

57 Pawlas, J. and Begtrup, M. (2002) *Org. Lett.*, **4**, 2687–90.

58 Kawabata, H., Nishino, T., Nishiyama, Y. and Sonoda, N. (2002) *Tetrahedron Lett.* **43**, 4911–13.

59 Sapountzis, I., Lin, W., Fischer, M. and Knochel, P. (2004) *Angew. Chem. Int. Ed.*, **43**, 4364–6.

60 Kitamura, T., Aoki, Y., Isshiki, S., Wasai, K. and Fujiwara, Y. (2006) *Tetrahedron Lett.*, **47**, 1709–12.

61 Bailly, F., Cottet, F., Schlosser, M. and S. (2005), 791–7.

62 Schlosser, M. and Castagnetti, E. (2001) *Eur. J. Org. Chem.*, 3991–7.

63 Masson, E. and Schlosser, M. (2005) *Eur. J. Org. Chem.*, 4401–5.

64 Biland-Thommen, A.S., Raju, G.S., Blagg, J., White, A.J.P. and Barrett, A. G.M. (2004) *Tetrahedron Lett.*, **45**, 3181–4.

65 Morton, G.E. and Barrett, A.G.M. (2005) *J. Org. Chem.*, **70**, 3525–9.

66 Morton, G.E. and Barrett, A.G.M. (2006) *Org. Lett.*, **8**, 2859–61.

67 Sygula, A., Sygula, R. and Rabideau, P.W. (2005) *Org. Lett.*, **7**, 4999–5001.

68 (a) Kaelin, D.E.Jr, , Lopez, O.D. and Martin, S.F. (2001) *J. Am. Chem. Soc.*, **123**, 6937–8;
(b) Kaelin, D.E.Jr, , Sparks, S.M., Plake, H.R. and Martin, S.F. (2003) *J. Am. Chem. Soc.*, **125**, 12994–5.

69 (a) Apsel, B., Bender, J.A., Escobar, M., Kaelin, D.E.Jr, , Lopez, O.D. and Martin, S.F. (2003) *Tetrahedron Lett.*, **44**, 1075–7;
(b) Martin, S.F. (2003) *Pure Appl. Chem*, **75**, 63–70.

70 Chen, C.-L., Sparks, S.M. and Martin, S.F. (2006) *J. Am. Chem. Soc.*, **128**, 13696–7.

71 Sörgel, S., Azap, C. and Reißig, H.-U. (2006) *Eur. J. Org. Chem.*, 4405–18.

72 Caster, K.C., Keck, C.G. and Walls, R.D. (2001) *J. Org. Chem.*, **66**, 2932–6.

73 Romero, C., Peña, D., Pérez, D. and Guitián, E. (2006) *Chem. Eur. J.*, **12**, 5677–84.

74 Chan, S.-H., Yick, C.-Y. and Wong, H. N.C. (2002) *Tetrahedron*, **58**, 9413–22. ,

75 Compain, C., Donnadieu, B. and Mathey, F. (2005) *Organometallics*, **24**, 1762–5.

76 Wood, T.K., Piers, W.E., Keay, B.A. and Parvez, M. (2006) *Org. Lett.*, **8**, 2875–8.

77 Fuchs, E., Keller, M. and Breit, B. (2006) *Chem. Eur. J.*, **12**, 6930–9.

78 Xie, C. and Zhang, Y. (2007) *Org. Lett.*, **9**, 781–4.

79 Schuster, I.I., Craciun, L., Ho, D.M. and Pascal, R.A., Jr (2002) *Tetrahedron*, **58**, 8875–82.

80 Duong, H.M., Bendikov, M., Steiger, D., Zhang, Q., Sonmez, G., Yamada, J. and Wudl, F. (2003) *Org. Lett.*, **5**, 4433–6.

81 Coe, J.W., Wirtz, M.C., Bashore, C.G. and Candler, J. (2004) *Org. Lett.*, **6**, 1589–92.

82 (a) Chen, Y.-L., Zhang, H.-K., Wong, W.-Y. and Lee, A.W.M. (2002) *Tetrahedron Lett.*, **43**, 2259–62;
(b) Chen, Y.-L., Sun, J.-Q., Wei, X., Wong, W.-Y. and Lee, A.W.M. (2004) *J. Org. Chem.*, **69**, 7190–7.

83 (a) Westcott, S.A., Kakkar, A.K., Stringer, G., Taylor, N.J. and Marder, T.B. (1990) *J. Organomet. Chem.*, **394**, 777–94;

(b) Wang, B., Xu, S. and Zhou, X. (1997) *J. Organomet. Chem.*, **540**, 101–4.
84 Wang, B., Mu, B., Chen, D., Xu, S. and Zhou, X. (2004) *Organometallics*, **23**, 6225–30.
85 Rybáčková, M., Bělohradský, M., Holý, P., Pohl, R. and Závada, J. (2006) *Synthesis*, 2039–42.
86 Wang, D.Z., Katz, T.J., Golen, J. and Rheingold, A.L. (2004) *J. Org. Chem.*, **69**, 7769–71.
87 Battiste, M.A., Duan, J.-X., Zhai, Y.-A., Ghiviriga, I., Abboud, K.A. and Dolbier, W.R. Jr (2003) *J. Org. Chem.*, **68**, 3078–83.
88 Dolbier, W.R., Jr, Zhai, Y.-A., Wheelus, W., Battiste, M.A., Ghiviriga, I. and Bartberger, M.D. (2007) *J. Org. Chem.*, **72**, 550–8.
89 Real, M.M., Pérez Sestelo, J. and Sarandeses, L.A. (2002) *Tetrahedron Lett.*, **43**, 9111–14.
90 Benzyne was generated by thermal decomposition of benzenediazonium 2-carboxylate. For a reference, see: Logullo, F.M., Seitz, A.H. and Friedman, L. (1973) *Org. Synth. Coll.*, **5**, 54–9.
91 Dockendorff, C., Sahli, S., Olsen, M., Milhau, L. and Lautens, M. (2005) *J. Am. Chem. Soc.*, **127**, 15028–9.
92 Thasana, N., Pisutjaroenpong, S. and Ruchirawat, S. (2006) *Synlett*, 1080–4.
93 Hayes, M.E., Shinokubo, H. and Danheiser, R.L. (2005) *Org. Lett.*, **7**, 3917–20.
94 Stevens, R.V. and Bisacchi, G.S. (1982) *J. Org. Chem.*, **47**, 2393–6.
95 Maurin, P., Ibrahim-Ouali, M. and Santelli, M. (2001) *Tetrahedron Lett.*, **42**, 8147–9.
96 Hamura, T., Hosoya, T., Yamaguchi, H., Kuriyama, Y., Tanabe, M., Miyamoto, M., Yasui, Y., Matsumoto, T. and Suzuki, K. (2002) *Helv. Chim. Acta*, **85**, 3589–604.
97 Hamura, T., Ibusuki, Y., Sato, K., Matsumoto, T., Osamura, Y. and Suzuki, K. (2003) *Org. Lett.*, **5**, 3551–4.
98 Hamura, T., Ibusuki, Y., Uekusa, H., Matsumoto, T. and Suzuki, K. (2006) *J. Am. Chem. Soc.*, **128**, 3534–5.
99 Hamura, T., Ibusuki, Y., Uekusa, H., Matsumoto, T., Siegel, J.S., Baldridge, K.K. and Suzuki, K. (2006) *J. Am. Chem. Soc.*, **128**, 10032–3.
100 Hamura, T., Arisawa, T., Matsumoto, T. and Suzuki, K. (2006) *Angew. Chem. Int. Ed.*, **45**, 6842–4.
101 Yoshida, H., Watanabe, M., Fukushima, H., Ohshita, J. and Kunai, A. (2004) *Org. Lett.*, **6**, 4049–51.
102 Shoji, Y., Hari, Y. and Aoyama, T. (2004) *Tetrahedron Lett.*, **45**, 1769–71.
103 Jin, T. and Yamamoto, Y. (2007) *Angew. Chem. Int. Ed.*, **46**, 3323–5.
104 Liu, Z., Shi, F., Martinez, P.D.G., Raminelli, C. and Larock, R.C. (2008) *J. Org. Chem.*, **73**, 219–26.
105 Raminelli, C., Liu, Z. and Larock, R.C. (2006) *J. Org. Chem.*, **71**, 4689–91.
106 Jayanth, T.T., Jeganmohan, M., Cheng, M.-J., Chu, S.-Y. and Cheng, C.-H. (2006) *J. Am. Chem. Soc.*, **128**, 2232–3.
107 Baigrie, B., Cadogan, J.I.G., Mitchell, J.R., Robertson, A.K. and Sharp, J.T. (1972) *J. Chem. Soc. Perkin Trans.* **1**, 2563–7.
108 Cram, D.J. and Day, A.C. (1966) *J. Org. Chem.*, **31**, 1227–32.
109 Rao, U.N. and Biehl, E. (2002) *J. Org. Chem.*, **67**, 3409–11.
110 For the Pd-catalyzed cycloaddition reactions of arynes, see: Guitián, E., Pérez, D. and Peña, D. (2005) *Top. Organomet. Chem.*, **14**, 109–46.
111 (a) Bennett, M.A. and Wenger, E. (1996) *Organometallics*, **15**, 5536–41;
(b) Bennett, M.A. and Wenger, E. (1995) *Organometallics*, **14**, 1267–77.
112 Peña, D., Escudero, S., Pérez, D., Guitián, E. and Castedo, L. (1998) *Angew. Chem. Int. Ed.*, **37**, 2659–61.
113 Peña, D., Pérez, D., Guitián, E. and Castedo, L. (1999) *Org. Lett.*, **1**, 1555–7.
114 Iglesias, B., Cobas, A., Pérez, D., Guitián, E. and Vollhardt, K.P.C. (2004) *Org. Lett.*, **6**, 3557–60.
115 Peña, D., Pérez, D., Guitián, E. and Castedo, L. (1999) *J. Am. Chem. Soc.*, **121**, 5827–8.
116 Peña, D., Pérez, D., Guitián, E. and Castedo, L. (2000) *Synlett*, 1061–3.
117 Peña, D., Pérez, D., Guitián, E. and Castedo, L. (2000) *J. Org. Chem.*, **65**, 6944–50.
118 Radhakrishnan, K.V., Yoshikawa, E. and Yamamoto, Y. (1999) *Tetrahedron Lett.*, **40**, 7533–5.

119 Yoshikawa, E., Radhakrishnan, K.V. and Yamamoto, Y. (2000) *J. Am. Chem. Soc.*, **122**, 7280–6.

120 Peña, D., Pérez, D., Guitián, E. and Castedo, L. (2003) *Eur. J. Org. Chem.*, 1238–43.

121 (a) Sato, Y., Tamura, T. and Mori, M. (2004) *Angew. Chem. Int. Ed.*, **43**, 2436–40;
(b) Sato, Y., Tamura, T., Kinbara, A. and Mori, M. (2007) *Adv. Synth. Catal.*, **349**, 647–61.

122 Caeiro, J., Peña, D., Cobas, A., Pérez, D. and Guitián, E. (2006) *Adv. Synth. Catal.*, **348**, 2466–74.

123 Deaton, K.R. and Gin, M.S. (2003) *Org. Lett.*, **5**, 2477–80.

124 Hsieh, J.-C. and Cheng, C.-H. (2005) *Chem. Commun.*, 2459–61.

125 Hsieh, J.-C., Rayabarapu, D.K. and Cheng, C.-H. (2004) *Chem. Commun.*, 532–3.

126 Jayanth, T.T., Jeganmohan, M. and Cheng, C.-H. (2004) *J. Org. Chem.*, **69**, 8445–50.

127 Quintana, I., Boersma, A.J., Peña, D., Pérez, D. and Guitián, E. (2006) *Org. Lett.*, **8**, 3347–9.

128 Yoshikawa, E. and Yamamoto, Y. (2000) *Angew. Chem. Int. Ed.*, **39**, 173–5.

129 Zhang, X. and Larock, R.C. (2005) *Org. Lett.*, **7**, 3973–6.

130 (a) Liu, Z., Zhang, X. and Larock, R.C. (2005) *J. Am. Chem. Soc.*, **127**, 15716–17;
(b) Liu, Z. and Larock, R.C. (2007) *J. Org. Chem.*, **72**, 223–32.

131 Jayanth, T.T. and Cheng, C.-H. (2006) *Chem. Commun.*, 894–6.

132 Bhuvaneswari, S., Jeganmohan, M. and Cheng, C.-H. (2006) *Org. Lett.*, **8**, 5581–4.

133 Liu, Z. and Larock, R.C. (2007) *Angew. Chem. Int. Ed.*, **46**, 2535–8.

134 Bennett, M.A., Hockless, D.C.R., Humphrey, M.G., Schultz, M. and Wenger, E. (1996) *Organometallics*, **15**, 928–33.

135 Chatani, N., Kamitani, A., Oshita, M., Fukumoto, Y. and Murai, S. (2001) *J. Am. Chem. Soc.*, **123**, 12686–7.

136 Dong, C.-G. and Hu, Q.-S. (2006) *Org. Lett.*, **8**, 5057–60.

137 Asao, N. and Sato, K. (2006) *Org. Lett.*, **8**, 5361–3.

138 Wender, P.A., Paxton, T.J. and Williams, T.J. (2006) *J. Am. Chem. Soc.*, **128**, 14814–5.

139 Yoshida, H., Honda, Y., Shirakawa, E. and Hiyama, T. (2001) *Chem. Commun.*, 1880–1.

140 (a) Yoshida, H., Ikadai, J., Shudo, M., Ohshita, J. and Kunai, A. (2003) *J. Am. Chem. Soc.*, **125**, 6638–9;
(b) Yoshida, H., Ikadai, J., Shudo, M., Ohshita, J. and Kunai, A. (2005) *Organometallics*, **24**, 156–62.

141 Yoshida, H., Tanino, K., Ohshita, J. and Kunai, A. (2004) *Angew. Chem. Int. Ed.*, **43**, 5052–5.

142 Yoshida, H., Tanino, K., Ohshita, J. and Kunai, A. (2005) *Chem. Commun.*, 5678–80.

143 Yoshikawa, E., Radhakrishnan, K.V. and Yamamoto, Y. (2000) *Tetrahedron Lett.*, **41**, 729–31.

144 Jeganmohan, M. and Cheng, C.-H. (2004) *Org. Lett.*, **6**, 2821–4.

145 Jeganmohan, M. and Cheng, C.-H. (2005) *Synthesis*, 1693–7.

146 Jayanth, T.T., Jeganmohan, M. and Cheng, C.-H. (2005) *Org. Lett.*, **7**, 2921–4.

147 (a) Henderson, J.L., Edwards, A.S. and Greaney, M.F. (2006) *J. Am. Chem. Soc.*, **128**, 7426–7;
(b) Henderson, J.L., Edwards, A.S. and Greaney, M.F. (2007) *Org. Lett.*, **9**, 5589–92.

13
Radical-Based Arylation Methods

Santiago E. Vaillard, Birte Schulte and Armido Studer

13.1
Introduction

Most radical reactions can be conducted under mild conditions. In contrast to ionic reactions and many transition metal-mediated processes, most of the functional groups are tolerated under radical conditions. Moreover, radical reactions can be performed in various solvents; even water is tolerated as a reaction medium. These facts, among others, make radical processes highly useful for arylations. Radical arylations can be performed using $S_{RN}1$-type reactions, by homolytic aromatic substitutions, and by reactions of aryl radicals with various radical acceptors. In this chapter we first focus on $S_{RN}1$-type reactions [1], and later concentrate on homolytic aromatic substitutions. Unfortunately, due to limitations of space, we cannot provide a comprehensive overview on this topic; hence, for further information the reader is referred to some excellent reviews on this issue [2].

13.2
$S_{RN}1$-Type Radical Arylations

Since the pioneering studies of Bunnett [3], the scope of the unimolecular radical nucleophilic substitution ($S_{RN}1$) reaction has increased considerably, and today this approach is well established for the formation of aryl–carbon and aryl–heteroatom bonds. The $S_{RN}1$ reaction is a chain process which includes radicals and radical anions as intermediates; the reaction mechanism is depicted in Scheme 13.1 [1].

Electron transfer (ET) to the radical precursor, such as an aryl halide, from a suitable electron source (step 1 in Scheme 13.1) provides a radical anion which subsequently fragments to the corresponding aryl radical and X^- (step 2). The coupling of the aryl radical with a nucleophile then generates a radical anion which acts as an electron-transfer reagent finally providing the arylation product after ET to the radical precursor (steps 3 and 4).

Modern Arylation Methods. Edited by Lutz Ackermann

ISBN: 978-3-527-31937-4

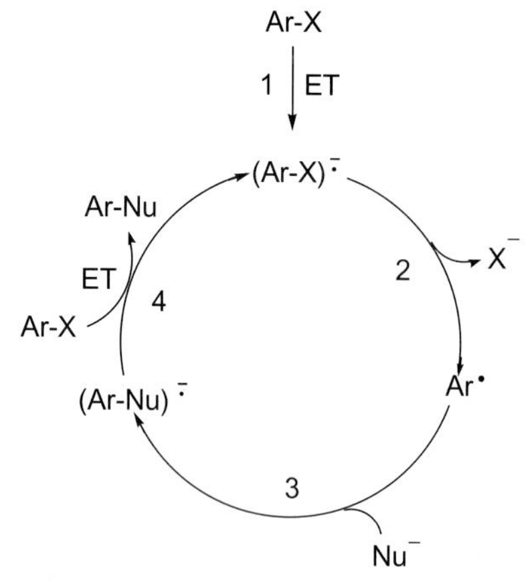

Scheme 13.1 Mechanism of the $S_{RN}1$ arylation.

Although some spontaneous or thermal $S_{RN}1$ reactions are known, in most of the cases initiation is necessary to trigger the process [4]. The most general method to initiate the chain reaction is the photoinduced ET from the nucleophile to the substrate (for a detailed discussion, see Chapter 14) [1]. Other known methods of promoting $S_{RN}1$ reactions are the use of: (i) alkali metals in liquid ammonia; (ii) ferrous salts; or (iii) electrochemistry.

Substituted arenes and heteroarenes bearing a suitable leaving group are good substrates for $S_{RN}1$ reactions. Halides are used as leaving groups in most of the cases, although other leaving groups, such as $(EtO)_2P(O)O^-$, RS^-, $PhSO^-$, $PhSO_2^-$, $PhSe^-$, PhS_2, N_2, and NMe_3, can also be employed.

Various carbon- and heteroatom-centered nucleophiles have been successfully used in the $S_{RN}1$ arylation. Hence, carbanions readily obtained by deprotonation can be employed as nucleophiles. The reaction with phenolates leads to the formation of carbon–carbon bonds (*ortho*-arylation) [5], while aryl–heteroatom bond formation occurs when using Sn, P, As, Sb, S, Se and Te-centered anions. $S_{RN}1$ reactions are not sensitive to steric hindrance, and good yields are usually obtained for *ortho*-substituted aryl radicals [6]. Many functional groups such as OR, SAr, CF_3, NH_2, SO_2R and CN are tolerated; however, the nitro group as aryl substituent leads to chain termination [7].

13.2.1
Intermolecular $S_{RN}1$ Reactions

Enolates derived from acyclic and cyclic aliphatic ketones can easily be arylated by the $S_{RN}1$ process. These reactions are usually conducted in dimethylsulfoxide (DMSO) or liquid ammonia as solvent. As an example, the photoinduced arylation of the enolate derived from 2-acetyl-*N*-methyl pyrrole with iodobenzene in

DMSO affording the corresponding phenylated product is depicted in Equation 13.1 [8].

(13.1)

The arylation and heteroarylation of various ketone enolates can also be achieved in DMSO in the presence of $FeCl_2$ or $FeBr_2$ as chain initiators. Moreover, it has been shown that SmI_2 promotes the reaction of $^{-}CH_2COPh$ with Ar-I in DMSO [9, 10].

Biologically important heteroarenes are readily prepared by the reaction of ketone, ester or amide enolates with *ortho*-substituted aryl halides [11]. The *ortho*-substituent can subsequently be used for further synthetic manipulations. For example, the reaction of *o*-iodo or *o*-bromoaniline with enolates generated from aliphatic or aromatic ketones under $S_{RN}1$ conditions provides 2,3-disubstituted indoles in moderate to excellent yields (Equation 13.2) [12, 13]. 2,3-Disubstituted isoquinolin-1-ones (Equation 13.3) and isoquinolines are readily prepared using *o*-iodobenzamide and *o*-iodobenzylamine as radical precursors by the same approach [14].

(13.2)

X = I, Br
R^1 = alkyl, aryl, heteroaryl
R^2 = alkyl, aryl, heteroaryl

(13.3)

The *ortho*-arylation reaction of phenolates with aryldiazonium sulfides as radical precursors represents an efficient process for the synthesis of benzopyranones and related compounds (Equation 13.4). The intermediately formed $S_{RN}1$ products are readily transformed *in situ* to the corresponding benzopyranones under mild acidic reaction conditions [15].

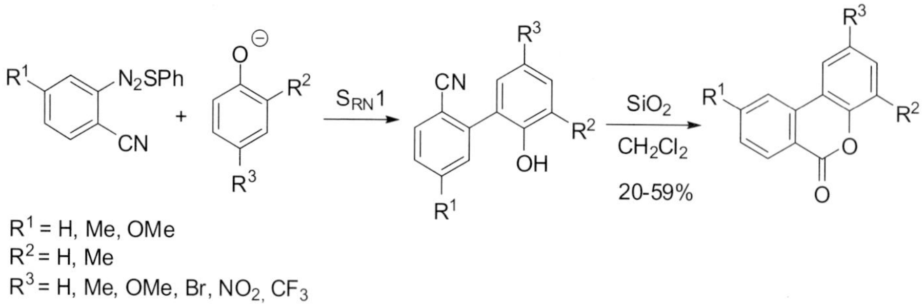

R^1 = H, Me, OMe
R^2 = H, Me
R^3 = H, Me, OMe, Br, NO_2, CF_3

(13.4)

Many heteroatom-centered nucleophiles can be arylated by the $S_{RN}1$ mechanism. For example, Ar–Cl and $^+NMe_3Ar$ react with Me_3Sn^- in liquid ammonia to provide the corresponding stannylated arenes in high yields [16, 17]. Di- and tri-stannylated aromatic compounds can be prepared using di- and tri-chlorobenzenes as radical precursors [16]. The fact that Ar–Cl reacts easily with Me_3Sn^- by $S_{RN}1$-type chemistry and that Ar–I bonds are much more reactive than Ar–Cl bonds in the Stille reaction allowed for the development of an attractive iterative approach to the synthesis of poly-aryl compounds, as shown in Scheme 13.2.

$(EtO)_2P^-$ and Ph_2P^- are good nucleophiles in aromatic $S_{RN}1$ reactions and react with aryl halides to yield the substitution products with acceptable to excellent yields. Moreover, Ar–I reacts efficiently with PhS^-, substituted thiophenolates and

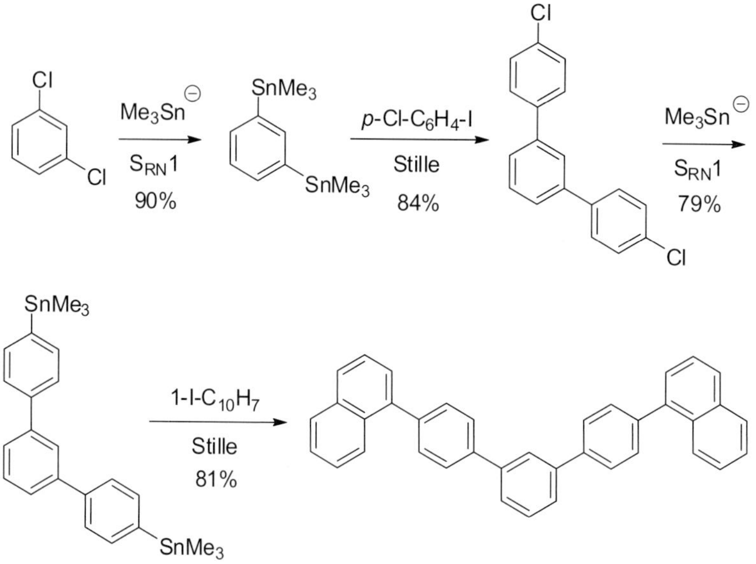

Scheme 13.2 Iterative $S_{RN}1$ and Stille-type arylations.

heteroaryl thiolates under photoinitiation in liquid ammonia to give diarylsulfides [18]. Recently, deprotonated thiourea has been successfully used as the *S*-nucleophile in $S_{RN}1$ reactions, affording the corresponding *S*-arylated compounds which are readily transformed to aryl methyl sulfides, diaryl sulfides, diaryl disulfides and aryl thiols in good yields (50–80%) [19]. As mentioned in Section 13.1, efficient Se–aryl- and Te–aryl-bond formation can also be achieved via $S_{RN}1$ chemistry [20].

13.2.2
Intramolecular $S_{RN}1$ Reactions

Although the intramolecular $S_{RN}1$ reaction has been studied to a lesser extent as compared to the intermolecular process, it has been shown to represent a highly useful route for the synthesis of various heterocyclic compounds. For example, the $S_{RN}1$ cyclization of *N*-alkyl-*N*-acyl-*o*-haloanilines and *N*-acyl-*N*-methyl-*o*-chlorobenzylamines (Equation 13.5) gives *N*-alkylindol-2-one and 1,4-dihydro-*2H*-isoquinolin-3-ones in moderate to good yields [21]. Iodoketone **1** reacts under $S_{RN}1$ conditions to provide the complex heterocyclic compound **2** with excellent yield (Equation 13.6) [22].

Cl N R O (R = H, Me, Et) — $h\nu$, $^{\ominus}NH_2/NH_3$, 54-62% → R O N (13.5)

O O N I O OMe **1** — $h\nu$, NH_3, KO*t*Bu, 94% → O O H N O OMe **2** (13.6)

An intramolecular $S_{RN}1$ reaction was used as key step in the synthesis of the alkaloid *O*-demethyleupoularamine **3**. The intermediately formed $S_{RN}1$ product is readily transformed by photochemical oxidative biaryl formation and subsequent methylation to the target compound **3** in good yield (Equation 13.7) [23]. The $S_{RN}1$ cyclization of 1-(2-bromobenzyl)-1,2,3,4-tetrahydroisoquinolin-7-ol derivatives **4** was recently applied to the synthesis of aporphine alkaloids **5** (Equation 13.8) [24] and, by using the same approach, homoaporphine alkaloids can also be synthesized.

(13.7)

(13.8)

R = SO_2Ar, COMe, CO_2Me, Me, H
R^1= F, Cl, OMe, H
(n = 1,2)

13.3
Homolytic Aromatic Substitutions

13.3.1
Intramolecular Homolytic Aromatic Substitutions

13.3.1.1 Arylations Using Nucleophilic C-Centered Radicals

When Pschorr reported more than a century ago on the first intramolecular homolytic aromatic substitution [25], he showed that biaryls could be readily prepared by intramolecular homolytic aromatic substitution using reactive aryl radicals and arenes as radical acceptors. The aryl radicals were generated by treatment of arenediazonium salts with copper(I) ions. Today, this reaction and related processes are referred to as Pschorr reactions. It was later found that radical biaryl synthesis could be conducted without copper salts by photochemical or thermal generation of the aryl radical from the corresponding diazonium salt [26]. Moreover, the reduction of aryl diazonium salts offers another route to generate reactive aryl radicals. Hence, electrochemistry [27], titanium(III) ions [28], Fe(II)-salts [29], tetrathiafulvalene [30] and iodide [31] have each been used successfully for the reduction of diazonium salts to generate the corresponding aryl radicals [32]. As an example, the iodide-induced cyclization of diazonium salt **6** to phenanthrene derivative **7** is presented in Scheme 13.3 [31]. For further information on the

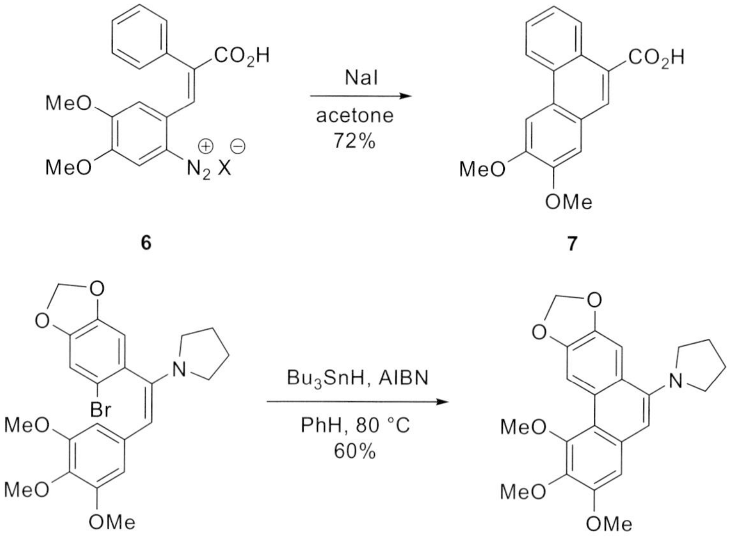

Scheme 13.3 Biaryl synthesis by homolytic aromatic substitution using reactive aryl radicals (AIBN = α,α′-azobisisobutyrodinitrile).

application of the Pschorr and related reactions in synthetic organic chemistry, the reader is referred elsewhere [26a].

Perhaps the most important contribution to radical biaryl synthesis was the discovery, made independently by Narasimhan [33] and Togo [34], which showed that aryl radicals generated from aryl bromides using tributyltin hydride, undergo intramolecular homolytic aromatic substitution reactions to form the corresponding biaryls. Hence, readily available and stable aryl bromides (and, of course, also aryl iodides) can be used as precursors in radical biaryl synthesis in combination with the ubiquitous Bu_3SnH as a reducing reagent. As an example, the pioneering tin hydride-mediated cyclization of aryl bromide **8** to give the phenanthrene derivative **9** is presented in Scheme 13.3 [33].

During the past years this method has been used successfully in modern biaryl synthesis, with several groups having shown its potential for $C(sp^2)$–$C(sp^2)$-bond formation [35]. Heteroarenes such as substituted pyrroles, indoles, pyridones and imidazoles can act as radical acceptors in these processes [36]. In addition, aryl bromides, chlorides and iodides have been used as substrates in electrochemically induced radical biaryl synthesis [37]. In a series of elegant studies, Harrowven recently applied the tin hydride-mediated radical biaryl synthesis to the preparation of helicenes [38]. These reactions comprise a double $C(sp^2)$–$C(sp^2)$-bond formation; an example – the impressive transformation of diiodide **10** to [7] helicene **11** – is presented in Scheme 13.4.

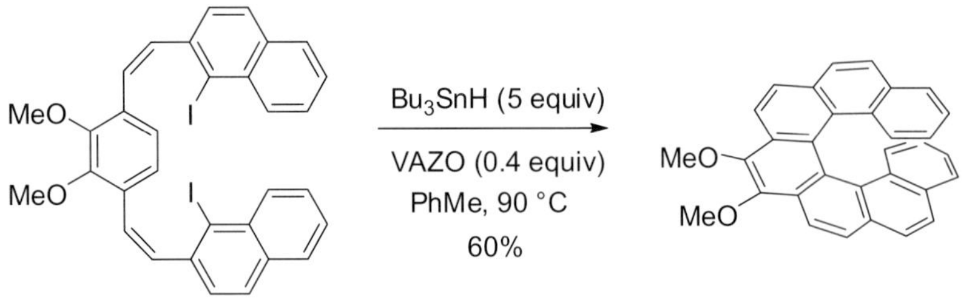

MeO
MeO
I
I
Bu_3SnH (5 equiv)
VAZO (0.4 equiv)
PhMe, 90 °C
60%
MeO
MeO
10
11

Scheme 13.4 Synthesis of a [7]helicene by double homolytic aromatic substitution (VAZO = 1,1′-azobis(cyclohexanecarbonitrile)).

The tin hydride-mediated C(sp^2)–C(sp^2)-bond formation has successfully been applied as key step in the synthesis of various natural products. *Amaryllidacaea* alkaloids [39], glaucine [40], cryptopleurine [41] and an anticancer benzo[*c*]phenanthridine alkaloid [42] were each successfully prepared using this approach.

Tin-mediated homolytic aromatic substitutions also occur with alkenyl radicals as reacting intermediates. For example, Curran introduced [4+1] annulations incorporating aromatic substitution reactions with alkenyl radicals to synthesize the core structure of various camptothecin derivatives [43]. For this, the alkenyl radicals can be readily generated from alkynes by radical addition reactions [44, 45]. For example, aryl radical **12**, generated from the corresponding iodide or bromide, was allowed to react with phenyl isonitrile to afford imidoyl radical **13**, which further reacts in a 5-*exo-dig* process to alkenyl radical **14** (Scheme 13.5) [43a, b]. The alkenyl radical **14** then reacts in a 1,6-cyclization followed by oxidation to the tetracycle **15**. There is some evidence [45] that the homolytic aromatic substitution can also occur by an initial *ipso* attack to afford the spiro radical **16**, followed by opening of this cyclohexadienyl radical to an iminyl radical **17**, reclosure, and finally oxidation. An interesting cascade reaction comprising an intramolecular alkenyl radical addition onto an arene was disclosed by Bowman [46]. The reaction of selenide **18** with tin hydride under radical conditions provided elipticine **19** in a cascade reaction, and in a good yield (Scheme 13.5).

Along with reactive aryl and alkenyl radicals, acyl radicals – readily generated from the corresponding acylselenides – undergo efficient homolytic aromatic substitutions in the presence of tin hydride or $(Me_3Si)_3SiH$ [47]. Arenes, as well as heteroarenes, can be used as acceptors. As an example, the highly efficient radical transformation of acylselenide **20** to heteroarene **21**, which readily isomerizes to phenol **22** is depicted in Scheme 13.6 [48].

Interestingly, the intramolecular radical addition onto arenes [49] and heteroarenes [50–54] can also be performed with primary and secondary alkyl radicals. As for the aryl radicals discussed above, the alkyl radicals used in these arylations are usually generated from the corresponding bromides or iodides using Bu_3SnH and a radical initiator. Bowman has carefully studied the intramolecular radical

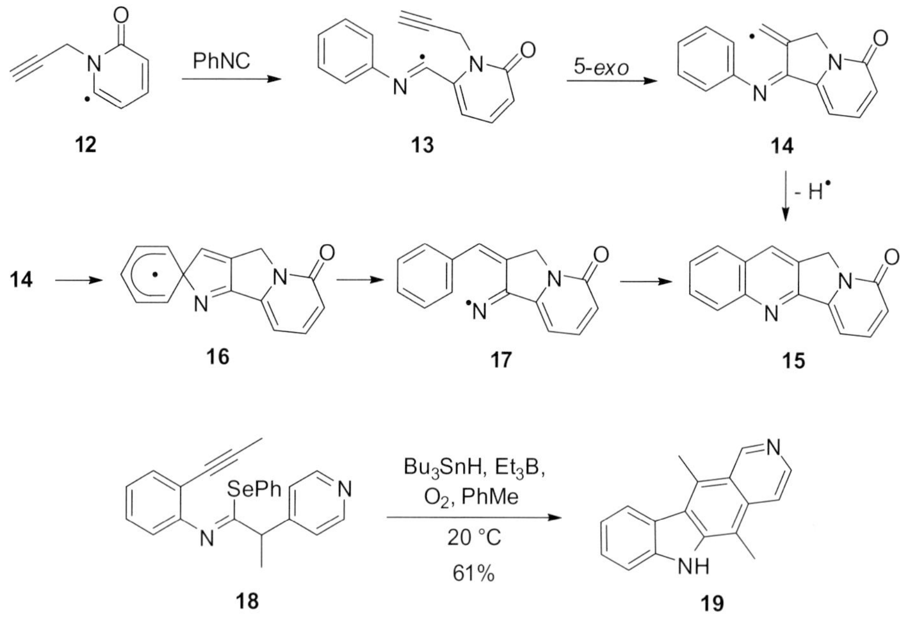

PhNC

5-*exo*

- H•

12 13 14 16 17 15

Bu_3SnH, Et_3B, O_2, PhMe

20 °C

61%

18 19

Scheme 13.5 Homolytic aromatic substitution using vinyl radicals.

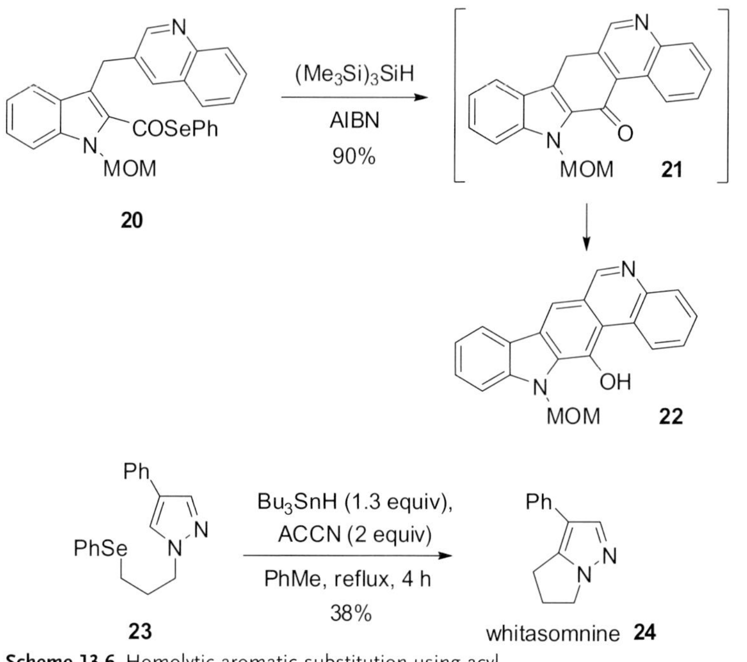

Scheme 13.6 Homolytic aromatic substitution using acyl and alkyl radicals (ACCN = 1,1′-(*Z*)-diazene-1, 2-diyldicyclohexanecarbonitrile).

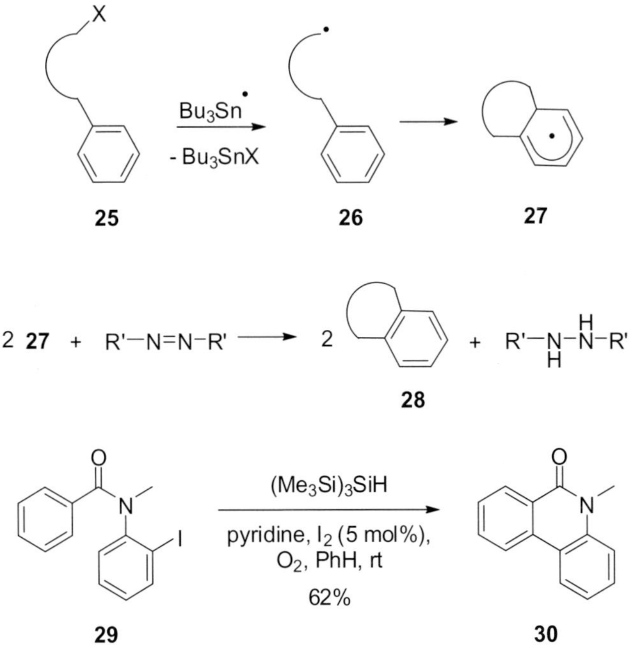

Scheme 13.7 Mechanism of the homolytic aromatic substitution under tin hydride/AIBN conditions. Dioxygen as oxidant of the cyclohexadienyl radical using $(Me_3Si)_3SiH$ as radical reducing reagent.

alkylation of differently substituted imidazoles and pyrroles [50], and achieved good results for imidazoles bearing electron-withdrawing substituents, such as the formyl or the methoxycarbonyl group. Phenyl groups at the pyrrole core also accelerate the addition reaction. As an example, whitasomnine **24** was successfully synthesized from selenide **23** using this approach [50b] (Scheme 13.6). Along with the activation of the heteroarene towards radical addition, these substituents further increase the regioselectivity of the homolytic substitution. Indoles have also been transformed to the corresponding tricyclic systems using the same method [51]. Moreover, pyridinium salts [52] and triazoles [53] were successfully alkylated under radical conditions with tin hydride as the mediator.

Furthermore, iminyl radicals have been used in intramolecular homolytic aromatic substitutions. The iminyl radicals can be generated *in situ* by *C*-radical addition onto nitriles [55] or they may be generated directly from the corresponding oximes [56] and hydrazones [57].

All of these tin hydride-mediated intramolecular arylations involve an initial radical generation by reaction of $Bu_3Sn\cdot$ with precursor **25** to afford the corresponding *C*-centered radical **26**, followed by an intramolecular addition onto an arene to give a stable cyclohexadienyl radical **27** which is then oxidized (!) to the product **28** (Scheme 13.7). It should be noted that this oxidation must occur under reducing tin hydride conditions. When debate centered on the mechanism of the homolytic aromatic substitution, a number of hypotheses was proposed. Thus, in a carefully devised study Beckwith, Bowman and Storey showed that the *pseudo*-$S_{RN}1$-type reaction often discussed in the literature was not compatible with their recent findings [58]. Rather, it was shown that the radical initiator (in most cases AIBN) was responsible for oxidation of the cyclohexadienyl radical **27**, as had been suggested initially by Curran [59]. Indeed, generally large amounts of initiator are necessary to obtain acceptable yields. Very recently, Curran also showed that oxidation of the cyclohexadienyl radical is best achieved using small amounts of dioxygen. Hence, those reactions generally conducted at high temperatures can also be realized at ambient temperatures, as was shown for the transformation of iodide **29** into lactam **30** [60]. Here, dioxygen acts as both the initiator and as the oxidant of the intermediate cyclohexadienyl radical. It is also likely that the hydroperoxyl radical formed after oxidation of the cyclohexadienyl radical can also further act as an oxidant to form HOOH. Of course, the same mechanism also operates for intermolecular homolytic substitutions (see Section 13.3.2).

Along with tin hydride-mediated arylations, electrochemical [61], photochemical [62] and iodine transfer methods [63] have each been successfully used for the generation of alkyl radicals in homolytic aromatic substitutions. Moreover, nucleophilic α-oxyalkyl radicals, prepared from the corresponding aldehydes by treatment with either samarium diiodide [64] or stannyl radicals [65], reacted with arenes to afford the homolytic substitution products. In a series of reports, Zard highlighted the potential of xanthates as alkyl radical precursors in homolytic aromatic substitutions [66].

13.3.1.2 Arylations Using Electrophilic C-Centered Radicals

Compared to the intramolecular aromatic alkylation with nucleophilic radicals, the analogous process with electrophilic radicals is far less common. Often, $Mn(OAc)_3$ is used to oxidatively generate the electrophilic radicals [67]. Citterio conducted a careful study of the $Mn(OAc)_3$-mediated intramolecular homolytic aromatic substitution of various dialkyl malonates [68, 69], and showed the reaction to be well suited for the formation of five- (see **31**), six- (see **32**) and seven-membered benzanellated rings (see **33**). The arylation with electrophilic *C*-radicals can also be performed on heteroarenes, as shown in the synthesis of **34** [70]. For cyclizations forming a six-membered ring, high yields were obtained in the alkylation of both electron-rich and electron-poor arenes. However, formation of the seven-membered ring occurred only with electron-rich arenes. Cerium(IV) ammonium nitrate [71] and iron(III) perchlorate [72] have also been used as oxidants in similar reactions.

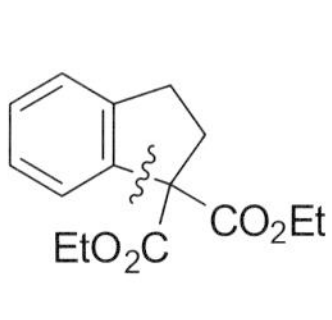

31 39% 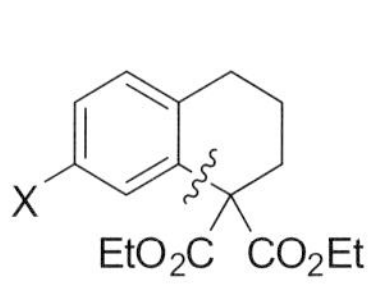 X = H, OMe, NO_2, NHCOMe **32** 80-88% **33** 70% 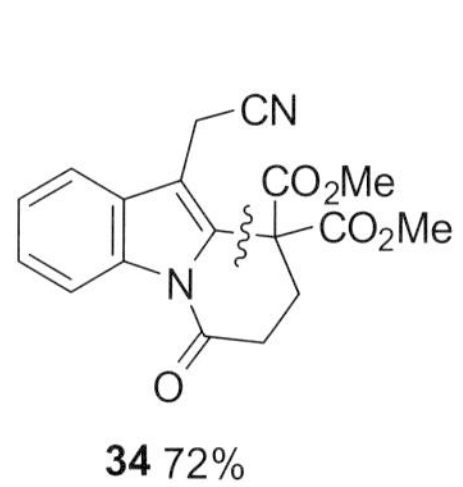 **34** 72%

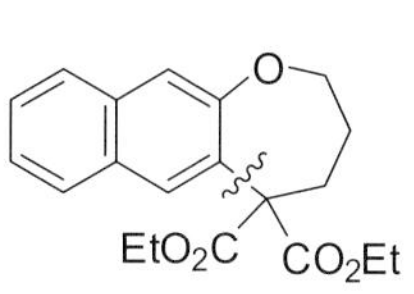

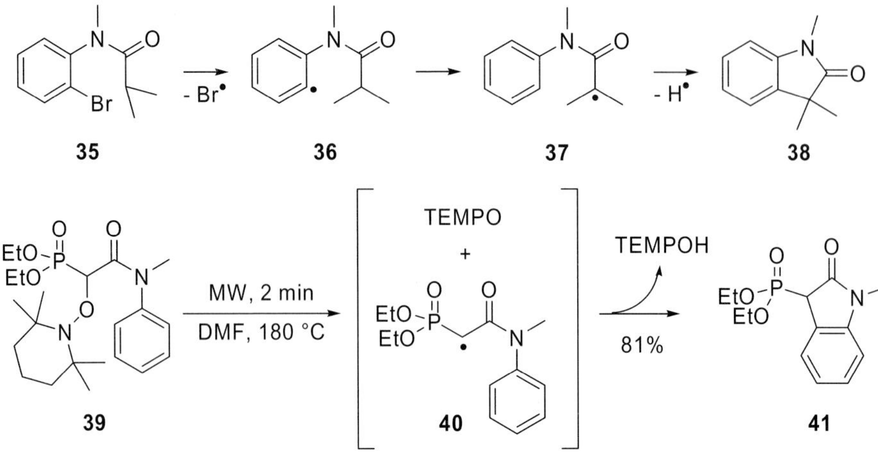

Scheme 13.8 1,5-H-Transfer for the generation of electrophilic C-radicals and TEMPO-mediated arylation.

Beckwith showed that the *N*-(*o*-bromophenyl)amide **35** could be transformed into the corresponding oxindole **38** (70%) at high temperatures, using Bu_3SnH via a tandem radical translocation of the initially formed aryl radical **36** to form **37** with subsequent intramolecular homolytic substitution (Scheme 13.8) [73]. Details of tin-free versions of this process have recently been published [74], where diethylphosphine oxide was used as a tin hydride substitute. Electrophilic *C*-radicals used in homolytic aromatic substitutions can also be generated directly from the corresponding selenides or bromides [75], or by radical addition reactions onto olefins bearing electron-withdrawing substituents [76]. Moreover, electrophilic *C*-radicals can also be generated via thermal C–O-bond homolysis of TEMPO-derived alkoxyamines and used in tin-free homolytic aromatic substitutions (TEMPO = 2,2,6,6-tetramethylpiperidine-1-oxyl radical) [77]. Hence, the heating of alkoxyamine **39** under microwave (MW) irradiation affords the corresponding *C*-radical **40** which then undergoes homolytic aromatic substitution to give oxindole **41**. It should be noted that, within these reactions, TEMPO acts as an oxidant (Scheme 13.8).

13.3.1.3 Radical Aryl Migration Reactions

In recent years, radical aryl migrations have received increased attention within the synthetic organic chemistry community. Yet, these reactions are also found as key steps in complex natural product synthesis [78]. For example, the neophyl rearrangement–which is the 1,2-phenyl migration of the neophyl radical **42** to form the tertiary radical **44** (probably via spirocyclohexadienyl radical **43**)–was discovered by Urry and Kharasch more than 60 years ago (Scheme 13.9) [79], since which time numerous reports on neophyl-type rearrangements have been presented [80]. However, despite these efforts the postulated intermediate **43** has not yet been identified [81]. The slow neophyl rearrangement ($k = 762\,s^{-1}$ at 25 °C, [82]) can be used as a radical clock [83]. The 1,2-aryl migration can also occur from *C*- to

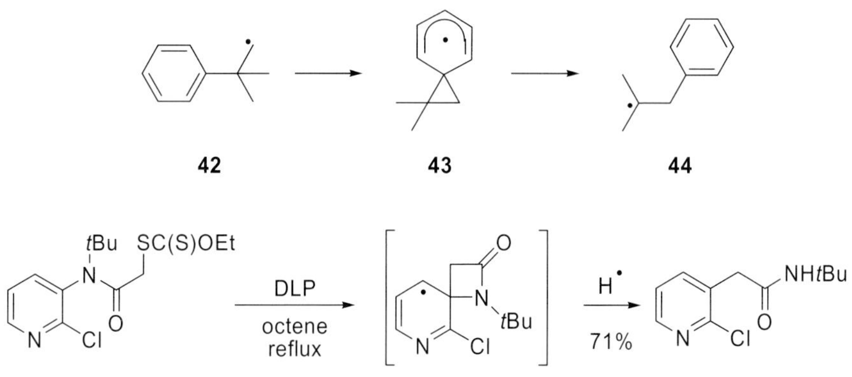

42 43 44

45 46 47

DLP octene reflux

H• 71%

Scheme 13.9 Radical 1,2- and 1,3-aryl migration reactions (DLP = dilauroyl peroxide).

O-centered radicals, and subsequently is referred to as the *O*-neophyl rearrangement [84].

Although a 1,3-aryl migration from carbon to *C*-centered radicals is unknown, Zard recently discovered the first radical 1,3-aryl migration from *N*- to *C*-centered radicals (radical Smiles rearrangement) [85]. The treatment of xanthogenate **45** with dilauroyl peroxide in refluxing octene provided the 1,3-aryl migration product **47** in 71% yield (the radical **46** is thought to be an intermediate in this interesting reaction). The first report on a 1,4-aryl migration between two carbon atoms appeared in 1956 [86], although in most cases the 1,4-aryl migration has only been observed as a side reaction. Nevertheless, some reports have been made on synthetically useful 1,4-aryl migrations [87, 88]. For example, Sherburn disclosed an interesting approach to podophyllotoxin which comprised a radical 1,4-aryl migration from oxygen to carbon [88a]. Reaction of the thionocarbonate **48** with a silyl radical generated the tertiary radical **49a** which subsequently cyclized in a 5-*exo* reaction to give **49b** (Scheme 13.10). Stereoselective 1,4-aryl migration leads to the alkoxyl radical **49c** which, after reduction and hydrolysis, eventually provides lactone **50**. The complex cascade process occurred in 40% yield. Radical 1,4-aryl migrations from sulfur to carbon are discussed below.

Whilst radical 1,5-aryl migration from carbon to carbon has been successfully used for the preparation of various biaryls [89], aryl migrations from heavier Group IV elements are also known. As an example, Wilt showed both the 1,4- and 1,5-phenyl migrations from silicon to carbon to be rather efficient processes [90], and also proved that the radical aryl migration between carbon and silicon could be reversible in the systems studied. Similar conclusions have been drawn by Sakurai and Hosomi, who investigated the 1,5-phenyl migration from carbon to silyl radicals [91]. Later, Studer investigated the stereoselective 1,5-aryl migration from silicon to secondary *C*-centered radicals [92]. As an example, the reaction of iodide **51** under radical reaction conditions provided the corresponding aryl migration product **52** in 70% yield with a 10 : 1 diastereoselectivity (Scheme 13.11). The cyclohexadienyl radical **53**, with the methyl substituents in *pseudo*-equatorial

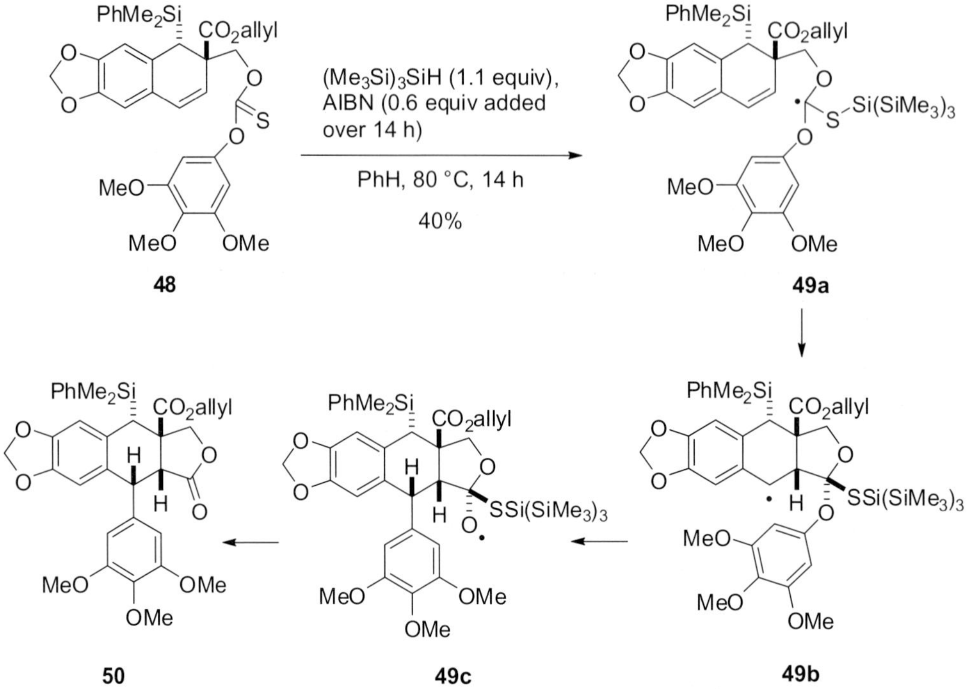

Scheme 13.10 A complex cascade reaction comprising a radical 1,4-aryl migration.

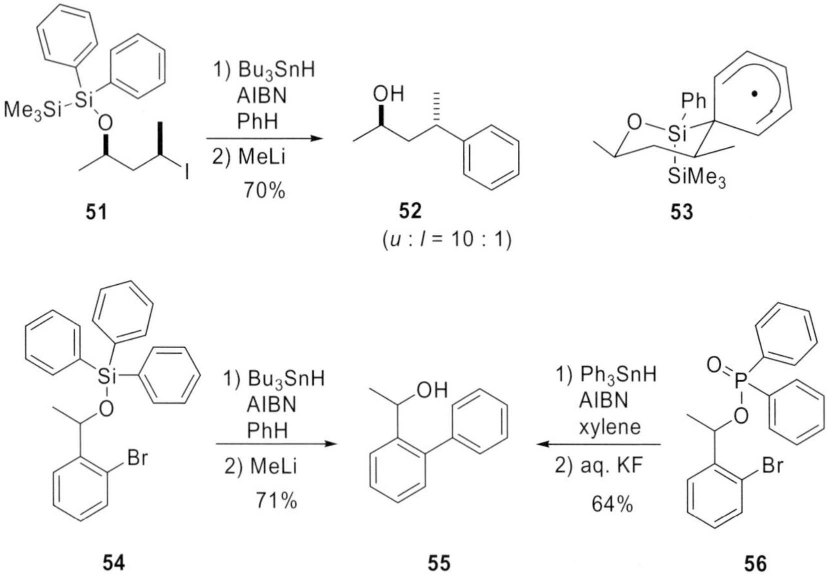

Scheme 13.11 Aryl migrations from silicon or phosphorus to carbon.

position, was suggested as an intermediate in the formation of the major isomer. Functionalized aryl groups can also be transferred by this method [93]. The analogous 1,4-aryl migration from silicon to secondary *C*-radicals was also reported to occur with excellent diastereoselectivity [93], and a radical aryl migration from silicon to carbon can also be used for the preparation of biaryls [94]. Thus, the transformation of bromide **54** under radical reaction conditions, followed by desilylation, afforded biaryl **55** in 71% yield. The same product could be obtained by aryl migration from phosphorus (see **56**) to the appropriate aryl radical [95].

The 1,4-aryl migration from sulfur in sulfonamides to *C*-centered radicals was first investigated by Speckamp during the 1970s [96], when iodides of type **57** were reacted with Bu_3SnH under radical conditions to afford the corresponding arylated products **59** (Scheme 13.12). In these processes, the initially formed primary alkyl radical was seen to react at the *ipso* position to form cyclohexadienyl radical **58**. Rearomatization, followed by reduction and SO_2 extrusion, finally led to the amine **59**. Both, electron-poor and electron-rich arenes can be transferred using this method, although unfortunately the homolytic substitution product **60**, deriving from the initial *ortho* attack, was formed as a side product. In a later study, Motherwell developed a new biaryl synthesis using both 1,4- and 1,5-aryl migrations from sulfur to aryl radicals [97a], and showed that readily available arenesulfonates, as well as arenesulfonamides, could be used as the starting materials in these reactions (**61** $\rightarrow$ **62**). As in the Speckamp studies with primary alkyl radicals, the aryl radicals **61** were shown to react also by *ortho* attack to afford the side product **63**; however, by judicious choice of the migrating aryl group it was possible to completely suppress the *ortho* attack [97d].

Studer used the radical aryl migration from sulfur in sulfonates to secondary *C*-radicals for the stereoselective $C(sp^2)$–$C(sp^3)$-bond formation [98]. In this way, various aryl groups were transferred with high selectivities and in high yields, as shown for the transformation of arylsulfonates **64** to the corresponding alcohols **65** (Scheme 13.12). The analogous aryl migrations in sulfonamides turned out to be far less efficient, however [99]. Furthermore, $C(sp^2)$–$C(sp^3)$-bond-forming processes using radical *ipso* substitutions not comprising aryl migrations were developed by Caddick [100] and Bowman [101]. Here, sulfone-substituted imidazoles and indoles were shown easily to undergo intramolecular *ipso* substitution with primary alkyl radicals and expulsion of the sulfonyl moiety. Even a methoxyl radical could act as a leaving group in these radical intramolecular *ipso*-substitution reactions [102].

13.3.2 Intermolecular Homolytic Aromatic Substitutions

13.3.2.1 Arylation with Nucleophilic C-Centered Radicals

The reaction of a nucleophilic alkyl radical $R^\bullet$ with benzene affords the σ-complex **66**, which may be readily oxidized to cation **67** (Scheme 13.13). The homolytic aromatic substitution product **68** is eventually formed by deprotonation. However, if the reaction is performed under nonoxidizing reaction conditions, the rather

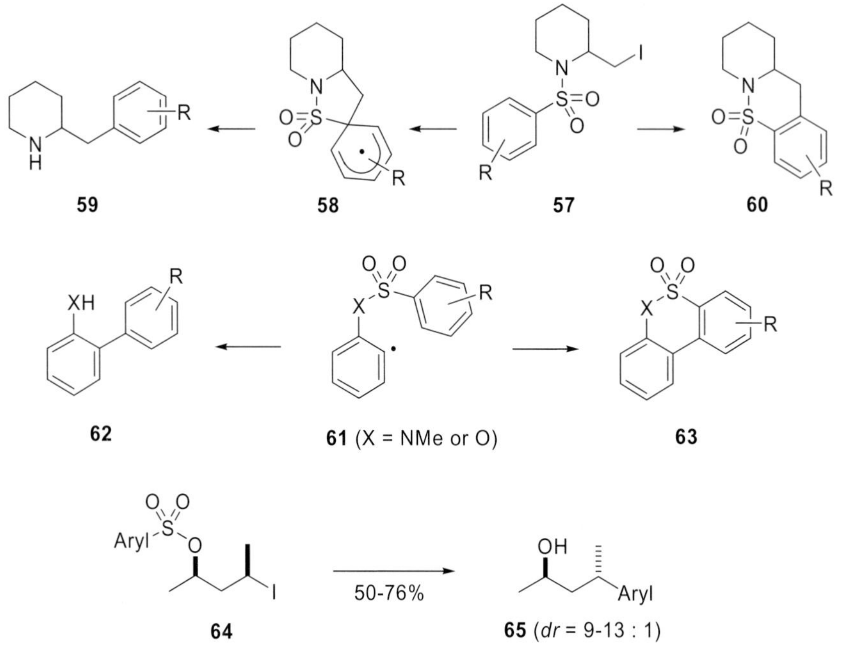

Aryl = C_6H_5, 4-F-C_6H_4, 4-MeO-C_6H_4, 2-thienyl, 5-Me_2N-$C_{10}H_6$

Scheme 13.12 Aryl migrations from sulfur to C-centered radicals.

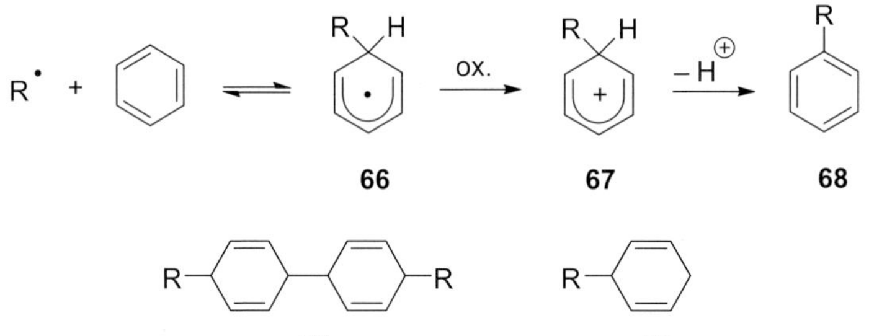

Scheme 13.13 General scheme for intermolecular homolytic aromatic substitution reactions.

stable longlived cyclohexadienyl radical **66** can either dimerize to give **69** or disproportionate to form cyclohexadiene **70** and arene **68**. Moreover, H-abstraction from cyclohexadienyl radical **66** directly leads to arene **68**.

Side reactions are often observed in these intermolecular radical arylations, one major limitation being the slow reaction of the alkyl radical with the arene. The *tert*-butyl radical addition onto benzene occurs at 79 °C with a rate constant of $3.8 \times 10^2\, M^{-1}\, s^{-1}$ [103], which is far below the rate constant of an efficient radical reaction. Consequently, side reactions – including rearrangement, dimerization and

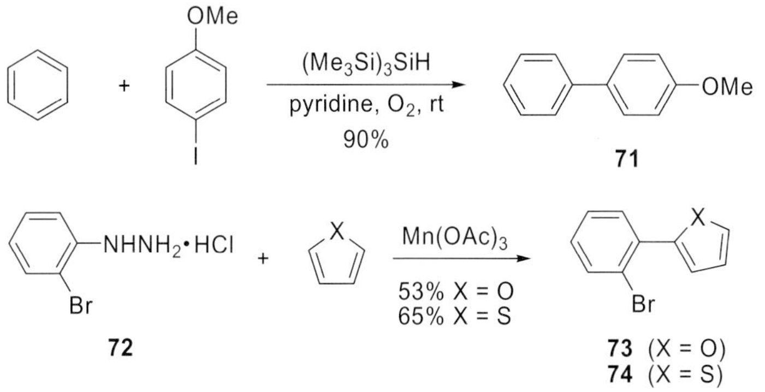

Scheme 13.14 Homolytic aromatic substitutions using reactive aryl radicals.

disproportionation, as well as halogen and hydrogen abstraction – may compete with the desired reaction. In addition, the reactivity of the homolytic aromatic substitution product **68** towards further alkyl radical addition is only slightly diminished as compared to the reactivity of the starting benzene. Consequently, overalkylation represents a serious problem.

Whilst the homolytic aromatic substitution with differently substituted benzene derivatives has been carefully studied by several groups [2], unfortunately a low regioselectivity is obtained for alkyl radical addition to substituted benzene derivatives [104]. However, high selectivities can be obtained for the addition of nucleophilic radicals with benzene derivatives bearing electron-withdrawing substituents, while good regioselectivities were achieved in the reaction with heteroarenes, such as thiophene, furan and thiazole [105]. A base-promoted homolytic aromatic substitution of electron-deficient arenes was developed in the Russell laboratory. Here, alkylmercury halides [106] or alkyl halides [107] were reacted with arenes in the presence of 1,4-diazabicyclo[2,2,2]octane (DABCO) to afford the corresponding alkylated arenes. The reactions of aryl radicals with arenes under reductive reaction conditions have also been investigated. In order to achieve acceptable yields, the arene acceptor must be used in large excess, which generally means as a solvent. The tin hydride-mediated addition of aryl iodides onto pyridine was recently disclosed [108]; this reaction can also be accomplished using silanes as reagents. Curran showed that the aryl radical addition to benzene occurred with a high yield when using $(Me_3Si)_3SiH$ as a radical reducing reagent, and aryl iodides in the presence of oxygen and pyridine. For example, the biphenyl derivative **71** was obtained in excellent yield using this method, starting from *p*-methoxy-iodobenzene at ambient temperature (Scheme 13.14) [60]. Without oxygen and pyridine, however, higher temperatures are necessary and lower yields are obtained [109]. Aryl radicals can also be generated using the $Mn(OAc)_3$-mediated oxidation of the corresponding aryl hydrazines and, if the reaction is conducted in an aromatic solvent, then a homolytic aromatic substitution will occur. An example of this procedure – the reaction of hydrazine hydrochloride **72** with $Mn(OAc)_3$ in

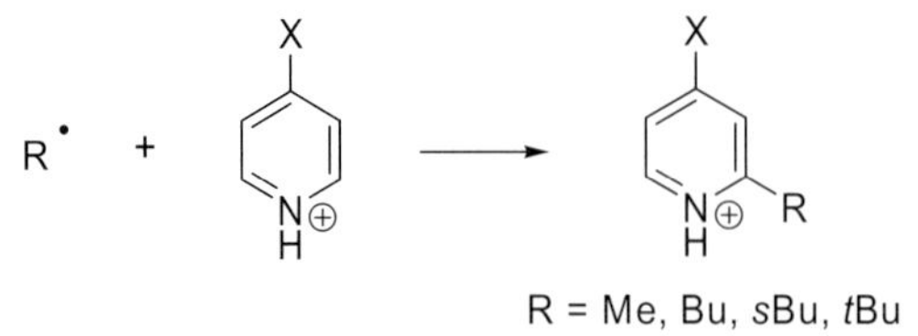

R• + 4-X-pyridinium → 2-R-4-X-pyridinium

R = Me, Bu, *s*Bu, *t*Bu

X	Me•	Bu•	*s*Bu•	*t*Bu•
CN	12.5	20.3	259.0	1890
COMe	3.6	5.6	55.6	144
Cl	2.4	–	–	11.1
H	1	1	1	1
Me	0.5	0.3	0.3	0.15
OMe	0.3	0.1	0.02	0.005

Scheme 13.15 Relative rates for the alkylation of protonated *para*-substituted pyridines with nucleophilic radicals.

furan or thiofuran as solvent to provide the arylation products **73** and **74**–is depicted in Scheme 13.14 [110].

The radical alkylation of protonated heteroaromatic compounds–the so-called Minisci reaction–has been intensively investigated [2e, g, 111]. Protonated heteroarenes are electron-deficient substrates, which react with nucleophilic radicals with high regioselectivity to yield the corresponding homolytic aromatic substitution products. For *para*-substituted pyridine derivatives the reaction occurs with complete regioselectivity at the 2-position, whereas for nonprotonated pyridines, arylations occur with low regioselectivity and in low yields.

The rates of radical additions to protonated heteroarenes correlate with the nucleophilicities of the attacking radicals [112]; for example, electrophilic radicals such as $^{\bullet}CH_2CO_2H$, $^{\bullet}CH_2CN$ and $^{\bullet}CH_2NO_2$ do not react with protonated pyridines. Furthermore, the reactivity towards aromatic substitution depends on the electrophilicity of the arene moiety, with the highest rates being observed for addition to the electron-poor 4-cyanopyridinium salts (Scheme 13.15). Similar reactions with the *para*-methoxy derivative may be up to 3.5×10^5 times slower [112, 113].

These reactivity trends show clearly that polar effects are involved in these radical substitution reactions. The transition state is thought to involve a charge-transfer process from the radical (electron donor) to the pyridinium ion (electron acceptor) [114]. In this respect, frontier molecular orbital (FMO) theory [115] has been applied to explain the reactivity differences that have been observed upon varying the substituents at the pyridinium ion and upon altering the nucleophilicity of the attacking radical. FMO can also be used to explain the regioselectivities obtained in these homolytic aromatic substitutions. Here, the LUMO of the substituted pyridinium cation has the highest coefficients at carbon atoms 2 and 4

[116], while the dominant interaction of the radical addition was between the SOMO of the nucleophilic radical and the LUMO of the protonated heteroarene. As the 4-position is blocked in the *para*-substituted systems presented in Scheme 13.15, addition would occur regioselectively at the 2-position. For the unsubstituted pyridinium cations, however, the reaction occurs at the 2- and 4-position [117]. The Minisci reaction has been used successfully for the alkylation of various heteroarenes, including lepidine, pyrazine, quinoline and quinoxaline [2e,g, 111]. Compounds such as alkanes, alkenes, carboxylic acids, esters, amides, amines, alcohols, ethers, aldehydes, ketones and halides (among others) have been used as radical precursors in the Minisci reaction [118]. An overview of the different methods that have been applied to generate alkyl radicals for these processes is provided in Ref. [111b].

13.3.2.2 **Arylation with Electrophilic C-Centered Radicals**

Heiba *et al.* were the first to report an example of a homolytic aromatic substitution using an electrophilic *C*-radical, approximately 40 years ago. Here, toluene was reported to react with acetic acid in the presence of either $Pb(OAc)_4$ [119a] or $Mn(OAc)_3$ [119b] to afford tolylacetic acid as a regioisomeric mixture, along with various side products. In this process, $^{\bullet}CH_2CO_2H$, which adds to toluene, is generated via the oxidation of acetic acid. Later, acetone was shown to be oxidized by $Mn(OAc)_3$ to the corresponding electrophilic *C*-radical, which could react in a homolytic aromatic substitution with substituted arenes to afford the corresponding arylation products [120, 121]. As expected, the reaction outcome depended on the nucleophilicity of the arene and, due to polar effects, better results were obtained with electron-rich arenes. The nitromethylation of arenes could also be achieved under similar conditions [122]. It transpired that the 'electrophilic' homolytic aromatic substitution using $Mn(OAc)_3$ as oxidant is a general process which can be performed with methylene derivatives bearing acidifying electron-withdrawing substituents. Most of the reactions of this type have been conducted with dialkyl malonates [123a], and an interesting example that documents the efficiency of such processes is depicted in Scheme 13.16. Here, intermolecular malonyl radical addition onto the indole core of **75** and oxidation leads to malonate **76**, which in turn undergoes intramolecular $Mn(OAc)_3$-mediated homolytic aromatic substitution to give **77** [124a]. Cerium (IV) ammonium nitrate has also been used successfully as an oxidant in aromatic malonylations [123d, e].

More recently, Baciocchi [125c] and Byers [125d, e] showed that α-iodoesters, α-iodomalonates and α-iodocarbonitriles could each be used as radical precursors in the homolytic aromatic substitution of various heteroarenes. The perfluoroalkylation of arenes is also possible using the same method [125f, g]. Electrophilic *C*-radicals for aromatic substitutions can also be generated from the corresponding xanthates by a xanthogenate group transfer [126]. As an example, the reaction of *N*-methyl-2-formylpyrrole with xanthate **78** in the presence of dilauroyl peroxide (DLP) to afford arylation product **79** is shown in Scheme 13.16. Upon thermal decomposition, DLP delivers alkyl radicals which can add to **78** to give, after fragmentation, the desired electrophilic *C*-radical. The addition of this *C*-radical onto

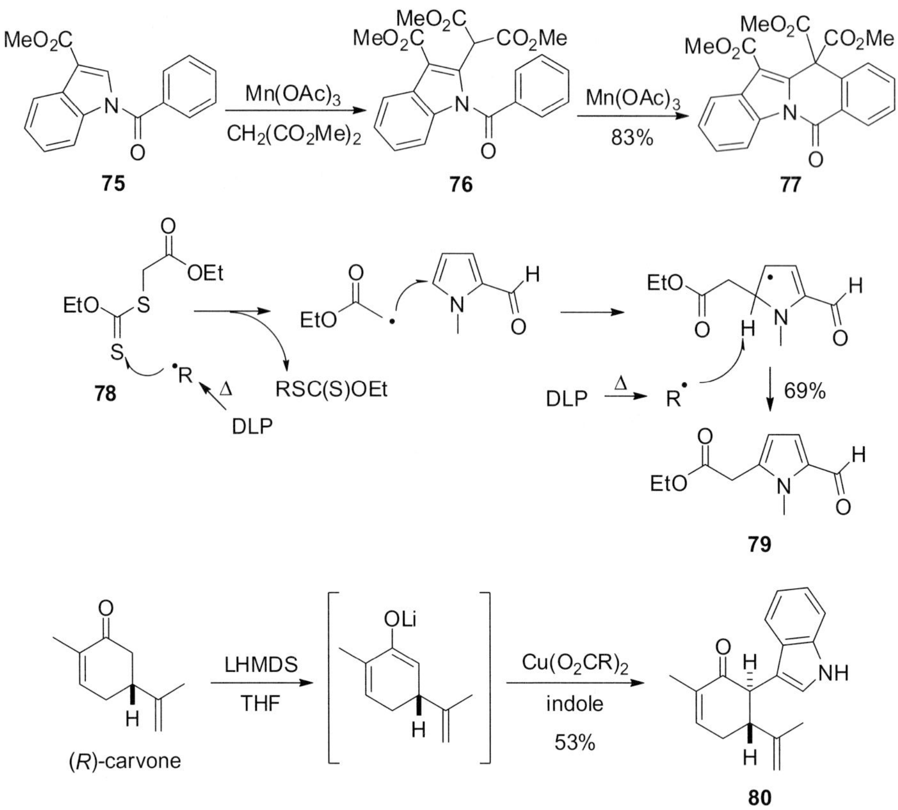

Scheme 13.16 Homolytic aromatic substitution using electrophilic C-radicals.

N-methyl-2-formylpyrrole and subsequent H-abstraction by radical R· eventually delivers the substituted pyrrole **79**.

Baran has recently shown that electrophilic radicals generated from Li-enolates upon oxidation with various metal salts (Fe- and Cu-salts) react efficiently with indoles and pyrroles to afford the corresponding arylation products [127]. For example, the deprotonation of (*R*)-carvone with LHMDS [$LiN(SiMe_3)_2$] and subsequent oxidation of the enolate with copper(II) 2-ethylhexanoate [$Cu(O_2CR)_2$] in the presence of indole provides the arylation product **80** in 53% yield as a single diastereoisomer (Scheme 13.16). This key arylation step was used for the synthesis of hepalindoles [127a], (*S*)-ketorolac [127b] and for a highly impressive synthesis of (+)-ambiguine H, without the use of any protecting groups [127c].

Homolytic aromatic substitution can also be performed with electrophilic heteroatom-centered radicals. The radical amination of arenes with electrophilic dialkyl aminyl radical cations represents a valuable method for the preparation of aniline derivatives [2e, 111a]. The radical cations (R_2HN^+) are generally prepared from the corresponding *N*-chloroamines in an acidic medium, using catalytic amounts of a metal salt [Fe(II)-, Ti(III)-, Cu(I)- and Cr(II)-salts]. The reaction functions best for electron-rich arenes such as phenols and aryl ethers, although in the amination of monosubstituted benzene derivatives a mixture of the *ortho* and *para* isomers is formed, with the *para*-compound as the major product. For further information on this reaction, the reader is referred elsewhere [2a, 111a]. Recently, radical phosphonations of arenes and heteroarenes were reported [128, 129]. As an example, activated furans such as **81** undergo regioselective phosphonation with dimethyl phosphate in the presence of $Mn(OAc)_3$ (Equation 13.9) [129], with phosphonate **82** being obtained in excellent yield (95%). Moreover, pyrroles and thiazoles can also be phosphonated under these conditions.

81 + $HP(O)(OMe)_2$ $\xrightarrow[\text{AcOH, 80 °C, 3 h, 95\%}]{Mn(OAc)_3 \cdot H_2O}$ **82** (13.9)

13.3.2.3 Intermolecular ipso-Substitutions

In recent years, intermolecular homolytic *ipso* substitution reactions have received very little attention and are, consequently, currently not of synthetic importance [130]. These reactions occur only in selected cases and, in general, the attacking radical must be nucleophilic and the arene acceptors electrophilic. An additional very important issue is the ability of the group which is being replaced to act as a radical leaving group. Of course, the radical leaving group at the arene exerts steric repulsion, and therefore the radical attack often occurs at an unsubstituted position. Thus, the intermolecular homolytic aromatic substitution will compete with the *ipso* attack and, in most cases, will be the exclusive reaction pathway. For example, the reaction of thiophene **83** with the nucleophilic 1-adamantyl radical afforded the *ipso* product **85** in 45% yield, whereas the analogous reaction with the methyl radical gave selectively the homolytic aromatic substitution product **84** in 35% yield (Scheme 13.17) [130]. In the case of the nucleophilic 1-adamantyl radical, the regioselectivity of the attack is controlled by polar effects, whereas for the methyl radical (where polar effects are less important) the stability of the intermediate σ-complex is responsible for the regioselectivity of the attack [130]. A similar change in chemoselectivity upon changing the attacking radical (1-adamantyl versus methyl radical) was observed in the reaction with 1,3,5-trinitrobenzene [131].

Alkyldenitration by the 1-adamantyl radical in *para*-substituted nitrobenzenes (p-X-C_6H_4-NO_2) is an efficient process only for electron-poor arenes. Thus, whilst for anisole (X = OMe), benzene (X = H) and toluene (X = Me) derivatives no *ipso* products were observed, the analogous reaction with nitrobenzenes bearing electron-withdrawing substituents (X = NO_2, CN, SO_2R, CO_2R, COMe, CHO) afforded the corresponding alkyldenitration products in 45–60% yield [132]. The effect of the leaving group on the *ipso* substitution was studied by reaction of the 1-adamantyl radical with differently substituted benzothiazole derivatives **86** to give

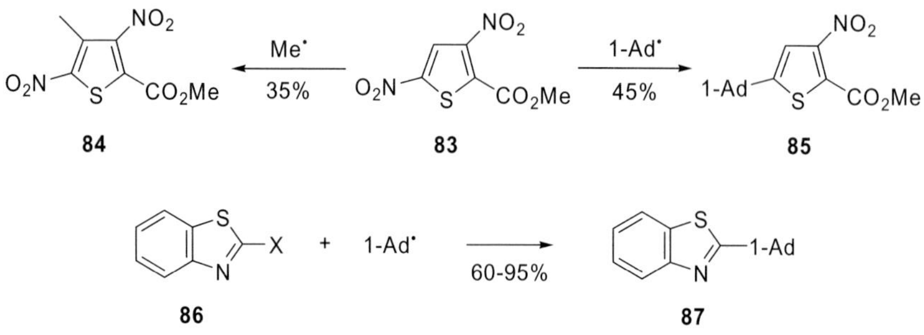

O_2N–(thiophene)–CO_2Me **84** ←(Me•, 35%)– **83** –(1-Ad•, 45%)→ **85**

86 (X) + 1-Ad• →(60-95%) **87** (1-Ad)

X = NO_2, SO_2Ph, SOPh, COMe   1-Ad = 1-adamantyl

Scheme 13.17 Intermolecular *ipso* substitutions.

heteroarene **87** (Scheme 13.17) [133]. While the best results were obtained for the alkyldenitration (X = NO_2, 95%), the phenylsulfonyl, phenylsulfinyl and acyl radicals were also seen to be good leaving groups in these *ipso* substitutions (X = SO_2Ph, 80%; X = SOPh, 80%; X = COMe, 60%). Halides [134], the methylsulfanyl and the methoxy group were not efficiently replaced by the adamantyl group. More recently, the *ipso* substitution with the phenylsulfonyl radical as a leaving group was applied to the preparation of various stannylated heterocycles (stannyldesulfonylation) [135].

13.4 Arylations Using Aryl Radicals

13.4.1 Additions onto Olefins: Meerwein Arylation

The cupric halide-catalyzed reaction of an aryl diazonium salt with an olefin–known as the Meerwein arylation–is a powerful and well-developed method for the arylation of olefins [136, 137]. Good results have been obtained for olefins bearing electron-withdrawing groups such as a carbonyl, a cyano, an aryl or an alkenyl group. Depending on the nature of the olefin and the diazonium salt, the reaction can lead either to the oxidation product **88** or to the Cl-atom transfer product **89** (Equation 13.10) [138].

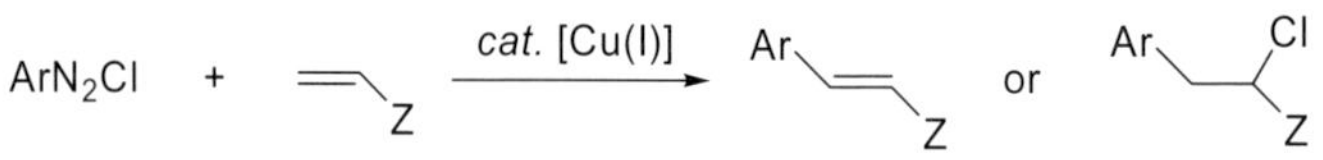

$$ArN_2Cl + CH_2{=}CH{-}Z \xrightarrow{cat.\ [Cu(I)]} Ar{-}CH{=}CH{-}Z\ (\mathbf{88}) \text{ or } Ar{-}CH_2{-}CHCl{-}Z\ (\mathbf{89}) \quad (13.10)$$

Z = electron-withdrawing group

Today, it is widely accepted that the Cu(I)-salt which serves as the catalytically active species is formed by the reduction of a Cu(II)-salt, either by the solvent or

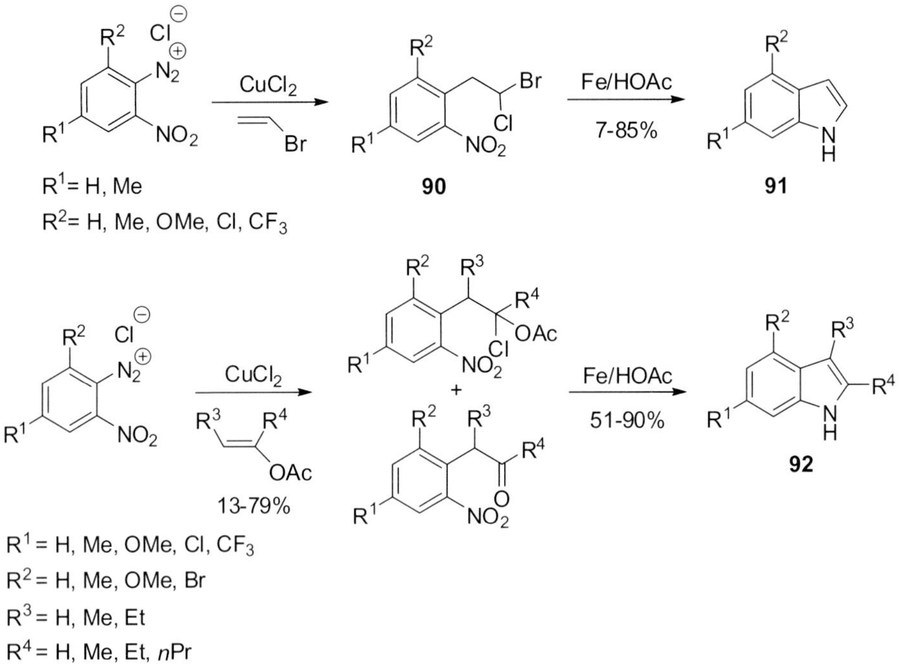

R^1 = H, Me
R^2 = H, Me, OMe, Cl, CF_3

R^1 = H, Me, OMe, Cl, CF_3
R^2 = H, Me, OMe, Br
R^3 = H, Me, Et
R^4 = H, Me, Et, *n*Pr

Scheme 13.18 Indole synthesis using a Meerwein arylation.

by impurities present in the reaction mixture [139]. Electron transfer from a Cu(I)-salt (CuX) to the aryl diazonium salt (ArN_2X) leads to CuX_2 and the corresponding aryl radical, which then adds to the double bond to give an alkyl radical. Halogen abstraction of the alkyl radical from CuX_2 provides the arylated compound **89** and CuX. Olefin **88** is then formed by the elimination of HX from **89**.

Nonconjugated olefins, acetylenes, styrenes, α,β-unsaturated aldehydes, α,β-unsaturated ketones, α,β-unsaturated acids and quinones are all arylated under Meerwein conditions. Conjugated dienes and ene-ynes can be also used as substrates.

As an example, an indole synthesis using a Meerwein arylation as the key step is presented in Scheme 13.18. The radical addition of substituted *o*-nitrophenyl diazonium chlorides to vinyl bromide under typical Meerwein conditions provides chlorides **90**. Then, reduction of the nitro group followed by cylization and aromatization, eventually give the corresponding disubstituted indoles **91**. Tetrasubstituted indoles **92** were prepared in moderate to good yields using a similar approach [140].

The synthesis of substituted quinolones [141], benzothiophenes [142] and benzylthiazolidine-2,4-diones [143] has also been achieved using the Meerwein reaction. Ti(III)chloride has been shown to be suitable for the generation of aryl radicals from diazonium salts [144]. Interestingly, a metal-free version of the Meerwein arylation has recently been applied to the synthesis of α-arylmethyl ketones (Equation 13.11) which, importantly, could easily be run on a multi-kilogram scale [145]. Although no mechanistic studies were provided in the original report, an $S_{RN}1$-type reaction appeared to be feasible.

OAc, Me; KOAc, acetone/water, 39-76%; N_2BF_4, R^1, R^2 (13.11)

R^1 = H, CF_3, Me, OCF_3, OMe, NO_2, Cl
R^2 = H, NO_2, CN, CO_2Me, Cl

The carbodiazenylation of olefins is related to the Meerwein arylation. In this case, the aryl radical addition intermediates are not halogenated (as in the original Meerwein protocol) but are trapped with a diazonium salt so as to provide the corresponding radical cation. Reduction finally leads to aryl diazenes. Fe(II)-salts, Ti(III)chloride and Cu(I)-salts or mixtures of these reducing reagents have been used to conduct carbodiazenylations, some reported examples of which are presented in Equation 13.12 [146].

2 R^2, R^1, NH_2; $NaNO_2$; 2 R^2, R^1, N_2X; R^3, R^4; Ti(III) or Ti(III)/Fe(II), $MeOH/H_2O$, - N_2, 47-80%; R^1, R^2, R^3, R^4, N=N, R^1, R^2

R^1 = H, Cl, CO_2Me, CF_3, OMe
R^2 = H, CO_2H, OMe
R^3 = CH_2OAc, CH_2CN, $(CH_2)_2COMe$, CH_2OH, $(CH_2)_2OH$
R^4 = H, Me

(13.12)

13.4.2 Cyclizations Using Aryl Radicals

The cyclization of an aryl radical is a useful process for the preparation of (hetero)cyclic compounds such as tetrahydrobenzofurans, chromanes, indanes, indolines and tetrahydroquinolines [147]. The synthetic strategy involves the generation of an aryl radical which subsequently adds to an unsaturated moiety, most usually in a 5-*exo* or a 6-*exo trig* process (Scheme 13.19). After ring closure, the intermediate cyclized radical **93** can either be reduced (pathway A) or it can react further with a nucleophile ($S_{RN}1$-type process, pathway C) or with another radical acceptor to yield, after reduction, compounds of type **94** (pathway B).

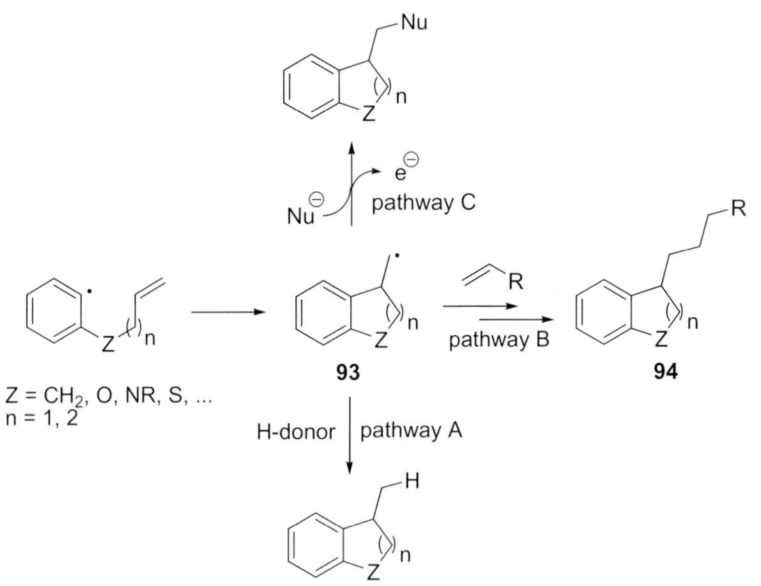

Scheme 13.19 Radical cyclizations using aryl radicals.

Along with aryl iodides and aryl bromides, aryl diazonium salts have been used successfully as aryl radical precursors in these processes. Alkenyl-substituted aryl diazonium tetrafluoroborates were shown to be efficient substrates for the synthesis of substituted dihydrobenzofurans and related compounds [148, 149]. Aryl radical generation was seen to occur via electron transfer to the aryl iodide, aryl bromide or aryl diazonium salt, in a Sandmeyer-type reaction [150, 151], with the reactions generally having high yields. It was also shown that the radicals obtained after aryl radical cyclization could be used for further bond-forming processes. Here, the thiolate and thioacetate anions reacted via the $S_{RN}1$ substitution mechanism to deliver the corresponding *S*-substituted products (pathway C). Reductive and oxidative termination of the radical chain process was also possible (for reduction, see pathway A in Scheme 13.19) [152, 153]. Although various environmentally benign hydrogen donors such as $(Me_3Si)_3SiH$ and *N*-ethyl piperidine hypophosphite can be used to mediate these reductive radical chain reactions, the more generally applicable (but toxic) Bu_3SnH was the most efficient radical reducing reagent, and therefore was commonly used in these cyclization reactions. Reductive aryl radical cyclizations have been successfully applied in natural product synthesis [154]. For example, (±)-vindoline (key steps shown in Scheme 13.20), (±)-horsfiline and (±)-aspidoespermidine can each be prepared using a 5-*exo* aryl radical cyclization as a key step [155–157]. Iodine abstraction by a silyl radical from the aryl iodide **95** generates radical **96**, which then undergoes 5-*exo* cyclization to **97**. Addition onto the azide functionality followed by N_2-extrusion leads to the aminyl radical **98**, which is eventually reduced to afford **99**.

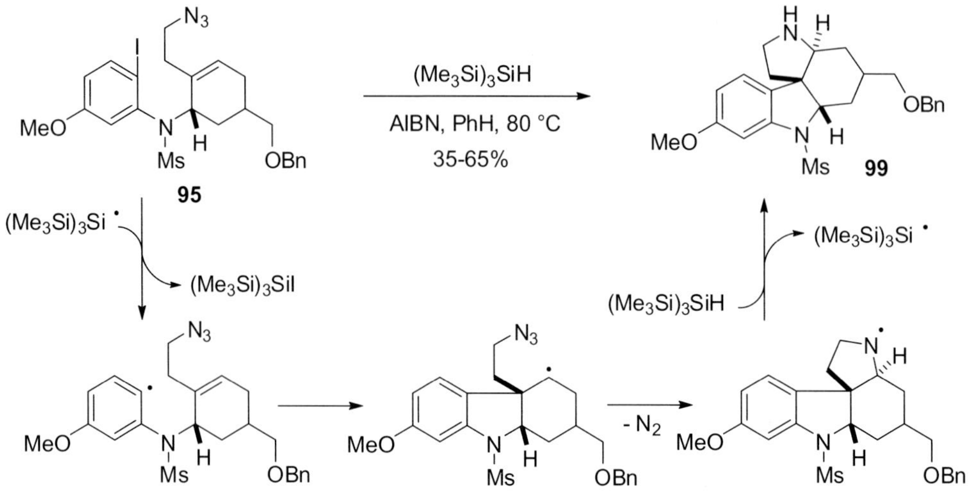

Scheme 13.20 Radical cascade reaction initiated by an aryl radical cyclization.

Aryl radicals can also be generated from the corresponding aryl iodides with SmI_2. A radical/polar crossover reaction with cyclobutenyl-substituted aryl iodides as starting materials was used for the synthesis of the BCD ring system of penitrems (Equation 13.13). After 6-*exo* cyclization, the intermediate alkyl radical is reduced with SmI_2 to the organosamarium compound **100** which finally reacts with acetone to provide, after hydrolysis, alcohol **101** as a valuable building block for the synthesis of penitrems [158].

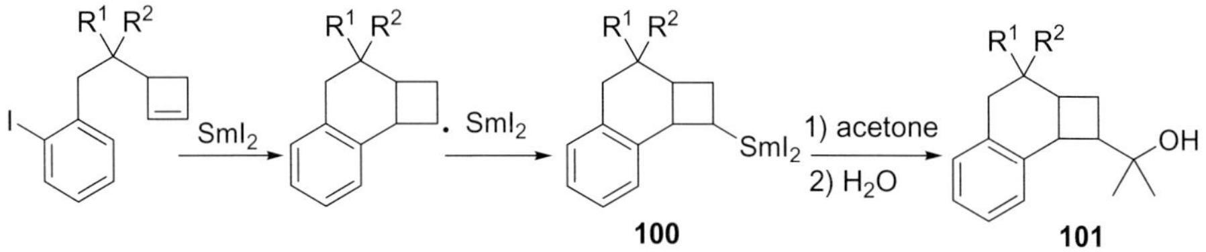

(13.13)

Atropisomers of *ortho*- and *N*-substituted *o*-iodoacrylamides, *N*-allyl-*N*-*o*-iodoarylamides are stable at ambient temperature with a barrier of interconversion of approximately 20–30 kcal mol^{-1}. Enantioenriched, axially chiral aryl iodides reacted at ambient temperature under reductive reaction conditions to provide the corresponding 5-*exo* cyclization products. It has been shown by Curran that the axial chirality can be transferred into centrochirality, and hence 5-*exo* aryl radical cyclization occurs more rapidly than does racemization of the axially chiral aryl radical. An example is presented in Equation 13.14; the transfer of chirality in these processes is usually around 90% [159, 160].

Bu_3SnH, Et_3B/O_2, PhH, 20 °C

M:*P* = 83:17 85% *R*:*S* = 76:24 (13.14)

The intramolecular homolytic substitution (S_Hi) reaction of aryl radicals at sulfur, selenium and tellurium has been intensively used for the synthesis of S-, Se- and Te-containing heterocycles. A typical example is the transformation of aryl iodide **102** to benzothiophene **103** under reductive radical conditions (Equation 13.15) [161]; here, the homolytic substitution is efficient if the leaving radical R is stabilized. This approach has been used for the selective deoxygenation of the hydroxy function of the 6-hydroxymethyl group in mannose derivatives [162]. The leaving radical R can also be a *P*-centered radical, as shown recently [163]. Interestingly, fused benzosultines and cyclic sulfinamides can be prepared in good yields via homolytic substitution at sulfur (Equation 13.16) [164].

102 - I• R• **103** (13.15)

Bu_3SnH, AIBN, PhH, 80 °C, 31-100% (13.16)

Z = O, NH, NMe
X = Br, I
R^2 = *p*-Tol, *o*-NO_2-C_6H_4, ^{t}Bu
R^1 = H, OMe, F, ...
n = 1,2

Various benzoselenophenes **106** were successfully prepared using the S_Hi reaction at selenium. Selenides **104**, readily prepared from the corresponding epoxides, react under standard radical conditions by homolytic substitution at selenium to the heterocycles **105**. Water elimination finally provides the benzoselenophenes **106** in high yields (Equation 13.17) [165, 166]. In analogy, Te-containing heterocycles can also be prepared [166, 167].

104 → 105 → 106

$(Me_3Si)_3SiH$, AIBN, PhH, 80 °C, 80-86%

R^1, R^2 = H, Me, Ph

(13.17)

13.4.3 Phosphonylation of Aryl Radicals

The homolytic substitution at trivalent phosphorus has recently been disclosed as a new method for the metal-free C–P bond formation. Oshima [168] has shown that *in situ*-generated tetraphenylbisphosphane (X = PPh_2) (Equation 13.18) reacts under standard conditions with aryl radicals to afford, after oxidation, the corresponding aryl diphenylphosphane sulfides.

Ph_2P-X; 1) V-40, PhH, 80 °C; 2) S_8 or H_2O_2 → P(Z)Ph_2

X = PPh_2, $SnMe_3$; Z = S, O

R = 2-OMe, 4-Cl, 4-CF_3, ...

(13.18)

More recently, Studer [169] has replaced Ph_2P-PPh_2 by the readily available trimethyltindiphenylphosphane (X = $SnMe_3$) as radical acceptor (Equation 13.18). The reaction of various aryl halides under radical reaction conditions produced, after oxidation, the corresponding aryl diphenylphosphane oxides with good yields.

13.5 Conclusions

The pioneering studies into radical arylation, which were conducted several decades ago, have paved the way to modern radical arylation chemistry. In particular, an improved knowledge on kinetic data has allowed the planning of complex cascade processes that comprise radical arylation reactions capable of being used also in the synthesis of complex natural products. Despite these recent efforts, a large number of qualitative and quantitative investigations remain to be conducted, especially in the area of stereoselective radical arylations. In conclusion, radical chemistry can today be considered as a valuable, highly promising method that complements ionic- and transition metal-mediated arylation reactions. Indeed, based on the large number of research groups currently engaged in this field, it is inevitable that many interesting new developments will emerge in the near future.

Abbreviations

ACCN	1,1′-(*Z*)-diazene-1,2-diyldicyclohexanecarbonitrile
AIBN	α,α′-azobisisobutyrodinitrile
DABCO	1,4-diazabicyclo[2,2,2]octane
DLP	dilauroyl peroxide
DMF	dimethylformamide
DMSO	dimethylsulfoxide
ET	electron transfer
FMO	frontier molecular orbital (theory)
LDA	lithium diisopropylamide
LHMDS	lithium bis(trimethylsilyl)amide
LUMO	lowest unoccupied molecular orbital
MOM	methoxymethyl
MW	microwave
SOMO	singly occupied molecular orbital
TEMPO	2,2,6,6-tetramethylpiperidine-1-oxyl radical
TMSCl	trimethylchlorosilane
V-40	1,1′-azobis(cyclohexane-1-carbonitrile)
VAZO	1,1′-azobis(cyclohexanecarbonitrile)

References

1 For reviews on the aromatic $S_{RN}1$ reactions, see: (a) Rossi, R.A. (2005) *Synthetic Organic Photochemistry*, Vol. 2 (eds A. Griesbeck and J. Mattay), Marcel Dekker, New York, p. 495;
(b) Rossi, R.A., Pierini, A.B. and Peñéñory, A.B. (2003) *Chem. Rev.*, **103**, 71;
(c) Rossi, R.A. Pierini, A.B. and Santiago, A.N. (1999) *Organic Reactions* (eds L.A. Paquette and R. Bittman), John Wiley & Sons, Ltd, p. 1.

2 (a) Williams, G.H. (1960) *Homolytic Aromatic Substitution*, Pergamon Press, Oxford;
(b) Hey, D.H. (1967) *Advances in Free Radical Chemistry*, Vol. 2 (ed. G.H. Williams), Logos Press, London, p. 47;
(c) Perkins, M.J. (1973) *Free Radicals*, Vol. 2 (ed. J.K. Kochi), John Wiley & Sons, Inc., New York, p. 231;
(d) Tiecco, M. (1975) *Free Radical Reactions* (ed. W.A. Waters), Butterworths, London, p. 25;
(e) Minisci, F. (1976) *Topics in Current Chemistry*, Vol. 62, Springer-Verlag, Berlin, p. 1;
(f) Tiecco, M. and Testaferri, L. (1983) *Reactive Intermediates*, Vol. 3 (ed. R.A. Abramovitch), Plenum Press, New York, p. 61;
(g) Minisci, F., Vismara, E. and Fontana, F. (1989) *Heterocycles*, **28**, 489;
(h) Studer, A. and Bossart, M. (2001) *Radicals in Organic Synthesis*, Vol. 2 (eds

P. Renaud and M.P. Sibi), Wiley-VCH Verlag GmbH, Weinheim, p. 62;
(i) Bowmann, W.R. and Storey, J.M.D. (2007) *Chem. Soc. Rev.*, **36**, 1803.

3 Kim, J.K. and Bunnett, J.F. (1970) *J. Am. Chem. Soc.*, **92**, 7463.

4 For examples of spontaneous $S_{RN}1$ reactions, see: (a) Dell'Erba, C., Novi, M., Petrillo, G. and Tavani, C. (1993) *Tetrahedron*, **49**, 235;
(b) Dell'Erba, C., Novi, M., Petrillo, G. and Tavani, C. (1992) *Tetrahedron*, **48**, 325;
(c) Dell'Erba, C., Novi, M., Petrillo, G. and Tavani, C. (1991) *Tetrahedron*, **47**, 333;
(d) Oostvee, E.A. and van der Plas, H.C., 1979. *Recl. Trav. Chim. Pays-Bas.*, **98**, 441;
(e) Carver, D.R., Komin, A.P., Hubbard, J.S. and Wolfe, J.F. (1981) *J. Org. Chem.*, **46**, 294.

5 Beugelmans, R. and Chastanet, J. (1993) *Tetrahedron*, **49**, 7883.

6 (a) Bunnett, J.F., Mitchel, E. and Galli, C. (1985) *Tetrahedron*, **41**, 4119;
(b) Bunnett, J.F. and Sundberg, J.E. (1975) *Chem. Pharm. Bull.*, **23**, 2620.

7 (a) Compton, R.G., Dryfe, R.A.W. and Fisher, A.C. (1994) *J. Chem. Soc., Perkin Trans.*, **2**, 1581;
(b) Compton, R.G., Dryfe, R.A.W., Eklund, J.C., Page, S.D., Hirst, J., Nei, L., Fleet, G.W.J., Hsia, K.Y., Bethell, D. and Martingale, L.J. (1995) *J. Chem. Soc., Perkin Trans.*, **2**, 1673.

8 Baumgartner, M.T., Pierini, A.B. and Rossi, R.A. (1999) *J. Org. Chem.*, **64**, 6487.

9 (a) Lukach, A.E., Morris, D.G., Santiago, A.N. and Rossi, R.A. (1995) *J. Org. Chem.*, **60**, 1000;
(b) Santiago, A.N., Takeuchi, K., Ohga, Y., Nishida, M. and Rossi, R.A. (1991) *J. Org. Chem.*, **56**, 1581;
(c) Galli, C., Gentili, P. and Guarnieri, A. (1997) *Gazz. Chim. Ital.*, **127**, 159;
(d) Galli, C., Gentili, P. and Guarnieri, A. (1995) *Gazz. Chim. Ital.*, **125**, 409.

10 Nazareno, M.A. and Rossi, R.A. (1994) *Tetrahedron Lett.*, **35**, 5185.

11 Rossi, R.A. and Baumgartner, M.T. (1999) *Target in Hetercyclic Systems: Chemistry and Properties*, Vol. 3 (eds O.A. Attanasi and D. Spinelli), Soc. Chimica Italiana, p. 215.

12 (a) Beugelmans, R. and Roussi, G. (1981) *Tetrahedron*, **37**, 393;
(b) Beugelmans, R. and Roussi, G. (1979) *J. Chem. Soc., Chem. Commun.*, 950.

13 Barolo, S.M., Lukach, A.E. and Rossi, R.A. (2003) *J. Org. Chem.*, **68**, 2807.

14 (a) Beugelmans, R. and Bois-Choussy, M. (1981) *Synthesis*, 729;
(b) Guastavino, J.F., Barolo, S.M. and Rossi, R.A. (2006) *Eur. J. Org. Chem.*, 3898.

15 Petrillo, G., Novi, M., Dell'Erba, C. and Tavani, C. (1991) *Tetrahedron*, **47**, 9297.

16 (a) Yammal, C.C., Podestá, J.C. and Rossi, R.A. (1992) *J. Org. Chem.*, **57**, 5720;
(b) Córsico, E.F. and Rossi, R.A. (2000) *Synlett*, 227.

17 (a) Córsico, E.F. and Rossi, R.A. (2002) *J. Org. Chem.*, **67**, 3311;
(b) Chopa, A.B., Lockhart, M.T. and Silbestri, G. (2001) *Organometallics*, **20**, 3358.

18 (a) Bunnett, J.F. and Creary, X. (1974) *J. Org. Chem.*, **39**, 3173;
(b) Bunnett, J.F. and Creary, X. (1975) *J. Org. Chem.*, **40**, 3740;
(c) Pierini, A.B., Baumgartner, M.T. and Rossi, R.A. (1991) *J. Org. Chem.*, **56**, 580;
(d) Baumgartner, M.T., Pierini, A.B. and Rossi, R.A. (1993) *J. Org. Chem.*, **58**, 2593;
(e) Beugelmans, R. and Chbani, M. (1995) *Bull. Soc. Chim. Fr.*, **132**, 290;
(f) Beugelmans, R. and Bois-Choussy, M. (1986) *Tetrahedron*, **42**, 1381.

19 Argüello, J.E., Schmidt, L.C. and Peñéñory, A.B. (2003) *Org. Lett.*, **5**, 4133.

20 Pierini, A.B., Peñéñory, A.B. and Rossi, R.A. (1984) *J. Org. Chem.*, **49**, 486.

21 (a) Goehring, R.R., Sachdeva, Y.P., Pisipati, J.S., Sleevi, M.C. and Wolfe, J.F. (1985) *J. Am. Chem. Soc.*, **107**, 435;
(b) Wolfe, J.F., Sleevi, M.C. and Goehring, R.R. (1980) *J. Am. Chem. Soc.*, **102**, 3646;
(c) Wu, G.S., Tao, T., Cao, J.J. and Wei, X.L. (1992) *Acta Chim. Sin.*, **50**, 614.

22 (a) Semmelhack, M.F., Chong, B.P., Stauffer, R.D., Rogerson, T.D., Chong, A. and Jones, L.D. (1975) *J. Am. Chem. Soc.*, **97**, 2507;

(b) Semmelhack, M.F., Stauffer, R.D. and Rogerson, T.D. (1973) *Tetrahedron Lett.*, **14**, 4519.

23 Goehring, R.R. (1992) *Tetrahedron Lett.*, **33**, 6045.

24 Barolo, S.M., Teng, X., Cunny, G.D. and Rossi, R.A. (2006) *J. Org. Chem.*, **71**, 8493.

25 Pschorr, R. (1896) *Ber. Dtsch. Chem. Ges.*, **29**, 496.

26 (a) Abramovitch, R.A. (1967) *Advances in Free Radical Chemistry*, Vol. 2 (ed. G.H. Williams), Logos Press, London, p. 87;
(b) Sainsbury, M. (1980) *Tetrahedron*, **36**, 3327.

27 Elofson, R.M. and Gadallah, F.F. (1971) *J. Org. Chem.*, **36**, 1769.

28 Caronna, T., Ferrario, F. and Servi, S. (1979) *Tetrahedron Lett.*, **20**, 657.

29 Wassmundt, F.W. and Kiesman, W.F. (1995) *J. Org. Chem.*, **60**, 196.

30 Lampard, C., Murphy, J.A., Rasheed, F., Lewis, N., Hursthouse, M.B. and Hibbs, D.E. (1994) *Tetrahedron Lett.*, **35**, 8675.

31 Chauncy, B. and Gellert, E. (1969) *Aust. J. Chem.*, **22**, 993.

32 For further methods to decompose diazonium salts to aryl radicals, see: Galli, C. (1988) *Chem. Rev.*, **88**, 765.

33 Narasimhan, N.S. and Aidhen, I.S. (1988) *Tetrahedron Lett.*, **29**, 2987.

34 Togo, H. and Kikuchi, O. (1988) *Tetrahedron Lett.*, **29**, 4133.

35 (a) Estévez, J.C., Villaverde, M.C., Estévez, R.J. and Castedo, L. (1993) *Tetrahedron*, **49**, 2783;
(b) Estévez, J.C., Villaverde, M.C., Estévez, R.J. and Castedo, L. (1995) *Tetrahedron*, **51**, 4075;
(c) Tsuge, O., Hatta, T. and Tsuchiyama, H. (1998) *Chem. Lett.*, **27**, 155;
(d) Orito, K., Uchiito, S., Satoh, Y., Tatsuzawa, T., Harada, R. and Tokuda, M. (2000) *Org. Lett.*, **2**, 307;
(e) Martínez, E., Estévez, J.C., Estévez, R.J. and Castedo, L. (2001) *Tetrahedron*, **57**, 1973;
(f) Ganguly, A.K., Wang, C.H., David, M., Bartner, P. and Chan, T.M. (2002) *Tetrahedron Lett.*, **43**, 6865;
(g) Harrowven, D.C., Helias, N.L., Moseley, J.D., Blumire, N.J. and Flanagan, S.R. (2003) *Chem. Commun.*, 2658;
(h) Zhang, W. and Pugh, G. (2003) *Tetrahedron*, **59**, 3009;
(i) Alcaide, B., Almendros, P., Pardo, C., Rodríguez-Vicente, A. and Pilar Ruiz, M. (2005) *Tetrahedron*, **61**, 7894;
(j) Clyne, M.A. and Aldabbagh, F. (2006) *Org. Biomol. Chem.*, 268.

36 (a) Kraus, G.A. and Kim, H. (1993) *Synth. Commun.*, **23**, 55;
(b) Suzuki, F. and Kuroda, T. (1993) *J. Heterocyclic Chem.*, **30**, 811;
(c) Antonio, Y., De, M.E. La Cruz, E., Galeazzi, A., Guzman, B.L., Bray, R., Greenhouse, L.J., Kurz, D.A., Lustig, M. L., Maddox, J.M. and Muchowski (1994) *Can. J. Chem.*, **72**, 15;
(d) Comins, D.L., Hong, H. and Jianhua, G. (1994) *Tetrahedron Lett.*, **35**, 5331;
(e) Ho, T.C.T. and Jones, K. (1997) *Tetrahedron*, **53**, 8287;
(f) Harrowven, D.C., Sutton, B.J. and Coulton, S. (2002) *Tetrahedron*, **58**, 3387;
(g) Harrowven, D.C., Sutton, B.J. and Coulton, S. (2003) *Org. Biomol. Chem.*, 4047;
(h) Flanagan, S.R., Harrowven, D.C. and Bradley, M. (2003) *Tetrahedron Lett.*, **44**, 1795;
(i) Ganguly, A.K., Wang, C.H., Chan, T.M., Ing, Y.H. and Buevich, A.V. (2004) *Tetrahedron Lett.*, **45**, 883;
(j) Allin, S.M., Bowman, W.R., Elsegood, M.R.J., McKee, V., Karim, R. and Rahman, S.S. (2005) *Tetrahedron*, **61**, 2689.

37 Donnelly, S., Grimshaw, J. and Trocha-Grimshaw, J. (1994) *J. Chem. Soc. Chem. Commun.*, 2171 and references cited therein.

38 (a) Harrowven, D.C., Nunn, M.I.T. and Fenwick, D.R. (2002) *Tetrahedron Lett.*, **43**, 7345;
(b) Harrowven, D.C., Guy, I.L. and Nanson, L. (2006) *Angew. Chem. Int. Ed.*, **45**, 2242.

39 Rosa, A.M., Lobo, A.M., Branco, P.S., Prabhakar, S. and Sá-da-Costa, M. (1997) *Tetrahedron*, **53**, 299.

40 (a) Estévez, J.C., Villaverde, M.C., Estévez, R.J. and Castedo, L. (1994) *Tetrahedron*, **50**, 2107;
(b) Comins, D.L., Thakker, P.M., Baevsky, M.F. and Badawi, M.M. (1997) *Tetrahedron*, **53**, 16327.

41 Suzuki, H., Aoyagi, S. and Kibayashi, C. (1995) *Tetrahedron Lett.*, **36**, 935.

42 Nakanishi, T., Suzuki, M., Mashiba, A., Ishikawa, K. and Yokotsuka, T. (1998) *J. Org. Chem.*, **63**, 4235.

43 (a) Curran, D.P. and Liu, H. (1992) *J. Am. Chem. Soc.*, **114**, 5863;
(b) Curran, D.P., Liu, H., Josien, H. and Ko, S.B. (1996) *Tetrahedron*, **52**, 11385;
(c) Josien, H., Ko, S.B., Bom, D. and Curran, D.P. (1998) *Chem. Eur. J.*, **4**, 67 and references cited therein;
(d) review on isonitriles acting as radical acceptors: Nanni, D. (2001) *Radicals in Organic Synthesis*, Vol. 2 (eds P. Renaud and M.P. Sibi), Wiley-VCH Verlag GmbH, Weinheim, p. 44.

44 (a) Nanni, D., Pareschi, P., Rizzoli, C., Sgarabotto, P. and Tundo, A. (1995) *Tetrahedron*, **51**, 9045 and references cited therein;
(b) Leardini, R., Nanni, D., Pareschi, P., Tundo, A. and Zanardi, G. (1997) *J. Org. Chem.*, **62**, 8394;
(c) Camaggi, C.M., Leardini, R., Nanni, D. and Zanardi, G. (1998) *Tetrahedron*, **54**, 5587;
(d) Nanni, D., Calestani, G., Leardini, R. and Zanardi, G. (2000) *Eur. J. Org. Chem.*, 707;
(e) Benati, L., Calestani, G., Leardini, R., Minozzi, M., Nanni, D., Spagnolo, P., Strazzari, S. and Zanardi, G. (2003) *J. Org. Chem.*, **68**, 3454;
(f) Du, W. and Curran, D.P. (2003) *Org. Lett.*, **5**, 1765;
(g) Rashatasakhon, P., Ozdemir, A.D., Willis, J. and Padwa, A. (2004) *Org. Lett.*, **6**, 917;
(h) Marion, F., Coulomb, J., Servais, A., Courillon, C., Fensterbank, L. and Malacria, M. (2006) *Tetrahedron*, **62**, 3856.

45 Nanni, D., Pareschi, P. and Tundo, A. (1996) *Tetrahedron Lett.*, **37**, 9337.

46 Pedersen, J.M., Bowman, W.R., Elsegood, M.R.J., Fletcher, A.J. and Lovell, P.J. (2005) *J. Org. Chem.*, **70**, 10615.

47 (a) Fontana, F., Minisci, F., Barbosa, M.C.N. and Vismara, E. (1991) *J. Org. Chem.*, **56**, 2866;
(b) Miranda, L.D., Cruz-Almanza, R., Pavón, M., Alva, E. and Muchowski, J.M. (1999) *Tetrahedron Lett.*, **40**, 7153;
(c) Doll, M.K.H. (1999) *J. Org. Chem.*, **64**, 1372;
(d) Motherwell, W.B. and Vázquez, S. (2000) *Tetrahedron Lett.*, **41**, 9667;
(e) Allin, S.M., Barton, W.R.S., Bowman, W.R. and McInally, T. (2001) *Tetrahedron Lett.*, **42**, 7887;
(f) Bennasar, M.L., Roca, T. and Ferrando, F. (2004) *Tetrahedron Lett.*, **45**, 5605;
(g) Bennasar, M.L., Roca, T. and Ferrando, F. (2005) *J. Org. Chem.*, **70**, 9077;
(h) Bowmann, W.R., Elsegood, M.R.J., Stein, T. and Weaver, G.W. (2007) *Org. Biomol. Chem.*, 103.

48 Bennasar, M.L., Roca, T. and Ferrando, F. (2006) *Org. Lett.*, **8**, 561.

49 Ishibashi, H., Nakamura, N., Ito, K., Kitayama, S. and Ikeda, M. (1990) *Heterocycles*, **31**, 1781.

50 (a) Aldabbagh, F., Bowman, W.R., Mann, E. and Slawin, A.M.Z. (1999) *Tetrahedron*, **55**, 8111;
(b) Allin, S.M., Barton, W.R.S., Bowman, W.R. and McInally, T. (2002) *Tetrahedron Lett.*, **43**, 4191.

51 (a) Moody, C.J. and Norton, C.L. (1997) *J. Chem. Soc., Perkin Trans.*, **1**, 2639;
(b) Van de Poël, H., Guillaumet, G. and Viaud-Massuard, M.C. (2002) *Tetrahedron Lett.*, **43**, 1205.

52 Murphy, J.A. and Sherburn, M.S. (1991) *Tetrahedron*, **47**, 4077.

53 Marco-Contelles, J. and Rodríguez-Fernández, M. (2000) *Tetrahedron Lett.*, **41**, 381.

54 (a) Pavé, G., Usse-Versluys, S., Viaud-Massuard, M.C. and Guillaumet, G. (2003) *Org. Lett.*, **5**, 4253;
(b) Huang, X. and Yu, L. (2005) *Synlett*, 2953.

55 Servais, A., Azzouz, M., Lopes, D., Courillon, C. and Malacria, M. (2007) *Angew. Chem. Int. Ed.*, **46**, 576 and references cited therein.

56 (a) Sakuragi, H., Ishikawa, S., Nishimura, T., Yoshida, M., Inamoto, N. and Tokumaru, K. (1976) *Bull. Chem. Soc. Jpn*, **49**, 1949;
(b) Forrester, A.R., Gill, M., Sadd, J.S. and Thomson, R.H. (1979) *J. Chem. Soc., Perkin Trans.*, **1**, 612;
(c) Alonso, R., Campos, P.J., García, B. and Rodríguez, M.A. (2006) *Org. Lett.*, **8**, 3521.

57 McNab, H. (1984) *J. Chem. Soc., Perkin Trans.*, **1**, 377.

58 Beckwith, A.L.J., Bowry, V.W., Bowman, W.R., Mann, E., Parr, J. and Storey, J.M.D. (2003) *Angew. Chem. Int. Ed.*, **43**, 95.

59 (a) Curran, D.P. and Liu, H. (1994) *J. Chem. Soc., Perkin Trans.*, **1**, 1377;
(b) Curran, D.P., Yu, H. and Liu, H. (1994) *Tetrahedron*, **50**, 7343.

60 Curran, D.P. and Keller, A.I. (2006) *J. Am. Chem. Soc.*, **128**, 13706.

61 Ozaki, S., Mitoh, S. and Ohmori, H. (1996) *Chem. Pharm. Bull.*, **44**, 2020.

62 (a) Ruchkina, E.L., Blake, A.J. and Mascal, M. (1999) *Tetrahedron Lett*, **40**, 8443;
(b) photochemical initiation: Taniguchi, T., Iwasaki, K., Uchiyama, M., Tamura, O. and Ishibashi, H. (2005) *Org. Lett.*, **7**, 4389.

63 (a) Artis, D.R., Cho, I.S., Jaime-Figueroa, S. and Muchowski, J.M. (1994) *J. Org. Chem.*, **59**, 2456;
(b) Menes-Arzate, M., Martínez, R., Cruz-Almanza, R., Muchowski, J.M., Osornio, Y.M. and Miranda, L.D. (2004) *J. Org. Chem.*, **69**, 4001.

64 Schmalz, H.G., Siegel, S. and Bats, J.W. (1995) *Angew. Chem. Int. Ed. Engl.*, **34**, 2383.

65 Sugawara, T., Otter, B.A. and Ueda, T. (1988) *Tetrahedron Lett.*, **29**, 75.

66 (a) Kaoudi, T., Quiclet-Sire, B., Seguin, S. and Zard, S.Z. (2000) *Angew. Chem. Int. Ed.*, **39**, 731;
(b) Quiclet-Sire, B., Sortais, B. and Zard, S.Z. (2002) *Chem. Commun.*, 1692;
(c) Binot, G. and Zard, S.Z. (2005) *Tetrahedron Lett.*, **46**, 7503;
(d) Bacque, E., Qacemi, M. El and Zard, S.Z. (2004) *Org. Lett.*, **6**, 3671.

67 Snider, B.B. (1996) *Chem. Rev.*, **96**, 339.

68 (a) Citterio, A., Fancelli, D., Finzi, C., Pesce, L. and Santi, R. (1989) *J. Org. Chem.*, **54**, 2713;
(b) see also: Citterio, A., Sebastiano, R. and Nicolini, M. (1993) *Tetrahedron*, **49**, 7743;
(c) García Ruano, J. L. and Rumbero, A. (1999) *Tetrahedron: Asymmetry*, **10**, 4427.

69 (a) Wang, S.F., Chuang, C.P. and Lee, W.H. (1999) *Tetrahedron*, **55**, 6109;
(b) Chen, H.L., Lin, C.Y., Cheng, Y.C., Tsai, A.I. and Chuang, C.P. (2005) *Synthesis*, 977.

70 Magolan, J. and Kerr, M.A. (2006) *Org. Lett.*, **8**, 4561.

71 Citterio, A., Pesce, L., Sebastiano, R. and Santi, R. (1990) *Synthesis*, 142.

72 Citterio, A., Cerati, A., Sebastiano, R., Finzi, C. and Santi, R. (1989) *Tetrahedron Lett.*, **30**, 1289.

73 (a) Beckwith, A.L.J. and Storey, J.M.D. (1995) *J. Chem. Soc., Chem. Commun.*, 977;
(b) See also: Storey, J.M.D. and Ladwa, M.M. (2006) *Tetrahedron Lett.*, **47**, 381.

74 Khan, T.A., Tripoli, R., Crawford, J.J., Martin, C.G. and Murphy, J.A. (2003) *Org. Lett.*, **5**, 2971.

75 (a) Murphy, J.A., Tripoli, R., Khan, T.A. and Mali, U.W. (2005) *Org. Lett.*, **7**, 3287;
(b) Bremner, J.B. and Sengpracha, W. (2005) *Tetrahedron*, **61**, 941.

76 (a) Araneo, S., Fontana, F., Minisci, F., Recupero, F. and Serri, A. (1995) *Tetrahedron Lett.*, **36**, 4307;
(b) See also: Bertrand, S., Hoffmann, N., Pete, J.P. and Bulach, V. (1999) *Chem. Commun.*, 2291.

77 Teichert, A., Jantos, K., Harms, K. and Studer, A. (2004) *Org. Lett.*, **6**, 3477.

78 Studer, A. and Bossart, M. (2001) *Tetrahedron*, **57**, 9649.

79 Urry, W.H. and Kharasch, M.S. (1944) *J. Am. Chem. Soc.*, **66**, 1438.

80 (a) Beckwith, A.L.J. and Ingold, K.U. (1980) *Rearrangements in Ground and Excited States* (ed. P. de Mayo), Academic Press, New York, p. 170;
(b) Freidlina, R.Kh. and Terent'ev, A.B. (1980) *Advances in Free Radical Chemistry*, Vol. 6 (ed. G.H. Williams), Heyden & Son, London, p. 32.

81 Effio, A., Griller, D., Ingold, K.U., Scaiano, J.C. and Sheng, S.J. (1980) *J. Am. Chem. Soc.*, **102**, 6063.

82 (a) Lindsay, D.A., Lusztyk, J. and Ingold, K.U. (1984) *J. Am. Chem. Soc.*, **106**, 7087;
(b) See also: Leardini, R., Nanni, D., Pedulli, G.F., Tundo, A., Zanardi, G., Foresti, E. and Palmieri, P. (1989) *J. Am. Chem. Soc.*, **111**, 7723.

83 (a) Griller, D. and Ingold, K.U. (1980) *Acc. Chem. Res.*, **13**, 317;
(b) Newcomb, M. (1993) *Tetrahedron*, **49**, 1151;
(c) Newcomb, M. (2001) *Radicals in Organic Synthesis*, Vol. 1 (eds P. Renaud and M.P. Sibi), Wiley-VCH Verlag GmbH, Weinheim, p.317.

84 (a) Ohno, A., Kito, N. and Ohnishi, Y. (1971) *Bull. Chem. Soc. Jpn*, **44**, 467;
(b) Cadogan, J.I.G., Husband, J.B. and McNab, H. (1983) *J. Chem. Soc., Perkin Trans.*, **2**, 697;
(c) Antunes, C.S.A., Bietti, M., Ercolani, G., Lanzalunga, O. and Salamone, M. (2005) *J. Org. Chem.*, **70**, 3884;
(d) Bietti, M. and Salamone, M. (2005) *J. Org. Chem.*, **70**, 10603;
(e) Ingold, K.U., Smeu, M. and DiLabio, G.A. (2006) *J. Org. Chem.*, **71**, 9906.

85 (a) Bacqué, E., El Qacemi, M. and Zard, S.Z. (2005) *Org. Lett.*, **7**, 3817;
(b) for aryl migrations from nitrogen to C-centered radicals, see: Hey, D.H. and Moynehan, T.M. (1959) *J. Chem. Soc.*, 1563;
(c) Benati, L., Spagnolo, P., Tundo, A. and Zanardi, G. (1979) *J. Chem. Soc., Chem. Commun.*, 141;
(d) Grimshaw, J. and Haslett, R.J. (1980) *J. Chem. Soc., Perkin Trans.*, **1**, 657;
(e) Lee, E., Whang, H.S. and Chung, C.K. (1995) *Tetrahedron Lett.*, **36**, 913.

86 Winstein, S., Heck, R., Lapporte, S. and Baird, R. (1956) *Experientia*, **12**, 138.

87 (a) Benati, L., Capella, L., Montevecchi, P.C. and Spagnolo, P. (1994) *J. Org. Chem.*, **59**, 2818;
(b) Montevecchi, P.C. and Navacchia, M.L. (1998) *J. Org. Chem.*, **63**, 537;
(c) Amii, H., Kondo, S. and Uneyama, K. (1998) *Chem. Commun.*, 1845;
(d) Tada, M., Shijima, H. and Nakamura, M. (2003) *Org. Biomol. Chem.*, **1**, 2499;
(e) Gheorghe, A., Quiclet-Sire, B., Villa, X. and Zard, S.Z. (2005) *Org. Lett.*, **7**, 1653;
(f) Palframan, M.J., Tchabanenko, K. and Robertson, J. (2006) *Tetrahedron Lett.*, **47**, 8423;
(g) Gandon, L.A., Russell, A.G., Güveli, T., Brodwolf, A.E., Kariuki, B.M., Spencer, N. and Snaith, J.S. (2006) *J. Org. Chem.*, **71**, 5198;
(h) Winkler, J.D. and Lee, E.C.Y. (2006) *J. Am. Chem. Soc.*, **128**, 9040.

88 (a) Reynolds, A.J., Scott, A.J., Turner, C.I. and Sherburn, M.S. (2003) *J. Am. Chem. Soc.*, **125**, 12108;
(b) for other 1,4-aryl migrations from oxygen to C-centered radicals, see: Bachi, M.D. and Bosch, E. (1989) *J. Org. Chem.*, **54**, 1234;
(c) Lee, E., Lee, C., Tae, J.S., Whang, H.S. and Li, K.S. (1993) *Tetrahedron Lett.*, **34**, 2343;
(d) Rosa, A.M., Lobo, A.M., Branco, P.S. and Prabhakar, S. (1997) *Tetrahedron*, **53**, 285;
(e) Crich, D. and Hwang, J.T. (1998) *J. Org. Chem.*, **63**, 2765;
(f) Fischer, J., Reynolds, A.J., Sharp, L.A. and Sherburn, M.S. (2004) *Org. Lett.*, **6**, 1345.

89 (a) Giraud, L., Lacôte, E. and Renaud, P. (1997) *Helv. Chim. Acta*, **80**, 2148;
(b) Alcaide, B. and Rodríguez-Vicente, A. (1998) *Tetrahedron Lett.*, **39**, 6589.

90 Wilt, J.W., Chwang, W.K., Dockus, C.F. and Tomiuk, N.M. (1978) *J. Am. Chem. Soc.*, **100**, 5534.

91 Sakurai, H. and Hosomi, A. (1970) *J. Am. Chem. Soc.*, **92**, 7507.

92 Studer, A., Bossart, M. and Steen, H. (1998) *Tetrahedron Lett.*, **39**, 8829.

93 Amrein, S., Bossart, M., Vasella, T. and Studer, A. (2000) *J. Org. Chem.*, **65**, 4281.

94 Studer, A., Bossart, M. and Vasella, T. (2000) *Org. Lett.*, **2**, 985.

95 Clive, D.L.J. and Kang, S. (2000) *Tetrahedron Lett.*, **41**, 1315.

96 (a) Loven, R. and Speckamp, W.N. (1972) *Tetrahedron Lett.*, **13**, 1567;
(b) Köhler, H.J. and Speckamp, W.N. (1980) *J. Chem. Soc., Chem. Commun.*, 142 and references cited therein.

97 (a) Motherwell, W.B. and Pennell, A. M.K. (1991) *J. Chem. Soc., Chem. Commun.*, 877;
(b) da Mata, M.L.E.N., Motherwell, W.B. and Ujjainwalla, F. (1997) *Tetrahedron Lett*, **38**, 137;
(c) da Mata, M.L.E.N., Motherwell, W.B. and Ujjainwalla, F. (1997) *Tetrahedron Lett*, **38**, 141;
(d) Bonfand, E., Forslund, L., Motherwell, W.B. and Vázquez, S. (2000) *Synlett*, 475.

98 Studer, A. and Bossart, M. (1998) *Chem. Commun.*, 2127.

99 Bossart, M., Fässler, R., Schoenberger, J. and Studer, A. (2002) *Eur. J. Org. Chem.*, 2742.

100 (a) Caddick, S., Aboutayab, K. and West, R. (1993) *Synlett*, 231;
(b) Caddick, S., Shering, C.L. and Wadman, S.N. (2000) *Tetrahedron*, **56**, 465.

101 (a) Aldabbagh, F. and Bowman, W.R. (1999) *Tetrahedron*, **55**, 4109;
(b) thiyl radicals can also act as leaving groups, see: Allin, S.M., Bowman, W.R., Karim, R. and Rahman, S.S. (2006) *Tetrahedron*, **62**, 4306.

102 Zhang, W. and Pugh, G. (2001) *Tetrahedron Lett.*, **42**, 5613.

103 Citterio, A., Minisci, F., Porta, O. and Sesana, G. (1977) *J. Am. Chem. Soc.*, **99**, 7960.

104 Tiecco, M. and Testaferri, L. (1983) *Reactive Intermediates*, Vol. 3 (ed. R.A. Abramovitch), Plenum Press, New York, p. 72;

105 (a) Hutton, J. and Waters, W.A. (1965) *J. Chem. Soc.*, 4253;
(b) Rudqvist, U. and Torssell, K. (1971) *Acta Chem. Scand.*, **25**, 2183;
(c) Baule, M., Vernin, G., Dou, H.J.M. and Metzger, J. (1971) *Bull. Soc. Chim. Fr.*, 2083;
(d) Janda, M., Šrogl, J. Stibor, I., Němec, M. and Vopatrná, P. (1973) *Tetrahedron Lett.*, **14**, 637.

106 (a) Russell, G.A., Chen, P., Kim, B.H. and Rajaratnam, R. (1997) *J. Am. Chem. Soc.*, **119**, 8795;
(b) Kim, B.H., Lee, Y.S., Lee, D.B., Jeon, I., Jun, Y.M., Baik, W. and Russell, G.A. (1998) *J. Chem. Res. Synop.*, 826.

107 Wang, C., Russell, G.A. and Trahanovsky, W.S. (1998) *J. Org. Chem.*, **63**, 9956.

108 McLoughlin, P.T.F., Clyne, M.A. and Aldabbagh, F. (2004) *Tetrahedron*, **60**, 8065.

109 Núñez, A., Sánchez, A., Burgos, C. and Alvarez-Builla, J. (2004) *Tetrahedron*, **60**, 6217.

110 Demir, A.S., Reis, Ö. and Emrullaho lu, M. (2002) *Tetrahedron*, **58**, 8055.

111 (a) Minisci, F. (1973) *Synthesis*, 1;
(b) Minisci, F., Fontana, F. and Vismara, E. (1990) *J. Heterocyclic Chem.*, **27**, 79.

112 Citterio, A., Minisci, F. and Franchi, V. (1980) *J. Org. Chem.*, **45**, 4752.

113 Minisci, F., Mondelli, R., Gardini, G.P. and Porta, O. (1972) *Tetrahedron*, **28**, 2403.

114 (a) Minisci, F., Galli, R., Malatesta, V. and Caronna, T. (1970) *Tetrahedron*, **26**, 4083;
(b) Clerici, A., Minisci, F. and Porta, O. (1974) *Tetrahedron*, **30**, 4201.

115 Fleming, I. (1976) *Frontier Orbitals and Organic Chemical Reactions*, John Wiley & Sons, Ltd, Chichester.

116 Klopman, G. (1968) *J. Am. Chem. Soc.*, **90**, 223.

117 Minisci, F., Vismara, E., Fontana, F., Morini, G., Serravelle, M. and Giordano, C. (1987) *J. Org. Chem.*, **52**, 730.

118 Recent examples: (a) Minisci, F., Recupero, F., Punta, C., Gambarotti, C., Antonietti, F., Fontana, F. and Pedulli, G.F. (2002) *Chem. Commun.*, 2496;
(b) Crowden, C.J. (2003) *Org. Lett.*, **5**, 4497;
(c) Minisci, F., Porta, O., Recupero, F., Punta, C., Gambarotti, C., Pruna, B., Pierini, M. and Fontana, F. (2004) *Synlett*, 874.

119 (a) Heiba, E.I., Dessau, R.M. and Koehl, W.J., Jr (1968) *J. Am. Chem. Soc.*, **90**, 1082;
(b) Heiba, E.I., Dessau, R.M. and Koehl, W.J., Jr (1969) *J. Am. Chem. Soc.*, **91**, 138;
(c) Heiba, E.I. and Dessau, R.M. (1971) *J. Am. Chem. Soc.*, **93**, 995.

120 Vinogradov, M.G., Verenchikov, S.P., Fedorova, T.M. and Nikishin, G.I. (1975) *J. Org. Chem. (USSR)*, **11**, 937.

121 Kurz, M.E., Baru, V. and Nguyen, P.N. (1984) *J. Org. Chem.*, **49**, 1603.

122 Kurz, M.E. and Chen, T.Y.R. (1978) *J. Org. Chem.*, **43**, 239.

123 (a) Citterio, A., Santi, R., Fiorani, T. and Strologo, S. (1989) *J. Org. Chem.*, **54**, 2703;
(b) Cho, I.S. and Muchowski, J.M. (1991) *Synthesis*, 567;
(c) Artis, D.R., Cho, I.S. and Muchowski, J.M. (1992) *Can. J. Chem.*, **70**, 1838;
(d) Baciocchi, E. and Muraglia, E. (1993) *J. Org. Chem.*, **58**, 7610;
(e) Baciocchi, E., Dell'Aira, D. and Ruzziconi, R. (1986) *Tetrahedron Lett.*, **27**, 2763;
(f) see also: Weinstock, L.M., Corley, E., Abramson, N.L., King, A.O. and Karady, S. (1988) *Heterocycles*, **27**, 2627.

124 (a) Chuang, C.P. and Wang, S.F. (1996) *Heterocycles*, **43**, 2215;
(b) Wang, S.F. and Chuang, C.P. (1997) *Heterocycles*, **45**, 347.

125 (a) Baciocchi, E., Muraglia, E. and Sleiter, C. (1992) *J. Org. Chem.*, **57**, 6817;
(b) Baciocchi, E. and Muraglia, E. (1993) *Tetrahedron Lett.*, **34**, 5015;
(c) Baciocchi, E., Muraglia, E. and Villani, C. (1994) *Synlett*, 821;
(d) Byers, J.H., Campbell, J.E., Knapp, F.H. and Thissell, J.G. (1999) *Tetrahedron Lett.*, **40**, 2677;
(e) Byers, J.H., Duff, M.P. and Woo, G.W. (2003) *Tetrahedron Lett.*, **44**, 6853;
(f) Bravo, A., Biørsvik, H.R., Fontana, F., Liguori, L., Mele, A. and Minisci, F. (1997) *J. Org. Chem.*, **62**, 7128 and references cited therein;
(g) Murakami, S., Kim, S., Ishii, H. and Fuchigami, T. (2004) *Synlett*, 815.

126 Osornio, Y.M., Cruz-Almanza, R., Jiménez-Montano, V. and Miranda, L.D. (2003) *Chem. Commun.*, 2316.

127 (a) Baran, P.S. and Richter, J.M. (2004) *J. Am. Chem. Soc.*, **126**, 7450;
(b) Baran, P.S., Richter, J.M. and Lin, D.W. (2005) *Angew. Chem.*, **117**, 615;
(c) Baran, P.S., Maimone, T.J. and Richter, J.M. (2007) *Nature*, **446**, 404.

128 Kagayama, T., Nakano, A., Sakaguchi, S. and Ishii, Y. (2006) *Org. Lett.*, **8**, 407.

129 Mu, X.J., Zou, J.P., Qian, Q.F. and Zhang, W. (2006) *Org. Lett.*, **8**, 5291.

130 (a) Tiecco, M. (1980) *Acc. Chem. Res.*, **13**, 51;
(b) Tiecco, M. (1981) *Pure Appl. Chem.*, **53**, 239.

131 Testaferri, L., Tiecco, M. and Tingoli, M. (1979) *J. Chem. Soc., Perkin Trans.*, **2**, 469.

132 Testaferri, L., Tiecco, M., Tingoli, M., Fiorentino, M. and Troisi, L. (1978) *J. Chem. Soc., Chem. Commun.*, 93.

133 Fiorentino, M., Testaferri, L., Tiecco, M. and Troisi, L. (1977) *J. Chem. Soc., Chem. Commun.*, 316.

134 Halides as leaving groups: Traynham, J.G. (1979) *Chem. Rev.*, **79**, 323.

135 Aboutayab, K., Caddick, S., Jenkins, K., Joshi, S. and Kahn, S. (1996) *Tetrahedron*, **52**, 11329.

136 Meerwein, H., Buchner, E. and van Emster, K. (1939) *J. Prakt. Chem.*, **152**, 237.

137 For a review on the Meerwein arylation, see: Rondestvedt, C.S., Jr (1976) *Org. React.*, **24**, 225.

138 For a review on radical reactions using aryldiazonium salts see: Galli, C. (1988) *Chem. Rev.*, **88**, 765.

139 (a) Kochi, J.K. (1955) *J. Am. Chem. Soc.*, **77**, 5090;
(b) Kochi, J.K. (1955) *J. Am. Chem. Soc.*, **77**, 5274.

140 Raucher, S. and Koolpe, G.A. (1983) *J. Org. Chem.*, **48**, 2066.

141 Theodoridis, G. and Malamas, P. (1991) *J. Heterocycl. Chem.*, **28**, 849.

142 Obushak, N.D., Matiichuk, V.S. and Martyak, R.L. (2003) *Heterocycl. Compd.*, **39**, 878.

143 (a) Takashi, S., Katsutoshi, M., Takeo, H., Yoshitaka, M. and Yutaka, K. (1983) *Chem. Pharm. Bull.*, **31**, 560;
(b) Yu, M., Kanji, M., Hitoshi, I., Chitoshi, H., Satoru, O. and Takashi, S. (1991) *Chem. Pharm. Bull.*, **39**, 1440.

144 Citterio, A. and Vismara, E. (1980) *Synthesis*, 291.

145 Molinaro, C., Mowat, J., Gosselin, F., O'Shea, P.D., Marcoux, J.F., Angelaud, R. and Davies, I.W. (2007) *J. Org. Chem.*, **72**, 1856.

146 Heinrich, M.R., Blank, O. and Wetzel, A. (2007) *J. Org. Chem.*, **72**, 476.

147 For recent surveys on the synthesis of five- and six-membered heterocyclic rings

by radical cyclization, see: (a) Majumdar, K.C., Basu, P.K. and Mukhopadhyay, P.P. (2005) *Tetrahedron*, **61**, 10603;
(b) Majumdar, K.C., Basu, P.K. and Chattopadhyay, S.K. (2007) *Tetrahedron*, **63**, 793.

148 Meijs, G.F. and Beckwith, A.L.J. (1986) *J. Am. Chem. Soc.*, **108**, 5890.

149 Beckwith, A.L.J. and Meijs, G.F. (1987) *J. Org. Chem.*, **52**, 1922.

150 Petrillo, G., Novi, M., Garbarino, G. and Filiberti, M. (1988) *Tetrahedron Lett.*, **29**, 4185.

151 Vaillard, S.E., Postigo, A. and Rossi, R.A. (2002) *J. Org. Chem.*, **67**, 8500.

152 Patel, V.F., Pattenden, G. and Russell, J.J. (1986) *Tetrahedron Lett.*, **27**, 2303.

153 Togo, H. and Kikuchi, O. (1988) *Tetrahedron Lett.*, **29**, 4133.

154 Bowman, W.R., Fletcher, A.J. and Potts, G.B.S. (2002) *J. Chem. Soc., Perkin Trans.*, **1**, 2747.

155 Zhou, S.Z., Bommezijn, S. and Murphy, J.A. (2002) *Org. Lett.*, **4**, 443.

156 Patro, B. and Murphy, J.A. (2000) *Org. Lett.*, **2**, 3599.

157 Lizos, D., Tripoli, R. and Murphy, J.A. (2001) *Chem. Commun.*, 2732.

158 Rivkin, A., Nagashima, T. and Curran, D.P. (2003) *Org. Lett*, **5**, 419.

159 Petit, M., Lapierre, A.J.B. and Curran, D.P. (2005) *J. Am. Chem. Soc.*, **127**, 14994.

160 Petit, M., Geib, S.J. and Curran, D.P. (2004) *Tetrahedron*, **60**, 7543.

161 Ooi, T., Furuya, M., Sakai, D. and Maruoka, K. (2001) *Adv. Synth. Catal.*, **343**, 166.

162 (a) Crich, D. and Yao, Q. (2004) *J. Am. Chem. Soc.*, **126**, 8232;
(b) Crich, D. and Yao, Q. (2003) *Org. Lett.*, **5**, 2189.

163 Carta, P., Puljic, N., Robert, C., Dhimane, A.L., Festerbank, L., Lacôte, E. and Malacria, M. (2007) *Org. Lett.*, **9**, 1061.

164 Coulomb, J., Certal, V., Festerbank, L., Lacôte, E. and Malacria, M. (2006) *Angew. Chem. Int. Ed.*, **45**, 633.

165 Schiesser, C.H. and Sutej, K. (1992) *Tetrahedron Lett.*, **33**, 5137.

166 Engman, L., Laws, M.J., Malmström, J., Schiesser, C.H. and Zugaro, L.M. (1999) *J. Org. Chem.*, **64**, 6764.

167 Laws, M.J. and Schiesser, C.H. (1997) *Tetrahedron Lett.*, **38**, 8429.

168 Sato, A., Yorimitsu, H. and Oshima, K. (2006) *J. Am. Chem. Soc.*, **128**, 4240.

169 Vaillard, S.E., Mück-Lichtenfeld, C., Grimme, S. and Studer, A. (2007) *Angew. Chem. Int. Ed.*, **46**, 6533.

14
Photochemical Arylation Reactions

Valentina Dichiarante, Maurizio Fagnoni and Angelo Albini

14.1
Introduction

Photochemical methods [1–3], in particular those which are aimed at carbon–carbon bond formation, occupy a rather small place among the tools of the synthetic organic chemist. It is the belief of the present authors that this situation is not due to these techniques having a limited potential, but rather to the limited interest that has been shown in this field. Today, whereas other methods are more widely investigated and produce a proportionally large output, photochemical methods tend to remain confined to a few dedicated laboratories.

Photoinduced arylations are an exemplary case. Although, initially, the emergence of the $S_{RN}1$ reaction [4–6] roused much interest, this has not been followed up as widely as might have been expected. Over the years many problems with photoinduced arylation have been identified but the solutions to these problems, or the introduction of improvements, might have met with a greater success, had the method been pursued by a greater number of laboratories. Nevertheless, new classes of photoinduced arylations have been reported, notably in relation to methods employing radical ions and, more recently, via aryl cations. Consequently, today's panorama of photoinduced arylation is sufficiently rich and varied as to deserve presentation among other arylation methods. In this chapter we describe the details of various types of photoarylation, and also review the relevant literature of the past decade. Where available, previous studies are referred to through review articles, and the literature as a whole is reviewed up to mid 2007.

Electronic excitation opens several possible reaction paths for arylations, as summarized in Scheme 14.1:

- Homolysis (path *a*): With simple benzene derivatives the energy of the excited states (certainly of the singlet, possibly also of the triplet), is of the same order as that of many covalent bonds. Thus, phenyl halides are cleaved, albeit with varying efficiency, and generate the phenyl radical that may give the arylation product via reaction with a radical trapping reagent T (for a detailed discussion

Modern Arylation Methods. Edited by Lutz Ackermann

ISBN: 978-3-527-31937-4

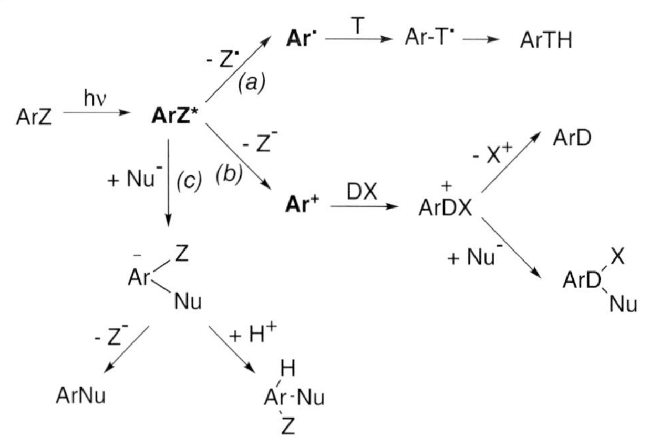

Scheme 14.1 Possible reaction paths in photoarylation reactions.

on radical arylation methods, see Chapter 13) [7]. The efficiency depends on the excited state involved (singlet or triplet) and on the halogen present. There is, however, no simple relationship, because heavier atoms not only result in weaker bonds but also shorten the lifetime of the excited states, thus diminishing the likelihood of fragmentation. With naphthalenes and polycondensed aromatics, the excited states are lower in energy and fragmentation becomes less likely.

- Heterolysis (path *b*): The above energetic considerations are also of relevance here, even if aryl cations have been considered as a paradigmatic example of highly reactive species. In polar media, heterolysis has been observed for a–perhaps unexpectedly–large range of compounds, such as phenyl chlorides, fluorides, triflates, mesylates and phosphates, provided that they bear an electron-donating substituent. The corresponding cation is subsequently trapped by a donor D-X [8, 9].

- Photosubstitution (path *c*): Arenes, particularly when bearing an electron-withdrawing substituent, are electrophiles in the excited state and may react with a nucleophile Nu^- forming a new bond via a $S_N2(Ar^*)$ reaction path [10, 11].

- Single electron transfer: this is an unusual process in ground-state chemistry, but is quite common in photochemistry; it leads to various, well recognized transformations, as summarized in Scheme 14.2 [12, 13].

Electron transfer (ET) may involve the initial formation of an electron donor–acceptor complex (ArZ–DX) in the ground state, in which case a new absorption band appears and a selective photoexcitation at a specific wavelength hν′ is possible (in Scheme 14.2). Either way, the chemistry following ET depends on which of the radical ions is more reactive. A quite general process is the $S_{RN}1$ reaction (path *d* in Scheme 14.2), during which an aromatic radical anion $ArZ^{\bullet -}$ (usually from an

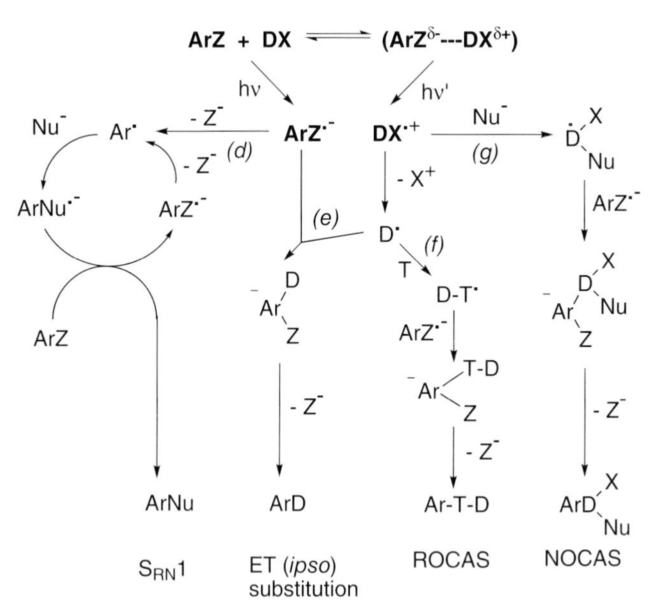

Scheme 14.2 Arylations via photoinduced electron transfer (PET) reactions.

aryl halide) is formed and cleaves to give an aryl radical. Here, the strength of the C–Z bond is an important parameter, and often the reaction is successfully performed with aryl iodides. Trapping by a nucleophile Nu^- (that usually acts as the donor in the initiating step), such as an enolate anion, and electron transfer from the starting material give rise to a chain process leading to the corresponding arylated product [4–6].

In the other transformations, the aromatic radical anion $ArZ^{\bullet -}$ is more stable (e.g. when generated from an aromatic nitrile), and it is thus the coformed radical cation $DX^{\bullet +}$ that reacts. Unimolecular fragmentation of the latter, which is a rapid process with various aliphatic derivatives such as acetals or silanes, leads to radical $D^\bullet$, which couples with the aryl radical anion. The resulting adduct anion $D–Ar–Z^-$ loses Z^-, resulting in overall *ipso*-substitution by an alkyl group (Scheme 14.2, path *e*). A synthetically useful reaction involves trapping of the alkyl radical D by a trapping reagent T (typically an electrophilic alkene), coupling of the radical adduct D–T with aryl radical anion $ArZ^{\bullet -}$, and a final aromatization to form three-component adducts T–Ar–D [the ROCAS (Radical Olefin Combination Aromatic Substitution) process], as shown in Scheme 14.2, path *f*.

When the radical cation $DX^{\bullet +}$ does not undergo fragmentation, as is typical when arising from an olefin, an alternative is found in the attack by a nucleophile Nu^-. The resulting radical X–D–Nu couples with the aryl radical anion $ArZ^{\bullet -}$ and regains aromaticity, yielding again a three-components adduct X–ArD–Nu [the NOCAS (Nucleophile Olefin Combination Aromatic Substitution) process, as shown in Scheme 14.2, path g [14, 15]. In Scheme 14.2, the aromatic compound

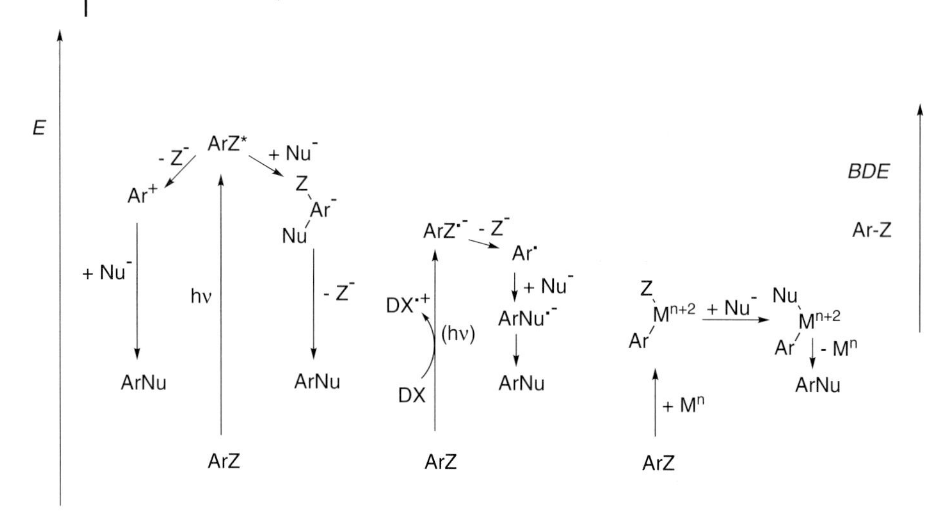

Scheme 14.3 Comparison of the energy of intermediates involved in arylation reactions.

is reduced in the electron-transfer process. However, the chemical reactivities are reversed when an electron-rich arene and an electron-deficient alkene are used. Here, the radical anion of the latter is protonated and the radical couples with the radical cation of the arene [the EOCAS (Electrophile Olefin Combination Aromatic Substitution) process] [15, 16].

In summary, photochemical methods exploit the high energy of the adsorbed photon in various ways – that is, either via reaction of the electronically excited state itself, whether involving the usual addition–elimination mechanism of arenes or unimolecular fragmentations, or via electron transfer followed by fragmentation of one of the charged radical ions.

One aspect worth noting here is the energy of the key intermediate (Scheme 14.3). Electronically excited states are rather high in energy, and when the subsequent chemical reactions proceed directly from them, high-energy intermediates are accessible. As an example, phenyl radicals or phenyl cations are formed by the cleavage of an Ar—Z bond (Scheme 14.3, left part) and even relatively strong bonds, such as C—F bonds, may fragment under these reaction conditions (see path *b* in Scheme 14.1).

However, excited states represent short-lived intermediates. When their chemical reaction is too slow, electron transfer to or from a suitable additive may prevent chemically unproductive decay to the ground state. Thereby, a radical cation or radical anion is formed, the energy of which is necessarily lower than that of the excited state (Scheme 14.3, central part). Therefore, the energy of the cleavable bond has a more important role in this process, and often such reactions are satisfactory with aryl iodides, but not with aryl chlorides.

In any case, the large amount of energy and the change in electronic structure induced by light absorption make excited states highly reactive intermediates under conditions as mild as those of the best catalytic reactions. It should be further noted here that photochemical arylations via aryl cation bear some mechanistic resemblance to metal-catalyzed arylations. In fact, the photogenerated Ar^+ cation reacts similarly to the complexed cation $Ar–M^+$ (Scheme 14.3, right part).

As the key intermediates are produced photochemically under mild reaction conditions, this can make it easier to control the subsequent chemistry, such as the trapping of such intermediates, and also make photochemical reactions experimentally less demanding than comparable, metal-catalyzed processes. In fact, this is another aspect that should encourage further work in this area of research [8, 9, 17]. The photochemical method is thus potentially complementary to metal-catalyses, as it allows a direct access to high-energy intermediates. However, in thermal metal-catalyses the interaction with the metal allows the possibility of fine-tuning the chemical reactivity due to ligand–metal interactions (Scheme 14.3). The extensive exploration of metal catalysis for arylation has led to a variety of new procedures, the wide selectivity of which has not yet been achieved by the much less-studied photochemical methods.

14.2 Photochemical Formation of Aryl—C Bonds

14.2.1 Intermolecular Formation of Aryl—Alkyl Bonds

Photochemistry offers a convenient access to the formation of aryl—alkyl bonds under mild reaction conditions. This can be accomplished either via fragmentation of an aromatic derivative to produce a trappable aryl radical or aryl cation as intermediate, or via the activation of an aliphatic component that then reacts with the arene derivative.

A classical example of a thermal method via radicals is the Meerwein arylation of alkenes, which is based on the copper-mediated reduction of aryldiazonium salts, causing dinitrogen loss and attack of the resulting phenyl radical onto an electrophilic alkene (see Section 13.4.1). The reaction can be also photochemically induced (path *a*, Scheme 14.1), for example through the photodecomposition of phenyl iodides. However, these compounds absorb only weakly in the near ultraviolet (UV) and the reaction is synthetically less practical.

Initially, the heterolytic fragmentation of substituted aromatics (path *b*, Scheme 14.1) was considered seldom to occur. However, recent studies have shown this to be the main photochemical process from phenyl fluorides, chlorides, mesylates, triflates and phosphates [18] bearing an amino, alkoxy, hydroxy or thioalkoxy [19] group in *ortho, meta* (despite the stronger competition by solvolysis in this case [20]) or *para* position in a polar medium, such that a S_N1 reaction results [8, 9, 21]. It should be noted here, that acetonitrile is sufficiently activating for substituted anilines, while methanol or trifluoroethanol are required for substituted anisoles, phenols and thioanisoles. Under these reaction conditions the corresponding *triplet* aryl cation is formed; this is a decisive feature, as these species exhibit a

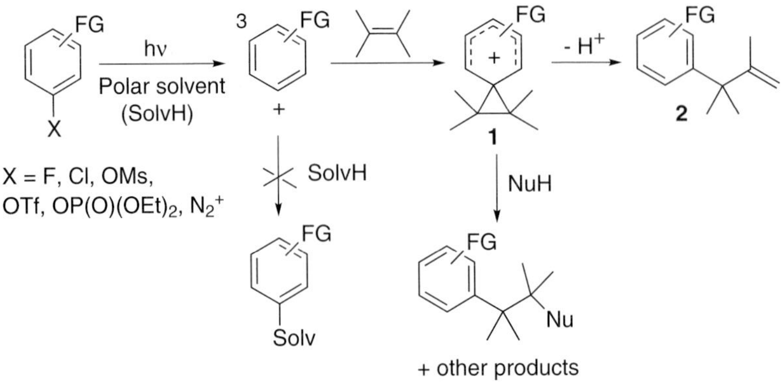

Scheme 14.4 Generation and reactivity of an aryl cation Ar^+.

high reactivity towards π-nucleophiles but are much less reactive with respect to n-nucleophiles. As a result, the irradiation can be carried out in an ion-stabilizing solvent (e.g. an alcohol) that make the fragmentation efficient but does not compete as a nucleophile for the aryl cation. These cations are rather selectively trapped by alkenes, alkynes or arenes, provided that a sufficiently high concentration of the π-nucleophile is used (0.5 to 1 *M*). The ensuing intersystem crossing in the thus-formed adduct cation leads to a singlet state cation, that is phenonium ion **1**, and hence to the 'common' carbocation chemistry, giving the final products via deprotonation or addition of a nucleophile NuH (Scheme 14.4).

Phenyl cations are likewise generated by irradiation of the corresponding diazonium salts. In this case, no activating group is required, but the substituents play an important role. Thus, the cation is formed selectively in the desired triplet state with electron-withdrawing substituted derivatives, whereas in the other cases sensitization by benzophenone or xanthone is required [22].

As an example, the products obtained from the irradiation of 4-chloroaniline (Scheme 14.4; FG = NH_2, X = Cl) in the presence of an olefin, such as 2,3-dimethyl-2-butene, arise from the adduct cation **1**. The addition of chloride or an alcohol, used as solvent, produces a β-chloroethylaniline or a β-alkoxyethylaniline, respectively, while deprotonation yields an allylaniline **2** [21, 23].

The product distribution can be somewhat complicated, particularly with unsymmetrically substituted olefins, because of the different viable reaction paths (as illustrated above with the competition between deprotonation and nucleophilic addition), as well as potential cationic (Wagner–Meerwein) rearrangements [24] of the adduct cation. The significance of these processes largely varies with the given reagents and reaction conditions. Nevertheless, good yields of a single product are obtained in a number of cases. Some typical examples are provided in Scheme 14.5. Thus, allylbenzenes are obtained exclusively by irradiation in the presence of allyltrimethylsilane, as shown in the synthesis of safrole (**3**) [25]. Further, a conjugated olefin or arylalkynes, such as 4-(1-hexynyl) anisole (**4**), were accessed from 1,1-diphenylethylene [26] or terminal alkynes [27], respectively (Scheme

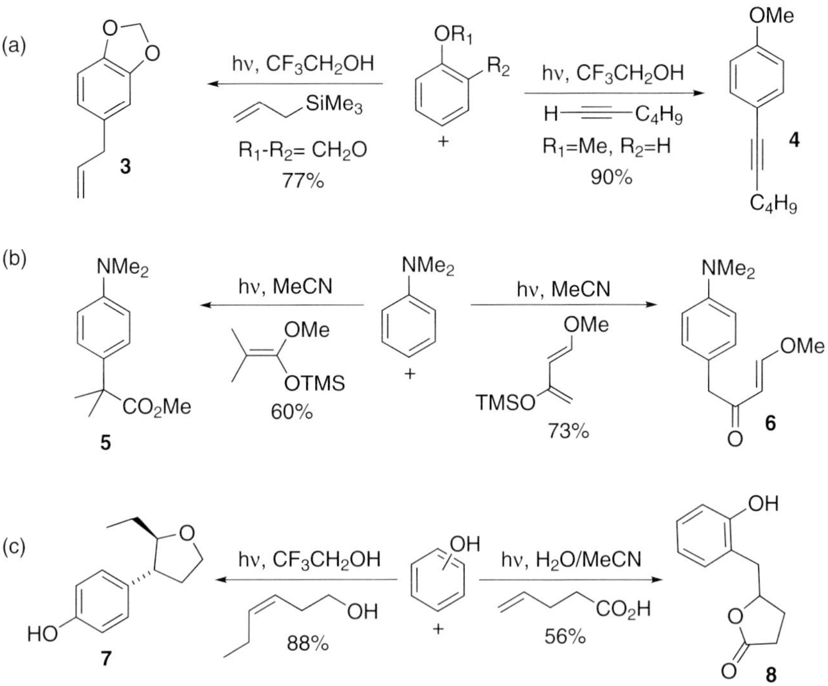

Scheme 14.5 Intermolecular arylations via photogenerated phenyl cations.

14.5a). Acyclic dienes are selectively arylated at a single alkene moiety, while the cyclic dienes form an additional C–C bond (see below) [28]. Enol ethers and analogous reagents have been proven quite useful, and phenylacetates or propionates (e.g. **5**), including common anti-inflammatory drugs, have been obtained by trapping of the cation by ketene silyl acetals. Aryl methyl ketones (e.g. **6**) have similarly been isolated using silyl enol ethers or the Danishefsky diene [29] (Scheme 14.5b). When the trapping reagents contain a nucleophilic functional group, cyclizations can take place at the cation adduct stage. This functional group can be a C–C double bond, as in the above case of cyclic dienes (e.g. 1,5-cyclo-octadiene) that provides arylated bicyclo[3.3.0]octanes [28]. Furthermore, a heteroatom functionality can also be used such that, when employing 3-alkenols or 4-alkenoic acids aryl-substituted tetrahydrofurans (e.g. **7**) or benzyl-substituted γ-lactones (e.g. **8**) are obtained respectively (Scheme 14.5c) [30].

These photochemical reactions with olefins can be considered a cationic analogue of the Meerwein arylation that occurs with nucleophilic rather than with electrophilic alkenes. The rapid cleavage of excited aryl halides and esters in polar solvent and the efficient trapping of the formed aryl cation render these arylations normally less-sensitive towards dissolved oxygen, in contrast to many other photochemical reactions. These characteristics, along with the mild reaction conditions and the simple experimental set-up, make the photochemical method a complementary and valuable alternative to metal-mediated or -catalyzed reactions.

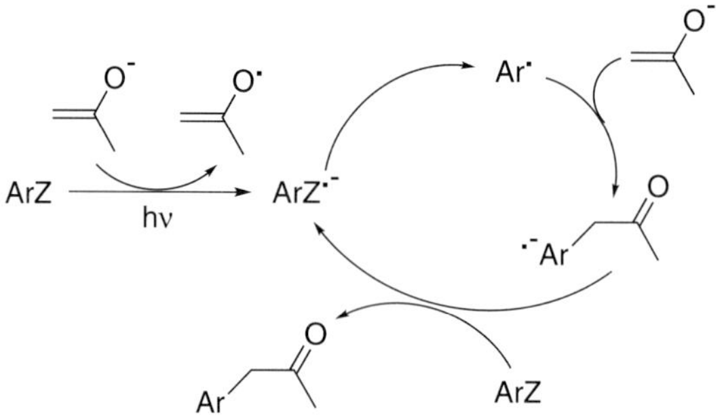

Scheme 14.6 Synthesis of aryl methyl ketones via $ArS_{RN}1$ reaction.

As mentioned above, radical alkylations may be achieved through the homolysis of an C(sp^2)—Z bond upon direct irradiation (path *a*, Scheme 14.1; see also Scheme 14.18 for an example). A much more convenient method, however, is the $S_{RN}1$ process (path *d*, Scheme 14.2). In this case, it is the aromatic radical anion arising from photoinduced electron transfer (PET) that cleaves and gives the radical. The process then proceeds via a radical-chain mechanism, provided that the fragmentation of the C—Z bond is efficient, and therefore depends on the strength of that bond. It is for this reason that iodides are often used successfully in this situation. Due to the chain mechanism only a weak illumination is required, and in some cases the ambient light in the research laboratory will be sufficient. In some systems, the initial ET step is sufficiently rapid so as to allow the reaction to occur thermally, without photochemical initiation. Enolates are widely used both as a trapping reagent for the aryl radical and as an initiating donor, and a large variety of aryl methyl ketones has been prepared by using this method (Scheme 14.6). Although some failures have been reported, this procedure has met with considerable success, and its interest is reflected by the production of numerous reviews on the subject [4–6].

Only selected recent examples will be presented here (Scheme 14.7) [31, 32]. (Hetero)aromatic ketones have been prepared in good yields most often using enolates of methyl aryl ketones and aryl halides (Scheme 14.7a), which makes this arylation a valuable methodology for the preparation of this class of compound. When a stereogenic center is formed during the transformation, diastereoselectivities of between 31% and 98% can be achieved with chiral, enantiomerically enriched phenylacetamide enolates [33]. Interestingly, with phenyl-substituted enolates arylation takes place to some extent at the arene. In fact, it has been found that the regioselectivity of the arylation can be controlled through the choice of the cation. Thus, selective arylation either at the α-position (**9**) or at the arene in *para*-position (**10**) was accomplished by using either the dilithium or the dipotassium dianion of phenylacetic acid, respectively (Scheme 14.7b) [34]. Substitution/cyclocondensation reaction sequences have been largely applied to phenyl halides bearing suitable *ortho*-substituents for the formation of heterocycles. This approach has been used for syntheses of indoles from *o*-iodoanilines, and similarly for the

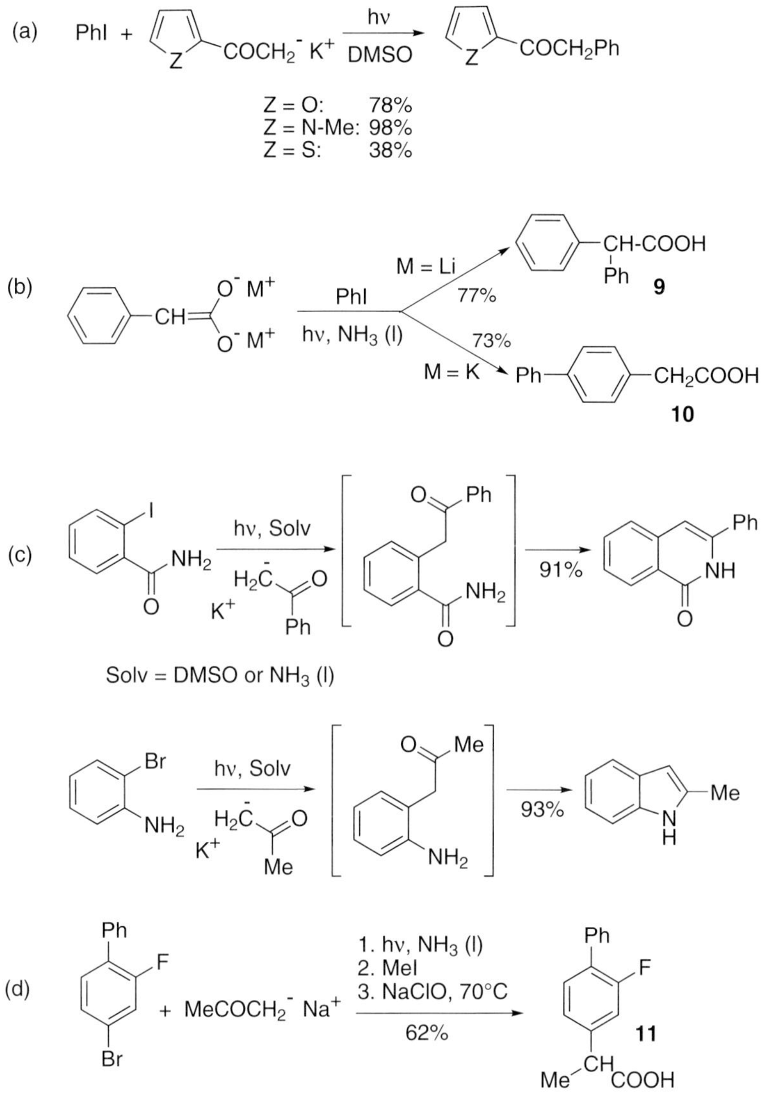

(a) PhI + [2-(Z-heteroaryl)]$COCH_2^-$ K^+ —hν, DMSO→ [2-(Z-heteroaryl)]$COCH_2Ph$

Z = O: 78%
Z = N-Me: 98%
Z = S: 38%

(b) PhCH=C(O^- M^+)$_2$ —PhI, hν, NH_3 (l)→ M = Li, 77%: PhCH(Ph)-COOH **9**; M = K, 73%: Ph–C_6H_4–CH_2COOH **10**

(c) 2-iodobenzamide (NH_2) + $H_2\bar{C}$C(O)Ph K^+ —hν, Solv→ [2-(PhCOCH$_2$)benzamide] —91%→ 3-phenylisoquinolinone (Ph, NH, O)

Solv = DMSO or NH_3 (l)

2-bromoaniline + $H_2\bar{C}$C(O)Me K^+ —hν, Solv→ [2-(MeCOCH$_2$)aniline] —93%→ 2-methylindole (Me, N–H)

(d) 4-bromo-2-fluorobiphenyl (Ph, F, Br) + $MeCOCH_2^-$ Na^+ —1. hν, NH_3 (l); 2. MeI; 3. NaClO, 70°C; 62%→ **11** (Ph, F, Me–CH–COOH)

Scheme 14.7 Formation of Ar–C bonds via photostimulated $ArS_{RN}1$ reaction.

preparation of benzofurans, quinolines, isoquinolines, phenanthridines, benzazepines and benzoxazine, as well as further annulated aza-heterocycles (Scheme 14.7c) [35–38]. The basic reaction conditions of these syntheses can be exploited for further alkylation reactions, as in the photostimulated reaction of 4-bromo-2-fluorobiphenyl with the acetone enolate in liquid ammonia. Here, the substitution product was methylated by methyl iodide to give 3-(3′fluorobiphenyl)-2-butanone, which subsequently afforded the drug flurbiprofen (**11**) by hypochlorite oxidation (Scheme 14.7d) [39].

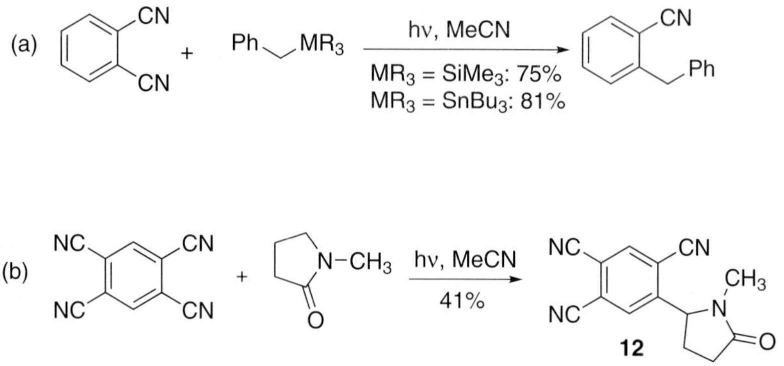

Scheme 14.8 Photoalkylation of aromatic nitriles.

As mentioned in Section 14.1, electron-withdrawing substituted arenes are good acceptors in the excited state, and often undergo electron-transfer reactions. The most typical transformation is that of aromatic nitriles which, in the presence of suitable donors, undergo an *ipso*-substitution reaction of the cyano group by an alkyl group. The process involves PET, fragmentation of the donor radical cation, coupling between the resulting radical and the acceptor radical anion and cyanide elimination allowing rearomatization (see Scheme 14.2, path *e*) [10]. This can be considered as a mild metal-free photochemical activation of aromatic nitriles [40] that is otherwise carried out thermally using nickel catalysts [41, 42]. Furthermore, this is an interesting example of a reaction, during which a Ar—C bond is formed at the expense of another Ar—C bond [40, 42].

Two examples belonging to this class of reaction are shown in Scheme 14.8. In the first transformation, *o*-dicyanobenzene was benzylated to form *o*-benzylbenzonitriles when irradiated in the presence of either benzyltrimethylsilane (75% yield) [43] or benzyltributylstannane (81%) [44] via the benzyl radical formed by fragmentation of the organometallic radical cations. The reaction led to a clean monobenzylation and the formation of disubstituted products could be prevented, because the reduction potential of the monobenzylated benzonitrile in the excited state is not sufficient to allow any further oxidation of either the silane or the stannane (Scheme 14.8a).

Recently, substituted amides and lactams, such as dimethylformamide (DMF) or 2-pyrrolidones, were used as electron donors in the photoalkylation of 1,2,4,5-tetracyanobenzene (TCB) [45]. The success of the reaction was ascribed to the high reduction potential of TCB in the excited state that made the initial step – that is, the oxidation of amides [$E_{1/2}^{ox}$ DMF = 2.29 V versus standard calomel electrode (SCE)] – possible. It should be noted here that the ensuing deprotonation step was found to be chemoselective when using *N*-methylpyrrolidone (NMP), as illustrated in Scheme 14.8b. Accordingly, deprotonation of the lactam radical cation intermediate occurred exclusively at the methylene, and not from the methyl group. Tricyano benzene **12** was thus isolated in 41% yield.

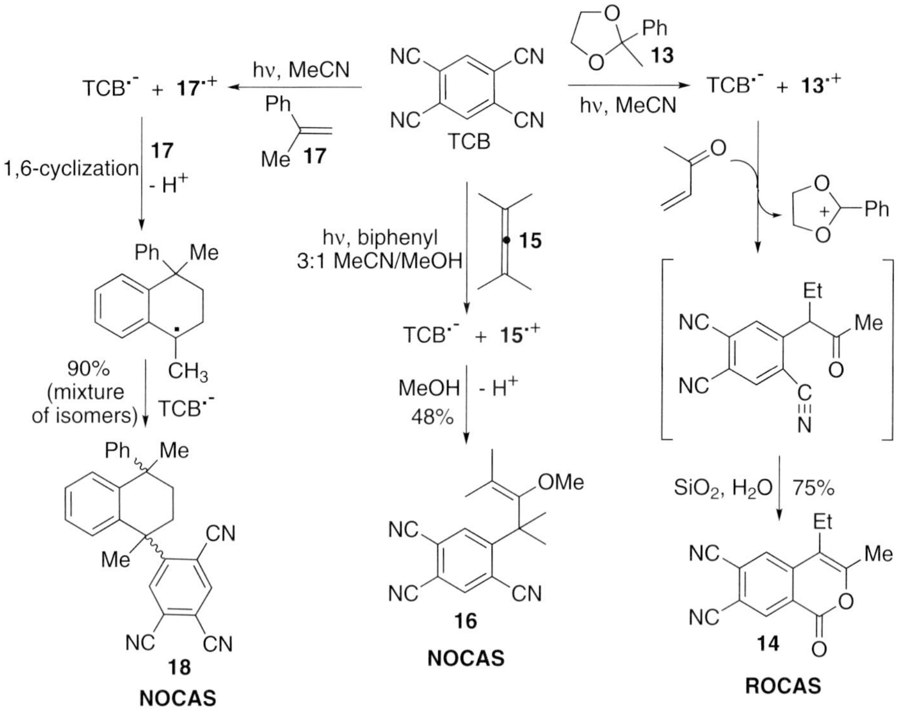

Scheme 14.9 Different pathways in the PET functionalization of tetracyanobenzene (TCB).

Heteroarenes have been photochemically functionalized by PET reactions forming new C–C bonds both in an inter- and intramolecular fashion via a similar mechanism [46]. The heteroarenes could serve both as electron donor (e.g. pyrroles or indoles) or electron acceptor (e.g. cyanopyridines or cyanopyrazines). Again, fragmentation of the radical cation, coupled with the radical anion and loss of the anion, led to overall *ipso*-substitution. In addition to the cyano group, halides could also function as leaving groups, such that in some cases an attack at an unsubstituted position took place [46].

N-Alkylacridinium ions and other cationic heteroarenes likewise underwent photoalkylation, but these processes were reductive additions, rather than photosubstitution, yielding acridanes [46]. In these cases, both tetra-alkylstannanes and the more environmentally benign 4-alkyl-1-benzyl-1,4-dihydronicotinamides have recently been used as a source of alkyl radicals [47].

More elaborate functionalizations could be accomplished when the donor or the radical cation underwent more complex reactions. As an example, the above synthesis of alkyltricyanobenzenes could be varied into a ROCAS reaction (Scheme 14.9, right part) [48, 49]. Thus, photolysis of a mixture of TCB as acceptor, 2-methyl-2-phenyl-1,3-dioxolane (**13**) as donor and an electron-poor olefin formed a 3-aryl substituted 2-pentanone, which cyclized during purification on silica gel to afford isocoumarine **14** in 75% yield [48]. This three-component reaction is based on the initial formation of radical cation $\mathbf{13}^{\bullet+}$ and fragmentation of this intermediate, generating a methyl radical, which in turn added to the enone (see also Scheme 14.2, path *f*). It should be noted that, in this reaction, two C—C bonds were cleaved, while two C—C bonds were formed in a one-pot procedure.

A further variation of these functionalizations of cyanoarenes is the NOCAS process [14, 15]. As shown in Scheme 14.2, path g, this involves the addition of a nucleophile (which is often the solvent) to the donor radical cation. The thus-formed neutral radical adds to the acceptor radical anion, while rearomatization by the loss of an anion leads again to an overall *ipso*-substitution. Allenes could be used as the donors in these reactions, as shown recently by Arnold [50]. Accordingly, the irradiation of TCB in the presence of tetramethylallene (**15**) in a 3:1 MeCN/MeOH mixture afforded 1:1:1 arene–allene–methanol adduct **16** in 48% yield (Scheme 14.9, central part). Interestingly, the addition of methanol took place exclusively at the central allene carbon, while aromatic substitution occurred through the terminal carbons. ω-Alkenols, in which an *O*-nucleophile and an easily oxidized moiety are both present, could also be used. In the latter case, the initial ET was followed by a cyclization, yielding aryl-substituted tetrahydrofurans or tetrahydropyrans as the final products via a tandem Ar–C, C—O bond formation [51].

Ammonia or amines were likewise used as nucleophiles in place of alcohols [52]. Here, PET between an arene S (e.g. triphenylbenzenes) and 1,4-dicyanobenzene (DCB) led to the arene radical cation $S^{\bullet+}$ and the radical anion $DCB^{\bullet-}$. Secondary ET from a donor (D) to $S^{\bullet+}$ produced the radical cation of the former ($D^{\bullet+}$) that added ammonia. After deprotonation, coupling of the resulting neutral radical with $DCB^{\bullet-}$ and cyanide loss led to the corresponding NOCAS product. Suitable donors were arylcyclopropanes, quadricyclane and dienes that gave 4-(1-aryl-3-aminopropyl)benzonitriles, 4-[7-3-aminonorborn-5-enyl)-benzonitrile [53] and 4-(4-aminobut-2-enyl)-benzonitriles, respectively [54].

A variation of the reaction involved the use of the alkene itself as nucleophile. In this case, a radical cation dimer was formed by attack of the alkene radical cation by the neutral alkene, forming a distonic radical cation (Scheme 14.9, left part). With α-methylstyrene (**17**) as the alkene, a cyclization took place and the neutral radical resulting from the ensuing deprotonation coupled with the radical anion of the acceptor (in this case TCB), leading to the NOCAS adduct **18** as a diastereoisomeric mixture in overall 90% yield [55]. The irradiation of aromatic nitriles in the presence of alkenes may lead to different products, particularly when carried out in an apolar medium. As an example, 1,4-dicyanobenzene gave isoquinolines by a [4+2]-cycloaddition with a cyano group through irradiation in the presence of diphenylethylenes in benzene via a polar exciplex [56].

Few examples have been reported involving the complementary photo-EOCAS reaction, which corresponds to path g in Scheme 14.2, but with reversed charges [15]. Typical reactions involved a methoxyarene (e.g. 1,4-dimethoxybenzene or 1-methoxynaphthalene) as donors and an electron-poor olefin (e.g. acrylonitrile) as acceptors [16]. As an example, the photochemically produced radical anion of the acrylonitrile was protonated by MeOH as solvent, and then coupled with the aryl ether radical cation to form 2-aryl-propionitriles. With regards to the donor, the reaction bore some similarity to a Friedel–Crafts alkylation, whilst with regards to the alkene it resulted in arylation at the α-position of the α,β-unsaturated nitrile, rather than at the more commonly arylated β-position.

14.2.2 Cyanations

Another important class of reactions involves the introduction of a cyano group by substitution in an Ar–Z precursor. In fact, novel pathways leading to aromatic nitriles – for example, photosubstitution reactions – are desirable in view of the many applications of aryl cyanides as agrochemicals and pharmaceuticals. Today, the classical copper(I)-mediated Rosenmund–von Braun and Sandmeyer reactions, from aryl halides and aryldiazonium salts respectively, have been supplanted by reactions which employ palladium- or copper-catalysis [57]. The rather common use of excess cyanide anion may lead to a deactivation of the catalyst, and affect to a remarkable extent each of the key steps of the catalytic cycle [58a]. Although the use of complex iron cyanide has been shown to offer an effective solution to this limitation [58b,c], photocyanation provides an equally useful alternative [10]. At present, these reactions have received only limited attention, but good chemical yields have been reported in a number of cases [10]. The advantages of this method are the mildness of the reaction conditions employed (e.g. a low reaction temperature), the use of water as cosolvent, and a wide variety of leaving groups that could be displaced, as illustrated in Scheme 14.10.

The reaction took place either via the ArS_N1 or via the ArS_N2 mechanism (see Scheme 14.1, paths *b,c*), and occurred successfully via the first mechanism when using aryl halides or esters with electron-donating substituents. Among halides, both aryl chlorides and fluorides underwent photosubstitution when irradiated in an aqueous MeCN solution of KCN (Scheme 14.10, left part) [59]. It should be noted that, in transition metal-catalyzed reactions, the substitution of a chloro- by a cyano-group occurs only under relatively harsh reaction conditions, whereas such a process does not take place at all with aryl fluorides [57]. In the case of aryl esters the photoinduced cyanation occurred conveniently. As esters are easily

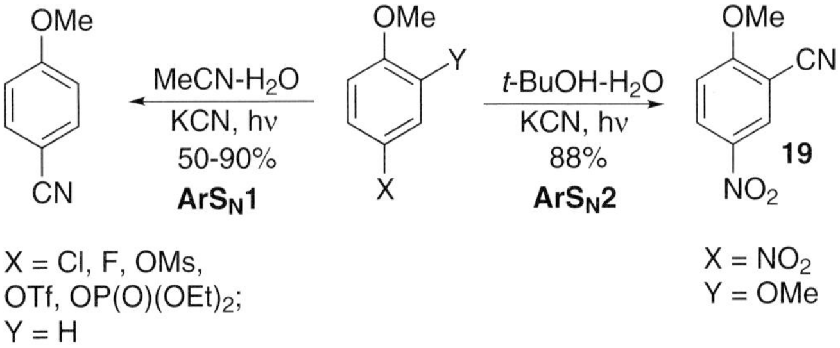

OMe, CN ← $MeCN\text{-}H_2O$, KCN, hν, 50-90%, **ArS_N1** — OMe, Y, X — $t\text{-}BuOH\text{-}H_2O$, KCN, hν, 88%, **$ArS_N2$** → OMe, CN, NO_2 **19**

X = Cl, F, OMs, OTf, $OP(O)(OEt)_2$; Y = H

X = NO_2; Y = OMe

Scheme 14.10 Photocyanation of arenes.

available from the corresponding phenols, this reaction could be viewed as a two-step strategy for the substitution of a hydroxy-group by a cyano-group [59].

In the case of ArS_N2 reactions, a number of processes involving this mechanism have been reported, although most have not been sufficiently selective as to be preparatively useful [10]. However, both methoxy- and nitro-groups acted as leaving groups under photochemical reaction conditions, and when both were present, one of these could be chemoselectively substituted, as shown by the synthesis of 2-methoxy-5-nitrobenzonitrile (**19**) through the irradiation of 4-nitroveratrole in the presence of KCN [60] (Scheme 14.10, right part). Phenyl sulfones, sulfonamides and sulfoxides could be also used as starting materials for the photochemical synthesis of benzonitriles [61]. Another possibility would be to introduce a cyano group by means of a $S_{RN}1$ reaction, thus involving the addition of cyanide to an aryl radical generated by the decomposition of phenylazosulfides [62].

An attack at an unsubstituted position was obtained via the addition of cyanide to the radical cation of an arene, and subsequent rearomatization. A recent example of this strategy is provided by the photocyanation of pyrene in the presence of DCB and NaCN in an oil-in-water emulsion system to give 1-cyanopyrene in 83% yield [63]. The direct formation of a σ-complex from the triplet was envisaged in the photocyanation of 2-halo-4-nitroanisoles in a acetonitrile:water solution (1:2), where the relative rates for the attack by cyanide on the halogenated compounds were in the ratio 27:2:2:1 for F, Cl, Br and I, respectively [64].

14.2.3 Intramolecular Formation of Aryl—Alkyl Bonds

Aryl—alkyl bonds are photochemically formed also in an intramolecular fashion, in most cases through a photostimulated $ArS_{RN}1$ reaction, starting from aryl halides bearing a nucleophilic moiety in *ortho* position to the halogen [4–6].

Thereby, 3-substituted 2,3-dihydrobenzofurans, 1,2-dihydronaphtho(2,1-*b*)furans and *N*-substituted 2,3-dihydro-1*H*-indoles were synthesized in very good yields in liquid ammonia, starting from aryl allyl ethers or amines using Me_3Sn^-, Ph_2P^- or $^-CH_2NO_2$ anions as nucleophiles (Scheme 14.11, path *a*, n = 1). The key step is the selective 5-*exo* ring closure by the phenyl radical onto the olefin moiety, which was followed by addition of the nucleophiles to give the substituted heterocycles. When using the nitromethane anion, two C—C bonds were formed in this reaction [65, 66]. The $S_{RN}1$ reaction with ring closure was again found useful for the preparation of valuable derivatives such as 4-substituted chromanes and 4-substituted benzo[*f*]chromanes in good yields (Scheme 14.11, path *a*, n = 2) [66]. The radical site in the cyclized radical intermediate could be reduced, rather than functionalized, as illustrated in path *b* of Scheme 14.11. This tandem cyclization/reduction sequence could be successfully achieved using cyclohexadienyl anion **20** as the hydrogen-donor, which is itself available from nontoxic, reusable and inexpensive ethyl benzoate [67].

The function of electron donor in the initial PET step could be fulfilled by a suitable additive. Thus, 1-allyloxy-2-iodobenzene (**21**), upon photolysis in the presence of 1,3-dimethyl-2-phenylbenzimidazoline (DMPBI) as donor in DMF, gave

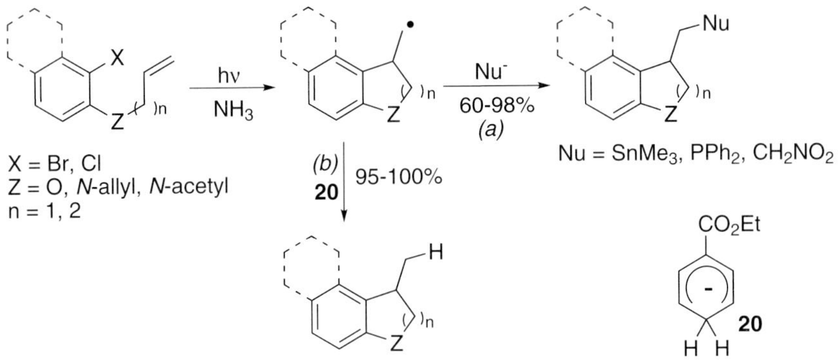
X
hν
NH3
)n
Z
Nu⁻
60-98%
(a)
Nu
Nu = SnMe3, PPh2, CH2NO2
X = Br, Cl
Z = O, N-allyl, N-acetyl
n = 1, 2
(b)
20
95-100%
H
CO2Et
20
H H

Scheme 14.11 Photoinduced synthesis of benzoannulated heterocycles.

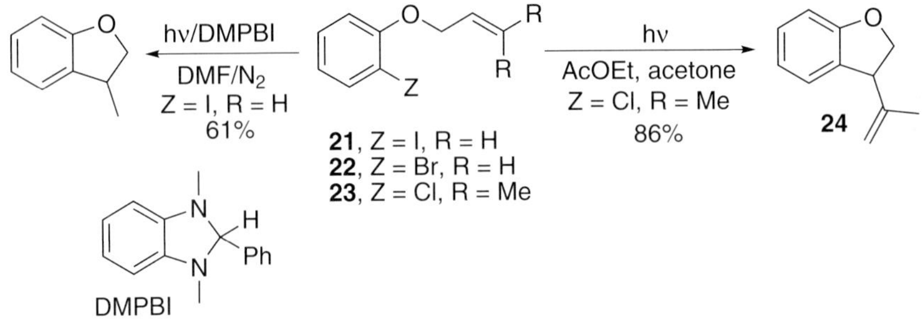
hν/DMPBI
DMF/N2
Z = I, R = H
61%
21, Z = I, R = H
22, Z = Br, R = H
23, Z = Cl, R = Me
hν
AcOEt, acetone
Z = Cl, R = Me
86%
24
H
Ph
DMPBI

Scheme 14.12 Synthesis of dihydrobenzofurans via intramolecular formation of C(sp^2)—C bonds.

3-methyl-dihydro-benzofuran in 61% yield through cleavage of the C(sp^2)—I bond and cyclization of the aryl radical (Scheme 14.12, left part) [68]. The reaction was not successful when starting from the less easily reduced 1-allyloxy-2-bromo derivative (**22**). Of note here was the fact that 3-isopropenyl-2,3-dihydrobenzofuran (**24**) was obtained in a high yield of 86%, starting from chloroaryl ether **23** (Scheme 14.12, right part) [69]. In this case, however, a S_N1 process was involved and the key intermediate was the aryl cation formed by photoheterolysis of the C(sp^2)—Cl bond, rather than an aryl radical as in the previous cases.

A further alternative involved the homolytic cleavage of an aliphatic C(sp^3)—halide bond, for example in α-chloroacetamides, rather than an aryl C(sp^3)—halide bond. The radical thus formed was exploited for the synthesis of a potential precursor to vinblastine-type dimeric indole alkaloids such as 20-de-ethylcatharanthine through a radical cyclization step onto an indole skeleton [70].

14.2.4
Intermolecular Formation of Aryl—Aryl Bonds

Benzene is a suitable trapping reagent for aryl cations, and the ArS_N1 reaction presented in Section 14.2.1 can be conveniently applied to the synthesis of biphenyls. The cations were formed by irradiation of aryl halides [71, 72] or esters [18]

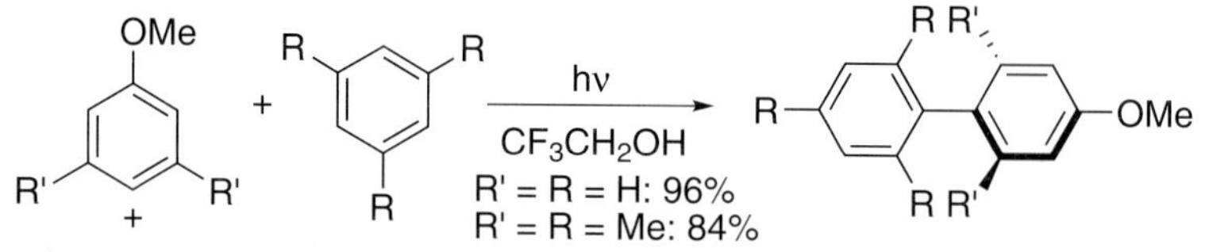

Scheme 14.13 Synthesis of sterically congested biphenyls via photo-S_N1 reaction.

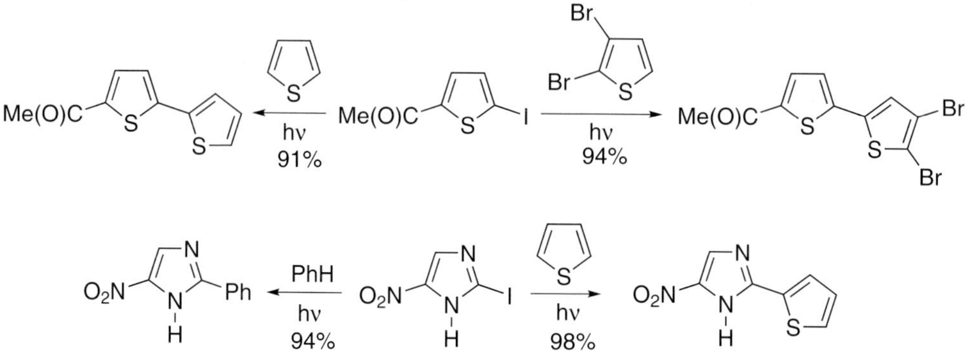

Scheme 14.14 Photoinduced cross-couplings for the synthesis of biheterocycles.

with electron-donating substituents or through the irradiation of phenyldiazonium salts, which may form ground-state complexes with many arenes [22, 73]. This method offers an alternative to the traditional procedures used to prepare biphenyls, such as the Ullmann coupling, or to modern transition metal catalysis. Contrary to most catalytic procedures, the reaction is a direct electrophilic substitution at an unsubstituted position, and does not require the presence of an activating group on the arene to be arylated. The high reactivity of the phenyl cation makes the process successful even when highly hindered biphenyls are formed. Thus, the yields of isolated products remain high when substituents are present in proximity to the reacting site, and tetra-*ortho*-substituted biphenyls were obtained without problems using this method (Scheme 14.13) [74].

Five-membered heterocycles were also conveniently arylated by photochemically generated aryl cations [75]. Likewise, the $S_{RN}1$ process has been widely applied to the synthesis of biaryls by coupling of an aryl halide and an electron-rich arene, such as phenolates or anilides [76–78]. A similar mechanism is also involved in the photochemical synthesis of heterobiaryl derivatives by the irradiation of five-membered heterocyclic halides with electron-withdrawing groups, such as thiophene, furan or pyrrole bearing in position C-5 an iodo- or bromo-substituent, and in position C-2 an aldehyde, a ketone or a nitro group (Scheme 14.14) [79, 80]. Cleavage of the halide appears to involve the assistance of the aromatic ring, and led to the production of an aryl–iodine complex along with the key intermediate, the heteroaryl radical [81].

14.2.5
Intramolecular Formation of Aryl—Aryl Bonds

Ultraviolet irradiation of 1,2-diarylethenes leading to *E–Z* isomerization and conrotatory cyclization to dihydrophenanthrenes represents a useful method for the formation of Ar—Ar bonds in an intramolecular fashion [82–84]. When this electrocyclic reaction is carried out in the presence of an oxidant, tricyclic arenes are formed, such as phenanthrene derivatives (Scheme 14.15, path *a*). Attempted aromatization may lead to lower yields when using easily oxidized electron-rich aromatics such as pyrroles. However, the introduction of an electron-withdrawing tosyl group in the starting 1,2-diarylethenes overcomes this problem, and tosylphenanthrene derivatives can be isolated in good yields upon irradiation in the presence of catalytic amounts of iodine as oxidant [85]. This novel photocyclization of tosylstilbenoids has been used in the preparation of some cyclopropaindolone analogues of the DNA-alkylating unit of the antitumor compound CC-1065 [86, 87].

The tosyl group could alternatively be eliminated during the reaction when a base, such as 1,8-diazabicyclo[5.4.0]undec-7-ene (DBU), was present, as is the case for the photolysis of 1,2-diaryl-1-tosylstilbenes that led to phenanthrenes or their heterocyclic analogues in one step (Scheme 14.15, path *b*). These results were consistent with a mechanism involving a base-induced elimination of *p*-toluenesulfinic acid from 9-tosyl-4a,4b-dihydrophenanthrene as intermediate, formed by photochemical cyclization of the starting ethene [88].

Photoinduced electrocyclization reactions were likewise adopted for the synthesis of 3,4-dihydroquinolin-2(1*H*)-ones from *N*-substituted prop-2-enoyl anilides in good yields (Scheme 14.16). The reaction was carried out in MeCN under Ar using a high-pressure Hg lamp (Pyrex filter) at ambient temperature). Both, *N*-(4′-acylphenyl) and *N*-(2′-acylphenyl) derivatives were used. In the latter case a thermal [1,5] acyl shift occurred as the last step to give photoproducts **25** [89, 90].

The photostimulated intramolecular *ortho*-arylation reaction of bromoarenes linked to *N*-substituted tetrahydroisoquinolines in liquid ammonia was used for

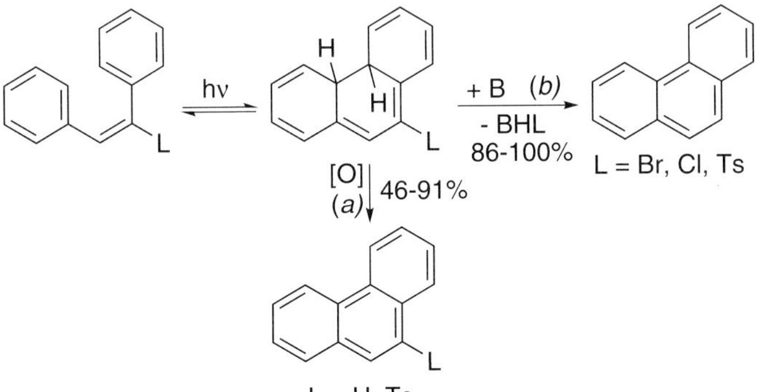

Scheme 14.15 Aryl—aryl bonds formations via photoelectrocyclic reactions.

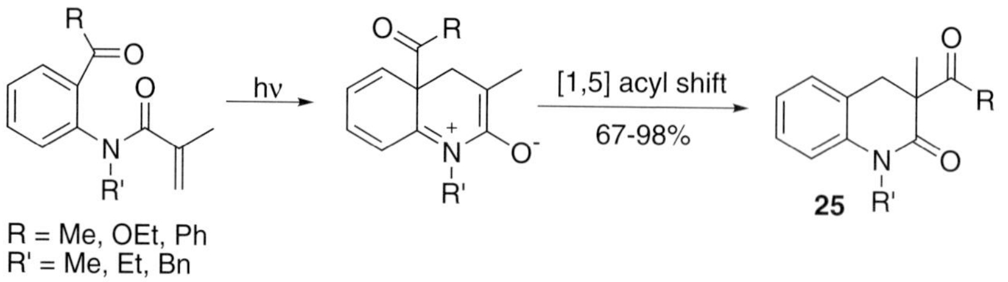
hν
[1,5] acyl shift
67-98%
25
R = Me, OEt, Ph
R' = Me, Et, Bn

Scheme 14.16 Photochemical approach to the formation of the dihydroquinoline skeleton.

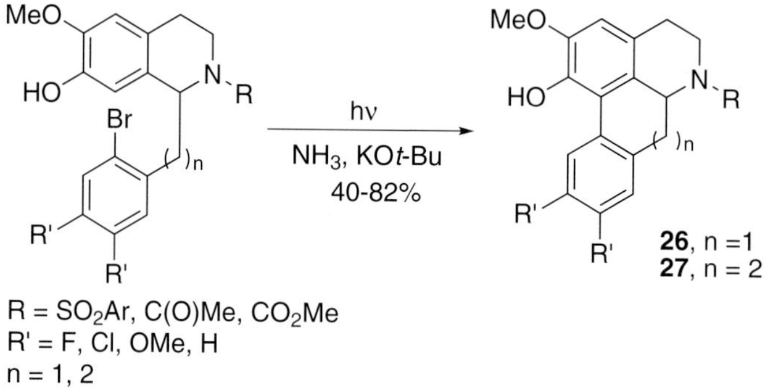
hν
NH3, KOt-Bu
40-82%
26, n =1
27, n = 2
R = SO2Ar, C(O)Me, CO2Me
R' = F, Cl, OMe, H
n = 1, 2

Scheme 14.17 Photoinduced synthesis of aporphine and homoaporphine derivatives.

the synthesis of the aporphine alkaloid skeleton **26** (54–82% yield; Scheme 14.17, n = 1) via a $S_{RN}1$ process. Recently, this strategy was extended to the synthesis of the analogous homoaporphine derivatives (**27**; Scheme 14.17, n = 2). Tetrahydroisoquinoline precursors having electron-withdrawing groups on nitrogen, such as amides, sulfonamides and carbamates, gave the corresponding homoaporphine in satisfactory yields (40%). However, when basic secondary or tertiary amines were present, cyclization took place in low yields [91].

The use of Bu_3SnH and AIBN has become commonplace in the synthesis of annulated arenes by the intramolecular aromatic substitution via aryl and heteroaryl radicals (see Chapter 13). Nevertheless, a simpler and more environmentally friendly protocol should be the direct generation of the σ-aryl or heteroaryl radicals by UV-induced homolytic cleavage of an aryl halide with subsequent intramolecular attack onto an arene in proximity. Accordingly, tricyclic [2,1-*a*] fused heterocycles were regioselectively formed in high yields from *N*-(2-arylethyl)-2-iodoimidazoles upon irradiation in acetonitrile (Scheme 14.18) [92].

14.3
Photochemical Formation of Aryl—N Bonds

The formation of aromatic C—N bonds is one of the key steps in the preparation of compounds of interest in the pharmaceutical and agrochemical industries. The

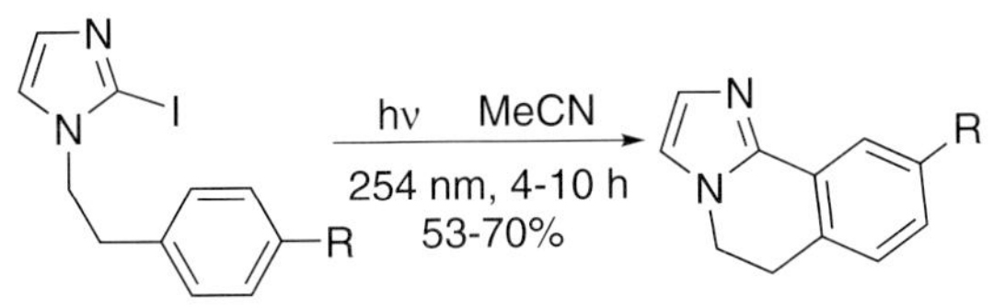

hν MeCN
254 nm, 4-10 h
53-70%

R = H, Cl, CF_3

Scheme 14.18 Synthesis of fused heterocycles via photogenerated aryl radicals.

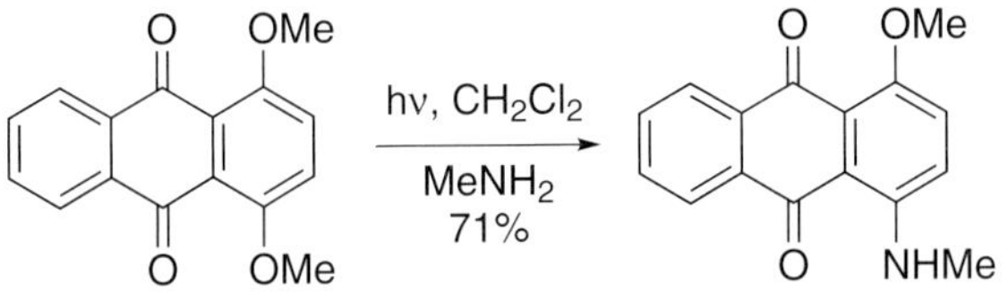

hν, CH_2Cl_2
$MeNH_2$
71%

Scheme 14.19 Photoamination of an anthraquinone derivative.

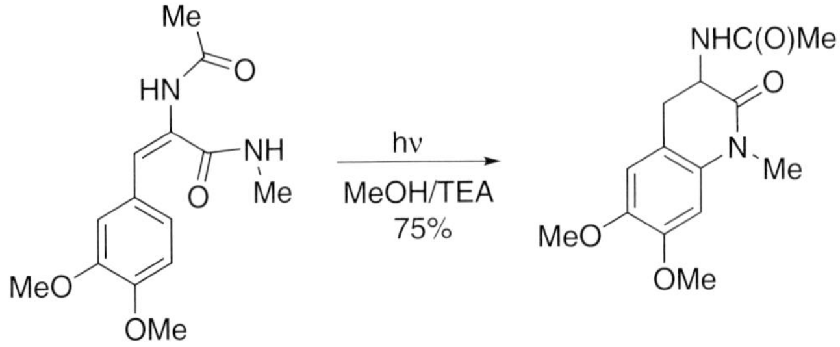

hν
MeOH/TEA
75%

Scheme 14.20 PET-mediated formation of $C(sp^2)$—N bonds.

copper-catalyzed (see Chapter 4) and palladium-catalyzed (see Chapter 3) aromatic amination of aryl halides or sulfonates is the method of choice for the formation of the $C(sp^2)$—N bonds [93–95]. The aspects of the photochemical versions of this reaction shown here are mainly based on the use of amines as nucleophiles in photosubstitution reactions [10] as illustrated in Scheme 14.19 [96].

Intramolecular nucleophilic photosubstitutions have also been reported. The most commonly found example is the photo-Smiles rearrangement, which is the conversion of aryl-ω-aminoalkylethers into the corresponding substituted anilines upon irradiation [10, 97, 98]. Recently, the photochemistry of (*Z*)-*N*-acyldehydroarylalaninamides in methanol in the presence of a base, such as DBU or triethylamine (TEA), has been reported and found to produce substituted dihydroquinolinones in high yields (Scheme 14.20) [99–101]. The reaction is initiated by ET from the amine to the excited state of the amide. Competitive side reactions were also observed, thereby lowering the yields of the cyclized products.

14.4 Conclusions

In this chapter we have shown that, today, a number of well-recognized photochemical arylation methods are available for the organic synthetic chemist which provide scope, yields and selectivities that often are comparable to those of thermal – usually metal-catalyzed – methods. Moreover, the experimental approach is quite simple in practice, requiring only the addition of a typically inexpensive, phosphor-coated lamp to the normal laboratory equipment. In this way, the sophisticated reaction conditions required for some catalytic reactions are avoided, as is the use of toxic and/or dangerous reagents. Taken together, these characteristics should lead to the photochemical, 'green chemistry', approach becoming more common among synthetic chemistry laboratories [17, 102].

Abbreviations

DBU	1,8-diazabicyclo[5.4.0]undec-7-ene
DCB	1,4-dicyanobenzene
DMF	*N,N*-dimethylformamide
DMPBI	1,3-dimethyl-2-phenylbenzimidazoline
EOCAS	electrophile olefin combination aromatic substitution
ET	electron transfer
NMP	*N*-methylpyrrolidinone
NOCAS	nucleophile olefin combination aromatic substitution
PET	photoinduced electron transfer
ROCAS	radical olefin combination aromatic substitution
TCB	1,2,4,5-tetracyanobenzene
TEA	triethylamine
UV	ultraviolet

References

1 Mattay, J. and Griesbeck, A.G. (eds) (1994) *Photochemical Key Steps in Organic Synthesis*, Wiley-VCH Verlag GmbH, New York.

2 Griesbeck, A.G. and Mattay, J. (eds) (2005) *Synthetic Organic Photochemistry*, Dekker, New York.

3 Albini, A., Fagnoni, M. and Mella, M. (2000) *Pure Appl. Chem.*, **72**, 1321–6.

4 Rossi, R.A. and Peñéñory, A.B. (2006) *Curr. Org. Synth.*, **3**, 121–58.

5 Rossi, R.A. and Peñéñory, A.B. (2004) The photostimulated $S_{RN}1$ process: reaction of haloarenes with carbanions, in *Handbook of Organic Photochemistry and Photobiology*, 2nd edn (eds W. Horspool and F. Lenci), CRC Press, Boca Raton, pp. 47/1–47/24.

6 Rossi, R.A. (2005) Photoinduced aromatic nucleophile substitution reaction, in *Molecular and Supramolecular Photochemistry*, Vol. 12 (eds A.G. Griesbeck and J. Mattay), Dekker, New York, pp. 495–527.

7 Kessar, S.V. and Mankotia, S. (1995) Photocyclization of haloarenes, in

Handbook of Organic Photochemistry and Photobiology (eds W.H. Horspool and P.-S. Song), CRC Press, New York, pp. 1218–28.
8 Fagnoni, M. and Albini, A. (2005) *Acc. Chem. Res.*, **38**, 713–21.
9 Fagnoni, M. (2006) *Lett. Org. Chem.*, **3**, 253–9.
10 Fagnoni, M. and Albini, A. (2006) Photonucleophilic substitution reactions, in *Molecular and Supramolecular Photochemistry*, Vol. 14 (eds V. Ramamurthy and K. Schanze), Dekker, New York, pp. 131–77.
11 Karapire, C. and Icli, S. (2004) Photochemical aromatic substitution, in *Handbook of Organic Photochemistry and Photobiology*, 2nd edn (eds W. Horspool and F. Lenci), CRC Press, Boca Raton, pp. 37/1–37/14.
12 Mella, M., Fagnoni, M., Freccero, M., Fasani, E. and Albini, A. (1998) *Chem. Soc. Rev.*, **27**, 81–9.
13 Albini, A., Fasani, E. and Mella, M. (1993) *Top. Curr. Chem.*, **168**, 143–73.
14 Mangion, D. and Arnold, D.R. (2002) *Acc. Chem. Res.*, **35**, 297–304.
15 Mangion, D. and Arnold, D.R. (2004) The Photochemical Nucleophile-Olefin Combination, Aromatic Substitution (Photo-NOCAS) reaction, in *Handbook of Organic Photochemistry and Photobiology*, 2nd edn (eds W. Horspool and F. Lenci), CRC Press, Boca Raton, pp. 40/1–40/17.
16 Mangion, D., Frizzle, M., Arnold, D.R. and Cameron, T.S. (2001) *Synthesis*, 1215–22.
17 Protti, S., Dondi, D., Fagnoni, M. and Albini, A. (2007) *Pure Appl. Chem.*, **79**, 1929–38.
18 De Carolis, M., Protti, S., Fagnoni, M. and Albini, A. (2005) *Angew. Chemie Int. Ed.*, **44**, 1232–6.
19 Lazzaroni, S., Dondi, D., Fagnoni, M. and Albini, A. (2007) *Eur. J. Org. Chem.*, 4360–5.
20 Dichiarante, V., Dondi, D., Protti, S., Fagnoni, M. and Albini, A. (2007) *J. Am. Chem. Soc.*, **129**, 5605–11. Correction: 11662.
21 (a) Guizzardi, B., Mella, M., Fagnoni, M., Freccero, M. and Albini, A. (2001) *J. Org. Chem.*, **66**, 6353–63; (b) Dichiarante, V. and Fagnoni, M. (2008) *Synlett*, 787–800.
22 Milanesi, S., Fagnoni, M. and Albini, A. (2005) *J. Org. Chem.*, **70**, 603–10.
23 Guizzardi, B., Mella, M., Fagnoni, M. and Albini, A. (2003) *J. Org. Chem.*, **68**, 1067–74.
24 Mella, M., Fagnoni, M. and Albini, A. (2004) *Org. Biomol. Chem.*, **2**, 3490–5.
25 Protti, S., Fagnoni, M. and Albini, A. (2005) *Org. Biomol. Chem.*, **3**, 2868–71.
26 Fagnoni, M., Mella, M. and Albini, A. (1999) *Org. Lett.*, **1**, 1299–301.
27 Protti, S., Fagnoni, M. and Albini, A. (2005) *Angew. Chemie Int. Ed.*, **44**, 5675–8.
28 Guizzardi, B., Mella, M., Fagnoni, M. and Albini, A. (2003) *Chem. Eur. J.*, **9**, 1549–55.
29 Fraboni, A., Fagnoni, M. and Albini, A. (2003) *J. Org. Chem.*, **68**, 4886–93.
30 Protti, S., Fagnoni, M. and Albini, A. (2006) *J. Am. Chem. Soc.*, **128**, 10670–1.
31 Baumgartner, M.T., Gallego, M.H. and Pierini, A.B. (1998) *J. Org. Chem.*, **63**, 6394–7.
32 Baumgartner, M.T., Pierini, A.B. and Rossi, R.A. (1999) *J. Org. Chem.*, **64**, 6487–9.
33 Baumgartner, M.T., Lotz, G.A. and Palacios, S.M. (2004) *Chirality*, **16**, 212–19.
34 Nwokogu, G.C., Wong, J.-W., Greenwood, T.D. and Wolfe, J.F. (2000) *Org. Lett.*, **2**, 2643–6.
35 Barolo, S.M., Lukach, A.E. and Rossi, R. A. (2003) *J. Org. Chem.*, **68**, 2807–11.
36 Greenwood, W.J., Jr, Layman, T.D., Downey, A.L. and Wolfe, J.F. (2005) *J. Org. Chem.*, **70**, 9147–55.
37 Barolo, S.M., Rosales, C., Guìo, J.E.A. and Rossi, R.A. (2006) *J. Heterocycl. Chem.*, **43**, 695–9.
38 Guastavino, J.F., Barolo, S.M. and Rossi, R.A. (2006) *Eur. J. Org. Chem.*, 3898–902.
39 Ferrayoli, C.G., Palacios, S.M. and Alonso, R.A. (1995) *J. Chem. Soc. Perkin Trans.*, **1**, 1635–8.
40 For some examples see: Frolov, A.N. (1998) *Russ. J. Org. Chem.*, **34**, 139–61.
41 Miller, J.A. (2001) *Tetrahedron Lett.*, **42**, 6991–3.
42 Nakao, Y., Oda, S. and Hiyama, T. (2004) *J. Am. Chem. Soc.*, **126**, 13904–5.

43 Mizuno, K., Ikeda, M. and Otsuji, Y. (1985) *Tetrahedron Lett.*, **26**, 461–4.

44 Mizuno, K., Nakanishi, K. and Otsuji, Y. (1988) *Chem. Lett.*, **17**, 1833–6.

45 Tsuji, M., Higashiyama, K., Yamauchi, T., Kubo, H. and Ohmiya, S. (2001) *Heterocycles*, **54**, 1027–32.

46 Fagnoni, M. (2003) *Heterocycles*, **60**, 1921–58.

47 Fukuzumi, S., Ohkubo, K., Suenobu, T., Kato, K., Fujitsuka, M. and Ito, O. (2001) *J. Am. Chem. Soc.*, **123**, 8459–67.

48 Mella, M., Fagnoni, M. and Albini, A. (1994) *J. Org. Chem.*, **59**, 5614–22.

49 Fagnoni, M., Mella, M. and Albini, A. (1995) *J. Am. Chem. Soc.*, **117**, 7877–81.

50 Mangion, D., Arnold, D.R., Cameron, T.S. and Robertson, K.N. (2001) *J. Chem. Soc. Perkin Trans.*, **2**, 48–60.

51 McManus, K.A. and Arnold, D.R. (1995) *Can. J. Chem.*, **73**, 2158–69.

52 Yasuda, M., Shiragami, T., Matsumoto, J., Yamashita, T. and Shima, K. (2006) Photoamination with Ammonia and Amines, in *Molecular and Supramolecular Photochemistry*, Vol. 14 (eds V. Ramamurthy and K. Schanze), Dekker, New York, pp. 207–53.

53 Yasuda, M., Kojima, R., Tsutsui, H., Utsunomiya, D., Ishii, K., Jinnouchi, K., Shiragami, T. and Yamashita, T. (2003) *J. Org. Chem.*, **68**, 7618–24.

54 Yamashita, T., Itagawa, J., Sakamoto, D., Nakagawa, Y., Matsumoto, J. and Shiragami, T., Yasuda, M. (2007) *Tetrahedron*, **63**, 374–80.

55 Zhang, M., Lu, Z.-F., Liu, Y., Grampp, G. Hu, H.-W. and Xu, J.-H. (2006) *Tetrahedron*, **62**, 5663–74.

56 Ishii, H., Imai, Y., Hirano, T., Maki, S., Niwa, H. and Ohashi, M. (2000) *Tetrahedron Lett.*, **41**, 6467–71.

57 (a) Sundermeier, M., Zapf, A. and Beller, M. (2003) *Eur. J. Inorg. Chem.*, 3513–26;
(b) Schareina, T., Zapf, A., Mägerlein, W., Müller, N. and Beller, M. (2007) *Chem. Eur. J.*, **13**, 6249–54.

58 (a) Dobbs, K.D., Marshall, W.J. and Grushin, V.V. (2007) *J. Am. Chem. Soc.*, **129**, 30–1;
(b) Schareina, T., Zapf, A., Mägerlein, W., Müller, N. and Beller, M. (2007) *Tetrahedron Lett.*, **48**, 1087–90;
(c) Cheng, Y., Duan, Z., Li, T. and Wu, Y. (2007) *Synlett*, 543–6.

59 Dichiarante, V., Fagnoni, M. and Albini, A. (2006) *Chem. Commun.*, 3001–3.

60 Kuzmič, P. and Souček, M. (1986) *Collect. Czech. Chem. Commun.*, **51**, 358–67.

61 El'tsov, A.V., Kul'bitskaya, O.V., Smirnov, E.V. and Frolov, A.N. (1973) *Russ. J. Org. Chem.*, **9**, 2542–5.

62 See for example: Dell'Erba, C., Houmam, A., Novi, M., Petrillo, G. and Pinson, J. (1993) *J. Org. Chem.*, **58**, 2670–7.

63 Kitagawa, F., Murase, M. and Kitamura, N. (2002) *J. Org. Chem.*, **67**, 2524–31.

64 Wubbels, G.G., Johnson, K.M. and Babcock, T.A. (2007) *Org. Lett.*, **9**, 2803–6.

65 Vaillard, S.E., Postigo, A. and Rossi, R.A. (2002) *J. Org. Chem.*, **67**, 8500–6.

66 Bardagì, J.I., Vaillard, S.E. and Rossi, R. A. (2007) *ARKIVOC*, **4**, 73–83.

67 Vaillard, S.E., Postigo, A., Rossi, R.A. (2004) *J. Org. Chem.*, **69**, 2037–41.

68 Hasegawa, E., Yoneoka, A., Suzuki, K., Kato, T., Kitazume, T. and Yanagi, K. (1999) *Tetrahedron*, **55**, 12957–68.

69 Dichiarante, V., Fagnoni, M., Mella, M. and Albini, A. (2006) *Chem. Eur. J.*, **12**, 3905–15.

70 Sundberg, R.J. and Bloom, J.D. (1980) *J. Org. Chem.*, **45**, 3382–7.

71 Mella, M., Coppo, P., Guizzardi, B., Fagnoni, M., Freccero, M. and Albini, A. (2001) *J. Org. Chem.*, **66**, 6344–52.

72 Protti, S., Fagnoni, M., Mella, M. and Albini, A. (2004) *J. Org. Chem.*, **69**, 3465–73.

73 Kosynkin, D., Bockman, T.M. and Kochi, J.K. (1997) *J. Am. Chem. Soc.*, **119**, 4846–55.

74 Dichiarante, V., Fagnoni, M. and Albini, A. (2007) *Angew. Chemie Int. Ed.*, **46**, 6495–8.

75 Guizzardi, B., Mella, M., Fagnoni, M. and Albini, A. (2000) *Tetrahedron*, **56**, 9383–9.

76 Pierini, A.B., Baumgartner, M.T. and Rossi, R.A. (1987) *Tetrahedron Lett.*, **28**, 4653–6.

77 Baumgartner, M.T., Pierini, A.B. and Rossi, R.A. (1992) *Tetrahedron Lett.*, **33**, 2323–6.

78 Tempesti, T.C., Pierini, A.B. and Baumgartner, M.T. (2005) *J. Org. Chem.*, **70**, 6508–11.

79 D'Auria, M., De Luca, E., Mauriello, G. and Racioppi, R. (1998) *J. Chem. Soc., Perkin Trans.*, **1**, 271–4.

80 D'Auria, M., D'Amico, A., D'Onofrio, F. and Piancatelli, G. (1987) *J. Chem. Soc., Perkin Trans.*, **1**, 1777–80.

81 Elisei, F., Latterini, L., Aloisi, G.G. and D'Auria, M. (1995) *J. Phys. Chem.*, **99**, 5365–72.

82 Gilbert, A. (2004) Cyclization of stilbene and its derivatives, in *Handbook of Organic Photochemistry and Photobiology*, 2nd edn (eds W. Horspool and F. Lenci), CRC Press, Boca Raton, pp. 41/1–41/11.

83 Hoffmann, N. (2004) Synthesis of heterocycles by photocyclization of arenes, in *Handbook of Organic Photochemistry and Photobiology*, 2nd edn (eds W. Horspool and F. Lenci), CRC Press, Boca Raton, pp. 34/1–34/20.

84 Matsuda, K. and Irie, M. (2006) *Chem. Lett.*, **35**, 1204–9.

85 Antelo, B., Castedo, L., Delamano, J., Gómez, A., López, C. and Tojo, G. (1996) *J. Org. Chem.*, **61**, 1188–9.

86 Enjo, J., Castedo, L. and Tojo, G. (2001) *Org. Lett.*, **3**, 1343–5.

87 Castedo, L., Delamano, J., Enjo, J., Fernández, J., Grávalos, D.G., Leis, R., López, C., Marcos, C.F., Ríos, A. and Tojo, G. (2001) *J. Am. Chem. Soc.*, **123**, 5102–3.

88 Almeida, J.F., Castedo, L., Fernández, D., Neo, A.G., Romero, V. and Tojo, G. (2003) *Org. Lett.*, **5**, 4939–41.

89 Nishio, T., Tabata, M., Koyama, H. and Sakamoto, M. (2005) *Helv. Chim. Acta*, **88**, 78–86.

90 Nishio, T., Koyama, H., Sasaki, D. and Sakamoto, M. (2005) *Helv. Chim. Acta*, **88**, 996–1003.

91 Barolo, S.M., Teng, X., Cuny, G.D. and Rossi, R.A. (2006) *J. Org. Chem.*, **71**, 8493–9.

92 Clyne, M.A. and Aldabbagh, F. (2006) *Org. Biomol. Chem.*, **4**, 268–77.

93 (a) Buchwald, S.L., Mauger, C., Mignani, G. and Scholz, U. (2006) *Adv. Synth. Catal.*, **348**, 23–39;
(b) Schlummer, B. and Scholz, U. (2004) *Adv. Synth. Catal.*, **346**, 1599–626.

94 Jiang, L. and Buchwald, S.L. (2004) Palladium-catalyzed aromatic carbon-nitrogen bond formation, in *Metal-Catalyzed Cross-Coupling Reactions*, 2nd edn (eds A. De Meijere and F. Diederich), Wiley-VCH, Weinheim, pp. 699–760.

95 Hartwig, J.F. (2002) Palladium-catalyzed amination of aryl halides and sulfonates, in *Modern Arene Chemistry* (ed. D. Astruc), Wiley-VCH Verlag GmbH, Weinheim, Germany, pp. 107–68.

96 Green-Buckley, G. and Griffiths, J. (1984) *J. Photochem.*, **27**, 119–21.

97 Nakagaki, R. and Mutai, K. (1996) *Bull. Chem. Soc. Jpn*, **69**, 261–74.

98 Wubbels, G.G., Ota, N. and Crosier, M.L. (2005) *Org. Lett.*, **7**, 4741–4.

99 Maekawa, K., Igarashi, T., Kubo, K. and Sakurai, T. (2001) *Tetrahedron*, **57**, 5515–26.

100 Maekawa, K., Shinozuka, A., Naito, M., Igarashi, T. and Sakurai, T. (2004) *Tetrahedron*, **60**, 10293–304.

101 Maekawa, K., Fujita, K., Iizuka, K., Igarashi, T. and Sakurai, T. (2005) *Heterocycles*, **65**, 117–31.

102 Albini, A. and Fagnoni, M. (2004) *Green Chem.*, **6**, 1–6.

Index

Modern Arylation Methods. Edited by Lutz Ackermann

ISBN: 978-3-527-31937-4

b